Lecture Notes in Computer Science 4947

Commenced Publication in 1973
Founding and Former Series Editors:
Gerhard Goos, Juris Hartmanis, and Jan van Leeuwen

T0181380

Jayant R. Haritsa Ramamohanarao Kotagiri
Vikram Pudi (Eds.)

Database Systems
for Advanced Applications

13th International Conference, DASFAA 2008
New Delhi, India, March 19-21, 2008
Proceedings

 Springer

Volume Editors

Jayant R. Haritsa
Indian Institute of Science
Supercomputer Education and Research Centre
Bangalore 560012, India
E-mail: haritsa@dsl.serc.iisc.ernet.in

Ramamohanarao Kotagiri
The University of Melbourne
Department of Computer Science and Software Engineering
Victoria 3010, Australia
E-mail: kotagiri@unimelb.edu.au

Vikram Pudi
International Institute of Information Technology
Gachibowli, Hyderabad 500032, India
E-mail: vikram@iiit.ac.in

Library of Congress Control Number: 2008922313

CR Subject Classification (1998): H.2, H.3, H.4, H.5, J.1

LNCS Sublibrary: SL 3 – Information Systems and Application, incl. Internet/Web
and HCI

ISSN	0302-9743
ISBN-10	3-540-78567-1 Springer Berlin Heidelberg New York
ISBN-13	978-3-540-78567-5 Springer Berlin Heidelberg New York

Springer is a part of Springer Science+Business Media

springer.com

© Springer-Verlag Berlin Heidelberg 2008
Printed in Germany

Typesetting: Camera-ready by author, data conversion by Scientific Publishing Services, Chennai, India
Printed on acid-free paper   SPIN: 12238484   06/3180   5 4 3 2 1 0

Foreword

Welcome to the proceedings of the 13th International Conference on Database Systems for Advanced Applications (DASFAA 2008) which was held in New Delhi, India. DASFAA 2008 continued the tradition of providing an international forum for technical discussion among researchers, developers and users of database systems from academia, business and industry. Organizing DASFAA 2008 was a very rewarding experience; it gave us an excellent opportunity to work with many fine colleagues both within and outside India.

We would like to thank Jayant Haritsa and Ramamohanarao Kotagiri for putting together a world-class Program Committee. The committee worked very hard to bring a high-quality technical program to the conference. DASFAA 2008 also included an industrial track co-chaired by Anand Deshpande and Takeshi Fukuda.

The conference also featured three tutorials: (1) Preference Query Formulation and Processing: Ranking and Skyline Query Approaches, by Seung-Won Huang and Wolf-Tilo Balke, (2) Stream Processing: Going Beyond Database Management Systems, by Sharma Chakravarthy, and (3) The Semantic Web: Semantics for Data and Services on the Web, by Vipul Kashyap and Christoph Bussler. We would like to thank S. Sudarshan and Kian-lee Tan for their effort in organizing the tutorials, Srinath Srinivasa and Wookey Lee for the panels, Prasan Roy and Anthony Tung for the demos, and Sanjay Chawla and Chee-Yong Chan for the workshops.

This conference would not have been possible without the support of many other colleagues: S.K. Gupta (Honorary Conference Chair), Mukul Joshi and Takahiro Hara (Publicity Chairs), Vikram Pudi (Publication Chair), Rajeev Gupta (Website Chair), J.P. Gupta (Organization Chair), Naveen Kumar and Neelima Gupta (Local Arrangement Chairs).

We greatly appreciate the support of our sponsors without which a high-quality conference like DASFAA would not have been accessible to a large number of attendees. Finally, our thanks to the DASFAA Steering Committee, especially its Chair, Kyu-Young Whang, for his constant encouragement and support.

March 2008

Mukesh Mohania
Krithi Ramamritham

Message from the Program Committee Chairs

The 13th International Conference on Database Systems for Advanced Applications (DASFAA 2008) was held in New Delhi, the historically-rich national capital of India, from March 19 to 21, 2008. DASFAA is an annual international database conference, located in the Asia-Pacific region, showcasing state-of-the-art research and development activities in database systems and their applications. It provides a forum for technical presentations and discussions among database researchers, developers and users from academia, business and industry.

The Call for Papers attracted 173 research and industry submissions from 30 countries, spanning the globe, reflecting DASFAA's truly international appeal. Each paper was rigorously reviewed by three members of the Program Committee, which included 100-plus database experts from 20 countries. After sustained discussions on the conference bulletin board, 27 full papers and 30 short papers were finally selected for presentation at the conference and are featured in this proceedings volume.

The chosen papers cover the entire spectrum of database research ranging from core database engines to middleware to applications. Featured topics include query and transaction processing, spatial and temporal databases, mobile and distributed databases, XML, data mining and warehousing, data streams and P2P networks, and industrial applications. This rich mixture of papers made the conference appealing to a wide audience and laid the foundation for a synergistic exchange of views between researchers and practitioners.

The paper titled "Bulk-Loading the ND-Tree in Non-ordered Discrete Spaces," by Seok, Qian, Zhu, Oswald and Pramanik from the USA, was selected for the Best Paper award by a committee chaired by Theo Haerder of the University of Kaiserslautern, Germany. A cleverly designed technique is presented in this paper for efficiently bulk-loading index structures on attribute domains, such as genomic sequences, that have a finite set of discrete values and no natural ordering among these values.

Three tutorials from leading international experts that explore new horizons of database technology extending to intelligent query responses, stream processing and the Semantic Web were also featured in the program. A set of eight demos of prototype systems, spanning data mining, data warehousing, Semantic Web and temporal data, showcased compelling applications of novel technologies. A thought-provoking panel on the future prospects of data mining research provided a forum for generating discussion and audience participation. Finally, there were two sessions of invited talks from industry experts, delivering a ringside view of database technologies and challenges.

The conference was privileged to host keynote addresses by Beng-Chin Ooi of the National University of Singapore, Surajit Chaudhuri of Microsoft Research, and Anand Deshpande of Persistent Systems, representing academia, industrial research and commercial industry, respectively. Together, they provided a holistic perspective on the world of databases from diverse angles of community storage, autonomic systems and technology transfer, especially useful for research students scouting around for dissertation topics.

We thank all the session chairs, tutorial speakers, authors, panelists and participants for ensuring a lively intellectual atmosphere and exchange of views. We also thank all members of the Dasfaa Steering Committee, the Organizing Committee and the Program Committee for generously contributing their time and expertise, ensuring a high-quality program of lasting value to the conference attendees. A special note of thanks to Kyu-young Whang for expertly guiding us through the entire conference process, and to the PC members for their painstaking reviews and discussions.

In closing, we believe that all participants of the DASFAA 2008 conference in New Delhi found it to be a technically fruitful and culturally stimulating experience.

March 2008   Jayant R. Haritsa
                                                     Ramamohanarao Kotagiri

DASFAA 2008 Conference Organization

Honorary Chair

S.K. Gupta   IIT Delhi, India

Conference Chairs

Mukesh Mohania   IBM India Research, India
Krithi Ramamritham   IIT Bombay, India

Program Chairs

Jayant R. Haritsa   IISc Bangalore, India
Ramamohanarao Kotagiri   University of Melbourne, Australia

Industrial Chairs

Anand Deshpande   Persistent Systems, Pune, India
Takeshi Fukuda   IBM Research, Tokyo, Japan

Tutorial Chairs

S. Sudarshan   IIT Bombay, India
Kian-lee Tan   NUS Singapore

Workshop Chairs

Sanjay Chawla   University of Sydney, Australia
Chee-Yong Chan   NUS Singapore

Panel Chairs

Srinath Srinivasa   IIIT Bangalore, India
Wookey Lee   Inha University, Incheon, South Korea

Demo Chairs

Prasan Roy   Aster Data Systems, USA
Anthony Tung   NUS, Singapore

Publicity Chairs

Mukul Joshi   Great Software Lab, Pune, India
Takahiro Hara   Osaka University, Osaka, Japan

Publication Chair

Vikram Pudi   IIIT Hyderabad, India

Web-site Chair

Rajeev Gupta   IBM India Research Lab, New Delhi, India

Organization Chair

J.P. Gupta   JP Inst. of Technology, Noida, India

Local Arrangements Chairs

Naveen Kumar   University of Delhi, India
Neelima Gupta   University of Delhi, India

Regional Chairs

Asia   Mizuho Iwaihara   Kyoto University, Japan
Oceania   Millist Vincent   University of South Australia, Australia
Europe   Ladjel Bellatreche Poitiers University, France
Americas   Kalpdrum Passi   Laurentian University, Canada

Program Committee

Srinivas Aluru   Iowa State University, USA
Toshiyuki Amagasa   University of Tsukuba, Japan
Arvind Arasu   Microsoft Research, USA
Masatoshi Arikawa   University of Tokyo, Japan
Masayoshi Aritsugi   Kumamoto University, Japan
Vijay Atluri   Rutgers University, USA
James Bailey   University of Melbourne, Australia

Ladjel Bellatreche	LISI/ENSMA France
Sonia Berman	University of Cape Town, South Africa
Sourav Bhowmick	NTU Singapore
Stephane Bressan	NUS Singapore
K. Selcuk Candan	Arizona State University, USA
Barbara Catania	University of Genoa, Italy
Chee-Yong Chan	NUS Singapore
Sanjay Chawla	University of Sydney, Australia
Ying Chen	IBM Research, China
Gao Cong	Microsoft Research, China
Ernesto Damiani	University of Milan, Italy
Gautam Das	University of Texas-Arlington, USA
Gillian Dobbie	University of Auckland, New Zealand
Jianhua Feng	Tsinghua University, China
Hakan Ferhatosmanoglu	Ohio State University, USA
Elena Ferrari	University of Insubria, Italy
Dimitrios Gunopulos	University of California-Irvine, USA
Theo Haerder	University of Kaiserslautern, Germany
Takahiro Hara	Osaka University, Japan
Kenji Hatano	Doshisha University, Japan
Haibo Hu	Baptist University, Hong Kong, China
Seungwon Hwang	POSTECH, South Korea
Bala Iyer	IBM Silicon Valley Lab, USA
Panos Kalnis	NUS Singapore
Ibrahim Kamel	University of Sharjah, UAE
Kamal Karlapalem	IIIT Hyderabad, India
Raghav Kaushik	Microsoft Research, USA
Myoung-Ho Kim	KAIST, South Korea
Rajasekar Krishnamurthy	IBM Research, USA
A. Kumaran	Microsoft Research, India
Arnd Christian Konig	Microsoft Research, USA
Manolis Koubarakis	N & K University of Athens, Greece
P. Sreenivasa Kumar	IIT Madras, India
Laks V.S. Lakshmanan	University of British Columbia, Canada
Jae-Gil Lee	University of Illinois-Urbana/Champaign, USA
Sang-Goo Lee	Seoul National University, South Korea
Wang-Chien Lee	Penn State University, USA
Young-Koo Lee	Kyoung Hee University, South Korea
Jinyan Li	Inst. for Infocomm Research, Singapore
Qing Li	City University of Hong Kong, China
Ee-peng Lim	NTU Singapore
Xuemin Lin	University of New South Wales, Australia
Qiong Luo	HKUST, Hong Kong, China
Anna Maddalena	University of Genoa, Italy

Sanjay Madria	University of Missouri-Rolla, USA
Sharad Mehrotra	University of California-Irvine, USA
Sameep Mehta	IBM Research, India
Kamesh Munagala	Duke University, USA
Yang-Sae Moon	Kangwon National University, South Korea
Yasuhiko Morimoto	Hiroshima University, Japan
Atsuyuki Morishima	University of Tsukuba, Japan
Ekawit Nantajeewarawat	SIIT, Thammasat University, Thailand
Wolfgang Nejdl	University of Hannover, Germany
Hwee-Hwa Pang	SMU Singapore
Sanghyun Park	Yonsei University, South Korea
Srinivasan Parthasarathy	Ohio State University, USA
Guenther Pernul	University of Regensburg, Germany
Radha Krishna Pisipati	IDRBT Hyderabad, India
Evaggelia Pitoura	University of Ioannina, Greece
Sunil Prabhakar	Purdue University, USA
Sujeet Pradhan	Kurashiki University, Japan
Vikram Pudi	IIIT Hyderabad, India
Weining Qian	Fudan University, China
Rajugan Rajagopalapillai	Curtin University of Technology, Australia
Prakash Ramanan	Wichita State University, USA
Ganesh Ramesh	Microsoft Corp., USA
Balaraman Ravindran	IIT Madras, India
Indrakshi Ray	Colorado State University, USA
P. Krishna Reddy	IIIT Hyderabad, India
Tore Risch	Uppsala University, Sweden
Simonas Saltenis	Aalborg University, Denmark
Sunita Sarawagi	IIT Bombay, India
Nandlal Sarda	IIT Bombay, India
Markus Schneider	University of Florida, USA
Heng Tao Shen	University of Queensland, Australia
Jialie Shen	SMU Singapore
Yanfeng Shu	CSIRO Tasmanian ICT Centre, Australia
Ambuj Singh	University of California-Santa Barbara, USA
Srinath Srinivasa	IIIT Bangalore, India
Kazutoshi Sumiya	University of Hyogo, Japan
Changjie Tang	Sichuan University, China
David Taniar	Monash University, Australia
Egemen Tanin	University of Melbourne, Australia
Yufei Tao	Chinese University of Hong Kong, China
Vicenc Torra	IIIA-CSIC Catalonia, Spain
Anthony Tung	NUS Singapore
Ozgur Ulusoy	Bilkent University, Turkey
Rachanee Ungrangsi	Shinawatra University, Thailand
Vasilis Vassalos	Athens University of Economics and Business, Greece

Sabrina De Capitani di Vimercati	University of Milan, Italy
Guoren Wang	Northeast University, China
Haixun Wang	IBM Research, USA
Wei Wang	University North Carolina-Chapel Hill, USA
Yan Wang	Macquarie University, Australia
Vilas Wuwongse	Asian Inst. of Tech., Thailand
Li Yang	Western Michigan University, USA
Jae-Soo Yoo	Chungbuk National University, South Korea
Jeffrey Xu Yu	Chinese University of Hong Kong, China
Mohammed Zaki	Rensselaer Polytechnic Inst., USA
Arkady Zaslavsky	Monash University, Australia
Rui Zhang	University of Melbourne, Australia
Qiankun Zhao	Penn State University, USA
Shuigeng Zhou	Fudan University, China
Xiaofang Zhou	University of Queensland, Australia

External Reviewers

Mohammed Eunus Ali	University of Melbourne, Australia
Ismail Sengor Altingovde	Bilkent University, Turkey
Fatih Altiparmak	Ohio State University, USA
Mafruz Z. Ashrafi	Inst. for Infocomm Research, Singapore
Alberto Belussi	University of Verona, Italy
Anirban Bhattacharya	Pelion, Australia
Arnab Bhattacharya	University of California - Santa Barbara, USA
Laurynas Biveinis	Aalborg University, Denmark
Petko Bogdanov	University of California - Santa Barbara, USA
Kyoung Soo Bok	KAIST, South Korea
Congxing Cai	Pennsylvania State University, USA
Vineet Chaoji	Rensselaer Polytechnic Inst., USA
M. A. Cheema	University of New South Wales, Australia
Ding Chen	National University of Singapore
Rinku Dewri	Colorado State University, USA
Philipp Dopichaj	University of Kaiserslautern, Germany
Stefan Durbeck	University of Regensburg, Germany
Ludwig Fuchs	University of Regensburg, Germany
Haris Georgiadis	Athens University of Economics and Business, Greece
Michael Gibas	Ohio State University, USA
Mohammad Al Hasan	Rensselaer Polytechnic Inst., USA
Hoyoung Jeung	University of Queensland, Australia
Cai Jing	Wuhan University of Technology, China
Hiroko Kinutani	University of Tokyo, Japan
R. Uday Kiran	IIIT Hyderabad, India
Jan Kolter	University of Regensburg, Germany

M. Kumaraswamy	IIIT Hyderabad, India
Nick Larusso	University of California - Santa Barbara, USA
Ken Lee	Pennsylvania State University, USA
Byoung Yeop Lee	Paejai University, South Korea
Vebjorn Ljosa	Broad Institute, USA
Bing-rong Lin	Pennsylvania State University, USA
Xingjie Lius	Pennsylvania State University, USA
Y. Luo	University of New South Wales, Australia
Matoula Magiridou	University of Athens, Greece
Krissada Maleewong	Shinawatra University, Thailand
Nikos Mamoulis	University of Hong Kong, China
Chris Mayfield	Purdue University, USA
Marco Mesiti	University of Milan, Italy
Iris Miliaraki	University of Athens, Greece
Kotaro Nakayama	Osaka University, Japan
Yuan Ni	National University of Singapore
Sarana Nutanong	University of Melbourne, Australia
Rifat Ozcan	Bilkent University, Turkey
Andrea Perego	University of Insubria, Italy
Paola Podesta	University of Genoa, Italy
Yinian Qi	Purdue University, USA
Lu Qin	Chinese University of Hong Kong, China
T. Raghunathan	IIIT Hyderabad, India
Sayan Ranu	University of California - Santa Barbara, USA
M. Venugopal Reddy	IIIT Hyderabad, India
Brian Ruttenberg	University of California - Santa Barbara, USA
Ahmet Sacan	Ohio State University, USA
Saeed Salem	Rensselaer Polytechnic Inst., USA
Rolf Schillinger	University of Regensburg, Germany
Christian Schlager	University of Regensburg, Germany
Rahul Shah	Louisiana State University, USA
Sarvjeet Singh	Purdue University, USA
Vishwakarma Singh	University of California - Santa Barbara, USA
Seok Il Song	Chungju National University, South Korea
Nobutaka Suzuki	University of Tsukuba, Japan
Yu Suzuki	Ritsumeikan University, Japan
Nan Tang	Chinese University of Hong Kong, China
Yuzhe Tang	Fudan University, China
Wee Hyong Tok	National University of Singapore
Yicheng Tu	Purdue University, USA
Muhammad Umer	University of Melbourne, Australia
Lucy Vanderwend	Microsoft Research, USA
Waraporn Viyanon	University of Missouri-Rolla, USA
Derry Wijaya	National University of Singapore
Xiaokui Xiao	Chinese University of Hong Kong, China

Sponsoring Institutions

Persistent Systems, India

Yahoo! India

Tata Consultancy Services, India

Great Software Laboratory, India

Google Inc., USA

IBM India

Satyam Computer Services, India

ARC Research Network in Enterprise
Information Infrastructure

Motorola India

Table of Contents

Invited Talks (Abstracts)

Part I: Full Papers

XML Schemas

Data Mining

Spatial Data

Indexes and Cubes

Data Streams

P2P and Transactions

XML Processing

Complex Pattern Processing

IR Techniques

Part II: Short Papers

Queries and Transactions

Data Mining

XML Databases

Data Warehouses and Industrial Applications

Mobile and Distributed Data

Part III: Demonstrations Track

Part IV: Panel

Storage and Index Design for Community Systems

Beng-Chin Ooi

National University of Singapore
ooibc@comp.nus.edu.sg

In recent years, we have witnessed a rapid growth in the number of web services. The popularity of these services has driven much activity in digital information publishing and searching. While these systems vary in the services they provide, they share the same operation mode - users publish data items such as URLs, photos and advertisements; in addition, users collaboratively contribute descriptions such as tags and attributes which are used by the system to organize the published data to facilitate searching (by browsing or querying the tags). Such collaborative but unsophisticated way of organizing information with user-created metadata is known as folksonomy, and such systems are sometimes also referred to as collaborative tagging systems. Besides simple tags, richer and flexible data structures could be used to describe a published item to provide more powerful expressiveness to the user. Some systems allow users to define their own attributes, and to describe their published objects with variable number of attributes. The freedom from a strict syntax of the published data items is very convenient to users. However, it is a challenging task to design efficient and scalable mechanisms to organize, classify, and index the data items with variable schemas and topics to facilitate searching.

There have also been recent efforts in creating community portals to provide relevant and available information for specific applications, and also systems for archiving information published in the web. These systems pose new challenges to the design of storage system and indexing structures that must be robust and scalable. In this talk, I shall speculate on some of these challenges and discuss some of our efforts in this area.

J.R. Haritsa, R. Kotagiri, and V. Pudi (Eds.): DASFAA 2008, LNCS 4947, p. 1, 2008.
© Springer-Verlag Berlin Heidelberg 2008

Self-tuning Database Systems: Past, Present and Future

Surajit Chaudhuri

Microsoft Research, USA
surajitc@microsoft.com

Although the goal of making database systems self-tuning remains elusive, there has been progress in this field in specific areas. We review some of these advances and also comment on directions that appear to be promising. In our view, the current architecture of the database systems makes building self-tuning database systems especially challenging. Therefore, it may be worth rethinking some of the key aspects of today's systems. Finally, we share our thoughts on opportunities and open issues, both in the context of self-tuning database systems, and the database field at large.

J.R. Haritsa, R. Kotagiri, and V. Pudi (Eds.): DASFAA 2008, LNCS 4947, p. 2, 2008.
© Springer-Verlag Berlin Heidelberg 2008

The Business of Managing Data: Implications for Research

Anand Deshpande

Persistent Systems, India
anand@pspl.co.in

Since 1990, Persistent Systems has been in the business of building systems to manage data. In this talk, based on my experience of working with companies building data management products, I propose to provide an overview of how the data business has evolved over the last two decades. I will attempt to connect research trends from database conferences during this period and explore the commercial impact of hot research topics. It would be interesting to explore why certain technologies were not successful in the commercial world and how certain technologies were commercially available before they became hot research topics. Through examples, I will try to chart the timeline of how technologies have moved from research projects to start-ups to products in the market.

J.R. Haritsa, R. Kotagiri, and V. Pudi (Eds.): DASFAA 2008, LNCS 4947, p. 3, 2008.
© Springer-Verlag Berlin Heidelberg 2008

Holistic Constraint-Preserving Transformation from Relational Schema into XML Schema

Rui Zhou, Chengfei Liu, and Jianxin Li

Faculty of Information and Communication Technologies
Swinburne University of Technology
Melbourne, VIC3122, Australia
{rzhou,cliu,jili}@ict.swin.edu.au

Abstract. In this paper, we propose a holistic scheme of transforming a relational schema into an XML Schema with integrity constraints preserved. This scheme facilitates constructing a schema for the published XML views of relational data. With this schema, users are able to issue qualified queries against XML views, and discover update anomalies in advance before propagating the view updates into relational database. Compared to the previous work which splits the transformation process into two steps, we establish a holistic solution to directly transform a relational schema into an XML Schema without building a reference graph. We achieve this by classifying the underlying relations in a more concise and effective way, and applying the converting rules wisely. The rules are also devised to be more compact and less complicated in contrast to those in our previous work. Finally, we manage to crack another hard nut which was seldom touched before, i.e. converting circularly referenced relations into recursive XML Schema.

1 Introduction

While XML has become the standard format for publishing and exchanging data on the Internet, most business data are still stored and maintained in relational database. One way to narrow the gap is to provide a virtual XML view[1] for relational data via an XML schema which can be transformed from the relational database schema [2]. Therefore, not only will users be able to issue qualified XML queries conforming to this schema, but invalid updates on the view can also be detected in advance by preserved constraints buried in the transformed schema without resorting to the underlying relational database. As a result, quality transformation from relational schema to XML schema is strongly in demand.

To clarify this schema conversion problem, we differentiate it from its well-studied counterpart named schema matching in data integration area by similar formalisms as in [3]. Given a source schema φ and a target schema Φ, the *schema matching* problem finds a "mapping" that relates elements in φ to ones in Φ. As to the *schema conversion* problem, only a source schema φ is given and the

---

[1] Part of the view may be materialized according to the literature [1].

J.R. Haritsa, R. Kotagiri, and V. Pudi (Eds.): DASFAA 2008, LNCS 4947, pp. 4–18, 2008.

goal is to find a target schema Φ that is equivalent to φ. Obviously, our aim is to investigate the schema conversion problem.

Previous renowned systems [1, 4] for publishing XML views from relational database laid more emphases on the structural aspect, but failed to recognize the importance of preserving constraints into the converted schema. This disadvantage turns evident when researchers focus on the XML view update problem [5, 6], i.e. whether the updates on XML views can be propagated into underlying relational database without violating buried constraints in relational data. Constraints-based transformation is first put forward in [7], where an algorithm called CoT utilized inclusion dependencies to nestedly structure the transformed XML schema. Such constraints can be acquired from database through ODBC/JDBC interface or provided by human experts who are familiar with the semantics of the relational schema. Liu et al. [8] advanced the transformation by taking into account more integrity constraints, such as primary keys, not-null and unique, and they employed XML Schema [9] to be the actual XML schema due to its powerful expressiveness. However, transformation process was split into two steps in [8] by firstly transforming each relation table into an XML Schema fragment and then restructuring the fragments in a nested way with a reference graph. The reference graph is used to capture reference relationships between relational tables, similar to IND-graph in [7].

In this paper, we propose a holistic scheme to transform a relational schema into an XML Schema. We achieve to directly produce a transformed XML Schema by evading the second step (restructuring the intermediate mapped XML Schema fragments). The basic idea is to devise a set of mapping rules to transform relational schema and integrity constraints into corresponding XML Schema expressions. And then to develop an algorithm to apply the rules in a wise way so that the target XML Schema can be incrementally constructed without any element being restructured or repositioned once the element is mapped into the target schema. Compared with the previous work [8], we highlight the contributions of this paper as follows:

- We introduce a more clear and concise classification of relational tables, and show that the classification serves as a base for holistic transformation algorithm.
- We achieve the conversion with less number of rules and the rules are less complicated as well.
- We devise a holistic schema transformation algorithm with the set of rules.
- Circularly referenced relations are converted into recursive XML Schema, which is seldom touched in previous work.

The rest of the paper is organized as follows. In Section 2, we will give out an overview about XML Schema, the advantage of XML Schema over DTD and five criteria of obtaining a quality transformed schema. In Section 3, after laying a foundation of our holistic transformation scheme by introducing a more concise, complete and effective categorization of relations, we will deliberately address a set of rules for converting relational schema into XML Schema. The holistic algorithm will be followed in Section 4 to illustrate how to combine the rules together, and moreover to utilize them in an efficient way. Related work is given in Section 5. Finally, we draw a conclusion and propose some future work in Section 6.

2   XML Schema

As a W3C XML Language, XML Schema [9] defines the structure, content and semantics of XML documents. Contrast with its counterpart DTD, XML Schema is superior in that (1) XML Schema supports powerful data typing with a rich set of built-in data types. Users are allowed to derive their own simple types by restriction and complex types by both restriction and extension. While in DTD, users are restrained from defining their own types, and are provided with only a limited number of built-in types for describing attributes. (2) XML Schema provides comprehensive support for representing integrity constraints such as id/idref, key/keyref, unique, fine grained cardinalities, etc. While DTD only provides limited support, such as id/idref. (3) Apart from the sequence and selection compositors for group elements, XML Schema also supports other compositors such as set. (4) XML Schema has the same syntax as XML, which allows the schema itself be processed by the same tools that manage the XML documents it describes. (5) Namespaces are well supported in XML Schema, but not in DTD. Although DTD is still used for simple application, XML Schema has becoming a dominant XML schema language and a W3C Recommendation.

In order to preserve the constraints from relational schema into XML schema, it is obviously advantageous to adopt XML Schema as the target schema rather than DTD. The expressiveness of XML Schema will be observed in Section 3 where we detailedly discuss how to transform a relational schema into an XML schema.

Before starting discussion about transformation strategy, it is crucial to clarify several significant criteria in designing a quality target schema. The criteria are also proposed in our previous work [10].

- Information preservation - it is fundamental that the target XML schema should entirely preserve structural and semantic information of the application.
- Highly nested structure - nesting is important in XML documents because it allows navigation of the paths in the document tree to be processed efficiently.
- No redundancy - there is no data redundancy in the XML views that conform to the target XML schema, and thus no inconsistency will be introduced while updating the XML views.
- Consideration of dominant associations - the structure of XML document should be accommodated with dominant associations so that queries can be processed efficiently.
- Reversibility of design - the original design can be achieved from the target XML schema, which is fundamentally important to data integration.

Our goal is to devise a mapping scheme to transform a relational schema to an XML schema while considering all the above criteria.

3   Mapping Scheme and Rules

In this section, we will deliberately address the mapping scheme. Firstly, we classify the relations into three categories, and show the new classification is more clear, concise and effective than the one in our previous work [8]. Then we introduce how to utilize XML Schema to describe multi-attribute values in relational schema. Finally, as the most significant part in this paper, we generate a set of rules to convert a relational schema into an XML schema while preserving all the integrity constraints, such as primary keys, foreign keys, null/not-null, unique characteristics.

3.1   Categorizing Relations

Before introducing our new categorization, we briefly review the previous one [8], in which all the relations are classified into four categories based on different types of primary keys. (1) *regular relation*: A regular relation is a relation where the primary key contains no foreign keys. (2) *component relation*: A component relation is a relation where the primary key contains one foreign key. This foreign key references another relation which we call the parent relation of the component relation. The other part of the primary key serves as a local identifier under the parent relation. The component relation is used to represent a component or a multivalued attribute of its parent relation. (3) *supplementary relation*: A supplementary relation is a relation where the whole primary key is also a foreign key which references another relation. The supplementary relation is used to supplement another relation or to represent a subclass for transforming a generalization hierarchy from a conceptual schema. (4) *association relation*: An association relation is a relation where the primary key contains more than one foreign key, each of which references a participant relation of the association.

In this paper, we classify relations into three categories according to the number of foreign keys in the relation tables. (1) *base relation*: a relation with no foreign keys; (2) *single-related relation*: a relation with only one foreign key; (3) *multi-related relation*: a relation with more than one foreign keys. Similar to previous work, we name the referenced relation of a single-related relation as parent relation, the referenced relations of a multi-related relation as participant relations respectively. The new classification is superior than the previous one in that:

- The current classification is more clear, concise without taking into account the containment relationship between primary keys and foreign keys.
- The three classified categories are complete and disjoint, which means any relation could fall into one and only one category. While in the previous classification, supplementary relation belongs to component relation according

to our new definition. Conceptually speaking, component relation and supplementary relation reflect different kinds of relationship between entities. However, on a mapping basis, we found they could be equally treated when transformed into XML Schema, because both of them can be nested under their referenced relation.

– With the help of the new classification, we are able to devise a holistic schema mapping strategy to produce a nested XML Schema structure by treating the three types of relations differently. This is a big improvement, since we used to separate the transforming process into two steps by firstly mapping the relations into XML Schema fragments and then restructuring the fragments to be as nestable as possible.

We assume the source relations are already in Boyce-Codd normal form (BCNF). This guarantees that no redundancy exists in the relational schema. The assumption is reasonable, since most relational schemas are well-designed in commercial relational database. At least, we can normalize the flat relation into BCNF. It is necessary for the source schema to be redundancy-free when we want to convert it into a redundancy-free target schema.

3.2 Expressing Multi-attribute Values

Keys can identify entity instance in relations. It associates with several kinds of integrity constraints and is an important component in relational schema. We aim to map the keys into XML Schema so that no related constraints will be lost. Before this step, we firstly begin with how to express multi-valued attribute.

In a relation, a key may be single-valued or multi-valued. We regard this property to be orthogonal to the types of primary/foreign keys, say both primary key and foreign key can be single-valued or multi-valued. Previous work [8] enumerated the combinations of them when conducting rules to transform a relational schema into an XML Schema. However, we find the complicated enumeration can be evaded by separating multi-valued mapping from primary/foreign key mapping. And this separation will ultimately produce more concise rules than our previous work.

XML Schema supports two mechanisms to represent identity and reference, id/idref and key/keyref. The former supports dereference function in path expressions in most XML query language, such as XQuery. However, it suffers two fatal disadvantages. (1) It could only be applied to a single element/attribute. (2) It also has a problem in precisely representing a reference, i.e. no restriction is given to protect an idref element/attribute from referencing unexpected element. The latter accommodates with reference to multiple elements/attributes, and hence supports mapping of multi-attribute keys. Although, there are still no support on dereference for key/keyref in XQuery. We believe the XQuery draft can be extended to achieve it just like the addition of XQuery Update [11] to support *transform queries* [12]. Therefore, we adopt key/keyref to map multi-attribute keys into XML Schema.

```
<xsd:selector xpath="//R" />
<xsd:field xpath="@A1" />
... ...
<xsd:field xpath="@An" />
```

Let a set of attribute in relation R be A_1, A_2, ..., A_n $(1 \leq n)$, we use a selector and n fields to map multi-attribute value as a whole.

The above structure will be frequently used in the following part of the paper to describe constraints residing in relational schema including primary key, foreign key and unique property. There are slight amendments for different constraints.

3.3 Constraint-Preserving Transformation Rules

In this section, we will comprehensively give out the transformation rules. We will firstly discuss how to map three categories of relations into XML Schema. Dominate application is considered when processing multi-related relations. Furthermore, with the goal of preserving constraints, we go on to map primary keys and foreign keys into XML Schema expressions. Finally, other constraints, such as unique, null/not-null, are also taken into account to make the mapping complete.

Mapping relations. We assume the relational schema is in BCNF and has been normalized into a set of relations, and thus we build specific mapping rules for each type of relations introduced in Section 3.1.

Rule 1. Schema Root - For a relational schema φ, a root element named *schemaXML* is created correspondingly for the target XML Schema.

```
<?xml version="1.0"?>
<xsd:schema xmlns:xsd="http://www.w3.org/2001/XMLSchema"
    targetNamespace="targetNamespaceURI"
    xmlns="targetNamespaceURI" elementFormDefault="qualified">
  <xsd:element name="schemaXML">
    <xsd:complexType>
      <xsd:sequence>
        <!--mapped relational schema is here -->
      <xsd:sequence>
    </xsd:complexType>
  </xsd:element>
</xsd>
```

This rule is required for converting any relational schema. A relational schema may consist of a number of relations while an XML Schema is only one single document. Since XML document including XML Schema conforms to a tree structure, it is suitable to pick the root element to take the name of the corresponding relational schema.

Rule 2. Base Relation - For each base relation named R in relational schema φ, a corresponding element named R is created as a child of the root *schemaXML* element.

```
<xsd:element name="R">
  <xsd:complexType>
    <!-- details of attributes in R -->
  </xsd:complexType>
</xsd:element>
```

Rule 3. Single-related Relation - For a single-related relation named R in relational schema φ, let the relation R referenced be R', an element with the name R will be created and placed as a child of R' (no matter which type of relation R' is).

As to Rule 2, compared to our previous work [8], we place base relations rather than regular relations as direct children of the root element, because base relations don't contain foreign keys, while regular relations may, as long as the foreign keys are not included in their primary keys. In Rule 3, for the nesting purpose, it is proper to put single-related relations (i.e. relations with only one foreign key) under their referenced relations so that the foreign key constraints can be implicitly preserved. XML Schema can also be built in a hierarchical manner by this means.

Before giving Rule 4, we introduce the concept of dominant association. We borrow a similar idea from [10].

Definition 1. Let R be a multi-related relation, $R_1, R_2, ..., R_n$ be its participant relations. A dominant association is a pair (R, R_i) $(i \in [1, n])$, where R_i is the most frequently queried relation together with R among all the participant relations. We call R_i the *dominant participant relation* of R and all the other participant relations *ordinary participant relations*.

In real application, dominant associations can be identified by query statistics. On a special case, if more than one participant relations are qualified to be the dominant participant relation, we randomly choose one as the result.

Rule 4. Multi-related Relation - For a multi-related relation named R in relational schema φ, let its dominant participant relation be R'. Similar to Rule 3, an element with the name R will be created and placed as a child of R'.

We choose to nest a multi-related relation under its dominant participant relation in the transformed schema, because by this means navigation of path expressions could be processed efficiently, since the multi-related relation seems to be most relevant with the dominant participant relation.

Rule 5. Attribute - For each attribute A in relation R, let the name and type of the attribute be $N(A)$ and $T(A)$, an "attribute" element is created under the corresponding element of R in XML Schema taking the name value $N(A)$ and type value $T(A)$.

```
<xsd:attribute name="N(A)" type="xsd:T(A)" />
```

Mapping Keys. Keys are the most important integrity constraints in relational database. Unlike previous work, we separate mapping keys apart from mapping relations to produce more compact rules. Keys are expressed identically regardless of relation type (base, single-related, multi-related) or key cardinality (single valued, multiple valued).

Rule 6. Primary Key - Let the primary key of a relation denoted as R be $(PK_1, PK_2, ..., PK_n)$, where $1 \leq n$. A "key" element is created for relation R with a selector to select the XPath of relation R and several fields to identify attributes $PK_1, PK_2, ..., PK_n$. The name of the "key" element should be unique within the namespace.

```
<xsd:key name="R_PK">
  <xsd:selector xpath="//R" />
  <xsd:field xpath="@PK1" />
  ... ...
  <xsd:field xpath="@PKn" />
</xsd:key>
```

Primary keys are mapped into XML Schema accompanied with every relation, while foreign keys are only required for multi-related relations in the mapping process, because foreign key constraint has been captured in parent/child relationship in XML Schema when placing a relation under its parent relation (for single-related relation) or dominant participant relation (for multi-related relation).

Rule 7. Foreign Key - Let the ordinary foreign keys of relation R come from a set of relations $R_1, R_2, ..., R_m$ $(1 \leq m)$. To map these foreign keys, a "keyref" element is created to capture the reference relationship from each ordinary participant relation of R. Without loss of generality, we take R_1 as an example to illustrate the foreign key mapping result. Assume the foreign key in R came from R_1 is $(FK_1, FK_2, ..., FK_n)$ $(1 \leq n)$, then we have the following mapping fragment in XML Schema. Mapping foreign keys from other relations is in a similar way.

```
<xsd:keyref name="R1_FK" refer="R1_PK">
  <xsd:selector xpath="//R">
  <xsd:field xpath="@FK1" />
  ... ...
  <xsd:field xpath="@FKn" />
</xsd:keyref>
```

Mapping Other Constraints. In addition to integrity constraints on keys, we further map two other constraints – not-null and unique – into XML Schema. We aim to preserve as many constraints as possible.

Rule 8. Null and Not-Null - For each attribute A with not-null constraints, we add attribute declaration *used= "required"* into the mapped attribute element, otherwise we add *used= "optional"*.

```
<xsd:attribute name="N(A)" type="xsd:T(A)" used="required" />
```

Rule 9. Unique - For each unique constraint defined on attributes $A_1, A_2, ..., A_n$ in relation R, a "unique" element is created with a selector to select the element for R and several fields to identify $A_1, A_2, ..., A_n$. The name of the "unique" element should be unique within the namespace.

```
<xsd:unique name="UniqueR">
  <xsd:selector xpath="//R">
  <xsd:field xpath="@A1" />
  ... ...
  <xsd:field xpath="@An" />
</xsd:unique >
```

4   Holistic Transformation Algorithm

In this section, we will introduce our holistic transformation algorithm. Previous work transformed the relational schema into XML Schema in two steps. In the first step, relations are converted into an intermediate XML Schema, then the schema will be reconstructed into the ultimate XML Schema. Moreover, in the second step, we need to build a IND-graph [3] or a reference graph [8] from relational schema to capture the reference relationships between different relations in the relational schema in order to assemble the mapped XML Schema fragments into a reasonable tree pattern. However, in the novel holistic transforming solution, we achieved to evade the second step, and directly build the XML Schema in a nested manner. The suitable, complete and concise categorization of relations in Section 3.1 serves as the foundation for our algorithm. After resolving circularly referenced relations, we will present our holistic transformation algorithm and give a transformation example.

4.1   Resolving Circularly Referenced Relations

Recursive characteristic of XML data is more expressive in some situation, such as nested "subsection" elements in an XML document about a book. It simplifies schema definition, but brings in too much complexity in many other aspects, eg. evaluating XPath queries, schema matching etc. Researchers often regard it as a hard nut, and are unwilling to touch it. However, for certain scenarios, particularly speaking in our problem, if a set of relational tables are circularly referenced, it will be proper to adopt the recursive nature of XML Schema to capture the circular relationship. We divide circularly referenced relations into two types.

For the first type, *self-referenced relation* is defined as a relation, which has a foreign key referencing itself. One typical example is that a course relation may have an attribute called prerequisite course referencing itself.

Rule 10. Self-referenced Relation - Given a self-referenced relation R, after modifying the self-referenced attribute to be a non-key attribute, R turns out to be a *normal relation* (one of base, single related, multi-related relations). Rules 2 - 4 can be applied accordingly. Finally, the following is added into the mapped schema element to restore the self-reference relationship.

```
<xsd:element name="R" type="T(R)" />
```

The other type of circularly referenced relations is *relation circle*, defined as a set of relations R_1, R_2, ..., R_n $(2 \leq n)$, where R_i is referenced by R_{i+1} $(1 \leq i \leq n)$

and R_n is referenced by R_1. Before addressing the mapping rule, we introduce a definition about dominant relation.

Definition 2. Let R_1, R_2, ..., R_n ($2 \leq n$) be a relation circle, the *dominant relation* is defined as a relation, which is most frequently queried among all the relations in the relation circle.

Rule 11. Relation Circle - Without loss of generality, let R_1 be the dominant relation of relation circle R_1, R_2, ..., R_n, break the reference from R_1 to R_n and map the relation set one by one with Rule 3, finally add the following element as a child element to R_n.

```
<xsd:element name="R1" type="T(R1)"/>
```

The aim of adopting dominant relation is similar as dominant participant relation in Section 3.3. The converted XML Schema is supposed to support applications efficiently.

4.2   Transformation Algorithm

Since a number of rules have been given in the previous section, another problem arises as how to apply the rules in an efficient manner in the transforming process. The sequence of mapping different relations results in different performance. For example, intuitively a base relation should be mapped before a single-related relation, otherwise the mapping of a single-relation may be blocked and postponed until its parent relation (which may probably be base relation) is mapped into the XML Schema. The above criteria is captured in our algorithm.

The algorithm is given in Algorithm 1, in which we aim to incrementally transform relational tables into XML Schema. The core idea lies: After a relation r is mapped into the XML Schema, single-related relations referencing r could be transformed next, so could a multi-related relation whose dominant participant relation is r. Furthermore, after these relations are added into XML Schema. r will become useless in further transformation process, since all the relations r could capture have been included in the XML Schema. We don't need to consider r any more.

In the algorithm, we first perform an initialization by marking dominant relations for relation circles and dominant participant relations for multi-related relations. After breaking the circles for circularly referenced relations, we are able to perform the rules devised in Section 3.3. ΔS is used to accommodate useful relations which may further induce other relations by its primary key. After mapping base relations line 5-9, we obtain a initial ΔS and the to-be-mapped tables will reduce to $S - \Delta S$. Line 10-22 is the core part of this holistic incremental algorithm. The meaning has been explained in the previous paragraph. Rule 5-6, 8-9 could be well employed together with Rule 2, 3 and 4. And there is also a little trick for Rule 7. Assume a multi-related relation is to be transformed into XML Schema, while not every its participant relation is already in the XML Schema (although the dominant one is), in such a case, foreign keys cannot be mapped into the target XML Schema until every participant relation has been transformed. Luckily, we can use a callback function to resolve the problem.

Algorithm 1. Holistic Transformation Algorithm

Input: A relational schema in terms of a set of relations S
Output: A transformed XML Schema with constraints preserved T

1: InitMarkDominant(S);
   {marking dominant relations for relation circles, dominant participant relations for multi-related relations}
2: BreakCircle(S);
   {using Rule 10 and 11 to break circles for circularly referenced relations}
3: CreateRoot();
   {apply Rule 1 to create the root element for empty XML Schema T}
4: $\Delta S = \phi$;
   {initialize a empty set}
5: **for all** base relation $r_i \in$ relational schema S **do**
6:    apply Rule 2, 5-9 to create a child element of XML Schema root;
7:    $\Delta S = \Delta S \cup \{r_i\}$;
8:    $S = S - \{r_i\}$;
9: **end for**
10: **for all** relation $r_i \in \Delta S$ **do**
11:    **for all** single-related relation $s_i \in S$ referencing r_i **do**
12:       apply Rule 3, 5-9 to create an element in XML Schema;
13:       $\Delta S = \Delta S \cup \{s_i\}$;
14:       $S = S - \{s_i\}$;
15:    **end for**
16:    **for all** multi-related relation $t_i \in S$ referencing r_i with dominant participant relation **do**
17:       apply Rule 4, 5-9 to create an element in XML Schema;
18:       $\Delta S = \Delta S \cup \{t_i\}$;
19:       $S = S - \{t_i\}$;
20:    **end for**
21:    $\Delta S = \Delta S - \{r_i\}$;
22: **end for**
23: return T;

4.3   Transformation Example

We now give an example to illustrate transformation from relational schema into XML Schema. We adopt the following relational schema, where primary keys are underlined, foreign keys are *italic*, a not-null attribute is followed by "/N".

> student(<u>sno</u>, sname/N)
> course(<u>cno</u>, cname/N, *preqcno*, *dno*)
> dept(<u>dno</u>, dname/N)
> takeCourse(*<u>sno</u>*, *<u>cno</u>*, score)

In this relational schema, several features of relational schema are reflected. *preqcno* is the prerequisite course standing for self-referenced relationship. "student" and "dept" are two base relations, "takeCourse" is a multi-related relation and "course" can be treated as a single-related relation after removing self-reference relationship. The ultimate transformation result is shown in

```
<?xml version="1.0"?>
<xsd:schema xmlns:xsd="http://www.w3.org/2001/XMLSchema"
    targetNamespace="targetNamespaceURI"
    xmlns="targetNamespaceURI" elementFormDefault="qualified">
  <xsd:element name="schemaXML">
    <xsd:complexType>
      <xsd:sequence>
        <xsd:element name="student">
          <xsd:complexType>
            <xsd:attribute name="sno" type="xsd:string"/>
            <xsd:attribute name="sname" type="xsd:string" used="required"/>
            <xsd:element name="takeCourse">
              <xsd:complexType>
                <xsd:attribute name="cno" type="xsd:string"/>
                <xsd:attribute name="score" type="xsd:integer" used="optional"/>
              </xsd:complexType>
            </xsd:element>
          </xsd:complexType>
        </xsd:element>
        <xsd:element name="dept">
          <xsd:complexType>
            <xsd:attribute name="dno" type="xsd:string"/>
            <xsd:attribute name="dname" type="xsd:string" used="required"/>
            <xsd:element name="course" >
              <xsd:complexType type="courseType">
                <xsd:attribute name="cno" type="xsd:string" />
                <xsd:attribute name="cname" type="xsd:string" used="required"/>
                <xsd:element name="course" type="courseType" />
              </xsd:complexType>
            </xsd:element>
          </xsd:complexType>
        </xsd:element>
      </xsd:sequence>
    </xsd:complexType>
  </xsd:element>
  <xsd:key name="student_PK">
    <xsd:selector xpath="//student" /><xsd:field xpath="@sno" />
  </xsd:key>
  <xsd:key name="course_PK">
    <xsd:selector xpath="//course" /><xsd:field xpath="@cno" />
  </xsd:key>
  <xsd:key name="dept_PK">
    <xsd:selector xpath="//dept" /><xsd:field xpath="@dno" />
  </xsd:key>
  <xsd:key name="takeCourse_PK">
    <xsd:selector xpath="//takeCourse" />
    <xsd:field xpath="../@sno" />
    <xsd:field xpath="@cno" />
  </xsd:key>
  <xsd:keyref name="takeCourse_FK" refer="course_PK">
    <xsd:selector xpath="//course"><xsd:field xpath="@cno" />
  </xsd:keyref>
</xsd>
```

Fig. 1. The Target XML Schema

Fig. 1. Recursive reference is firstly broken, and then base relations are transformed into XML Schema. Then "course" and "takeCourse" are converted in a following way. In this example, "student" is considered to be the dominant relation for "takeCourse".

5 Related Work

Publishing XML views have been extensively studied during the past few years and several systems are built. SilkRoute [1] defines XML views through a relational-to-XML transformation language called RXL. Issued XML queries are combined with the views by a query composer and the combined RXL queries are then translated into the corresponding SQL queries. XPERANTO [4] takes a similar approach, but focused more on constructing XML views inside the relational engine. Unfortunately, for both systems, users are not able to find integrity constraints buried in relational schema from the published XML views. On the contrary, users should be aware of the constraints in the XML schema against which they are going to issue queries. Benedikt et al. [13] proposed a DTD-directed publishing scheme base on *attribute translation grammars* (ATGs). An ATG extends a DTD by associating semantic rules via SQL queries. With the extended DTD, data are extracted from relational database and reconstructed into an XML document. The difference between their work and ours is that they leverage a predefined domain DTD (as XML view schema) which is universally recognized, while we aim to transform the relational schema into a new XML Schema.

There are also several works on schema transformation. An early work in transforming relational schema to XML schema is DB2XML [14]. DB2XML uses a simple algorithm to map flat relational model to flat XML model in almost one-to-one manner. DTD is used as the target XML schema. Followed works [15,16] mostly focus on structural aspect and haven't taken deep thought about preserving constraints. Lee at al. [7] presented two algorithms NeT and CoT. NeT derives nested structures from flat relations by repeatedly applying the nest operator on tuples of each relation. The resulting nested structures may be useless since the derivation is not at the type level. CoT considers inclusion dependencies as constraints to generate a more intuitive XML Schema. XViews [17] constructs a graph based on primary key/foreign key relationship and generates candidate views by choosing the node with either maximum in-degree or zero in-degree as the root element. The candidate XML views generated may be highly nested. However, all of the above approaches fail to fully consider the preservation of integrity constraints, and some of them also suffer from certain level of data redundancy. The most similar work attempted to achieve the same goal as this paper is [8], where mapping process is divided into two steps. The classified relations and derived rules are also not as compact as ours. Furthermore, by using holistic constraint-preserving mapping, we simplified the problem without sacrificing efficiency. Schema transformation is also studied in data integration area, such as X-Ray [18].

6 Conclusions and Future Work

In this paper, we have addressed a holistic scheme of transforming a relational schema into an XML Schema with integrity constraints preserved. This scheme facilitates constructing a schema for the published XML views of relational data. With this schema, users will be able to issue qualified queries against XML views, and moreover update anomalies can also be detected in advance before propagating the updates into relational database. Compared to the previous work which splits the transformation process into two steps, we established a holistic solution to directly transform a relational schema into an XML Schema without building a reference graph. We achieved this by classifying the underlying relations in a more concise and effective way, and applying the mapping rules wisely. The rules are also refined to be more compact and less complicated in contrast to those previous ones.

A promising future work will be to make the transformed XML Schema more robust. Specifically speaking, current XML Schema is intended for dominant application, because we have nested a multi-related relation under its dominant participant relation. It may be inefficient to validate a XPath query which reaches a multi-related relation via an ordinary participant relation. Furthermore, for circularly referenced relations, the dominant relation is used to decide the root of the XML schema, this approach would work for a single chain of circularly referenced relations, while it is not clear how this approach could be extended for more complex relational schema. The last significant future work is to build a system and evaluate the algorithm performance in real applications.

Acknowledgments. This work was supported by the Australian Research Council Discovery Project under the grant number DP0878405.

References

1. Fernandez, M.F., Kadiyska, Y., Suciu, D., Morishima, A., Tan, W.C.: Silkroute: A framework for publishing relational data in XML. ACM Trans. Database Syst. 27(4), 438–493 (2002)
2. Liu, C., Vincent, M.W., Liu, J., Guo, M.: A virtual XML database engine for relational databases. In: Bellahsène, Z., Chaudhri, A.B., Rahm, E., Rys, M., Unland, R. (eds.) XSym 2003. LNCS, vol. 2824, pp. 37–51. Springer, Heidelberg (2003)
3. Lee, D., Mani, M., Chu, W.W.: Schema conversion methods between XML and relational models. In: Knowledge Transformation for the Semantic Web, pp. 1–17 (2003)
4. Shanmugasundaram, J., Shekita, E.J., Barr, R., Carey, M.J., Lindsay, B.G., Pirahesh, H., Reinwald, B.: Efficiently publishing relational data as XML documents. VLDB J. 10(2-3), 133–154 (2001)
5. Wang, L., Rundensteiner, E.A., Mani, M.: Updating XML views published over relational databases: Towards the existence of a correct update mapping. Data Knowl. Eng. 58(3), 263–298 (2006)
6. Braganholo, V.P., Davidson, S.B., Heuser, C.A.: Pataxo: A framework to allow updates through XMLviews. ACM Trans. Database Syst. 31(3), 839–886 (2006)

7. Lee, D., Mani, M., Chiu, F., Chu, W.W.: Net & cot: translating relational schemas to XML schemas using semantic constraints. In: CIKM, pp. 282–291. ACM, New York (2002)
8. Liu, C., Vincent, M.W., Liu, J.: Constraint preserving transformation from relational schema to XML schema. World Wide Web 9(1), 93–110 (2006)
9. Consortium, W.W.W.: Xml schema part 0, 1, 2. In: W3C Candidate Recommendation, http://www.w3.org/XML/schema/
10. Liu, C., Li, J.: Designing quality xml schemas from e-r diagrams. In: Yu, J.X., Kitsuregawa, M., Leong, H.-V. (eds.) WAIM 2006. LNCS, vol. 4016, pp. 508–519. Springer, Heidelberg (2006)
11. Chamberlin, D., Florescu, D., Robie, J.: Xquery update. In: W3C working draft (July 2006), http://www.w3.org/TR/xqupdate/
12. Fan, W., Cong, G., Bohannon, P.: Querying xml with update syntax. In: Chan, C.Y., Ooi, B.C., Zhou, A. (eds.) SIGMOD Conference, pp. 293–304. ACM, New York (2007)
13. Benedikt, M., Chan, C.Y., Fan, W., Rastogi, R., Zheng, S., Zhou, A.: Dtd-directed publishing with attribute translation grammars. In: VLDB, pp. 838–849. Morgan Kaufmann, San Francisco (2002)
14. Turau, V.: Making Legacy Data Accessible for XML Applications (1999), http://www.informatik.fh-wiesbaden.de/~turau/veroeff.html
15. Deutsch, A., Fernández, M.F., Suciu, D.: Storing semistructured data with stored. In: Delis, A., Faloutsos, C., Ghandeharizadeh, S. (eds.) SIGMOD 1999, Proceedings ACM SIGMOD International Conference on Management of Data, June 1-3, 1999, pp. 431–442. ACM Press, USA (1999)
16. Tatarinov, I., Viglas, S.D., Beyer, K., Shanmugasundaram, J., Shekita, E., Zhang, C.: Storing and querying ordered xml using a relational database system. In: SIGMOD 2002: Proceedings of the 2002 ACM SIGMOD international conference on Management of data, pp. 204–215. ACM Press, New York (2002)
17. Baru, C.K.: XViews: XML views of relational schemas. In: DEXA Workshop, pp. 700–705 (1999)
18. Kappel, G., Kapsammer, E., Retschitzegger, W.: Integrating XML and relational database systems. World Wide Web 7(4), 343–384 (2004)

An Optimized Two-Step Solution for Updating XML Views

Ling Wang, Ming Jiang, Elke A. Rundensteiner, and Murali Mani

Worcester Polytechnic Institute, Worcester, MA 01609, USA

Abstract. View updating is a long standing difficult problem. Given a view defined over base data sources and a view update, there are several different updates
over the base data sources, called *translations*, that perform the update. A translation is said to be *correct* if it performs the update and at the same time does not
update any portion of the view not specified in the update (no *view side-effects*).
The view update problem finds a correct translation for a given view update if
one exists. In the relational scenario, previous research has attempted to study
the view update problem either by utilizing only the schema knowledge, or by
directly examining the base data. While utilizing only the schema knowledge is
very efficient, we are not guaranteed to find a correct translation even if one exists. On the other hand, examining the base data is guaranteed to find a correct
translation if one exists, but is very time-consuming. The view update problem is
even more complex in the XML context due to the nested hierarchical structure
of XML and the restructuring capabilities of the XQUERY view specification. In
this paper we propose a schema-centric framework, named HUX, for efficiently
updating XML views specified over relational databases. HUX is complete (always finds a correct translation if one exists) and is efficient. The efficiency of
HUX is achieved as follows. Given a view update, HUX first exploits the schema
to determine whether there will never be a correct translation, or there will always
be a correct translation. Only if the update cannot be classified using the schema,
HUX will examine the base data to determine if there is a correct translation.
This data-level checking is further optimized in HUX, by exploiting the schema
knowledge extracted in the first step to significantly prune the space of potential
translations that is explored. Experiments illustrate the performance benefits of
HUX over previous solutions.

1 Introduction

Both XML-relational systems [6,20] and native XML systems [14] support creating
XML wrapper views and querying against them. However, update operations against
such virtual XML views are not yet supported in most cases. While several research
projects [4,22] began to explore this XML view updating problem, they typically pick
one of the translations, even if it is not correct (in other words, they might pick a translation that causes view side-effects). Allowing side-effects to occur is not practical in
most applications. The user would like the updated view to reflect changes as expected,
instead of bearing additional changes in the view.

One approach to solve this problem is to determine the view side-effects caused by
each translation. Here, for each translation, compare the view before the update and

J.R. Haritsa, R. Kotagiri, and V. Pudi (Eds.): DASFAA 2008, LNCS 4947, pp. 19–34, 2008.
© Springer-Verlag Berlin Heidelberg 2008

the view after the update as in [21]. If a side-effect is detected, then the translation is rejected. However, this could be very expensive, as the space of possible translations could be very large, and also computing the before and after image for each translation could also be very time consuming.

Some other approaches such as [16] allow the user to determine if the side-effects caused by a translation are acceptable. In other words, the users participate in setting rules for pruning down the candidate translations. However, in most cases users do not have such knowledge, and simply want the view updated exactly as expected. This is the same scenario that we assume in this paper. Therefore we employ the strategy that a translation will be directly rejected if any view side-effects are detected.

To identify the space of possible translations, the concepts of provenance [5] and lineage [11] are used, where lineage refers to descriptions of the origins of each piece of data in a view. In [11], the authors exhaustively examine every possible translation obtained from the lineage to determine if the translation causes any view side-effects. This is done by examining the actual instance data. This data-centric approach can be quite time consuming as shown in [5].

In stead of a data-centric approach, the schema of the base and the view could be used to determine a correct translation for a view update, if one exists. Such schema centric approaches are proposed for Select-Project-Join views in the relational context in [13,15]. In [4], the authors map an XML view to a set of relational views and therefore transform XML view update problems back to relational view update problems. Such a transformation, however, is not sufficient to detect all the possible view side-effects, as we will discuss in the following two sections. A pure schema-based approach is efficient, but rather restrictive. The disadvantage of a pure schema-based approach is that, for many view update cases, it is not possible to determine the translatability (whether there exists a correct translation) by only examining the schema.

Now, with the following three examples, let us examine how the above data-based or schema-based approach can be used in XML view updates scenario. In these examples, a view update can be classified as **translatable**, which indicates there exist a side-effect free translation, or otherwise **untranslatable** using either schema or data knowledge. Also we assume that when we delete an element in the XML view, we will delete all its descendant elements as well.

1.1 Motivating Examples

Fig. 1(a) shows a running example of a relational database for a course registration system. An XML view in Fig. 1(c) is defined by the *view query* in Fig. 1(b). The following examples illustrate cases of classifying updates as translatable or untranslatable. XML update language from [22] or update primitives from [4] can be used to define update operations. For simplicity, in the examples below we only use a delete primitive with the format *(delete* nodeID*)*, where *nodeID* is the abbreviated identifier of the element to be deleted[1]. For example, C1 is the first *Course* element, P1.TA1 is the first *TA* element of the first *Professor* element. We use *Professor.t*$_1$ to indicate the first tuple of base relation *Professor*.

---

[1] Note that the view here is still *virtual*. In reality, this nodeID is achieved by specifying conditions in the update query.

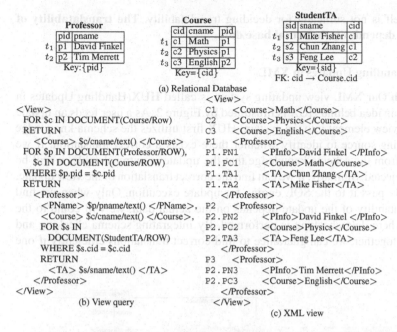

(a) Relational Database

<View>
FOR $c IN DOCUMENT(Course/Row)
RETURN
    <Course> $c/cname/text() </Course>
FOR $p IN DOCUMENT(Professor/ROW),
    $c IN DOCUMENT(Course/ROW)
WHERE $p.pid = $c.pid
RETURN
    <Professor>
        <PName> $p/pname/text() </PName>,
        <Course> $c/cname/text() </Course>,
    FOR $s IN
        DOCUMENT(StudentTA/ROW)
    WHERE $s.cid = $c.cid
    RETURN
        <TA> $s/sname/text() </TA>
    </Professor>
</View>

(b) View query

```
<View>
C1      <Course>Math</Course>
C2      <Course>Physics</Course>
C3      <Course>English</Course>
P1      <Professor>
P1.PN1      <PInfo>David Finkel </PInfo>
P1.PC1      <Course>Math</Course>
P1.TA1      <TA>Chun Zhang</TA>
P1.TA2      <TA>Mike Fisher</TA>
        </Professor>
P2      <Professor>
P2.PN2      <PInfo>David Finkel </PInfo>
P2.PC2      <Course>Physics</Course>
P2.TA3      <TA>Feng Lee</TA>
        </Professor>
P3      <Professor>
P2.PN3      <PInfo>Tim Merrett</PInfo>
P2.PC3      <Course>English</Course>
        </Professor>
</View>
```

(c) XML view

Fig. 1. Running Example of the Course Registration System

Example 1. Update u_1 ={delete P1.TA1} over the XML view in Fig. 1(c) deletes the student "Chun Zhang". We can delete *StudentTA.t_2* to achieve this without causing any view side-effect. We can determine that any TA view element can be deleted by deleting the corresponding tuple in the studentTA relation, by utilizing only the schema knowledge (as we will study in Section 4). **The schema knowledge is sufficient to determine that this update is translatable.**

Example 2. Consider the update u_2 = {delete P2.PC2}.

The appearance of the view element P2.PC2 is determined by two tuples: *Professor.t_1* and *Course.t_2*. There are three translations for this view update: T_1={delete *Professor.t_1*}, T_2={delete *Course.t_2*} and T_3={delete *Professor.t_1*, delete *Course.t_2*}. All three translations are incorrect and cause view side-effects - translation T1 will delete the parent Professor view element (P2), translation T2 will delete one of the Course view elements (C2), translation T3 will delete both P2 and C2. We can determine that there is no correct translation for deleting any Course view element by utilizing only the schema knowledge (studied in Section 4). **The schema knowledge is sufficient to determine that this update is untranslatable.**

Example 3. Consider the update u_3 = {delete P3}. This can be achieved by deleting *Professor.t_2*}. Let us consider the update u_4 = {delete P2}. The appearance of the view element P2 is determined by two tuples: *Professor.t_1* and *Course.t_2*. However, we cannot find any correct translation for this view update: if we delete *Course.t_2*, view element $C2$ will also get deleted; if we delete *Professor.t_1*, view element $P1$ will also get deleted. The difference between u_3 and u_4 indicates that sometimes the schema knowledge itself is not sufficient for deciding translatability. **The translatability of these updates depends on the actual base data.**

1.2 HUX: Handling Updates in XML

Our Approach Our XML view updating system is called **HUX**(Handling Updates in XML). The main idea behind HUX is illustrated by Figure 2. As a user feeds an update over an XML view element into the system, HUX first utilizes the schema knowledge of the underlying source to identify whether there exists a correct translation. If we can conclude from the schema knowledge that the update is untranslatable, it will be immediately rejected. Similarly, if we can find the correct translation of the view update, we will directly pass it to the SQL engine for update execution. Only when we find that the translatability of the update depends on the actual data, we will come to the second step, where data checking is performed. By integrating schema checking and data checking together, we can guarantee to find correct translations efficiently if one exits.

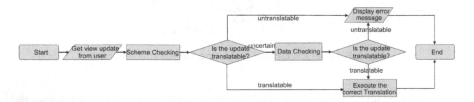

Fig. 2. Flowchart of HUX

Let us illustrate HUX for the motivating examples. During the schema level check, update u_1 is classified as translatable. We translate this update by deleting the corresponding tuple in the $StudentTA$ relation. Update u_2 will be found to be untranslatable by the schema-level check and is directly rejected. Updates u_3 and u_4 cannot be classified as translatable or untranslatable by the schema-level check. Therefore we proceed to the data-level check, where we find that u_3 is translatable and u_4 is not. As the correct translation of u_3 has also been identified, it will be updated and u_4 will be rejected.

In addition, we can further optimize data-checking step by utilizing the extracted schema knowledge to prune down the search space of finding correct translations. For example, consider Example 3. When updating a Professor view element, there are three possible ways: deleting the corresponding tuple in Professor table, deleting the corresponding tuple in Course table or deleting both of them. However, as we analyzed in the schema-checking step, each Course tuple also contributes to the existence of a Course view element. Therefore we cannot delete a Professor view element by deleting its Course tuple as a Course view element will get deleted as view side-effects. Therefore, on the data level, we only need to check the Professor base tuple. Details of how to maximally utilize schema knowledge in the data-checking are presented in Section 5.

XML-Specific Challenges. Note the above integration of two approaches can be applied to all data models. In this paper, we examine the scenario where the view update

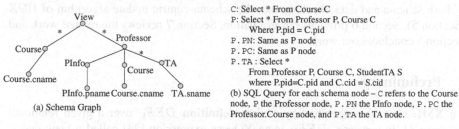

C: Select * From Course C
P: Select * From Professor P, Course C
      Where P.pid = C.pid
P.PN: Same as P node
P.PC: Same as P node
P.TA : Select *
      From Professor P, Course C, StudentTA S
      where P.pid=C.pid and C.cid = S.cid
(b) SQL Query for each schema node – C refers to the Course
node, P the Professor node, P.PN the PInfo node, P.PC the
Professor.Course node, and P.TA the TA node.

Fig. 3. Schema graph of the XML view in Figure 1

is against an XML view built over relational databases. For studying this problem, we will try to extend relational view update solutions to handle this scenario for the following two reasons:

1. There has been lot of work on relational view update problems, including both schema approach [1,10,17,15,13] and data approach [12].
2. We can treat XML views as a "composition" of a set of relational views [20,4]. Here, each node in the schema graph of the view (Fig. 3) can be considered as generated by a relational view, with an associated SQL query. The set of instances of a schema node is therefore given by this SQL query.

Intuitively, an update over a certain XML view schema node can be treated as an update over its relational mapping view. This in turn can be handled as relational view update problem. However, a simple transformation of XML view update problem into relational view update problem will not suffice. The relational view update problem considers side-effects only on elements with the same view query as the element to be updated. However, an XML view has many different view schema nodes, with different mapping SQL queries, and we need to examine the side-effects on all these elements. In the following sections, we will highlight these differences and explain how they are handled in our HUX approach.

Contributions. We make the following contributions in this paper. (1) We propose the first pure data-driven strategy for XML view updating, which guarantees that all updates are correctly classified. (2) We also propose a schema-driven update translatability reasoning strategy, which uses schema knowledge including now both keys and foreign keys to efficiently filter out untranslatable updates and identify translatable updates when possible. (3) We design an interleaved strategy that optimally combines both schema and data knowledge into one update algorithm, which performs a complete classification in polynomial time for the core subset of XQuery views[2] (4) We have implemented the algorithms, along with respective optimization techniques in a working system called HUX. We report experiments assessing its performance and usefulness.

Outline. Section 2 formally defines the view updating problem and reviews the clean source theory. The pure data-driven side-effect checking strategy is described in Section 3, while the schema-driven one is described in Section 4. We combine the power

---

[2] The complexity analysis is omitted for space considerations.

of both schema and data checking into the schema-centric update algorithm of HUX (Section 5). Section 6 provides our evaluation, Section 7 reviews the related work and Section 8 concludes our work.

2 Preliminary

An **XML view** V is specified by a **view definition** DEF_V over a given relational database D. In our case, DEF_V is an XQuery expression [24] called a *view query*. Let $\mho$ be the domain of update operations over the view. Let $u \in \mho$ be an update on the view V. An *insertion* adds while a *deletion* removes an element from the XML view. A *replacement* replaces an existing view element with a new one.

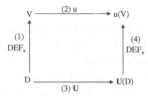

Fig. 4. Correct Translation of View Updates

Definition 1. *A relational update sequence U on a relational database D is a* **correct translation** *of an update u on the view V iff* $u(DEF_V(D)) = DEF_V(U(D))$.

A correct translation is shown in Fig. 4 holds. Intuitively, a correct translation exactly performs the view update and nothing else, namely, without view side-effects.

Clean Extended Source Theory. The *clean source theory* [13], has been widely used as theoretical foundation for the relational view update problem [12,5]. Let $R_1, R_2, ...,$ R_n be the set of relations referenced by the SQL query Q of a view schema node v. Informally a view element e's **generator** $g(e)$ is $\{R_1^*, R_2^*, ..., R_n^*\}$, where $R_i^* \in R_i (i = 1..n)$ contains exactly the tuple in R_i used to derive e. For example, the generator of the *Professor* view P1 in Fig. 1(c) is $g(\text{P3}) = \{Professor.t_2, Course.t_3\}$. The definition of the generator follows [13], also called *Data lineage* [12] and *Why Provenance* [5].

Further, each R_i^* is a **source** of the view element, denoted by s. For example, there are two possible sources of P3, namely, $s_1 = \{Professor.t_2\}$, $s_2 = \{Course.t_3\}$. A **clean source** is a source of an element used only by this particular element and no other one. For instance, s_1 is a clean source of $P3$, but s_2 is not since s_2 is also part of the generator of C3.

Given an element e and its source s. Let E be the set of tuples t_j in database D, which directly or indirectly refer to a tuple in the source s through foreign key constraint(s). In other words, E is the additional set of tuples that will get deleted when we delete s. Now, $extend(s) = s \cup E$ is called the **extended source** of s. For example, extend($Professor.t_1$)={$Professor.t_1$, $StudentTA.t_1$, $StudentTA.t_2$}.

As concluded in [13,25], an update translation is correct if and only if it deletes or inserts a clean extended source of the view tuple. Intuitively, it means that the update operation only affects the "private space" of the given view element and will not cause any view side-effect.

However, the update translatability checking based on the clean source theory above must examine the actual base data. Also, the number of potential translations of a given update can be large [5]. Therefore we propose instead to use the schema knowledge to filter out the problematic updates whenever possible. This prunes the search space in terms of candidates we must consider. We thus introduce a set of corresponding schema-level concepts as below.

e	A view element		v	A view schema node
$g(e)$	The generator of e		$G(v)$	The schema-level Generator of v
s	The source of e		S	The schema-level Source of v
$extend(s)$	The extended source of s		$Extend(S)$	The schema-level Extended Source of S

(a) Main Concepts at the Data-level	(b) Main Concepts at the Schema-level

Fig. 5. Main Concepts Used for View Updates

Given a view element e and its schema node v. *Schema-level generator* $G(v)$ indicates the set of relations from which the generator $g(e)$ is extracted. Similarly, S denotes the set of relations the source s derived from, named *schema-level source*. For example, $G(\text{P})=\{Professor, Course\}$. Schema level sources include $S_1=\{Professor\}$, $S_2=\{Course\}$. Note that $S \subseteq Extend(S) \subseteq D$.

XML View Elements Classification. Given an XML view update on element e, a correct translation for this update can update any descendant element of e in addition to updating e. However, this translation should *not* affect any of the non-descendant elements of e. We classify these non-descendant elements into three groups as shown in Figure 6. **Group-NonDesc** includes view elements whose schema nodes are non-descendant ones of v. **Group-Self** includes those whose schema node is v. **Group-Desc** includes those whose schema nodes are descendants of v.

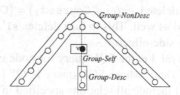

Fig. 6. Schema Tree Structure

For example, let the view element e be P1 from Fig 1. Then Group-NonDesc includes C1, C2, C3. Group-Self includes P2, P3. Group-Desc includes P2.PN2, P2.PC2, P2.TA3, P3.PN3, P3.PC3. For updating a view element e, if there is a translation that performs the update without affecting any element of any of the three groups, then this is a correct translation of the given update.

3   Data-Driven View Updating

Using clean source theory, most commercial relational data base systems [21,2,8] and some research prototypes [5,12] directly issue SQL queries over the base data to identify view side-effects. If any clean source (exclusive data lineage [12]) is found to exist, then this source can be a correct translation. Below we extend this approach to find a clean extended source for updating elements in an XML view.

Given the generator $g(e) = \{R_1^*, R_2^*, ..., R_n^*\}$ of a view element e of a schema node v. Intuitively, deleting any R_i^* from the generator will certainly delete the element e. However, when R_i^* get deleted, $extend(R_i^*)$ (all the base tuples that directly or indirectly refer to it) will also get deleted. Thus if any $t' \in extend(R_i^*)$ also belongs to $g(e')$, where e' is another view element other than e or its descendants, e' will also get deleted. This is a view side-effect and implies R_i^* cannot be a correct translation. Therefore, if R_i^* is a correct translation, all the view elements should remain unchanged after deleting $extend(R_i^*)$, except e and its descendants.

Let D and D' be relational database instances before and after deleting R_i^* [3]. Also, Q^v is the SQL query for a schema node v in view schema graph. Then $Q^v(D)$ and $Q^v(D')$ stand for sets of view elements that correspond to v before and after deleting $extend(R_i^*)$ respectively. Thus, $Q^v(D) - Q^v(D')$, **denoted as** $\lambda(v)$, are view elements that are deleted by deleting R_i^*. The following three rules detect side effects in each of the three groups of nodes (Fig. 6). The rules are self-explanatory and we skip examples for the first two rules for space considerations.

Rule 1. *Consider Group-NonDesc node v'. Deleting R_i^* from $g(e)$ will delete the element e without causing side-effect on any element e' of v' if $\lambda(v') = \emptyset$.*

Rule 2. *Consider Group-Self node v'. Deleting R_i^* from $g(e)$ will delete the element e without causing side-effect on any element e' of v' if $\lambda(v') = e$.*

For a schema node v' in Group-Desc, e' of v' will get deleted if e' is a descendant of e. For example, consider $c1$ in Figure 7(c). Its generator is $\{\text{Course}.t_1\}$. Let R_i^* be $\text{Course}.t_1$. When R_i^* gets deleted, $extend(\text{Course}.t_1) = \{\text{Course}.t_1, \text{StudentTA}.t_1,$ $\text{StudentTA}.t_2\}$ gets deleted as well. This in turn deletes $s1'$, $s1''$, $s2'$ and $s2''$ in Figure 7(c), which causes view side-effects.

Let us make it more general. Let the SQL query for node v be $Q^v(R_1, R_2, \ldots, R_n)$. Let v' be a node in Group-Desc, whose SQL query is $Q^{v'}(R_1, R_2, \ldots, R_n, S_1, S_2, \ldots, S_m)$ (note that $Q^{v'}$ will include all relations specified in Q^v). For an element e of node v, let $g(e) = \{R_1^*, R_2^*, \ldots, R_n^*\}$. The elements of v' that are descendants of e are given by $Q^{v'}(R_1^*, R_2^*, \ldots, R_n^*, S_1, S_2, \ldots, S_m)$. To determine if deleting R_i^* causes side effects on elements of v', we need to check whether the elements of v' that are deleted are those that are descendants of e, as stated below.

Rule 3. *Consider Group-Desc node v'. Deleting R_i^* will delete e without causing side-effects on elements of node v' if $\lambda(v') = Q^{v'}(R_1^*, R_2^*, \ldots, R_n^*, S_1, S_2, \ldots, S_m)$.*

---

[3] As $extend(R_i^*)$ gets deleted when deleting R_i^*, terms "deleting R_i^*" and "deleting $extend(R_i^*)$" are used interchangeably.

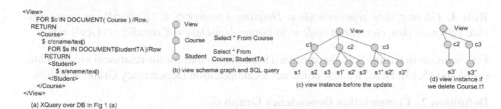

(a) XQuery over DB in Fig 1 (a)

Fig. 7. Examples for Group-Desc

Theorem 1. *Consider a source R_i^* of e. For $\forall v$, where v is a schema node in the view, if there are no side effects caused by deleting R_i^* according to Rules 1 - 3, then R_i^* is a correct translation for deleting e.*

4 Schema-Driven View Updating

In the previous section, we described the approach of identifying side-effects by *examining the actual base data*. This approach is *correct* and *complete*, meaning we can always reject all untranslatable updates and identify all translatable updates. However, this data examination step could be quite expensive. We now propose a more effective solution based on schema knowledge. Using schema knowledge only, we will classify an update as *untranslatable* (Example 2), *translatable* (Example 1) and *uncertain* (Example 3).

On the schema level, when we classify an update as "translatable", it implies for all the database instances, there exists a correct translation for this update. Just like in Section 3, we will set three rules to check if the given user view update can achieve this in Section 4.1. Similarly, when we classify an update as "untranslatable", it implies for all the database instances, any translation of this update will always cause side-effects. We will discuss this in Section 4.2. For those updates that we cannot guarantee it is translatable or untranslatable, we will examine the actual base data to classify it.

4.1 Schema-Level Translatable Updates

To classify an update as always translatable (Example 1), we have to check whether a clean source always exists for any update on the schema node. The following rules are used to identify whether it is possible to delete a source without ever causing any side-effects on Group-NonDesc, Group-Self and Group-Desc nodes. Note as we need to guarantee this update is translatable for all database instances, the following rules tend to be conservative, which means they may be too strict for some specific instances. For a specific database instance, we will not reject an view update if none of its translations satisfy all these rules, we only claim that we haven't found a correct translation.

Similar to the idea in data-checking in Section 3, we will consider every Source $S \in G(v)$ as a translation, as deleting the generator $R_i^* \in S$ will delete e. For each S, we use the following rules to check if deleting tuples from R_i will cause side-effects on nodes in different groups. We again describe these rules without examples as they are self-explanatory and for space considerations.

Rule 4. *Given a view schema node v. Deleting a source $S \in G(v)$ will not cause any side-effect on view element of node v' in Group-NonDesc if Extend(S) $\cap$ $G(v') = \emptyset$.*

For elements in Group-Self, the approach is similar to that for the relational view update problem [15,13]. Here, we first construct a Computation Dependency Graph as below.

Definition 2. Computation Dependency Graph $\mathcal{G}_C$.

1. *Given a view schema node v computed by SQL query Q^v. Let $R_1, R_2, \ldots, R_n$ be relations referenced by Q^v. Each R_i, $1 \le i \le n$ forms a node in $\mathcal{G}_C$.*
2. *Let R_i, R_j be two nodes ($R_i \neq R_j$). There is an edge $R_i \to R_j$ if Q has a join condition of the form $R_i.a = R_j.b$ and $R_j.b$ is UNIQUE in R_j.*
3. *If Q has a join condition $R_i.a = R_j.b$ where $R_i.a$ is UNIQUE for R_i and $R_j.b$ is UNIQUE for R_j, then there are two edges $R_i \to R_j$ and also $R_j \to R_i$.*

Fig. 8 shows the computation dependency graph for P and $P.TA$ nodes in Fig 3.

Fig. 8. (a) $\mathcal{G}_C$ of P-node and (b) $\mathcal{G}_C$ of P.TA-node

Rule 5. *Given a view schema node v and its computation dependency graph $\mathcal{G}_C$. Deleting a source S of v will not cause side-effect in any view element of Group-Self if the corresponding node of S in $\mathcal{G}_C$ can reach all other nodes.*

For a node v' in Group-Desc, we check that the view query for v' captures all the key/foreign key constraints between any relation in $G(v')$ - $G(v)$ and the Source S of v. We say that a query captures a key/foreign key constraint, fk references k, if the query includes a predicate of the form k=fk. Intuitively, this ensures that when a tuple of S is deleted, the only view elements that are affected are the descendants of e, the view element to be deleted.

Rule 6. *Deleting a source S of v will not cause any view side-effect in any view element of node v' in Group-Desc if for any $R_j \in G(v') - G(v)$, key/foriegn key constraints between R_j and S are captured in the query for v'.*

Theorem 2. *For a given view update u, if a translation U does not cause side-effects on Group-NonDesc nodes (by Rule 4, on Group-Self nodes (by Rule 5) and on Group-Desc nodes (by Rule 6), then U is a correct translation.*

4.2 Schema-Level Untranslatable Updates

Schema-level untranslatable updates are determined by the rule below. If the generator of a view schema node v is the same as that of its parent node, then it is guaranteed that deletion of any source of an element of v will cause side-effects on its parent element. See Example 2 in Section 1.1.

Rule 7. *Given a view schema node v, and its parent schema node v_p, a source S of v will cause side-effects on v_p if $S \in G(v_p)$.*

Theorem 3. *Given a view schema node v, and its parent schema node v_p, if $G(v) = G(v_p)$, then there is no correct translation for updating any element of v.*

4.3 Schema-Level Uncertain Updates

If for a view schema node v, we cannot determine at the schema-level whether an update of an element of v is translatable or untranslatable, we say that update of this node is schema-level uncertain.

5 Schema-Centric XML View Updating Algorithm

Given a user view update, after applying rules defined in Section 4, we may not still be able to classify it as translatable or not. In such cases, we can use rules defined in Section 3 to determine which translations are correct. With those rules, we can always classify an update as translatable or untranslatable. Further, the observations made during the schema-checking can be used to optimize the data-checking.

For example, consider the updates u_3 and u_4 in Example 3 in Section 1. Here, v is P in the schema graph shown in Figure 3. There are two relations in $G(v)$: $Professor$ and $Course$. After applying the schema checking rules on these two translations separately, we cannot find a correct translation, and these updates are classified as "uncertain". Now we perform the data-checking. However, the schema-checking reveals that $Course$ is not a correct translation, and hence this need not be checked. Further, while checking $Professor$, the schema-checking reveals that there will be no side-effects on $P.TA$, and hence side-effects on $P.TA$ need not be examined during data-checking.

In summary, when doing data-checking, schema-checking knowledge can help us in two ways:

1. Prune down translations. Those translations identified as incorrect on the schema-level need not be considered at the data-level.
2. Prune down view schema nodes. Given a translation R_i and a schema node v' in Group-NonDesc, Group-Self or Group-Desc, if R_i will never cause any side-effects on v', there is no need to check side-effects on v' on the data-level.

In order to pass the observations from the schema-level checking to the data-level checking, we introduce a data structure $SS(v)$ (Search Space), that keeps track of the observations made with regards to the update of an element of v. The column names are all the relations in $G(v)$ (they are the different possible translations). The row names are nodes in the schema graph of the view (the nodes for which we need to determine whether there will be side-effects). Consider the view defined in Figure 1 in Section 1. The initial Search Space (before any rules are applied) consists of view schema nodes $P, P.PN, P.PC, P.TA, C$ are shown in Figure 9.

During the schema-level checking, if we determine that a relation R_i in $G(v)$ does not cause side-effects on a view schema node v', we denote it as $"\sqrt{}"$ in the cell (R_i, v').

SS(C) translations view schema nodes	Course
P	
P.PN	
P.PC	
P.TA	
C	

SS(P), SS(P.PN), SS(P.PC) translations view schema nodes	Professor	Course
P		
P.PN		
P.PC		
P.TA		
C		

SS(P.TA) translations view schema nodes	Professor	Course	StudentTA
P			
P.PN			
P.PC			
P.TA			
C			

Fig. 9. Initial Search Space for view schema nodes in Figure 3

On the other hand, if we determine that R_i will cause side-effects on v', we denote it as "×" in the cell (R_i, v'). If the schema-level check cannot provide any guarantees as to whether there will be side-effects on v' or not, then the cell is left blank. Let us illustrate how the schema level checking proceeds for our example in Fig. 1.

First Step: Schema-Checking. After applying Rule 7 on each relation in $G(v)$, $SS(P)$ and $SS(C)$ remain unchanged. For the rest of view schema nodes, their search spaces are updated as in Figure 10. This also implies we should not consider $Professor$ for deleting view elements of $P.PN$, $P.PC$ and $P.TA$ nodes. Now, we can classify any update on $P.PC$ and $P.PN$ as untranslatable, as every one of their Sources will cause side-effects.

SS(P.PN), SS(P.PC) translations view schema nodes	Professor	Course
P	X	X
P.PN		
P.PC		
P.TA		
C		

SS(P.TA) translations view schema nodes	Professor	Course	StudentTA
P	X	X	
P.PN			
P.PC			
P.TA			
C			

Fig. 10. Search Spaces after applying Rule 7

Now, we utilize rules in Section 4.1. Currently, we need to check translation $Course$ for C, $Professor$ for P and $StudentTA$ for $P.PA$. After applying the three rules, updated search spaces are shown in Figure 11. We classify any update on $P.TA$ as translatable, and the correct translation is to delete from $StudentTA$.

SS(P) translations view schema nodes	Professor	Course
P		
P.PN	v	
P.PC	v	
P.TA	v	
C	v	X

SS(C) translations view schema nodes	Course
P	
P.PN	
P.PC	
P.TA	
C	v

SS(P.TA) translations view schema nodes	Professor	Course	StudentTA
P	X		v
P.PN		X	v
P.PC		X	v
P.TA			v
C			v

Fig. 11. Search Spaces after applying rues in Section 4.1

Second Step: Data-Checking. Updates on P and C nodes are classified as uncertain by the schema-level checking, therefore given a user update, we need to do a data-level check. However, for update of an element of node P, we need to check whether the source from *Professor* table will cause any side-effect on elements of node P, which implies we need to use Rule 2 only. Now, for updates u_3 and u_4 in Example 3, we find that u_3 is translatable, and the correct translation is delete $\{Professor.t_2\}$; u_4 is untranslatable.

6 Evaluation

We conducted experiments to address the performance impact of our system. The test system used is a dual Intel(R) PentiumIII 1GHz processor, 1G memory, running SuSe Linux and Oracle 10g. The relational database is built using TPC-H benchmark [23]. Two views are used in our experiments, with their view schemas shown below.

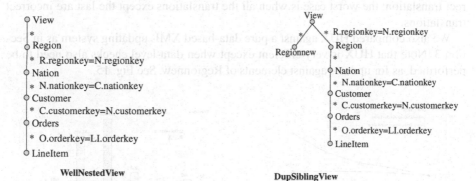

WellNestedView DupSiblingView

The cost of the schema-level marking and checking as the view query size increases is shown in Fig. 12. The view L uses only the lineitem table; the view LO uses the lineitem and orders table, and so on, and the views are WellNestedViews. The schema-level marking is done at compile time, and the time for this marking increases with the view query size. However, given an update, the check involves only checking the mark for that node and hence it is constant, and negligible.

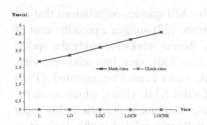

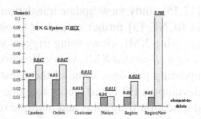

Fig. 12. Schema-Level Checking Performance

Fig. 13. HUX vs. Non-Guaranteed System

Run-time Overhead of HUX. Let us compare the run-time overhead of HUX. The run-time costs of HUX include: check the compile-time mark, do a run-time check if needed, find the correct translation and perform it (if translatable). We compared HUX against a Non-Guaranteed (NG) System that arbitrarily chooses a translation. The comparison using the DupSiblingView is shown in Fig. 13. Updates on Lineitem, Orders, Customer and Region are found to be translatable at the schema-level. For effective comparison, we assume that the NG system magically finds the correct translation. Note the small overhead for HUX. Nation is determined to be schema-level untranslatable. Updates on RegionKey require a data-level check and hence is expensive.

HUX vs. Data-based View Update System. If schema-level check can classify an update as translatable or untranslatable, then HUX is very efficient as compared to data-based systems. Fig 14 shows that the time taken at run-time for HUX stays constant. However, the data level check is much more expensive for different view query sizes. The best case for data level check is when the first translation checked is a correct translation; the worst case is when all the translations except the last are incorrect translations.

We also compared HUX against a pure data-based XML updating system as in Section 3. Note that HUX is very efficient except when data-level checks also need to be performed, as for updates against elements of Regionnew. See Fig. 15.

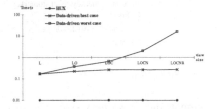

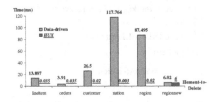

Fig. 14. HUX vs. relational view update system

Fig. 15. HUX vs. Pure data-based view update system

7   Related Work

[17,15] study view update translation mechanism for SPJ queries on relations that are in BCNF. [3] further extends it to object-based views. [22] studies execution cost of updating XML views using triggers versus indices. Recent works [4] study the update over *well-nested* XML views. However, as authors map XML view into relational view updating problem, some of the constraints in XML views cannot be captured. [19,7] develop an ER based theory to guide the design of valid XML views, which avoid the duplication from joins and multiple references to the relations. Our work in this paper is *orthogonal* to these works by addressing new challenges related to the decision of translation existence instead of assuming a view to be well-nested or valid.

We have studied schema-based approaches for checking the existence of a correct translation [25]. This work combines the schema-based approaches with data-based approaches efficiently to perform more updates. The concept of generator that we used is borrowed from [13], and is similar to the concepts of *data provenance* [5] or *lineage* [12]. In a recent work [18], the authors propose an approach to store auxiliary data for relational view updates; in their framework all updates are translatable. Another recent work [9] studies the problem of updating XML views materialized as relations. However, they do not consider performing the updates against the original relations from which the XML view was published as studied in this work.

8 Conclusion

In this paper, we have proposed an efficient solution for the XML view update problem. A progressive translatability checking approach is used to guarantee that only translatable updates are fed into the actual translation system to obtain the corresponding SQL statements. Our solution is *efficient* since we perform schema-level (thus very inexpensive) checks first, while utilizing the data-level checking only at the last step. Even during this last step, we utilize the schema knowledge to effectively reduce the search space for finding the translation. Our experiments illustrate the benefits of our approach. Our approach can be applied by any existing view update system for analyzing the translatability of a given update before its translation is attempted.

References

1. Bancilhon, F., Spyratos, N.: Update Semantics of Relational Views. ACM Transactions on Database Systems, 557–575 (December 1981)
2. Banerjee, S., Krishnamurthy, V., Krishnaprasad, M., Murthy, R.: Oracle8i - The XML Enabled Data Management System. In: ICDE, pp. 561–568 (2000)
3. Barsalou, T., Siambela, N., Keller, A.M., Wiederhold, G.: Updating Relational Databases through Object-Based Views. In: SIGMOD, pp. 248–257 (1991)
4. Braganholo, V.P., Davidson, S.B., Heuser, C.A.: From XML view updates to relational view updates: old solutions to a new problem. In: VLDB, pp. 276–287 (2004)
5. Buneman, P., Khanna, S., Tan, W.-C.: Why and where: A characterization of data provenance. In: Van den Bussche, J., Vianu, V. (eds.) ICDT 2001. LNCS, vol. 1973, Springer, Heidelberg (2000)
6. Carey, M.J., Kiernan, J., Shanmugasundaram, J., Shekita, E.J., Subramanian, S.N.: XPERANTO: Middleware for Publishing Object-Relational Data as XML Documents. The VLDB Journal, 646–648 (2000)
7. Chen, Y.B., Ling, T.W., Lee, M.-L.: Designing Valid XML Views. In: Spaccapietra, S., March, S.T., Kambayashi, Y. (eds.) ER 2002. LNCS, vol. 2503, pp. 463–478. Springer, Heidelberg (2002)
8. Cheng, J.M., Xu, J.: XML and DB2. In: ICDE, pp. 569–573 (2000)
9. Choi, B., Cong, G., Fan, W., Viglas, S.: Updating Recursive XML Views of Relations. In: ICDE (2007)
10. Cosmadakis, S.S., Papadimitriou, C.H.: Updates of Relational Views. Journal of the Association for Computing Machinery, 742–760 (October 1984)
11. Cui, Y., Widom, J.: Run-time translation of view tuple deletions using data lineage. In: Technique Report, Stanford University (June 2001)
12. Cui, Y., Widom, J., Wienner, J.L.: Tracing the lineage of view data in a warehousing environment. ACM Transactions on Database Systems 25(2), 179–227 (2000)
13. Dayal, U., Bernstein, P.A.: On the Correct Translation of Update Operations on Relational Views. ACM Transactions on Database Systems 7(3), 381–416 (1982)

14. Jagadish, H., Al-Khalifa, S., Lakshmanan, L., Nierman, A., Paparizos, S., Patel, J., Srivastava, D., Wu, Y.: Timber: A native XML database. In: Bressan, S., Chaudhri, A.B., Li Lee, M., Yu, J.X., Lacroix, Z. (eds.) CAiSE 2002 and VLDB 2002. LNCS, vol. 2590, Springer, Heidelberg (2003)

15. Keller, A.M.: Algorithms for Translating View Updates to Database Updates for View Involving Selections, Projections and Joins. In: Fourth ACM SIGACT-SIGMOD Symposium on Principles of Database Systems, pp. 154–163 (1985)

16. Keller, A.M.: Choosing a view update translator by dialog at view definition time. In: Chu, W.W., Gardarin, G., Ohsuga, S., Kambayashi, Y. (eds.) VLDB 1986 Twelfth International Conference on Very Large Data Bases, Kyoto, Japan, August 25-28, 1986, pp. 467–474. Morgan Kaufmann, San Francisco (1986)

17. Keller, A.M.: The Role of Semantics in Translating View Updates. IEEE Transactions on Computers 19(1), 63–73 (1986)

18. Kotidis, Y., Srivastava, D., Velegrakis, Y.: Updates Through Views: A New Hope. In: VLDB (2006)

19. Ling, T.W., Lee, M.-L.: A Theory for Entity-Relationship View Updates. In: Pernul, G., Tjoa, A.M. (eds.) ER 1992. LNCS, vol. 645, pp. 262–279. Springer, Heidelberg (1992)

20. Fernandez, M., et al.: SilkRoute: A Framework for Publishing Relational Data in XML. ACM Transactions on Database Systems 27(4), 438–493 (2002)

21. Rys, M.: Bringing the Internet to Your Database: Using SQL Server 2000 and XML to Build Loosely-Coupled Systems. In: VLDB, pp. 465–472 (2001)

22. Tatarinov, I., Ives, Z.G., Halevy, A.Y., Weld, D.S.: Updating XML. In: SIGMOD, May 2001, pp. 413–424 (2001)

23. TPCH. TPC Benchmark H (TPC-H),
    http://www.tpc.org/information/benchmarks.asp

24. W3C. XQuery 1.0 Formal Semantics (June 2003),
    http://www.w3.org/TR/query-semantics/

25. Wang, L., Rundensteiner, E.A., Mani, M.: Updating XML Views Published Over Relational Databases: Towards the Existence of a Correct Update Mapping. DKE Journal (2006)

Even an Ant Can Create an XSD*

Ondřej Vošta, Irena Mlýnková, and Jaroslav Pokorný

Charles University, Faculty of Mathematics and Physics,
Department of Software Engineering,
Malostranské nám. 25, 118 00 Prague 1, Czech Republic
ondra.vosta@centrum.cz, irena.mlynkova@mff.cuni.cz,
jaroslav.pokorny@mff.cuni.cz

Abstract. The XML has undoubtedly become a standard for data representation and manipulation. But most of XML documents are still created without the respective description of its structure, i.e. an XML schema. Hence, in this paper we focus on the problem of automatic inferring of an XML schema for a given sample set of XML documents. In particular, we focus on new features of XML Schema language and we propose an algorithm which is an improvement of a combination of verified approaches that is, at the same time, enough general and can be further enhanced. Using a set of experiments we illustrate the behavior of the algorithm on both real-world and artificial XML data.

1   Introduction

Without any doubt the XML [17] is currently a de-facto standard for data representation. Its popularity is given by the fact that it is well-defined, easy-to-use, and, at the same time, enough powerful. To enable users to specify own allowed structure of XML documents, so-called *XML schema*, the W3C[1] has proposed two languages – DTD [17] and XML Schema [32,16]. The former one is directly part of XML specification and due to its simplicity it is one of the most popular formats for schema specification. The latter language was proposed later, in reaction to the lack of constructs of DTD. The key emphasis is put on simple types, object-oriented features (such as user-defined data types, inheritance, substitutability, etc.), and reusability of parts of a schema or whole schemes. On the other hand, statistical analyses of real-world XML data show that a significant portion of real XML documents (52% [26] of randomly crawled or 7.4% [27] of semi-automatically collected[2]) still have no schema at all. What is more, XML Schema definitions (XSDs) are used even less (only for 0.09% [26] of randomly crawled or 38% [27] of semi-automatically collected XML documents) and even if they are used, they often (in 85% of cases [14]) define so-called *local tree grammars* [29], i.e. languages that can be defined using DTD as well.

---

* This work was supported in part by Czech Science Foundation (GAČR), grant number 201/06/0756.
[1] http://www.w3.org/
[2] Data collected with the interference of a human operator.

J.R. Haritsa, R. Kotagiri, and V. Pudi (Eds.): DASFAA 2008, LNCS 4947, pp. 35–50, 2008.

In reaction to this situation a new research area of automatic construction of an XML schema has opened. The key aim is to create an XML schema for the given sample set of XML documents that is neither too general, nor too restrictive. It means that the set of document instances of the inferred schema is not too broad in comparison with the provided set of sample data but, also, it is not equivalent to the sample set. Currently there are several proposals of respective algorithms (see Section 2), but there is still a space for further improvements. Primarily, almost all of the existing approaches focus on construction of regular expressions of DTD (though sometimes expressed in XML Schema language) which are relatively simple. Hence, our key aim is to focus on new constructs of XML Schema language, such as, e.g., unordered sequences of elements or elements having the same name but different structure, that enable to create more realistic schemes. For this purpose we propose an algorithm which is an improvement of a combination of verified approaches that is, at the same time, enough general and can be easily further enhanced. Such algorithm will enable to increase popularity and exploitation of XML Schema which fails especially in complexity of schema definition. Having an automatic generator of XSDs, a user is not forced to specify the whole schema manually, since the inferred one can serve, at least, as a good initial draft and thus eases the definition.

The paper is structured as follows: Section 2 overviews existing works focussing on automatic construction of XML schemes. Section 3 introduces the proposed algorithm in detail and Section 4 discusses the results of experimental testing. Finally, Section 5 provides conclusions and outlines possible future work.

2   Related Work

The existing solutions to the problem of automatic construction of an XML schema can be classified according to several criteria. Probably the most interesting one is the type of the result and the way it is constructed, where we can distinguish heuristic methods and methods based on inferring of a grammar.

Heuristic approaches [28, 33, 21] are based on experience with manual construction of schemes. Their results do not belong to any class of grammars and they are based on generalization of a trivial schema (similar to the classical *dataguide* [23]) using a set of predefined heuristic rules, such as, e.g., "if there are more than three occurrences of an element, it is probable that it can occur arbitrary times". These techniques can be further divided into methods which generalize the trivial schema until a satisfactory solution is reached [28, 33] and methods which generate a huge number of candidates and then choose the best one [21]. While in the first case the methods are threatened by a wrong step which can cause generation of a suboptimal schema, in the latter case they have to cope with space overhead and specify a reasonable function for evaluation quality of the candidates. A special type of heuristic methods are so-called *merging state algorithms* [33]. They are based on the idea of searching a space of all possible generalizations, i.e. XML schemes, of the given XML documents represented using a prefix tree automaton. By merging its states they construct the optimal solution. In fact, since the space is theoretically infinite, only a reasonable subspace of possible solutions is searched using various heuristics.

On the other hand, methods based on *inferring of a grammar* [19, 11] are based on theory of languages and grammars and thus ensure a certain degree of quality of the result. They are based on the idea that we can view an XML schema as a grammar and an XML document valid against the schema as a word generated by the grammar. Although grammars accepting XML documents are in general context-free [13], the problem can be reduced to inferring of a set of regular expressions, each for a single element (and its subelements). But, since according to Gold's theorem [22] regular languages are not identifiable only from positive examples (i.e. sample XML documents which should conform to the resulting schema), the existing methods exploit various other information such as, e.g., predefined maximum number of nodes of the target automaton representing the schema, user interaction specifying (in)correctness of a partial result, restriction to an identifiable subclass of regular languages, etc.

Finally, let us mention that probably the first approach which focuses directly on true XML Schema constructs has only recently been published in [15]. The authors focus on the importance of context of elements for inferring of XSDs. They define a subclass of XSDs which can be learned from positive examples and focus especially on constructs which are used in real-world XML schemes.

3   Proposed Algorithm

As we have mentioned, almost all the described papers focus on constructs of DTD which are quite restricted. All the elements are defined on the same level thus it is impossible to define two elements having the same name but different structure. But this requirement can have quite reasonable usage due to homonymy of element names. For instance, we can have an XSD of a library where each book as well as author have name. In the former case it can be only a simple string, while in the latter case the name of a human can consist of a couple of elements each having its own semantics – see Fig. 1.

```
<book>                                 <author>
    <name>Sherlock Holmes</name>         <name>
</book>                                      <first>Arthur</first>
                                             <middle>Conan</middle>
                                             <last>Doyle</last>
                                         </name>
                                     </author>
```

Fig. 1. Elements with the same name but different structure

Another interesting XSD feature is element `all`, i.e. unordered sequence of subelements, which allows to use an arbitrary permutation of the specified elements in respective document instances. This operator does not increase the expressive power of DTD operators, but it significantly simplifies the notation of used regular expressions, in particular in case of data-centric documents.

Our proposal focuses mainly on these two constructs. It is inspired especially by papers [33] and [21]. From the former one we "borrow" the idea of the Ant Colony Optimization (ACO) heuristic [18] for searching the space of possible generalizations, from the latter one we use a modification of the MDL principle [24] of their evaluation. The ACO heuristic enables to join several approaches into a single one and thus to infer more natural XML schemes. The MDL (Minimum Description Length) principle enables their uniform evaluation.

The main body of the algorithm can be described as follows: Firstly, each document D from the input set I_D is transformed into tree T. The trees hence form a set of input *document trees* I_T. (Note that for simplicity we omit attributes that can be considered as special types of elements.)

Definition 1. *A document tree of a document D is a directed graph $T = (V_T, E_T)$, where V_T is a set of nodes corresponding to element instances in D, and E_T is a set of edges s.t. $\langle a, b \rangle \in E_T$ if b is a (direct) subelement of a in D.*

Now, for each distinct element name in I_T, we perform clustering of respective element instances (see Section 3.1) with regard to the above described XSD feature. Then, for each cluster we generalize the trivial schema accepting purely the items in the cluster (see Section 3.2). And finally, we rewrite all the generalized schemes into XSD syntax (see Section 3.3). All the three steps involve a kind of improvement of current approaches. In the first case the existing works simply group elements on the basis of their names, elements with the same name but different structure are not supported and, hence, the result can be misleading. We solve this problem using clustering of elements on the basis of their context and structure. In the second case we propose a general combination of the best of the existing approaches which exploit some of the pure XML Schema constructs and can be easily extended. And in the final step we are able to output XSDs involving not only DTD constructs but also some of the purely XML Schema ones.

3.1   Clustering of Elements

To cluster elements having the same name but different structure we need to specify the similarity measure and the clustering algorithm.

Similarity of Elements. As an XML element e can be viewed as a subtree T_e (in the following text denoted as an *element tree*) of corresponding document tree T, we use a modified idea of *tree edit distance*, where the similarity of trees T_e and T_f is expressed using the number of edit operations necessary to transform T_e into T_f (or vice versa). The key aspect is obviously the set of allowed edit operations. Consider two simple operations – adding and removal of a leaf node (and respective edge). As depicted in Fig. 2, such similarity is not suitable, e.g., for recursive elements. The example depicts two element trees of element a having subelement i having subelement j having subelement k which contains either subelement z or again i. With the two simple edit operations the edit

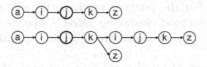

Fig. 2. Tree edit distance of recursive elements

distance would be 4, but since the elements have the same XML schema we would expect the optimal distance of 0.

Therefore, for our purpose we exploit a similarity measure defined in [30] which specifies more complex XML-aware tree edit operations involving operations on whole subtrees, each having its cost, as follows:

- *Insert* – a single node n is inserted to the position given by parent node p and ordinal number expressing its position among subelements of p
- *Delete* – a leaf node n is deleted
- *Relabel* – a node n is relabeled
- *InsertTree* – a whole subtree T is inserted to the position given by parent node p and ordinal number expressing position of its root node among subelements of p
- *DeleteTree* – a whole subtree rooted at node n is deleted

As it is obvious, for given trees T_e and T_f there are usually several possible transformation sequences for transforming T_e into T_f. A natural approach is to evaluate all the possibilities and to choose the one with the lowest cost. But such approach can be quite inefficient. Thus authors of the approach propose so-called *allowable sequences* of edit operations, which significantly reduce the set of possibilities and, at the same time, speed up their cost evaluation.

Definition 2. *A sequence of edit operations is* allowable *if it satisfies the following two conditions:*

1. *A tree T may be inserted only if T already occurs in the source tree T_e. A tree T may be deleted only if it occurs in the destination tree T_f.*
2. *A tree that has been inserted via the InsertTree operation may not subsequently have additional nodes inserted. A tree that has been deleted via the DeleteTree operation may not previously have had children nodes deleted.*

The first restriction forbids undesirable operations like, e.g., deleting whole T_e and inserting whole T_f, etc., whereas the second one enables to compute the costs of the operations efficiently. The evaluating algorithm is based on the idea of determining the minimum cost of each required insert of every subtree of T_e and delete of every subtree of T_f using a simple bottom-up procedure.

Hence, in the following text we assume that we have a function $dist(T_e, T_f)$ which expresses the edit distance of XML trees T_e and T_f.

Clustering Algorithm. For the purpose of clustering elements we use a modification of *mutual neighborhood clustering (MNC) algorithm* [25]. We start with initial clusters $c_1, c_2, ... c_K$ of elements given by the equivalence of their context.

Definition 3. *A context of an element e in a document tree $T = (V_T, E_T)$ is a concatenation of element names $e_0 e_1 ... e_N$, where e_0 is the root node of T, $e_N = e$, and $\forall i \in (1, N) : \langle e_{i-1}, e_i \rangle \in E_T$.*

Thus, the initial clustering is based on a natural assumption that elements having the same context (and element name) are likely to have the same schema definition. The initial clusters are then merged on the basis of element structure using the tree edit distance *dist*. Firstly, $\forall i \in (1, K)$ we determine a representative element r_i of cluster c_i. Then, for each pair of $\langle r_i, r_j \rangle$ s.t. $i \neq j; i, j \in (1, K)$ we determine tree edit distance $dist(T_{r_i}, T_{r_j})$ of the respective trees. The MNC algorithm is parameterized by three parameters – minimum distance $dist_{MIN}$, maximum distance $dist_{MAX}$, and factor F – and exploits the definition of *mutual neighborhood*:

Definition 4. *Let T_e and T_f be two element trees, where T_e is i-th closest neighbor of T_f and T_f is j-th closest neighbor of T_e. Then mutual neighborhood of T_e and T_f is defined as $MN(T_e, T_f) = i + j$.*

The MNC algorithm places two element trees T_{r_i} and T_{r_j} into the same group if $dist(T_{r_i}, T_{r_j}) \leqslant dist_{MIN}$ or $(dist(T_{r_i}, T_{r_j}) \leqslant dist_{MAX}$ and $MN(T_{r_i}, T_{r_j}) \leqslant F)$, resulting is a set of clusters $c_1, c_2, ..., c_L$ (where $L \leqslant K$) of elements grouped on the basis of their context and structure.

3.2 Schema Generalization

Now, for each cluster of elements $c_i; i \in (1, L)$ we infer an XML schema which "covers" all the instances in c_i and, at the same time, is still reasonably general. We speak about *generalization* of a trivial schema accepting purely the given set of elements. We can view the problem as a kind of optimization problem [12].

Definition 5. *A model $P = (S, \Omega, f)$ of a combinatorial optimization problem consists of a search space S of possible solutions to the problem (so-called feasible region), a set Ω of constraints over the solutions, and an objective function $f : S \to \mathbb{R}_0^+$ to be minimized.*

In our case S consists of all possible generalizations of the trivial schema. As it is obvious, S is theoretically infinite and thus, in fact, we can search only for a reasonable suboptimum. Hence, we use a modification of ACO heuristic [18]. Ω is given by the features of XML Schema language we are focussing on. In particular, apart from DTD concatenation (","), exclusive selection ("|"), and iteration ("?", "+", and "*"), we want to use also the `all` operator of XML Schema representing unordered concatenation (for simplicity denoted as "&"). And finally, to define f we exploit the MDL principle [24].

Ant Colony Optimization. The ACO heuristic is based on observations of nature, in particular the way ants exchange information they have learnt. A set of artificial "ants" $A = \{a_1, a_2, ..., a_{card(A)}\}$ search the space S trying to find the optimal solution $s_{opt} \in S$ s.t. $f(s_{opt}) \leqslant f(s); \forall s \in S$. In i-th iteration each $a \in A$ searches a subspace of S for a local suboptimum until it "dies" after performing a predefined amount of steps N_{ant}. While searching, an ant a spreads a certain amount of "pheromone", i.e. a positive feedback which denotes how good solution it has found so far. This information is exploited by ants from the following iterations to choose better search steps.

We modify the algorithm in two key aspects. Firstly, we change the heuristic on whose basis the ants search S to produce more natural schemes with the focus on XML Schema constructs. And secondly, we add a temporary negative feedback which enables to search a larger subspace of S. The idea is relatively simple – whenever an ant performs a single step, it spreads a reasonable negative feedback. The difference is that the positive feedback is assigned after i-th iteration is completed, i.e. all ants die, to influence the behavior of ants from $(i + 1)$-st iteration. The negative feedback is assigned immediately after a step is performed, i.e. it influences behavior of ants in i-th iteration and at the end of the iteration it is zeroed. The algorithm terminates either after a specified number of iterations N_{iter} or if $s'_{opt} \in S$ is reached s.t. $f(s'_{opt}) \leqslant T_{max}$, where T_{max} is a required threshold.

The obvious key aspect of the algorithm is one step of an ant. Each step consist of generating of a set of possible movements, their evaluation using f, and execution of one of the candidate steps. The executed step is selected randomly on the basis of probability given by f.

Generating a Set of Possible Movements. Each element in cluster c_i can be viewed as a simple *grammar production rule*, where the left-hand side contains element name and right-hand side contains the sequence of names of its (direct) subelements. For instance the sample elements `name` in Fig. 1 can be viewed as

    `name` → `#PCDATA`

    `name` → `first middle last`

Hence, we can represent the trivial schema accepting purely elements from c_i as a *prefix-tree automaton* and by merging its states we create its generalizations. Each possible merging of states represents a single step of an ant. To generate a set of possible movements, i.e. possible generalizations, we combine two existing methods – k,h-context and s,k-string – together with our own method for inferring of & operators.

The *k,h-context method* [11] specifies an identifiable subclass of regular languages which assumes that the context of elements is limited. Then merging states of an automaton is based on an assumption that two states t_x and t_y of the automaton are identical (and can be merged) if there exist two identical paths of length k terminating in t_x and t_y. In addition, also h preceding states in these paths are then identical.

The *s,k-string method* [33] is based on Nerod's equivalency of states of an automaton assuming that two states t_x and t_y are equivalent if sets of all paths leading from t_x and t_y to terminal state(s) are equivalent. But as such condition is hardly checked, we can restrain to k-strings, i.e. only paths of length of k or paths terminating in a terminal state. The respective equivalency of states then depends on equivalency of sets of outgoing k-strings. In addition, for easier processing we can consider only s most probable paths, i.e. we can ignore singular special cases.

Finally, the idea of *inferring of & operators*, i.e. identification of unordered sequences of elements, can be considered as a special kind of merging states of an automaton. It enables to replace a set of ordered sequences of elements with a single unordered sequence represented by the & operator. We describe the process of inferring the unordered sequences in Section 3.3 in detail.

Evaluation of Movements. The evaluation of moving from schema s_x to s_y, where $s_x, s_y \in S$ (i.e. the evaluation of merging states of the respective automaton) is defined as

$$mov(s_x, s_y) = f(s_x) - f(s_y) + pos(s_x, s_y) + neg(s_x, s_y)$$

where f is the objective function, $pos(s_x, s_y) \geqslant 0$ is the positive feedback of this step from previous iterations and $neg(s_x, s_y) \leqslant 0$ is the respective negative feedback. For the purpose of specification of f we exploit a modification of the MDL principle based on two observations [21]: A good schema should be enough general which is related to the low number of states of the automaton. On the other hand, it should preserve details which means that it enables to express document instances using short codes, since most of the information is carried by the schema itself and does not need to be encoded. Hence, we express the quality of a schema $s \in S$ described using a set of production rules $R_s = \{r_1, r_2, ..., r_{card(R_s)}\}$ as the sum of the size (in bits) of s and the size of codes of instances in cluster $c_i = \{e_1, e_2, ..., e_{card(c_i)}\}$ used for inferring of s.

Let O be the set of allowed operators and E the set of distinct element symbols in c_i. Then we can view right-hand side of each $r \in R_s$ as a word over $O \cup E$ and its code can be expressed as $|r| \cdot \lceil \log_2(card(O) + card(E)) \rceil$, where $|r|$ denotes length of word r. The size of code of a single element instance $e \in c_i$ is defined as the size of code of a sequence of production rules $R_e = \langle g_1, g_2, ..., g_{card(R_e)} \rangle$ necessary to convert the initial nonterminal to e using rules from R_s. Since we can represent the sequence R_e as a sequence of ordinal numbers of the rules in R_s, the size of the code of e is $card(R_e) \cdot \lceil \log_2(card(R_s)) \rceil$.

Note that since the ACO heuristic enables to use any inferring method to produce possible movements, i.e. schema generalizations, whereas the MDL principle does not consider the way they were inferred, the algorithm can easily be extended to other inferring methods as well as new constructs.

3.3   Inferring of Regular Expressions

The remaining open issue is how to infer the XSD, i.e. the set of regular expressions, from the given automaton. We exploit this information thrice – for evaluation of the objective function f, for generation of set of possible movements of an ant, and to output the resulting XSD. The automaton A can be represented as a directed graph, whose nodes correspond to states and edges labeled with symbols from the input alphabet E represent the transition function, i.e. an edge $\langle v_x, v_y \rangle$ labeled with symbol $e \in E$ denotes that the transition function from state v_x to state v_y on symbol e is defined. The task is to convert A to an equivalent regular expression. As for the DTD operators we use the well-known rules for transforming an automaton to a regular expression, similarly to [28, 33]. A brand new one approach we propose to identify a subgraph of A representing the all operator (& operator for simplicity).

In general the & operator can express the unordered sequence of regular expressions of any complexity such as, e.g., $(e_1|e_2) * \& e_3?\&(e_4, e_5, e_6)$, where $e_1, e_2, ..., e_6$ are element names. But, in general, the W3C recommendation of XML Schema language does not allow to specify so-called *nondeterministic data model*, i.e. a data model which cannot be matched without looking ahead. A simple example can be a regular expression $(e_1, e_2)|(e_1, e_3)$, where while reading the element e_1 we are not able to decide which of the alternatives to choose unless we read the following element. Hence, also the allowed complexity of unordered sequences is restricted. The former and currently recommended version 1.0 of XML Schema specification [32, 16] allows to specify an unordered sequence of elements, each with the allowed occurrence of $(0, 1)$, whereas the allowed occurrence of the unordered sequence itself is of $(0, 1)$ too. The latter version 1.1 [20, 31], currently in the status of a working draft, is similar but the allowed number of occurrence of items of the sequence is $(0, \infty)$. In our approach we focus on the more general possibility, since it will probable soon become a true recommendation.

First-Level Candidates. For the purpose of identification of subgraphs representing the allowed type of unordered sequences, we first define so-called *common ancestors* and *common descendants*.

Definition 6. *Let $G = (V, E)$ be a directed graph. A* common descendant *of a node $v \in V$ is a descendant $d \in V$ of v s.t. all paths traversing v traverse also d.*

Definition 7. *Let $G = (V, E)$ be a directed graph. A* common ancestor *of a node $v \in V$ is an ancestor $a \in V$ of v s.t. all paths traversing v traverse also a.*

Considering the example in Fig. 3 we can see that the common descendants of node 3 are nodes 6 and 7, whereas node 1 has no common descendants, since paths traversing node 1 terminate in nodes 7 and 9. Similarly, the common ancestor of node 6 is node 1. Note that in the former case there can exist paths which traverse d but not v (see node 3 and its common descendant 6), whereas in the latter case there can exist paths which traverse a but not v.

For the purpose of searching the unordered sequences we need to further restrict the Definition 7.

Definition 8. *Let $G = (V, E)$ be a directed graph. A* common ancestor *of a node $v \in V$ with regard to a node $u \in V$ is an ancestor $a \in V$ of v s.t. a is a common ancestor of each direct ancestor of v occurring on path from u to v.*

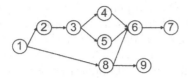
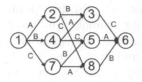

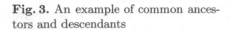

Fig. 3. An example of common ancestors and descendants

Fig. 4. Automaton P_3 – permutation of three items

For example, considering Fig. 3, the common ancestors of node 6 with regard to node 2 are nodes 2 and 3.

We denote the node v from Definition 6 or the node a from Definitions 7 and 8 as *input nodes* and their counterparts as *output nodes*. The set of nodes occurring on paths starting in an input node and terminating in an output node are called a *block*. Using the definitions we can now identify subgraphs which are considered as *first-level candidates* for unordered sequences. A node n_{in} is an input node of block representing a first-level candidate if

1. its out-degree is higher that 1,
2. the set of its common descendants is not empty, and
3. at least one of its common descendants, denoted as n_{out}, whose set of common ancestors with regard to n_{in} contains n_{in}.

The three conditions ensure that there are at least two paths leading from n_{in} representing at least two alternatives and that the block is *complete* meaning that there are no paths entering or leaving the block otherwise than using n_{in} or n_{out}. For example, considering Fig. 3, the only first-level candidate is subgraph consisting od nodes 3, 4, 5, 6.

Second-Level Candidates. Having a first-level candidate we need to check it for fulfilling conditions of an unordered sequence and hence being a *second-level candidate*. As we know from the specification of XML Schema, such unordered sequence can consist only of simple elements which can repeat arbitrarily. Hence, firstly, we can skip processing of first-level candidates which contain subgraphs representing other operators or repetitions of more complex expressions. For the purpose of further checking we exploit the idea of similarity of graphs: For each $n \in \mathbb{N}$ we know the structure of the automaton P_n which accepts each permutation of n items having all the states fully merged. (An example of automaton P_3 is depicted in Fig. 4 for three items A, B, C.) Thus the idea is to compare the similarity of the first-level candidates with P_n automatons. But the situation is more complicated, since the W3C recommendation allows optional and repeatable elements in the unordered sequences, i.e. the allowed number of occurrences can be also 0 or greater than 1. Together with the fact that the input elements on whose basis the automaton was built do not need to contain all possible permutations, the candidate graph can have much different structure than any P_n. And naturally, we cannot check the similarity with all P_n automatons.

To solve the problem of multiple occurrence of an element, we temporarily modify the candidate graphs by replacing each repeatable occurrence of an element e with auxiliary element e' with single occurrence. For the purpose of similarity evaluation, we can consider the modified graph without repetitions. The repetitions will influence the resulting regular expression, i.e. the respective operator will be added. Then, we can denote the maximum path length l_{max} in the candidate graph and hence denote the size of the permutation. And finally, since both the modified candidate graph and $P_{l_{max}}$ graph are always acyclic, we can evaluate their similarity using a classical edit-distance algorithm.

We can also observe, that the candidate graph must be always a subgraph of $P_{l_{max}}$, otherwise we can skip its processing. Hence, the problem of edit distance is highly simplified. We use the following types of edit operations:

- Adding an edge between two existing nodes, and
- Splitting an existing edge into two edges, i.e. adding a new node and an edge.

The first operation is obvious and corresponds to the operation of adding paths that represent permutations which were not present in the source data and its cost is > 0. In the latter case the operation corresponds to adding an item of a permutation which was not present in the source data and its cost is 0, but it influences the resulting regular expression similarly to the above described multiple occurrence. Naturally, only nodes and edges which are present in $P_{l_{max}}$ and not in the candidate graph or edges from n_{in} to n_{out} can be added. From all the possible edit sequences we choose the one with the lowest total cost [30].

Extension of Candidates. As we can see from the third condition which describes a first-level candidate, there can exist more candidates for the output node n_{out}. Hence, the remaining question is which of the candidates ought to be selected. In our proposal we use an approach which is, at the same time, a kind of a heuristic. The idea is based on the simple observation that each permutation of n items contains permutations of $n - 1$ items. It can be seen in Fig. 4, which contains not only the P_3 automaton for the three items A, B, and C, but also all three P_2 automatons for pairs A and B (see subgraphs on nodes 1, 2, 3, 4), B and C (see subgraph on nodes 2, 3, 6, 8), C and A (see subgraph on nodes 1, 4, 5, 7).

Hence, for each node of the graph we firstly determine the set of all nodes which fulfill the condition of first-level candidate and we sort them totally in ascending order using the size of the respective block. The ordering determines the order in which we check the conditions of second-level candidates and enables to exploit the knowledge of previously determined second-level candidates, i.e. subgraphs corresponding to permutations for subsets of the given items.

4   Experimental Implementation

For the purpose of experimental analysis of the proposed algorithm we have created a prototype implementation called *SchemaMiner*. Since our approach results from existing verified approaches [33, 21] and focuses on XSD constructs which were not considered yet, a comparison with any other method would either lead to similar results or would be quite unfair. Thus we rather compare our approach with existing real-world XSDs. From the available real-world XML documents we have selected subsets having an XSD and classified them on the basis of their structure. For each collection of documents we have inferred the schema using *SchemaMiner* and analyzed the results.

Since the real-world XML documents did not cover all the features of the proposed algorithm, we have also generated a set of artificial documents which enable to illustrate the remaining properties. Hence, a natural question may arise whether these constructs are worth considering since they do not occur in real applications quite often. As we have mentioned in the Introduction, the general problem of real-world XML data is that they exploit only a small part of all constructs allowed by the W3C specifications. On one hand, XML data processing approaches can exploit this knowledge and focus only on these common constructs. But, on the other hand, it is also necessary to propose methods which help users to use also more complex tools which enable to describe and process XML data more efficiently. In our case it means to find an XSD which describes the selected situation more precisely.

4.1   Real-World XML Documents

We have classified the used XML documents according to their structure into the following categories:

- **Category 1:** Documents having very simple and general structure of type $(e_1|e_2|...|e_n)*$. They do not exploit optional elements, deeper hierarchy of exclusive selections, or sequences of elements. (e.g. [1])
- **Category 2:** These documents exploit purely optional elements, sequences of elements, and repetitions. (e.g. [2,3])
- **Category 3:** These documents exploit all constructs of DTD, i.e. exclusive selections, repetitions, optional elements, and sequences of elements. (e.g. [4,5,6])
- **Category 4:** These documents have fairly regular structure suitable for storing into relational databases. The root element typically contains repetition of an element corresponding to n-tuple of simple data in fixed order. (e.g. [2])
- **Category 5:** These documents exploit the `all` construct of XSDs. (e.g. [7])
- **Category 6:** Since each XSD is at the same time an XML document and there exists an XSD of XML Schema language, we have included also XSDs, i.e. XML documents describing structure of other XML documents. (e.g. [8,9,10])

4.2   Results of Experiments

Comparing the difference between inferred and real-world XML schemes we have found several interesting observations. The most striking difference was inferring of less general schema in all the cases. This finding corresponds to the results of statistical analyses which show that XML schemes are usually too general [27]. Obviously, if a structural aspect is not involved in the sample XML documents, it can hardly be automatically generated. We can even consider this feature as an advantage to specify a more precise schema. On the other hand, another interesting difference was generalization of the inferred schema by setting the interval of repetition to $(0, \infty)$ instead of $(1, \infty)$, i.e. enabling to omit an element in case it should not be omitted. This situation occurs usually in case of repetition of a simple sequence of elements, in particular in category 4. But in this category such behavior is rather natural than harmful. And another significant difference was "inefficient" notation of the regular expressions, e.g., in case of expressions of the form $((e_1, e_2, e_3)|e_2)$ that could be rewritten to $(e_1?, e_2, e_3?)$, though this expression is not equivalent. This property is given by the features of the algorithm which directly transforms an automaton to a regular expression and indicates that additional optimization or user interaction would be useful.

Using category 5 we wanted to focus especially on the unordered sequences. But, unfortunately, the amount of real-world XML data was so small that the results were not usable. That is why we have decided to use artificial data (see Section 4.3). And a similar problem occurred in case of elements with the same name but different structure.

In case of the category 6, i.e. XSDs, we have divided the source data into two groups – a set of XSDs randomly downloaded from various sources and a set of XSDs inferred by the algorithm. The result in case of the first set was highly influenced by the variety of input data and the fact that XML Schema provides plenty of constructs which enable to specify the same structure in various ways. Nevertheless, the resulting schema was still recognizable as the XML Schema XSD. In case of the latter set of XSDs the result was naturally much better since the data came from the same source. An example is depicted in Fig. 5, where we can see the inferred XSD fragment of element choice of XML Schema.

```
<xs:complexType name="Txs:choice1" >
  <xs:choice minOccurs="1" maxOccurs="1">
    <xs:sequence minOccurs="1" maxOccurs="1">
      <xs:element name="xs:element" type="Txs:element1" minOccurs="1" maxOccurs="1"/>
      <xs:element name="xs:element" type="Txs:element1" minOccurs="0" maxOccurs="unbounded"/>
    </xs:sequence>
    <xs:sequence minOccurs="1" maxOccurs="1">
      <xs:element name="xs:sequence" type="Txs:sequence1" minOccurs="1" maxOccurs="1"/>
      <xs:element name="xs:sequence" type="Txs:sequence1" minOccurs="0" maxOccurs="unbounded"/>
    </xs:sequence>
    <xs:sequence minOccurs="1" maxOccurs="1">
      <xs:element name="xs:element"  type="Txs:element1"  minOccurs="1" maxOccurs="1"/>
      <xs:element name="xs:element"  type="Txs:element1"  minOccurs="0" maxOccurs="unbounded"/>
      <xs:element name="xs:sequence" type="Txs:sequence1" minOccurs="1" maxOccurs="1"/>
    </xs:sequence>
  </xs:choice>
</xs:complexType>
```

Fig. 5. An example of generated XSD

4.3   Artificial XML Documents

As we have mentioned, to analyze all the features of the algorithm we have prepared a set of artificial XML documents. In particular we have focused on the occurrence of unordered sequences of elements and the occurrence of elements with the same name but different structure.

Permutated Set. We have generated several sets of XML documents which differentiate in two aspects – the size of the permutated set and the percentage of permutations represented in the documents, i.e. having a set of n elements, the percentage of $n!$ of their possible orders.

According to our results the size of set has almost no impact on the resulting schema, whereas the percentage of permutations is crucial. Particular results are depicted in Table 1 for the size of the set of 3, 4, and 5 (for more than 5 items the results were almost the same) and the percentage of permutations of 10, 20, ..., 100%. Value no denotes that the permutation operator did not occur in the resulting schema; value partly denotes that it occurred, but not for the whole set; value yes denotes that it occurred correctly.

Table 1. Influence of percentage of permutations

Size of the set	Percentage of permutations									
	10%	20%	30%	40%	50%	60%	70%	80%	90%	100%
3	no	partly	no	partly	no	no	partly	yes	yes	yes
4	no	yes	partly	yes	partly	partly	partly	yes	yes	yes
5	no	no	partly	partly	partly	partly	partly	yes	yes	yes

Elements with Different Content. Similarly we have created XML documents containing an element with the same name but different structure in various contexts. The key parameter of the testing sets was the percentage of the same subelements within the element. The experiments showed that the algorithm behaves according to the expectations and joins two elements if they "overlap" in more than 50% of their content.

5   Conclusion

The aim of this paper was to propose an algorithm for automatic construction of an XML schema for a given sample set of XML documents exploiting new constructs of XML Schema language. In particular we have focussed on unordered sequences allowed by the all element and the ability to specify elements having the same name but different content. We have proposed a hybrid algorithm that combines several approaches and can be easily further enhanced. Our main motivation was to increase the exploitation of XSDs providing a reasonable draft of an XML schema that can be further improved manually by a user, if necessary.

Our future work will focus on further exploitation of other XML Schema constructs, such as, e.g., element groups, attribute groups, or inheritance, i.e. reusability of parts of a schema that can increase naturalness of the result. This idea is highly connected with our second future improvement that will focus on user interaction. This way we can ensure more realistic and suitable results than using a purely automatic reverse engineering. Even our current approach can be extended using user interaction in several steps. The first one is the process of clustering, where a user can specify new clusters, e.g., on the basis of semantics of elements and attributes. A second example of exploitation of user-provided information can be a set of negative examples, i.e. XML documents which do not conform to the target XML schema. Such documents would influence steps of ants, in particular the objective function and, hence, enable to create better result. And, finally, the user interaction can be exploited directly for specifying steps of ants, enabling to find the optimal solution more efficiently.

References

1. Available at: http://arthursclassicnovels.com/
2. Available at: http://www.cs.wisc.edu/niagara/data.html
3. Available at: http://research.imb.uq.edu.au/rnadb/
4. Available at: http://www.assortedthoughts.com/downloads.php
5. Available at: http://www.ibiblio.org/bosak/
6. Available at: http://oval.mitre.org/oval/download/datafiles.html
7. Available at: http://www.rcsb.org/pdb/uniformity/
8. Available at: http://www.eecs.umich.edu/db/mbench/
9. Available at: http://arabidopsis.info/bioinformatics/narraysxml/
10. Available at: http://db.uwaterloo.ca/ddbms/projects/xbench/index.html
11. Ahonen, H.: Generating Grammars for Structured Documents Using Grammatical Inference Methods. Report A-1996-4, Dep. of Computer Science, University of Helsinki (1996)
12. Bartak, R.: On-Line Guide to Constraint Programming (1998), http://kti.mff.cuni.cz/~bartak/constraints/
13. Berstel, J., Boasson, L.: XML Grammars. In: Nielsen, M., Rovan, B. (eds.) MFCS 2000. LNCS, vol. 1893, pp. 182–191. Springer, Heidelberg (2000)
14. Bex, G.J., Neven, F., Van den Bussche, J.: DTDs versus XML Schema: a Practical Study. In: WebDB 2004: Proc. of the 7th Int. Workshop on the Web and Databases, New York, NY, USA, pp. 79–84. ACM Press, New York (2004)
15. Bex, G.J., Neven, F., Vansummeren, S.: XML Schema Definitions from XML Data. In: VLDB 2007: Proc. of the 33rd Int. Conf. on Very Large Data Bases, Vienna, Austria, pp. 998–1009. ACM Press, New York (2007)
16. Biron, P.V., Malhotra, A.: XML Schema Part 2: Datatypes, 2nd edn. W3C (2004), http://www.w3.org/TR/xmlschema-2/
17. Bray, T., Paoli, J., Sperberg-McQueen, C.M., Maler, E., Yergeau, F.: Extensible Markup Language (XML) 1.0, 4th edn. W3C (2006)
18. Dorigo, M., Birattari, M., Stutzle, T.: Ant Colony Optimization – Artificial Ants as a Computational Intelligence Technique. Technical Report TR/IRIDIA/2006-023, IRIDIA, Bruxelles, Belgium (2006)

19. Fernau, H.: Learning XML Grammars. In: Perner, P. (ed.) MLDM 2001. LNCS (LNAI), vol. 2123, pp. 73–87. Springer, Heidelberg (2001)
20. Gao, S., Sperberg-McQueen, C.M., Thompson, H.S.: XML Schema Definition Language (XSDL) 1.1 Part 1: Structures. W3C (2007), http://www.w3.org/TR/xmlschema11-1/
21. Garofalakis, M., Gionis, A., Rastogi, R., Seshadri, S., Shim, K.: XTRACT: a System for Extracting Document Type Descriptors from XML Documents. In: SIGMOD 2000: Proc. of the 2000 ACM SIGMOD Int. Conf. on Management of Data, pp. 165–176. ACM Press, New York (2000)
22. Gold, E.M.: Language Identification in the Limit. Information and Control 10(5), 447–474 (1967)
23. Goldman, R., Widom, J.: DataGuides: Enabling Query Formulation and Optimization in Semistructured Databases. In: VLDB 1997: Proc. of the 23rd Int. Conf. on Very Large Data Bases, pp. 436–445. Morgan Kaufmann, San Francisco (1997)
24. Grunwald, P.D.: A Tutorial Introduction to the Minimum Description Principle (2005), http://homepages.cwi.nl/~pdg/ftp/mdlintro.pdf
25. Jain, A.K., Dubes, R.C.: Algorithms for Clustering Data. Prentice Hall College Div., Englewood Cliffs (1988)
26. Mignet, L., Barbosa, D., Veltri, P.: The XML Web: a First Study. In: WWW 2003: Proc. of the 12th Int. Conf. on World Wide Web, vol. 2, pp. 500–510. ACM Press, New York (2003)
27. Mlynkova, I., Toman, K., Pokorny, J.: Statistical Analysis of Real XML Data Collections. In: COMAD 2006: Proc. of the 13th Int. Conf. on Management of Data, pp. 20–31. Tata McGraw-Hill Publishing Company Limited, New York (2006)
28. Moh, C.-H., Lim, E.-P., Ng, W.-K.: Re-engineering Structures from Web Documents. In: DL 2000: Proc. of the 5th ACM Conf. on Digital Libraries, pp. 67–76. ACM Press, New York (2000)
29. Murata, M., Lee, D., Mani, M.: Taxonomy of XML Schema Languages Using Formal Language Theory. ACM Trans. Inter. Tech. 5(4), 660–704 (2005)
30. Nierman, A., Jagadish, H.V.: Evaluating Structural Similarity in XML Documents. In: WebDB 2002: Proc. of the 5th Int. Workshop on the Web and Databases, Madison, Wisconsin, USA, pp. 61–66. ACM Press, New York (2002)
31. Peterson, D., Biron, P.V., Malhotra, A., Sperberg-McQueen, C.M.: XML Schema 1.1 Part 2: Datatypes. W3C (2006), http://www.w3.org/TR/xmlschema11-2/
32. Thompson, H.S., Beech, D., Maloney, M., Mendelsohn, N.: XML Schema Part 1: Structures, 2nd edn., W3C (2004), http://www.w3.org/TR/xmlschema-1/
33. Wong, R.K., Sankey, J.: On Structural Inference for XML Data. Technical Report UNSW-CSE-TR-0313, School of Computer Science, The University of New South Wales (2003)

A User Driven Data Mining Process Model and Learning System

Esther Ge, Richi Nayak, Yue Xu, and Yuefeng Li

CRC for Construction Innovations
Faculty of Information technology
Queensland University of Technology, Brisbane, Australia
r.nayak@qut.edu.au

Abstract. This paper deals with the problem of using the data mining models in a real-world situation where the user can not provide all the inputs with which the predictive model is built. A learning system framework, Query Based Learning System (QBLS), is developed for improving the performance of the predictive models in practice where not all inputs are available for querying to the system. The automatic feature selection algorithm called Query Based Feature Selection (QBFS) is developed for selecting features to obtain a balance between the relative minimum subset of features and the relative maximum classification accuracy. Performance of the QBLS system and the QBFS algorithm is successfully demonstrated with a real-world application.

Keywords: Data Mining, Learning System, predictive model, lifetime prediction, corrosion prediction, feature selection, civil engineering.

1 Introduction

Data Mining (DM) has been driven by the need to solve practical problems since its inception [20]. In order to achieve a greater usability of DM models, there are three main phases in the lifecycle of a data mining project: (1) training of the model, (2) evaluation (or testing) of the model and (3) using the final trained model in practice. The third phase is usually carried out by business managers or a typical user of the system. A number of Knowledge Discovery and Data Mining (KDDM) process models have been established to organize the lifecycle of a DM project within a common framework. The existing KDDM process models end up with the deployment phase in which rules and patterns inferred from the trained models are utilized in decision making [6]. These process models, however, do not consider the utilization of trained model as a prediction tool in real use for prediction purpose.

DM has been successfully applied in many areas such as marketing, medical and financial [13]. Civil engineering is also one of the areas where a variety of successful real-world data mining applications are reported in building construction [5, 9, 12, 14, 17-19, 21-23, 26]. One such application is metallic corrosion prediction in buildings. These applications can be classified into two main categories: 1) building the predictive models using various data mining techniques [5, 9, 12, 17, 18, 21-23, 26]; and 2) improving the prediction accuracy using new hybrid methods [14, 19]. All of

J.R. Haritsa, R. Kotagiri, and V. Pudi (Eds.): DASFAA 2008, LNCS 4947, pp. 51–66, 2008.

these predictive models assume that the inputs that users will provide in *using* the model are the same as the input features used in *training* the models. However, if users have information of limited inputs only, the predicted results will not be as good as they were during the training and evaluation phases of the data mining system. In other words, the performance of the predictive model degrades due to the absence of many input values. A major problem that still needs to be solved is how to select appropriate features to build the model for a real situation when users have information on limited inputs only.

A considerable body of research has emerged to address the problems of selecting the relevant features from a set of features. The existing feature selection algorithms can be grouped into two broad categories: filters and wrappers [2, 15, 27] for finding the best subset for maximizing classification accuracy [7] [11] [16] [1]. Filters are independent of the inductive algorithm and evaluate features based on the general characteristics of the data, whereas wrappers evaluate features using the inductive algorithm that is finally employed for learning as well. However, these algorithms do not suffice to address our problem. These algorithms can not be used to remove relevant features while the classification accuracy is still acceptable and, in particular, it is known that users in practice would not be familiar with these features while using the system in predictive modeling.

This paper deals with the problem of using the data mining models in a real-world situation where the user can not provide all the inputs with which the model is built. We have developed a learning system framework, called as Query Based Learning System (QBLS), for improving the performance of the predictive models in practice where not all inputs are available for querying to the system. The automatic feature selection algorithm, called as Query Based Feature Selection (QBFS) is developed for selecting features to obtain a balance between the relative minimum subset of features and the relative maximum classification accuracy. This algorithm is evaluated on a number of synthetic data sets and a real-world application of the lifetime prediction of metallic components in buildings.

2 Query Based Learning System

The proposed Query Based Learning System (QBLS) (presented in Figure 1) is a data mining process model based on the most popular industry standard model, CRISP-DM (Cross Industry Standard Process for Data Mining)[6]. The QBLS model consists of nine phases structured as sequences of predefined steps. Three procedures that are different from the CRISP-DM are highlighted. These three procedures - Query Based Feature Selection (QBFS), Results Post-processing and Model in Use - are critical for the success of the proposed QBLS model. The QBFS is separated from the data pre-processing step as it has the involvement of users or domain experts and hence is different from the usual feature selection. The basic idea of the QBFS is to select a minimum subset of relevant features with which the predictive model provides an acceptable performance, as well as, to make the selected features available to users when the model is used in practice. Section 3 will discuss this further.

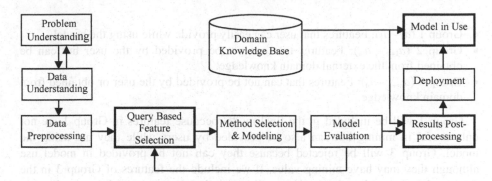

Fig. 1. Query based learning system

In the proposed process model QBLS, a domain knowledge base assists to deal with the vague queries in the "Model in Use" phase and with eliminating illogical outcomes in post-processing. Some features included in the final model may not be directly provided by users but can be inferred by the domain knowledge base. For example, "annual rainfall" is an important factor in determining the service life of building components in civil engineering. While using the DM model to predict the service life of a building component, the user will most likely provide the location and material as an input. The user may not be aware of the exact value of rainfall in the area. However, a domain knowledge base will have such information. This information can now be treated as one of the input values for the model. Furthermore, the domain knowledge base can be used in reinforcing the outputs inferred by the predictive model. Since the real-world DM models are for solving practical problems, the final result is critical to users. However, mining errors are inevitable even for a perfect model. The domain knowledge base is used to confirm that the results predicted by the data mining system do abide by the rules of the domain and/or domain experts. For example, it is domain knowledge in civil engineering that (1) a roof in a severe marine location will not last longer than one in a benign environment, and (2) a stainless steel roof should last longer than one with galvanized steel. Such in-built rules are checked to ensure the correctness of the results processed by models.

3   Query Based Feature Selection (QBFS)

The basic idea of QBFS is to obtain the minimum subset of features that will produce an accurate predictor (numerical prediction) by improving the performance of the QBLS when used in practice where not all inputs are available for querying.

3.1   Algorithms

The first step of QBFS involves removing the features with no mining value such as identification features. The remaining features, $A = \{a_1, a_2, ..., a_k, a_{k+1}, ..., a_m, a_{m+1}, ..., a_n\}$ are clustered into three groups according to their easy availability to users.

- Group 1 (a_1 - a_k): Features that user can easily provide while using the model.
- Group 2 (a_{k+1} - a_m): Features that can not be provided by the user but can be obtained from the external domain knowledge.
- Group 3 (a_{m+1} - a_n): Features that can not be provided by the user or obtained from domain knowledge.

Group 1 will be included in the final model because features in Group 1 are not only useful in mining but can also be provided by users while they are using the model. Group 3 will be rejected because they can not be provided in model use although they may have mining value. If we include the features of Group 3 in the final model, their values in new data will be missing. As a result, the generalization accuracy will decrease. A decision has to be made for features in Group 2, as they can not be provided by users but they can be obtained from external domain knowledge. If we include all the features of Group 2, the measurements to obtain some of these values may be too complex and computationally expensive. If we exclude those features, the performance of the model may not be accepted by users.

For QBFS to be commonly used or to be used in applications in which the expertise is not available to guide the categorisation of features into three groups, the size of Group 1 and Group 3 is reduced to zero so that all available features become the Group 2 members and can now be tested for their relevancy to the target feature. The three variations of QBFS - QBFS-F, QBFS-W, and QBFS-FW (described below) - are applied to the features of Group 2 for selecting a minimum subset.

QBFS-F. QBFS-F is a correlation-based filter algorithm based on the concept that a feature is good if it is highly correlated to the target feature but not highly correlated to any of the other features [28]. Features in Group 1 are already chosen so they become a starting subset to remove the features of Group 2. A Group 2 feature is removed if the level of correlation between any of the Group 1's features and the Group 2 feature is high enough to cause redundancy. Moreover, if the features in Group 2 are highly correlated to each other, one of them, which is more redundant to Group 1's features, is removed. The Pearson's correlation (R) is used to estimate the level of correlation between two features as it is a symmetrical measure and there is no notion of one feature being the "class". When two features (X and Y) are continuous, the standard linear (Pearson's) correlation is calculated (Equation 1) where $\bar{x}_i$ is the mean of X and $\bar{y}_i$ is the mean of Y features.

$$R_{XY} = \frac{\sum_i (x_i - \bar{x}_i)(y_i - \bar{y}_i)}{\sqrt{\sum_i (x_i - \bar{x}_i)^2} \sqrt{\sum_i (y_i - \bar{y}_i)^2}} \tag{1}$$

When one feature is continuous and the other is discrete, a weighted Pearson's correlation is calculated (Equation 2) where p is the prior probability that X takes value x_i and X_{bi} is a binary feature that takes value 1 when X has value x_i and 0 otherwise.

$$R_{XY} = \sum_{i=1}^{k} p(X = x_i) R_{X_{bi}Y} \qquad (2)$$

When both features are discrete, all weighted correlations are calculated for all combinations [11] (Equation 3).

$$R_{XY} = \sum_{i=1}^{k} \sum_{j=1}^{l} p(X = x_i, Y = y_j) R_{X_{bi}Y_{bj}} \qquad (3)$$

The QBFS-F algorithm as shown in Figure 2 consists of two major parts. The first part removes features in Group 2 that are highly correlated to features in Group 1 (Step 3) and the second part removes features in Group 2 that are highly correlated to each other (Step 6). The correlation level between a feature a_i and a class C, denoted as Ra_i,c is used as reference to determine whether the feature is highly correlated. For each feature a_i in Group 2, if its correlation to any features a_j in Group 1 (Ra_j,a_i) is equal to or greater than Ra_i,c, it will be removed. The remaining features in Group 2 are sorted in ascending order based on their Ra_j,a_i values so the first one is the least correlated to any features in Group 1. If any two features are highly correlated to each other and one of them needs to be removed, the one with bigger Ra_j,a_i values is removed. The bigger Ra_j,a_i value indicates the higher possibility of this feature to be redundant to Group 1's features.

If the size of group 1 is zero, the algorithm is simplified as only the second part. In such a case, the features in Group 2 arc sorted in descending order based on the Ra_i,c values. If any two features are highly correlated to each other, the one with smaller Ra_i,c value is removed.

QBFS-W. As the QBFS-F algorithm does not take into account the inductive algorithms, the features it has chosen might not be appropriate for a learning method. The proposed wrapper algorithm, called as QBFS-W involves an inductive algorithm as the evaluation function. As a wrapper method includes an iterative step for retraining the model, the run time for QBFS-W is inevitably long. The forward selection heuristic search strategy [2, 7, 8] is employed to reduce the time complexity. Forward selection starts from the empty set, and each iteration generates new subsets by adding a feature selected by some evaluation function [7].

The algorithm (Figure 3) begins a search at the Group 1's features and adds the p 2's features one at a time. If the size of Group 1 is zero, it begins the search with the empty set of features. Each time a feature with the best performance is chosen and then this feature is removed from Group 2. The loop is terminated when an acceptable model with the performance $\geq \delta$ is built with the minimum set of features or all Group 2's features are chosen. The stopping criterion δ is a predefined threshold ($0 < \delta < 1$) using Correlation Coefficient (CC) as performance measures. A small δ is associated with a high probability of removing relevant features. This parameter is fine-tuned empirically to provide good performance. The worst case is that in which all features in Group 2 are included in the feature set. Suppose the size of Group 2 is N, the search space of this algorithm is $O(N!)$ for the worst case.

Input:
  D: the whole data set
  A_1: the features of Group 1 $\{a_1, ..., a_k\}$
  A_2: the features of Group 2 $\{a_{k+1}, ..., a_m\}$
  C: the target feature
Output:
  S_{min}: the minimum subset

1. **For** $i = k+1$ **to** m
      Calculate $Ra_i,_c$ for a_i
   End
2. $A_2' = \emptyset$
3. **For** $i = k+1$ **to** m
      For $j = 1$ **to** k
           Calculate $Ra_j,_{a_i}$ for a_i
           If $Ra_j,_{a_i} \geq Ra_i,_c$
                Remove a_i
                Exit For
           End
      End
      If a_i is not removed
           Append a_i to A_2'
      Keep the biggest $Ra_j,_{a_i}$ for a_i
   End
4. Sort A_2' in ascending values of $Ra_j,_{a_i}$
5. $a_p \leftarrow$ get the first element of A_2'
6. **Do begin**
      $a_q \leftarrow$ get the next element of A_2' following a_p
      Do begin
           Calculate $Ra_p,_{a_q}$
           If $Ra_p,_{a_q} \geq Ra_q,_c$
                Remove a_q from A_2'
           $a_q \leftarrow$ get the next element of A_2' following a_q
      Until $a_q == NULL$
      $a_p \leftarrow$ get the next element of A_2' following a_p
   Until $a_p == NULL$
7. $S_{min} = A_2'$

Fig. 2. QBFS-F algorithm

QBFS-FW. Wrappers and filters can complement each other to propose a better method, in that filters search through the feature space efficiently while the wrappers provide good accuracy [29]. The algorithms combining filters and wrappers usually choose some best subsets using a goodness measure and then exploit cross validation to decide a final best subset across different cardinalities [28]. The proposed QBFS-FW algorithm (shown in Figure 4) tries to combine the advantages of both methods.

QBFS-FW first calculates the correlation level (Ra_i,c) between each feature a_i in Group 2 and target feature C and sorts Group 2's features in descending order based on their Ra_i,c so that the most relevant feature is positioned at the beginning of the list. Then it employs backward elimination [7, 8] to remove the Group 2's features one at a time. The algorithm attempts to keep the features that are strongly relevant to the target feature in the selected subset. It also reduces the time complexity to be linear. Since the time for calculating Ra_i,c can be ignored compared to the model training time, the search space of this algorithm is $O(N)$ for the worst case where N is the size of Group 2.

Input:
    D: the whole data set
    A_1: the features of Group 1 $\{a_1, ..., a_k\}$
    A_2: the features of Group 2 $\{a_{k+1}, ..., a_m\}$
    C: the target feature
    δ: a predefined threshold
Output:
    S_{min}: the minimum subset

1. $S_{min} = A_1$
2. Train the model with S_{min} and keep the performance P
3. If $P \geq \delta$, **Return** S_{min}
4. $Q = A_2$
5. While $Q \neq \phi$
    1) For each $q \in Q$
        Set $S' \leftarrow \{q\} \cup S_{min}$

        Train the model with S' and note the performance P
    2) Set $S_{min} \leftarrow \{q^*\} \cup S_{min}$ where q^* corresponds to the best P obtained in step 5.1
    3) If $P \geq \delta$, **Return** S_{min}
        Else
            Set $Q \leftarrow Q \setminus \{q^*\}$
        End

Fig. 3. QBFS-W algorithm

3.2 Evaluation

The proposed algorithms are evaluated in terms of number of selected features and prediction accuracy. The experiments were performed on a real-world dataset for predicting lifetime of metallic components. Other datasets from the UCI collection [3] are also used in experiments for comparison purposes. The model tree algorithm (M5) [24] is chosen as inductive algorithms for QBFS-W and QBFS-FW algorithm. Two representative feature selection methods - CFS [11] and ReliefF [16] - are chosen for comparison as they are leading algorithms. For ReliefF, we set $m = 250$ (the number of instances sampled), $k = 10$ (number of nearest neighbors) for discrete class data and

200 for numeric class data, $\sigma = 20$ (a parameter that controls the influence of nearest neighbors) and $\delta = 0.01$ (a threshold by which features can be discarded). Tenfold cross validation (10-CV) was used throughout the experiments.

Input:
  D: the whole data set
  A_1: the features of Group 1 $\{a_1, ..., a_k\}$
  A_2: the features of Group 2 $\{a_{k+1}, ..., a_m\}$
  C: the target feature
  δ: a predefined threshold
Output:
  S_{min}: the minimum subset

1. $S_{min} = A_1 + A_2$
2. **For** $i = k+1$ **to** m
      Calculate Ra_i,c for a_i
   End
3. Sort A_2 in descending values of Ra_i,c
4. $a_p \leftarrow$ get the last element of A_2
5. **Do begin**
      Set $S' \leftarrow S_{min} \setminus \{ a_p \}$
      Train the model with S'
      If the performance $P \geq \delta$, remove a_p from S_{min}
      $a_p \leftarrow$ get the next element of A_2 preceding a_p
   Until $a_p == NULL$ or $P < \delta$
6. **Return** S_{min}

Fig. 4. QBFS-FW algorithm

The Real-world Dataset. The objective is to predict the service life of metallic components in Queensland school buildings based on the information from multiple sources. The details of these input datasets are presented in Table 1. The datasets include four different sources of service life information from the Delphi Survey, Holistic- I, -II and –III, where Holistic-III was divided into two parts in terms of different target features. The multiple sources are independent but complementary to each other. These sources can not be combined and the models are required to be constructed independently from each of them. Holistic-I, -II and -III relate to different component types with different materials while Delphi contains all component types with all materials. Each data source contains completely different features in which some can not be provided by users or obtained from domain knowledge. Features of each data source are divided into three groups (as defined in section 3.1) after the consultation with domain experts before applying QBFS.

For each of the data sources, we run all the three algorithms of QBFS, CFS and ReliefF. Table 2 presents the total number of features selected by the three algorithms for each of the datasets and their selected feature ID. The results in Table 2 reveal that on an average, both QBFS and CFS reduce more than half of the features while

Table 1. Details of real-world datasets

Dataset	I	F	BC	BM	T
Delphi Survey	683	10	Roofs / Gutters / Others	Galvanized Steel / Zincalume / Colorbond / Others	Mean
Holistic-I	9640	11	Gutters	Galvanized Steel / Zincalume	MLannual
Holistic-II	4780	20	Gutters	Colorbond	Life of gutter at 600um
Holistic-III	1297	18	Roofs	Galvanized Steel / Zincalume	Zincalume Life / Galvanized Life

I: number of instances; F: number of features; BC: building component;
BM: building material; T: target feature.

Table 2. Total number and feature ID of selected features on real-world datasets

Data Set	QBFS			CFS	ReliefF	Full Set
	QBFS-F	QBFS-W	QBFS-FW			
Delphi Survey	5 (1,2,4,5,6)	5 (1,2,4,5,6)	5 (1,2,4,5,6)	3 (2,4,6)	1 (4)	5
Holistic-I	6 (2,3,6,8,9,10)	5 (2,3,8,9,10)	5 (2,3,8,9,10)	4 (3,6,9,10)	4 (2,3,6,10)	6
Holistic-II	4 (2,3,7,12)	5 (2,3,7,12,14)	5 (2,3,7,12,14)	3 (2,20,21)	10 (2,3,4,13, 14,15,18-21)	14
Holistic-III_Ga	3 (3,4,9)	3 (3,4,9)	3 (3,4,9)	6 (3,9,10,11,13,15)	10 (3,4,6,7,10, 11,12,13,15,16)	12
Holistic-III_Zi	3 (3,4,9)	3 (3,4,9)	3 (3,4,9)	4 (3,8,9,14)	5 (3,4,8,9,14)	6
Average	4.2	4.2	4.2	4	6	8.6

ReliefF selects more features. There are some overlapping between the selected features by QBFS, CFS and ReliefF while for QBFS-F, QBFS-W and QBFS-FW, the selected features are almost the same.

In order to examine the effect of these selections on accuracy, we apply M5 on the selected features. Table 3 shows the Correlation Coefficient (CC) using M5 on 10-CV and a specified test set. This specified test set only includes those features whose values can be provided by users and leaving other features as missing values. For example, CFS chose 4 features (Longitude, ZincalumeMassLoss, Marine and N) for Holistic-III_Zi to build the model. When users utilise this model, they only can input "Longitude" and "Marine", leaving "ZincalumeMassLoss" and "N" as missing values. Therefore, the values of "ZincalumeMassLoss" and "N" in the specified test set for CFS should be omitted. Such test set reflects predictive performance of the models on unseen cases in a practical scenario.

Based on the CC on 10-CV of each algorithm, we observe that the learning accuracy is very close to when using the full set. This indicates the ability of QBFS, CFS and ReliefF to identify redundant features. On the other hand, based on the CC

Table 3. Correlation Coefficient (CC) of M5 on selected features of real-world datasets

Dataset	QBFS-F		QBFS-W/FW		CFS		ReliefF		Full Set	
	10-CV	test	10-CV	test	10-CV	test	10-CV	test	10-CV	test
Delphi Survey	0.9198	0.9016	0.9198	0.9016	0.8577	0.8850	0.6812	0.6581	0.9198	0.9016
Holistic-I	0.9790	0.9746	0.9764	0.9672	0.9081	0.9568	0.6543	0.7709	0.9790	0.9746
Holistic-II	0.8421	0.8796	0.9973	0.9978	0.9994	0.3037	1	-0.7067	1	-0.8349
Holistic-III _Ga	0.9393	0.9321	0.9393	0.9321	0.9948	-0.633	0.9913	0.3403	0.9883	0.4756
Holistic-III _Zi	0.8801	0.9449	0.8801	0.9449	0.9969	0	0.9971	0	0.9971	0
Average	**0.9121**	**0.9266**	**0.9426**	**0.9487**	**0.9514**	**0.3025**	**0.8648**	**0.2125**	**0.9768**	**0.3034**

* 10-CV denotes CC on 10-CV, test denotes CC on a specified test set.

on the specified test set of each algorithm, we find that only the proposed QBFS algorithms maintain the accuracy while the CFS and ReliefF algorithms perform very poorly in a practical situation. Even in some cases (e.g. CFS and ReliefF for Holistic-III_Zi and ReliefF for Holistic-II), the accuracy is reduced to zero or minus. This indicates that the good generalization accuracy can be obtained in real situations with the use of QBFS, while the existing algorithms can not deal with the problem of selecting appropriate features for building the model in a real-world situation in which the user can not provide all the information with which the model is being trained.

The UCI Datasets. The details of the datasets taken from the UCI repository are presented in Table 4. Since we do not have domain experts in these datasets to guide the categorization of three groups, the size of Group 1 and Group 3 are reduced to zero. All available features are kept in Group 2.

Table 5 shows the total number of features selected by the algorithms for each dataset and their selected feature ID. Table 6 presents the prediction accuracy using M5 on 10-CV. From Table 5, we observe that on average, both of the QBFS-F and QBFS-W algorithms reduce more than half of the features (especially for QBFS-W, which reduces features to 2.5) while CFS and ReliefF select more features (4.75 and 4.25). From Table 6, we observe that the CC of all algorithms on average is very close to the CC when calculating with the full set. That means all algorithms can maintain the accuracy after feature reduction. QBFS-FW has achieved 0.9198 of CC in spite of using 4.5 features only. On the contrary, CFS only achieves 0.9175 of CC in spite of using 4.75 features and ReliefF achieves 0.8796 of CC in spite of using 4.25 features. QBFS-F and QBFS-W select relatively fewer features and achieves relatively higher accuracy. The above experimental results suggest that the QBFS algorithm is a practical solution to the domain driven data mining problems. Moreover, for all datasets, QBFS performs comparably with two representative feature selection algorithms CFS and ReliefF.

Table 4. Details of UCI datasets

ID	Dataset	Instances	Features	Target Feature
1	autoMPG	398	9	mpg
2	CPU	209	10	PRP
3	housing	506	14	medv
4	servo	167	5	class

Table 5. Total number and feature ID of selected features on UCI datasets

Dataset	QBFS			CFS	ReliefF	Full Set
	QBFS-F	QBFS-W	QBFS-FW			
autoMPG	3 (5,6,7)	2 (3,7)	5 (2,3,4,5,7)	7 (2-8)	6 (2,3,4,5,7,8)	7
CPU	5 (4,5,6,7,8)	2 (5,6)	3 (4,5,6)	5 (4,5,6,7,8)	5 (4,5,6,7,8)	7
housing	4 (4,6,11,13)	3 (6,9,13)	6	4 (4,6,11,13)	5 (1,4,6,11,13)	13
servo	3 (1,2,3)	3 (1,2,3)	4 (1,2,3,4)	3 (1,2,3)	1 (3)	4
Average	**3.75**	**2.5**	**4.5**	**4.75**	**4.25**	**7.75**

Table 6. Correlation Coefficient (CC) of M5 on selected features of UCI datasets

Dataset	QBFS-F	QBFS-W	QBFS-FW	CFS	ReliefF	Full Set
autoMPG	0.9213	0.9033	0.9228	0.9258	0.9230	0.9258
CPU	0.9420	0.9149	0.9248	0.9420	0.9420	0.9242
Housing	0.8767	0.8756	0.8961	0.8767	0.8977	0.9131
Servo	0.9254	0.9254	0.9356	0.9254	0.7558	0.9356
Average	**0.9164**	**0.9048**	**0.9198**	**0.9175**	**0.8796**	**0.9247**

4 Implementation of QBLS

In this section, we deploy the Query Based Learning System in a real-world application of the lifetime prediction of metallic components in buildings.

4.1 Overview of the System

The proposed QBLS system (illustrated in Figure 5) basically consists of three main parts: feature selection, predictors and domain knowledge. The QBFS algorithm is first applied to the datasets to select a minimum subset of features which can be provided by users. A data mining method (we discuss the method selection in next section) is applied on selected features to build the predictors for all of the datasets. The predictors are used to carry out prediction for user input queries. The domain knowledge base consists of three parts: salt deposition knowledge, rainfall knowledge and generalized rules extracted from domain expert opinions. Because the features

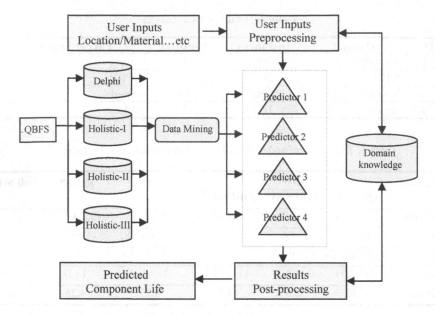

Fig. 5. Overview of the system

selected to build the predictors include features of "Salt Deposition" and "Rainfall Annual", the salt deposition and rainfall database is included in the knowledge base, which is for pre-processing user inputs. Generalized rules are used in post-processing the predicted results, for example, solving the inconsistency in predicted results.

4.2 Representation of Knowledge Base

The knowledge base is implemented as relational database in which the knowledge is represented as items in the database. Examples of the salt deposition knowledge and rainfall knowledge in the generated knowledge base are presented in Tables 7 and 8 respectively. There are total 18 generalized rules in the generated knowledge base and some of them are presented in Table 9.

As the location (longitude and latitude) that users input in the queries may not exactly match the salt deposition and rainfall knowledge, a similarity principle is employed to obtain the value of salt deposition and rainfall. The similarity principle means that the nearest geographic location will have the most similar value for salt deposition and annual rainfall. Once the user inputs longitude and latitude, the system finds the nearest location from the knowledge base and then gets the value of salt deposition and rainfall. These values can then be treated as user inputs for the predictors.

In terms of the predicted results, the system checks them with the generalised rules. If the component, material and environment are matched and the predicted service life is in the range, the results are considered reasonable and presented to users. Otherwise

Table 7. Salt Deposition Knowledge

XLong	YLat	Salt Deposition
151.986	-28.0373	3.80842

Table 8. Rainfall Knowledge

XLong	YLat	Rain Annual (mm)
151.986	-28.0373	1595

Table 9. Generalised Rules

Component	Environment	Material	Min (years)	Max (years)
Gutters	Marine	Galvanised Steel	5	15
Gutters	Benign	Colorbond	20	50

the system gives a message that the result does not abide by the generalised rules with presenting the values instead of removing this unreasonable result.

4.3 Method Selection and Modeling

There are various data mining methods like Naïve Bayes, K-Nearest Neighbors (K-NN), Decision Tree (DT) [28] and Neural Network (NN) that can be considered to do prediction tasks. Ge et al. [10] have reported the detailed method to choose the best data mining algorithm for building predictors. The results show that the DT and naive bayes methods on the discretised output perform poorly [10]. Three best methods are M5 for Delphi Survey, KNN for Holistic-I and NN and M5 for Holistic-II and III. Considering the balance between accuracy and comprehensibility of predictors, M5 is chosen as the learning method in QBLS.

Since the notion of QBLS is user query based, we combine the model-based learning (M5) with the instance-based learning [25] to improve the performance. This method first uses the instance-based approach to find a set of instances similar to the target instance. Then the class values of similar instances are adjusted using the value predicted by the model tree before they are combined. We use the KNN ($k=3$) for the instance-based method.

Therefore, the final predictors are built using M5+KNN on the features selected by QBFS. The performance of the predictors is presented graphically in Figures 7 and 8. The performance of this M5+KNN combined model is compared with the M5 model and the ensemble model with bagging [4]. Figures 7 and 8 show that the better correlation coefficient and lower mean absolute error are obtained by combining the M5 and KNN learning methods. The method seems to provide significant improvement for relatively weaker models like Holistic-II and Holistic-III_Zi, whereas the improvement for the near-perfect models such as Holistic-I, is not so obvious. The combined M5+KNN model also outperforms the ensemble model with bagging.

D: Delphi H-I: Holistic-I H-II: Holistic-II H-III_G: Holistic-III for Galvanized Steel
H-III_Z: Holistic-III for Zincalume

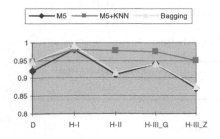 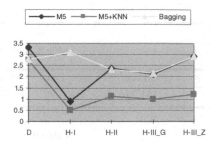

Fig. 7. Correlation Coefficient of M5, M5+ KNN and bagging

Fig. 8. Mean Absolute Error of M5, M5+ KNN and bagging

5 Conclusions

This paper develops a new learning system framework, called as Query Based Learning System (QBLS) for improving the performance of the predictive models in practice where not all inputs are available for querying to the system. This paper also presents a new feature selection algorithm, called as Query Based Feature Selection (QBFS) for selecting the features according to the interest of domain expert or user, while maintaining the accuracy of the predictive model. The available features can all be grouped into three types: 1) that can be provided by users 2) that can be obtained from domain knowledge 3) neither 1) nor 2). External domain knowledge is successfully used for dealing with incomplete and vague queries in pre-processing. It is also used in dealing with inconsistency of the predicted service life from different predictors in post-processing. This novel use of domain knowledge improves the prediction accuracy when users can not provide all inputs.

The proposed Query Based Feature Selection algorithm is compared with two representative feature selection methods CFS and ReliefF. The results prove that QBFS outperforms the existing methods in selecting the features of domain-driven and expert-driven data mining problems. The use of query based feature selection procedure in QBLS indicates that feature selection is not only used for removing irrelevant features but also greatly assists in choosing features according to the use in practice. If a feature, which belongs to type 3 that is it can not be available for user, is included in the final model, the performance of model reduces significantly in using the model. Therefore, such features, even if they are useful in mining, should be rejected. The proposed Query Based feature selection may result in some useful features being rejected. This may reduce the performance of the predictive models. We show that the integrated method combining M5 (model-based) and KNN (instance-based) is successfully applied in such cases for improving performance.

Acknowledgement. Authors would sincerely like to thank the WEKA developers and owners to make use of WEKA codes in this project. We would like to thank CRC-CI to provide us the financial support, and to Penny Corrigan and Michael Ball to provide us the data and feedback during intermittent progress.

References

1. Almuallim, H., Dietterich, T.G.: Learning Boolean Concepts in the Presence of Many Irrelevant Features. Artificial Intelligence 69(1-2), 279–305 (1994)
2. Bengio, S.: Statistical Machine Learning from Data Feature Selection, Matigny, Switzerland (2006)
3. Blake, C., Keogh, E., Merz, C.J.: UCI Repository of Machine Learning Data Bases: Irvine, CA: University of California, Department of Information and Computer Science (1998)
4. Breiman, L.: Bagging Predictors. Machine Learning 24(2), 123–140 (1996)
5. Brence, J.R., Brown, D.E.: Data mining corrosion from eddy current non-destructive tests. Computers & Industrial Engineering 43(4), 821–840 (2002)
6. CRISP-DM, Cross Industry Standard Process for Data Mining (2003)
7. Dash, M., Liu, H.: Feature selection for classification. Intelligent Data Analysis 1, 131–156 (1997)
8. Devijver, P.A., Kittler, J.: Pattern Recognition: A Statistical Approach. Prentice-Hall, Englewood Cliffs (1982)
9. Furuta, H., Deguchi, T., Kushida, M.: Neural network analysis of structural damage due to corrosion. In: Proceedings of ISUMA - NAFIPS 1995 The Third International Symposium on Uncertainty Modeling and Analysis and Annual Conference of the North American Fuzzy Information Processing Society (1995)
10. Ge, E., Nayak, R., Xu, Y., Li, Y.: Data Mining for Lifetime Prediction of Metallic Components, (AusDM 2006). In: The Australasian Data Mining Conference, Sydney (2006)
11. Hall, M.A.: Correlation-based Feature Selection for Discrete and Numeric Class Machine Learning. In: Proc. 17th International Conf. on Machine Learning, Morgan Kaufmann, San Francisco (2000)
12. KamrunNahar, M., Urquidi-Macdonald, M.: Data mining of experimental corrosion data using Neural Network. In: 208th Meeting of the Electrochemical Society, October 16-21, 2005, Electrochemical Society Inc, Los Angeles, CA, United States (2005)
13. Kantardzic, M., Zurada, J.: Next Generation of Data-Mining Applications. Wiley-IEEE Press (2005)
14. Kessler, W., Kessler, R.W., Kraus, M., Kubler, R., Weinberger, K.: Improved prediction of the corrosion behaviour of car body steel using a Kohonen self organising map. In: Advances in Neural Networks for Control and Systems, IEE Colloquium (1994)
15. Kohavi, R., John, G.H.: Wrappers for Feature Subset Selection. Artificial Intelligence 97(1-2), 273–324 (1997)
16. Kononenko, I.: Estimating Attributes: Analysis and Extensions of RELIEF. In: European Conference on Machine Learning (1994)
17. Leu, S.-S., Chen, C.-N., Chang, S.-L.: Data mining for tunnel support stability: neural network approach. Automation in Construction 10(4), 429–441 (2001)
18. Melhem, H.G., Cheng, Y.: Prediction of remaining service life of bridge decks using machine learning. Journal of Computing in Civil Engineering 17(1), 1–9 (2003)
19. Melhem, H.G., Cheng, Y., Kossler, D., Scherschligt, D.: Wrapper Methods for Inductive Learning: Example Application to Bridge Decks. Journal of Computing in Civil Engineering 17(1), 46–57 (2003)

20. Melli, G., Zaiane, O.R., Kitts, B.: Introduction to the special issue on successful real-world data mining applications. SIGKDD Explor. Newsl. 8(1), 1–2 (2006)
21. Mita, A., Hagiwara, H.: Damage Diagnosis of a Building Structure Using Support Vector Machine and Modal Frequency Patterns. In: Smart Structures and Materials 2003: Smart Systems and Nondestructive Evaluation for Civil Infrastructures, March 3-6, 2003. 2003, The International Society for Optical Engineering, San Diego, CA, United States (2003)
22. Morcous, G., Rivard, H., Hanna, A.M., Asce, F.: Modeling Bridge Deterioration Using Case-based Reasoning. Journal of Infrastructure Systems 8(3), 86–95 (2002)
23. Morcous, G., Rivard, H., Hanna, A.M.: Case-Based Reasoning System for Modeling Infrastructure Deterioration. Journal of Computing in Civil Engineering 16(2), 104–114 (2002)
24. Quinlan, J.R.: Learning with Continuous Classes. In: 5th Australian Joint Conference on Artificial Intelligence (1992)
25. Quinlan, J.R.: Combining instance-based and model-based learning. In: Proceedings of the Tenth International Conference on Machine Learning, Morgan Kaufmann, Amherst, Massachusetts (1993)
26. Skomorokhov, A.O.: A knowledge discovery method - APL implementation and application. In: Proceedings of the APL Berlin 2000 Conference, Berlin, Germany, July 24-27, 2000, Association for Computing Machinery (2000)
27. Tang, W., Mao, K.: Feature Selection Algorithm for Data with Both Nominal and Continuous Features. In: Ho, T.-B., Cheung, D., Liu, H. (eds.) PAKDD 2005. LNCS (LNAI), vol. 3518, pp. 683–688. Springer, Heidelberg (2005)
28. Yu, L., Liu, H.: Feature Selection for High-Dimensional Data: A Fast Correlation-Based Filter Solution. In: Proceedings of The Twentieth International Conference on Machine Leaning (ICML 2003), Washington, D.C. (2003)
29. Zexuan, Z., Ong, Y.-S., Dash, M.: Wrapper-Filter Feature Selection Algorithm Using A Memetic Framework. IEEE Transactions on System, Man, and Cybernetics, Part B

Efficient Mining of Recurrent Rules from a Sequence Database

David Lo[1], Siau-Cheng Khoo[1], and Chao Liu[2]

[1] Department of Computer Science, National University of Singapore
[2] Department of Computer Science, University of Illinois at Urbana-Champaign
dlo@comp.nus.edu.sg, khoosc@comp.nus.edu.sg, chaoliu@cs.uiuc.edu

Abstract. We study a novel problem of mining significant recurrent rules from a sequence database. Recurrent rules have the form "whenever a series of precedent events occurs, eventually a series of consequent events occurs". Recurrent rules are intuitive and characterize behaviors in many domains. An example is in the domain of software specifications, in which the rules capture a family of program properties beneficial to program verification and bug detection. Recurrent rules generalize existing work on sequential and episode rules by considering repeated occurrences of premise and consequent events within a sequence and across multiple sequences, and by removing the "window" barrier. Bridging the gap between mined rules and program specifications, we formalize our rules in linear temporal logic. We introduce and apply a novel notion of rule redundancy to ensure efficient mining of a compact representative set of rules. Performance studies on benchmark datasets and a case study on an industrial system have been performed to show the scalability and utility of our approach.

1 Introduction

The information age has caused an explosive growth in the amount of data produced. Mining for knowledge from data has been shown useful for many purposes [12] ranging from finance, advertising, bio-informatics and recently software engineering [17,20]. Addressing the same issue of knowledge discovery from data, we study the problem of mining recurrent rules, each having the following form:

"Whenever a series of precedent events occurs, eventually another series of consequent events occurs"

The above rule is intuitive and represent an important form of knowledge characterizing the behaviors of many systems appearing in various domains. Examples of rules in this format include:

1. Resource Locking Protocol: Whenever a lock is acquired, eventually it is released.
2. Internet Banking: Whenever a connection to a bank server is made and an authentication is completed and money transfer command is issued, eventually money is transferred and a receipt is displayed.

J.R. Haritsa, R. Kotagiri, and V. Pudi (Eds.): DASFAA 2008, LNCS 4947, pp. 67–83, 2008.

3. Network Protocol: Whenever an HDLC connection is made and an acknowledgement is received, eventually a disconnection message is sent and an acknowledgement is received.

Zooming into the domain of software specification and verification, recurrent rules correspond to a family of program properties useful for program verification (*c.f.* [9]). The first example given above corresponds to a program property. Research in software verification addresses rigorous approaches to check the correctness of a software system with respect to a formal specification which often corresponds to a set of program properties (*c.f.*, [26,6]). However, specification might often be outdated or missing due to software evolution, reluctance in writing formal specification and short-time-to-market cycle of software development (*c.f.*, [8,3,5]). Recovering or mining specifications expressed as rules and automata has been a recent interest in software engineering and programming language domain [27,29,3,19]. However, recent approaches on mining rules as specification [27,29] has only focused on two-event rules due to the exponential complexity associated with mining rules of arbitrary length.

To address the above issue, in this paper we propose a novel extension of work on pattern mining, in particular sequential pattern mining [2] and episode mining [22]. Sequential pattern mining first addressed by Agrawal and Srikant in [2] discovers patterns that are supported by a *significant number of sequences*. A pattern is supported by a sequence if the former is a sub-sequence of the later. On the other hand, Mannila *et al.* perform episode mining to discover *frequent episodes within a sequence of events* [22]. An episode is supported by a window if it is a sub-sequence of the series of events appearing in the window. Garriga later extends Mannila *et al.*'s work to replace a fixed-window size with a gap constraint between one event to the next in an episode [11]. In both cases, an episode is defined as a series of events occurring *relatively close* to one another (*i.e.* they occur in the same window). Episode mining focuses on mining from a single sequence of events.

Rules can be formed from both sequential patterns and episodes as proposed in [24,22]. Different from a pattern, a rule expresses a *constraint* involving its premise (*i.e.*, pre-condition) and consequent (*i.e.*, post-condition). These constraints are needed for potential uses of rules in filtering erroneous sequences, detecting outliers, etc.

However, rules from sequential patterns and episodes have different semantics from recurrent rules. A sequential rule *pre* → *post* states: "whenever a sequence is a super-sequence of *pre* it will also be a super-sequence of *pre* concatenated with *post*". An episode rule *pre* → *post* states: "whenever a window is a super-sequence of *pre* it will also be a super-sequence of *pre* concatenated with *post*". Recurrent rules generalize sequential rules where for each rule, *multiple* occurrences of the rule's premise and consequent both *within a sequence and across multiple sequences* are considered. Recurrent rules generalize episode rules by allowing precedent and consequent events to be separated by an *arbitrary* number of events in a *sequence database*. Also, a set of sequences rather than a single sequence is considered during mining.

These generalizations are needed in many application areas, and an example is in mining program properties from execution traces. Because of loops and recursions, an execution trace can contain repeated occurrences of a particular property. Also, program properties are often inferred from a set of traces instead of a single trace (*c.f* [3,19]). Finally, important patterns for verification, such as, lock acquire and release or stream open and close (*c.f* [29]) often have their events occur at some arbitrary distance away from one another in a program trace. Hence, there is a need to "break" the "window barrier" or "gap constraints" in order to capture program properties of interest.

Our goal is to mine a set of rules satisfying given support and confidence thresholds. To reduce the number of mined rules and improve efficiency, we define a novel notion of *rule redundancy* and devise search space pruning strategies to detect redundant rules "early" in the mining process. The final output is a set of non-redundant rules satisfying the given support and confidence thresholds.

In order to bridge the gap between mined rules and program specifications, we formalize our rules using linear temporal logic (LTL) – a widely used formalism in program verification [6]. By mapping rules to LTL expressions, these rules can be directly consumed by existing program verifiers.

We carry out a performance study on several standard benchmark datasets to demonstrate the effectiveness of our search space pruning strategies. We also perform a case study on traces of JBoss Application Server – the most widely used J2EE server – to illustrate the usefulness of our technique in recovering the specifications that a software system obeys.

The contributions of this work are as follows:

1. We present a novel notion of recurrent rules along with its mining algorithm.
2. We introduce and utilize the definition of redundant rules to reduce the number of mined rules.
3. We employ two "apriori"-like properties and "early" detection of redundant rules effective in aiding the scalability of rule mining.
4. We bridge the gap between data mining and program verification by translating mined rules to useful LTL expressions.
5. We show a case study on the utility of our technique in recovering specifications of a large industrial programs.

The outline of this paper is as follows. In Section 2, we discuss related work. Section 3 contains important background information on LTL formalizing our definition of recurrent rules. Section 4 presents the principles behind mining recurrent rules and the pruning strategies employed. Section 5 presents our algorithm. Section 6 describes the study conducted to evaluate the performance of our mining framework and the benefits of various pruning strategies. Section 7 describes our case study, and Section 8 concludes this paper and presents some future work.

2 Related Work

Two related research threads are sequential pattern mining(*e.g.*, [2,28,25,24]) and episode mining (*e.g.*, [22,11]). This work can be viewed as an extension of both sequential rules and episode rules. Differences between our work and the above have been discussed in the introduction section.

In the area of specification mining, a number of studies on mining software temporal properties have been performed [29,27,3,19]. Most of them mine an automata (*e.g.*, [3,19]) and hence are very different from our work. Of the most relevance is the work on mining rule-based specification [29,27], where the rules have a similar semantics as ours but are limited to two-event rules (*e.g.*, ⟨*lock*⟩ → ⟨*unlock*⟩). Their algorithms do not scale for mining multi-event rules since they first list all possible two-event rules and then check the significance of each rules. For rules of arbitrary lengths, the number of possible rules is arbitrarily large. Our work generalizes their work by mining a complete set of rules of arbitrary lengths that satisfy given support and confidence thresholds. To enable efficient mining, we devise a number of search space pruning strategies.

In [20], we proposed iterative patterns to discover software specifications, which are defined based on the semantics of Message Sequence Charts (MSC) [15]. Different from [20], this work is based on a different formalism, namely the semantics of Linear Temporal Logic (LTL). LTL has a wider application area than MSC, ranging from software engineering [29] to security & privacy [4]. In the software domain, LTL (but not MSC) is one of the most widely-used formalism for program verification (*i.e.*, ensuring correctness of a software system) [6]. There are many standard verification tools readily taking software properties expressed in LTL as inputs. Since the underlying target formalisms and semantics are different, both the search space pruning strategies and the mining algorithm are very different from our previous work in [20].

3 Preliminaries

This section introduces preliminaries on LTL and its verification which dictate the semantics of recurrent rules. Also, notations used in this paper are described.

Linear-time Temporal Logic Our mined rules can be expressed in Linear Temporal Logic (LTL) [14]. LTL is a logic that works on possible program paths. A possible program path corresponds to a program trace. A path can be considered as a series of events, where an event is a method invocation. For example, (file_open, file_read, file_write, file_close), is a 4-event path.

There are a number of LTL operators, among which we are only interested in the operators 'G','F' and 'X'. The operator 'G' specifies that *globally* at every point in time a certain property holds. The operator 'F' specifies that a property holds either at that point in time or *finally (eventually)* it holds. The operator 'X' specifies that a property holds at the *next* event. Some examples are listed in Table 1.

Our mined rules state whenever a series of precedent events occurs eventually another series of consequent events also occurs. A mined rule denoted as *pre* →

Table 1. LTL Expressions and their Meanings

$F(unlock)$
    Meaning: Eventually *unlock* is called

$XF(unlock)$
    Meaning: From the next event onwards, eventually *unlock* is called

---

$G(lock \rightarrow XF(unlock))$
    Meaning: Globally whenever *lock* is called, then from the next event onwards, eventually *unlock* is called

---

$G(main \rightarrow XG(lock \rightarrow (\rightarrow XF(unlock \rightarrow XF(end)))))$
    Meaning: Globally whenever *main* followed by *lock* are called, then from the next event onwards, eventually *unlock* followed by *end* are called

Table 2. Rules and their LTL Equivalences

Notation	LTL Notation
$a \rightarrow b$	$G(a \rightarrow XFb)$
$\langle a, b \rangle \rightarrow c$	$G(a \rightarrow XG(b \rightarrow XFc))$
$a \rightarrow \langle b, c \rangle$	$G(a \rightarrow XF(b \wedge XFc))$
$\langle a, b \rangle \rightarrow \langle c, d \rangle$	$G(a \rightarrow XG(b \rightarrow XF(c \wedge XFd)))$

post, can be mapped to its corresponding LTL expression. Examples of such correspondences are shown in Table 2. Note that although the operator 'X' might seem redundant, it is needed to specify rules such as $\langle a \rangle \rightarrow \langle b, b \rangle$ where the 'b's refer to *different occurrences of 'b'*. The set of LTL expressions minable by our mining framework is represented in the Backus-Naur Form (BNF) as follows:

$$rules := G(prepost)$$
$$prepost := event \rightarrow post | event \rightarrow XG(prepost)$$
$$post := XF(event) | XF(event \wedge XF(post))$$

Checking/Verifying LTL Expressions. LTL expressions are originally developed for checking software systems expressed in the form of automata [13] (a transition system with start and end nodes). There are existing tools converting code to an automata (*e.g.*, [7]). Given an automata and an LTL property one can check for its satisfaction through a well-known technique of model checking [6].

    Consider the example in Figure 1, the pseudo-code on the left corresponds to the automaton on the right. Given the property $\langle main, lock \rangle \rightarrow \langle unlock, end \rangle$, a model checking tool (*c.f*, [6]) will ensure that for all states in the model where *lock* preceded by a *main* occurs (marked by the red dashed arrows), eventually (whichever path is taken) *unlock and* then eventually *end* can be reached. For the above example, the property is violated. The *lock* immediately before *end* is not followed by an *unlock*. Note however, the property $\langle main, lock, use \rangle \rightarrow \langle unlock, end \rangle$ is satisfied. This is the case since the *lock* immediately before *end* is not followed by a *use*, i.e., the pre-condition of the rule is not satisfied and the rule vacuously holds.

    In this paper, we map this to sequences. We consider a sequence as a form of automata (a linear one). An event is mapped to a state. A mined rule (or property)

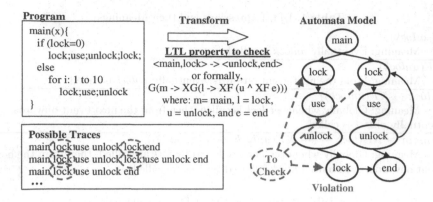

Fig. 1. Code -> Automata -> Verification

$pre \rightarrow post$ with a perfect confidence (*i.e.*, confidence=1) states that in the sequences from all states where the *pre* holds eventually *post* occurs. In the above example, for all points in the sequence (*i.e.*, temporal points) where $\langle main, lock \rangle$ occurs (marked with dashed red circle), one need to check whether eventually $\langle unlock, end \rangle$ occurs. Based on the definition of LTL properties and how they are verified, our technique analyzes sequences and captures *strong* LTL expressions that satisfy given support and confidence thresholds.

Basic Notations. Let I be a set of distinct events considered. The input to our mining framework is a sequence database referred to as $SeqDB$. Each sequence is an ordered list of events, and is denoted as $\langle e_1, e_2, \ldots, e_{end} \rangle$ where $e_i \in I$.

We define a pattern P to be a series of events. We use $last(P)$ to denote the last event of P. A pattern $P_1 + + P_2$ denotes the concatenation of patterns P_1 and P_2. A pattern P_1 ($\langle e_1, e_2, \ldots, e_n \rangle$) is considered a *subsequence* of another pattern P_2 ($\langle f_1, f_2, \ldots, f_m \rangle$) if there exists integers $1 \leq i_1 < i_2 < \ldots < i_n \leq m$ such that $e_1 = f_{i_1}, e_2 = f_{i_2}, \cdots, e_n = f_{i_n}$ (denoted as $P_1 \sqsubseteq P_2$).

4   Generation of Recurrent Rules

Each recurrent rule has the form $P_1 \rightarrow P_2$, where P_1 and P_2 are two series of events. P_1 is referred to as the *premise* or *pre-condition* of the rule, while P_2 is referred to as the *consequent* or *post-condition* of the rule. The rules correspond to temporal constraints expressible in LTL notations. Some examples are shown in Table 2. We use the sample database in Table 3 as our running example.

Table 3. Example Database – $DBEX$

Seq ID.	Sequence
S1	$\langle a, b, e, a, b, c \rangle$
S2	$\langle a, c, b, e, a, e, b, c \rangle$

4.1   Concepts and Definitions

Mined rules are formalized as Linear Temporal Logic(LTL) expressions with the format: G(... → XF...). The semantics of LTL and its verification technique described in Section 3 will dictate the semantics of recurrent rules described here. Noting the meaning of the temporal operators illustrated in Table 1, to be precise, a recurrent rule expresses:

"Whenever a series of events *has just occurred at a point in time (i.e. a temporal point)*, eventually another series of events occurs"

From the above definition, to generate recurrent rules, we need to "peek" at interesting temporal points and "see" what series of events are likely to occur next. We will first formalize the notion of temporal points and occurrences.

Definition 1 (Temporal Points). *Consider a sequence S of the form $\langle a_1, a_2, \ldots, a_{end} \rangle$. All events in S are indexed by their position in S, starting at 1 (e.g., a_j is indexed by j). These positions are called* temporal points *in S. For a temporal point j in S, the prefix $\langle a_1, \ldots, a_j \rangle$ is called the j-prefix of S.*

Definition 2 (Occurrences & Instances). *Given a pattern P and a sequence S, the* occurrences *of P in S is defined by a set of temporal points T in S such that for each $j \in T$, the j-prefix of S is a super-sequence of P and $last(P)$ is indexed by j. The set of* instances *of pattern P in S is defined as the set of j-prefixes of S, for each $j \in T$.*

Example. Consider a pattern P $\langle a, b \rangle$ and the sequence $S1$ in the example database (*i.e.*, $\langle a, b, e, a, b, c \rangle$). The *occurrences* of P in $S1$ form the set of temporal points $\{2,5\}$, and the corresponding set of *instances* are $\{\langle a, b \rangle, \langle a, b, e, a, b \rangle\}$.

We can then define a new type of database projection to capture events occurring after each temporal point. The following are two different types of projections and their associated support notions.

Definition 3 (Projected & Sup). *A database* projected *on a pattern p is defined as:*

$SeqDB_P = \{(j, sx) \mid$ *the jth sequence in SeqDB is s, where $s = px$++sx, and px is the minimum prefix of s containing $p\}$*

Given a pattern P_X, we define $sup(P_X, SeqDB)$ to be the size of $SeqDB_{P_X}$ (equivalently, the number of sequences in SeqDB containing P_X). Reference to the database is omitted if it is clear from the context.

Definition 4 (Projected-all & Sup-all). *A database* projected-all *on a pattern p is defined as:*
$SeqDB_P^{all} = \{(j, sx) \mid$ *the jth sequence in SeqDB is s, where $s = px$++sx, and px is an instance of p in s }*

Given a pattern P_X, we define $sup^{all}(P_X, SeqDB)$ to be the size of $SeqDB_{P_X}^{all}$. Reference to the database is omitted if it is clear from the context.

Definition 3 is a standard database projection (*c.f.* [28,25]) capturing events occurring after the first *temporal point*. Definition 4 is a new type of projection capturing events occurring after *each temporal point*.

Example. To illustrate the above concepts, we project and project-all the example database $DBEX$ with respect to $\langle a, b \rangle$. The results are shown in Table 4(a) & (b) respectively.

Table 4. $(a); DBEX_{\langle a,b \rangle}$ & $(b); DBEX^{all}_{\langle a,b \rangle}$

	Seq ID.	Sequence
(b)	$S1_1$	$\langle e, a, b, c \rangle$
	$S1_2$	$\langle c \rangle$
	$S2_1$	$\langle e, a, e, b, c \rangle$
	$S2_2$	$\langle c \rangle$

	Seq ID.	Sequence
(a)	S1	$\langle e, a, b, c \rangle$
	S2	$\langle e, a, e, b, c \rangle$

The two projection methods' associated notions of sup and sup^{all} are different. Specifically, sup^{all} reflects the number of occurrences of P_X in $SeqDB$ rather than the number of sequences in $SeqDB$ supporting P_X.

Example. Consider the example database, $sup(\langle a, b \rangle, DBEX) = |DBEX_{\langle a,b \rangle}| = 2$. On the other hand, $sup^{all}(\langle a, b \rangle, DBEX) = |DBEX^{all}_{\langle a,b \rangle}| = 4$.

From the above notions of temporal points, projected databases and pattern supports, we can define support and confidence of a recurrent rule.

Definition 5 ((S-/I-)Support & Confidence). *Consider a recurrent rule R_X ($pre_X \rightarrow post_X$). The [prefix-]sequence-support (s-support) of R_X is defined as the number of sequences in $SeqDB$ where pre_X occurs, which is equivalent to $sup(pre_X, SeqDB)$. The instance-support (i-support) of R_X is defined as the number of occurrences of pattern $pre_X + post_X$ in $SeqDB$, which is equivalent to $sup^{all}(pre_X + post_X, SeqDB)$. The confidence of R_X is defined as the likelihood of $post_X$ happening after pre_X. This is equivalent to the ratio of $sup(post_X, SeqDB^{all}_{pre_X})$ to the size of $SeqDB^{all}_{pre_X}$.*

Example. Consider $DBEX$ and a recurrent rule R_X, $\langle a, b \rangle \rightarrow \langle c \rangle$. From the database, the s-support of R_X is the number of sequences in $DBEX$ supporting (or is a super-sequence of) the rule's pre-condition – $\langle a, b \rangle$. There are 2 of them – see Table 4(a). Hence s-support of R_X is 2. The i-support of R_X is the number of *occurrences* of pattern $\langle a, b, c \rangle$ in $DBEX$ – i.e., the number of *temporal points* where $\langle a, b, c \rangle$ occurs. There are also 2 of them. Hence, i-support of R_X is 2. The confidence of the rule R_X ($\langle a, b \rangle \rightarrow \langle c \rangle$) is the likelihood of $\langle c \rangle$ occurring after each *temporal point* of $\langle a, b \rangle$. Referring to Table 4(b), we see that there is a $\langle c \rangle$ occurring after each temporal point of $\langle a, b \rangle$. Hence, the confidence of R_X is 1.

Strong rules to be mined must have their [prefix-] sequence-supports greater than the $min\_s\text{-}sup$ threshold, their instance-supports greater than the $min\_i\text{-}sup$ threshold, *and* their confidences greater the min_conf threshold.

In mining program properties, the confidence of a rule (or property), which is a measure of its certainty, matters the most (*c.f.*, [29]). Support values are considered to differentiate high confidence rules according to the frequency of their occurrences in the traces. Rules with confidences <100% are also of interest due to the imperfect trace collection and the presence of bugs and anomalies [29]. Similar to the assumption made by work in statistical debugging (*e.g.*, [10]), simply put, if a program behaves in one way 99% of the time, and the opposite 1% of the time, the latter is a possible bug. Hence, a high confidence and highly supported rule is a good candidate for bug detection using program verifiers.

We denote recurrent rules, expressible in LTL template $G(pre \rightarrow post)$, as $pre \rightarrow post$, where *pre* and *post* correspond to an event or a series of events. We added the notions of [prefix-] sequence-support, instance-support, and confidence to the rules. The formal notation of recurrent rules is defined below.

Definition 6 (Recurrent Rules). *A recurrent rule R_X is denoted by pre $\rightarrow$ post (s-sup,i-sup,conf). The series of events pre and post represents the rule pre- and post-condition and are denoted by R_X.Pre and R_X.Post respectively. The notions s-sup, i-sup and conf represent the sequence-support, instance-support and confidence of R_X respectively. They are denoted by s-sup(R_X), i-sup(R_X), and conf(R_X) respectively.*

Example. Consider $DBEX$ and the rule R_X, $\langle a, b \rangle \rightarrow \langle c \rangle$ shown in the previous example. It has s-support value of 2, i-support value of 2 and confidence of 1. It is denoted by $\langle a, b \rangle \rightarrow \langle c \rangle (2, 2, 1)$.

4.2 Apriori Properties and Non-redundancy

Apriori properties have been widely used to ensure efficiency of many pattern mining techniques (*e.g.*, [1,2]). Fortunately, recurrent rules obey the following apriori properties:

Theorem 1 (Apriori Property – S-Support). *If a rule $evs_P \rightarrow evs_C$ does not satisfy the min_s-sup threshold, neither will all rules $evs_Q \rightarrow evs_C$ where evs_Q is a super-sequence of evs_P.*

Theorem 2 (Apriori Property – Confidence). *If a rule $evs_P \rightarrow evs_C$ does not satisfy the min_conf threshold, neither will all rules $evs_P \rightarrow evs_D$ where evs_D is a super-sequence of evs_C.*

To reduce the number of rules and improve efficiency, we define a notion of rule redundancy defined based on *super-sequence relationship* among rules having the same support and confidence values. This is similar to the notion of *closed* patterns applied to sequential patterns [28,25].

Definition 7 (Rule Redundancy). *A rule R_X (pre$_X \rightarrow$post$_X$) is redundant if there is another rule R_Y (pre$_Y \rightarrow$post$_Y$) where:*
(1); R_X is a sub-sequence of R_Y (i.e., pre$_X$++post$_X \sqsubseteq$ pre$_Y$++post$_Y$)
(2); Both rules have the same support and confidence values
   Also, in the case that the concatenations are the same (i.e., pre$_X$++post$_X$ = pre$_Y$++post$_Y$), to break the tie, we call the one with the longer premise as being redundant (i.e., we wish to retain the rule with a shorter premise and longer consequent).

A simple approach to reduce the number of rules is to first mine a full set of rules and then remove redundant ones. However, this "late" removal of redundant rules is inefficient due to the exponential explosion of the number of intermediary rules that need to be checked for redundancy. To improve efficiency, it is therefore necessary to identify and prune a search space containing redundant rules "early" during the mining process. The following two theorems are used for 'early' pruning of redundant rules. The proofs are available in our technical report [21].

Theorem 3 (Pruning Redundant Pre-Conds). *Given two pre-conditions P_X and P_Y where $P_X \sqsubset P_Y$, if $SeqDB_{P_X} = SeqDB_{P_Y}$ then for all sequences of events post, rules $P_X \to post$ is rendered redundant by $P_Y \to post$ and can be pruned.*

Theorem 4 (Pruning Redundant Post-Conds). *Given two rules R_X (pre $\to$ P_X) and R_Y (pre $\to P_Y$) if $P_X \sqsubset P_Y$ and $(SeqDB_{pre}^{all})_{P_X} = (SeqDB_{pre}^{all})_{P_Y}$ then R_X is rendered redundant by R_Y and can be pruned.*

Utilizing Theorems 3 & 4, many redundant rules can be pruned 'early'. However, the theorems only provide sufficient conditions for the identification of redundant rules – there are redundant rules which are not identified by them. To remove remaining redundant rules, we perform a post-mining filtering step based on Definition 7.

Our approach to mining a set of non-redundant rules satisfying the support and confidence thresholds is as follows:

Step 1. Leveraging Theorems 1 & 3, we generate a *pruned* set of pre-conditions satisfying *min_s-sup*.
Step 2. For each pre-condition *pre*, we create a *projected-all* database $SeqDB_{pre}^{all}$.
Step 3. Leveraging Theorems 2 & 4, for each $SeqDB_{pre}^{all}$, we generate a *pruned* set containing such post-condition *post*, such that the rule *pre* $\to$ *post* satisfies *min_conf*.
Step 4. Checking the rules' instance-supports, we remove rules from step 3 that do not satisfy *min_i-sup*.
Step 5. Using Definition 7, we filter any remaining redundant rules.
In the next section, we describe our algorithm in detail.

5 Algorithm

In the previous section, the process of mining non-redundant rules has been divided into 5 steps. Steps 1 and 3 sketch how a pruned set of pre- and post- conditions are mined. The following paragraphs will elaborate them in more detail.

Before proceeding, we first describe a set of patterns called *projected database closed* (or LS-Set) first mentioned in [28]. A pattern is in the set if *there does not exist any super-sequence pattern having the same projected database*. Patterns having the same projected database must have the same support, but not vice versa. *Projected database closed* patterns is of special interest to us, as explained in the following paragraphs.

At step 1, a pruned set of pre-conditions is generated from the input database $SeqDB$. From Theorem 3, a pattern is in the pruned pre-condition set if *there does not exist any super-sequence pattern having the same projected database*. Comparing with the definition of *projected database closed* patterns in the previous paragraph, we note that this pruned set of pre-conditions corresponds to the *projected database closed set* (or LS-Set) mined from $SeqDB$.

At step 3, starting with a projected-all database $SeqDB_{pre}^{all}$, we generate a pruned set of post-conditions. From Theorem 4, a pattern is in the pruned post-condition set if *there does not exist any super-sequence pattern having the same projected database*. Again, this set of pruned post-condition corresponds to the *projected database closed set* (or LS-Set) mined from $SeqDB_{pre}^{all}$.

Our mining algorithm (NR^3-Miner: Non-Redundant Recurrent Rule Miner) is shown in Figure 2. First, a pruned set of pre-conditions satisfying the minimum sequence-support threshold (*i.e.*, $min\_s\text{-}sup$) is mined using an LS-Set miner modified from BIDE [25], the state-of-the-art closed sequential pattern miner.[1] Next, for each pre-condition mined, a database projected-all on it is formed. Consequently, another LS-Set Generator is run on each projected-all database to mine the set of post-conditions of the corresponding candidate rules having enough sequence-support and confidence values. Next, each candidate rule is further checked for the

---

Procedure Mine Non-Redundant Recurrent Rules
Inputs: $SeqDB$: Sequence DB; $min\_s\text{-}sup$: Min. S-Support Thresh.;
$min\_i\text{-}sup$: Min. I-Support Thresh.; min_conf : Min. Conf. Thresh.
Output:
Rules: Non Redundant Set of Recurrent Rules
Method:
1: Let $PreCond =$ Generate an LS-Set from $SeqDB$ with the threshold
              set at $min\_s\text{-}sup$
2: For every *pre* in *PreCond*
3:    Let $SeqDB_{pre}^{all} = SeqDB$ projected-all by pattern *pre*
4:    Let $bthd = min\_conf \times |SeqDB_{pre}^{all}|$
5:    Let $PostCond =$ Generate an LS-Set from $SeqDB_{pre}^{all}$ with the
              threshold set at $bthd$
6:    For every *post* in *PostCond*
7:      Add (*pre* $\rightarrow$ *post*) to *Rules*
8: For every *rx* in *Rules*
9:    If (i-sup(*rx*) $< min\_i\text{-}sup$)
10:     Remove *rx* from *Rules*
11:   If (*rx* is redundant according to Def. 7)
12:     Remove *rx* from *Rules*
13: Output *Rules*

Fig. 2. Mining Algorithm – NR^3-Miner

---

[1] BIDE, in effect prunes all search sub-spaces containing patterns not in LS-Set. To mine LS-Set using BIDE, we keep the search space pruning strategy but remove the closure check. The details are available in our technical report [21].

satisfaction of the minimum instance-support threshold (*i.e.*, *min_i-sup*). Provided that the pattern *pre++post* is in the pruned set of pre-conditions computed in the first step of the mining process (*i.e.*, line 1 of the algorithm), the instance-support of a rule *pre → post* has been computed during the second step of the mining process (*i.e.*, line 3 of the algorithm). Otherwise, an additional database scan need to be made to compute the rule's instance-support value. Finally, a filtering step to remove any remaining redundant rules based on Definition 7 is performed. To perform the final filtering step scalably, each remaining rule is first hashed based on its support and confidence values. Only rules falling into the same hash bucket need to be checked for super-sequence relationship.

The algorithm can be adapted easily to generate a full set of recurrent rules. This is performed to serve as a point of reference for investigating the benefit of the early identification and pruning of redundant rules. To generate the full set we can simply: (1); Generate a full set of pre- and post- conditions of rules satisfying the s-support and confidence thresholds at lines 1 and 5 of the algorithm respectively (we use PrefixSpan [23] for this purpose), and (2); Skip the final redundancy filtering step (*i.e.*, lines 11-12 of the algorithm in Figure 2).

6   Performance Evaluation

Experiments have been performed on both synthetic and real datasets on low support thresholds to evaluate the *scalability of our mining framework*. The lower the thresholds, the more difficult it is to mine the rules. Our algorithms are the *first* algorithms mining recurrent rules, hence we compare and contrast the runtime required and the number of rules mined when full and non-redundant sets of recurrent rules are mined to evaluate the *effectiveness of our pruning strategies*.

Datasets. We use 2 datasets in our experiments: one synthetic and another real. Synthetic data generator provided by IBM was used with modification to ensure generation of single-event sequences (*i.e.*, all transactions are of size 1). We also experimented on a click stream dataset (*i.e.*, Gazelle dataset) from KDD Cup 2000 [16]. It contains 23639 sequences with an average length of 3 and a maximum length of 651.

Environment and Configuration. All experiments were performed on a Pentium M 1.6GHz IBM X41 tablet PC with 1.5GB main memory, running Windows XP Tablet PC Edition 2005. Algorithms were written using Visual C#.Net.

Experiment Methodology & Presentation. Experiments were performed by varying *min_s-sup*, *min_i-sup* & *min_conf* thresholds. The results are plotted as line graphs. 'Full' and 'NR' correspond to the full set and non-redundant set of rules respectively. The x-axis of the graph corresponds to the thresholds used while the y-axis represents the runtime required, or the number of mined rules.

For the experiment with the Gazelle dataset, only the results for mining 'NR' rules are plotted. The 'Full' set is not mine-able even at the highest *min_s-sup* threshold shown in Figure 5 & 6 – our attempt produced a gigantic 51 GB file before we had to stop the process.

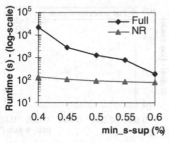
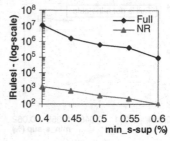

Fig. 3. Varying *min_s-sup* (at *min_conf*=50%,*min_i-sup*=1) for D5C20N10S20 dataset

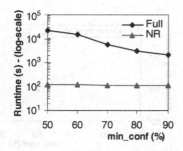
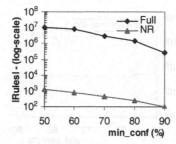

Fig. 4. Varying *min_conf* (at *min_s-sup*=0.4%,*min_i-sup*=1) for D5C20N10S20 dataset

Synthetic Dataset Result. The experiment results for the synthetic dataset are shown in Figure 3 & 4. We produce a synthetic dataset by running the IBM synthetic data generator with the following parameter setting: D (number of sequences - in 1000s) = 5, C (average sequence length) = 20, N (number of unique events - in 1000s) = 10 and S (average number of events in maximal sequences) = 20.

Comparing the results of mining a non-redundant set with that of mining a full set of rules, we note that for the non-redundant set both the runtime and the number of mined rules were reduced by a large amount: up to *147 times less* for the runtime, and *8500 times less* for the number of mined rules.

Both the runtime and the number of mined rules are significantly increased when the *min_s-sup* threshold is lowered. When a full set of rules is mined, lowering the *min_conf* threshold from 90% to 50% significantly increases the runtime and the number of mined rules. Reducing the confidence threshold has less effect when non-redundant rules are mined.

Varying the i-support threshold does not affect the runtime because we do not have any apriori property involving the instance support of mined rules. Hence, i-support threshold is not used to prune the search space. However, the i-support threshold still affects the number of mined rules: the number decreases as the threshold increases. Due to the space limitation, experimental results on varying the i-support threshold is moved to the technical report [21].

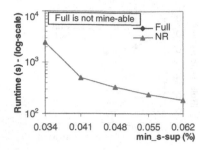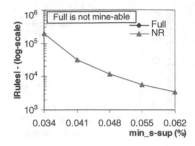

Fig. 5. Varying $min\_s\text{-}sup$ (at min_conf=50%, min_i-sup=1) for Gazelle dataset

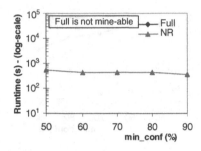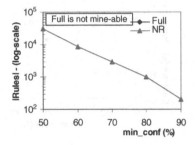

Fig. 6. Varying min_conf (at min_s-sup=0.041%, min_i-sup=1) for Gazelle dataset

Gazelle Dataset Result. The experiment results of mining non-redundant rules from the Gazelle dataset are shown in Figure 5 & 6. The runtime is significantly increased when the $min\_s\text{-}sup$ threshold is lowered. Lowering the min_conf threshold does not affect the runtime much. However, we note that the number of mined rules sharply reduces when the min_conf threshold is increased from 50% to 90%. The results also show that we can efficiently mine recurrent rules from real data set even at a low $min\_s\text{-}sup$ support of 0.034%.

Summary. The experiment results show the effectiveness of our pruning strategies in reducing both the runtime and the number of mined rules. A non-redundant set of recurrent rules can be mined efficiently from both real and synthetic datasets even at low $min\_s\text{-}sup$ and $min\_i\text{-}sup$ thresholds. We did not experiment with low min_conf thresholds as we believe the usefulness of low confidence rules (if any) is minimal.

7   Case Study

A case study was performed on the security component of JBoss Application Server (JBoss AS). JBoss AS is the most widely used J2EE application server. It contains over 100,000 lines of code and comments. The purpose of this study is to show the usefulness of the mined rules to describe the behavior of a real software system.

Premise	➤	Consequent
XLoginConfImpl.getConfEntry()		ClientLoginModule.initialize()
AuthenticationInfo.getName()		ClientLoginModule.login()
		ClientLoginModule.commit()
		SecAssocActs.setPrincipalInfo()
		SetPrincipalInfoAction.run()
		SecAssocActs.pushSubjectContext()
		SubjectThreadLocalStack.push()
		SimplePrincipal.toString()
		SecAssoc.getPrincipal()
		SecAssoc.getCredential()
		SecAssoc.getPrincipal()
		SecAssoc.getCredential()

Fig. 7. A Rule from JBoss-Security

We instrumented the security component of JBoss-AS using JBoss-AOP and generated traces by running the test suite that comes with the JBoss-AS distribution. In particular, we ran regression tests testing Enterprise Java Bean (EJB) security implementation of JBoss-AS. Twenty-three traces of a total size of 4115 events, with 60 unique events, were generated. Running the algorithm with the minimum support and confidence thresholds set at 15 and 90% respectively, ten non-redundant rules were mined. The algorithm completed within three seconds.

A sample of the mined rules (with abbreviated class and method names) is shown in Figure 7. The rule, read from top to bottom, left to right, describes authentication using Java Authentication and Authorization Service (JAAS) for EJB within JBoss-AS. When authentication scenario starts, first configuration information is checked to determine authentication service availability – this is described by the premise of the rule. This is followed by: invocations of actual authentication events, binding of principal information to the subject being authenticated, and utilizations of subject's principal and credential information in performing further actions – these are described by the consequent of the rule.

8   Conclusion and Future Work

In this paper, we proposed a novel framework to mine *recurrent rules* from a sequence database. Recurrent rules have the form "whenever a series of precedent events occurs, eventually a series of consequent events occurs". Recurrent rules are intuitive and characterize behaviors in many domains. Support and confidence values are attached to recurrent rules to distinguish significant ones. Two apriori properties pertaining to the sequence-support and confidence values of rules have been used to prune the search space of possible rules. Also, we have proposed a novel definition of rule redundancy. Employing "early" pruning of redundant rules has further improved the efficiency of the mining process and reduced the number of mined rules. Our performance study shows the effectiveness of our pruning strategies in reducing runtime (up to 147 times less) and in removing redundant rules (up to 8500 times less). Non-redundant recurrent rules can be efficiently mined even at low support thresholds by our proposed mining framework. A case study on JBoss Application Server shows the applicability of our rules in mining program properties.

As future work, we plan to apply our technique to analyze other real-life datasets from various domains not restricted to software data. A comparative study to compare recurrent rules with other forms of software specifications mined from execution traces [3,19,20] is another future work. Improving scalability of the mining process further and extending to mining from stream data are other possible future work.

Acknowledgement. We thank the anonymous reviewers for their valuable comments and advice. We thank judges of SIGPLAN Symposium on Programming Language, Design and Implementation 2007 student research competition for their advice and encouragement [18]. Our thanks to Jianyong Wang and Jiawei Han for the source code of BIDE. Also, we thank Jiawei Han and his group at UIUC for the binary of PrefixSpan bundled in the data mining package ILLIMINE. We wish to thank Blue Martini Software for contributing the KDD Cup 2000 data.

References

1. Agrawal, R., Srikant, R.: Fast algorithms for mining association rules. In: VLDB (1994)
2. Agrawal, R., Srikant, R.: Mining sequential patterns. In: ICDE (1995)
3. Ammons, G., Bodik, R., Larus, J.R.: Mining specification. In: SIGPLAN-SIGACT POPL (2002)
4. Barth, A., Datta, A., Mitchell, J.C., Nissenbaum, H.: Privacy and contextual integrity: Framework and applications. In: S&P (2006)
5. Capilla, R., Duenas, J.C.: Light-weight product-lines for evolution and maintenance of web sites. In: CSMR (2003)
6. Clarke, E.M., Grumberg, O., Peled, D.A.: Model Checking. MIT Press, Cambridge (1999)
7. Corbett, J., Dwyer, M., Hatcliff, J., Pasareanu, C., Robby,, Laubach, S., Zheng, H.-J.: Bandera: extracting finite-state models from java source code. In: ICSE (2000)
8. Deelstra, S., Sinnema, M., Bosch, J.: Experiences in software product families: Problems and issues during product derivation. In: Nord, R.L. (ed.) SPLC 2004. LNCS, vol. 3154, Springer, Heidelberg (2004)
9. Dwyer, M., Avrunin, G., Corbett, J.: Patterns in property specifications for finite-state verification. In: ICSE (1999)
10. Engler, D.R., Chen, D.Y., Chou, A.: Bugs as inconsistent behavior: A general approach to inferring errors in systems code. In: SOSP (2001)
11. Garriga, G.C.: Discovering unbounded episodes in sequential data. In: Lavrač, N., Gamberger, D., Todorovski, L., Blockeel, H. (eds.) PKDD 2003. LNCS (LNAI), vol. 2838, Springer, Heidelberg (2003)
12. Han, J., Kamber, M.: Data Mining Concepts and Techniques, 2nd edn. Morgan Kaufmann, San Francisco (2006)
13. Hopcroft, J.E., Motwani, R., Ullman, J.D.: Introduction to Automata Thoery, Language, and Computation. Addison-Wesley, Reading (2001)

14. Huth, M., Ryan, M.: Logic in Computer Science. Cambridge (2004)
15. ITU-T. ITU-T Recommendation Z.120: Message Sequence Chart (MSC) (1999)
16. Kohavi, R., Brodley, C., Frasca, B., Mason, L., Zheng, Z.: KDD-Cup 2000 organizers report: Peeling the onion. SIGKDD Explorations 2, 86–98 (2000)
17. Liu, C., Lian, Z., Han, J.: How bayesians debug. In: Perner, P. (ed.) ICDM 2006. LNCS (LNAI), vol. 4065, Springer, Heidelberg (2006)
18. Lo, D.: A sound and complete specification miner. In: SIGPLAN PLDI Student Research Competition (awarded 2nd position) (2007),
    http://www.acm.org/src/winners.html
19. Lo, D., Khoo, S.-C.: SMArTIC: Toward building an accurate, robust and scalable specification miner. In: SIGSOFT FSE (2006)
20. Lo, D., Khoo, S.-C., Liu, C.: Efficient mining of iterative patterns for software specification discovery. In: SIGKDD (2007)
21. Lo, D., Khoo, S.-C., Liu, C.: Mining recurrent rules from sequence database. In SoC-NUS Technical Report TR12/07 (2007)
22. Mannila, H., Toivonen, H., Verkamo, A.I.: Discovery of frequent episodes in event sequences. DMKD 1, 259–289 (1997)
23. Pei, J., Han, J., Mortazavi-Asl, B., Pinto, H., Chen, Q., Dayal, U., Hsu, M.-C.: Prefixspan: Mining sequential patterns efficiently by prefix-projected pattern growth. In: ICDE (2001)
24. Spiliopoulou, M.: Managing interesting rules in sequence mining. In: Żytkow, J.M., Rauch, J. (eds.) PKDD 1999. LNCS (LNAI), vol. 1704, Springer, Heidelberg (1999)
25. Wang, J., Han, J.: BIDE: Efficient mining of frequent closed sequences. In: ICDE (2004)
26. Wing, J.M.: A specifier's introduction to formal methods. IEEE Computer 23, 8–24 (1990)
27. Weimer, W., Necula, G.: Mining temporal specifications for error detection. In: Halbwachs, N., Zuck, L.D. (eds.) TACAS 2005. LNCS, vol. 3440, Springer, Heidelberg (2005)
28. Yan, X., Han, J., Afhar, R.: CloSpan: Mining closed sequential patterns in large datasets. In: SDM (2003)
29. Yang, J., Evans, D., Bhardwaj, D., Bhat, T., Das, M.: Perracotta: Mining temporal API rules from imperfect traces. In: ICSE (2006)

Uniqueness Mining

Rohit Paravastu, Hanuma Kumar, and Vikram Pudi

IIIT-H,Gachibowli,Hyderabad 500032,India
prohit@students.iiit.ac.in,
hanuma@students.iiit.ac.in,
vikram@iiit.ac.in
http://www.iiit.net/ṽikram

Abstract. In this paper we consider the problem of extracting the special properties of any given record in a dataset. We are interested in determining what makes a given record unique or different from the majority of the records in a dataset. In the real world, records typically represent objects or people and it is often worthwhile to know what special properties are present in each object or person, so that we can make the best use of them. This problem has not been considered earlier in the research literature. We approach this problem using ideas from clustering, attribute oriented induction (AOI) and frequent itemset mining. Most of the time consuming work is done in a preprocessing stage and the online computation of the uniqueness of a given record is instantaneous.

1 Introduction

In this paper we consider the problem of extracting the special properties of any given record in a dataset. We are interested in determining what makes a given record unique or different from the majority of the records in a dataset. In the real world, records typically represent objects or people and it is often worthwhile to know what special properties are present in each object or person, so that we can make the best use of them. This problem has not been considered earlier in the research literature.

At first glance it may be tempting to dismiss this problem as another avatar of outlier mining. However, in the latter, we are interested in extracting records that are significantly different from the majority; here, we are interested in extracting the properties that make any given record seem like an outlier.

There are several scenarios where this type of mining can be useful:

1. Deans of universities and principals of schools generally send letters to parents of students describing their child's performance. It may be easy to identify students who got the top rank in each subject. However,it will be nice to be able to identify a student who is the only one who got more than (or less than!) 70% in some subjects A, B and C. If some unique property of each student were mentioned, it would be interesting to the readers of their reports.

J.R. Haritsa, R. Kotagiri, and V. Pudi (Eds.): DASFAA 2008, LNCS 4947, pp. 84–94, 2008.

2. A student comes to a faculty member in a university asking for a project. The best outcome can be expected if a project is chosen that makes use of the special skills of this student. One way to identify these special skills would be to mine the database of student marks in various subjects, to find unique combinations of subjects in which this student performs well.
3. The hiring wing of a company would be interested to know what are the special skills of applicants. Many online job portals require applicants to enter their individual skills, academic performance, etc. in digital form. The resulting database can be mined to obtain the speciality of each applicant.
4. Special skills of people working in a company can be used to assign them to various projects.
5. Consider an exam paper having 10 questions. Rather than deciding the weightage of marks for each question,grades to students can be given depending on what unique combinations of questions they answered correctly. For example, if everyone answered questions 1,2 and 3 correctly, these questions wouldn't count much in the final grade. If some student was the only one to answer questions 5 and 6, then it indicates that this student deserves a good grade.

We approach the uniqueness mining problem using ideas from clustering, attribute oriented induction (AOI) and frequent itemset mining. Most of the time-consuming work is done in a preprocessing stage and the online computation of the uniqueness of a given record is instantaneous. We experimentally evaluated our approach on a synthetically generated dataset, by adding records that are known to contain very different properties, and showing that our system is able to identify them.

The remainder of the paper is organized as follows: In Section 2 we formulate the uniqueness mining problem. In Section 3 we present our uniqueness mining algorithm.Then, in Section 5 we experimentally evaluate our algorithm and show the results. Related work that seems connected to the problem definition or the algorithm is reviewed in Section 4.Finally, in Section 6, we summarize the conclusions of our study and identify future work.

2   The Uniqueness Mining Problem

In this section we formulate the uniqueness mining problem. We provide a general definition of the uniqueness of objects and show how it applies to records in relational datasets.

Intuitively, we say that an object is unique if it has an important property that is absent in similar objects. With this intuition in place, we tentatively define the notion of uniqueness as follows. This will be refined later by placing restrictions on what properties are applicable in the definition.

Definition 1 (unique object). *An object x in a dataset D is said to be unique with respect to a boolean property P that is computable for all objects in D and a threshold η in the range $(0, 1)$ iff $P(x) = True$ and $|\ \{\ y : y \in D \wedge P(y) = True\ \}\ | < \eta\ |\ D\ |$*

When applied to relational datasets, D consists of the set of records in the dataset and P is a boolean property defined on the attributes of D. Without loss of generality,we treat attributes as being either categorical or numerical.As an example consider a students database. Then P could be "Is the first language of this student $=$ English??" or "Is the marks in Physics of this student in the range (60,70)??".

In general, any "where" clause of an SQL query can be treated as a boolean property. However, we restrict ourselves in this paper to properties of the form: (categorical-attribute $=$ value) and (numerical-attribute in range (value1, value2)). We also consider combinations of several categorical and numerical attributes along with their corresponding values and ranges.

Notice that if an object $x \in D$ is unique with respect to a property P, then all objects $y \in D$ for which $P(y) = True$ must also be unique with respect to P. For convenience, we define the notion of uniqueness of a property as follows:

Definition 2 (Unique Property). *A boolean property P that is computable for all objects in D is unique with respect to a threshold η in the range $(0, 1)$ iff $P(x) = True$ and $| \{ y: y \in D \wedge P(y) = True \} | < \eta(D)$, where x is a record in the dataset D.*

Too many unique objects problem: One shortcoming of Definition 1 is that it does not consider the "structure" of the boolean property being considered. In the case of relational data, the structure consists of attributes and their values or ranges of values. As a consequence of ignoring the structure, too many unique objects may be obtained as illustrated below.

Suppose our student database consists of 100 students and $\eta = 10\%$. Also, suppose that there are 10 first languages in the dataset. In a random dataset, on an average for 5 of these languages, there will be slightly less than 10 students and for the remaining 5, there will be slightly more than 10 students. It is thus *normal* that according to our definition, about 50% of the students will be considered unique with respect to their respective first language!

Clearly, our definition of uniqueness fails to satisfy our intuitive notion of uniqueness. Half of the records cannot be considered unique with respect to a single factor. If any definition of uniqueness results in a *large* number of records being labelled as unique, then the definition has failed its purpose. In order to overcome this condition, we need to neglect sets of properties that apply to a large number of records. We refer to such properties as *trivially unique*. Before we formally define such properties, we first need to develop the notion of a *sibling property*:

Definition 3 (Sibling Property). *For any boolean property P applicable on objects in a dataset D, the set of all specializations of a generalization of P are said to be sibling properties of P.*

By a generalization of property P, we mean any property obtained by relaxing one of the requirements of P. By a specialization of property P, we mean any property obtained by adding one more requirement on P. Notice that any boolean property P is a sibling of itself.

In the case of relational data, boolean properties may be of the form: (attribute = value) or (attribute in range (value1,value2)). A *generalization* of such a property would be to leave the value or range unspecified. This can be represented by the form: (attribute = ?) or (attribute in range (?,?)). The specializations of this generalization would correspond to all possible values or ranges that can be substituted for the '?' marks.

Thus, in the above example, the generalization for (first-language = English) would be (first-language = ?) and its specializations would include all possible first-languages in the domain. Alternatively, another kind of useful specialization is to add additional requirements on the values of attributes. For example, (first-language = English and physics-marks in (60,70)) is a specialization of (first-language = English).

Definition 4 (Trivially Unique Property). *If according to definition 1 , the total number of unique objects with respect to (D, P', η) over each sibling P' of some boolean property P,is large $(\geq \eta \mid D \mid)$,then every unique P' is trivially unique.*

In the above example, this definition simply states that if the total number of unique students with respect to all first-languages exceeds a threshold, then every unique first-language is trivially unique.

Thus to fix the above mentioned problem of too many unique records, the boolean properties considered in Definition 1 must not be trivially unique.

We also note that the uniqueness threshold η would in general be inversely proportional to the number of siblings of the property being considered. Hence, we set $\eta = \sigma/N_P$ where σ is a user-specified threshold in the range $(0,1)$ and N_P is the number of siblings of the property P being considered.

The uniqueness mining problem that we consider in this paper is for relational datasets and is to extract for a given record all non-trivially unique boolean properties of the form (categorical-attribute = value) and (numerical-attribute in range (value1,value2)). If there is no single attribute that makes this record unique, we consider combinations of attributes along with their corresponding values and ranges.

3   The Uniqueness Mining Algorithm

In this section we describe our algorithm to mine the uniqueness of records in a relational dataset. Our task is to extract for a given record all unique boolean properties of the form (categorical-attribute = value) and (numerical-attribute in range (value1,value2)). If there is no single attribute that makes this record unique, we consider combinations of attributes.

To enable efficient algorithms to be built for uniqueness mining, we first identify the following lemmas. Their proofs follow easily from the definitions of uniqueness given in the previous section.

Lemma 1. *If a property P is unique in a dataset D, then any specializations of P must also be unique.*

Lemma 2. *If a Property P is trivially unique in a dataset D, then any specialization of P must also be trivially unique.*

Lemma 3. *If a Property P is unique in a dataset D, then any specialization of P must be trivially unique.*

The last lemma follows from the first lemma. If all specializations of P are unique, then these specializations together (which are all mutual siblings) will result in all records in the dataset being unique.

Due to these lemmas, we can follow a level-wise strategy of mining unique properties, starting with properties which are based on a single attribute (referred to as singletons) and then moving on to combinations of these singletons of length 2, 3, etc. At each level, we need not consider any properties that are known to be unique from the previous level, as by the second and third lemmas above, these will only result in trivially unique properties. The reader will surely notice that this processing is similar to that of the Apriori [1] algorithm of frequent itemset mining.

3.1   Preprocessing: Identifying Singletons

In order to allow an interactive user experience while mining uniqueness of records in a dataset, we do most of the time-consuming work in a preprocessing phase. The pseudo-code for this phase is shown in Figure 1 and explained below.

```
Preprocess(D, μ, O)
Input:Dataset D,Threshold μ,Ontology O
1.            for each categorical attribute A:
2.                    if | DistinctValues(A) |> μ | D |
3.                        generalize(A, O, D)
4.                        compute and store frequency of values in A
5.                else:
6.                        Prune(A)//ignore A from now on
7.            For each numerical attribute A:
8.                cluster the values of attribute A
9.                Compute and store the frequency of each cluster.
```

Fig. 1. Preprocessing Phase

During the preprocessing stage, we do the following:

1. We apply ideas from Attribute Oriented Induction (AOI) [6] to do the following: Generalize categorical attributes that have too many distinct values (e.g. addresses of people) using an ontology if available (lines 2-3 of Figure 1), or else to remove such attributes (e.g. names of people) that have too many distinct values (lines 5-6).

2. Determine ranges of numerical attributes for the purpose of defining boolean properties. We do this by applying any standard clustering algorithm on each such attribute (lines 7-9).

At the end of preprocessing, each value of unpruned categorical attributes and each range of numerical attributes correspond to singletons. After preprocessing, we identify combinations of these singletons that are unique. This is explained in the next subsection.

3.2 Determining Uniqueness of Records

In this phase, the user provides an input record x and desires to see all the ways in which x is unique. The algorithm then proceeds to first identify if x is unique with respect to singleton properties extracted during the preprocessing phase. If that fails, then combinations of singleton properties are considered. The detailed pseudo-code is shown in figure 2 .

$Uniqueness(x, D, \sigma, \mu)$	
Input : $Record x, Dataset D, Thresholds \sigma, \mu$	
1	k=1
2	Candidates = $\{\{A\}$: for each unpruned attribute A $\}$
3	while $\mid Candidates \mid > 0$
4	$NonUniqProp = \phi$
5	for each candidate $P \in Candidates$:
6	$result = IsUnique(x, P)$
7	If $result \neq 1$//not unique
8	$NonUniqProp = NonUniqProp \cup \{P\}$
9	k=k+1
10	Candidates = Combine(NonUniqProp,k)

Fig. 2. Levelwise Uniqueness Mining

The above algorithm works in an iterative manner, starting with a set of candidate properties. Note that by "property", here we mean an attribute or a set of attributes. Implicitly, the values or ranges corresponding to these attributes are obtained from the attribute values of x. The initial candidates correspond to all single attributes that have not been pruned in the preprocessing phase (lines 1-2 of figure 2). The given record x has a value for each categorical attribute and lies in a range of values for each numerical attribute. Hence every attribute corresponds to a unique singleton property extracted during the preprocessing phase.

The algorithm proceeds to check for each attribute, if x is unique with respect to that attribute (lines 5-6). The details of the IsUnique function used here is shown later in Figure 3 . If the attribute is not found to be unique, it is added to a list NonUniqProp (lines 7-8). This list is used to generate candidates for the next iteration (line 10).

The Combine function used to generate candidates operates as follows: It extends each entry in NonUniqProp with one more attribute, such that all immediate subsets of the new entry are already present in NonUniqProp. In a sense, it works similar to the AprioriGen function of [1]. We now describe the details of the IsUnique function used in line 6 above. Its pseudo-code is shown in Figure 3 .

```
IsUnique(x, P)
Input:Record x, Property P
1        count =| {y : y ∈ D ∧ P(y) = P(x)} |
2        η = σ/N_P
3        if count < η | D |:
4                if no.of unique records w.r.t P < μ | D |:
5                        Print x is unique w.r.t P  //non-trivially unique.
6                Return 1
7        else:
8                return 0
```

Fig. 3. Is record unique w.r.t. Property?

In this Pseudo-code ,note that the property P represents a set of attributes. Implicitly, the required values or ranges corresponding to these attributes are obtained from the attribute values of x. The function initially determines the number of records in D that have the same attribute values and ranges as x (line 1 of Figure 3). Next, the function determines if the property P is unique or not by comparing this number with a threshold (line 2).

Recall from Section 2 that in the computation of the threshold, N_P is the number of siblings of the property P being considered. To illustrate: For a singleton numeric attribute,N_P would be the number of ranges that this attribute has been divided into during the preprocessing phase. For a combination of numeric attributes, N_P would be the product of number of ranges of each attribute.

Next, the function verifies whether P is trivially unique.It does this by computing the total number of records that are unique with respect to P and checking if this number is less than a threshold (line 3). If the property is found to be non-trivially unique, the property is output as being so (line 4). If the property is determined not to be unique, the function returns 0 (line 6-7).

Also,if the number of attributes are too high, we incorporate the IsUnique() function in figure 3 and Combine() function in figure 2 into preprocessing stage. Thus, we store each and every unique properties of all records in preprocessing stage itself.So,The uniqueness of any record can be obtained instantaneously irrespective of the number of attributes.

4   Related Work

The uniqueness mining problem introduced in this paper is related to several other areas including outlier mining,subspace clustering and frequent itemset mining.At first glance it may be tempting to dismiss this problem as another avatar of outlier mining [2, 3]. However, in the latter, we are interested in extracting records that are significantly different from the majority; here, we are interested in extracting the properties that make any given record seem like an outlier.It is possible to extract outliers using our approach in the following way: If a record is an outlier its unique properties would be very "dominant". It is likely to have unique properties that are very general. It is also likely that these properties can be extracted in our approach using very lenient thresholds.It is not possible to use general-clustering based outlier detection methods to determine unique records. This is because the clusters obtained using general clustering algorithms on n-dimensional data objects is hard to characterize.In our approach, we apply clustering algorithms seperately on each dimension - this is used only to identify good ranges for each dimension.Recent and upcoming works in subspace clustering [4] may eventually resolve the above problem. These studies aim to identify clusters in sub-spaces of dimensions, rather than on all dimensions at once. If this is possible efficiently,then records present in very small subspace clusters can be treated as unique. However, the work in this paper is different in the sense that it aims to output the unique properties of *any* input record.

5 Experiments

In this section, we evaluate the Uniqueness Mining algorithm using synthetic relational datasets.In these experiments we attempt to demonstrate that the algorithm and framework are useful in that the unique properties of records are output. We do this by adding handmade records to the dataset whose properties are significantly different from the remaining records. We then show that our algorithm is able to identify the unique properties of the hand-made records accurately. An important point is that the output of our algorithm is small enough for a human to parse. This is due to the pruning of properties that are trivially unique.Next, we also demonstrate the scalability of our approach by measuring its response times on two synthetic datasets of very different sizes and a real dataset. In all cases, the response times of our approach were instantaneous.

The experiments were run on a computer with a 2.6 GHz Intel Pentium IV Processor and a RAM of 512 MB and the data was stored in a 40 GB DDR HDD.All the algorithms have been implemented in Python(version 2.4) and the data was stored in a MySQL relational database server (version 5.0.22).

We generated datasets that replicate a student information database.The attributes used were name, gender, major,Date of birth(DOB),telephone, city of residence, GPA,major course Grade. In our experiments we fix $\mu = 10\%$ and $\sigma = 10\%$.

During the preprocessing stage, the numerical attribute GPA was clustered using the CURE clustering algorithm [5]. Similarly, other attributes were clustered as described in the algorithm and the frequencies of clusters were stored in a database.

Experiment 1

In this experiment we verified the accuracy of the algorithm in finding already known hand-made unique tuples.We introduced 30 hand-made records in a 1000 record dataset. We illustrate our output using one such record (Gender:"F", Major:"ECE", Residence:"City X, Country Y",GPA:6.13, Major course grade: A). This record was created uniquely in the combination (GPA, major course grade)-usually there are very few people with the given combination of values for GPA and major course grade.

In the preprocessing phase, the attributes 'name', 'phone' and 'DOB' were pruned by keeping $\mu = 10\%$. There were 1000, 874 and 1000 distinct values in these attributes, respectively.This clearly shows that even though a record may be unique in these attributes it would be trivially unique and uninteresting. Thus the attribute elimination methods in the preprocessing stage prove to be very useful.

When we run the algorithm with the $\sigma = 10\%$,we obtained the tuple as unique in the subspace ('GPA', 'Major course grade'), as expected. We also obtained that the record is unique in the subspace ('GPA','Residence'). The record did not contain any special properties based on a single attribute.

Further details of the statistics of the database with respect to this record are as follows:

The cluster that contains GPA=6.13 contained 46 records. The cluster with Grade=A consists of 95 records which are too many when compared to the threshold values.But the combination contains only 1 point while the minimum threshold value evaluates to approximately 2 in this case. This shows that although there are many average and intelligent students, there are a few people that have an average overall GPA and also have a very good grade in their major course.

This may be very useful for the hiring wing of a company who are looking for students who have got good grades in their major course but not based only on their GPA. Also,this might be very useful for those unique students who are normally rejected based on their GPAs.

Additionally, we come to know that there is only one person in 130 students from City X in the GPA range 6.04-6.28 showing that the student is unique in the subspace ('gpa','residence').

Experiment 2

In this experiment we deal with the scalability of the algorithm. We created a synthetic data set of the same format as in the above experiment consisting of 1 Million records. The dataset in the previous experiment had only 1000 records.

The complexity of the preprocessing stage is proportional to the complexity of the clustering algorithms and the speed of query retrieval of the relational database servers.The algorithm we use for clustering numerical attributes is CURE. So the preprocessing stage would be of the order of $O(n^2 log(n))$.

In this dataset, only 'Name' and 'Phone' are removed during preprocessing and the clusters of other attributes are stored in the database. We ran the level-wise algorithm on the resulting database containing a million records and 6 attributes. The algorithm was executed in 26 seconds on an average.

One interesting result was obtained when the input record was (gender:'F', major:'CIVIL', DOB:1-11-1986, Residence:'City X', GPA:6.23, majorcourseGrade :'A').The uniqueness of this record was found in the subspaces (GPA, Major course grade); (DOB,residence,major);(DOB,residence,gender); (DOB,major, gender).

In this case out of a cluster of 100949 people having 'A' in Major course grade and 45981 people having GPA range of (6.34 - 6.04), only 2 people lie in the intersection of both the clusters making them very special indeed!

Experiment 3

Unlike the earlier experiments, where we have dealt with the scalability and accuracy of the algorithm with respect to datasets with less number of attributes and fairly large number of records, this experiment verifies the accuracy and scalability of our algorithm using large number of attributes.

For this experiment,we have used a dataset containing the answers of students in a competitive exam.There are 50 attributes in the dataset each representing a question in the exam.The value of each attribute can be either 'yes' or 'no' , 'yes' if the answer is correct , 'no' if it isn't correct.Totally,there are 1000 attributes in the dataset.

The intuitive approach of our algorithm is to find unique candidates in these exams.Normally,the candidates are assessed based on their final marks only, but this algorithm provides another scope of ranking the candidates.

We can find some unique students from the total set who have answered a set of answers correctly,while many other students have answered them incorrectly.These unique students can be given some bonus marks based on the toughness of the questions.This can ensure that talented students do not fail to qualify in competitive exams, just because they have wasted some extra time answering tough questions.

For this,The threshold value σ is set to 0.1.

For the given threshold value,we found a student having the following unique set of correct answers.

{'ques32': 'yes', 'ques36': 'yes', 'ques21': 'yes', 'ques4': 'yes','ques42':'yes'}

Normally,the student got 28 marks out of 50.But,he was able to answer these questions which others were not able to answer ,wasting some valuable time.So, he may be given some bonus marks as he had answered some relatively tough questions.

6   Conclusions

In this paper we introduced the problem of Uniquess Mining. We are interested in determining what makes a given record unique or different from the majority of the records in a dataset. We have discussed several examples of applications where this may be useful. Our solution to this problem uses ideas from clustering, attribute oriented induction (AOI) and frequent itemset mining. Most of the time-consuming work is done in a preprocessing stage and the online computation of the uniqueness of a given record is instantaneous. We demonstrated the accuracy and scalability of our approach using synthetic datasets.

Future work includes designing faster algorithms. This is necessary for applications where there are a large number of attributes. Also, in many applications, there may be a notion of different weights to attributes - some attributes may be more important in defining uniqueness. Further, numerical attributes such as GPA or salary usually have a bias - higher (or lower) values typically mean that the record is better in some sense. This may be useful in defining or ranking the unique properties of records.

References

1. Agrawal, R., Srikant, R.: Fast algorithms for mining association rules. In: Proc. of Intl. Conf. on Very Large Databases (VLDB) (September 1994)
2. Agrawal, C.C., Yu, P.S.: Outlier detection for high dimensional data. In: Proc. of ACM SIGMOD Intl.Conf. on Management of Data 2001 (2001)
3. Ng, R.T., Breunig, M.M., Kriegel, H.-P., sander, J.: Identifying density based local outliers. In: Proc. of ACM SIGMOD Intl. Conf. on Management of Data (2000)
4. Knorr, E.M., Ng, R.T.: Finding Intensional Knowledge of Distance-Based Outliers. In: Proc. of VLDB 1999 (1999)
5. Gunopulos, D., Agrawal, R., Gehrke, J., Raghavan, P.: Automatic subspace clustering of high dimensional data for data mining applications. In: Proc. of ACM SIGMOD Intl.Conf. on Management of Data (1998)
6. Sudipto Guha, R.R., Shim, K.: An efficient clustering algorithm for large databases. In: Proc. of ACM SIGMOD Intl. Conf. on Management of Data 1998 (1998)
7. Cercone, N., Cai, Y., Han, J.: Attribute-oriented induction in relational databases. In: Knowledge Discovery in Databases, AAAI/MIT Press (1991)

Discovering Spatial Interaction Patterns

Chang Sheng, Wynne Hsu, Mong Li Lee, and Anthony K.H. Tung

Department of Computer Science,National University of Singapore, Singapore
{shengcha,whsu,leeml,atung}@comp.nus.edu.sg

Abstract. Advances in sensing and satellite technologies and the growth of Internet have resulted in the easy accessibility of vast amount of spatial data. Extracting useful knowledge from these data is an important and challenging task, in particular, finding interaction among spatial features. Existing works typically adopt a grid-like approach to transform the continuous spatial space to a discrete space. In this paper, we propose to model the spatial features in a continuous space through the use of influence functions. For each feature type, we build an influence map that captures the distribution of the feature instances. Superimposing the influence maps allows the interaction of the feature types to be quickly determined. Experiments on both synthetic and real world datasets indicate that the proposed approach is scalable and is able to discover patterns that have been missed by existing methods.

1 Introduction

Among various spatial analysis tasks, finding nearby relation between spatial features is an important problem with many scientific applications. For example, in epidemiology studies, dengue fever and Aedes mosquito tend to exhibit spatial closeness; while in location-based services, it is beneficial to find E-services that are located together. In view of this, there has been sustained interest in developing techniques to discover spatial association and/or collocation patterns. The definition of these patterns is based on a **binary** notion of proximity where a distance threshold is set and objects are either close or far away based on the threshold. Interesting patterns are then defined to be a set of features that are frequently close to each other.

A classic framework discover spatial collocation patterns is the reference feature centric model [3] which converts the spatial relationship to transactions, and utilizes association rule mining techniques to discover patterns. The work in [2,4] propose an event centric model whose statistical foundation is based on Ripley's K function [5]. However, the main drawback is its requirement for expensive and frequent instance spatial-join operations during the mining process. Many methods have been proposed to reduce the number of spatial join operations, e.g., grid partitioning approach [11], summary structure [9] and join-less approaches [10]. However, these mining algorithms remain expensive, particularly, on the large databases. Further, they are also sensitive to the distance threshold.

Fig. 1 shows a sample dataset of objects[1] with three feature types f_1, f_2, f_3, denoted by the symbols □, △, and ○, respectively. The dotted circles define the object's

---

[1] Without special presentation, objects refer to the spatial instances in dataset in the rest paper.

J.R. Haritsa, R. Kotagiri, and V. Pudi (Eds.): DASFAA 2008, LNCS 4947, pp. 95–109, 2008.

distance threshold in the context of collocation mining. We observe that many of the $\triangle$ objects have $\square$ objects within their distance thresholds. In other words, $\{\square, \triangle\}$ is a collocation pattern. Yet, a closer examination reveals that there are many $\bigcirc$ objects that are just outside the distance threshold of the $\triangle$ objects. However, $\{\triangle, \bigcirc\}$ is not a collocation pattern due to the artificial boundary imposed by the distance threshold. Suppose the $\triangle$ denotes the victims of an infectious disease and $\square$ and $\bigcirc$ denotes two possible agents/carriers of the infectious disease. Missing out the $\{\triangle, \bigcirc\}$ pattern may result in not being able to locate the true agent/carrier of the infectious disease. To remove this artificial boundary, we propose to model the features in a continuous space through the use of influence functions.

In this paper, we introduce the notion of *Spatial Interaction Patterns* (SIPs) to capture the interactions among the set of features. Such patterns are sets of binary features that are spatially correlated based on the distributions of their instances. For each feature type, we build an approximate influence map with certain error control that captures the influence distribution of the feature instances by iterative approach. Superimposing the influence maps allows the spatial closeness of the feature types to be easily determined. This approach allows us to avoid the costly spatial joins. We carried out experiments on both synthetic and real world datasets to demonstrate that the proposed approach is scalable and is able to discover patterns that have been missed by existing methods.

ID	X	Y	FID
001	22.8	27.5	f_1
002	32.2	60.8	f_1
003	75.1	60.4	f_1
004	70.3	72.5	f_1
005	32.4	17.7	f_2
006	25.1	40.2	f_2
007	42.4	67.0	f_2
008	61.0	61.2	f_2
009	45.2	33.0	f_3
010	54.9	37.0	f_3
011	40.1	48.2	f_3

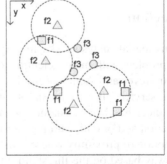

(a) Dataset   (b) Instance distribution

Fig. 1. Some instances and their relation

2   Influence Model

In this section, we introduce the notations used in defining the influence function to capture the degree of affinity between two spatial objects. We extend the notations to the influence maps of features, i.e., object groups, and infer some useful properties.

2.1   Influence Function of Objects

When two spatial objects are near each other, they exert an influence on each other. This degree of influence is represented as a radial basis kernel function in the form of either Gaussian, Epanechnikov, Biweight or Triangle function [7]. In this work, we select the Gaussian function to be the kernel due to its unique range $(-\infty, \infty)$ which can naturally model the distribution of influence. Let $k_i(\cdot)$ denotes the kernel function on the i-th dimension. We can easily generalize the influence function of one object in the d-dimensional space.

Definition 1. *Assuming that each dimension has identical Gaussian kernel $\mathcal{N}(0, \sigma_k^2)$. Given a d-dimensional object $o = (h_1, h_2, \ldots, h_d)$, its **influence** to the neighboring point $p = (p_1, p_2, \ldots, p_d)$ in the d-dimensional space, is the product of influence on each dimension:*

$$inf(o,p) = \prod_{i=1}^{d} \frac{1}{\sqrt{2\pi\sigma_k^2}} exp\{-\frac{(h_i - p_i)^2}{2\sigma_k^2}\} = (2\pi\sigma_k^2)^{-d/2} exp\{-\frac{\sum\limits_{i=1}^{d}(h_i - p_i)^2}{2\sigma_k^2}\}$$

(1)

The "points" in Definition 1 could be any d-dimensional vector in a d-dimensional space. In the case of a 2D spatial object o on the x-y plane, its influence distribution is a bivariate Gaussian function whose mean is the observed position of o and standard deviation is σ_k. If the kernel on each dimension has the same standard deviation (i.e., $\sigma_x = \sigma_y$), the influence distribution is circular in shape, otherwise, it is an ellipse. We call this bell-like 3D shape an **influence unit**. Note that over 95% influence of an influence unit is captured within a rectangular region of mean $\pm 3\sigma$ along the x-y plane.

2.2 Influence Map of Features

Similar to the kernel density estimator [7], the influence map of a feature on a 2D plane is the normalized summation of all the instances' influence units of this feature[2].

Definition 2. *Given a set of spatial objects $\{o_1, o_2, \ldots, o_n\}$ with feature f on a spatial plane $\mathcal{P}$, and the influence function $inf(\cdot)$. The **influence** of feature f on one position $p \in \mathcal{P}$, denoted by $S(f, p)$, is*

$$S(f,p) = \frac{1}{n} \sum_{i=1}^{n} inf(o_i, p).$$

(2)

*We use $S(f)$ to denote the **influence map** of feature f.*

The volume of $S(f)$ can be computed as the integral of the influence of all points in $\mathcal{P}$. This leads to Lemma 1.

Lemma 1. *The volume of an influence map $S(f)$ is 1.*

PROOF: Assume the feature f contains n objects $\{o_1, o_2, \ldots, o_n\}$.
Volume of $S(f) = \int_{p \in \mathcal{P}} S(f, p) dp$

---

[2] Nevertheless, the smoothing techniques well-developed in kernel density estimator are beyond the scope of this paper, because influence maps are only utilized to model the distribution of influences of objects instead of objects themselves

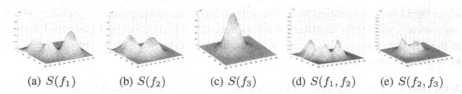

(a) $S(f_1)$ (b) $S(f_2)$ (c) $S(f_3)$ (d) $S(f_1, f_2)$ (e) $S(f_2, f_3)$

Fig. 2. Examples of influence maps and their interaction

$= \int_{p \in \mathcal{P}} \frac{1}{n} \sum_{i=1}^{n} inf(o_i, p)dp$

$= \frac{1}{n} \int_{p \in \mathcal{P}} inf(o_1, p)dp + \ldots + \int_{p \in P} inf(o_n, p)dp$

$= \frac{1}{n} \times n \times (Volume\ of\ influence\ unit)$

$= 1.$ □

Definition 3. *Given two influence maps $S(f_1)$ and $S(f_2)$ with respect to feature f_1 and f_2 and the spatial plane $\mathcal{P}$, the* **influence of a feature pair** *$\{f_1, f_2\}$ on a position $p \in \mathcal{P}$, denoted as $S(f_1, f_2, p)$, is $min\{S(f_1, p), S(f_2, p)\}$. We use $S(f_1, f_2)$ to denote the* **influence map of the feature pair** *$\{f_1, f_2\}$ on plane $\mathcal{P}$.*

Definition 4. *The interaction between a pair of features $\{f_i, f_j\}$ is measured as the volume of the influence map $S(f_i, f_j)$. We call this measure the* **Interaction (I)** *of feature pair $\{f_i, f_j\}$ and is denoted as $I(f_i, f_j) = \int_{p \in P} S(f_i, f_j, p)dp$.*

From Lemma 1, we infer that $0 \leq I \leq 1$. The interaction between a feature and itself is 1, i.e., $I(f_i, f_i) = 1$. Hence, $I(f_i, f_j) = 1$ indicates that the objects of feature f_i and f_j have the same influence distribution. On the other hand, $I(f_i, f_j) = 0$ implies that the data distributions of feature f_i and f_j are far apart from each other. Clearly, $I(f_i, f_j) = I(f_j, f_i)$ since both $Euclidean()$ and $min()$ subfunctions are symmetric.

Fig. 2 shows the influence maps of features f_1, f_2 and f_3 in Fig. 1, as well as the feature pairs $\{f_1, f_2\}$ and $\{f_2, f_3\}$. Definition 3 and Definition 4 can be extended to three or more features. For three features, we have $I(f_1, f_2, f_3) = \int_{p \in \mathcal{P}} min\{S(f_1, p), S(f_2, p), S(f_3, p)\}dp$. The Interaction measure (I) is used to determine the significance of a spatial interaction pattern. This measure indicates how much a feature is affected by the interaction from the other features in a feature set.

Lemma 2. *(Apriori) The Interaction measure (I) is monotonically non-increasing with the increase in the number of features.*

PROOF: Let us assume that a feature set P_n consists of n features, $f_0, f_1, \ldots, f_{n-1}$, and the interaction of n features, $f_0, f_1, \ldots, f_{n-1}$, to be $I(f_0, f_1, \ldots, f_{n-1})$. According to Definition 3, $I(f_0, f_1, \ldots, f_{n-1}) = \int_{p \in \mathcal{P}} min\{NS(f_0, p), S(f_1, p), \ldots, S(f_{n-1}, p)\}dp$. Then for a longer feature set $P_{n+1} = P \bigcup \{f_n\}$, we have

$I(f_0, f_1, \ldots, f_{n-1}, f_n)$

$= \int_{p \in \mathcal{P}} min\{S(f_0, p), S(f_1, p), \ldots, S(f_{n-1}, p), S(f_n, p)\}dp$

$= \int_{p \in \mathcal{P}} min\{min\{S(f_0, p), \ldots, S(f_{n-1}, p)\}, S(f_n, p)\}dp$

$\leq \int_{p \in \mathcal{P}} min\{S(f_0, p), S(f_1, p), \ldots, S(f_{n-1}, p)\}\}dp$

$= I(f_0, f_1, \ldots, f_{n-1}).$ □

3   Mining Spatial Interaction Patterns

Here, we present the algorithm PROBER (sPatial inteRactiOn Based pattErns mineR) to find the interaction patterns in spatial databases.

Definition 5. *Given a spatial database containing a feature set* $\mathcal{F}$ *and a threshold* min_I, *a* **Spatial Interaction Pattern** *(SIP) is the set of features* $\{f_1, f_2, \ldots, f_k\} \subseteq$ $\mathcal{F}$ *and* $I(f_1, f_2, \ldots, f_k) \geq min\_I$.

If the interaction of a SIP is greater than a predefined threshold min_I, this SIP is a *valid* or *frequent* SIP. The *maximal* SIP set is the smallest set of valid SIPs that covers all valid SIPs. For example, given the valid SIPs $\{\{f_1, f_2\}, \{f_2, f_3\}, \{f_1, f_3\}, \{f_3, f_4\},$ $\{f_1, f_2, f_3\}\}$, the maximal SIP set is $\{\{f_3, f_4\}, \{f_1, f_2, f_3\}\}$. Each member in maximal SIP set is called a maximal SIP.

The problem of mining spatial interaction patterns is defined as follows. *Given a spatial database containing m features and n instances, as well as the minimal interaction measure* min_I, *our goal is to find the set of all valid spatial interaction patterns.*

Mining of SIP is computationally expensive due to two reasons. First, the comparison of continuous spaces is computationally infinite. Second, the enumeration of all candidate patterns is exponential. We consider the first problem in Section 3.1 and examine the second problem in Section 3.2. In Section 3.3, we present the PROBER algorithm and analyze its complexity.

3.1   Uniform Sampling Approximation

In theory, the influence map of one feature is continuous. In order to facilitate mutual comparison in practice, continuous influence maps must be approximated to discrete space by sampling. Uniform sampling are utilized in case of no prior knowledge about underlying distribution. We use a *progressive refinement approach* to build the approximated influence maps such that the approximation error falls within a pre-specified error bound. Assume that the target geographical plane is a square of length L. Given a resolution R, we divide the plane into $\frac{L}{R} \times \frac{L}{R}$ cells. For each cell, we use its center to approximate the influences exerted on this cell by other objects in the neighbouring cells. As long as R is sufficiently small, our model will provide a good approximation. Let us denote an influence map and its refined approximation as $\widehat{S}$ and $\widehat{S'}$ repsectively. We define the **Influence Error (IErr)** as follows.

Definition 6. *The difference in the two influence maps,* $\widehat{S}$, *and* $\widehat{S'}$ *is given by:*

$$IErr(\widehat{S'}, \widehat{S}) = \frac{1}{|\widehat{S'}|} \times \sum_{every\ cell \in \widehat{S'}} \frac{\| cell_{\widehat{S'}} - cell_{\widehat{S}} \|}{MAX(cell_{\widehat{S'}}, cell_{\widehat{S}})} \tag{3}$$

where $|\widehat{S'}|$ *denotes the number of cells in* $\widehat{S'}$, $cell_{\widehat{S'}}$ *is a single cell at* $\widehat{S'}$, *and* $cell_{\widehat{S}}$ *is the cell which covers the corresponding cells in* $\widehat{S'}$.

For example, Fig. 3 shows $\widehat{S}$ and $\widehat{S'}$, respectively. In this example, the influence difference of $\widehat{S}$ and two cells in $\widehat{S'}$, is the volume of the shaded areas in the right part of Fig. 3. $IErr = \frac{1}{4} \times (\frac{|60-50|}{60} + \frac{|65-50|}{65} + \frac{|45-50|}{50} + \frac{|40-50|}{50}) \approx \frac{1}{4} \times 0.69 = 0.17$.

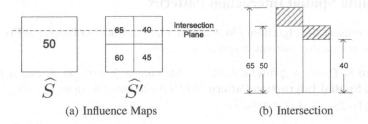

| (a) Influence Maps | (b) Intersection |

Fig. 3. An example to compute IErr

Algorithm 1. BuildApproSpace

---

input : Dataset O, kernel deviation σ, inital resolution r and error bound
  min_err
output: Approximate influence map $\widehat{S}(\cdot)$
1 Initialize two approximation spaces $S_b = \widehat{S(\cdot)}|_r$ and $S_f = \widehat{S(\cdot)}|_{r/2}$;
2 **while** $(IErr(S_f, S_b) > min\_err)$ **do**
3 $\quad$ $S_b \leftarrow S_f$;
4 $\quad$ $r \leftarrow \frac{r}{2}$;
5 $\quad$ Initialize $S_f = \widehat{S(\cdot)}|_{r/2}$;
6 return S_f;

Lemma 3. *Let θ denote $\frac{R}{\sigma}$. The error bound $IErr$ between $\widehat{S_R}$, and $\widehat{S_{R/2}}$, is*

$$IErr(\widehat{S_R}, \widehat{S_{R/2}}) \leq \frac{1 - e^{-\frac{R^2}{16\sigma^2}} e^{-\frac{kR}{4\sigma}} + e^{\frac{kR}{4\sigma}} - e^{\frac{R^2}{16\sigma^2}}}{2}, \qquad (4)$$

where $0 \leq k \leq 3$.

The proof for this lemma is omitted due to space limitation. Interested reader please refer to given in [8] for details.

We now present the algorithm *BuildApproSpace* that progressively refine the influence map until the IErr is within the specified error bound. Let $O = \{o_1, o_2, \ldots, o_n\}$ be a set of objects of one particular feature. Algorithm 1 give the sketch. Line 1 builds two spaces, S_f and S_b. S_f is implemented as a $\frac{L}{r} \times \frac{L}{r}$ matrix while S_b is a $\frac{2L}{r} \times \frac{2L}{r}$ matrix, where L is the plane width. For each object $o_i \in O$, we superimpose a minimal bounding rectangle (MBR) of side 6σ onto the two spaces, centering at the position of o_i. This results in the updates of element values on the two spaces respectively.

To compare the two matrices S_f and S_b, for each element of S_b, we find the corresponding four elements in matrix S_f and obtain the absolute difference among them. Line 2 computes the approximate error within each cell of S_b individually as given by Definition 6.

If the approximation error is greater than the user specified parameter min_err, we initialize a new matrix at half the resolution. We compute the approximation error of this finer resolution space. This process repeats until the error is less than min_err, and S_f is the final approximate influence map.

3.2   Pattern Growth and Pruning

Lemma 2 indicates that the interaction measure satisfies the downward closure property. In other words, a candidate pattern is valid only if all of its subpatterns are valid. This property allows many SIPs to be pruned during the mining process. Even then, the number of valid SIPs can still be very large. This is because for a valid SIP of n features, there are $(2^n - 1)$ subsets of features that are also valid SIPs. However, majority of these valid SIPs are redundant as their interactions can be inferred from their supersets. Avoiding the generation of these redundant SIPs can significantly improve mining efficiency and reduce memory space requirements. Current state-of-the-art maximal pattern mining algorithms [1,6] use a search tree structure to facilitate depth-first search to find frequent itemsets, but they cannot deal with maximal interaction pattern mining problem directly.

Motivated by the idea of maximal pattern mining algorithms, we employ a depth-first search with "look ahead". A tree structure called *interaction tree* is used to facilitate the mining process. In order to construct this interaction tree, we first obtain an *interaction graph* where each feature is a node in the interaction graph and two nodes are connected by an edge if the two features form an SIP. The transformation from the interaction graph to an interaction tree is as follows. A root node is first created at level 0. At level 1, we create a node for each feature as a child of the root, and the order of children follows the lexicographical order. In the subsequent levels, for each node u at level k ($k > 1$) and for each right sibling v of u, if (u, v) is connected in the interaction graph (namely $\{u, v\}$ is an interaction pattern), we create a child node for u with the same label of v. For each node, we could enumerate one candidate pattern, with prefix feature set of its parent node by concatenating the feature in this node. For example, we construct the tree shown in Fig. 4(b) based on Fig. 4(a).

Note that if a pattern p is not a valid SIP, then any longer pattern that contains p cannot be a valid SIP. This allows us to effectively prune off many unnecessary computations. Further, the structure of the interaction tree always forces the evaluation of the longest patterns first. This implies that if the longest pattern is a valid SIP, then we do not need to evaluate any of its sub-patterns.

The evaluation of SIP is performed by algorithm *EvalSIP* (see Algorithm 2). Given a feature node f_n in the interaction tree, line 1 obtains the parent node of f_n. Line 2 forms a candidate pattern P_{cand} by backtracking from the current node to the root of interaction tree. As an example, for the feature node f_6 at level 3 in Fig. 4(b), we can backtrack from f_6 to form a candidate pattern $\{f_1, f_3, f_6\}$ with prefix $\{f_1, f_3\}$.

Since the computation of influence map interaction is expensive, it makes sense to delay this computation as long as possible. Here, we note that if there is one superset of P_{cand} in C, this computation can be delayed. Line 4 sets this node to be a *delay* node and Line 5 propagates this to its children nodes. Lines 7-13 recover the influence map of the prefix. For example, suppose we already have a maximal SIP $\{f_1, f_2, f_4, f_5\}$ in maximal SIP set C. The candidate patterns $\{f_1, f_4\}$ and $\{f_1, f_4, f_5\}$ can be exempted from the evaluation as they are both subpatterns of $\{f_1, f_2, f_4, f_5\} \in C$. Here f_4 and

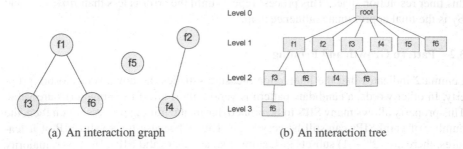

(a) An interaction graph   (b) An interaction tree

Fig. 4. Data structure for mining maximal SIPs

f_5 are marked *delay* nodes in interaction tree. If we need to evaluate another candidate pattern $\{f_1, f_4, f_5, f_6\}$, the interaction computation of its prefixes $\{f_1, f_4\}$ and $\{f_1, f_4, f_5\}$ becomes necessary. So the influence map computation will start from f_4 via f_5 along the path to f_6. The pseudocode of this operation is given in Lines 7-13. Finally, Line 13 obtains the final influence map of prefix $parS$.

Lines 14-18 compute the interaction between the influence maps of the prefix node and the current node in a depth-first manner. If the interaction is no less than the threshold min_I, this candidate pattern is valid, and it will be added to the maximal pattern set C in Line 15.

3.3   Algorithm PROBER

With the necessary computing components, we now present the Algorithm PROBER to mine spatial interaction patterns. The algorithm incorporates the pattern enumeration technique into the mining process. Algorithm PROBER takes as input the spatial database D, the influence error threshold min_err, the interaction measure threshold min_I, and outputs the set of maximal SIPs.

Line 1 finds the feature set from the dataset. Lines 2-4 build the approximate influence maps for each feature, by calling Algorithm *BuildApproSpace*. To facilitate the next mining phase, the approximate influence maps of all features are required to be superimposed using the same resolution. Therefore, the halt condition (Line 2 in Algorithm 1) is modified to be "If the maximal $IErr(S_b, S_f)$ of all features is greater than min_err, then do the next iteration".

Lines 6-10 discover the interaction in all feature pairs combination. In particular, Line 7 computes the interaction of the one feature pair by taking the minimal value between each element pair and summing up all the minimal values. If the measure is greater than min_I, this feature pair is considered to be a SIP.

Line 11 builds the *interaction tree* using the set of interaction pairs found previously. Line 13 invokes Algorithm *EvalSIP* to recursively visit the necessary feature nodes in *interaction tree*, starting from the root node of tree. Finally, Line 14 returns the maximal pattern set C.

Consider the example in Fig. 1, assume that σ=10 and min_I= 0.3, Table 1 shows the mining process of PROBER. The mining stops at level 2 because the pattern $\{f_1, f_3\}$ is

---

Algorithm 2. $EvalSIP(f_n, min\_I, C)$

---

1 $parNode \leftarrow$ the parent node of f_n;
2 backtracking f_n till root to form a candidate P_{cand};
3 **if** P_{cand} *has a superpattern in* C **then**
4 | set f_n to be a *delay* node;
5 | call $EvalSIP(f_n's\ child\ node, min\_I, C)$;

6 **else**
7 | $parS$ = a unit matrix;
8 | **while** $parNode$ *is a delay node* **do**
9 | | $parS = Interaction(parS, S(parNode))$;
10 | | $parNode \leftarrow$ the parent node of $parNode$;

11 | $parS2 \leftarrow$ the influence map of $parNode$;
12 | $parS = Interaction(parS, parS2)$;
13 | $fS = Interaction(parS, S(f_n))$;
14 | **if** $fS > min\_I$ **then**
15 | | add P_{cand} to C;
16 | | call $EvalSIP(f_n's\ child\ node, min\_I, C)$;

17 | **else**
18 | | call $EvalSIP(f_n's\ sibling\ node, min\_I, C)$;

---

not a valid SIP, hence the pattern $\{f_1, f_2, f_3\}$ is pruned. As a result, the maximal SIPs are $\{\{f_1, f_2\}$ and $\{f_2, f_3\}\}$.

Complexity Analysis. Let $\theta = \sigma/R$ where R is the final resolution after multiple iterations. To build the influence map, an influence unit of range $6\sigma \times 6\sigma$ is computed for each instance in the database, thus it needs $(6\sigma/R)^2 = (6/\theta)^2$ distance computations. The overall computational complexity to build the influence maps is $O(n(6/\theta)^2)$, where n is the database size. Since there is one influence map matrix for each feature, the space complexity is $O(f(\frac{L}{R})^2)$ where L is the plane length, f is the feature number.

The PROBER algorithm, in the worst case, could generate $\binom{f}{\lfloor \frac{f}{2} \rfloor}$ maximal SIPs, and each of them requires $\lfloor \frac{f}{2} \rfloor$ influence map comparison along the path from root to the leaf node of the maximal SIP. Each influence map comparison requires the computation of complexity $O((\frac{L}{R})^2)$. Hence, the overall computational complexity in mining phase, is $O((\binom{f}{\lfloor \frac{f}{2} \rfloor}) \times \lfloor \frac{f}{2} \rfloor \times (\frac{L}{R})^2)$. The space complexity includes the space to store emphinteraction tree and influence matrixes. The *interaction tree* has at most 2^f nodes, and each node contains an influence matrix. Since the space required for influence matrix dominates, the worst space requirement is $O((2^f - f) \times (\frac{L}{R})^2)$.

In summary, the computational complexity of PROBER is $O(n(6/\theta)^2 + (\binom{f}{\lfloor \frac{f}{2} \rfloor}) \times \lfloor \frac{f}{2} \rfloor \times (\frac{L}{R})^2)$, and the space complexity of PROBER is $O(2^f(\frac{L}{R})^2)$.

Algorithm 3. PROBER

---

   input : D: the spatial database, σ: kernel deviation, min_err: influence error
            threshold, min_I: interaction threshold;
   output: C: the set of SIPs;
1 Let RF to be all features in D;
2 /*Phase 1: impose approximate influence map*/;
3 **for** *each feature* $f_i \in RF$ **do**
4   call $BuildApproSpace(\cdot)$ to build the influence map;
5 /*Phase 2: build interaction tree*/;
6 $C = \emptyset$;
7 Impose an ordering on RF;
8 **for** *each feature pair* (f_i, f_j), *where* $f_i \prec f_j$ **do**
9   evaluate feature pair (f_i, f_j);
10   **if** $I(f_i, f_j) \geq min\_I$ **then** add pattern $\{f_i, f_j\}$ to C;
11 build the interaction tree T_{col} based on C;
12 /*Phase 3: mine maximal SIP*/;
13 call $EvalSIP(T_{col}.root, min\_I, C)$;
14 return C;

Table 1. Mining SIPs by influence model

level=1		
f_1	$S(f_1)$ (see Fig. 2(a))	$I(f_1)$=1
f_2	$S(f_2)$ (see Fig. 2(b))	$I(f_2)$=1
f_3	$S(f_3)$ (see Fig. 2(c))	$I(f_3)$=1
level=2		
$\{f_1, f_2\}$	$S(f_1, f_2)$ (see Fig. 2(d))	$I(f_1, f_2)$=0.58
$\{f_1, f_3\}$	$S(f_1, f_3)$	$I(f_1, f_3)$=0.27
$\{f_2, f_3\}$	$S(f_2, f_3)$ (see Fig. 2(e))	$I(f_2, f_3)$=0.38

4   Experiment Study

In this section, we present the results of experiments to evaluate the performance of
PROBER. We also compare PROBER with existing collocation algorithms FastMiner
[11] and TopologyMiner [9]. In order to ensure meaningful comparison, we generate
two versions of PROBER, PROBER_ALL to discover all patterns, and PROBER_MAX
to discover only maximal patterns. Note that while comparing the effectiveness of
PROBER with FastMiner and TopologyMiner may not be appropriate due to the different interesting measures used, however, we could treat FastMiner and TopologyMiner
as good baselines with respect to both effectiveness and scalability issues. Table 2
shows the parameter counterparts between influence model and distance model. In the
following experiments, we assign identical values to the parameter counterparts, e.g.
$\sigma = d = 50$ and $min\_I = min\_prev = 0.4$.

Table 2. Parameter counterparts

Distance Model	Influence Model
distance threshold: d	influence deviation: σ
minimal prevalence: min_prev	minimal interaction: min_I

Synthetic Datasets: We extend the synthetic data generator in [9] to generate the synthetic spatial databases with Gaussian noise. All the data are distributed on the plane of 8192×8192. The synthetic datasets are named using the format "Data-(m)-(n)-(d)-(N)" where m is the number of SIP features, n is the number of noise feature, d is the distance threshold and N is the number of instances in the dataset. For each object, we assign a Gaussian noise of 0 mean and σ_e, say 5, deviation on each dimension.

Real-life Datasets: The real-life dataset used in our experiments is the *DCW environmental data*. We downloaded 8 layers of Minnesota state from Digital Chart of the World[3] (DCW). Each layer is regarded as a feature (see Table 3 for details) in our experiments and the Minnesota state is mapped to a 8192×8192 plane.

Table 3. DCW dataset feature description

FID	Name	# of Points	FID	Name	# of Points
f_0	Populated Place	517	f_4	Hypsography Supplemental	687
f_1	Drainage	6	f_5	Land Cover	28
f_2	Drainage Supplemental	1338	f_6	Aeronautical	86
f_3	Hypsography	72	f_7	Cultural Landmarks	103

All the algorithms are implemented in C++. The experiments were carried out on a Pentium 4 3Ghz PC with 1GB of memory, running Windows XP.

4.1 Performance of Influence Map Approximation

We evaluate the convergence of $IErr$ and the runtime of the $BuildApproSpace$ algorithm on both synthetic data Data-6-2-100-50k and the real-life DCW dataset. We assign the initial resolution $r = 256$ and $min\_err = 0.05$. We set $\sigma = 100$ for Data-6-2-100-50k and $\sigma = 50$ for DCW data. Table 4 and 5 show the results where each row gives the influence error $IErr$ for each iteration. We observe that the $IErr$ converges rapidly and falls below min_error after limited (e.g. 3 or 4) iterations.

The time taken for each iteration is shown in Fig. 5. The runtime increases exponentially for the synthetic dataset; while the increase in runtime is as not as significant for the smaller DCW dataset. This is expected because a finer resolution will result in an increase in both time and space complexity as analyzed in Section 3.3.

---

[3] http://www.maproom.psu.edu/dcw

Table 4. Convergence on DCW data

Iteration	f_0	f_1	f_2	f_3	f_4	f_5	f_6	f_7	MAX
1	0.16	0.02	0.15	0.13	0.16	0.04	0.17	0.17	0.17
2	0.10	0.01	0.09	0.08	0.10	0.03	0.11	0.11	0.11
3	0.06	0.01	0.05	0.05	0.06	0.02	0.06	0.07	0.07
4	0.03	0.00	0.03	0.03	0.03	0.01	0.04	0.04	0.04

Table 5. Convergence on Data-6-2-100-50k

Iteration	f_0	f_1	f_2	f_3	f_4	f_5	f_6	f_7	MAX
1	0.17	0.17	0.17	0.17	0.17	0.17	0.25	0.24	0.25
2	0.10	0.10	0.10	0.10	0.10	0.10	0.15	0.14	0.15
3	0.05	0.06	0.06	0.06	0.06	0.06	0.08	0.08	0.08
4	0.03	0.03	0.03	0.03	0.03	0.03	0.04	0.04	0.04

4.2 Effectiveness Study

In this set of experiments, we show that PROBER_ALL algorithm is effective in eliminating the effect of the artificial boundaries imposed by distance models such as FastMiner and TopologyMiner. As the results of FastMiner and TopologyMiner are similar, we only compare PROBER_ALL with TopologyMiner in our experiment.

To model the artificial boundary effect, we apply to each object some guassian noise with mean 0 and standard deviation that varies from 1 to 30. We take note of the number of interaction patterns discovered. The results are shown in Fig. 6. We observe that when the standard deviation is small (less than 10), both PROBER_ALL and TopologyMiner discover the same number of patterns. This is expected because the small standard deviation does not cause much impact on the boundary imposed by the distance model. However, once the standard deviation exceeds 10, we observe that the number of patterns discoverd by TopologyMiner falls sharply. This is because many more objects may happen to lie just outside the distance threshold imposed by the distance model. As a result, many interaction patterns are missed.

Next, we apply PROBER_ALL on the DCW environment data. We set the interaction threshold min_I to be 0.4. Tables 6 show the mining results. We observe that regardless of how the σ varies, the patterns discovered by PROBER_ALL are always a superset of those found by the two distance model-based techniques, FastMiner and TopologyMiner. In particular, when $\sigma = 200$, we find that {populated place, drainage supplemental, hypsography supplemental} and {populated place, aeronautical, cultural landmarks} are missed by the distance model-based techniques but are discovered as SIPs.

4.3 Scalability

In this set of experiments, we demonstrate the scalability of both PROBER_ALL and PROBER_MAX. We set the number of features to 10 (including non-noise and noise

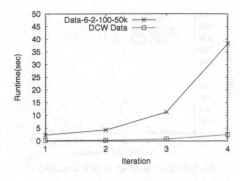

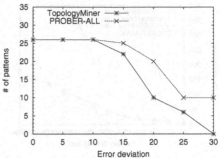

Fig. 5. Runtime of the *BuildApproSpace* Algorithm

Fig. 6. Effectiveness study

Table 6. Comparison of patterns on DCW dataset

d/σ	Distance Model	Influence Model
50	NA	$\{f_0, f_2\}, \{f_2, f_4\}$
100	$\{f_0, f_2\}, \{f_2, f_4\}$	$\{f_0, f_2\}, \{f_0, f_4\}, \{f_0, f_6\}, \{f_2, f_4\}$
150	$\{f_0, f_2\}, \{f_0, f_4\}, \{f_2, f_4\}$	$\{f_0, f_2\}, \{f_0, f_3\},\{f_0, f_4\}, \{f_0, f_6\},\{f_0, f_7\},$ $\{f_2, f_3\}, \{f_2, f_4\}, \{f_2, f_6\}, \{f_3, f_4\},\{f_3, f_7\},$ $\{f_4, f_6\},\{f_4, f_7\}, \{f_6, f_7\}$
200	$\{f_0, f_2, f_4\}, \{f_3, f_4\}$	$\{f_0, f_2, f_4\}, \{f_0, f_2, f_6\}, \{f_0, f_6, f_7\}, \{f_0, f_3\},$ $\{f_2, f_3\}, \{f_2, f_7\}, \{f_3, f_4\},\{f_3, f_6\},\{f_3, f_7\},$ $\{f_4, f_6\},\{f_4, f_7\}$

features), and generate twelve datasets Data-8-2-50-{20k, 40k, 60k, 80k, 100k, 200k, ..., 800k}. We compare the performance of PROBER with FastMiner and TopologyMiner by varying the total number of instances. Fig. 7(a) shows that both FastMiner and TopologyMiner increase exponentially as the number of instances increases while PROBER shows a linear increase. This is expected because the time complexity for the distance model is polynomial in the number of instances while that of PROBER_ALL and PROBER_MAX is linear. In addition, PROBER_MAX runs slightly faster than PROBER_ALL in terms of runtime, because PROBER_MAX only detects the maximal patterns thus reducing the mining cost.

We also set the database size at 20k instances and generate eight datasets Data-{4,6,8,10,12,14,16,18}-0-50-20k to evaluate the three algorithms. The results are shown in Fig. 7(b). Both FastMiner and TopologyMiner do not scale well with respect to the number of SIP features. TopologyMiner allows pattern growth in a depth-first manner, but the extraction of the projected databases requires much time and space. Algorithm PROBER_MAX shows the best scalability compared to the other algorithms, although it slows down when the number of features exceeds 16. This is because of the large number of SIP features which results in the exponential growth of its *interaction tree*.

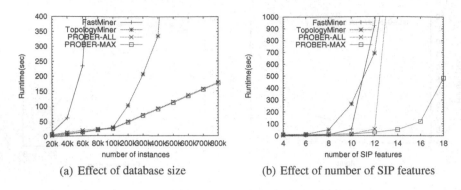

(a) Effect of database size   (b) Effect of number of SIP features

Fig. 7. Scalability study

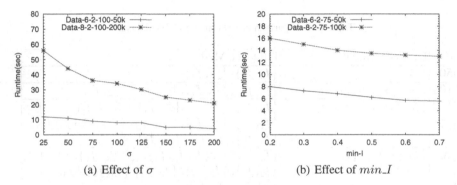

(a) Effect of σ (b) Effect of min_I

Fig. 8. Sensitivity study

4.4   Sensitivity

In the final set of experiments, we examine the effect of two parameters, σ and min_I, on the performance of PROBER.

Effect of σ. We first evaluate the effect of σ on PROBER_MAX. The two datasets used in this experiment are Data-6-2-100-50k and Data-8-2-100-200k. Fig. 8(a) gives the results. PROBER_MAX run faster as σ increases. This is expected due to two reasons: (1) The time complexity of PROBER are inversely related to the σ; (2) A larger σ value implies a smoother distribution of the influence maps, hence requiring fewer iterations to construct the approximated spaces.

Effect of min_I. We evaluate PROBER_MAX on two dataset Data-6-2-75-50k and Data-8-2-75-100k with min_I varying between 0.2 to 0.7. Fig. 8(b), shows that the runtime of PROBER_MAX is independent of min_I. This is reasonable as the cost of building the approximate influence maps dominates in PROBER_MAX. On the other hand, the mining cost is decreasing as min_I increases (see Fig. 8(b)). This is because fewer patterns become frequent as min_I increases.

5   Conclusion

In this paper, we have introduced an influence model to discover spatial interaction patterns from spatial databases. Compared to the distance model used in existing works, the influence model consider the spatial affinity in terms of continuous functions instead of discrete functions. This leads to more meaningful mining results. Another advantage of the influence model is its ability to avoid expensive join operations that are traditionally required to discover the relationship among spatial instances. By analyzing the bounds for computational error, we have also designed an approximation mining algorithm PROBER to find such patterns. The experiment results on both synthetic and real-life datasets demonstrated that PROBER is both effective and scalable.

We observe that some features exhibit interactions not on the whole plane but only in some localized regions. Our future work will investigate the utilization of non-uniform sampling approach to build compact influence maps to mine such local interaction patterns.

References

1. Gouda, K., Zaki, M.J.: Efficiently mining maximal frequent itemsets. In: ICDM, pp. 163–170 (2001)
2. Huang, Y., Shekhar, S., Xiong, H.: Discovering colocation patterns from spatial data sets: a general approach. IEEE Transactions on Knowledge and Data Engineering 16(12), 1472 (2004)
3. Koperski, K., Han, J.: Discovery of spatial association rules in geographic information databases. In: Egenhofer, M.J., Herring, J.R. (eds.) SSD 1995. LNCS, vol. 951, pp. 47–66. Springer, Heidelberg (1995)
4. Morimoto, Y.: Mining frequent neighboring class sets in spatial databases. In: ACM SIGKDD, pp. 353–358 (2001)
5. Ripley, B.D.: Spatial Statistics. Wiley, Chichester (1981)
6. Bayardo, J. R.J.: Efficiently mining long patterns from databases. In: SIGMOD 1998: Proceedings of the 1998 ACM SIGMOD international conference on Management of data, pp. 85–93. ACM Press, New York (1998)
7. Scott, D.W.: Multivariate Density Estimation: Theory, Practice, and Visualization. Wiley, Chichester (1992)
8. Sheng, C., Hsu, W., Lee, M.L.: Discovering spatial interaction patterns. Technique Report TRC6/07. National University of Singapore (June 2007)
9. Wang, J., Hsu, W., Lee, M.L.: A framework for mining topological patterns in spatio-temporal databases. In: ACM CIKM 2005, pp. 429–436. ACM Press, New York (2005)
10. Yoo, J.S., Shekhar, S., Celik, M.: A join-less approach for co-location pattern mining: A summary of results. In: ICDM, pp. 813–816 (2005)
11. Zhang, X., Mamoulis, N., Cheung, D.W., Shou, Y.: Fast mining of spatial collocations. In: ACM SIGKDD 2004, pp. 384–393. ACM Press, New York (2004)

Topological Relationships between Map Geometries

Mark McKenney and Markus Schneider*

University of Florida, Department of Computer and Information Sciences
{mm7,mschneid}@cise.ufl.edu

Abstract. The importance of topological relationships between spatial objects is recognized in many disciplines. In the field of spatial databases, topological relationships have played an important role, providing mechanisms for spatial constraint specification and spatial indexing techniques. The use of spatial data in a map form has become popular, resulting in spatial data models such as *topologies* or, more generally, *map geometries*, which model collections of spatial objects that satisfy certain topological constraints. However, the topological relationships between map geometries remain unexplored. In this paper, we identify the set of valid topological relationships between map geometries, and then provide a mechanism by which they can be directly implemented on top of existing systems by using topological relationships between regions.

1 Introduction

The study of topological relationships between objects in space has lately received much attention from a variety of research disciplines including robotics, geography, cartography, artificial intelligence, cognitive science, computer vision, image databases, spatial database systems, and geographical information systems (GIS). In the areas of databases and GIS, the motivation for formally defining topological relationships between spatial objects has been driven by a need for querying mechanisms that can filter these objects in spatial selections or spatial joins based on their topological relationships to each other, as well as a need for appropriate mechanisms to aid in spatial data analysis and spatial constraint specification.

A current trend in spatial databases and GIS is the use of *topologically integrated spatial data models*, or what are often referred to as *topologies*. We will refer to these models as *map geometries* to avoid confusion with other terms. A map geometry, in this sense, is a collection of spatial objects that satisfy certain topological constraints; specifically, spatial data objects are only allowed to *meet* or be *disjoint*. Typically, spatial data arranged in a map geometry can be viewed from two levels: the object level, in which spatial objects are allowed to overlap, and the topological level, in which *topological primitives* consisting of polygons,

---

* This work was partially supported by the National Science Foundation under grant number NSF-CAREER-IIS-0347574.

J.R. Haritsa, R. Kotagiri, and V. Pudi (Eds.): DASFAA 2008, LNCS 4947, pp. 110–125, 2008.

lines, and nodes form a map geometry in the plane. We use the term map geometry to refer to the topological level. For instance, consider a region representing the United States, and a region representing a high temperature zone that intersects both the US and the Pacific Ocean. These two objects arranged in a map geometry would result in three topological primitives, the polygon consisting of the intersection of both regions, and the polygons representing the remainder of the US and the high temperature zone.

Currently, map geometries are used in spatial databases to enforce topological constraints between points, lines, and regions, and ensure spatial data quality standards when manipulating them. However, there is currently no method of performing topological data analysis between map geometries. For instance, consider a user who creates a map geometry modelling a swamp that is broken into regions based on pollution levels. For example, a section that is not very polluted will be represented as a region, as will be an area that is highly polluted. Once the map geometry is created, it would be interesting to query the database to find related map geometries. For instance, if someone has already completed a survey of a more specific part of the swamp, there may be map geometry in the database that consists of a more detailed view of a region in the original map geometry. Furthermore, map geometries may exist of the same swamp, but broken into regions differently than the original map. Such map geometries may offer information as to pollution sources. Finally, it may be useful to discover which other map geometries overlap the original one, or are adjacent to it.

Although topological queries over map geometries allow many useful queries, we currently do not know the possible topological relationships between map geometries. Because map geometries are more general than complex regions, it follows that there are more possible relationships between them than there are between complex regions. Furthermore, it is unclear how such topological relationships can be implemented. Existing techniques of computing topological relationships could be extended in order to compute the new relationships, but this requires modification of database internals. If the topological relationships between map geometries could be computed based upon topological relationships between their components (i.e., their topological primitives) that correspond to existing spatial data types, then they could be directly used by existing spatial database systems. Therefore, the first goal of this paper is to discover the complete set of valid topological relationships between map geometries. We use the spatial data type of *spatial partitions* to represent map geometries, and derive the topological relationships based on their definition. Note that spatial partitions only represent map geometries consisting of regions. We leave the treatment of map geometries containing point and line features to future work. The second goal of this paper is to characterize topological relationships between map geometries based on their components. In this case, the components of spatial partitions are complex regions. Thus, by characterizing the new topological relationships as thus, we provide a method to directly use them in spatial databases in which topological predicates between complex regions are already implemented.

The remainder of this paper is structured as follows: in Section 2, we examine research related to maps, spatial objects, and topological relationships. We then present the definition for the type of spatial partitions in Section 3. The topological relationships between map geometries are derived and presented in Section 4. We then characterize the topological relationships between map geometries using topological relationships between complex regions, and show to integrate them into existing databases in Section 5. Finally, we draw some conclusions and consider topics for future work in Section 6.

2   Related Work

The work in this paper deals with defining topological relationships between map geometries modeled by spatial partitions. The research related to this falls into two general categories: research exploring spatial data types and map geometries in general, and research into the topological relationships between spatial data types. The spatial data types that have received the most attention in the spatial database and GIS literature are types for simple and complex points, lines, and regions. Simple lines are continuous, one-dimensional features embedded in the plane with two endpoints; simple regions are two dimensional point sets that are topologically equivalent to a closed disc. Increased application requirements and a lack of closure properties of the simple spatial types lead to the development of the complex spatial types. In [13], the authors formally define complex data types, such as complex points (a single point object consisting of a collection of simple points), complex lines (which can represent networks such as river systems), and complex regions that are made up of multiple faces and holes (i.e., a region representing Italy, its islands, and the hole representing the Vatican).

Map geometries have been studied extensively in the literature. In [6,7,10,16], a map is not defined as a data type itself, but as a collection of spatial regions that satisfy some topological constraints. Because these map types are essentially collections of more basic spatial types, it is unclear how topological constraints can be enforced. Other approaches to defining map geometries center around raster or tessellation models [11,15]. However, these lack generality in that they are restricted to the tessellation scheme in use. Topologies, such as those used in GIS and spatial systems, provide the most general map geometries available [8,9]; however, these approaches all center around *implementation models*, and do not provide a formal mathematical model upon which we can base our work. This paper is based on the formal model of spatial partitions presented in [5] in which map geometries are defined as a complete partition of the Euclidean plane into regions which are complex regions such that each region has a unique *label* and regions are allowed to share common boundaries or be disjoint. This model, formally presented in Section 3, has a formal, mathematical definition and provides very few restrictions as to its generality.

Topological relationships indicate qualitative properties of the positions of spatial objects that are preserved under continuous transformations such as translation, rotation, and scaling. Quantitative measures such as distance or

$$\begin{pmatrix} A^\circ \cap B^\circ \neq \varnothing & A^\circ \cap \partial B \neq \varnothing & A^\circ \cap B^- \neq \varnothing \\ \partial A \cap B^\circ \neq \varnothing & \partial A \cap \partial B \neq \varnothing & \partial A \cap B^- \neq \varnothing \\ A^- \cap B^\circ \neq \varnothing & A^- \cap \partial B \neq \varnothing & A^- \cap B^- \neq \varnothing \end{pmatrix}$$

Fig. 1. The 9-intersection matrix for spatial objects A and B

size measurements are deliberately excluded in favor of modeling notions such as connectivity, adjacency, disjointedness, inclusion, and exclusion. Attempts to model and rigorously define the relationships between certain types of spatial objects have lead to the development of two popular approaches: the *9-intersection model* [3], which is developed based on point set theory and point set topology, and the *RCC model* [12], which utilizes spatial logic. The 9-intersection model characterizes the topological relationship between two spatial objects by evaluating the non-emptiness of the intersection between all combinations of the interior ($^\circ$), boundary (∂) and exterior ($^-$) of the objects involved. A unique 3×3 matrix, termed the *9-intersection matrix* (9IM), with values filled as illustrated in Figure 1 describes the topological relationship between a pair of objects:

Various models of topological predicates using both *component derivations*, in which relationships are derived based on the interactions of all components of spatial objects, and *composite derivations*, in which relationships model the global interaction of two objects, exist in the literature. Examples of component derivations can be found in [4,1]. In [4], the authors define topological relationships between regions with holes in which each of the relationships between all faces and holes are calculated. Given two regions, R and S, containing m and n holes respectively, a total of $(n + m + 2)^2$ topological predicates are possible. It is shown that this number can be reduced $mn + m + n + 1$; however, the total number of predicates between two objects depends on the number of holes the objects contain. Similarly, in [1], predicates between complex regions without holes are defined based on the topological relationship of each face within one region with all other faces of the same region, all faces of the other region, and the entire complex regions themselves. Given regions S and R with m and n faces respectively, a matrix is constructed with $(m + n + 2)^2$ entries that represent the topological relationships between S and R and each of their faces.

The most basic example of a composite derivation model (in which the global interaction of two spatial objects is modeled) is the derivation of topological predicates between simple spatial objects in [3]. This model has been used as the basis for modeling topological relationships between object components in the component models discussed above. In [13], the authors apply an extended 9-intersection model to point sets belonging to complex points, lines, and regions. Based on this application, the authors are able to construct a composite derivation model for complex data types and derive a complete and finite set of topological predicates between them, thus resolving the main drawback of the component derivation model. We use a composite derivation model in this paper.

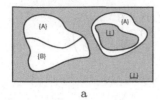
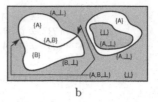

a   b

Fig. 2. Figure *a* shows a spatial partition with two regions and annotated with region labels. Figure *b* shows the same spatial partition with its region and boundary labels (arrows are used to indicate points and are not part of the spatial partition). Note that labels are modeled as sets of attributes in spatial partitions.

3   Data Model: Spatial Partitions

In this section, we review the definition of spatial partitions upon which we base our model of topological relationships between map geometries. A spatial partition, in two dimensions, is a subdivision of the plane into pairwise disjoint *regions* such that each region is associated with a *label* or *attribute* having simple or complex structure, and these regions are separated from each other by *boundaries*. The label of a region describes the thematic data associated with the region. All points within the spatial partition that have an identical label are part of the same region. Topological relationships are implicitly modeled among the regions in a spatial partition. For instance, neglecting common boundaries, the regions of a partition are always disjoint; this property causes spatial partitions to have a rather simple structure. Note that the *exterior* of a spatial partition (i.e., the unbounded face) is always labeled with the ⊥ symbol. Figure 2a depicts an example spatial partition consisting of two regions.

We stated above that each region in a spatial partition is associated with a single attribute or label. A spatial partition is modeled by mapping Euclidean space to such labels. Labels themselves are modeled as sets of attributes. The regions of the spatial partition are then defined as consisting of all points which contain an identical label. Adjacent regions each have different labels in their interior, but their common boundary is assigned the label containing the labels of both adjacent regions. Note that labels are not relevent in determining topological relationships, but are required by the type definition of spatial partitions. Figure 2b shows an example spatial partition complete with boundary labels.

3.1   Notation

We now briefly summarize the mathematical notation used throughout the following sections. The application of a function $f : A \to B$ to a set of values $S \subseteq A$ is defined as $f(S) := \{f(x) | x \in S\} \subseteq B$. In some cases we know that $f(S)$ returns a singleton set, in which case we write $f[S]$ to denote the single element, i.e. $f(S) = \{y\} \implies f[S] = y$. The inverse function $f^{-1} : B \to 2^A$ of f is defined as $f^{-1}(y) := \{x \in A | f(x) = y\}$. It is important to note that f^{-1} is a total function and that f^{-1} applied to a set yields a set of sets. We define the range function of a function $f : A \to B$ that returns the set of all elements that f returns for an input set A as $rng(f) := f(A)$.

Let (X, T) be a topological space [2] with topology $T \subseteq 2^x$, and let $S \subseteq X$. The *interior* of S, denoted by $S°$, is defined as the union of all open sets that are contained in S. The *closure* of S, denoted by $\overline{S}$ is defined as the intersection of all closed sets that contain S. The *exterior* of S is given by $S^- := (X - S)°$, and the *boundary* or *frontier* of S is defined as $\partial S := \overline{S} \cap \overline{X - S}$. An open set is *regular* if $A = \overline{A}°$ [14]. In this paper, we deal with the topological space $\mathbb{R}^2$.

A *partition* of a set S, in set theory, is a complete decomposition of the set S into non-empty, disjoint subsets $\{S_i | i \in I\}$, called blocks: (i) $\forall i \in I : S_i \neq \emptyset$, (ii) $\bigcup_{i \in I} S_i = S$, and (iii) $\forall i, j \in I, i \neq j : S_i \cap S_j = \emptyset$, where I is an index set used to name different blocks. A partition can equivalently be regarded as a total and surjective function $f : S \to I$. However, a spatial partition cannot be defined simply as a set-theoretic partition of the plane, that is, as a partition of $\mathbb{R}^2$ or as a function $f : \mathbb{R}^2 \to I$, for two reasons: first, f cannot be assumed to be total in general, and second, f cannot be uniquely defined on the borders between adjacent subsets of $\mathbb{R}^2$.

3.2   The Definition of Spatial Partitions

In [5], spatial partitions have been defined in several steps. First a *spatial mapping* of type A is a total function $\pi : \mathbb{R}^2 \to 2^A$. The existence of an undefined element $\perp_A$ is required to represent undefined labels (i.e., the exterior of a partition). Definition 1 identifies the different components of a partition within a spatial mapping. The labels on the borders of regions are modeled using the power set 2^A; a *border* of π (Definition 1(ii)) is a block that is mapped to a subset of A containing two or more elements, as opposed to a *region* of π (Definition 1(i)) which is a block mapped to a singleton set. The *interior* of π (Definition 1(iii)) is defined as the union of π's regions. The *boundary* of π (Definition 1(iv)) is defined as the union of π's borders. The *exterior* of π (Definition 1(v)) is the block mapped $\perp_A$. It is also useful to note the *exterior boundary* of a spatial partition as the union of all borders that carry the label $\perp_A$. As an example, let π be the spatial partition in Figure 2 of type $X = \{A, B, \perp\}$. In this case, $rng(\pi) = \{\{A\}, \{B\}, \{\perp\}, \{A, B\}, \{A, \perp\}, \{B, \perp\}, \{A, B, \perp\}\}$. Therefore, the regions of π are the blocks labeled $\{A\}$, $\{B\}$, and $\{\perp\}$ and the boundaries are the blocks labeled $\{A, B\}$, $\{A, \perp\}$, $\{B, \perp\}$, and $\{A, B, \perp\}$. Figure 3 provides a pictorial example of the interior, exterior, and boundary of a more complex example map (note that the borders and boundary consist of the same points, but the boundary is a single point set whereas the borders are a set of point sets).

Definition 1. Let π be a spatial mapping of type A
(i) $\rho(\pi) := \pi^{-1}(rng(\pi) \cap \{X \in 2^A | \ |X| = 1\})$ *(regions)*
(ii) $\omega(\pi) := \pi^{-1}(rng(\pi) \cap \{X \in 2^A | \ |X| > 1\})$ *(borders)*
(iii) $\pi° := \bigcup_{r \in \rho(\pi) | \pi[r] \neq \{\perp_A\}} r$ *(interior)*
(iv) $\partial \pi := \bigcup_{b \in \omega(\pi)} b$ *(boundary)*
(v) $\pi^- := \pi^{-1}(\{\perp_A\})$ *(exterior)*

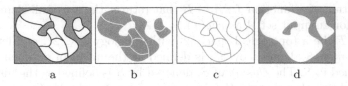

Fig. 3. Figure a shows a spatial partition π with two disconnected faces, one containing a hole. The interior (π°), boundary ($\partial\pi$), and exterior (π^-) of the partition are shown if Figures b c, and d, respectively. Note that the labels have been omitted in order to emphasize the components of the spatial partition.

A *spatial partition* of type A is then defined as a spatial mapping of type A whose regions are regular open sets [14] and whose borders are labeled with the union of labels of all adjacent regions. Since we are using spatial partitions to model map geometries, we use the terms 'spatial partition' and 'map geometry' interchangeably for the remainder of the paper.

Definition 2. A *spatial partition* of type A is a spatial mapping π of type A with:

    (i) $\forall r \in \rho(\pi) : r = \overline{r}^\circ$

    (ii) $\forall b \in \omega(\pi) : \pi[b] = \{\pi[[r]] | r \in \rho(\pi) \wedge b \subseteq \partial r\}$

4 Topological Relationships

In this section, we describe a method for deriving the topological relationships between a given pair of map geometries. We begin by describing various approaches to the problem, then outline our chosen method, and finally derive the actual relationships based on this method.

The goal of this paper is to define a complete, finite set of topological predicates between map geometries; therefore, we employ a method similar to that found in [13], in which the 9-intersection model is extended to describe complex points, lines, and regions. In Section 3, we defined a point set topological model of map geometries in which we identified the interior, exterior, and boundary point sets belonging to maps. Based on this model, we extend the 9-intersection model to apply to the point sets belonging to map objects. However, due to the spatial features of map geometries, the embedding space ($\mathbb{R}^2$), and the interaction of map geometries with the embedding space, some topological configurations are impossible and must be excluded. Therefore, we must identify topological constraints that must be satisfied in order for a given topological configuration to be valid. Furthermore, we must identify these constraints such that all invalid topological configurations are excluded, and the complete set of valid configurations remains. We achieve this through a proof technique called *Proof-by-Constraint-and-Drawing*, in which we begin with the total set of 512 possible 9-intersection matrices, and determine the set of valid configurations by first providing a collection of topological constraint rules that invalidate impossible topological configurations, and second, validating all matrices that satisfy *all* constraint rules by providing a prototypical spatial configuration (i.e., the configurations can be drawn in the embedding space). Completeness is achieved because all topological configurations are either eliminated by constraint rules, or are proven to be possible through the drawing of a prototype. The remainder of this section contains the constraints, and the prototypical drawings of map geometries are shown in Table 1.

We identify eight constraint rules that 9IMs for map geometries must satisfy in order to be valid. Each constraint rule is first written in sentences and then expressed mathematically. Some mathematical expressions are written in two equivalent expressions so that they may be applied to the 9-intersection matrix more easily. Following each rule is the rationale explaining why the rule is correct. In the following, let π and σ be two spatial partitions.

Lemma 1. *Each component of a map geometry intersects at least one component of the other map geometry:*

$$(\forall C_\pi \in \{\pi^\circ, \partial\pi, \pi^-\} : C_\pi \cap \sigma^\circ \neq \varnothing \vee C_\pi \cap \partial\sigma \neq \varnothing \vee C_\pi \cap \sigma^- \neq \varnothing)$$
$$\wedge (\forall C_\sigma \in \{\sigma^\circ, \partial\sigma, \sigma^-\} : C_\sigma \cap \pi^\circ \neq \varnothing \vee C_\sigma \cap \partial\pi \neq \varnothing \vee C_\sigma \cap \pi^- \neq \varnothing)$$

Proof. Because spatial mappings are defined as total functions, it follows that $\pi^\circ \cup \partial\pi \cup \pi^- = \mathbb{R}^2$ and that $\sigma^\circ \cup \partial\sigma \cup \sigma^- = \mathbb{R}^2$. Thus, each part of π must intersect at least one part of σ, and vice versa. ☐

Lemma 2. *The exteriors of two map geometries always intersect:*

$$\pi^- \cap \sigma^- \neq \varnothing$$

Proof. The closure of each region in a map geomtry corresponds to a complex region as defined in [13]. Since complex regions are closed under the union operation, it follows that the union of all regions that compose a map geometry is a complex region, whose boundary is defined by a Jordan curve. Therefore, every spatial partition has an exterior. Furthermore, in [5], the authors prove that spatial partitions are closed under intersection. Thus, the intersection of any two spatial partitions is a spatial partition that has an exterior. Therefore, the exteriors of any two spatial partitions intersect, since their intersection contains an exterior. ☐

Lemma 3. *If the boundary of a map geometry intersects the interior of another map geometry, then their interiors intersect:*

$$((\partial\pi \cap \sigma^\circ \neq \varnothing \Rightarrow \pi^\circ \cap \sigma^\circ \neq \varnothing) \wedge (\pi^\circ \cap \partial\sigma \neq \varnothing \Rightarrow \pi^\circ \cap \sigma^\circ \neq \varnothing))$$
$$\Leftrightarrow ((\partial\pi \cap \sigma^\circ = \varnothing \vee \pi^\circ \cap \sigma^\circ \neq \varnothing) \wedge (\pi^\circ \cap \partial\sigma = \varnothing \vee \pi^\circ \cap \sigma^\circ \neq \varnothing))$$

Proof. Assume that a boundary b of partition π intersects the interior of partition σ but their interiors do not intersect. In order for this to be true, the label of the regions on either side of b must be labeled with the empty label. According to the definition of spatial partitions, a boundary separates two regions with different labels; thus, this is impossible and we have a proof by contradiction. ☐

Lemma 4. *If the boundary of a map geometry intersects the exterior of a second map geometry, then the interior of the first map geometry intersects the exterior of the second:*

$$((\partial\pi \cap \sigma^- \neq \varnothing \Rightarrow \pi° \cap \sigma^- \neq \varnothing) \wedge (\pi^- \cap \partial\sigma \neq \varnothing \Rightarrow \pi^- \cap \sigma° \neq \varnothing))$$
$$\Leftrightarrow ((\partial\pi \cap \sigma^- = \varnothing \vee \pi° \cap \sigma^- \neq \varnothing) \wedge (\pi^- \cap \partial\sigma = \varnothing \vee \pi^- \cap \sigma° \neq \varnothing))$$

Proof. This proof is similar to the previous proof. Assume that the boundary b of partition π intersects the exterior of partition σ but the interior of π does not intersect the exterior of σ. In order for this to be true, the label of the regions on either side of b must be labeled with the empty label. According to the definition of spatial partitions, a boundary separates two regions with different labels; thus, this is impossible and we have a proof by contradiction.   □

Lemma 5. *If the boundaries of two map geometries are equivalent, then their interiors intersect:*

$$(\partial\pi = \partial\sigma \Rightarrow \pi° \cap \sigma° \neq \varnothing) \Leftrightarrow (c \Rightarrow d) \Leftrightarrow (\neg c \vee d) \text{ where}$$
$$c = \partial\pi \cap \partial\sigma \neq \varnothing \wedge \pi° \cap \partial\sigma = \varnothing \wedge \partial\pi \cap \sigma° = \varnothing$$
$$\wedge \partial\pi \cap \sigma^- = \varnothing \wedge \pi^- \cap \partial\sigma = \varnothing$$
$$d = \pi° \cap \sigma° \neq \varnothing$$

Proof. Assume that two spatial partitions have an identical boundary, but their interiors do not intersect. The only configuration which can accommodate this situation is if one spatial partition's interior is equivalent to the exterior of the other spatial partition. However, according to Lemma 2, the exteriors of two partitions always intersect. If a partition's interior is equivalent to another partition's exterior, then their exteriors would not intersect. Therefore, this configuration is not possible, and the interiors of two partitions with equivalent boundaries must intersect.   □

Lemma 6. *If the boundary of a map geometry is completely contained in the interior of a second map geometry, then the boundary and interior of the second map geometry must intersect the exterior of the first, and vice versa:*

$$(\partial\pi \subset \sigma° \Rightarrow \pi^- \cap \partial\sigma \neq \varnothing \wedge \pi^- \cap \sigma° \neq \varnothing) \Leftrightarrow (\neg c \vee d) \text{ where}$$
$$c = \partial\pi \cap \sigma° \neq \varnothing \wedge \partial\pi \cap \partial\sigma = \varnothing \wedge \partial\pi \cap \sigma^- = \varnothing$$
$$d = \pi^- \cap \partial\sigma \neq \varnothing \wedge \pi^- \cap \sigma° \neq \varnothing$$

Proof. If the boundary of spatial partition π is completely contained in the interior of spatial partition σ, it follows from the Jordan Curve Theorem that the boundary of σ is completely contained in the exterior of π. By Lemma 4, it then follows that the interior of σ intersects the exterior of π.   □

Lemma 7. *If the boundary of one map geometry is completely contained in the interior of a second map geometry, and the boundary of the second map geometry is completely contained in the exterior of the first, then the interior of the first map geometry cannot intersect the exterior of the second and the interior of the second map geometry must intersect the exterior of the first and vice versa:*

$$((\partial\pi \subset \sigma^\circ \;\wedge\; \pi^- \supset \partial\sigma) \Rightarrow (\pi^\circ \cap \sigma^- = \varnothing \;\wedge\; \pi^- \cap \sigma^\circ \neq \varnothing))$$
$$\Leftrightarrow (c \Rightarrow d) \Leftrightarrow (\neg c \;\vee\; d) \text{ where}$$
$$c = \partial\pi \cap \sigma^\circ \neq \varnothing \;\wedge\; \partial\pi \cap \partial\sigma = \varnothing \;\wedge\; \partial\pi \cap \sigma^- = \varnothing$$
$$\wedge\; \pi^\circ \cap \partial\sigma = \varnothing \;\wedge\; \pi^- \cap \partial\sigma \neq \varnothing$$
$$d = \pi^\circ \cap \sigma^- = \varnothing \;\wedge\; \pi^- \cap \sigma^\circ \neq \varnothing$$

Proof. We construct this proof in two parts. According to Lemma 6, if $\partial\pi \subseteq \sigma^\circ$, then $\pi^- \cap \sigma^\circ \neq \varnothing$. Now we must prove that π° cannot intersect σ^-. Since $\partial\pi \subset \sigma^\circ$, it follows that π° intersects σ°. Therefore, the only configuration where $\pi^\circ \cap \sigma^- \neq \varnothing$ can occur is if σ contains a hole that is contained by π. However, in order for this configuration to exist, the $\partial\sigma$ would have to intersect the interior or the boundary of π. Since the lemma specifies the situation where $\pi^- \supset \partial\sigma$, this configuration cannot exist; thus, the interior of π cannot intersect the exterior of σ. $\qquad\square$

Lemma 8. *If the boundary of a map geometry is completely contained in the exterior of a second map geometry and the boundary of the second map geometry is completely contained in the exterior of the first, then the interiors of the map geometries cannot intersect:*

$$((\partial\pi \subset \sigma^- \;\wedge\; \pi^- \supset \partial\sigma) \Rightarrow (\pi^\circ \cap \sigma^\circ = \varnothing))$$
$$\Leftrightarrow (c \Rightarrow d) \Leftrightarrow (\neg c \;\vee\; d) \text{ where}$$
$$c = \partial\pi \cap \sigma^\circ = \varnothing \;\wedge\; \partial\pi \cap \partial\sigma = \varnothing \;\wedge\; \partial\pi \cap \sigma^- \neq \varnothing$$
$$\wedge\; \pi^\circ \cap \partial\sigma = \varnothing \;\wedge\; \pi^- \cap \partial\sigma \neq \varnothing$$
$$d = \pi^\circ \cap \sigma^\circ = \varnothing$$

Proof. The lemma states that the interiors of two disjoint maps do not intersect. Without loss of generality, consider two map geometries that each consist of a single region. We can consider these map geometries as complex region objects. If two complex regions are disjoint, then their interiors do not intersect. We can reduce any arbitrary map to a complex region by computing the spatial union of its regions. It follows that because the interiors of two disjoint regions do not intersect, the interiors of two disjoint maps do not intersect. $\qquad\square$

Using a simple program to apply these eight constraint rules reduces the 512 possible matrices to 49 valid matrices that represent topological relationships between two maps geometries. The matrices and their validating prototypes are depicted in Table 1. Finally, we summarize our results as follows:

Theorem 1. *Based on the 9-intersection model for spatial partitions, 49 different topological relationships exist between two map geometries.*

Proof. The argumentation is based on the *Proof-by-Constraint-and-Drawing* method. The constraint rules, whose correctness has been proven in Lemmas 1 to 8, reduce the 512 possible 9-intersection matrices to 49 matrices. The ability to draw prototypes of the corresponding 49 topological configurations in Table 1 proves that the constraint rules are complete. $\qquad\square$

Table 1. The valid matrices and protoypical drawings representing the possible topological relationships between maps. Each drawing shows the interaction of two maps, one map is medium-grey and has a dashed boundary, the other is light-grey and has a dotted boundary. Overlapping map interiors are dark-grey, and overlapping boundaries are drawn as a solid line. For reference, the figure for matrix 41 shows two disjoint maps and the figure for matrix 1 shows two equal maps.

Matrix 1	Matrix 2	Matrix 3	Matrix 4	Matrix 5
$\begin{pmatrix} 1\,0\,0 \\ 0\,1\,0 \\ 0\,0\,1 \end{pmatrix}$	$\begin{pmatrix} 1\,1\,0 \\ 0\,1\,0 \\ 0\,0\,1 \end{pmatrix}$	$\begin{pmatrix} 1\,0\,1 \\ 0\,1\,0 \\ 0\,0\,1 \end{pmatrix}$	$\begin{pmatrix} 1\,1\,1 \\ 0\,1\,0 \\ 0\,0\,1 \end{pmatrix}$	$\begin{pmatrix} 1\,0\,0 \\ 1\,1\,0 \\ 0\,0\,1 \end{pmatrix}$
Matrix 6	Matrix 7	Matrix 8	Matrix 9	Matrix 10
$\begin{pmatrix} 1\,1\,0 \\ 1\,1\,0 \\ 0\,0\,1 \end{pmatrix}$	$\begin{pmatrix} 1\,0\,1 \\ 1\,1\,0 \\ 0\,0\,1 \end{pmatrix}$	$\begin{pmatrix} 1\,1\,1 \\ 1\,1\,0 \\ 0\,0\,1 \end{pmatrix}$	$\begin{pmatrix} 1\,1\,1 \\ 0\,0\,1 \\ 0\,0\,1 \end{pmatrix}$	$\begin{pmatrix} 1\,1\,1 \\ 1\,0\,1 \\ 0\,0\,1 \end{pmatrix}$
Matrix 11	Matrix 12	Matrix 13	Matrix 14	Matrix 15
$\begin{pmatrix} 1\,0\,1 \\ 0\,1\,1 \\ 0\,0\,1 \end{pmatrix}$	$\begin{pmatrix} 1\,1\,1 \\ 0\,1\,1 \\ 0\,0\,1 \end{pmatrix}$	$\begin{pmatrix} 1\,0\,1 \\ 1\,1\,1 \\ 0\,0\,1 \end{pmatrix}$	$\begin{pmatrix} 1\,1\,1 \\ 1\,1\,1 \\ 0\,0\,1 \end{pmatrix}$	$\begin{pmatrix} 1\,0\,0 \\ 0\,1\,0 \\ 1\,0\,1 \end{pmatrix}$
Matrix 16	Matrix 17	Matrix 18	Matrix 19	Matrix 20
$\begin{pmatrix} 1\,1\,0 \\ 0\,1\,0 \\ 1\,0\,1 \end{pmatrix}$	$\begin{pmatrix} 1\,0\,1 \\ 0\,1\,0 \\ 1\,0\,1 \end{pmatrix}$	$\begin{pmatrix} 1\,1\,1 \\ 0\,1\,0 \\ 1\,0\,1 \end{pmatrix}$	$\begin{pmatrix} 1\,0\,0 \\ 1\,1\,0 \\ 1\,0\,1 \end{pmatrix}$	$\begin{pmatrix} 1\,1\,0 \\ 1\,1\,0 \\ 1\,0\,1 \end{pmatrix}$
Matrix 21	Matrix 22	Matrix 23	Matrix 24	Matrix 25
$\begin{pmatrix} 1\,0\,1 \\ 1\,1\,0 \\ 1\,0\,1 \end{pmatrix}$	$\begin{pmatrix} 1\,1\,1 \\ 1\,1\,0 \\ 1\,0\,1 \end{pmatrix}$	$\begin{pmatrix} 1\,1\,1 \\ 1\,0\,1 \\ 1\,0\,1 \end{pmatrix}$	$\begin{pmatrix} 0\,0\,1 \\ 0\,1\,1 \\ 1\,0\,1 \end{pmatrix}$	$\begin{pmatrix} 1\,0\,1 \\ 0\,1\,1 \\ 1\,0\,1 \end{pmatrix}$

Table 1. *(Continued)*

Matrix 26	Matrix 27	Matrix 28	Matrix 29	Matrix 30
$\begin{pmatrix} 1 & 1 & 1 \\ 0 & 1 & 1 \\ 1 & 0 & 1 \end{pmatrix}$	$\begin{pmatrix} 1 & 0 & 1 \\ 1 & 1 & 1 \\ 1 & 0 & 1 \end{pmatrix}$	$\begin{pmatrix} 1 & 1 & 1 \\ 1 & 1 & 1 \\ 1 & 0 & 1 \end{pmatrix}$	$\begin{pmatrix} 1 & 0 & 0 \\ 1 & 0 & 0 \\ 1 & 1 & 1 \end{pmatrix}$	$\begin{pmatrix} 1 & 1 & 0 \\ 1 & 0 & 0 \\ 1 & 1 & 1 \end{pmatrix}$
Matrix 31	Matrix 32	Matrix 33	Matrix 34	Matrix 35
$\begin{pmatrix} 1 & 1 & 1 \\ 1 & 0 & 0 \\ 1 & 1 & 1 \end{pmatrix}$	$\begin{pmatrix} 1 & 0 & 0 \\ 0 & 1 & 0 \\ 1 & 1 & 1 \end{pmatrix}$	$\begin{pmatrix} 1 & 1 & 0 \\ 0 & 1 & 0 \\ 1 & 1 & 1 \end{pmatrix}$	$\begin{pmatrix} 0 & 0 & 1 \\ 0 & 1 & 0 \\ 1 & 1 & 1 \end{pmatrix}$	$\begin{pmatrix} 1 & 0 & 1 \\ 0 & 1 & 0 \\ 1 & 1 & 1 \end{pmatrix}$
Matrix 36	Matrix 37	Matrix 38	Matrix 39	Matrix 40
$\begin{pmatrix} 1 & 1 & 1 \\ 0 & 1 & 0 \\ 1 & 1 & 1 \end{pmatrix}$	$\begin{pmatrix} 1 & 0 & 0 \\ 1 & 1 & 0 \\ 1 & 1 & 1 \end{pmatrix}$	$\begin{pmatrix} 1 & 1 & 0 \\ 1 & 1 & 0 \\ 1 & 1 & 1 \end{pmatrix}$	$\begin{pmatrix} 1 & 0 & 1 \\ 1 & 1 & 0 \\ 1 & 1 & 1 \end{pmatrix}$	$\begin{pmatrix} 1 & 1 & 1 \\ 1 & 1 & 0 \\ 1 & 1 & 1 \end{pmatrix}$
Matrix 41	Matrix 42	Matrix 43	Matrix 44	Matrix 45
$\begin{pmatrix} 0 & 0 & 1 \\ 0 & 0 & 1 \\ 1 & 1 & 1 \end{pmatrix}$	$\begin{pmatrix} 1 & 1 & 1 \\ 0 & 0 & 1 \\ 1 & 1 & 1 \end{pmatrix}$	$\begin{pmatrix} 1 & 0 & 1 \\ 1 & 0 & 1 \\ 1 & 1 & 1 \end{pmatrix}$	$\begin{pmatrix} 1 & 1 & 1 \\ 1 & 0 & 1 \\ 1 & 1 & 1 \end{pmatrix}$	$\begin{pmatrix} 0 & 0 & 1 \\ 0 & 1 & 1 \\ 1 & 1 & 1 \end{pmatrix}$
Matrix 46	Matrix 47	Matrix 48	Matrix 49	
$\begin{pmatrix} 1 & 0 & 1 \\ 0 & 1 & 1 \\ 1 & 1 & 1 \end{pmatrix}$	$\begin{pmatrix} 1 & 1 & 1 \\ 0 & 1 & 1 \\ 1 & 1 & 1 \end{pmatrix}$	$\begin{pmatrix} 1 & 0 & 1 \\ 1 & 1 & 1 \\ 1 & 1 & 1 \end{pmatrix}$	$\begin{pmatrix} 1 & 1 & 1 \\ 1 & 1 & 1 \\ 1 & 1 & 1 \end{pmatrix}$	

5   Computing Topological Relationships

Map geometries are currently implemented in many spatial systems as GIS topologies, but no mechanism to compute topological relationships between pairs of map geometries exists. However, map geometries, as considered in this paper, are composed of complex regions, and the topological predicates between complex regions are implemented in many spatial systems. Therefore, we now derive a method to determine the topological relationship between two map geometries based on the topological relationships between their component regions.

Recall that topological relationships based on the 9-intersection model are each defined by a unique 9IM. Given two map geometries, our approach is to derive the values in the 9IM representing their topological relationship by examining the *local* interactions between their component regions. We begin by discussing the properties of the 9IM, and then use these properties to show how we can derive a 9IM representing the topological relationship between two map geometries from the 9IMs representing the relationships between their component regions.

We begin by making two observations about the expression of local interactions between the component regions of map geometries in the 9IM that represents the geometries' relationship. The first observation is that if the interiors of two regions intersect, then the interiors of the map geometries intersect. This is because the presence of additional regions in a map geometry cannot cause two intersecting regions to no longer intersect. This observation also holds for interior/boundary intersections, and boundary/boundary intersections between regions composing map geometries. Therefore, if two map geometries contain respective regions such that those regions' interiors or boundaries intersect each other, then the 9IM for those regions will have the corresponding entry set to *true*.

The first observation does not hold for local interactions involving the exteriors of regions composing two map geometries. Consider Figure 4. This figure depicts two map geometries, one with a solid boundary (which we name A), and a second one with a dashed boundary (which we name B) that is completely contained in the closure of the first. Note that if we examine the topological relationship between the leftmost regions in each geometry, it is clear that interior of the region from B intersects the exterior of the region from A. However, this does not occur globally due to the other regions contained in the geometries. However, we make the observation that the union of all regions that compose a map geometry is a complex region whose exterior is equivalent to the exterior of the map geometry. Therefore, a map geometry's interior or boundary intersects the exterior of a second map geometry if the interior or boundary of a single region in the first map geometry intersects the exterior of the union of all regions in the second. Therefore, we can discover the values in the 9IM representing the topological relationship between two map geometries by examining the interactions of those geometries' component regions and their unions. Given a map geometry π a we define the set of all complex regions in π as $R(\pi) = \{\overline{r} | r \in \rho(\pi)\}$ (recall that the regions in the set $\rho(\pi)$ are open point sets, and complex regions are defined as closed point sets):

Definition 3. *Let A and B be map geometries and $U_r = \bigcup_{r \in R(A)} r$ and $U_s = \bigcup_{s \in R(B)} s$. Let $M_{(A,B)}{}^{\circ\circ}$ be the entry in the matrix M representing a topological relationship between A and B corresponding to the intersection of the interiors*

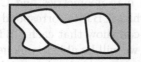

Fig. 4. Two example map geometries shown individually and overlaid

of A and B, etc. The entries in the 9IM representing the topological relationship between two map geometries can be defined as follows:

$$M_{(A,B)}{}^{\circ\circ}= \exists r \in R(A), s \in R(B)|r^\circ \cap s^\circ \neq \varnothing$$
$$M_{(A,B)}{}^\circ\partial= \exists r \in R(A), s \in R(B)|r^\circ \cap \partial s \neq \varnothing$$
$$M_{(A,B)}\partial^\circ= \exists r \in R(A), s \in R(B)|\partial r \cap s^\circ \neq \varnothing$$
$$M_{(A,B)}\partial\partial= \exists r \in R(A), s \in R(B)|\partial r \cap \partial s \neq \varnothing$$
$$M_{(A,B)}{}^{-\circ}= \exists s \in R(B)|U_r^- \cap s^\circ \neq \varnothing$$
$$M_{(A,B)}{}^-\partial= \exists s \in R(B)|U_r^- \cap \partial s \neq \varnothing$$
$$M_{(A,B)}{}^{\circ-}= \exists r \in R(A)|r^\circ \cap U_s^- \neq \varnothing$$
$$M_{(A,B)}\partial^-= \exists r \in R(A)|\partial r \cap U_s^- \neq \varnothing$$

At this point we are able to characterize a 9IM representing the topological relationship between two map geometries based on intersections of the components of the regions that make up the map geometries. However, our goal is to go a step further and characterize such a relationship based on the topological predicates between complex regions that are currently implemented in spatial systems. In order to achieve this, we represent a topological relationship between map geometries as an ordered triple of sets of topological predicates between complex regions, which we denote a *component based topological relationship* (CBTR). This triple consists of a set of topological predicates between the component regions in two map geometries (from which we can directly identify whether the interiors, boundaries, or interiors and boundaries of the map geometries intersect), a set of topological predicates between the regions of the first map geometry and union of the regions in the second (to determine whether the interior and boundary of the first intersect the exterior of the second), and a set of topological predicates between the regions of the second map geometry and the union of the regions in the first.

Definition 4. *Let P_{CR} be the set of topological predicates between complex regions. A component based topological relationship between map geometries A and B is an ordered triple $CBTR(A, B) = (P, N_1, N_2)$ of sets of topological predicates between complex regions such that:*
$$P = \{p \in P_{CR}|r \in R(A) \ \wedge \ s \in R(B) \ \wedge \ p(r, s)\}$$
$$N_1 = \{p \in P_{CR}|r \in R(A) \ \wedge \ U_s = \bigcup_{s \in R(B)} s \ \wedge \ p(r, U_s)\}$$
$$N_2 = \{p \in P_{CR}|U_r = \bigcup_{r \in R(A)} r \ \wedge \ s \in R(B) \ \wedge \ p(U_r, s)\}$$

Definition 4 allows us to identify a CBTR given two map geometries. However, we still cannot determine the topological relationship between two map geometries given their CBTR. In order to do this, we must define a correspondence between

9IMs representing topological relationships between map geometries and CB-TRs. Given such a correspondence, we can identify which CBTRs correspond to a particular 9IM, and vice versa. Furthermore, if we can show that each CBTR corresponds to a single 9IM for map geometries, then we will be able to uniquely identify a topological relationship between two map geometries by determining their CBTR.

In order to find the possible CBTRs that correspond to a particular 9IM between map geometries R, we find all possible values for each set in the triple (P, N_1, N_2). We then find all combinations of these values that form a valid CBTR. To find all possible values of the set P, we take the powerset of the set of 9IMs representing topological relationships between complex regions. We then keep only the sets in the power set such that (i) for each interaction between interiors, boundaries, or interior and boundary in R with a value of *true*, at least one 9IM exists in the set that has a corresponding entry set to *true*, and (ii), for each interaction between interiors, boundaries, or interior and boundary in R with a value of *false*, all 9IMs in the set have a corresponding entry of *false*. This follows directly from the observations made about the intersections of interiors and boundaries among the regions that make up a map geometry. The set of possible values for N_1 and N_2 are computed identically, except entries corresponding to interactions involving the exterior of a map geometry are used. A CBTR that corresponds to the topological relationship R is then a triple (P, N_1, N_2) consisting of any combination of the computed values for each set. Because Definition 3 defines each entry in a 9IM based on an equivalence to information found in the CBTR, it follows that each CBTR corresponds to a single topological relationship between map geometries. Therefore, we are able to uniquely represent a topological relationship between map geometries as a CBTR, which consists of topological relationships between complex regions. To use this in a spatial database, one must compute the CBTR for two given map geometries, and then use the rules in Definition 3 to construct the 9IM that represents their topological relationship.

6   Conclusions and Future Work

In this paper, we have defined and provided examples of the 49 topological relationships between map geometries modeled as spatial partitions. These relationships can be used to pose topological queries over map geometries in spatial systems. Furthermore, we have shown how these topological relationships can be computed using the existing topological relationships between complex regions that are currently implemented in many spatial systems. For future work, we plan to investigate a map geometry querying mechanism. Because map geometries are more complex than other spatial types, it is not yet clear if the traditional spatial querying mechanisms are adequate for handling general queries over map geometries.

References

1. Clementini, E., Di Felice, P., Califano, G.: Composite Regions in Topological Queries. Information Systems 20, 579–594 (1995)
2. Dugundi, J.: Topology. Allyn and Bacon (1966)
3. Egenhofer, M.J., Herring, J.: Categorizing Binary Topological Relations Between Regions, Lines, and Points in Geographic Databases. Technical report, National Center for Geographic Information and Analysis, University of California, Santa Barbara (1990)
4. Egenhofer, M.J., Clementini, E., Di Felice, P.: Topological Relations between Regions with Holes. Int. Journal of Geographical Information Systems 8, 128–142 (1994)
5. Erwig, M., Schneider, M.: Partition and Conquer. In: Frank, A.U. (ed.) COSIT 1997. LNCS, vol. 1329, pp. 389–408. Springer, Heidelberg (1997)
6. Güting, R.H.: Geo-relational algebra: A model and query language for geometric database systems. In: Schek, H.J., Saltor, F., Ramos, I., Alonso, G. (eds.) EDBT 1998. LNCS, vol. 1377, pp. 506–527. Springer, Heidelberg (1998)
7. Güting, R.H., Schneider, M.: Realm-Based Spatial Data Types: The ROSE Algebra. VLDB Journal 4, 100–143 (1995)
8. Herring, J.: TIGRIS: Topologically Integrated Geographic Information Systems. In: 8th International Symposium on Computer Assisted Cartography, pp. 282–291 (1987)
9. Hoel, E.G., Menon, S., Morehouse, S.: Building a Robust Relational Implementation of Topology. In: Hadzilacos, T., Manolopoulos, Y., Roddick, J.F., Theodoridis, Y. (eds.) SSTD 2003. LNCS, vol. 2750, pp. 508–524. Springer, Heidelberg (2003)
10. Huang, Z., Svensson, P., Hauska, H.: Solving spatial analysis problems with geosal, a spatial query language. In: Proceedings of the 6th Int. Working Conf. on Scientific and Statistical Database Management, pp. 1–17. Institut f. Wissenschaftliches Rechnen Eidgenoessische Technische Hochschule Zürich (1992)
11. Ledoux, H., Gold, C.: A Voronoi-Based Map Algebra. In: Int. Symp. on Spatial Data Handling (July 2006)
12. Randell, D.A., Cui, Z., Cohn, A.: A Spatial Logic Based on Regions and Connection. In: International Conference on Principles of Knowledge Representation and Reasoning, pp. 165–176 (1992)
13. Schneider, M., Behr, T.: Topological Relationships between Complex Spatial Objects. ACM Trans. on Database Systems (TODS) 31(1), 39–81 (2006)
14. Tilove, R.B.: Set Membership Classification: A Unified Approach to Geometric Intersection Problems. IEEE Trans. on Computers C-29, 874–883 (1980)
15. Tomlin, C.D.: Geographic Information Systems and Cartographic Modelling. Prentice-Hall, Englewood Cliffs (1990)
16. Voisard, A., David, B.: Mapping conceptual geographic models onto DBMS data models. Technical Report TR-97-005, Berkeley, CA (1997)

MBR Models for Uncertainty Regions of Moving Objects

Shayma Alkobaisi[1], Wan D. Bae[1], Seon Ho Kim[1],
and Byunggu Yu[2]

[1] Department of Computer Science, University of Denver, USA
{salkobai,wbae,seonkim}@cs.du.edu
[2] Computer Science and Information Technology, University of the District of
Columbia, USA
byu@udc.edu

Abstract. The increase in the advanced location based services such
as traffic coordination and management necessitates the need for advanced models tracking the positions of Moving Objects (MOs) like
vehicles. Computers cannot continuously update locations of MOs because of computational overhead, which limits the accuracy of evaluating MOs' positions. Due to the uncertain nature of such positions,
efficiently managing and quantifying the uncertainty regions of MOs are
needed in order to improve query response time. These regions can be
rather irregular which makes them unsuitable for indexing. This paper
presents *Minimum Bounding Rectangles* (MBR) approximations for
three uncertainty region models, namely, the *Cylinder Model* (CM), the
Funnel Model of *Degree* 1 (FMD_1) and the *Funnel Model* of *Degree*
2 (FMD_2). We also propose an estimation of the MBR of FMD_2 that
achieves a good balance between computation time and selectivity (false-
hits). Extensive experiments on both synthetic and real spatio-temporal
datasets showed an order of magnitude improvement of the estimated
model over the other modeling methods in terms of the number of MBRs
retrieved during query process, which directly corresponds to the number
of physical page accesses.

1   Introduction

Applications based on large sets of Moving Objects (MOs) continue to grow,
and so does the demand for efficient data management and query processing
systems supporting MOs. Although there are many different types of MOs, this
paper focuses on physical MOs such as vehicles, that can continuously move in
geographic space. Application examples include traffic control, transportation
management and mobile networks, which require an efficient on-line database
management system that is able to store, update, and query a large number of
continuously moving data objects.

J.R. Haritsa, R. Kotagiri, and V. Pudi (Eds.): DASFAA 2008, LNCS 4947, pp. 126–140, 2008.
© Springer-Verlag Berlin Heidelberg 2008

Many major challenges for developing such a system are due to the following fact: the actual data object has one or more properties (e.g., geographic location and velocity) that can continuously change over time, however, it is only possible to discretely record the properties. All the missing or non-recorded states collectively form the uncertainty of the object's history [7,8,11,10]. This challenging aspect of MO management is much more pronounced in practice, especially when infrequent updates, message transmission delays (e.g., SMS or SMTP), more dynamic object movements (e.g., unmanned airplanes), and the scalability to larger data sets are considered.

As technology advances, more sophisticated location reporting devices are able to report not only locations but also some higher order derivatives, such as velocity and acceleration at the device level, that can be used to more accurately model the movement of objects [11]. To be practically viable though, each uncertainty model must be paired with an appropriate approximation that can be used for indexing. This is due to the modern two-phase query processing system consisting of the refinement phase and the filtering phase. The refinement phase requires an uncertainty model that minimally covers all possible missing states (minimally conservative). The filtering step is represented by an access method whose index is built on an approximation of the refinement phase's uncertainty model. For example, minimum bounding rectangles or parametric boundary polynomials [2] that cover more detailed uncertainty regions of the underlying uncertainty model can be indexed for efficient filtering. The minimality (or selectivity) of the approximation determines the false-hit ratio of the filtering phase, which translates into the cost of the refinement step; the minimality of the uncertainty model determines the final false-hit ratio of the query.

This paper considers the following three uncertainty models for continuously moving objects: the *Cylinder Model* (*CM*) [7,8], the *Funnel Model* of *Degree* 1 (*FMD_1*) [10], and the *Funnel Model* of *Degree* 2 (*FMD_2*, a.k.a. the tornado model [11]). To quantify an uncertainty region, CM takes only the recorded positions (e.g., 2D location) at a given time as input and considers the maximum possible displacement at all times. FMD_1 takes the positions as CM does, however, FMD_1 considers the maximum displacement as a linear function of time. FMD_2 takes velocities as additional input and considers acceleration of moving objects. It calculates the maximum possible displacement using non-linear functions of time. The derived regions of these models can be rather irregular which makes them unsuitable for indexing. This paper proposes an MBR approximation for each of these three uncertainty models. Moreover, this paper proposes a simplified FMD_2, called the *Estimated Funnel Model* of *Degree* 2 (*EFMD_2*). Then the paper shows how $EFMD_2$ achieves a balance between selectivity (false hits) and computational cost. The selectivity aspects of CM, FMD_1 and $EFMD_2$ are compared in the MBR space [3]. To the best of our knowledge, no previous work has attempted to approximate the uncertainty regions discussed above by providing corresponding MBR calculations.

2 Related Work

To accommodate the uncertainty issue, several application-specific trajectory models have been proposed. One model is that, at any point in time, the location of an object is within a certain distance d, of its last reported location. Given a point in time, the uncertainty region is a circle centered at a reported point with a radius d, bounding all possible locations of the object [9]. Another model assumes that an object always moves along straight lines (linear routes). The location of the object at any point in time is within a certain interval along the line of movement [9]. Some other models assume that the object travels with a known velocity along a straight line, but can deviate from this path by a certain distance [8,5]. These models presented in [7,8] as a spatio-temporal uncertainty model (referred to in this paper as the *Cylinder Model*) produce 3-dimensional cylindrical uncertainty regions representing the past uncertainties of trajectories. The parametric polynomial trajectory model [2] is not for representing unknown states but for approximating known locations of a trajectory to make the index size smaller, assuming that all points of the trajectory are given.

Another study in [4] proved that when the maximum velocity of an object is known, all possible locations of the object during the time interval between two consecutive observations (reported states) lie on an error ellipse. A complete trajectory of any object is obtained by using linear interpolation between two adjacent states. By using the error ellipse, the authors demonstrate how to process uncertainty range queries for trajectories. The error ellipse defined and proved in [4] is the projection of a three-dimensional spatio-temporal uncertainty region onto the two-dimensional data space. Similarly, [1] represents the set of all possible locations based on the intersection of two half cones that constrain the maximum deviation from two known locations. It also introduces multiple granularities to provide multiple views of a moving object.

A recent uncertainty model reported in [10] formally defines both past and future spatio-temporal uncertainties of trajectories of any dimensionality. In this model, the uncertainty of the object between two consecutive reported points is defined to be the overlap between the two funnels. In this paper, we call this model the *Funnel Model* of *Degree* 1 (FMD_1). A non-linear extension of the funnel model, *Funnel Model* of *Degree* 2 (FMD_2), also known as the *Tornado Uncertainty Model* (TUM), was proposed in [11]. This higher degree model reduces the size of the uncertainty region by taking into account the temporally-varying higher order derivatives, such as velocity and acceleration.

3 MBR Models

This section quantifies the MBR approximation of the uncertainty regions generated by four models: CM, FMD_1, FMD_2 and $EFMD_2$. The notations of Table 1 will be used for the rest of this paper. We will use the notations CM, FMD_1, FMD_2 and $EFMD_2$ to refer to both, the uncertainty models and the corresponding MBR models throughout this paper.

Table 1. Notations

Notation	Meaning
P_{1i}	reported position of point 1 in the ith dimension
P_{2i}	reported position of point 2 in the ith dimension
t_1	time instance when P_1 was reported
t_2	time instance when P_2 was reported
t	any time instance between t_1 and t_2 inclusively
T	time interval between t_2 and t_1, $T = t_2 - t_1$
V_{1i}	velocity vector at P_1
V_{2i}	velocity vector at P_2
d	space dimensionality (time dimension is not included)
e	instrument and measurement error
M_v	maximum velocity of an object
M_a	maximum acceleration of an object
MD	maximum displacement of an object
MBR_{CM}	MBR of CM
MBR_{FMD_1}	MBR of FMD_1
MBR_{FMD_2}	MBR of FMD_2
MBR_{EFMD_2}	MBR of $EFMD_2$
low_i	lower bound of an MBR in the ith dimension
$high_i$	upper bound of an MBR in the ith dimension

3.1 Linear MBR Models

Cylinder Model (CM) The *Cylinder Model* is a simple uncertainty model that represents the uncertainty region as a cylindrical body [8]. Any two adjacent reported points, P_1 and P_2 of a trajectory segment, are associated with a circle that has a radius equal to an uncertainty threshold r (see Figure 1). The value of r represents the maximum possible displacement (MD) from the reported point including the instrument and measurement error, e.

CM quantifies the maximum displacement using the maximum velocity M_v of the object and the time interval between P_1 and P_2 such that $MD = M_v \cdot (t_2 - t_1)$.

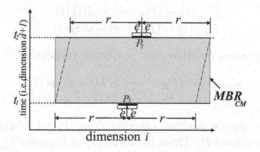

Fig. 1. MBR estimation of the Cylinder Model (MBR_{CM})

CM has the following properties: The two cross-sections at t_1 and at t_2 are hyper-circles that are perpendicular to dimension d+1 (i.e., the time dimension) and centered at, respectively, P_1 and P_2, with their radius: $r = e + MD$. Then, the two points that define MBR_{CM} in the ith dimension can be determined as follows:

$$low_i(MBR_{CM}) = min\{P_{1i}, P_{2i}\} - r$$
$$high_i(MBR_{CM}) = max\{P_{1i}, P_{2i}\} + r \qquad (1)$$

Note that the calculation of the MBR of the CM is simple and straightforward with little computational overhead.

Funnel Model of Degree 1 (FMD_1). While CM provides a simple and fast estimation of MBRs, its MBR includes large areas (volumes in 3D) that cannot be reached by the moving objects even in the worst case. This section introduces the *Funnel Model of Degree* 1 (FMD_1) which effectively excludes some areas of MBR_{CM}.

FMD_1 uses the maximum velocity M_v of the moving object to calculate the maximum displacement. However, as displayed in Figure 2, the maximum displacement, MD, is a function of time t in FMD_1. Moreover, the maximum displacement from both directions are considered in the calculation of an MBR. Each direction defines a funnel (similar to the cone in [1]) that has its top as a circle with radius equal to e centered at one of the reported points and its base as a circle that is perpendicular to the time dimension at the other reported point. The overlapping region of the two funnels generated between two adjacent reported positions defines the uncertainty region of FMD_1 (see Figure 2). The boundary of the uncertainty region is the maximum possible deviation of an object travelling between P_1 and P_2 during T. The maximum displacement at a given time t is $disp(t) = M_v \cdot t$, and the maximum displacement MD is defined as $MD = M_v \cdot (t_2 - t_1)$.

To calculate MBR_{FMD_1}, we define the minimum and maximum future and past locations of the moving object at a given time t along dimension i as follows: The minimum, maximum possible future positions, respectively:

$$f^1_{min}(t) = (P_{1i} - e) - disp(t),$$
$$f^1_{max}(t) = (P_{1i} + e) + disp(t)$$

The minimum, maximum possible past positions, respectively:

$$p^1_{min}(t) = (P_{2i} - e - MD) + disp(t),$$
$$p^1_{max}(t) = (P_{2i} + e + MD) - disp(t)$$

The "f" stands for future position starting from P_1 while the "p" stands for past position starting from P_2. Then, the cross point between $f^1_{min}(t)$ and $p^1_{min}(t)$ defines the theoretical lower bound of the MBR in the negative direction between P_1 and P_2. Similarly, the cross point between $f^1_{max}(t)$ and $p^1_{max}(t)$ defines the theoretical upper bound of the MBR in the positive direction. To find the two

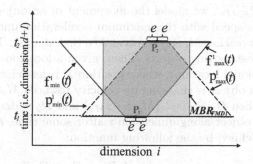

Fig. 2. MBR estimation of the Funnel Model of Degree 1 (MBR_{FMD_1})

cross points, one can solve $f^1_{min}(t) = p^1_{min}(t)$ and $f^1_{max}(t) = p^1_{max}(t)$ for t to obtain:

$$low_i(MBR_{FMD_1}) = P_{1i} - e - \frac{P_{1i} - P_{2i} + MD}{2}$$

$$high_i(MBR_{FMD_1}) = P_{1i} + e + \frac{P_{2i} - P_{1i} + MD}{2} \tag{2}$$

Note that the calculation of MBR_{FMD_1} is also simple and straightforward with little computational overhead.

3.2 Non Linear MBR Models

Funnel Model of Degree 2 (FMD_2). In defining their MBRs, both CM and FMD_1, assume that a moving object can instantly reach the maximum velocity from the current velocity. However, moving objects in reality move with momentum, i.e., they need some time to change their velocities. Thus, many moving objects in a lot of cases (e.g. vehicles) move with a certain acceleration that is bounded by a maximum value. This provides the idea of a 2nd degree uncertainty model called the *Funnel Model* of *Degree* 2 (FMD_2). FMD_2 uses both the maximum velocity M_v and the maximum acceleration M_a of the moving object to calculate the maximum displacement, taking into consideration both directions: from P_1 to P_2 using V_1 as an initial velocity and from P_2 to P_1 using V_2 as an initial velocity (see Figure 3).

Let $displ_1$ and $displ_2$ be, respectively, the first-degree and second-degree displacement functions defined as follows:

$$displ_1(V, t) = V \cdot t$$

$$displ_2(V, a, t) = \int_0^t (V + a \cdot x)dx \approx V \cdot t + (a/2) \cdot t^2$$

where V is velocity, a is acceleration and t is time. In addition, let t_{Mv} be the amount of time required to reach the maximum velocity M_v, given an initial velocity I_v and a maximum acceleration M_a, then we have:

$$t_{Mv} = (M_v - I_v)/M_a, \text{ when } M_v > I_v \text{ (where } I_v = V_1 \text{ for } P_1 \text{ and } I_v = V_2 \text{ for } P_2)$$

In defining MBR_{FMD_2}, we model the movement of an object as follows: an object accelerates its speed with the maximum acceleration until it reaches M_v (i.e., $displ_2$). Once it reaches M_v, it travels at Mv (i.e., $displ_1$). This is a realistic approximation of most moving objects. Then, given a location-time $< P, V, t >$ (i.e., a $d \geq 1$ dimensional location), where position P is associated with a time t and a velocity V, an object is changing its velocity towards M_v at a rate of M_a along dimension i when $D_1 = displ_1(M_v, t - t_{Mv})$ and $D_2 = displ_2(V, M_a, t_{Mv})$. The position of the object along dimension i after some time, t, from the start time t_s (i.e., t_1) is defined by the following function:

$$f\_pos(P_{1i}, V_{1i}, M_v, M_a, t, i) = \begin{cases} P_{1i} + D_2 + D_1 & \text{if } t_{Mv} < t \\ P_{1i} + D_2 & \text{otherwise} \end{cases}$$

Similarly, the position of the object before some time, t, from the start time t_s (i.e., t_2) is defined as follows:

$$p\_pos(P_{2i}, V_{2i}, M_v, M_a, t, i) = \begin{cases} P_{2i} - D_2 - D_1 & \text{if } t_{Mv} < t \\ P_{2i} - D_2 & \text{otherwise} \end{cases}$$

The first case (i.e., $t_{Mv} < t$) in the above functions is the case when the object reaches the maximum velocity M_v in a time period that is less than t. Hence, the position of the object is evaluated by a curve from t_s to t_{Mv} applying the maximum acceleration and by a linear function from t_{Mv} to t. However, the second case is the case when the object cannot reach M_v between t_s and t, thus the position of the object is only calculated using a non-linear function that uses the maximum acceleration M_a. When the range of velocity and acceleration of the object is $[-Mv, +Mv]$ and $[-Ma, +Ma]$, respectively, FMD_2 defines the future and past maximum displacements of the object along dimension i as follows:

The minimum, maximum possible future positions, respectively:

$$f^2_{min}(t) = f\_pos(P_{1i}, V_{1i}, -Mv, -Ma, t, i),$$
$$f^2_{max}(t) = f\_pos(P_{1i}, V_{1i}, +Mv, +Ma, t, i)$$

The minimum, maximum possible past positions, respectively:

$$p^2_{min}(t) = p\_pos(P_{2i}, V_{2i}, +Mv, +Ma, t, i),$$
$$p^2_{max}(t) = p\_pos(P_{2i}, V_{2i}, -Mv, -Ma, t, i)$$

Figure 3 shows the output of the above functions between P_1 and P_2. The funnel formed between $f^2_{min}(t)$ and $f^2_{max}(t)$ corresponds to all possible displacements from P_1 during T and the funnel formed between $p^2_{min}(t)$ and $p^2_{max}(t)$ corresponds to all possible displacements from P_2 during T. The area formed by the overlapping regions of the two funnels is the uncertainty region generated by FMD_2. To calculate MBR_{FMD_2}, one needs to determine the lower and upper bounds of the uncertainty region in every space dimension. In most cases, the lower bound can be the intersection of $f^2_{min}(t)$ and $p^2_{min}(t)$. The upper bound

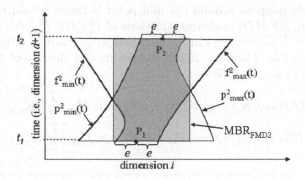

Fig. 3. MBR estimation of the Funnel Model of Degree 2 (MBR_{FMD_2})

can be the intersection of $f^2_{max}(t)$ and $p^2_{max}(t)$. Hence, the following set of equations need to be solved for t to find the two cross points: $f^2_{min}(t) = p^2_{min}(t)$ for low_i and $f^2_{max}(t) = p^2_{max}(t)$ for $high_i$. However, in some cases, low_i can be still greater than $P_{1i} - e$ or $P_{2i} - e$ (similarly, $high_i$ can be smaller than $P_{1i} + e$ or $P_{2i} + e$). Thus, MBR_{FMD_2} is defined as:

$$low_i(MBR_{FMD_2}) = min\{P_{1i} - e, P_{2i} - e, low_i\}$$
$$high_i(MBR_{FMD_2}) = max\{P_{1i} + e, P_{2i} + e, high_i\} \qquad (3)$$

To calculate the two cross points, one needs to solve a set of quadratic equations. The number of equations that need to be solved for each dimension is eight. There are four possible cases of intersections between $f^2_{min}(t)$ and p^2_{min} for one dimension: a non-linear (curve) part intersecting a non-linear part, a non-linear part intersecting a linear part, a linear part intersecting a non-linear part and finally a linear part intersecting a linear part. Due to the number of equations involved, the computation time of MBR_{FMD_2} is significantly high (as demonstrated in Section 4.2).

Estimated Funnel Model of Degree 2 ($EFMD_2$). FMD_2 more accurately estimates the uncertainty region of a moving object so that MBR_{FMD_2} is expected to be smaller than both MBR_{CM} and MBR_{FMD_1}. However, unlike the other models, the calculation of MBR_{FMD_2} is rather CPU time intensive, which may restrict the use of FMD_2. Practical database management systems need a fast MBR quantification to support large amount of updates. In order to take advantage of the small size of MBR_{FMD_2} and to satisfy the low computation overhead requirement, we propose an estimated model of FMD_2, which provides a balance between modeling power and computational efficiency. We refer to this estimated MBR model as the *Estimated Funnel Model of Degree 2 ($EFMD_2$)*.

The main idea of $EFMD_2$ is to avoid the calculation of the exact cross points while keeping the 2*nd* degree model that presents the smallest uncertainty region. $EFMD_2$ is based on the idea that the maximum displacement from P_1

to P_2 most likely happens around the mid point between t_1 and t_2. Based on that observation, $EFMD_2$ evaluates equations of $f^2_{min}(t)$, $f^2_{max}(t)$, $p^2_{min}(t)$ and $p^2_{max}(t)$ at $t_{mid} = \frac{t_1+t_2}{2}$, setting t in the functions to $\frac{T}{2}$, where T is the time interval between t_1 and t_2. Then, by slightly overestimating the size of MBR_{FMD_2}, we define MBR_{EFMD_2} as follows:

$$low_i(MBR_{EFMD_2}) = min\{P_{1i} - e, P_{2i} - e, f^2_{min}(\frac{T}{2}), p^2_{min}(\frac{T}{2})\}$$

$$high_i(MBR_{EFMD_2}) = max\{P_{1i} + e, P_{2i} + e, f^2_{max}(\frac{T}{2}), p^2_{max}(\frac{T}{2})\} \qquad (4)$$

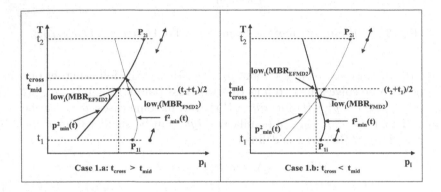

Fig. 4. Case2 a, b: calculating $low_i(MBR_{EFMD2})$

Theorem 1: MBR_{EFMD_2} guarantees to contain MBR_{FMD_2}.

Proof: We will show the proof for $low_i(MBR_{EFMD_2})$ and it is clear that the argument is similar for $high_i(MBR_{EFMD_2})$. Note that the functions f^2_{min} and p^2_{min} consist of at most two parts; a monotonically increasing part and a monotonically decreasing part. Let t_{cross} be the time instant when $f^2_{min} = p^2_{min}$ and P_{cross} be the intersection point of the two functions. Based on that observation we have two cases:

Case 1: P_{cross} is at the increasing part of any of the two functions. Then, P_{cross} is greater than or equal to either $P_{1i} - e$ or $P_{2i} - e$. Hence, $low_i(MBR_{EFMD_2}) \le P_{cross}$.

Case 2: P_{cross} is the intersection of the decreasing parts of both functions. Then, P_{cross} is in between $f^2_{min}(\frac{T}{2})$ and $p^2_{min}(\frac{T}{2})$ as follows:

Case 2 (a): If $t_{cross} > t_{mid}$, then $p^2_{min}(\frac{T}{2}) < p^2_{min}(t_2 - t_{cross})$ and $f^2_{min}(t_{cross} - t_1) < f^2_{min}(\frac{T}{2})$. We know that $p^2_{min}(t_2 - t_{cross}) = f^2_{min}(t_{cross} - t_1)$.

Therefore, $p^2_{min}(\frac{T}{2}) < f^2_{min}(t_{cross} - t_1) < f^2_{min}(\frac{T}{2})$. Hence, $low_i(MBR_{EFMD_2}) = p^2_{min}(\frac{T}{2})$ which covers $low_i(MBR_{FMD_2})$ (Figure 4).

Case 2 (b): If $t_{cross} < t_{mid}$, then $p^2_{min}(\frac{T}{2}) > p^2_{min}(t_2 - t_{cross})$ and $f^2_{min}(t_{cross} - t_1) > f^2_{min}(\frac{T}{2})$. Similarly, we know that $p^2_{min}(t_2 - t_{cross}) = f^2_{min}(t_{cross} - t_1)$.

Therefore, $f_{min}^2(\frac{T}{2}) < f_{min}^2(t_{cross} - t_1) < p_{min}^2(\frac{T}{2})$. Hence, $low_i(MBR_{EFMD_2})$ $= f_{min}^2(\frac{T}{2})$ which covers $low_i(MBR_{FMD_2})$ (Figure 4).

Case 2 (c): If $t_{cross} = t_{mid}$, then $low_i(MBR_{EFMD_2}) = low_i(MBR_{FMD_2})$.

4   Experimental Evaluation

In this section, we evaluate the proposed MBR models for CM, FMD_1, FMD_2 and $EFMD_2$ for both synthetic and real data sets. All velocities in this section are in $meters/second$ (m/s), all accelerations are in $meters/scond^2$ (m/s^2) and all volumes are in square meters times seconds ($meters^2 \cdot second$). We performed all our experiments on an Intel based computer running MS XP operating system, 1.66 GHz CPU, 1GB main memory space, using Cygwin/Java tools.

4.1   Datasets and Experimental Methodology

In all experiments, we assumed vehicles as moving objects. However, the proposed MBR calculation models can be applied to any moving objects. Our synthetic datasets were generated using the "Generate_Spatio_Temporal_Data" ($GSTD$) algorithm [6] with various parameter sets such as varied velocities and different directional movements (see Table 2). On top of $GSTD$, we added a module to calculate the velocity values at each location. Each group consisted of five independent datasets (datasets $1 - 5$ in group 1 and datasets $6 - 10$ in group 2). Each dataset in group 1 was generated by 200 objects moving towards Northeast with a rather high average velocity. Each dataset in group 2 was generated by 200 objects moving towards East with a lower average velocity than the datasets in group 1. For the datasets in group 1, we varied the average velocity between 17.69 m/s and 45.99 m/s, and between 12.52 m/s and 32.51 m/s for the datasets in group 2. As a specific example, dataset 11 was generated by 200 objects moving with the velocity in the x direction greater than the velocity in the y direction with an average velocity of 17.76 m/s. Each object in the eleven synthetic datasets reported its position and velocity every second for an hour. The real data set was collected using a GPS device while driving a car in the city of San Diego in California, U.S.A. The actual position and velocity were reported every one second and the average velocity was 11.44 m/s.

Table 2 shows the average velocity of the moving objects, the maximum recorded velocity and the maximum acceleration of the moving object. The last two columns show the maximum velocity and maximum acceleration values that were used in the calculation of the MBRs for each model.

Our evaluation of the models are based on two measurements. First, we quantified the volume of MBRs using each model. Then we calculated the average percentage reduction in the volume of MBRs. This indicates how effectively the MBR can be defined in each model. Second, given a range query, we measured the number of overlapping MBRs. This indicates how efficiently a range query can be evaluated using each model. While varying the query size, range queries

Table 2. Synthetic and real data sets and system parameters

datasets		reported records			parameters	
		AVG Vel.	MAX Vel.	MAX Acc.	M_v	M_a
synthetic	Group 1	17.69 - 45.99	21.21 - 49.49	7.09 - 7.13	55	8
	Group 2	12.52 - 32.51	15.00 - 35.00	5.02 - 5.04	55	8
	Dataset 11	17.76	20.61	6.41	55	8
real	San Diego	11.44	36.25	6.09	38.89	6.5

Table 3. FMD_2 and $EFMD_2$ comparison using synthetic and real datasets

Datasets		# MBRs/second		AVG MBR volume		AVG No. Intersections	
		FMD_2	$EFMD_2$	FMD_2	$EFMD_2$	FMD_2	$EFMD_2$
synthetic	Dataset 11	777.73	77822.18	179516.71	213190.98	2351.75	2616.25
real	San Diego	132.77	46920.12	111147.29	144018.71	33.5	42.25

were generated randomly by choosing a random point in the universe, then appropriate x, y extents (query area) and t extent (query time) were added to that point to create a random query region (volume) in the MBR universe. The MBRs of each model for both the synthetic and real datasets were calculated using time interval TI (time interval of MBRs) equal to 5, 10, 15 and 20 seconds.

4.2 FMD_2 and $EFMD_2$ Comparison

We analyze the performance of FMD_2 and $EFMD_2$ and show the tradeoff between the average computation time (CPU time) required to compute an MBR and the query performance as the number of intersections resulted by random range queries. Table 3 shows the number of MBRs that can be calculated in one second by FMD_2 and $EFMD_2$, the average MBR volumes and the average number of intersections with 1000 random range queries with area equal to 0.004% of the universe area. The time extent of the queries was 8 minutes. The average represents the overall average that was taken over different time intervals (T.I = 5, 10, 15, 25). The average time to calculate a single MBR_{FMD_2} for the real dataset was 7.532 milliseconds, while 0.022 milliseconds was the time needed to compute MBR_{EFMD_2}. Using the synthetic dataset, 1.285 milliseconds and 0.013 milliseconds were needed to calculate MBR_{FMD_2} and MBR_{EFMD_2}, respectively. The average reduction in the CPU time to calculate MBR_{EFMD_2} over MBR_{FMD_2} was around 99% for both datasets. The third and forth columns of Table 3 show the average number of MBRs that can be calculated per second using FMD_2 and $EFMD_2$ respectively. The next comparison was the average MBR volume of each of the two models using different time intervals. The values in columns five and six show the overall average of the MBR volume of MBR_{FMD_2} and MBR_{EFMD_2}, respectively. The last two columns show the average number of intersections resulted by the random queries. Based on our experimental results, varying the query size, FMD_2 provided average percentage reduction of $10\% - 25\%$ in the number of intersections over $EFMD_2$.

To be able to compare the two models, the applications that will make use of the models need to be addressed. In cases where the applications are retrieval query intensive and responding time is a critical issue, then FMD_2 would result in faster query answers than $EFMD_2$ by reducing the number of false-hits. On the other hand, when dealing with update intensive applications where very large number of moving objects change their locations very frequently, $EFMD_2$ can be a better choice, since it can handel more than two orders of magnitude larger number of moving objects.

4.3   MBR Model Evaluation

As shown in the previous section, it is more applicable to use $EFMD_2$ since it can handel much larger number of MBRs for the same amount of time compared to FMD_2 with out having to compensate for the 2nd degree model efficiency (i.e., average MBR volume). For this reason, we decide to compare $EFMD_2$ rather than FMD_2 with the other two proposed models in this section.

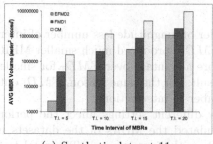
(a) Synthetic dataset 11

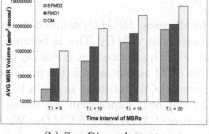
(b) San Diego dataset

Fig. 5. Average MBR volume of each model

Synthetic Data Results. Figure 5 (a) shows the average volume of the MBRs generated by each model for synthetic dataset 11. The x-axis represents the time interval (TI) used to calculate the MBRs. The y-axis (logarithmic scale) represents the average volume of the MBRs. Regardless of the TI value, FMD_1 resulted in much smaller MBRs than CM. $EFMD_2$ resulted in even much smaller MBRs compared to FMD_1. The larger the TI value is, the less advantage we gain from $EFMD_2$ compared to FMD_1. When TI is large, all calculated MBRs are very large because the maximum velocity is assumed during most of the time interval (T) between any two reported points, regardless of the model.

Next, we generated and evaluated 4000 random queries to synthetic dataset 11. We varied the query area between 0.004% and 0.05% of the area of the universe and varied the time extent of the query between 2 minutes and 8 minutes. Figure 6 (a) shows the average number of intersecting MBRs per 1000 queries that each model resulted in while varying TI.

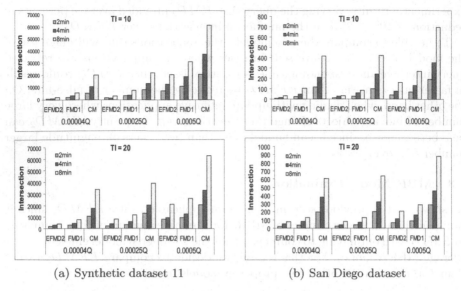

(a) Synthetic dataset 11   (b) San Diego dataset

Fig. 6. Average nuber of intersections per 1000 query

In all cases, $EFMD_2$ resulted in an order of magnitude less number of intersections than FMD_1. This is because $EFMD_2$ produced much smaller MBRs than FMD_1. Notice that $EFMD_2$ has more advantage over FMD_1 for smaller values of TI as explained in the previous result. For the same reason, FMD_1 outperformed CM resulting in much less number of intersections.

In the next experiment we tested how different settings affect the performance of FMD_1 and $EFMD_2$. We calculated the MBRs of the datasets in group 1 and 2 using $EFMD_2$ and FMD_1 with time interval (TI) equal to 5 seconds. Then, range queries were conducted on the calculated MBRs. 4000 random range queries were generated each with an area of 0.01% of the total area of the universe, and the time extent of the query was 2 minutes.

Figure 7 (a) and (b) show the average percentage reductions in the MBR volumes and in the number of intersecting MBRs after evaluating 4000 queries on group 1 and group 2, respectively. The x-axis represents the average velocity of the moving objects and the y-axis represents the reduction rate ($EFMD_2$ to FMD_1). The top curves show the percentage reduction in the average MBR volume, and the bottom curves show the percentage reduction in the number of intersections with the range queries. This confirms that $EFMD_2$ greatly outperforms FMD_1 and shows that the reduction in the average volume of the MBRs directly resulted in the reduction of the number of intersecting MBRs. Another observation is that the performance gain of $EFMD_2$ decreased when the average velocity increased. This was more obvious when objects were moving faster as shown in Figure 7 (a) since the objects reach the maximum velocity in a short time. Thus, the difference in MBR sizes between the two models becomes less.

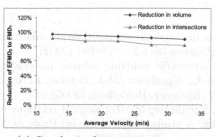
(a) Synthetic datasets in group 1

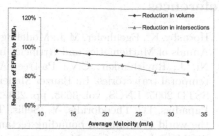
(b) Synthetic datasets in group 2

Fig. 7. Average reduction in MBR volume and number of intersections

Real Data Results. All observations on the synthetic dataset results hold with the real dataset. Figure 5 (b) shows the average volumes of the MBRs generated by each model for the San Diego dataset. Regardless of the TI value, FMD_1 resulted in much smaller MBRs than CM. Also, $EFMD_2$ resulted in even much smaller MBRs compared to FMD_1. In Figure 6 (b), we generated 4000 random queries. We varied the query area between 0.004% and 0.05% of the area of the universe and varied the time extent of the query between 2 minutes and 8 minutes. Figure 6 (b) shows the average number of intersections per 1000 query that each model resulted in when varying TI.

5   Conclusions

Due to the uncertain nature of locations between reported positions of moving objects, it is important to efficiently manage the uncertainty regions in order to improve query response time. Because it is not efficient to accommodate irregular uncertainty regions in indexing, we proposed *Minimum Bounding Rectangles* approximations for recently proposed uncertainty models, CM, FMD_1 and FMD_2. We also presented $EFMD_2$, an estimation of the MBR of FMD_2. MBR approximations of CM and FMD_1 are based on linear equations so that their computation requires minimal CPU overhead. However, the size of their MBRs were significantly larger than those of FMD_2 which are based on non-linear equations. To remedy the problem of the high computation time of FMD_2, while maintaining the advantage of smaller MBR size, we proposed an estimated model, $EFMD_2$, that achieves a good balance between accuracy and computation overhead. Experiments on synthetic and real datasets showed that $EFMD_2$ significantly outperformed MBR approximations of CM and FMD_1. The trade-off between FMD_2 and $EFMD_2$ was also demonstrated.

Acknowledgments

The authors thank Dr. Petr Vojtěchovský, Dr. Sada Narayanappa and Brandon Haenlein for their helpful suggestions during early stages of this paper.

References

1. Hornsby, K., Egenhofer, M.J.: Modeling moving objects over multiple granularities. Annals of Mathematics and Artificial Intelligence 36(1-2), 177–194 (2002)
2. Ni, J., Ravishankar, C.V.: Pa-tree: A parametric indexing scheme for spatio-temporal trajectories. In: Bauzer Medeiros, C., Egenhofer, M.J., Bertino, E. (eds.) SSTD 2005. LNCS, vol. 3633, pp. 254–272. Springer, Heidelberg (2005)
3. Papadias, D., Theodoridis, Y., Sellis, T., Egenhofer, M.: Topological relations in the world of minimum bounding rectangles: A study with r-trees. In: Proc. ACM SIGMOD Int. Conf. on Management of Data, pp. 92–103 (1995)
4. Pfoser, D., Jensen, C.S.: Capturing the uncertainty of moving-objects representations. In: Proceedings of Int. Conf. on Scientific and Statistical Database Management, pp. 123–132 (1999)
5. Sistla, P.A., Wolfson, O., Chanberlain, S., Dao, S.: Querying the uncertain position of moving objects. In: Etzion, O., Jajodia, S., Sripada, S. (eds.) Temporal Databases: Research and Practice, Dagstuhl Seminar 1997. LNCS, vol. 1399, pp. 310–337. Springer, Heidelberg (1998)
6. Theodoridis, Y., Silva, J.R.O., Nascimento, M.A.: On the generation of spatiotemporal datasets. In: Proceedings of Int. Symposium on Advances in Spatial Databases, pp. 147–164 (1999)
7. Trajcevski, G., Wolfson, O., Hinrichs, K., Chamberlain, S.: Managing uncertainty of moving objects databases. ACM Trans. on Databases Systems 29(3), 463–507 (2004)
8. Trajcevski, G., Wolfson, O., Zhang, F., Chamberlain, S.: The geometry of uncertainty in moving object databases. In: Jensen, C.S., Jeffery, K.G., Pokorný, J., Šaltenis, S., Bertino, E., Böhm, K., Jarke, M. (eds.) EDBT 2002. LNCS, vol. 2287, pp. 233–250. Springer, Heidelberg (2002)
9. Wolfson, O., Sistla, P.A., Chamberlain, S., Yesha, Y.: Updating and querying databases that track mobile units. Distributed and Parallel Databases 7(3), 257–387 (1999)
10. Yu, B.: A spatiotemporal uncertainty model of degree 1.5 for continuously changing data objects. In: Proceedings of ACM Int. Symposium on Applied Computing, Mobile Computing and Applications, pp. 1150–1155 (2006)
11. Yu, B., Kim, S.H., Alkobaisi, S., Bae, W.D., Bailey, T.: The tornado model: Uncertainty model for continuously changing data. In: Kotagiri, R., Radha Krishna, P., Mohania, M., Nantajeewarawat, E. (eds.) DASFAA 2007. LNCS, vol. 4443, pp. 624–636. Springer, Heidelberg (2007)

Summarization Graph Indexing: Beyond Frequent Structure-Based Approach

Lei Zou[1,*], Lei Chen[2], Huaming Zhang[3], Yansheng Lu[1], and Qiang Lou[4]

[1] Huazhong University of Science and Technology, Wuhan, China
{zoulei,lys}@mail.hust.edu.cn
[2] Hong Kong of Science and Technology, Hong Kong, China
leichen@cse.ust.hk
[3] The University of Alabama in Huntsville, Huntsville, AL, 35899, USA
hzhang@cs.uah.edu
[4] The Temple University, USA
qianglou@temple.edu

Abstract. Graph is an important data structure to model complex structural data, such as chemical compounds, proteins, and XML documents. Among many graph data-based applications, sub-graph search is a key problem, which is defined as *given a query Q, retrieving all graphs containing Q as a sub-graph in the graph database.* Most existing sub-graph search methods try to filter out false positives (graphs that are not possible in the results) as many as possible by indexing some frequent sub-structures in graph database, such as [20,22,4,23]. However, due to ignoring the relationships between sub-structures, these methods still admit a high percentage of false positives. In this paper, we propose a novel concept, *Summarization Graph,* which is a complete graph and captures most topology information of the original graph, such as sub-structures and their relationships. Based on Summarization Graphs, we convert the filtering problem into retrieving objects with set-valued attributes. Moreover, we build an efficient signature file-based index to improve the filtering process. We prove theoretically that the pruning power of our method is larger than existing structure-based approaches. Finally, we show by extensive experimental study on real and synthetic data sets that the size of candidate set generated by Summarization Graph-based approach is only about 50% of that left by existing graph indexing methods, and the total response time of our method is reduced 2-10 times.

1 Introduction

As a popular data structure, graphs have been used to model many complex data objects in real world, such as, chemical compounds [18] [11], proteins [2], entities in images [15], XML documents[21] and social networks [3]. Due to the wide usage of graphs, it is quite important to provide users to organize, access and analyze graph data efficiently. Therefore, graph database has attracted a lot of attentions from database and

---

* This work was done when the first author visited Hong Kong University of Science and Technology as a visiting scholar. The first author was partially supported by National Natural Science Foundation of China under Grant 70771043.

J.R. Haritsa, R. Kotagiri, and V. Pudi (Eds.): DASFAA 2008, LNCS 4947, pp. 141–155, 2008.
© Springer-Verlag Berlin Heidelberg 2008

data mining community, such as sub-graph search [16,20,9,22,8,6,4,23], frequent sub-graph mining [10,13,19], correlation sub-graph query [12] and so on. Among many graph-based applications, it is very important to retrieve related graphs containing a query graph from a large graph database efficiently. For example, given a large chemical compound database, a chemist wants to find all chemical compounds having a particular sub-structure.

In this paper, we focus on *sub-graph search*, which is defined as follows: *given a query graph Q, we need to find all data graphs G_i, where query graph Q is sub-graph isomorphism to data graph G_i.* Because sub-graph isomorphism is NP-complete [7], we have to employ filtering-and-verification framework to speed up the search efficiency. Specifically, we use an effective and efficient pruning strategy to filter out the false positives (graphs that are not possible in the results) as many as possible first (that is *filtering process*), then validate the remaining candidates by subgraph isomorphism checking (that is *verification process*). Therefore, the overall response time consists of filtering time and verification time. The state art of pruning strategy in the literature is frequent structures-based pruning. These methods apply data mining algorithms offline first to find frequent sub-structures (sub-graphs or sub-trees, we also call them *features*) [20,22,4,23] in graph database. Then some discriminate sub-structures are chosen as indexed features. After that, we build the inverted index for each feature, namely, the data graphs containing these features are linked to these features. The relationship between indexed features and data graphs is similar to the one between indexed words and documents in IR (Information Retrieve). To answer a sub-graph query Q, Q is represented as a set of features and all the graphs that may contain Q are retrieved by examining the inverted index. The rational behind this type of filtering strategy is that if some features of graph Q do not exist in a data graph G, G cannot contain Q as its sub-graph.

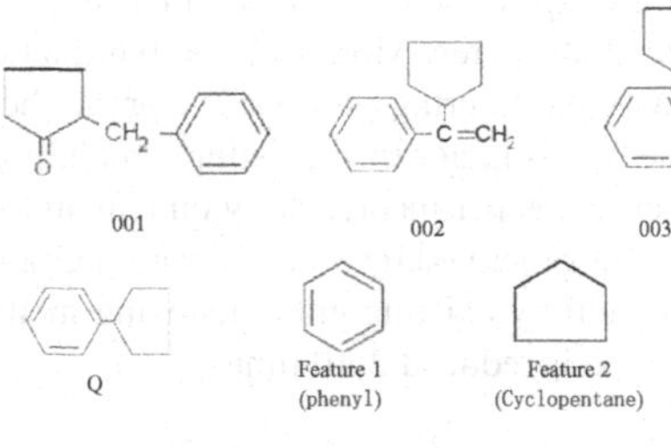

Fig. 1. Motivation Example

However, existing sub-structure based approaches treat each data graph as a bag of indexed features. They fail to capture the relationship among the features. As a consequence, many false positives still exist after filtering. Given a graph database with 3 data graphs 001−003 and a query Q in Fig. 1, assume that we choose feature 1 and 2 as indexed features. Since the query Q and data graphs 001−003 all contain both feature 1 and feature 2 respectively, after filtering, the candidates are still 001,002 and 003, which is exactly the whole database. *Can we reduce the candidate size based on the same feature 1 and 2 ?* Our answer is yes. Observed from Fig. 1, the length of the *shortest path* between feature 1 and feature 2 is 1 in query Q, but 2 in 001 and 002, and 1 in 003. It means that only graph 003 may be a correct answer, and we can filter out 001 and 002 safely. Considering both feature and pairwise feature relationship is primary motivation of this paper. However, simply checking feature first, then feature relationship is neither effective nor efficient, which will be discussed in Section 3.2.

In feature-based approaches, more indexed features means higher pruning power, but also means longer filtering time and larger space cost. On the other hand, less features lead to poor pruning power in the filtering process, which leads to bad performance in overall response time. Thus, there is a trade off between number of features and pruning power. In this paper, we propose a novel filtering method, which requires less features and provides higher pruning power, compared to the previous approaches, such as gIndex [20], FG-index [4]. Specifically, we encode both features and feature relationships of graph Q into Summarization Graph (i.e. $Sum(Q)$) and use Summarization Graph to conduct filtering. The contributions that we made in this paper are listed as follows:

1. We propose a novel concept, Summarization Graph, which is a complete graph and captures most topology information of the original graph, such as sub-structures and their relationships.
2. We encode summarization graph into a set, and we convert a filtering process into retrieving of objects with set-valued attributes to get candidate set, which can be solved efficiently by building an effective signature-based index.
3. Last but not least, through extensive experiments on a real dataset AIDS and synthetic datasets, we show that the candidate size of our methods is only about 50% of that achieved by existing approaches on average and the total searching time is reduced 2-10 times in our method.

The remaining of the paper is organized as follows. The related work is discussed in Section 2. Summarization Graph framework is proposed in Section 3. The performance study is reported in Section 4. Section 5 concludes the paper.

2   Related Work

Usually, to speed up the sub-graph search, we use the filtering-and-verification framework. First, we remove false positives (graphs that are not possible in the results) by *pruning strategy*; Then, we perform sub-graph isomorphism algorithm on each candidate to obtain the final results. Obviously, less candidates mean better search performance.

So far, many pruning strategies have been proposed [16,20,9,22,8,6,4,23]. The most popular approaches are *feature*-based. The approaches apply data mining techniques to extract some discriminate sub-structures as indexed features, then, build inverted index for each feature. Query graph Q is denoted as a set of features, and the pruning power always depends on selected features. By the inverted indexes, we can fix the complete set of candidates. Many algorithms have been proposed to improve the effectiveness of selected features, such as gIndex [20], TreePi [22], FG-Index [4] and Tree+δ [23]. In gIndex, Yan et al. propose a "discriminative ratio" for features. Only frequent and discriminative subgraphs are chosen as indexed features. In TreePi, due to manipulation efficiency of trees, Zhang et al. propose to use frequent and discriminate subtrees as indexed feature. In FG-Index [4], James et al. propose to use frequent sub-graphs as indexed features. In [23], Zhao et al. use "subtree" and a small number of discriminative sub-graphs as indexed features instead of sub-graphs in gIndex and FG-Index.

Different from the above approaches, in Closure-tree [9], He and Singh propose pseudo subgraph isomorphism by checking the existence of a semi-perfect matching from vertices in query graph to vertices a data graph (or graph closure). The major limitation of Closure-tree is the high cost in filtering phase, since Closure-tree needs to perform expensive structure comparison and maximal matching algorithm in filtering phase. There are some other interesting recent work in graph search problem, such as [6] [8]. In [6], Williams et al. propose to enumerate all connected induced subgraphs in the graph database, and organize them in a Graph Decomposition Index (GDI). This method cannot work well in a graph database with large-size graphs, due to combination explosion of enumerating all connected induced subgraphs. Jiang et al. propose gString [8] for compound database. It is not trivial to extend gString to graph database in other applications.

3   Summarization Graph

As mentioned in Section 1, feature-based indexing methods are not effective due to ignoring the relationship between features. However, it is not trivial to include the relationship into the indexes, checking feature first, then relationship is neither effective nor efficient, which will be discussed in Section 3.2.

In this section, we first give some preliminary background related to sub-graph search in Section 3.1. Then, we present a novel concept, *Summarization Graph*, which encodes the features as well as their relationships in Section 3.2. In Section 3.3, we present how Summarization Graph is used for filtering. Finally, in Section 3.4, we give an efficient signature file-based index structure to improve the filter efficiency.

3.1   Preliminary

Definition 1 (Graph Isomorphism). *Assume that we have two graphs $G_1 < V_1$, E_1, L_{1v}, L_{1e}, F_{1v}, $F_{1e} >$ and $G_2 < V_2$, E_2, L_{2v}, L_{2e}, F_{2v}, $F_{2e} >$. G_1 is graph isomorphism to G_2, if and only if there exists at least one bijective function $f : V_1 \rightarrow V_2$ such that: 1) for any edge $uv \in E_1$, there is an edge $f(u)f(v) \in E_2$; 2) $F_{1v}(u) = F_{2v}(f(u))$ and $F_{1v}(v) = F_{2v}(f(v))$; 3) $F_{1e}(uv) = F_{2e}(f(u)f(v))$.*

Definition 2 (Sub-graph Isomorphism). *Assume that we have two graphs G' and G, if G' is graph isomorphism to at least one sub-graph of G under bijective function f, G' is sub-graph isomorphism to G under injective function f.*

Definition 3 (Problem Definition). *Given a query graph Q, we need to find all data graphs G_i, where Q is sub-graph isomorphism to G_i in a large graph database.*

As discussed in Section 1, due to hardness of sub-graph isomorphism, sub-graph search should adopt *filtering-and-verification* framework. In feature-based approaches, given a query Q, we can fix the complete set of candidates by scanning the inverted index. In the filtering process, in order to find all features that are contained in the query Q, we need to perform sub-graph isomorphism. After filtering, for each candidate, we also need to perform sub-graph isomorphism to check whether it is a correct answer. Therefore, the cost of sub-graph search can be modeled below:

$$Cost = |N_f| * C_f + |N_c| * C_c \qquad (1)$$

Here, $|N_f|$ is the number of sub-graph isomorphism test in the filtering process. $|N_c|$ is the size of candidate size. C_f and C_c are average cost of sub-graph isomorphism in filtering and verification process respectively. Intuitively, if we choose more features, $|N_c|$ will be smaller. However, more indexed features also lead to more sub-graph isomorphism tests in the filtering process (namely, $|N_f|$ is large). Therefore, to improve sub-graph search performance, $|N_f|$ and $|N_c|$ should be both small. In this paper, we will show that our method can use less feature to obtain greater pruning power.

3.2 Summarization Graph

In this subsection, we first give some important definitions as the base of Summarization Graph. Then we illustrate, through an example, the reason not combining feature-index and feature relationship information directly. Finally, our Summarization Graph framework is discussed in details.

Definition 4. *Occurrence.* *Given two graphs G and S, if there are m different sub-graphs in G that are all isomorphic to the graph S, we call these sub-graphs as **Occurrences** of S in G, which is denoted as $O_i(G, S)$, where $i = 1...m$.*

For example, in Fig.1, there is only one occurrence of phenyl in 001, 002, and 003. Similarly, in Fig.2, there are two occurrences of triangle in 001 and 002 respectively.

Definition 5 (Distance between occurrences). *Given a graph G and two occurrences $O_i(G, S_1)$ and $O_j(G, S_2)$, where S_1 and S_2 are sub-graphs of G, $Dist(O_i(G, S_1), O_j(G, S_2))$ is the distance between two occurrences, which is defined as follows:*

1. *If $O_i(G, S_1)$ does not overlap with $O_j(G, S_2)$, then $Dist(O_i(G, S_1), O_j(G, S_2)) = Min(SDist(V_n, V_m))$, where $SDist(V_n, V_m)$ is the shortest path length between vertices V_n and V_m, and V_n, V_m are two vertices in $O_i(G, S_1)$ and $O_j(G, S_2)$ respectively.*
2. *If $O_i(G, S_1)$ overlaps with $O_j(G, S_2)$, then $Dist(O_i(G, S_1), O_j(G, S_2)) = 0 - ComVer$, where $ComVer$ is the number of common vertices in $O_i(G, S_1)$ and $O_j(G, S_2)$.*

For example, in Fig.1, the distance between two occurrences, feature 1 (phenyl) and feature 2 (cyclopentane), is 2 in 001 and 002, and 1 in 003. In graph 001 of Fig.2, the distance between two occurrences of triangle and square denoted by shade area is -1, since they share one vertex.

In Fig.2, given a query Q, 001 should be result, which contains Q. Obviously, it is not efficient to include the distance information into feature index by simply checking feature first then distances (more processing time in filter phase). Furthermore, the straightforward approach is not effective either (there are still many false positives left). For example, in Fig.2, F_1 (triangle) and F_2 (square) are chosen as indexed features. The distances between all occurrences of F_1 and F_2 are also recorded in Fig.2. There are two occurrences of F_1 and F_2 in 001 respectively, thus, there are four distances among these occurrences, and the same applies to 002. Unfortunately, as shown in Fig. 2, all distance information is the same between two graphs. Thus, simply plug in the distance information to feature-based approach may not help to improve the filtering power. In order to address the above issue, in this paper, we introduce a novel concept, *Summarization Graph*, which encodes both the features and their distance relationships together.

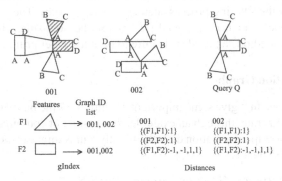

Fig. 2. Running Example

Fig.3 shows the steps to construct summarization graph. Given a graph 001, for each feature occurrence O_i, it is denoted as a vertex in the corresponding summarization graph, that is $Sum(001)$. The pair-wise distances among occurrences are recorded as edge weights in $Sum(001)$ in Fig.3. Then, we encode the edge weights into corresponding vertices in $Sum(001)$. Since $Sum(001)$ is a complete graph, we can denote it by the set of vertices in $Sum(001)$. The formal definition about summarization graph is given as follows:

Definition 6 (Summarization Graph). *Given a graph G and a set of its sub-graphs F_i, for every occurrence of F_i in G, it is denoted as a vertex in the Summarization Graph of G, i.e Sum(G). Each vertex label in Sum(G) is a set of Pairs. Each pair is denoted as <label, length>, which denotes the topology and distance information about this occurrence in G.*

To transform original graphs into Summarization Graph, 1) we should find some sub-graphs as features; 2) with these features, we transform all data graphs and the query graph into Summarization Graphs. Since feature selection approaches in the previous work are orthogonal to our methods, we omit the discussions about feature selection. Interested readers can refer to [20,22,4,23] for details.

Here, we discuss how to build a Summarization Graph for an original graph as follows. There are three sub-steps to generate Summarization Graph for an original graph: 1) find all occurrences of features. All occurrences form the vertexes of the Summarization Graph; 2) compute pair-wise distances between occurrences; 3) insert the distance

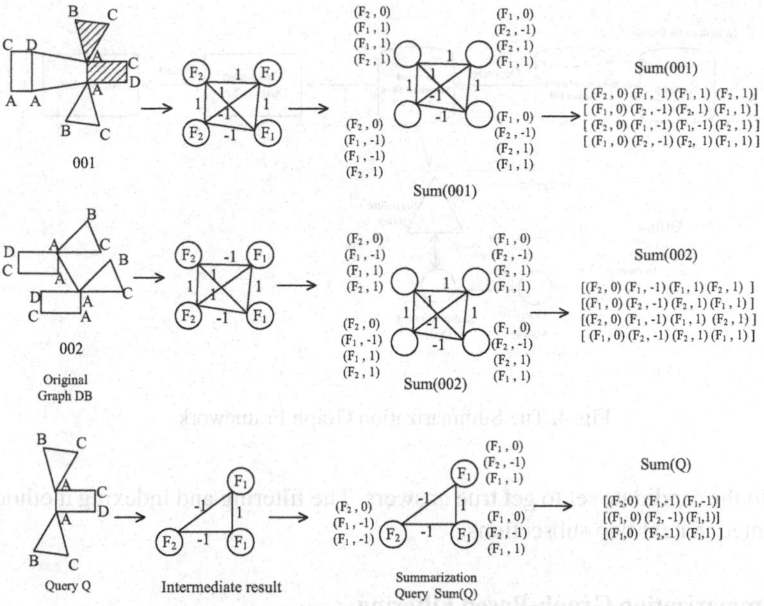

Fig. 3. Generating Summarization Graphs

information into vertexes of Summarization Graphs. The whole process of generating Summarization Graph is illustrated with an example in Fig.3. Notice that a Summarization Graph is a complete graph with unlabeled edges, which facilitate the latter process. In Fig.3, after sub-step 2, we get a complete graph with edge label (that is intermediate result). Then, we insert edge label, i.e. distances, into vertexes. Each vertex in the Summarization Graph is a set of Pair, and each pair is denoted as $< Label, Length >$. Take the up-right vertex in summarization graph 001 in Fig.3, i.e. Sum(001), for example. The pair set is $[(F_1, 0)(F_2, -1)(F_2, 1), (F_1, 1)]$. Since the corresponding vertex label in intermediate result is F_1, there is a Pair $(F_1, 0)$ in the vertex of Sum(001). In the intermediate result, there is an edge with distance 1 from the right-up vertex to another vertex with label F_2. So there is a Pair $(F_2, 1)$. Based on the similar process, we get another two Pairs, $(F_2, -1)$ and $(F_1, 1)$. Now, all together the four Pairs form the vertex label in the summarization graph.

With summarization graph, Fig.4 illustrates the overall framework of sub-graph search. In our framework, first, for all data graphs in the original graph database, we transform them into their corresponding Summarization Graphs. Then, Inverted Summarization Graph DB together with a signature file-based index is created for Summarization Graph DB.

In the online phase, when the query graph comes, 1) it is first transformed into its corresponding summarization graph, 2) and then the query summarization graph is searched in the signature file-based indexes and Inverted Summarization Graph DB to generate candidate set. 3) Finally, the sub-graph isomorphism checking algorithm is

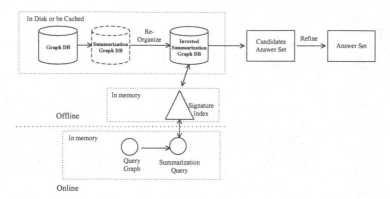

Fig. 4. The Summarization Graph Framework

applied to the candidate set to get true answers. The filtering and indexing methods will be presented in next two subsections.

3.3 Summarization Graph-Based Filtering

Definition 7 (Corresponding Vertexes). *Given two Summarization Graphs, $Sum(Q)$ and $Sum(G)$, for a vertex V_1 in $Sum(Q)$ and a vertex V_2 in Sum(G), if they satisfy the following conditions, we say that V_2 corresponds to V_1, where V_1 is a set of pairs $< F_i, L_i >$, and V_2 is also a set of pairs $< F_j, L_j >$: We can find an injective function from Pairs $< F_i, L_i >$ in V_1 to Pairs $< F_j, L_j >$ in V_2, where*
1) if $L_i \leq 0$, then $F_i = F_j$ and $L_i = L_j$
2) if $L_i > 0$, then $F_i = F_j$ and $0 < L_j \leq L_i$.

Lemma 1. *Given two Graphs Q and G, and their Summarization Graphs, Sum(Q) and Sum(G), respectively, if Q is a sub-graph of G, for each vertex in Sum(Q), there must exist a corresponding vertex in Sum(G).*

Proof. (Sketch:) 1) If Q is a sub-graph of G, according to Definition 2, we can find at least an injective function from Q to G. Each vertex and edge in Q has a unique corresponding image in G. Obviously, for each feature occurrence in Q, there is also a corresponding occurrence in G. According to the definition about summarization graph, for each vertex I in $Sum(Q)$, there must be a corresponding vertex J in $Sum(G)$. For any Pair $p < F_1, L_1 >$ in I:

a)if $L_1 = 0$: Because J is corresponding to I, so their feature labels should be same. So there must be a Pair q in $J < F_2, L_2 >$, where $L_2 = 0$ and $F_2 = F_1$.

b)if $L_1 < 0$: It means the occurrence I overlaps another occurrence of F_1 with $|L_1|$ vertexes in Q. Since for each occurrence of features in Q, there also exits a corresponding occurrence in G. If two occurrences overlap $|L_1|$ vertexes in Q, the corresponding occurrences also overlap $|L_1|$ vertexes in G. Therefore, there must be a Pair q in $J < F_2, L_2 >$, where $L_2 = L_1, F_2 = F_1$.

c) if $L_1 > 0$: It means that the minimum distance between the occurrence I and another occurrence of F_1 is L_1 in Q. According to graph theory, obviously, the minimum distance between these two corresponding occurrences in G cannot be larger than L_1, if Q is a sub-graph of G. Because, at least, the minimum path between these two occurrences in Q must be preserved in G.

In conclusion, Lemma 1 is correct.   □

Filtering Rule 1. Given two graphs Q and G, their corresponding Summarization Graphs are $Sum(Q)$ and $Sum(G)$ respectively. If there exists some vertex in $Sum(Q)$, it has no corresponding vertex in $Sum(G)$, Q cannot be a sub-graph of G.   □

According to Lemma 1, it is straightforward to know that the following theorem holds.

Theorem 1. *For sub-graph search, **Filtering Rule 1** satisfies no-negative requirement (no dismissals in query results).*

Since we consider both feature and pair-wise feature relationship together, it has greater filtering power than feature-based approaches. Thus, we have the following remark.

Remark 1. Based on the same selected feature set, Filtering Rule 1 has greater filtering power than feature-based methods.

001	002	Query Q
$[(F_2,0)(F_1, 1)(F_1, 1)(F_2, 1)]$	$[(F_2,0)(F_1,-1)(F_1,1)(F_2,1)]$	
$[(F_1,0)(F_2,-1)(F_2,1)(F_1,1)]$	$[(F_1,0)(F_2,-1)(F_2,1)(F_1,1)]$	$[(F_2,0)(F_1,-1)(F_1,-1)]$
$[(F_2,0)(F_1,-1)(F_1,-1)(F_2,1)]$	$[(F_2,0)(F_1,-1)(F_1,1)(F_2,1)]$	$[(F_1,0)(F_2,-1)(F_1,1)]$
$[(F_1,0)(F_2,-1)(F_2,1)(F_1,1)]$	$[(F_1,0)(F_2,-1)(F_2,1)(F_1,1)]$	$[(F_1,0)(F_2,-1)(F_1,1)]$

Fig. 5. Summarization Graph Database

Now, all data graphs in the graph database have been converted into their corresponding Summarization Graphs, which form the Summarization Graph Database, as shown in Fig.5. The query $Sum(Q)$ is also shown in Fig.5. We can sequentially scan the Summarization Graph Database to prune false alarms. In fact, pruning according to Filter Rule 1 is not time-consuming since it is a subset checking process. However, in order to avoid sequential scan in the Summarization Graph Database, we want to store the Summarization Graph Database like an inverted index, as shown in Fig.6. We call it *Inverted Summarization Graph Database*. In Fig.6, given a query Q, we convert it into its corresponding Summarization Graph, i.e Sum(Q), which is a set of vertexes. For each vertex in $Sum(Q)$, according to the Lemma 1, scanning the vertexes list on the left to find the corresponding vertexes in the Inverted Summarization Graph Database, we will get a graph list. The intersection of these graph lists will be candidate results for sub-graph search. In the running example, for a vertex $[(F_2,0),(F_1,-1),(F_1,-1)]$ in Sum(Q), we will get a graph list "001". For a vertex $[(F_1,0),(F_2,-1),(F_1,1)]$ in $Sum(Q)$, the graph list is "001,002". The intersection of graph lists is "001", which is the candidate.

Vertexes   Graph ID list

[(F_2 , 0) (F_1 , 1) (F_1 , 1) (F_2 , 1)]   →   001, ...

[(F_1 , 0) (F_2 , -1) (F_2 , 1) (F_1 , 1)]   →   001, 002...

[(F_2 , 0) (F_1 , -1) (F_1 , -1) (F_2 , 1)]   →   001, ...

[(F_2 , 0) (F_1 , -1) (F_1 , 1) (F_2 , 1)]   →   002, ...

... .

Fig. 6. Inverted Summarization Graph Database

3.4 Signature-Based Index

As discussed in the above subsection, using the Inverted Summarization Graph avoids the sequential scan in the whole database. However, we have to sequentially scan the vertexes list on the left of Fig.6. In order to avoid this, we build a signature file-based index for the Inverted Summarization Graph Database. In fact, the vertex list in the Inverted Summarization Graph Database is the objects with set-value attributes. Each vertex is an object and it has a set of attributes, i.e. Pairs. According to the methods in [17], we propose a hash function and build an efficient index for the Inverted Summarization Graph Database. For each pair $< F_i, L_i >$ of a vertex in summarization graph, if $L_i \leq 0$, we assign a distinct bit-string to the Pair. In Fig.7, on the left, we assign a distinct bit-string to each Pair. For example, for the vertex [(F_2,0)(F_1,-1)(F_1,1)(F_2, 1)], its hash value is (0000 0001) |(0100 0000) = (0100 0001) , where (0000 0001) is the bit-string of (F_2,0) and (0100 0000) is the bit-string of (F_1,-1). Using the hash function, each vertex in the Inverted Summarization Graph Database has a hash value, thus, we can index all vertexes by a hash table.

Given a query sub-graph Q, its Summarization Graph is $Sum(Q)$. Using the same hash function, for each vertex V_i in Sum(Q), it has a hash value, denoted by $Hash(V_i)$. Scanning the hash table, for each hash value H_i in the hash table, if ($H_i\&Hash(V_i)\neq Hash(V_i)$), the vertexes in the bucket cannot correspond to V_i in $Sum(Q)$.

Furthermore, we can also built a S-tree index [17] for the hash table, as shown in Fig.8. S-tree is a balanced tree, where each node has at least m children ($m \leq 2$), and at most M children ((M+1)/2 $\geq m$). Assume that the intermedin node (directory node) I in S-Tree has n child nodes C_i (i=1...n), we set $I[j] = (C_1[j]\vee...\vee C_i[j]...\vee C_n[j])$, where $I[j]$ is the j-th bit in I and $C_i[j]$ is the j-th bit in C_i. Given a vertex V_i in $Sum(Q)$, it has a hash value, denoted by $Hash(V_i)$. For an intermedin node I in S-tree, if ($I\&Hash(V_i)\neq Hash(V_i)$), all descendant nodes can be pruned. For example, given a vertex V [(F2,0) (F1,-1) (F1,-1)] in $Sum(Q)$ in Fig. 5, its hash code is $Hash(V)$=(0100 0001). Obviously, for the intermedin node I_1 in S-tree in Fig.8, $I_1\&Hash(V)\neq Hash(V)$). Therefore, all descendant nodes of I_1 cannot be corresponding vertex to V in $Sum(Q)$, and they can be pruned safely. Because of S-tree, we reduce the search space. Due to limited space, the more details about S-tree are omitted, and interested readers can refer to [17].

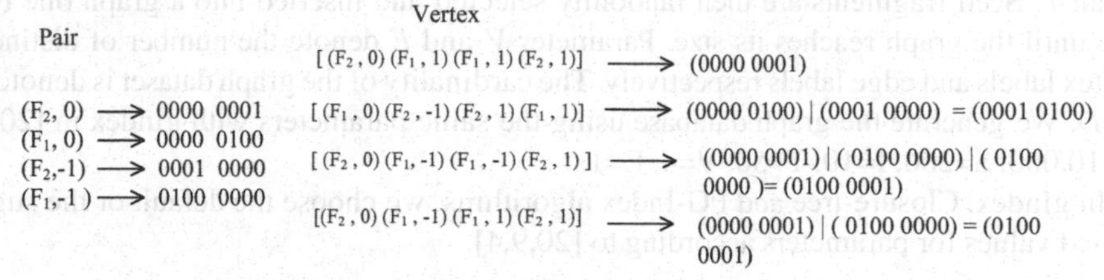

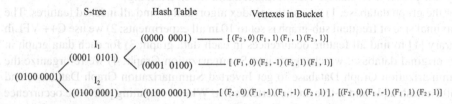

Fig. 7. Hash Function

Fig. 8. S-Tree

4 Experiments

In this section, we report our experiment results to evaluate our methods with existing approaches. In our experiment, some of state of art feature-based approaches, such as gIndex, Closure-Tree and FGIndex, are chosen to compare with our method. All experiments are done on a 1.7GHz Pentium 4 PC with 1 GB main memory, running RedHat Fedora Core 3. All algorithms are implemented in standard C++ with STL library support and compiled by g++ 3.4.2 compiler.

4.1 Datasets

1) **AIDS Antiviral Screen Dataset.** This dataset is available publicly on the website of the Developmental Therapeutics Program[1]. We generate 10,000 connected and labeled graphs from the molecule structures and omit Hydrogen atoms. The graphs have an average number of 24.80 vertices and 26.80 edges, and a maximum number of 214 vertices and 217 edges. A major portion of the vertices are C, O and N. The total number of distinct vertex labels is 62, and the total number of distinct edge labels is 3. We refer to this dataset as AIDS dataset. Each query set Qm has 1000 connected query graphs and query graphs in Qm are connected size-m graphs (the edge number in each query is m), which are extracted from some data graphs randomly, such as $Q4, Q8, Q12, Q16, Q20$ and $Q24$.

2) **Synthetic Dataset.** The synthetic dataset is generated by a synthetic graph generator provided by authors of [14]. The synthetic graph dataset is generated as follows: First, a set of S seed fragments are generated randomly, whose size is determined by a Poisson distribution with mean I. The size of each graph is a Poisson random variable with

---
[1] http://dtp.nci.nih.gov/

mean T. Seed fragments are then randomly selected and inserted into a graph one by one until the graph reaches its size. Parameter V and E denote the number of distinct vertex labels and edge labels respectively. The cardinality of the graph dataset is denoted by D. We generate the graph database using the same parameters with gIndex in [20]: D=10,000, S=200, I=10, T=50, V=4, E=1.

In gIndex, Closure-tree and FG-Index algorithms, we choose the default or the suggested values for parameters according to [20,9,4].

4.2 Methods

Off-line process: (Database Transformation and Index Building)
For the graph database, 1) we use the gIndex algorithm to find all indexed features. The maximal size of frequent sub-graph is set to 10 in all experiments; 2) we use C++ VFLib library [1] to find all feature occurrences in each data graph; 3) for each data graph in the original database, we change it into Summarization Graph; 4) we re-organize the Summarization Graph Database to get Inverted Summarization Graph Database, and build a hash table and S-tree on the vertexes list. When computing pair-wise occurrence distances, we first get all vertex-pair shortest paths in each data graph using Johnson algorithm [5], then get occurrence distances according to Definition 5.

On-line process: (Query Executing)
Given a query graph Q, we transform it into Summarization Graph, i.e. Sum(Q). For each vertex in Sum(Q), according to the S-tree and hash table, we find the corresponding vertexes in Inverted Summarization Graph Database. Then we get a graph list for each corresponding vertex. The intersection of these graph lists is the Candidate Answer Set. The above process is called getting candidate process. For each data graph in candidates, we have to do sub-graph isomorphic checking, which we call refining process.

4.3 Performance Studying

We evaluate offline and online performance of our method respectively, and compare them with other existing approaches. Notice that, the online performance is crucial for sub-graph search, such as pruning power and total response time.

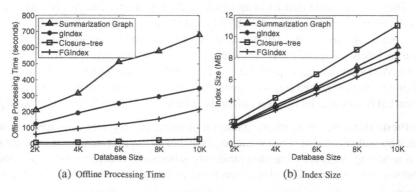

(a) Offline Processing Time   (b) Index Size

Fig. 9. Evaluate the Offline Performance in AIDS dataset

1) Offline Performance
Fig.9 shows the offline performance of Summarization Graph method. Compared with
feature-based methods, such as gIndex and FGIndex, our method needs to compute the
feature distances. Therefore, our offline processing time is larger than other approaches.
We encode both features and feature distances of graph G into $Sum(G)$. Even though
we utilize signature file to store $Sum(G)$, the space cost of our method is also a little
larger than gIndex and FGIndex, and smaller than Closure-Tree.

Observed from Fig.9, the offline performance of our method is not as good as other
approaches. However, in the online performance, our method outperform other methods

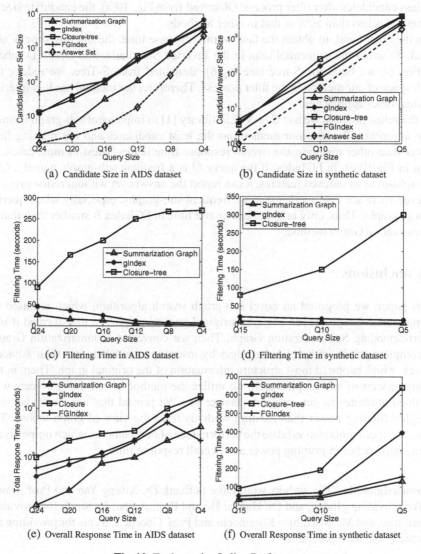

(a) Candidate Size in AIDS dataset   (b) Candidate Size in synthetic dataset

(c) Filtering Time in AIDS dataset   (d) Filtering Time in synthetic dataset

(e) Overall Response Time in AIDS dataset   (f) Overall Response Time in synthetic dataset

Fig. 10. Evaluate the Online Performance

orders of magnitude (see Fig. 10). Therefore, it is really worth to paying more time on off-line processing.

2) Online Performance

In Fig. 10, we evaluate the online performance our method in both AIDS and synthetic datasets, and compare it with other methods.

As we all know, since sub-graph isomorphism is NP-complete problem, all sub-graph methods adopt *filter-and-refine* framework. Obviously, the less candidates after filtering, the faster query response time. Therefore, pruning power in the filter process is critical to overall online performance. Compared with existing feature-based methods, we not only consider features but also the pairwise feature relationships. Therefore, we have less candidates after filter process. Observed from Fig. 10(a), the candidate size in our method is less than 50% of that in other methods.

On the other hand, to obtain the fast overall response time, the filtering time is also critical. To avoid the sequential scan in the Inverted Summarization Graph Database (see Fig. 6), we build the S-tree (see Fig.8). Benefited from S-Tree, we reduce the search space of our method in the filter process. Therefore, our method has less filtering time than other approaches.

In the refine process, we use C++ VFLib library [1] to implement sub-graph isomorphism algorithm. Because our method has the least candidates and our refining time is faster than other methods, the overall response time is the fastest in most cases, as shown in Fig 10(e). In FGIndex, if the query Q is a frequent sub-graph, namely, Q is isomorphism to an indexed features, it can report the answer set without refine process. However, there are few queries that are frequent sub-graphs, especially when queries are large graphs. Thus, only in $Q4$, the response time in FGIndex is smaller than that in Summarization Graph method.

5   Conclusions

In this paper, we proposed an novel sub-graph search algorithm, which is based on Summarization Graph model. For each original general graph, we transformed it into its corresponding Summarization Graph. Then we converted Summarization Graphs into complete graphs with unlabeled edges by inserting edge labels into their adjacent vertexes, which captured most structure information of the original graph. Then, in the filtering process of sub-graph search, we utilize the method of retrieving objects with set-valued attributes to get candidate answer set. We proved that our methods could get higher filtering power than existing methods from the view of graph theory. The extensive experiments also validate the performance of our methods, which outperforms existing approaches in pruning power and overall response time.

Acknowledgments. The authors would like to thank Dr. Xifeng Yan and Prof. Jiawei Han for providing gIndex, and Dr. Huahai He and Prof. Ambuj K. Singh for providing Closure-tree, and Mr. Michihiro Kuramochi and Prof. George Karypis for providing the synthetic graph data generator.

References

1. Available at: http://amalfi.dis.unina.it/graph
2. Berman, H.M., Westbrook, J., Feng, Z., Gilliland, G., Bhat, T.N., Weissig, H., Shindyalov, I.N., Bourne., P.E.: Frequent subtree mining - an overview. Nucleic Acids Research 23(10) (2000)
3. Cai, D., Shao, Z., He, X., Yan, X., Han, J.: Community mining from multi-relational networks. In: Jorge, A.M., Torgo, L., Brazdil, P.B., Camacho, R., Gama, J. (eds.) PKDD 2005. LNCS (LNAI), vol. 3721, Springer, Heidelberg (2005)
4. Cheng, J., Ke, Y., Ng, W., Lu, A.: fg-index: Towards verification-free query processing on graph databases. In: SIGMOD (2007)
5. Cormen, T.H., Leiserson, C.E., Rivest, R.L., Stein, C.: Introduction to Algorithm, 2nd edn. MIT Press, Cambridge (2000)
6. Huan, D.W.W.J., Wang, W.: Graph database indexing using structured graph decomposition. In: ICDE (2007)
7. Fortin, S.: The graph isomorphism problem. Department of Computing Science, University of Alberta (1996)
8. Jiang, P.Y.H., Wang, H., Zhou, S.: Gstring: A novel approach for efficient search in graph databases. In: ICDE (2007)
9. He, H., Singh, A.K.: Closure-tree: An index structure for graph queries. In: ICDE (2006)
10. Inokuchi, A., Washio, T., Motoda, H.: An apriori-based algorithm for mining frequent substructures from graph data. In: Zighed, A.D.A., Komorowski, J., Żytkow, J.M. (eds.) PKDD 2000. LNCS (LNAI), vol. 1910, Springer, Heidelberg (2000)
11. James, C.A., Weininger, D., Delany, J.: Daylight theory manual daylisght version 4.82. Daylight Chemical Information Systems, Inc. (2003)
12. Ke, Y., Cheng, J., Ng, W.: Correlation search in graph databases. In: SIGKDD (2007)
13. Kuramochi, M., Karypis, G.: Frequent subgraph discovery. In: ICDM (2001)
14. Kuramochi, M., Karypis, G.: Frequent subgraph discovery. In: ICDM (2001)
15. Petrakis, E.G.M., Faloutsos, C.: Similarity searching in medical image databases. IEEE Transactions on Knowledge and Data Enginnering 9(3) (1997)
16. Shasha, D., Wang, J.T.-L., Giugno, R.: Algorithmics and applications of tree and graph searching. In: PODS (2002)
17. Tousidou, E., Bozanis, P., Manolopoulos, Y.: Signature-based structures for objects with set-valued attributes. Inf. Syst. 27(2) (2002)
18. Willett., P.: Chemical similarity searching. J. Chem. Inf. Comput. Sci. 38(6) (1998)
19. Yan, X., Han., J.: Gspan: Graph-based substructure pattern mining. In: Proc. of Int. Conf. on Data Mining (2002)
20. Yan, X., Yu, P.S., Han, J.: Graph indexing: A frequent structure-based approach. In: SIGMOD (2004)
21. Zhang, N., Özsu, M.T., Ilyas, I.F., Aboulnaga, A.: Fix: Feature-based indexing technique for XML documents. In: VLDB (2006)
22. Zhang, S., Hu, M., Yang, J.: Treepi: A novel graph indexing method. In: ICDE (2007)
23. Zhao, P., Yu, J.X., Yu, P.S.: Graph Indexing: Tree + Delta >=Graph. In: VLDB (2007)

Bulk-Loading the ND-Tree in Non-ordered Discrete Data Spaces*

Hyun-Jeong Seok[1], Gang Qian[2], Qiang Zhu[1],
Alexander R. Oswald[2], and Sakti Pramanik[3]

[1] Department of Computer and Information Science,
The University of Michigan - Dearborn, Dearborn, MI 48128, USA
{hseok,qzhu}@umich.edu
[2] Department of Computer Science,
University of Central Oklahoma, Edmond, OK 73034, USA
{gqian,aoswald}@ucok.edu
[3] Department of Computer Science and Engineering,
Michigan State University, East Lansing, MI 48824, USA
pramanik@cse.msu.edu

Abstract. Applications demanding multidimensional index structures for performing efficient similarity queries often involve a large amount of data. The conventional tuple-loading approach to building such an index structure for a large data set is inefficient. To overcome the problem, a number of algorithms to bulk-load the index structures, like the R-tree, from scratch for large data sets in continuous data spaces have been proposed. However, many of them cannot be directly applied to a non-ordered discrete data space (NDDS) where data values on each dimension are discrete and have no natural ordering. No bulk-loading algorithm has been developed specifically for an index structure, such as the ND-tree, in an NDDS. In this paper, we present a bulk-loading algorithm, called the NDTBL, for the ND-tree in NDDSs. It adopts a special in-memory structure to efficiently construct the target ND-tree. It utilizes and extends some operations in the original ND-tree tuple-loading algorithm to exploit the properties of an NDDS in choosing and splitting data sets/nodes during the bulk-loading process. It also employs some strategies such as multi-way splitting and memory buffering to enhance efficiency. Our experimental studies show that the presented algorithm is quite promising in bulk-loading the ND-tree for large data sets in NDDSs.

Keywords: Multidimensional indexing, bulk-loading, non-ordered discrete data space, algorithm, similarity search.

---

* Research supported by the US National Science Foundation (under grants # IIS-0414576 and # IIS-0414594), the US National Institute of Health (under OK-INBRE Grant # 5P20-RR-016478), The University of Michigan, and Michigan State University.

J.R. Haritsa, R. Kotagiri, and V. Pudi (Eds.): DASFAA 2008, LNCS 4947, pp. 156–171, 2008.

1   Introduction

Multidimensional index structures such as the R-trees [3,4,12] and the K-D-B-tree [19] are vital to efficient evaluation of similarity queries in multidimensional data spaces. Applications requiring similarity queries often involve a large amount of data. As a result, how to rapidly bulk-load an index structure for a large data set from scratch has become an important research topic recently. Most research efforts done so far are for bulk-loading index structures in continuous data spaces (CDS). In this paper, our discussion focuses on bulk-loading an index tree for non-ordered discrete data spaces (NDDS).

An NDDS contains multidimensional vectors whose component values are discrete and have no natural ordering. Non-ordered discrete data domains such as gender and profession are very common in database applications. Lack of essential geometric concepts such as (hyper-)rectangle, edge length and region area raises challenges for developing an efficient index structure in an NDDS. An ND-tree, which utilizes special properties of an NDDS, was proposed recently to support efficient similarity queries in NDDSs [16,18]. A conventional tuple-loading (TL) algorithm was introduced to load vectors/tuples into the tree one by one. However, such a TL method may take too long when building an index for a large data set in an NDDS. In fact, many contemporary applications need to handle an increasingly large amount of data in NDDSs. For example, genome sequence databases (with non-ordered discrete letters 'a', 'g', 't', 'c') have been growing rapidly in size in the past decade. The size of the GenBank, a popular collection of all publicly available genome sequences, increased from 71 million residues (base pairs) and 55 thousand sequences in 1991, to more than 65 billion residues and 61 million sequences in 2006 [11]. Note that a genome sequence is typically broken into multiple fixed-length q-grams (vectors) in an NDDS when similarity searches are performed. Clearly, an efficient bulk-loading (BL) technique is required to effectively utilize the ND-tree in such applications.

A number of bulk-loading algorithms have been proposed for multidimensional index structures such as the R-tree and its variants in CDSs. The majority of them are sorting-based bulk-loading [5,9,10,14,15,20]. Some of these adopt the bottom-up approach, while the others employ the top-down approach. The former algorithms [9,14,20] typically sort all input vectors according to a chosen one-dimensional criterion first, place them into the leaves of the target tree in that order, and then build the tree level by level in the bottom-up fashion via recursively sorting the relevant MBRs at each level. The latter algorithms [10,15] typically partition the set of input vectors into K subsets of roughly equal size (where $K \leq$ the non-leaf node fan-out) based on one or more one-dimensional orderings of the input vectors, create a root to store the MBRs of the subsets, and then recursively construct subtrees for the subsets until each subset can fit in one leaf node. Unfortunately, these algorithms cannot be directly applied to an index structure in NDDSs where ordering as well as relevant geometric concepts such as centers and corners are lacking.

Another type of bulk-loading algorithms, termed the generic bulk-loading, has also been suggested [5,7,6,8]. The main characteristic of such algorithms is to bulk-load a target index structure T by utilizing some operations/interfaces (e.g., *ChooseSubtree* and *Split*) provided by the conventional TL (tuple-loading) algorithm for T. Hence, the generic bulk-loading algorithms can be used for a broad range of index structures that support the required operations. A prominent generic bulk-loading algorithm [6], denoted as GBLA in this paper, adopts a buffer-based approach. It employs external queues (buffers) associated with the internal nodes of a tree to temporarily block an inserted vector. Only when a buffer is full will its blocked vectors be forwarded to the buffer of a next-level node. It builds the target index tree level by level from the bottom up. The buffer-based bulk-loading approach is also used in the techniques presented in [1,13]. Some other generic bulk-loading algorithms employ a sample-based approach [7,8]. Vectors are randomly sampled from the input data set to build a seed index structure in the memory. The remaining vectors are then assigned to individual leaves of the seed structure. The leaves are processed in the same way recursively until the whole target structure is constructed. The effectiveness and efficiency of such sample-based bulk-loading techniques typically rely on the chosen samples. Generic bulk-loading algorithms are generally applicable to the ND-tree since the conventional TL algorithm for the ND-tree provides necessary operations.

In this paper, we propose a new algorithm, called NDTBL (the ND-Tree Bulk Loading), for bulk-loading the ND-tree in an NDDS. It was inspired by the above GBLA [7]. Although it also employs an intermediate tree structure and buffering strategies to speed up the bulk-loading process, it is significantly different from the GBLA in several ways: (1) a new in-memory buffer tree structure is adopted to more effectively utilize the available memory in guiding input vectors into their desired leaf nodes; (2) the non-leaf nodes of the in-memory buffer tree can be directly used in the target index tree to reduce node construction time; (3) a multi-way splitting based on the auxiliary tree technique for sorting discrete minimum bounding rectangles (DMBR) in an NDDS is employed to improve efficiency over the traditional two-way split; and (4) a unique adjusting process to ensure the target tree meeting all the ND-tree properties is applied when needed. Our experiments demonstrate that our algorithm is promising in bulk-loading the ND-tree in NDDSs, comparing to the conventional TL algorithm and the representative generic bulk-loading algorithm GBLA.

The rest of this paper is organized as follows. Section 2 introduces the essential concepts and notation. Section 3 discusses the details of NDTBL. Section 4 presents our experimental results. Section 5 concludes the paper.

2   Preliminaries

To understand our bulk-loading algorithm for the ND-tree in an NDDS, it is necessary to know the relevant concepts about an NDDS and the structure of the ND-tree, which were introduced in [16,17,18]. For completion, we briefly describe them in this section.

A d-*dimensional NDDS* Ω_d is defined as the Cartesian product of d alphabets: $\Omega_d = A_1 \times A_2 \times ... \times A_d$, where $A_i(1 \le i \le d)$ is the *alphabet* of the i-th dimension of Ω_d, consisting of a finite set of letters. There is no natural ordering among the letters. For simplicity, we assume A_i's are the same in this paper. As shown in [17], the discussion can be readily extended to NDDSs with different alphabets. $\alpha = (a_1, a_2, ..., a_d)$ (or '$a_1a_2...a_d$') is a vector in Ω_d, where $a_i \in A_i$ $(1 \le i \le d)$. A *discrete rectangle* R in Ω_d is defined as $R = S_1 \times S_2 \times ... \times S_d$, where $S_i \subseteq A_i(1 \le i \le d)$ is called the i-th *component set* of R. The *area of* R is defined as $|S_1| * |S_2| * ... * |S_d|$. The *overlap* of two discrete rectangles R and R' is $R \cap R' = (S_1 \cap S'_1) \times (S_2 \cap S'_2) \times ... \times (S_d \cap S'_d)$. For a given set SV of vectors, the *discrete minimum bounding rectangle (DMBR)* of SV is defined as the discrete rectangle whose i-th component set $(1 \le i \le d)$ consists of all letters appearing on the i-th dimension of the given vectors. The DMBR of a set of discrete rectangles can be defined similarly.

As discussed in [16,18], the Hamming distance is a suitable distance measure for NDDSs. The Hamming distance between two vectors gives the number of mismatching dimensions between them. Using the Hamming distance, a similarity (range) query is defined as follows: given a query vector α_q and a query range of Hamming distance r_q, find all the vectors whose Hamming distance to α_q is less than or equal to r_q.

The ND-tree based on the NDDS concepts was introduced in [16,18] to support efficient similarity queries in NDDSs. Its structure is outlined as follows.

The ND-tree is a disk-based balanced tree, whose structure has some similarities to that of the R-tree [12] in continuous data spaces. Let M and m $(2 \le m \le \lceil M/2 \rceil)$ be the maximum number and the minimum number of entries allowed in each node of an ND-tree, respectively. An ND-tree satisfies the following two requirements: (1) every non-leaf node has between m and M children unless it is the root (which may have a minimum of two children in this case); (2) every leaf node contains between m and M entries unless it is the root (which may have a minimum of one entry/vector in this case).

A leaf node in an ND-tree contains an array of entries of the form (op, key), where key is a vector in an NDDS Ω_d and op is a pointer to the object represented by key in the database. A non-leaf node N in an ND-tree contains an array of entries of the form $(cp, DMBR)$, where cp is a pointer to a child node N' of N in the tree and $DMBR$ is the discrete minimum bounding rectangle of N'. Since each leaf or non-leaf node is saved in one disk block, while their entry sizes are

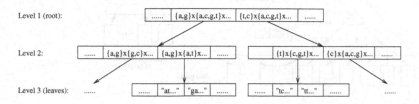

Fig. 1. An example of the ND-tree

different, M and m for a leaf node are usually different from those for a non-leaf node.

Figure 1 shows an example of the ND-tree for a genome sequence database with alphabet $\{a, g, t, c\}$ [16].

3 Bulk-Loading the ND-Tree

In this section, we introduce a bulk-loading algorithm for the ND-tree, which is inspired by the generic bulk-loading algorithm (GBLA) suggested by Bercken *et al.* in [6]. Although the two algorithms have some common ideas, they are significantly different in several ways including the intermediate tree structure, the memory utilization strategy, the target tree construction process, and the overflow node splitting approach.

3.1 GBLA and Shortcomings

Although no bulk-loading algorithm has been proposed specifically for the ND-tree so far, Bercken *et al.*'s GBLA can be applied to bulk-load the ND-tree due to its generic nature. GBLA can load any multidimensional index tree that provides the following operations: (1) *InsertIntoNode* to insert a vector/entry into a node of the tree, (2) *Split* to split an overflow node into two, and (3) *ChooseSubtree* to choose a subtree of a given tree node to accommodate an input vector/region, which are all provided by the conventional TL (tuple-loading) algorithm for the ND-tree [16,18].

The key idea of GBLA is to use an intermediate structure, called the buffer-tree (see Figure 2), to guide the input vectors to the desired leaf nodes of the target (index) tree that is being constructed. Once all the input vectors are loaded in their leaf nodes, GBLA starts to build another buffer-tree to insert the bounding rectangles/regions of these leaf nodes into their desired parent (non-leaf) nodes in the target tree. This process continues until the root of the target tree is generated.

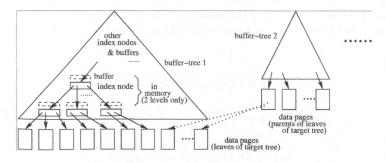

Fig. 2. Bulk-loading process of GBLA

Each node in the buffer-tree is called an index node. Each index node is associated with a buffer. Each buffer may consist of several pages. When input vectors/regions come to an index node N, they are accumulated in the buffer of N first. When the buffer is full, its stored vectors/regions are then pushed (cleared) into the desired child nodes determined by *ChooseSubtree*. This process is repeated until all stored vectors/regions reach their desired data pages (i.e., the nodes of the target tree). Splits are necessary when a data page or an index node is full. Since input vectors/regions are pushed into a data page or an index node in batches, the number of I/Os required for the same node/page is reduced during the bulk-loading.

During a bulk-loading, GBLA only keeps the following in memory: (1) the current index node N; (2) the last page of the buffer associated with N; and (3) the last pages of the buffers associated the child nodes of N if N is a non-leaf node of the buffer-tree, or the data pages pointed to by N if N is a leaf. All the other index nodes, buffer pages and data pages are kept on disk.

The way that GBLA makes use of all given memory space is to maximize the fan-out of each index node. A main drawback of this approach is following. Although an index node N in the buffer-tree is similar to a non-leaf node N' of the target tree, N cannot be directly used in the target tree since N and N' may have different fan-outs. This forces GBLA to generate the nodes of the target tree level by level, wasting much work done for constructing the buffer-tree. In addition, reading a large index node of the buffer-tree from disk to memory requires multiple I/Os.

3.2   Key Idea of a New Algorithm

To overcome the shortcomings of GBLA, we introduce a new bulk-loading algorithm NDTBL. The basic idea is following. Instead of using a separate buffer-tree, we directly buffer the non-leaf nodes of a target ND-tree T in memory (see Figure 3). In other words, the top portion (above the leaves) of T serves as an in-memory buffer tree BT during our bulk-loading. However, we only associate an auxiliary buffer (page) to each non-leaf node of T that is directly above the leaf nodes of T. This is because a non-leaf node needs a buffer only if its child nodes need to be read from disk so that multiple input vectors in the buffer can be pushed into a child node N when N is read in memory. We call an ND-tree with their non-leaf nodes buffered in memory a buffered ND-tree. Note that, when NDTBL splits an overflow node, it may split the node into more than two nodes to achieve high efficiency. Like the conventional TL algorithm of the ND-tree, NDTBL splits an overflow node by grouping/partitioning its data. Hence, it is a data-partitioning-based approach. Note that, when memory space is more than enough to keep all non-leaf nodes, the remaining memory space is used to cache as many leaf nodes as possible on the first-come first-served (FCFS) basis. Some cached leaf nodes are output to disk when memory space is needed for new non-leaf nodes. Hence only non-leaf nodes are guaranteed to be in memory.

When available memory is not enough to keep more non-leaf nodes during bulk-loading, NDTBL stops splitting the overflow leaf nodes, i.e., allowing

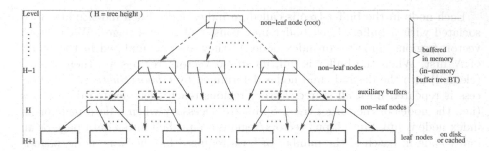

Fig. 3. Structure of a buffered ND-tree

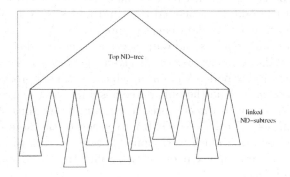

Fig. 4. An intermediate ND-tree resulting from linking subtrees

oversized leaf nodes. Once all input vectors are loaded in the tree, it then recursively builds a buffered ND-subtree for each oversized leaf node (see Figure 4). Since the subtrees may not have the same height and their root nodes may have less than m subtrees/children, some adjustments may be needed to make the final integrated tree meet all properties of an ND-tree.

Since the structure of a node in the in-memory buffer tree BT is the same as that of a non-leaf node in the target ND-tree T, BT can be directly output to disk as part of T (after appropriately mapping the memory pointers to the disk block pointers/numbers). If available memory is sufficiently large, causing no oversized leaf nodes, the target tree can be obtained by simply dumping the in-memory buffer tree into disk. Otherwise, the target ND-tree is built by parts with some possible adjustments done at the end. Since we keep the fan-out of a non-leaf node small, comparing to GBLA, we can better narrow down the set of useful child nodes for a given group of input vectors.

The following subsections describe the details of NDTBL.

3.3 Main Procedure

The main control procedure of algorithm NDTBL is given as follows. It recursively calls itself to build the target ND-tree by parts when memory is insufficient.

ALGORITHM 1 : **NDTBL**
Input: a set SV of input vectors in a d-dimensional NDDS.
Output: an ND-tree for SV on disk.
Method:
1. $BufTree$ = BuildBufferedNDtree(SV);
2.   **if** no leaf node of $BufTree$ is oversized **then**
3.   output the target ND-tree $TgtTree$ represented by $BufTree$ to disk;
4.   **else**
5.   output the buffered non-leaf nodes to disk as the top
         part $TopTree$ of the target ND-tree;
6.   **for** each oversized leaf node N of $BufTree$ **do**
7.   let SV_1 be the set of input vectors in N;
8. $ND\text{-}subtree$ = NDTBL(SV_1);
9.   replace N by $ND\text{-}sutree$ in its parent node in $TopTree$;
10.   **end for**;
11.   let $TgtTree$ be the intermediate ND-tree obtained by
         linking all $ND\text{-}subtree$s to $TopTree$;
12.   **if** heights of the $ND\text{-}subtree$s are different **then**
13.   find the smallest height h for all $ND\text{-}subtree$s;
14.   **for** each parent P of an $ND\text{-}subtree$ with height $> h$
           in $TopTree$ **do**
15. $TgtTree$ = CutTree($TgtTree$, P, h);
16.   **end for**;
17.   **end if**;
18.   **if** exists an $ND\text{-}subtree$ ST having $< m$ subtrees **then**
19.   let h = height(ST) $- 1$;
20.   **for** each parent P of an $ND\text{-}subtree$ in $TopTree$ **do**
21. $TgtTree$ = CutTree($TgtTree$, P, h);
22.   **end for**;
23.   **end if**;
24. **end if**;
25. **return** $TgtTree$.

Algorithm NDTBL first invokes function BuildBufferedNDtree to build a
buffered ND-tree for the given input set (step 1). If no leaf node is oversized,
the target ND-tree has been successfully built from the buffered ND-tree (steps
2 - 3). If there is at least one oversized leaf node (step 4), NDTBL first outputs
the top portion of the current ND-tree to disk (step 5) and then recursively calls
itself to build an ND-subtree for the vectors in each oversized leaf node (steps 6
- 11). If the heights of the above ND-subtrees are different, NDTBL re-balances
the ND-tree by cutting the subtrees that are taller than others (steps 12 - 17).
Since each ND-subtree is built as an ND-tree for a subset of input vectors, it is
possible that its root has less than m children (entries). In this case, NDTBL cuts
every ND-subtree by one level so that all nodes of the resulting target ND-tree
meets the minimum space utilization requirement (steps 18 - 23).

3.4   Building Buffered ND-Tree

The following function creates a buffered ND-tree for a given set of input vectors.

FUNCTION 1 : **BuildBufferedNDtree**
Input: a set SV of input vectors.
Output: an ND-tree for SV with non-leaf nodes buffered in memory.

Method:
1. create a buffered root node RN with an empty leaf node;
2. let $H = 1$;
3. **while** there are more uninserted vectors in SV **do**
4.   fetch the next vector b from SV;
5. $nodelist$ = InsertVector(b, RN, H);
6.   **if** $|nodelist| > 1$ **then**
7.   create a new buffered root RN with entries for nodes
        in $nodelist$ as children;
8.   let $H = H + 1$;
9.   **end if**;
10. **end while**;
11. **return** RN.

Function BuildBufferedNDtree starts with a non-leaf root node with an empty leaf node (steps 1 - 2). Note that, if a buffered ND-tree returned by Build-BufferedNDtree has only one leaf node N, the target ND-tree output by NDTBL consists of N only (i.e., the buffered non-leaf node is removed). Otherwise, all non-leaf nodes of a buffered ND-tree are used for the target ND-tree. Build-BufferedNDtree inserts one input vector at a time into the buffered ND-tree by invoking function InsertVector (steps 4 - 5). After a new vector is inserted into the tree, the tree may experience a sequence of splits making the original root to be split into several nodes. In such a case, BuildBufferedNDtree creates a new root to accommodate the nodes (steps 6 - 9).

Function InsertVector inserts an input vector into a given buffered ND-tree/ subtree. To improve performance, it inserts input vectors into the relevant auxiliary buffers of the parents of the desired leaf nodes first. Once an auxiliary buffer is full, it then clears the buffer by moving its vectors into the desired leaf nodes in batch. Since multiple buffered vectors can be pushed into a leaf node at once, multiple splits could happen, leading to possibly multiple subtrees returned.

FUNCTION 2 : **InsertVector**
Input: vector b to be inserted into a subtree with root RN and height H.
Output: a list of root nodes of subtrees resulting from inserting vector b.
Method:
1. let resultnodes = { RN };
2. **if** the level of RN is H **do**
3.   insert b into the auxiliary buffer $AuxBuf$ for RN;
4.   **if** $AuxBuf$ is full **then**
5.   sort vectors in $AuxBuf$ according to the order of
        their desired leaf nodes in RN;
6.   **for** each group of vectors in $AuxBuf$ with the same
        desired leaf node number **do**
7.   insert the vectors into the desired leaf node N;
8.   **if** N overflows and there is enough memory **then**
9. $splitnodes$ = Multisplit(N);
10.   replace entry for N in its parent P with entries
           for nodes from $splitnodes$;
11.   **if** P overflows **then**
12. $splitnodes$ = Multisplit(P);
13. $resultnodes = (resultnodes - \{P\}) \cup splitnodes$;
14.   **end if**;

```
15.      end if;
16.    end for;
17.   end if;
18. else
19.   SN = ChooseSubtree(RN, b);
20.   tmplist = InsertVector(b, SN, H);
21.   if |tmplist| > 1 then
22.      replace entry for SN in RN with entries for nodes from tmplist;
23.      if RN overflows then
24.         splitnodes = Multisplit(RN);
25.         resultnodes = (resultnodes − {RN}) ∪ splitnodes;
26.      end if;
27.   end if;
28. end if;
29. return resultnodes.
```

If the given root RN is a parent of leaf nodes (i.e., at level H), InsertVector inserts the input vector into the auxiliary buffer associated with RN (steps 2 - 3). If the buffer is full, InsertVector clears it by moving all its vectors into their desired leaf nodes (steps 4 - 17). To avoid reading the same leaf node multiple times, InsertVector sorts the vectors first and moves all vectors for the same leaf node together (steps 5 - 7). If a leaf node overflows and there is enough memory, InsertVector will split the node (steps 8 - 10). A leaf node split could cause its parent to split (steps 11 - 14). Note that the parent P of a leaf node can be a new node generated from a previous split rather than the original given root RN. If there is no enough memory for splitting, the input vectors are put into their desired leaf nodes without splitting, resulting in oversized leaf nodes. If the given root RN is not at level H (i.e., a parent of leaf nodes), InsertVector chooses the best subtree to accommodate the input vector by utilizing operation ChooseSubtree from the conventional TL algorithm for the ND-tree (steps 19 - 20). The split of the root of the subtree may cause the given root RN to split (steps 21 - 26).

3.5 Multi-way Splitting

When a node N overflows, a conventional TL algorithm for an index tree (such as the one for the ND-tree) would split N into two nodes. As mentioned earlier, when clearing an auxiliary buffer during our bulk-loading, multiple vectors may be put into the same child node N. Splitting N into two after accommodating multiple vectors may not be sufficient since a new node may still overflow. Hence it may be necessary to split N into more than two nodes. One way to handle such a case is to split N into two whenever a new vector from the auxiliary buffer triggers an overflow for N, and the splitting process is immediately propagated to its parent if the parent also overflows. This approach is essentially applying a sequence of conventional two-way splits to N. To avoid propagating the splitting process to the parents multiple times, a more efficient way to handle the case is not splitting N until all relevant vectors from the auxiliary buffer are loaded into N. If N overflows, it is split into a set G of new nodes, where $|G| \geq 2$. This new multi-way splitting requires to extend the conventional TL two-way splitting. The splitting process may be propagated (only once) to the ancestors of N. Note that if a node at level H is split, the vectors in its associated auxiliary buffer are also split accordingly.

One approach employed by the conventional TL algorithm for the ND-tree to splitting an overflow node N with $M + 1$ entries into two nodes is to apply an auxiliary tree technique [16,18]. The key idea is as follows. For each dimension i $(1 \leq i \leq d)$, it sorts the entries in N into an ordered list L_i by building an auxiliary tree for the i-th component sets of the DMBRs for the entries. The auxiliary tree is built in such a way that overlap-free partitions for the i-th component sets can be exploited in L_i. For ordered list L_i, $M - 2m + 2$ candidate partitions are generated by placing the first j $(m \leq j \leq M - m + 1)$ entries in the first portion of a partition and the remaining entries in the second portion of the partition. The ND-tree TL algorithm then chooses a best partition among all candidate partitions from all dimensions according to several criteria, including minimizing overlap, maximizing span, centering split, and minimizing area. Node N is then split into two nodes N_0 and N_1 according to the best partition chosen in the above process. For the bulk-loading case, N may have $\geq M + 1$ entries, and N_0 or N_1 (or both) can still overflow (i.e., having $\geq M + 1$ entries). In this situation, the multi-way splitting applies the above splitting process recursively to the overflow node(s). Note that the overflow splitting process is not propagated to the ancestors until the split of N is completely done.

The details of the multi-way splitting function are described as follows.

FUNCTION 3 : **Multisplit**
Input: an overflow node N
Output: a list of nodes resulting from splitting N
Method:
1. apply the auxiliary tree technique to find a best partition
      to split node N into two nodes N_0 and N_1;
2. let $splitnodes = \{ N_0, N_1 \}$;
3. **if** N_0 overflows **then**
4. $tmplist = \text{Multisplit}(N_0)$;
5. $splitnodes = (splitnodes - \{N_0\}) \cup tmplist$;
6. **end if**;
8. **if** N_1 overflows **then**
9. $tmplist = \text{Multisplit}(N_1)$;
10. $splitnodes = (splitnodes - \{N_1\}) \cup tmplist$;
11. **end if**;
12. **return** $splitnodes$.

3.6   Adjust Irregular Tree

Algorithm NDTBL invokes the following function to cut the subtrees of a given node to a given height. It is needed when making proper adjustments to an intermediate ND-tree that does not meet all the ND-tree requirements.

FUNCTION 4 : **CutTree**
Input: (1) current target ND-tree $TgtTree$; (2) parent node P of subtrees that need to be cut; (3) desired subtree height h.
Output: adjusted target ND-tree.

Method:
1. let ST be the set of subtrees with height $> h$ in P;
2. **for** each T in ST **do**
3.   let $S = \{$ subtrees of height h in T $\}$;
4.   replace T in its parent PN with all subtrees from S;
5.   **if** PN overflows **then**
6. $splitnodes = $ Multisplit(PN);
7.   **if** PN is not the root of $TgtTree$ **then**
8.   replace entry for PN in its parent with entries for
                  nodes from $splitnodes$;
9.   **else**
10.   create a new root of $TgtTree$ with entries for nodes from $splitnodes$;
11.   **end if**
12.   the overflow splitting process can be propagated up
                  to the root of $TgtTree$ when needed;
13.   **end if**;
14. **end for**;
15. **return** $TgtTree$.

Function CutTree cuts each subtree T of a given node P to a given height h, and links the resulting subtrees to the parent PN of T (steps 1 - 4). Note that, since a split may have happened, PN may not be the same as the original parent P. If PN overflows, CutTree invokes Multisplit to split PN. This splitting process can be propagated up to the root of the given tree (steps 5 - 13).

4   Experimental Results

To evaluate the efficiency and effectiveness of NDTBL, we conducted extensive experiments. Typical results from the experiments are reported in this section.

Our experiments were conducted on a PC with Pentium D 3.40GHz CPU, 2GB memory and 400 GB hard disk. Performance evaluation was based on the number of disk I/Os with the disk block size set at 4 kilobytes. The available memory sizes used in the experiments were simulated based on the program configurations rather than physical RAM changes in hardware. The data sets used in the presented experimental results included both real genome sequence data and synthetic data. Genomic data was extracted from bacteria genome sequences of the GenBank, which were broken into q-grams/vectors of 25 characters long (25 dimensions). Synthetic data was randomly generated with 40 dimensions and an alphabet size of 10 on all dimensions. For comparison, we also implemented both the conventional TL (tuple-loading) algorithm for the ND-tree [16,18] and the representative generic bulk-loading algorithm GBLA [6]. All programs were implemented in C++ programming language. The minimum space utilization percentage for a disk block was set to 30%. According to [6], we set the size (disk block count) of the external buffer (pages on disk) of each index node of the buffer-tree in GBLA at half of the node fan-out, which was decided by the available memory size.

Figures 5 and 6 (logarithmic scale in base 10 for Y-axis) show the number of I/Os needed to construct ND-trees for data sets of different sizes using TL, GBLA, and NDTBL for genomic and synthetic data, respectively. The size of

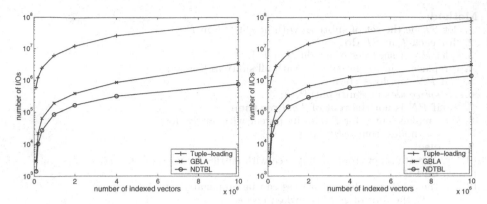

Fig. 5. Bulk-loading performance comparison for genomic data

Fig. 6. Bulk-loading performance comparison for synthetic data

memory available for the algorithms was fixed at 4 megabytes. From the figures, we can see that the bulk-loading algorithms significantly outperformed the conventional TL algorithm. For example, NDTBL was about 89 times faster than TL when loading 10 million genomic vectors in our experiments. Between the two bulk loading algorithms, GBLA was consistently slower than NDTBL, which showed that the strategies adopted by the latter, including avoiding the level-by-level construction process, applying the multi-way splitting and narrowing down search scopes, were effective. In fact, the performance improvement was increasingly larger as the database size increased. For example, NDTBL was about 4.5 times faster than GBLA when bulk-loading 10 million genomic vectors.

Since both NDTBL and GBLA employ a special in-memory intermediate (sub)tree structure for bulk-loading, the available memory size has a significant impact on their performance. Experiments were also conducted to study the effect of different memory sizes on the performance of NDTBL and GBLA. Table 1 shows the number of I/Os needed by these two algorithms to construct

Table 1. Effect of memory size on bulk-loading performance

Memory	4MB	8MB	16MB	32MB	64MB	128MB	256MB
GBLA/genomic	869793	815999	705982	612496	274381	100526	39027
NDTBL/genomic	319905	298272	270024	235886	182347	76698	35502
GBLA/synthetic	1265294	1187039	1026998	891003	399144	156235	72653
NDTBL/synthetic	585019	545591	493984	441445	339919	147864	68552

the ND-trees for a genomic data set (4 million vectors) and a synthetic data set (4 million vectors) under different sizes of available memory. From the experimental results, we can see that NDTBL was consistently more efficient than GBLA. When the memory size was small comparing to the database size, the performance of NDTBL was significantly better than that of GBLA. On the

Table 2. Query performance comparison for genomic data

key#	$r_q = 1$			$r_q = 2$			$r_q = 3$		
	io TL	io GBLA	io NDTBL	io TL	io GBLA	io NDTBL	io TL	io GBLA	io NDTBL
100000	16.0	15.9	15.9	63.8	63.7	63.6	184.3	184.0	183.6
200000	18.2	18.0	18.0	79.9	79.3	79.0	249.8	248.0	247.2
400000	20.1	19.8	19.6	96.5	95.4	94.4	327.6	323.6	320.4
1000000	22.7	22.6	22.3	121.8	121.1	119.8	451.4	449.0	444.0
2000000	26.6	26.3	26.2	145.2	143.2	142.9	572.8	565.0	563.8
4000000	29.8	29.7	29.4	172.0	169.3	167.9	724.9	715.3	709.2
10000000	33.5	33.3	33.0	210.1	209.5	207.7	958.2	955.1	946.4

Table 3. Query performance comparison for synthetic data

key#	$r_q = 1$			$r_q = 2$			$r_q = 3$		
	io TL	io GBLA	io NDTBL	io TL	io GBLA	io NDTBL	io TL	io GBLA	io NDTBL
100000	19.6	19.5	19.2	78.0	77.3	76.0	228.3	228.3	225.1
200000	22.4	23.4	22.8	97.2	98.3	98.1	305.9	305.9	299.2
400000	25.3	24.9	24.1	119.1	119.0	116.8	403.3	403.3	391.6
1000000	29.6	30.3	30.2	154.6	157.3	155.2	578.3	578.3	573.2
2000000	32.7	31.9	31.6	183.7	191.1	188.7	734.9	734.9	729.6
4000000	37.5	35.8	35.6	217.9	192.7	190.5	921.2	921.5	910.9
10000000	42.6	41.2	40.0	264.1	258.4	236.0	1188.2	1162.7	1137.4

other hand, when the memory was very large so that almost the entire ND-tree could be fit in it, the performance of two algorithms became close. However, since GBLA still needs to construct the target ND-tree level by level in such a case, its performance is still worse than that of the NDTBL. In real applications such as genome sequence searching, since the available memory size is usually small comparing to the huge database size, NDTBL has a significant performance benefit. In other words, for a fixed memory size, the larger the database size is, the more performance benefit the NDTBL can provide.

To evaluate the effectiveness of NDTBL, we compared the quality of the ND-trees constructed by all algorithms. The quality of an ND-tree was measured by its query performance and space utilization. Tables 2 and 3 show query performance of the ND-trees constructed by TL, GBLA, and NDTBL for genomic and synthetic data, respectively. These trees are the same as those presented in Figures 5 and 6. Query performance was measured based on the average number of I/Os for executing 100 random range queries at Hamming distances 1, 2 and 3. The results show that the ND-trees constructed by NDTBL have comparable performance as those constructed by TL and GBLA.

Table 4 shows the space utilization of the same set of ND-trees for genomic and synthetic data. From the table, we can see that the space utilization of NDTBL was also similar to that of TL and GBLA.

Table 4. Space utilization comparison

key#	genomic data			synthetic data		
	ut% TL	ut% GBLA	ut% NDTBL	ut% TL	ut% GBLA	ut% NDTBL
100000	69.7	69.8	70.0	62.5	62.5	63.6
200000	69.0	69.5	70.0	62.1	62.1	64.5
400000	69.0	69.9	70.3	62.3	62.3	64.6
1000000	72.4	72.8	73.9	65.7	65.7	67.6
2000000	71.9	72.2	73.3	65.7	66.0	68.0
4000000	70.9	71.6	72.4	65.7	67.8	68.2
10000000	68.7	68.9	69.7	81.1	81.5	82.6

Besides the experiments reported above, we have also conducted experiments with data sets of various alphabet sizes and dimensionalities. The results were similar. Due to the space limitation, they are not included in this paper.

5   Conclusions

There is an increasing demand for applications such as genome sequence searching that involve similarity queries on large data sets in NDDSs. Index structures such as the ND-tree [16,18] are crucial to achieving efficient evaluation of similarity queries in NDDSs. Although many bulk-loading techniques have been proposed to construct index trees in CDSs in the literature, no bulk-loading technique has been developed specifically for NDDSs. In this paper, we present a new algorithm NDTBL to bulk-load the ND-tree for large data sets in NDDSs.

The algorithm employs a special intermediate tree structure with related buffering strategies (e.g., keeping non-leaf nodes in memory, providing auxiliary buffers for parents of leaves, FCFS caching for leaves, and sorting buffered vectors before clearing, etc.) to build a target ND-tree by parts using available memory space. It applies a unique adjustment processing to ensure the properties of the ND-tree when needed. It also adopts a multi-way splitting method to split an overflow node into multiple nodes rather than always two nodes to improve efficiency. The auxiliary tree technique from the original ND-tree TL algorithm [16,18] is utilized to sort discrete rectangles/vectors based on the properties of an NDDS during the multi-way splitting.

Our experimental results demonstrate that the proposed algorithm to bulk-load an ND-tree significantly outperforms the conventional TL algorithm and the generic bulk-loading algorithm GBLA in [6], especially when being used for large data sets with limited available memory. The target ND-trees obtained from all the algorithms have comparable searching performance and space utilization.

Our future work includes studying bulk-loading techniques exploiting more characteristics of an NDDS and developing bulk-load methods for the space-partitioning-based index tree, the NSP-tree [17], in NDDSs.

References

1. Arge, L., Hinrichs, K., Vahrenhold, J., Viter, J.S.: Efficient Bulk Operations on Dynamic R-trees. In: Goodrich, M.T., McGeoch, C.C. (eds.) ALENEX 1999. LNCS, vol. 1619, pp. 328–348. Springer, Heidelberg (1999)
2. Arge, L., Berg, M., Haverkort, H., Yi, K.: The Priority R-tree: a practically efficient and worst-case optimal R-tree. In: Proc. of SIGMOD, pp. 347–358 (2004)
3. Beckman, N., Kriegel, H., Schneider, R., Seeger, B.: The R*-tree: an efficient and robust access method for points and rectangles. In: Proc. of SIGMOD, pp. 322–331 (1990)
4. Berchtold, S., Keim, D.A., Kriegel, H.-P.: The X-tree: an index structure for high-dimensional data. In: Proc. of VLDB 1996, pp. 28–39 (1996)
5. Berchtold, S., Bohm, C., Kriegel, H.-P.: Improving the Query Performance of High-Dimensional Index Structures by Bulk-Load Operations. In: Schek, H.-J., Saltor, F., Ramos, I., Alonso, G. (eds.) EDBT 1998. LNCS, vol. 1377, pp. 216–230. Springer, Heidelberg (1998)
6. Bercken, J., Seeger, B., Widmayer, P.: A Generic Approach to Bulk Loading Multidimensional Index Structures. In: Proc. of VLDB, pp. 406–415 (1997)
7. Bercken, J., Seeger, B.,, B.: An Evaluation of Generic Bulk Loading Techniques. In: Proc. of VLDB, pp. 461–470 (2001)
8. Ciaccia, P., Patella, M.: Bulk loading the M-tree. In: Proc. of the 9th Australian Database Conference, pp. 15–26 (1998)
9. De Witt, D., Kabra, N., Luo, J., Patel, J., Yu, J.: Client-Server Paradise. In: Proc. of VLDB, pp. 558–569 (1994)
10. Garcia, Y., Lopez, M., Leutenegger, S.: A greedy algorithm for bulk loading R-trees. In: Proc. of ACM-GIS, pp. 02–07 (1998)
11. http://www.ncbi.nlm.nih.gov/Genbank/
12. Guttman, A.: R-trees: a dynamic index structure for spatial searching. In: Proc. of SIGMOD, pp. 47–57 (1984)
13. Jermaine, C., Datta, A., Omiecinski, E.: A novel index supporting high volumne data warehouse insertion. In: Proc. of VLDB, pp. 235–246 (1999)
14. Kamel, I., Faloutsos, C.: On packing R-trees. In: Proc. of CIKM, pp. 490–499 (1993)
15. Leutenegger, S., Edgington, J., Lopez, M.: STR: A Simple and Efficient Algorithm for R-Tree Packing. In: Proc. of ICDE, pp. 497–506 (1997)
16. Qian, G., Zhu, Q., Xue, Q., Pramanik, S.: The ND-Tree: a dynamic indexing technique for multidimensional non-ordered discrete data spaces. In: Aberer, K., Koubarakis, M., Kalogeraki, V. (eds.) VLDB 2003. LNCS, vol. 2944, pp. 620–631. Springer, Heidelberg (2004)
17. Qian, G., Zhu, Q., Xue, Q., Pramanik, S.: A Space-Partitioning-Based Indexing Method for Multidimensional Non-ordered Discrete Data Spaces. ACM TOIS 23, 79–110 (2006)
18. Qian, G., Zhu, Q., Xue, Q., Pramanik, S.: Dynamic Indexing for Multidimensional Non-ordered Discrete Data Spaces Using a Data-Partitioning Approach. ACM TODS 31, 439–484 (2006)
19. Robinson, J.T.: The K-D-B-tree: a search structure for large multidimensional dynamic indexes. In: Proc. of SIGMOD, pp. 10–18 (1981)
20. Roussopoulos, N., Leifker, D.: Direct spatial search on pictorial databases using packed R-trees. In: Proc. of SIGMOD, pp. 17–31 (1985)

An Incremental Maintenance Scheme of Data Cubes

Dong Jin, Tatsuo Tsuji, Takayuki Tsuchida, and Ken Higuchi

Graduate School of Engineering, University of Fukui
3-9-1 Bunkyo, Fukui-shi, 910-8507, Japan
{jindong,tsuji,tsuchida,higuchi}@pear.fuis.fukui-u.ac.jp

Abstract. Data cube construction is a commonly used operation in data warehouses. Because of the volume of data stored and analyzed in a data warehouse and the amount of computation involved in data cube construction, *incremental maintenance of data cube* is really effective. To maintain a data cube incrementally, previous methods were mainly for relational databases. In this paper, we employ an *extendible multidimensional array* model to maintain data cubes. Such an array enables incremental cube maintenance without relocating any data dumped at an earlier time, while computing the data cube efficiently by utilizing the fast random accessing capability of arrays. Our data cube scheme and related maintenance methods are presented in this paper, and cost analysis on our approach is shown and compared with existing methods.

1 Introduction

Analysis on large datasets is increasingly guiding business decisions. Retail chains, insurance companies, and telecommunication companies are some of the examples of organizations that have created very large datasets for their decision support systems. A system storing and managing such datasets is typically referred to as a data warehouse and the analysis performed is referred to as On Line Analytical Processing (OLAP). At the heart of all OLAP applications is the ability to simultaneously aggregate across many sets of dimensions. Jim Gray has proposed the cube operator for data cube[7]. Data cube provides users with aggregated results that are *group-bys* for all possible combinations of dimension attributes. When the number of dimension attributes is n, the data cube computes 2^n group-bys, each of which is called a *cuboid*.

As the computation of a data cube typically incurs a considerable query processing cost, it is usually precomputed and stored as materialized views in data warehouses. A data cube needs updating when the corresponding source relation changes. We can reflect changes in the source relation to the data cube by either recomputation or incremental maintenance. Here, the incremental maintenance of a data cube means the propagation of only its changes. When the amount of changes during the specified time period are much smaller than the size of the source relation, computing only the changes of the source relation and reflecting to the original data cube is usually much cheaper than recomputing from scratch. Thus, several methods that allow the incremental maintenance of a data cube have been proposed in the past. The most recent one as we are aware of is [9]. But these methods are all for relational model, i.e. no papers for MOLAP (Multidimensional OLAP) until now as far as we know.

J.R. Haritsa, R. Kotagiri, and V. Pudi (Eds.): DASFAA 2008, LNCS 4947, pp. 172–187, 2008.

In MOLAP systems, a snapshot of a relational table in a front-end OLTP database is taken and dumped into a fixed size multidimensional array periodically like in every week or month. At every dumping, a new empty fixed size array has to be prepared and the relational table will be dumped again from scratch. If the array dumped previously is intended to be used, all the elements in it should be relocated by using the corresponding address function of the new empty array, which also incurs huge cost.

In this paper, we use the extendible multidimensional array model proposed in [8] as a basis for incremental data cube maintenance in MOLAP. The array size can be extended dynamically in any directions during execution time [1][2][4]. While a *dynamic array* is newly allocated when required at the execution time, all the existing elements of an extendible array are used as they are without any relocation; only the extended part is dynamically allocated. For each record inserted after the latest dumping, its column values are inspected and the fact data are stored in the corresponding extendible array element. If a new column value is found, the corresponding dimension of the extendible array is extended by one, and the column value is mapped to the new subscript of the dimension. Thus incremental dumping is sufficient instead of wholly dumping a relational table.

To maintain a data cube incrementally, existing methods compute a *delta cube*, which represents the data cube consisting of the change of the original data cube. The incremental maintenance of a data cube is divided into two stages: *propagate* and *refresh* [6]. The propagate stage computes the change of a data cube from the changes of the source relation, i.e., constructing delta cube. Then, the refresh stage refreshes the original data cube by applying the computed change (delta cube) to it. In this paper, we address a number of data structure and algorithm issues for efficient incremental data cube maintenance using the extendible multidimensional array. We use a single extendible array to store a full data cube. We call the scheme as *single-array data cube scheme*. The main contributions of this paper can be summarized as follows.

(a) To avoid huge overhead in refresh stage, we propose shared dimension method for incremental data cube maintenance using the single-array data cube scheme.

(b) We propose to materialize only *base cuboid* of delta cube, in propagate stage. Thus the cost of the propagate stage is significantly reduced in our method compared with the previous methods.

(c) By partitioning the data cube based on the single-array data cube scheme, we present a subarray-based algorithm refreshing the original data cube by scanning the base cuboid of the delta cube only once with limited working memory usage.

2   Employing Extendible Array

The extendible multidimensional array used in this paper is proposed in [8]. It is based upon the index array model presented in [4]. An n dimensional extendible array A has a history counter h and three kinds of auxiliary table for each extendible dimension $i(i=1,...,n)$. See Fig. 1. These tables are *history table H_i*, *address table L_i*, and *coefficient table C_i*. The history tables memorize extension history. If the size of A is $[s_1, s_2,..., s_n]$ and the extended dimension is i, for an extension of A along dimension i, contiguous memory area that forms an $n-1$ dimensional subarray S of size $[s_1, s_2,...,s_{i-1}, s_{i+1},..., s_{n-1}, s_n]$ is dynamically allocated. Then the current history counter value is incremented by one, and it is memorized on H_i, also the first address of S is held on L_i. Since h increases monotonously, H_i is an ordered set of history values. Note that an extended subarray is one to one corresponding with its history value, so the subarray is uniquely identified by its history value.

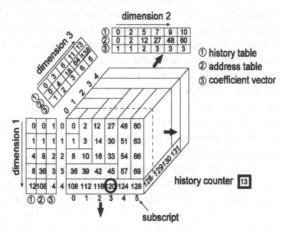

Fig. 1. A three dimensional extendible array

As is well known, element $(i_1, i_2, ..., i_{n-1})$ in an $n-1$ dimensional fixed size array of size $[s_1, s_2, ..., s_{n-1}]$ is allocated on memory using addressing function like:

$$f(i_1,..., i_{n-1})= s_2 s_3 ...s_{n-1}i_1+s_3 s_4 ...s_{n-1}i_2+ ...+s_{n-1}i_{n-2}+i_{n-1} \qquad (1)$$

We call $<s_2 s_3...s_n, s_3 s_4...s_n, ..., s_n>$ as a *coefficient vector*. Such a coefficient vector is computed at array extension and is held in a coefficient table. Using these three kinds of auxiliary tables, the address of array element $(i_1, i_2, ..., i_n)$ can be computed as follows.

(a) Compare $H_1 [i_1]$, $H_2 [i_2]$, ... , $H_{n-1} [i_n]$. If the largest value is $H_k [i_k]$, the subarray corresponding to the history value $H_k [i_k]$, which was extended along dimension k, is known to include the element.

(b) Using the coefficient vector memorized at $C_k[i_k]$, the offset of the element $(i_1, ..., i_{k-1}, i_{k+1}, ..., i_n)$ in the subarray is computed according to its addressing function in (1).

(c) $L_k [i_k]+$*the offset in* (b) is the address of the element.

Consider the element (4,3,0) in Fig.1. Compare the history values $H_1[4]=12$, $H_2[3]=7$ and $H_3[0]=0$. Since $H_1[4]>H_2[3]>H_3[0]$, it can be proved that (4,3,0) is involved in the subarray S corresponding to the history value $H_1[4]$ in the first dimension and the first address of S is found out in $L_1[4]$ =108. From the corresponding coefficient vector $C_1[4]$ =<4>, the offset of (4,3,0) from the first address of S is computed as 4×3+0=12 according to expression (1), the address of the element is determined as 108+12=120. Note that we can use such a simple computational scheme to access an extendible array element only at the cost of small auxiliary tables. The superiority of this scheme is shown in [4] compared with other schemes such as hashing [2].

3 Our Approach

In our approach, we use a single multidimensional array to store a full data cube[7]. Each dimension of the data cube corresponds to a dimension of the array with the same dimensionality as the data cube. Each dimension value of a cell of the data cube is uniquely mapped to a subscript value of the array. Note that special value *All* in each dimension of the data cube is always mapped to the first subscript value 0 in each dimension of the array. For concreteness, consider a 2-dimensional data cube, in which we have the dimensions *product* (*p*), *store* (*s*) and the "measure" (fact data value) *sales* (*m*). To get the cube we will compute *sales* grouped by all subsets of these two dimensions. That is, we will have *sales* by *product* and *store*; *sales* by *product*; *sales* by *store*; and overall *sales*. We can denote these group-bys as cuboid *ps*, *p*, *s*, and Φ, where Φ denotes the empty group-by. We call cuboid *ps* as *base cuboid* because other cuboids can be aggregated from it. Let Fig. 2(a) be the fact table of the data cube. Fig. 2(b) shows the realization of the 2-dimensional data cube using a single 2-dimensional array. Note that the *dimension value tables* are necessary to map the dimension values of the data cube to the corresponding array subscript values.

Obviously, we can retrieve any cuboid as needed by simply specifying corresponding array subscript values. For the above 2-dimensional data cube, see Fig. 2(b): Cuboid *ps* can be got by retrieving array element set $\{(x_p,x_s)|\ x_p{\neq}0,x_s{\neq}0\}$; Cuboid *p* by $\{(x_p,0)|\ x_p{\neq}0\}$; Cuboid *s* by $\{(0,x_s)|\ x_s{\neq}0\}$; Cuboid Φ by $(0,0)$. x_p denotes subscript value of dimension *p*, x_s denotes subscript value of dimension *s*.

The data cube cells is one-to-one correspondence to the array elements. So we may also call a data cube cell as an element of the data cube in the following. For example in Fig. 2(b), cube cell <Yplaza, Pen> can be also referred to as cube element (1, 1).

Now we implement the data cube with the extendible multidimensional array model presented in Section 2. Consider the example in Fig. 2(b). First, the array is empty, and cell <All, All> which represents overall *sales* with initial value 0 is added into the array. Then the fact data are loaded into the array one after another to build the base cuboid *ps* into the array; this causes extensions of the array. Then the cells in the cuboids other than base cuboid are computed from the base cuboid *ps* and added into the array. For example, we can compute the value of <Yplaza, All> as the sum of the values of <Yplaza, Pen> and <Yplaza, Glue> in the base cuboid. Refer to the result in Fig. 3. For simplicity of the figure, the address tables and the coefficient tables of the extendible array explained in Section 2 are omitted. We call such a data cube scheme as *single-array data cube scheme*.

We call the cells in the cuboids other than the base cuboid as *dependent cells* [17] because such cells can be computed from the cells of the base cuboid. For the same reason, we call the cuboids other than the base cuboid as *dependent cuboids*. Obviously, any dependent cell has at least one dimension value "*All*". Therefore in our single-array data cube scheme any array element having at least one zero subscript value is a dependent cell. Note that a subarray generally consists of base cuboid cell(s) and dependent cell(s). For example in Fig. 3, the subarray with history value 4 consists of two base cuboid cells <Yplaza, Glue> and <Genky, Glue>, and one dependent cell <All, Glue>.

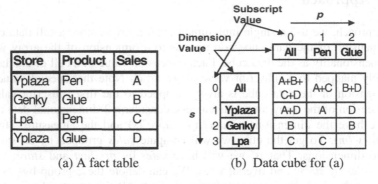

(a) A fact table   (b) Data cube for (a)

Fig. 2. A data cube using a single array

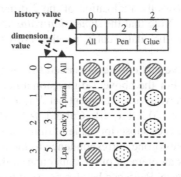

Fig. 3. Single-array data cube scheme

In the following we will use the single-array data cube scheme to maintain a data cube incrementally. The aggregate functions used in the data cube maintenance need to be *distributive* [7]. For simplicity, we only focus on SUM function in this paper. Furthermore, we assume that the change of the corresponding source relation involves only insertion. However, our approach can be easily extended to handle deletions and updates using the techniques provided in [6].

3.1 Shared Dimension Method

As we described in Section 1, the incremental maintenance of a data cube consists of the propagate stage and the refresh stage. The propagate stage computes the change of a data cube from the change of the source relation. Then, the refresh stage refreshes the data cube by applying the computed change to it. Let ΔF denote a set of newly inserted tuples into a fact table F. The propagate stage computes ΔQ which denotes the change of a data cube Q from ΔF. Take the 2-dimensional data cube Q in the above as an example, ΔQ can be computed using the following query:

```
ΔQ : SELECT p, s, SUM(m)
     FROM ΔF
     CUBE BY p, s
```

We call ΔQ as a *delta cube*. A delta cube represents the change of a data cube. The definition of ΔQ is almost the same as Q except that it is defined over ΔF instead of F. In this example, ΔQ computes four cuboids as Q. We call a cuboid in a delta cube as a delta cuboid, and denote delta cuboids in ΔQ as Δps, Δp, Δs and $\Delta \Phi$ which represent the change of cuboid ps, p, s and Φ in the original data cube Q respectively.

We can implement original data cube Q and delta data cube ΔQ as distinct extendible arrays. As ΔF is usually much smaller than F, the dimension sizes of the extendible array for ΔQ are supposed to be smaller than that for the original data cube Q. For example, the original data cube Q has six distinct values in a dimension, while the delta cube ΔQ has four distinct values in the dimension. See Fig. 4. They all have fewer distinct values than the dimension of the updated data cube Q' which has seven distinct values (a new dimension value 'F' is appended from ΔQ). In such a method, we can keep the array for ΔQ size as small as possible, but we need keep another dimension value table for ΔQ. Thus the same dimension value may have different subscript values between the arrays. Assume the first subscript value is 0. The dimension value 'H' in ΔQ has different subscript value with the one in Q and Q': 3 in ΔQ, 4 in Q and Q'. Thus in the refresh stage, each dimension value table should be checked to get the corresponding array elements updated. It will lead to huge overhead for large datasets.

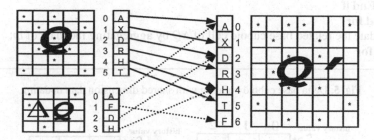

Fig. 4. Non-shared dimension method

To avoid such huge overhead, our approach uses the same dimension table for original data cube Q and delta cube ΔQ. For the example in Fig. 4, it means only the dimension value table for Q' will be used. So in the refresh stage, the dimension value tables need not to be checked because the corresponding array elements have the same subscript values in every array. We call such a method as *shared dimension method*.

To apply the shared dimension method into the extendible array model, the original and delta data cubes physically share one set of *dimension value tables, history tables,* and *coefficient tables,* while each data cube has independent set of subarrays to store fact data. The shared dimension data (*dimension value tables, history tables,* and *coefficient tables*) are updated together with building the delta base cuboid caused by source relation change ΔF. The algorithm of building delta base cuboid with shared dimension data updating is shown in Fig. 5.

For example, let the original fact table F consist of the first two tuples in Fig. 2(a), and the rest two tuples be ΔF, the change of the source relation. Fig. 6(a) shows the original data cube Q, Fig. 6(b) shows the resulted delta base cuboid in delta cube ΔQ with the updated shared dimension data. Note that the dimension data in Fig. 6(a) and 6(b) are physically in the same storage and shared between Q and ΔQ. In another words, updating of the shared dimension data in the delta base cuboid (Fig. 6(b)) is reflected in the shared dimension data in the original data cube (Fig. 6(a)).

Inputs:
   The changes of source relation (fact table) ΔF;
   Shared dimension data (*dimension value tables, history tables,*
                                   and *coefficient tables*);
Outputs:
   The delta data cube ΔQ with only delta base cuboid materialized;
   Updated shared dimension data;
Method:
  Initialize delta data cube ΔQ with the current shared dimension data;
  For each tuple tp in ΔF **do**
    For each dimension value v in tp **do**
      If v is not found in the corresponding shared dimension value table **then**
        update shared dimension data accordingly;
        allocate new subarray by extending along the dimension by one;
      End if
    End for
    update the related base cuboid cell of ΔQ by aggregating fact data of tp;
  End for

Fig. 5. Delta base cuboid building with shared dimension data updating

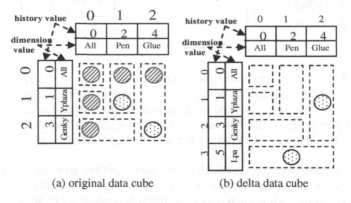

(a) original data cube   (b) delta data cube

Fig. 6. Original data cube and delta data cube with shared dimension data

3.2 Subarray-Based Method

Our approach materializes only one delta cuboid – base cuboid in the propagate stage by executing the algorithm in Fig. 5, so the cost of the propagate stage is significantly reduced. While in the refresh stage, the delta base cuboid is scanned by subarray to refresh the corresponding subarray in the original data cube in one pass. So we name such a method as *subarray-based method*.

Partitioning of Data Cube. In the single-array data cube scheme, a subarray is allocated for each distinct value of a dimension. Since a subarray is one-to-one corresponding with its history value as we noted in Section 2, a distinct dimension value is also one to one corresponding with the history value of its subarray. So, we can call the history value corresponding to a distinct dimension value v as the history value of v. Let $e=(v_1, v_2, v_3)$ be any base cuboid element in a 3-dimensional data cube, so $v_1, v_2, v_3 \neq 0$. We denote h_i as the history value of v_i ($i=1, 2, 3$). Without loss of generality, we assume the history value h_i satisfies $0<h_1<h_2<h_3$. Note that history value 0 in the following often denotes as h_0 for clarity. According to the semantics of the CUBE operator, there are $2^3-1=7$ dependent cells of e in the data cube. Table 1 shows the list of the base cuboid element e and its 7 dependent elements implemented by our data cube scheme.

Table 1. A base cuboid element and its dependent elements in a 3-dimensional data cube

Related Elements	Subarray history value	can be aggregated with
(v_1, v_2, v_3)	h_3	-
$(0, v_2, v_3)$		(v_1, v_2, v_3)
$(v_1, 0, v_3)$		(v_1, v_2, v_3)
$(0, 0, v_3)$		(v_1, v_2, v_3)
$(v_1, v_2, 0)$	h_2	(v_1, v_2, v_3)
$(0, v_2, 0)$		$(v_1, v_2, 0)$
$(v_1, 0, 0)$	h_1	$(v_1, v_2, 0)$
$(0, 0, 0)$	h_0	$(v_1, 0, 0)$

In Table 1, eight elements are partitioned into four groups according to the subarrays to which they belong. Each group is one-to-one corresponding with the history value of its corresponding subarray. So we call the group corresponding to history value h as *group h*. So (v_1, v_2, v_3), $(0, v_2, v_3)$, $(v_1, 0, v_3)$, and $(0, 0, v_3)$ is in group h_3, $(v_1, v_2, 0)$ and $(0, v_2, 0)$ is in group h_2, $(v_1, 0, 0)$ is in group h_1, and $(0, 0, 0)$ is in group h_0. In each group there is one and only one element that the rest elements can be aggregated with it. Such element will be called the *base element* of the group. Obviously (v_1, v_2, v_3), $(v_1, v_2, 0)$, $(v_1, 0, 0)$, and $(0, 0, 0)$ are the base elements of the group h_3, h_2, h_1, and h_0 respectively. They are in bold in Table 1. Furthermore, the base element of group h_{i-1} can be aggregated with the base element of group h_i along the extended dimension i ($i=1, 2, 3$). That is, $(v_1, v_2, 0)$ is the base element of group h_2 and can be aggregated with (v_1, v_2, v_3), the base element of group h_3 along the extended dimension 3; $(v_1, 0, 0)$ is the base element of group h_1 and can be aggregated with $(v_1, v_2, 0)$; $(0, 0, 0)$ is the base element of group h_0 and can be aggregated with $(v_1, 0, 0)$.

From the above 3-dimensional data cube we can generalize our data cube scheme for an n-dimensional data cube as follows. For any base cuboid element e in an n-dimensional data cube, there are 2^n-1 dependent cells of e. We can get these cells by substituting 0 for the n coordinate terms of e. The obtained 2^n elements (including e itself) can be partitioned into $n + 1$ groups according to their history values of the subarrays to which they belong. So *group* h_i is the group of cells corresponding to history value h_i $(i = 0, ..., n)$, where $0=h_0<h_1<h_2<...<h_n$. We can arrange the dimension values of e as $(v_1, v_2, ..., v_n)$, where the corresponding history value of v_i is $h_i(i = 1, ..., n)$. Then, all the elements in group h_i can be presented as n dimensional coordinate (*, ..., *, v_i, 0, ..., 0) where * represents either $v_j (j = 1, ..., i$-1) or 0. Element $(v_1, v_2, ..., v_i, 0, ..., 0)$ is the *base element of* group h_i, and the rest elements in the group can be aggregated with it. There are total 2^{i-1} elements in group h_i $(i > 0)$. In group h_0 there is always only one element, $(0, 0, ..., 0)$. Furthermore, the base element of group h_{i-1} can be aggregated with the base element of group h_i along the extended dimension i $(i = 1, ..., n)$. We will show later that we need to keep the intermediate result for the base element of group h_{i-1} by aggregating with the base element of group h_i along the extended dimension i.

Refreshing Scheme. As our subarray-based method only materializes the delta base cuboid in the propagate stage, for each element in the delta base cuboid we need updating the corresponding 2^n elements of the original cube in the refresh stage. We can separate the updating of the 2^n elements into $n+1$ groups. The elements in group h_n are refreshed together with the base cuboid element as they are in the same subarray. For the elements in the other n groups, we keep the intermediate results for the base elements of the groups until we refresh the corresponding subarrays whose history values are $h_i (i = 0, ..., n$-1). As we mentioned, the intermediate result for the base element of group h_{i-1} can be aggregated with the base element of group h_i along the extended dimension of i $(i = 1, ..., n)$. See the example in Table 1; for any element $e=(v_1, v_2, v_3)$ in the subarray of the delta data cube, its $2^3=8$ corresponding elements are updated in the subarrays corresponding to h_3, h_2, h_1, and h_0 of the original data cube. For the refreshment of the subarray corresponding to h_3, update (v_1, v_2, v_3), $(0, v_2, v_3)$, $(v_1, 0, v_3)$, and $(0, 0, v_3)$ and keep the intermediate result for $(v_1, v_2, 0)$ by aggregating (v_1, v_2, v_3) along dimension 3; for the refreshment of the subarray corresponding to h_2, update $(v_1, v_2, 0)$ and $(0, v_2, 0)$ and keep the intermediate result for $(v_1, 0, 0)$ by aggregating $(v_1, v_2, 0)$ along dimension 2; for the refreshment of the subarray corresponding to h_1, update $(v_1, 0, 0)$ and keep the intermediate result for $(0, 0, 0)$ by aggregating $(v_1, 0, 0)$ along dimension 1; finally for the refreshment of the subarray corresponding to h_0 only update $(0, 0, 0)$ without keeping further intermediate result.

To describe generally, for all the delta base cuboid elements of a subarray ΔS in the delta cube and all the intermediate result T for ΔS, our refreshing scheme based on the subarray-based method performs two things to refresh the corresponding subarray S in the original data cube; one is to update S with T and all the base cuboid elements in

ΔS; the other is to keep the further intermediate results by aggregating with T and all the base cuboid elements in ΔS along the extended dimension of ΔS. See the detail in our subarray-based refreshing algorithm in Fig. 7 and the result with our example in Table 2.

Inputs:
   The delta data cube ΔQ generated from the algorithm in Fig. 5;
   The original data cube Q;
Output:
   The updated data cube Q;
Method:
   For each history value h of a subarray in ΔQ from the max history value to 1 **do**
      let d = the extended dimension on history value h;
      If the intermediate result array Td does not exist **then**
         create n-1 dimensional array Td in memory;
      End if
      update Td by aggregating with $S[h]$ in ΔQ along dimension d;
      Let $Td'[h]$ = all the elements in Td' whose corresponding history value is h;
      update Td by aggregating with $Td'[h]$ along dimension d;
      refresh $S[h]$ in Q by aggregating with $Td'[h]$ and $S[h]$ in ΔQ;
   End for
   Refresh $S[0]$ in Q by aggregating with intermediate result array for $S[0]$

Fig. 7. Refreshing algorithm for subarray-based method

In the algorithm shown in Fig. 7, $S[h]$ denotes the subarray whose corresponding history value is h. As the intermediate result data are always aggregated along the subarray extended dimension, we can hold the intermediate result data for dependent cells in n arrays of n-1 dimensionality. We denote such data as Td which holds the intermediate result by aggregating with the subarrays extended on dimension d. Td' denotes all the intermediate result arrays except Td. Note that the intermediate results for the dependent cells in a subarray with history value h always come from the subarrays whose corresponding history values are larger than h. Thus, to get all the intermediate results, we must start refreshing from the subarray with the maximum history value.

Assume the example in Fig. 2(a) as ΔF. For simplicity, we assume the original data cube Q is empty. See the running result in Table 2. The intermediate result array Ts is generated on history value 5 and Tp on history value 4. Ts consists of the intermediate result for <All, Pen> and <All, Glue>; Tp consists of the intermediate result for <Yplaza, All>, <Genky, All>, and <All, All>. As there are only two intermediate result arrays Ts and Tp in this example, Ts is equivalent to Tp' and Tp equivalent to Ts'.

In order to avoid frequent accesses to the disks, the intermediate result arrays must be kept in main memory. If array extension is in round-robin manner for all dimensions just like in Fig. 3, it can be known that the total memory requirement for the intermediate result arrays is

$$M = \sum_{i=1}^{n} (\prod_{j=1, j\neq i}^{n} C_i) ,$$

where C_i is the cardinality of the i-th dimension in base cuboid ($1 \leq i \leq n$). Obviously M is much smaller than the size of the base cuboid if the dimension cardinalities are large enough.

It can be further proved that the total storage requirement in any array extension manner is bounded by M. In practical situation, the array extension may be not in round-robin manner and generally controllable. In the data cube maintenance for the real-world dataset, it is common that the valid elements of the delta cube are not uniformly distributed in the base cuboid. So the actual memory requirement can be much smaller than M. Furthermore we can refine our subarray-based algorithm to deallocate the memory for those intermediate results which are not needed in later computation. All these are worth further study with experiments on real-world datasets.

Table 2. Result of the refresh algorithm against ΔF in Fig. 2(a)

History value	Extended dimension	Aggregated elements in ΔQ & intermediate result	Updated intermediate result	Subarray elements refreshed
5	s	\<Lpa,Pen,C\>	\<All,Pen,C\> in Ts	\<Lpa,Pen,C\>,
			\<All,Glue,0\> in Ts	\<Lpa,All,C\>
4	p	\<Yplaza,Glue,D\>	\<Yplaza,All,D\> in Tp	\<Yplaza,Glue,D\>,
		\<Genky,Glue,B\>	\<Genky,All,B\> in Tp	\<Genky,Glue,B\>,
		\<All,Glue,0\> in Ts	\<All,All,0\> in Tp	\<All,Glue,B+D\>
3	s	\<Genky,All,B\> in Tp	\<All,All,B\>	\<Genky,All,B\>
2	p	\<Yplaza,Pen,A\>	\<Yplaza,All,A+D\>	\<Yplaza,Pen,A\>,
		\<All,Pen,C\> in Ts	\<All,All,B+C\>	\<All,Pen,A+C\>
1	s	\<Yplaza,All,A+D\> in Tp	\<All,All,A+B+C+D\>	\<Yplaza,All,A+D\>
0		\<All,All,A+B+C+D\> in Tp		\<All,All,A+B+C+D\>

4 Evaluation and Comparison

In this section, cost model for incremental data cube maintenance is developed. Based on the cost model, we make comparison among three methods: naïve method NV using all of 2^n delta cuboids, our subarray-based method SB and relational advanced method RA using $_nC_{n/2}$ delta cuboids [9]. Note that we only focus on the maintenance cost, but actually SB method also has the storage cost advantage over the other methods as it only materializes a single delta base cuboid in propagate stage.

4.1 Parameters

n: Number of dimensions of data cube Q

C_i: Number of dimension values (cardinality) of the i-th dimension in the base cuboid
($1 \le i \le n$)

ρ: Density of valid elements in the base cuboid, so it reflects the sparseness of the base cuboid

Nb: Total number of valid elements in the base cuboid. Obviously, $Nb = \rho \prod\limits_{i=1}^{n} C_i$.

Nd: Total number of valid elements in the dependent cuboids, so the total number of valid elements in Q is $Nb+Nd$.

4.2 Cost Model

We will use a 4-dimensional data cube for our cost evaluation and comparison. The 4 dimensions are denoted as a, b, c and d separately. Fig. 8 is the lattice diagram for the 4-dimensional data cube in analysis. We assume the values of some parameters of the data cube as follows:

$n=4$ $C_i=100$ (for all $1 \le i \le n$) $\rho=0.1, 0.05, 0.01$

We assume that the base cuboid is sparse which is common in practice (we can store only valid elements by some sparse array physical scheme like [8] [19]), while all the dependent cuboids are not sparse. It can be known that the total size of all the dependent cuboids Nd is $\prod\limits_{i=1}^{n}(C_i + 1) - \prod\limits_{i=1}^{n} C_i$.

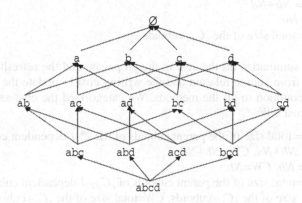

Fig. 8. Lattice diagram for a data cube

It is important for our evaluation that how many delta cuboids are materialized in these three methods. NV method materializes all the 2^n delta cuboids and get all the 2^n-1 dependent delta cuboids using *parent method* in [14]. SB method materializes only the base delta cuboid and uses it to refresh all the 2^n cuboids of the original data cube.

RA method is a little complex since it materializes $_nC_{n/2}$ delta cuboids to refresh the original data cube: We can assume the $_nC_{n/2}$ delta cuboids are *abcd, abc, abd, acd, bc, ad*, for more detail refer to [9]. Furthermore, we assume that any of the dependent cuboids are aggregated only from its parent cuboids in RA method. For example, cuboid *bc* can be only aggregated with cuboid *abc* or *bcd*.

We will denote the reading cost as *CR* and writing cost as *CW*. For simplicity, the total maintenance cost of the data cube is assumed to be the sum of *CR* and *CW*. We will use the simple linear cost model like [17] to measure *CR* and *CW*; the number of tuples read from or written to disk. The model was proved to be effective by experimental validation such as [17]. In Fig. 8 for example, if cuboid *ab* is computed from *abc*, *CR* is the size of cuboid *abc* and *CW* is the size of cuboid *ab*.

Propagate Cost. In the propagate stage, reading from source relation is common to all the methods. The difference is how many delta cuboids are materialized in the delta cube. So we only compare the reading cost *CR* to materialize dependent cuboids and writing cost *CW* to materialize cuboids in this stage:

NV method: *CR*=total size of the parent cuboids of the 2^n-1 dependent cuboids,
          CW=Nb+Nd
SB method: *CR=0, CW=Nb*
RA method: *CR*=total size of the parent cuboids of the $_nC_{n/2}$-1 dependent cuboids,
          CW=total size of the $_nC_{n/2}$ cuboids

Refresh Cost. In the refresh stage, writing cost to update the original data cube is common to all the methods. So we only compare the reading cost in this stage:

NV method: *CR= Nb+Nd*
SB method: *CR=Nb*
RA method: *CR*=total size of the $_nC_{n/2}$ cuboids

Total Cost. We summarize all the cost in the propagate and the refresh stage except the reading cost from source relation and the writing cost to update the original data cube which are common to all the methods. We categorized the cost as reading cost *CR* and writing cost *CW*:

NV method: *CR*= total size of the parent cuboids of the 2^n-1 dependent cuboids +
          Nb+Nd, CW=Nb+Nd
SB method: *CR= Nb, CW=Nb*
RA method: *CR*=total size of the parent cuboids of $_nC_{n/2}$-1 dependent cuboids + total
          size of the $_nC_{n/2}$ cuboids, *CW*=total size of the $_nC_{n/2}$ cuboids

The total cost of the three methods are compared and shown in Fig. 9 and Fig. 10 respectively on reading and writing cost.

To summarize the evaluation and comparison, our subarray-based method shows significant advantage over other methods; as density of the data cube (base cuboid) becomes smaller, the advantage becomes even larger. Tuples in multidimensional model consist of only fact data with position information (refer to sparse array physical storage scheme such as [8] [19]), which are generally much smaller than tuples in relational model. Thus it saves much more storage cost in multidimensional model. In another word, if we abstract the cost model in terms of storage pages accessed, our approach which is based on multidimensional model will show much more advantage over the other methods in relational model.

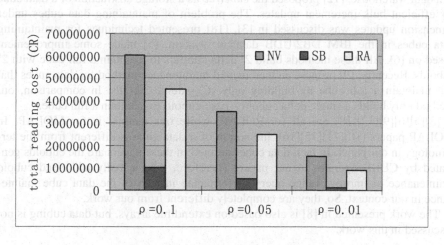

Fig. 9. Total reading cost comparison

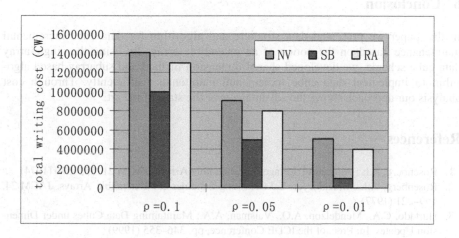

Fig. 10. Total writing cost comparison

5 Related Works

Since Jim Gray proposed the data cube operator, techniques for data cube construction and maintenance have been extensively studied. As far as we know, these papers include [9][13][14][17][19] mainly optimize the computation on cuboid level. So they usually implement an n-dimensional data cube into 2^n *nested relations* or arrays corresponding to the 2^n cuboids on data organization for relational and multidimensional databases. In this paper, we propose to organize all the 2^n cuboids of a data cube into only a single extendible array, thus provide opportunities to simplify the data cube management.

[6] is the first paper that addressed the issue of efficiently maintaining a data cube in a data warehouse. [12] proposed the *cubetree* as a storage abstraction of a data cube for efficient bulk increment updates. The problem of maintaining data cubes under dimension updates was discussed in [3]. [18] presented techniques for maintaining data cubes in the IBM DB2/UDB database system. [5] made some improvement based on [6]. All these methods build 2^n delta cuboids to maintain a full cube with 2^n cuboids. Recently, [9] propose an incremental maintenance method for data cubes that can maintain a full cube by building only $_nC_{n/2}$ delta cuboids. In comparison, our method only builds a single delta cuboid – base cuboid to maintain a full cube.

[3][5][6][9][12][18] are all for ROLAP, while our method is for MOLAP. In MOLAP papers [10][11][15][16], the notion of a data cube is different from the terminology in our paper. In fact, data cubes defined in these papers are the cuboids generated by CUBE operator in our paper. Therefore, they actually addressed cuboid maintenance to improve range query performance instead of the data cube maintenance in our context. So, they are completely different from our work.

The work presented in [8] is also based on extendible arrays, but data cubing is not discussed in this work.

6   Conclusion

In this paper we presented data structure and algorithm for data cube incremental maintenance based on the notion of an extendible array. By using the single-array data cube scheme, we developed shared dimension method and subarray-based algorithm to implement data cube incremental maintenance efficiently. Through cost analysis our approach shows the advantage over the state of the art.

References

1. Rosenberg, A.L.: Allocating Storage for Extendible Arrays. JACM 21, 652–670 (1974)
2. Rosenberg, A.L., Stockmeyer, L.J.: Hashing Schemes for Extendible Arrays. JACM 24, 199–221 (1977)
3. Hurtado, C.A., Mendelzon, A.O., Vaisman, A.A.: Maintaining Data Cubes under Dimension Updates. In: Proc. of the ICDE Conference, pp. 346–355 (1999)
4. Otoo, E.J., Merrett, T.H.: A Storage Scheme for Extendible Arrays. Computing 31, 1–9 (1983)
5. Li, H., Huang, H., Lin, Y.: DSD: Maintain Data Cubes More Efficiently. Fundam. Inform. 59(2-3), 173–190 (2004)
6. Mumick, I.S., Quass, D., Mumick, B.S.: Maintenance of Data Cubes and Summary Tables in a Warehouse. In: Proc. of the ACM SIGMOD Conference, pp. 100–111 (1997)
7. Gray, J., Bosworth, A., Layman, A., Pirahesh, H.: Data Cube: A Relational Aggregation Operator Generalizing Group-By, Cross-Tab, and Sub-Totals. In: Proc. of the ICDE Conference, pp. 152–159 (1996)

8. Hasan, K.M.A., Kuroda, M., Azuma, N., Tsuji, T., Higuchi, K.: An Extendible Array Based Implementation of Relational Tables for Multidimensional Databases. In: Tjoa, A.M., Trujillo, J. (eds.) DaWaK 2005. LNCS, vol. 3589, pp. 233–242. Springer, Heidelberg (2005)

9. Lee, K.Y., Kim, M.H.: Efficient Incremental Maintenance of Data Cubes. In: Proc. of the VLDB Conference, pp. 823–833 (2006)

10. Riedewald, M., Agrawal, D., Abbadi, A.E.: Flexible Data Cubes for Online Aggregation. In: Van den Bussche, J., Vianu, V. (eds.) ICDT 2001. LNCS, vol. 1973, pp. 159–173. Springer, Heidelberg (2001)

11. Riedewald, M., Agrawal, D., Abbadi, A.E., Pajarola, R.: Space-Efficient Data Cubes for Dynamic Environments. In: Kambayashi, Y., Mohania, M., Tjoa, A.M. (eds.) DaWaK 2000. LNCS, vol. 1874, pp. 24–33. Springer, Heidelberg (2000)

12. Roussopoulos, N., Kotidis, Y., Roussopoulos, M.: Cubetree: Organization of and Bulk Incremental Updates on the Data Cube. In: Proc. of the ACM SIGMOD Conference, pp. 89–99 (1997)

13. Jin, R., Yang, G., Vaidyanathan, K., Agrawal, G.: Communication and Memory Optimal Parallel Data Cube Construction. IEEE Transactions On Parallel and Distributed Systems 16(12), 1105–1119 (2005)

14. Agarwal, S., Agrawal, R., Deshpande, P.M., Gupta, A., Naughton, J.F., Ramakrishnan, R., Sarawagi, S.: On the Computation of Multidimensional Aggregates. In: Proc. of the VLDB Conference, pp. 506–521 (1996)

15. Geffner, S., Agrawal, D., Abbadi, A.E.: The Dynamic Data Cube. In: Zaniolo, C., Grust, T., Scholl, M.H., Lockemann, P.C. (eds.) EDBT 2000. LNCS, vol. 1777, pp. 237–253. Springer, Heidelberg (2000)

16. Geffner, S., Riedewald, M., Agrawal, D., Abbadi, A.E.: Data Cubes in Dynamic Environments. IEEE Data Eng. Bull. 22(4), 31–40 (1999)

17. Harinarayan, V., Rajaraman, A., Ullman, J.D.: Implementing Data Cubes Efficiently. In: Proc. of the ACM SIGMOD Conference, pp. 205–216 (1996)

18. Lehner, W., Sidle, R., Pirahesh, H., Cochrane, R.: Maintenance of Cube Automatic Summary Tables. In: Proc. of the ACM SIGMOD Conference, pp. 512–513 (2000)

19. Zhao, Y., Deshpande, P.M., Naughton, J.F.: An array based algorithm for simultaneous multidimensional aggregate. In: ACM SIGMOD, pp. 159–170 (1997)

A Data Partition Based Near Optimal Scheduling Algorithm for Wireless Multi-channel Data Broadcast*

Ping Yu, Weiwei Sun**, Yongrui Qin, Zhuoyao Zhang, and Bole Shi

Department of Computing & Information Technology, Fudan University
220 Handan Road, Shanghai, China
{yuping,wwsun,yrqin,zhangzhuoyao,bshi}@fudan.edu.cn

Abstract. Data broadcast is an efficient way to disseminate information to large numbers of users in wireless environments. The Square Root Rule (SRR) is the theoretical basis for the single channel broadcast scheduling. In this paper, we extend the SRR and propose the Multi-channel Square Root Rule (MSRR) for scheduling variable-length data with skewed access probabilities on variable-bandwidth channels. The theoretical optimal average access latency is also provided. However, this optimal value can not be achieved in reality. Based on MSRR, we provide a two-phase scheduling algorithm which achieves near optimal access latency. First data are partitioned and allocated to different channels according to MSRR. Second, different scheduling strategies are adopted on each channel according to the skewness of data subset allocated on that channel. Experiments show that the difference of average access latency between our results and the optimal value is below five percent in most situations.

Keywords: Wireless mobile environments, data broadcast, data partition, hybrid scheduling, multiple channels.

1 Introduction

With the rapid development of wireless communication and computer technologies, wireless mobile computing is becoming to reality. Millions of mobile users can access a large variety of information through mobile devices anywhere and anytime. Asymmetric is a main characteristic of wireless network, and the amount of concurrent user is unpredictable. Frequently demanded data (such as stock information, traffic conditions and weather information, etc.) can be disseminated periodically on the downlink channels to meet the needs of most mobile users. This method is known as *data broadcasting*, and has the advantage of *scalability*, *power conservation* and *efficient bandwidth usage*. A single broadcast of a data item can satisfy all the outstanding requests for that item

---

* This research is supported in part by the National Natural Science Foundation of China (NSFC) under grant 60503035 and the National High-Tech Research and Development Plan of China under Grant 2006AA01Z234.
** Corresponding author.

simultaneously. So broadcast can scale up to an arbitrary number of users [1]. As to power conservation, mobile devices consume more power energy in transmit mode than in receive mode, access information through only listening to the broadcast channels can reduce the energy consumption greatly.

Access efficiency and *power conservation* are two main aspects considered by many researchers in the broadcast area. By using *air index* [2,3,4], mobile devices can switch into energy-saving mode most of time thus reduce battery consumption furthermore. To improve the access efficiency, *scheduling* technologies are usually adopted. Data broadcast *scheduling* is to determine what is to be broadcast and when to by the server to minimize the average access time. When the user access pattern is uniform, flat scheduling is a naive method. But in most situations, access pattern is skewed. The Square Root Rule (SRR) [5,6] is a theoretical foundation for single channel skewed scheduling. But for multi-channel broadcast, to the best of our knowledge, there is still no similar precise rule. We focus on multi-channel scheduling problem in this paper, and our main contributions are:

(1) We propose the Multi-channel Square Root Rule (MSRR), prove the theoretical minimal access latency of multi-channel broadcast and provide the precise condition which can lead to the minimal access latency. Before our work, only heuristic properties and qualitative analysis were put forward.

(2) Based on MSRR, a two-phase multi-channel scheduling approach is proposed. First, partition data into several groups, each group broadcasts on one channel. Three data partition algorithms are provided, which can achieve near optimal access latency. Then, in the second phase, existing single channel scheduling algorithms can be applied.

(3) We observe that after data partition the access probability distribution on each channel is different, so a hybrid scheduling strategy is proposed. We use a statistics measure, *Kurtosis*, to evaluate the skewness of access probability on each channel after data partition. Then different single channel scheduling algorithm is applied on different channel according to the *Kurtosis* value of that channel. With this method, the average access time is improved further.

The remainder of this paper is organized as follows: In section 2, related work is reviewed. The theoretical model and MSRR are proposed in section 3. In section 4 we describe the two-phase scheduling, three partition algorithms and a hybrid scheduling algorithm for multi-channel broadcast are introduced. Experiments results are presented in section 5 which demonstrate the performance benefits of our techniques. We conclude the paper and discuss future work in section 6.

2   Related Work

Early data broadcast researches focused on problems on single channel. Based on the SRR, several scheduling algorithms were studied [5,6,7]. [5] proposed a naive way to implement SRR. [6] provided an on-line algorithm. A general algorithm was found in [7] which can be applied to non-uniform length data. [8]

introduced a hierarchical architecture (*broadcast disks*) to disseminate data with equal instance spacing. When the volume of data requested from broadcast channel becomes large, the access latency to user will increase. One of the solutions is to increase the physical bandwidth, for example, allowing the server to transmit over multiple disjoint physical channels [9]. There are also many situations that physical channels can not be coalesced into a lesser number of high-bandwidth channels [10,11] Thus new problems in multi-channel broadcast occur, which concern scheduling [10,12,13,9,14], indexing [4,11] and channel function assignment [15] etc.

There are two strategies to allocate data on multiple channels, *data partition* and *data replication* [16]. In data partition, data are assigned into disjoint sets and each set is broadcast in one channel. Data replication allows same data items appear on several channels. It depends on the user access ability to decide which strategy is adopted. If user can access only one channel, then data replication must be adopted; if user can access every channel, data partition is preferred.

Data partition based scheduling is a hot research area. Assume that there are N data items and K channels in a broadcast system. [10] proposed a dynamic scheduling approach based on the *temperature* of data items. Data items with higher access probability have higher temperature. But the paper did not give a precise definition of the temperature. A *Greedy* algorithm was studied in [12], data items are sorted by non-increasing access probabilities, each time one of the existing partitions is chosen to split into two which can achieve minimal access latency. The algorithm complexity is $O(N\log N)$. [13] studied a channel allocation tree which can be used to represent the data partition; their algorithm was named VF^K (Variant Fanout). The VF^K and *Greedy* algorithms have the same time complexity and scheduling result. All the above algorithms are based on data partition and adopt flat scheduling on each channel. Under this assumption, dynamic programming (DP) produces optimal scheduling. *Dichotomic* [9] is an efficient DP scheduling algorithm which reduces the time complexity from $O(N^2K)$ to $O(NK\log N)$. But scheduling skewed data in a flat way on each channel apparently can not achieve optimal access latency.

[14] studied a flexible data schedule generator for multi-channel broadcast systems. It assumed data partition and non-flat scheduling on each channel which is similar to our approach. Their method is an application of broadcast disks virtually. To obtain ideal access latency, very long broadcast period is needed which leads to long scheduling time and store space. [17] proposed a theoretical lower bound of average access latency for multi-channel scheduling and also a scheduling algorithm TOSA. But the lower bound is not accurate.

3   Theoretical Model

The *cyclic policy* is optimal as far as minimizing the average access latency is concerned in broadcast [5]. In this paper we assume adopting cyclic policy to broadcast data on each single channel and use notations described in Table 1.

3.1 Model for Single Channel

We assume that the database contains N data items, $d_1, d_2, \ldots, d_N$, and the access probability of each data item is known. We put forward some definitions first.

Broadcast cycle is a sequence in which each item is transmitted at least once. We use L to denote the *length of broadcast cycle*, $L = \sum_{i=1}^{N} f_i l_i$. An *instance* of a data item is an appearance of the data item in the broadcast. *Spacing* between two instances, s_{ij}, is the time from the beginning of transmitting the jth instance to the beginning of transmitting the $j+1$th instance (see Fig.1.).

Average access latency of data item d_i, denoted by t_i , is the time from the moment when a request is issued to the time user begins to receive the data. Thus $t_i = \sum_{j=1}^{f_i} \frac{s_{ij}^2}{2LB_S}$. [6] has proved that when the instances of each data item is equally spaced, t_i is minimized and $t_i = \frac{L}{2f_i B_S} = \frac{s_i}{2B_S}$.

Overall average access latency, denoted by AAL, is defined as the average waiting time over all items encountered by a user, thus

$$AAL = \sum_{i=1}^{N} t_i p_i = \frac{1}{2} \sum_{i=1}^{N} \left(\frac{L}{f_i B_S} \right) p_i \qquad (1)$$

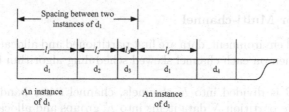

Fig. 1. Fragment of a broadcast cycle

Table 1. Summary of notations

Notation	Description
N	the number data items
K	the number of channels
B_S	the total bandwidth in the single channel system
B	the total bandwidth in the multi-channel system
d_i	the ith data item
p_i	the access probability of data item d_i
f_i	the broadcast frequency of data item d_i
l_i	the length of data item d_i
L	the length of broadcast cycle
s_{ij}	the spacing between the jth and $j+1$th instances of data item d_i
s_i	the average spacing between two continual instances of data item d_i
N_i	the number of data items on channel i
t_i	the average access latency of data item d_i
b_i	the bandwidth of channel i
p_{ij}	the access probability of the jth data item in channel i
l_{ij}	the length of the jth data item in channel i

Data broadcast scheduling is to determine the broadcast frequency of each data item and the spacing between continual instances of each data items to make AAL minimized. That is to solve the following optimization problem:

$$\min \; f(f_1, f_2, ...f_N) = \frac{1}{2} \sum_{i=1}^{N} \left(\frac{L}{f_i B_S} \times p_i \right)$$
$$s.t. \; \sum_{i=1}^{N} f_i l_i = L \qquad\qquad (Problem 1)$$

Vaidya and Hameed proposed the Square Root Rule to solve the problem.

Lemma 1 (Square Root Rule (SRR)[6]). *Given the access probability p_i of each data item d_i, the minimal overall average access latency is achieved when the following condition holds:*

$$f_i = \frac{L}{\sum_{j=1}^{N} \sqrt{p_j l_j}} \sqrt{\frac{p_i}{l_i}}, \qquad 1 \le i \le N, \qquad (2)$$

assuming that instances of each data item are equally spaced. The optimal AAL, denoted by AAL^{opt}, is:

$$AAL^{opt} = \frac{1}{2B_S} \left(\sum_{i=1}^{N} \sqrt{p_i l_i} \right)^2 \qquad (3)$$

3.2   Model for Multi-channel

In multi-channel environment, data are first partitioned and allocated to multiple channels, and then on each channel skewed scheduling algorithm based on SRR is adopted.

Bandwidth B is divided into K channels, channel i has bandwidth b_i, and $\sum_{i=1}^{K} b_i = B$. We partition N data items into K groups and allocate each group to a channel. The number of data items in channel i is N_i, $\sum_{i=1}^{K} N_i = N$. Data items in channel i have access probability $p_{i1}, p_{i2}, \ldots, p_{iN_i}$, and length $l_{i1}, l_{i2}, \ldots, l_{iN_i}$. The multi-channel overall average access latency, denoted by $MAAL$, is defined as

$$MAAL = \sum_{i=1}^{K} \frac{1}{2b_i} \left(\sum_{j=1}^{N_i} \sqrt{p_{ij} l_{ij}} \right)^2 \qquad (4)$$

assuming the scheduling on each channel obeys SRR.

Let $\sum_{j=1}^{N_i} \sqrt{p_{ij} l_{ij}} = x_i$ and $\sum_{i=1}^{N} \sqrt{p_i l_i} = m$, then the multi-channel broadcast scheduling is to solve the following optimization problem:

$$\min \; f(x_1, x_2, ...x_K) = \frac{1}{2} \sum_{i=1}^{K} \left(\frac{x_i^2}{b_i} \right)$$
$$s.t. \; \sum_{i=1}^{K} x_i = m \qquad\qquad (Problem 2)$$

Theorem 1 (Multi-channel Square Root Rule (MSRR)). *Given the access probability p_i of each item d_i, allocating these N data items disjointedly into K channels each with bandwidth b_i, the MAAL is minimized when the following condition holds:*

$$\sum_{j=1}^{N_i} \sqrt{p_{ij} l_{ij}} = \frac{b_i}{B} \times \sum_{i=1}^{N} \sqrt{p_i l_i}, \qquad 1 \le i \le K, \qquad (5)$$

assuming that data on each channel is scheduled based on SRR. The optimal MAAL, *denoted by* $MAAL^{opt}$, *is*:

$$MAAL^{opt} = \frac{\left(\sum_{i=1}^{N} \sqrt{p_i l_i}\right)^2}{2B} \qquad (6)$$

Proof. It is easily to prove that function f is convex, and the equality constraint is a linear function, so we can use *Lagrange Multiplier method* to solve above problem. Let μ be the *Lagrange Multiplier*, the corresponding *Lagrange function* is

$$L(x_1, x_2, ...x_K) = \frac{1}{2} \sum_{i=1}^{K} \frac{x_i^2}{b_i} - \mu \left(m - \sum_{i=1}^{K} x_i\right),$$

When the following conditions hold, the minimum of function f is obtained:

$$\begin{cases} \frac{\partial L}{\partial x_1} = \frac{x_1}{b_1} + \mu = 0 \\ ... \\ \frac{\partial L}{\partial x_K} = \frac{x_K}{b_K} + \mu = 0 \\ \frac{\partial L}{\partial \mu} = m - \sum_{i=1}^{K} x_i = 0 \end{cases}$$

Solving the above equations, we can get when $x_i = mb_i / \sum_{j=1}^{K} b_j = mb_i/B$, that is when $\sum_{j=1}^{N_i} \sqrt{p_{ij} l_{ij}} = \left(\sum_{i=1}^{N} \sqrt{p_i l_i}\right) b_i/B$, f achieves the optimum which is: $MAAL^{opt} = \sum_{i=1}^{K} \frac{1}{2} \frac{m^2 b_i}{B^2} = \frac{m^2}{2B} = \frac{\left(\sum_{i=1}^{N} \sqrt{p_i l_i}\right)^2}{2B}$. $\square$

The lower bound of $MAAL$ proposed in [17] is $\frac{K}{2} \times \frac{\left(\sum_{i=1}^{N} \sqrt{p_i l_i}\right)^2}{\left(\sum_{j=1}^{K} \sqrt{b_j}\right)^2}$ and the condition to achieve it is $\sum_{j=1}^{N_i} \sqrt{p_{ij} l_{ij}} = \frac{\sqrt{b_i}}{\sum_{j=1}^{K} \sqrt{b_j}} \times \sum_{i=1}^{N} \sqrt{p_i l_i}$, $1 \leq i \leq K$. It can be easily to prove that their lower bound is larger than our result when the bandwidth of channels is non-uniform.

4   Scheduling Algorithm

We regard $\sqrt{p_i l_i}$ as d_i's *temperature*, and the MSRR can be explained as follows: to obtain the minimal $MAAL$, the total temperature of all the data on one channel should be proportional to the channel's bandwidth.

According to MSRR, the broadcast scheduling can be executed in *two phases*:

Phase 1: data partition
Partition data items into K groups and allocate them to different channels according to MSRR. That is, make the total temperature of the data subset on each channel proportional to the corresponding channel's bandwidth.

Phase 2: single channel broadcast scheduling
Broadcast data items on each channel with single channel scheduling algorithms. For data items are partitioned disjointedly into different channels, they can be scheduled independently on each channel.

We describe the details in the following subsections.

4.1 Skewed Data Partition Algorithms (SP)

We define *channel temperature* of channel i as $(\sum_{i=1}^{N} \sqrt{p_i l_i}) \times b_i/B$. The problem to partition N data items into K channels is similar to the *bin-packing problem*. The *temperature* of each data item is similar to the *volume* of each parcel which will be put into the bins, and the *channel temperature* is similar to the *bin capacity*.

Besides the similarity between the two problems, there are some differences:

i) In bin-packing problem, the total volumes of all parcels in a bin can not be larger than the bin capacity, but in data partition, the total temperature of all items assigned to a channel can be larger than the channel temperature.

ii) In bin-packing problem, the objective is to minimize the number of bins used to contain the given parcels, but in data partition, the number of channels is given in advance.

For the bin-packing problem is combinatorial *NP-hard* problem, we propose some heuristic algorithms here.

Three SP Algorithms. The *input* of these three algorithms are N data items sequence $d_1, d_2, \ldots, d_N$ with their access probabilities, K channels with their bandwidths, and the *output* is the data assignment on each channel.

```
Algorithm 1. Data partition algorithm SP_NF (Next Fit)
1) Calculate channel i's temperature ch_temp[i]
2) Initialize each channel used capacity USED[i]=0
3) Initialize the current channel number k=1
4) for j=1 to N
5)      if USED[k]≥ch_temp[k]
6)          then k=k+1
7)      USED[k]=USED[k]+√(p_j l_j)
8)      Assign d_j to channel k
```

SP_NF assigns data items on channels in the given sequence. First put d_1 on channel 1; if d_i is the current item to assign, and the current channel is not full, then put d_i on the current channel. Otherwise assign it on the next channel. The complexity of SP_NF algorithm is $O(N)$.

Another $O(N)$ time complexity partition algorithm adopts *round robin* strategy, and is denoted by SP_RR. SP_RR assigns data items on channels in the given sequence. It put d_1 on channel 1,..., d_K on channel K, and then d_{K+1} in channel 1 if the channel is not full. If d_i is assigned to the current channel, then d_{i+1} is assigned to the next channel which is not full in a round robin sequence of all channels.

A more precise algorithm SP_FF is described as follows, which adopts first fit strategy.

```
Algorithm 2. Data Partition Algorithm SP_FF (First Fit)
1) Calculate channel i's temperature ch_temp[i]
2) Initial each channel unused capacity LEFT[i]=ch_temp[i]
```

```
3) for j=1 to N
4)     k=1
5)     while LEFT[k]<√(p_j l_j) & k<=K-1
6)         k=k+1
7)     LEFT[k] = LEFT[k]-√(p_j l_j)
8)     Assign d_j to channel k
```

SP_FF assigns data items on channels in the given sequence. First put d_1 on channel 1; if d_i is the current item to assign, then begin from channel 1 to find the first channel whose left capacity is not smaller than d_i's volume, and put d_i in that channel. The complexity of SP_FF algorithm is $O(N \log N)$.

Example 1. We use the same data items set in [17] to show the results of the above partition algorithms and the method TOSA in [17]. There are 8 uniform length data items and 2 uniform bandwidth channels in the system. Table 2 shows the access probabilities of all the data items. The temperature of every channel is $(\sum_{i=1}^{8} \sqrt{p_i})/2 = 1.1757$. The results of the partition algorithms are shown in Table 3 (the *MAAL* is calculated using formula (4) after partition), the $MAAL^{opt}=1.3822$. SP_FF achieves best *MAAL* (1.3827), smaller than that by TOSA.

Table 2. Access probabilities

d_i	d_1	d_2	d_3	d_4	d_5	d_6	d_7	d_8
p_i	0.5	0.2	0.1	0.1	0.07	0.01	0.01	0.01
$\sqrt{p_i}$	0.71	0.45	0.32	0.32	0.26	0.1	0.1	0.1

Table 3. Partition result of different algorithms

Algorithm	Time complexity	Partition result	MAAL
SP_NF	O(N)	Ch1: $d_1\ d_2\ d_3$ Ch2: $d_4\ d_5\ d_6\ d_7\ d_8$	1.4692
SP_FF	O(NlogN)	Ch1: $d_1\ d_2$ Ch2: $d_3\ d_4\ d_5\ d_6\ d_7\ d_8$	1.3827
SP_RR	O(N)	Ch1: $d_1\ d_3\ d_5$ Ch2: $d_2\ d_4\ d_6\ d_7\ d_8$	1.3948
TOSA	O(NlogN)	Ch1: $d_1\ d_4\ d_5$ Ch2: $d_2\ d_3\ d_6\ d_7\ d_8$	1.3948

SP Algorithms' Effect on Access Probability Distribution on Different Channels. In this subsection, we pay more attention to the effect of partition algorithms on the access probability distribution of data on different channels. We observe that after data partition, the access probability distribution on each channel is different and also different from the initial distribution before data partition.

For example, assuming that uniform-length date items are sorted in a descending order of their access probabilities, after partition with SP_NF, the distributions on some channels are almost flat and that on other channels are more skewed. But if we partition data using SP_RR, then each channel has skewed distribution. This phenomenon hints us that different scheduling strategies should be taken to adapt to the different distribution on different channel.

We use the *Kurtosis* metric in statistics to measure the skewness of the probability distribution on each channel. *Kurtosis* is the degree of peakedness of a distribution, and is defined as a normalized form of the fourth central moment of a distribution (we use *Kurt* for the abbreviation of *Kurtosis*):

$$Kurt = \frac{m_4}{\sigma^4} \quad (m_4 = \frac{\sum_{i=1}^{n} (x_i - \bar{x})^4}{n}, \ \sigma^2 = \frac{\sum_{i=1}^{n} (x_i - \bar{x})^2}{n}) \tag{7}$$

Here x_i denotes the probability of data item d_i on one channel, $\bar{x}$ is the mean of all the probabilities of the items on the channel, n is the number of data items on the channel. When $Kurt > 3$, the skewness on that channel is very high, otherwise the skewness is moderate or the distribution is almost flat. Table 6 in section 5.1 can be taken as an example.

4.2   Scheduling Algorithms for Single Channel

After data partition, we can schedule data on each channel using single channel scheduling algorithms.

Flat Scheduling (FS). The naive method is to broadcast each item one by one on the channel without considering their access probability. This method is *flat scheduling* (FS), and we use SPFS to denoting flat scheduling based on skewed partition. When the probability distribution is almost flat, FS performs well; but when the skewness is high, FS will lead to long average access latency.

Skewed Scheduling for Uniform-length Data. We propose a skewed scheduling algorithm (SS) here. The algorithm is based on SRR, similar to M.H. Ammar's algorithm [5]. But in their algorithm, data broadcast length L is an input parameter, if L is not chosen carefully, two unexpected cases will occur: the access latency will have a larger gap with the optimum or the broadcast length will be very long. In our algorithm, L is calculated within the algorithm, thus makes the AAL closer to the optimum.

Assuming N is the number of data on one channel here, and d_N is the item with smallest access probability, set its broadcast frequency to 1, from SRR we can get the formula to calculate L:

$$L = round(\sum_{i=1}^{N} \sqrt{p_i} / \sqrt{p_N}) \tag{8}$$

SS algorithm is designed based on SRR: first, calculating the *broadcast frequency* and *instance spacing* of each data item, then allocating them to the channel. It should achieve optimal average access latency theoretically. But it is inevitable to make some *round* and *floor* operation to make *broadcast cycle length*, the *frequency* and the *instance spacing* to be integer, some deviation arises. So SS algorithm can only achieve near optimal access latency. When the skewness is high or the number of items is large, the deviation is small relatively.

Example 2. Consider the data set in Table 2. The result of data partition using SP and scheduling using SS are shown in Table 4 (SPSS(NF) denotes using SP_NF partition algorithm and SS scheduling algorithm). SPSS(FF) achieves the best *MAAL* too because its partition result is the most precise one according to MSRR.

Table 4. Result of data partition and scheduling ($MAAL^{opt} = 1.3822$)

Algorithm	Partition result	L	Scheduling result	MAAL
SPSS(NF)	Ch1: d_1 d_2 d_3	5	d_1 d_2 d_1 d_3 d_1	1.5900
	Ch2: d_4 d_5 d_6 d_7 d_8	9	d_4 d_5 d_6 d_4 d_5 d_7 d_4 d_5 d_8	
SPSS(FF)	Ch1: d_1 d_2	3	d_1 d_2 d_1	1.4367
	Ch2: d_3 d_4 d_5 d_6 d_7 d_8	12	d_3 d_4 d_5 d_6 d_3 d_4 d_5 d_7 d_3 d_4 d_5 d_8	
SPSS(RR)	Ch1: d_1 d_3 d_5	5	d_1 d_3 d_1 d_5 d_1	1.4536
	Ch2: d_2 d_4 d_6 d_7 d_8	11	d_2 d_6 d_4 d_2 d_7 d_2 d_4 d_2 d_8 d_2 d_4	

Besides the SS algorithm, other skewed scheduling algorithms can be adopted in the second phase too, such as the Broadcast Disks method ([8]) and the method in [14].

Skewed Scheduling for Non-uniform-length Data. Hameed and Vaidya proposed a general skewed scheduling algorithm [7] for non-uniform-length data. The algorithm is an online algorithm, assuming the current time is T, it determines which item is chosen to broadcast at T. For each data item d_i, two variables, B_i and C_i, are maintained. B_i is the earliest time when next instance of d_i should begin transmission, and $C_i = B_i + s_i$. At the current time T, the algorithm first determines a set S of items whose $B_i \leq T$, chooses one item $d_j \in S$ which has minimal C_j, broadcasts d_j at time T and updates B_j and C_j as $B_j = C_j$, $C_j = B_j + s_j$ respectively. Then updates the current time $T = T + l_j$, and iterates the above steps. The average time complexity of each iteration is $O(\log N)$. This algorithm has an initial transient phase, after that, the schedule becomes cyclic. It can be used in the second scheduling phase for non-uniform-length data.

4.3   Hybrid Scheduling on Multiple Channels

As stated in 4.1, after data partition, access probability distribution on each channel will change. FS performs well in flat distribution; SS performs better in high skewed distribution. We can use *Kurtosis* metric to measure the skewness of each channel, and adopt different scheduling algorithm (flat or skewed scheduling) according to the *Kurtosis* value of each channel. This is the basic idea of hybrid scheduling algorithm with skewed partition (SPHS).

The *input* of SPHS is N data items sequence $d_1, d_2, \ldots, d_N$ with their access probabilities, K channels with their bandwidths; and the *output* is the broadcast sequences for each channel.

```
Algorithm 3. Hybrid Scheduling with Skewed Partition (SPHS)
1) Partition N data items into K channels using SP
2) for i=1 to K
3)      calculate Kurt[i] for each channel using formula (7)
4)      if Kurt[i]<=3
5)          then using flat scheduling for channel i
6)          else using skewed scheduling for channel i
```

Hybrid scheduling has the following advantages:

i) It benefits from both flat and skewed scheduling and can achieve better access latency than only using flat or skewed scheduling on all channels.

ii) Because some channels adopt flat scheduling, the scheduling is simple and the time complexity decreases.

iii) It is adaptive to different access probability distribution. No matter what partition algorithms or single channel scheduling algorithms are chosen in the two scheduling phases, hybrid scheduling can always achieve near optimal access latency.

5   Experiments

In this section we will first evaluate the performances of our methods for uniform-length items scheduling on uniform-bandwidth channels. Then the performances of data partition algorithms are compared for variable-length items scheduling on variable-bandwidth channels.

We adopt similar parameter ranges and control values to those used in other work [12,9,14,17]. Table 5 shows the parameters used in our experiments. Similar as in [17], the data item length and the channel bandwidth follow the *uniform distribution* between 1 and the maximum value (*MaxItemLength* and *MaxBandwidth* respectively). The unit of item length and channel bandwidth is logical unit. No specific physical unit is chosen because it does not affect the results of our experiments.

Table 5. Parameters in experiments

Description	Parameter	Value Range	Step	Control Value
Database size	N	500-10,000	1000	2500
Zipf skew	θ	0-1.5	0.1	80/20
Number of channels	K	1-10	1	4
Maximum item length	MaxItemLength	1-20		5
Maximum bandwidth	MaxBandwidth	1-20		5

We assume data items access probability follows *Zipf* distribution, the probability of accessing the i-th most popular object is: $p_i = \frac{(1/i)^\theta}{\sum_{i=1}^{N}(1/i)^\theta}$ $1 \le i \le N$.

When $\theta=0$, the distribution is flat and as θ becomes larger, the skewness become higher. The control value chosen for θ will produce 80/20 distribution; when $N=2500$ and $\theta=1.0$ *Zipf* distribution almost obey 80/20 rule, so we use 1.0 as control value.

To measure the access efficiency, we use the *MAAL* metric and calculate the *deviation percentage* compared with the $MAAL^{opt}$. Here *deviation percentage* is computed as $\frac{(MAAL - MAAL^{opt})}{MAAL^{opt}} \times 100\%$.

5.1 Scheduling for Uniform-Length Items on Uniform-Bandwidth Channels

MSRR is applicable in variable-length and variable-bandwidth situations. In this section, we assume the length of data items and the bandwidth of channels are uniform in order to compare our methods with others which are only applicable under the uniform environment.

The performances of hybrid (SPHS), flat (SPFS) and skewed (SPSS) scheduling algorithms are compared first. Then another skewed scheduling algorithm-FlexSched will be compared with SPHS. Two typical flat scheduling algorithms, Dichotomic and Greedy, will be compared with SPHS in the end of this subsection.

Performance of Hybrid Scheduling. After data partition, the real probability distribution on each channel is not the *Zipf* distribution as initial. Furthermore, the distribution on different channel is different.

When $N=2500$, $K=4$, and $\theta=1.0$, the *Kurtosis* of each channel after data partition is shown in Table 6. After data partition using SP_NF or SP_FF, the data on the first channel has high skewness, and the distribution on channel $2 \sim 4$ is almost flat. SP_RR makes each channel occupied by data with high skewness.

Table 6. Kurtosis of each channel after data partition

Channel	Kurtosis		
	SP_NF	SP_FF	SP_RR
1	75.61	75.60	524.28
2	2.95	2.97	422.66
3	2.19	2.18	334.63
4	1.99	1.99	269.8

Fig. 2 shows the performance of three scheduling strategies (flat, skewed and hybrid) with same partition method (SP_NF). When $\theta < 0.5$, FS performs best. When θ becomes larger, FS's performance decreases quickly. HS always outperforms SS, especially when the skewness is high. For example, when $\theta=0.9$, the difference between HS and SS is 3.6%. The reason is that when the channel's *Kurtosis* is less than 3, the distribution is almost flat. In this situation, SS will generate relatively larger deviation because of *round* operation and FS has better performance.

Comparison with Other Skewed Scheduling per Channel Algorithm.
FlexSched is a skewed partition and skewed scheduling algorithm. Its scheduling
strategy is a special case of broadcast disks method, for it may produce many
disks to achieve better access latency. When the skewness is low, FlexSched
performs a little better (less than 0.5%) (Fig. 3). When θ becomes larger, the
deviation percentage of FlexSched increases more quickly than SPHS. When
$\theta=1.4$ the difference is more than 3%. Although the difference of access latency
between the two algorithms is not large, their broadcast cycle length differs a
lot. When $\theta=1.0$, on channel 1, the broadcast cycle length for SPHS is 322, but
for FlexSched, the length is 970200, almost 3000 times larger than SPHS. The
long broadcast cycle will increase the scheduling time and storage space.

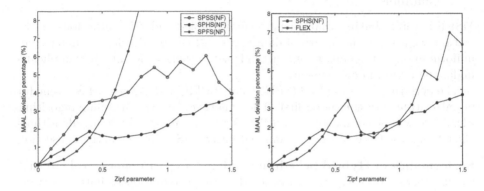

Fig. 2. Performance of scheduling strategies   **Fig. 3.** Comparison with flexSched

Comparison with Flat Scheduling per Channel Algorithms. We choose
SPHS(NF) to compare with two flat scheduling per channel algorithms, Dichotomic and Greedy. These two algorithms do partition data into different
channels, but the partition does not conform to MSRR; and in each channel,
they use flat scheduling.

(1) Effect of Database Size

Fig. 4 shows that as database size grows the *deviation percentage* of Dichotomic and Greedy grows quickly, but the variety of database size has little
effect on SPHS. With all the database size, SPHS outperforms the other two
algorithms. When $N=2500$, the *deviation percentage* of SPHS is 9.27% smaller
than Dichotomic and 11.43% smaller than Greedy. When $N=10000$, the difference is larger.

(2) Effect of Skewness

The *deviation percentage* of Dichotomic and Greedy increases quickly when
data skewness grows, SPHS performs well and the *deviation percentage* changes
a little while data skewness growing (Fig.5). When $\theta < 0.5$, Dichotomic and
Greedy perform a little better than SPHS (less than 1%); when $\theta >= 0.5$, SPHS
has obvious advantage. The reason for these results is that Dichotomic and
Greedy do not partition data complying with MSRR, and each channel adopts

flat scheduling. As database size and skewness grow, the deviation will become larger. SPHS partitions data according to MSRR, and at the most time the *deviation percentage* is about 2%.

(3) Effect of Number of Channel

In this experiment (Fig. 6), as the number of channels (K) grows, the *deviation percentage* of Dichotomic and Greedy decrease, but SPHS always performs better and steadily. The reason is that, when $K=1$, there is only one channel in the system, Dichotomic and Greedy turn to pure flat scheduling, thus behave poorly. As the number of channels grows, data items are partitioned into more groups. Although flat scheduling is taken for each group in Dichotomic and Greedy, each group has different AAL, thus to some extent, the skewness of the access probabilities is considered. So their performances get better as K increases. SPHS always adopts skewed scheduling in every channel, so it performs steadily and always better. When $K=4$, the *deviation percentage* of SPHS is 9.27% less than Dichomotic and 11.43% less than Greedy.

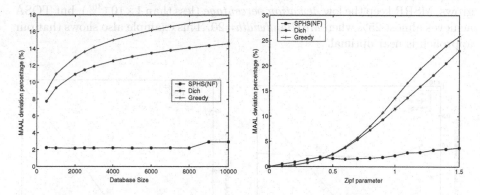

Fig. 4. Effect of database size   **Fig. 5.** Effect of skewness

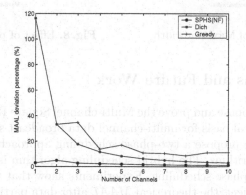

Fig. 6. Effect of number of channels

5.2   Data Partition for Variable-Length Items on Variable-Bandwidth Channels

In this subsection, we focus on the performance of data partition algorithms, and use formula (4) to calculate the $MAAL$ after the data partition. Fig. 7 shows that under all *MaxBandwidth*, SP_FF performs best, and the theoretical *deviation percentage* of all the three algorithms is below 10^{-3}%(without the deviation of scheduling). Among them, SP_FF performs best, because it has higher time complexity ($O(N\log N)$) than the other two ($O(N)$), thus its partition result is more close to the optimal result. SP_NF has similar performance as SP_RR, but using SP_RR, hot items have a chance to be dispersed to different channels. While using SP_NF, the hot items may cram into one or two channels, and the other channels are filled by large numbers of cold items.

Fig. 8 compares the partition criteria of MSRR with that of TOSA. We use the same partition algorithm SP_FF, but the *channel temperature* is calculated using different formulas. As we analyze in section 3.2, only when the bandwidth is uniform, TOSA has the same performance as MSRR, when the *MaxBandwidth* grows, MSRR keep the low *deviation percentage* (less than 1×10^{-3}%), but TOSA achieves almost 25% when *MaxBandwidth*=20. This example also shows that our approach is near optimal.

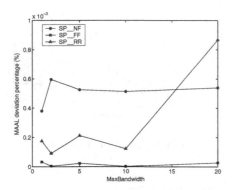

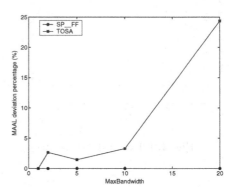

Fig. 7. Effect of MaxBandwidth   **Fig. 8.** Effect of partition criteria

6   Conclusions and Future Work

In this paper we propose and prove the Multi-channel Square Root Rule (MSRR) which is a theoretical basis for multi-channel data broadcast scheduling. On the basis of MSRR, we propose a two-phase scheduling approach and design three data partition algorithms. The hybrid scheduling algorithm is an integrated solution for the two-phase scheduling. Experiments show that the data partition algorithms are efficient, the theoretical $MAAL$ after data partition approximates the optimum (the difference is below 10^{-3}%). The hybrid scheduling algorithm also outperforms other data partition based multi-channel scheduling algorithms and the difference compared with the optimal access latency is below 5%.

Data partition is only one of data allocation strategies in multi-channel environment. We will study algorithms for other allocation strategies (such as data replication) and we are also very interested in the problems about indexing and channel function assignment on multiple channels.

References

1. Xu, J., Lee, D.L., Hu, Q., Lee, W.C.: Data Broadcast. In: Handbook of Wireless Networks and Mobile Computing, John Wiley, Chichester (2002)
2. Imielinski, T., Viswanathan, S., Badrinath, B.R.: Data on air: Organization and access. IEEE Trans. on Knowledge and Data Engineering 9(3), 353–372 (1997)
3. Xu, J., Lee, W.C., Tang, X.: Exponential index: A parameterized distributed indexing scheme for data on air. In: MobiSys. 2004, pp. 153–164 (2004)
4. Yao, Y., Tang, X., Lim, E., Sun, A.: An energy-efficient and access latency optimized indexing scheme for wireless data broadcast. IEEE Trans. on Knowledge and Data Engineering 18(8), 1111–1124 (2006)
5. Ammar, M.H., Wong, J.W.: The design of teletext broadcast cycles. Performance Evaluation 5(4), 235–242 (1985)
6. Vaidya, N.H., Hameed, S.: Scheduling data broadcast in asymmetric communication environments. ACM/Baltzer Journal of Wireless Networks (WINET) 5(3), 171–182 (1995)
7. Hameed, S., Vaidya, N.H.: Efficient algorithms for scheduling data broadcast. ACM/Baltzer Journal of Wireless Networks (WINET) 5(3), 183–193 (1999)
8. Acharya, S., Alonso, R., Franklin, M., Zdonik, S.: Broadcast disks: Data management for asymmetric communications environments. In: SIGMOD 1995, pp. 199–210 (1995)
9. Ardizzoni, E., Bertossi, A.A., Pinotti, M.C., Ramaprasad, S., Rizzi, R., Shashanka, M.V.S.: Optimal skewed data allocation on multiple channels with flat broadcast per channel. IEEE Trans. on Computer 54(5), 558–572 (2005)
10. Prabhakara, K., Hua, K., Oh, J.: Multi-level multi-channel air cache design for broadcasting in a mobile environment. In: ICDE 2000, pp. 167–176 (2000)
11. Yee, W., Navathe, S.: Efficient data access to multi-channel broadcast programs. In: CIKM 2003, pp. 153–160 (2003)
12. Yee, W.G., Navathe, S.B., Omiecinski, E., Jermaine, C.: Efficient data allocation over multiple channels at broadcast servers. IEEE Trans. on Computers, Special Issue on Mobility and Databases 51(10), 1231–1236 (2002)
13. Peng, W.C., Chen, M.S.: Efficient channel allocation tree generation for data broadcasting in a mobile computing environment. Wireless Networks 9, 117–129 (2003)
14. Hung, J.J., Seifert, A.: Flexsched: A parameterized data schedule generator for multi-channel broadcast systems. In: MDM 2006 (2006)
15. Huang, J.L., Peng, W.C., Chen, M.S.: Som: Dynamic push-pull channel allocation framework for mobile data broadcasting. IEEE Trans. on Mobile Computing 5(8), 974–990 (2006)
16. Leong, H.V., Si, A.: Data broadcasting strategies over multiple unreliable wireless channels. In: CIKM 1995 (1995)
17. Zheng, B.H., Wu, X., Jin, X., Lee, D.L.: Tosa: a near-optimal scheduling algorithm for multi-channel data broadcast. In: MDM 2005, pp. 29–37 (2005)

A Test Paradigm for Detecting Changes in Transactional Data Streams

Willie Ng and Manoranjan Dash

Centre for Advanced Information Systems,
Nanyang Technological University,
Singapore 639798
{ps7514253f,AsmDash}@ntu.edu.sg

Abstract. A pattern is considered useful if it can be used to help a person to achieve his goal. Mining data streams for useful patterns is important in many applications. However, data stream can change their behavior over time and, when significant change occurs, much harm is done to the mining result if it is not properly handled. In the past, there have been many studies mainly on adapting to changes in data streams. We contend that adapting to changes is simply not enough. The ability to detect and characterize change is also essential in many applications, for example intrusion detection, network traffic analysis, data streams from intensive care units etc. Detecting changes is nontrivial. In this paper, an online algorithm for change detection in utility mining is proposed. In order to provide a mechanism for making quantitative description of the detected change, we adopt the statistical test. We believe there is the opportunity for an immensely rewarding synergy between data mining and statistic. Different statistical significance tests are evaluated and our study shows that the Chi-square test is the most suitable for enumerated or count data (as is the case for high utility itemsets). We demonstrate the effectiveness of the proposed method by testing it on IBM QUEST market-basket data.

1 Introduction

Discovery of association rules is an interesting subfield of data mining and has seen a proliferation of techniques over the past several years [1]. This problem originates from the field of market basket analysis where the items are the products in a supermarket, and every transaction can be seen as a record of the purchases of an individual customer. An active and prominent aspect of data mining research is the measuring of "usefulness" of the discovered patterns. Unfortunately, traditional association rule mining (ARM) algorithms only consider if an item is absence or present in a transaction. The quantity of the item presents in a transaction has been tacitly ignored. Moreover, the external information such as the profit or cost of the item has not been considered as well. Usually, a large number of highly frequent itemsets are generated. They do not necessarily provide adequate answers for what the high utilities itemsets are. Recently, the term *utility mining* (UM) was suggested by [2] to address the shortcomings of ARM. In this context, utility refers to the measuring of how valuable an itemset is. The main goal is to mine high utility itemsets from a transactional dataset.

J.R. Haritsa, R. Kotagiri, and V. Pudi (Eds.): DASFAA 2008, LNCS 4947, pp. 204–219, 2008.

In this paper, we extend the challenge of UM from traditional static databases to data streams. The problem of mining frequent patterns in data streams has received a lot of attention [10]. For many recent applications such as trend learning, intrusion detection, fraud detection, etc, applying the concept of data stream might be more appropriate than normal databases. Unlike traditional static databases, data streams are continuous, unbounded and often arrive at high speed. Moreover, the underlying process that generates the data streams may evolve over time.

Suppose we employ a mining algorithm to create a model that describes certain aspects of the data stream. Assuming the current data generating process is stationary, it is possible to simply feed this algorithm with a "good" sample. In such a scenario, our main problem will be only to seek the best sampling technique such that a sample can be used to represent the unbounded data stream. We can then draw upon years of research in sampling where powerful techniques have already been implemented [14,15,16]. Unfortunately, such an assumption is impractical as data streams do change. Moreover, it is a non-trivial task to determine when change really occurs.

In mining frequent itemsets over stream data, past research mainly focus on adapting rather than detecting changes [27,28,29,30]. Maintaining an up to date model is important. However, we contend that it is also imperative to alarm the user when change is detected. A user may be interested to understand the nature of the change so that he can take the appropriate action against such change in the shortest possible time. Therefore, in this spirit, we investigate a feasible detection method that is capable to work with UM algorithms. In addition, to provide a mechanism for making quantitative description of the detected change, we adopt the statistical test. We believe there is the opportunity for an immensely rewarding synergy between data mining and statistic. More importantly, in this paper, the hypothesis testing tool is seen as a"verifier", while the UM algorithm is seen as a "discoverer". Any model that is generated by the UM algorithm will be compared with a reference model. A statistical test will determine if there is any discrepancy between the two models. Different statistical significance tests are evaluated and our study shows that the Chi-square test is the most suitable for enumerated data (as in this case for high utility itemsets).

1.1 Problem Statement

Our research problem is illustrated in Figure 1. We define the problem as follows: Consider that a data stream composed of two consecutive subparts ($Stream_A$ and $Stream_B$). The transition from $Stream_A$ to $Stream_B$ occurs at time t_{change}. Such scenario is particularly true in applications that have seasonal changes , e.g. from winter to spring. At certain fixed time interval, we conduct sampling. For example, we obtain the sample S_{A1} at time t_0 and t_1. To obtain S_{A1}, we can use some technique such as reservoir sampling. We can then mine S_{A1} and construct a reference model (high utility itemsets) that represents $Stream_A$. Assuming that the data are generated independently, the data distribution of S_{A1} and S_{A2} should be identical ($P(S_{A1}) \approx P(S_{A2})$). In this respect, whatever high utility itemsets found in S_{A1} is likely to be uncovered again in S_{A2}.

We let S_{A1}, the sample that is used to build the reference model, be the reference sample. Our objective will be to design a test algorithm that can determine if the present

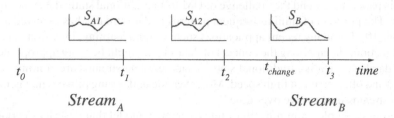

Fig. 1. Comparing sampling distribution

and past data stream are different. If the test indicates that they are different, the mining algorithm will be triggered to alarm the user. In this case, S_B will be the next reference sample and the monitoring process continues. All the samples will be processed in the main memory.

1.2 Preliminaries

For ease of exposition, this section provides the common notations that can be found in UM. We adopt the same definitions as in [2,3]. The problem can be formally stated as follows: Let $I = \{i_1, i_2, \ldots, i_m\}$ be a set of distinct items. Let D denote a database of transactions, where each transaction has a unique identifier (*tid*) and contains a set of items. The local transaction utility value, $lo(i_p, T_q)$, is the amount of item i_p in transaction T_q (Table 1). The external utility, $ex(i_p)$, is the value associated with i_p in Table 2. To measure the utility of i_p in transaction T_q, we simply compute the product of $lo(i_p, T_q)$ and $ex(i_p)$. For example, for $u(A, T_1) = 5 \times 2 = 10$.

An itemset X is a set of items such that $X \in (2^{|I|} - \{\emptyset\})$ where $2^{|I|}$ is the power set of I. An itemset with k items is called a k-itemset. We write "ABC" for the itemset $\{A, B, C\}$ when no ambiguity arises. The definition of utility of an itemset X, $u(X)$, is the sum of utilities of X in all the transactions containing X. For example, $u(AB) = u(A, T_1) + u(B, T_1) + u(A, T_2) + u(B, T_2) = 57$. An itemset X is considered a high utility itemset if and only if its utility is more than or equal to some minimum utility threshold (α), i.e. $u(X) \geq \alpha$. Thus, the job of UM is to mine for all the high utility itemsets in D. We denote by AHI the set of all high utility itemsets. In this paper, we employ the Two-Phase algorithm to efficiently mine high utility itemsets. A comprehensive description of this algorithm can be found in [3]. The rest of this paper is organized as follows: We explain the reason of using statistical test for change detection in Section 2. In section 3, we give a formal treatment of change detection by introducing the ACD[1]. Section 4 is devoted to the discussion of choosing the right statistical tool. In order to demonstrate the applicability of our methodology, Section 6 presents the preliminary experiment results. In Section 6, we provide a literature review on the works that are related to this paper. Section 7 gives the conclusions and future directions on our presented work.

---

[1] Algorithm for change detection.

Table 1. Transaction table. Each row is a transaction. The columns represent the amount of items in a particular transaction.

tid	A	B	C	D	E
T_1	5	1	3	1	1
T_2	1	2	1	0	0
T_3	0	4	1	0	2
T_4	20	0	3	2	4
T_5	0	2	0	1	6
T_6	12	0	1	0	7

Table 2. The external utility table

Item	Profit ($)
A	2
B	15
C	30
D	100
E	7

2   Change Detection

We would need a metric that can compare two models. In the case of UM, there will be two sets of AHI for the two models. The standard solution is to measure the symmetric difference [31] between these two sets of AHI:

$$S.D. = \frac{|(AHI_{cur} - AHI_{ref}) \cup (AHI_{ref} - AHI_{cur})|}{|AHI_{cur}| + |AHI_{ref}|}. \tag{1}$$

The smaller the size of $S.D.$, the greater is the similarity between AHI_{ref} and AHI_{cur} (and vice versa). Note, however, that while computing $S.D.$, the absolute utility value of each high utility itemset has been tacitly ignored.

We view the utility value of the itemsets as a potentially useful information. Consider a threshold $\alpha = 200$. Then, an itemset ABC having the utility value of 200 in the reference sample and 2000 in the current sample will be treated as high at the same time. Obviously, there is a significant difference but this will not be reflected in equation 1.

Another approach is to compare two models in a statistically meaningful way. In order to determine whether there are deviations between the current and the past data stream with respect to high utility itemsets, we construct two discrete distributions. Each bin in the distribution corresponds to an itemset. We want to know if the past and present data streams are related based on the distributions. However, it is practically impossible to establish a distribution that contains all the possible itemsets. For example, if I is a set of all distinct items, to build a histogram, one would required $2^{|I|} - 1$ number of bins! Since we are unable to build a distribution out of the data stream, we must base our conclusions on the models produced from the samples. Statistical tools for exploring such hypotheses are often called hypothesis testing.

The basic principle of the hypothesis testing is as follows. We start by defining two complementary hypotheses: the null hypothesis and the alternative hypothesis. In our case, the null hypothesis, denoted as H_0, might state that $Hist_{ref} = Hist_{cur}$, i.e. there has been no detectable change in the data stream, and the alternative hypothesis (H_1) might state that $Hist_{ref} \neq Hist_{cur}$, i.e. there is a detectable change. $Hist_{ref}$ denotes distribution representing the reference sample and $Hist_{cur}$ denotes distribution representing the current sample. Using the reference and current samples, we can compute a test statistic e.g., student-t statistic. The statistic value would vary from sample to sample. Based on the test statistic, the p-value is determined. The p-value represents the probability of having a sample as extreme as or more extreme than the current sample with respect to the reference sample. If the p-value is less than the level of significant, the null hypothesis is rejected, or else the null hypothesis is accepted [33].

3   Algorithm for Change Detection

In this section, the algorithm for change detection is describe. First, we determine the appropriate size of the sample. Second, we give the general overview of *ACD*.

3.1   Hoeffding Bound

Choosing a random sample of the data is one of the most natural ways of choosing a representative subset. The primary challenge in developing sampling-based algorithms stems from the fact that the utility value of an itemset in a sample almost always deviates from the utility value in the entire database. A wrong sample size may result in missing itemsets that have high utility value in the database but not in the sample and false itemsets that are of high value in the sample but not in the database.

Therefore, the determination of the sample size then becomes the critical factor in order to ensure that the outcomes of the mining process on the sample generalize well. Large deviation bounds such as the Hoeffding bound (sometimes called the additive Chernoff bounds) can often be used to compute a sample size n so that a given statistics on the sample is no more than ϵ away from the same statistics on the entire database, where ϵ is a tolerated error. As an example, if μ is the expected value of some amount in [0, 1] taken over all items in the database, and if $\hat{\mu}$ is a random variable denoting the value of that average over an observed sample, then the Hoeffding bound states that with probability at least $1 - \delta$,

$$|\mu - \hat{\mu}| \leq \epsilon. \tag{2}$$

The sample size must be at least n, where

$$n = \frac{1}{2\epsilon^2} \cdot \ln \frac{2}{\delta}. \tag{3}$$

This equation is very useful and it has been successfully applied in ARM algorithm [17]. Although the Hoeffding bound is often considered conservative, it gives user at least a certain degree of confidence. Note that demanding small error is expensive, due to the quadratic dependence on ϵ, but demanding high confidence is affordable, thanks to the logarithmic dependence on δ.

In this paper, we will use the Hoeffding bound to determine the sample size. For example, in UM, given $\epsilon = 0.01$ and $\delta = 0.001$, $n \approx 38,005$, which means that for itemset X, if we sample 38,005 transactions from a data set, then its true utility value $u(X)$ is beyond the range of $[u(X) - 0.01, u(X) + 0.01]$ with probability 0.001. In other words, $u(X)$ is within $\pm \epsilon$ of $u(X)$ with high confidence 0.999. It should be noted that this bound applies to experiments involving n distinct Bernoulli trials and thus it suits ARM perfectly. However, in UM, each item has amount not only to whether it is present or absent $(0, 1)$. The sample size is just a rough estimate. Fortunately, in our experiments, we did not encounter much problem. We leave a more comprehensive investigation of Hoeffding bound as a topic for future research.

3.2 Reservoir Sampling

The issue of how to maintain a sample of a specified size over data that arrives online has been studied in the past. The standard solution is to use reservoir sampling [20,32]. The technique of reservoir sampling is, in one sequential pass, to select a random sample of n transactions from a data set of N transactions where N is unknown and $n \ll N$.

Reservoir Sampling can be very efficient, with time complexity less than linear in the size of the stream. In [32], Vitter proposed the algorithm Z where the average run time is $O(n(1 + log(N/n)))$. The sample serves as a reservoir that buffers certain transactions from the data stream. New transactions appearing in the stream may be captured by the reservoir, whose limited capacity then forces an existing transaction to exit the reservoir. Variations on the algorithm allow it to idle for a period of time during which it only counts the number of transactions that have passed by. After a certain number of transactions are scanned, the algorithm can awaken to capture the next transaction from the stream.

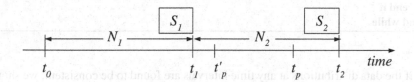

Fig. 2. A fragment of sequence

For our change detector, we would need to generate two sets of samples. It is not difficult to create two reservoir samples at separate time intervals. For example, from time 0 to t_1, we maintain a reservoir of size n and from t_1 to t_2, we maintain another reservoir of the same size (see Figure 2). The interval $[t_0, t_1]$ provides the range of transactions (N_1) for computing high utility itemsets. However, (N_1) can easily overwhelm the memory space. Even if it is possible to store them in the main memory, the time $(t_p - t_1)$ to compute all the transactions may be too expensive. To meet the real time requirement, we allow the reservoir sample to be buffered to the main memory at t_1 and processed such that the approximate high utility itemsets can be uncovered at t'_p where $t'_p - t_1 \ll t_p - t_1$.

3.3 Algorithm for Change Detection

Having determined the sample size, we shall proceed to explain our algorithm. The algorithm for detecting change in data steam is outlined in Algorithm 1. In the beginning, the first set of sample (S_1) will be loaded into the reference window $(window_{ref})$. Using Two-Phase algorithm [3], we can efficiently generate all the high utility itemsets, AHI_{ref}, from $window_{ref}$. In the next time interval, we update the current window, $window_{cur}$, using the new set of sample S_2. We then generate the AHI_{cur}. The absolute utility value of the high utility itemsets will be recorded. To test for discrepancy, we compare AHI_{ref} with AHI_{cur}. Note that when we conduct the statistical test, we are actually comparing the high utility itemsets based on their utility values. In an ideal

Algorithm 1. Algorithm for Change Detection

1: $T \leftarrow t_1$;
2: load S_1 to $window_{ref}$ at t_1;
3: $AHI_{ref} \leftarrow$ TwoPhase($window_{ref}$);
4: $i = 1$;
5: **while** not end of stream **do**
6:     $i + +$;
7:     load S_n to $window_{cur}$ at t_i;
8:     $AHI_{cur} \leftarrow$ TwoPhase($window_{cur}$);
9:     $AHI'_{ref} = scan(window_{ref}, AHI_{cur} - AHI_{ref}) + AHI_{ref}$;
10:     $AHI'_{cur} = scan(window_{cur}, AHI_{ref} - AHI_{cur}) + AHI_{cur}$;
11:     **if** $TEST(AHI'_{ref}, AHI'_{cur})$ = Reject **then**
12:         $T \leftarrow t_i$;
13:         Report change at time T;
14:         load S_n to $window_{ref}$;
15:         $AHI_{ref} = AHI_{cur}$;
16:     **end if**
17: **end while**

case, if the data distributions at any time intervals are found to be consistent, we should be able to find a match. That is, whatever high utility itemset discovered in $window_{ref}$ can also be discovered in $window_{cur}$, and vice versa. Therefore, each high utility itemset will contain two utility values (one from each window). However, there may be a small set of itemsets whose utility value are very close to the minimum utility threshold. Since we are dealing with random sampling, it is inevitable to encounter random fluctuation. As a consequence, even when there is no changes in the data stream, we might still have some itemsets that are present in AHI_{ref} but absent in AHI_{cur}, and vice versa.

It should be emphasized that our test does not end at simply comparing itemsets which are $AHI_{ref} \cap AHI_{cur}$. In fact, as long as an itemset is considered high utility in either $window_{ref}$ or $window_{cur}$, it has to be included in the test. We would like to look into why this itemset is considered high in one window but not in another. To include this itemset into the test, the absolute utility value must be obtained from both the windows. Since we already have one of its utility value through the Two Phase algorithm, what we need to do is to perform a single scan on the window that classified this itemset as low utility. For example, if itemset ABC is considered high utility in $window_{cur}$ but not in $window_{ref}$, we will do a scan on $window_{ref}$ to compute its utility value. That means, at any time there are two samples residing in the memory. In this paper, we define $AHI_{cur} - AHI_{ref}$ as a set of itemsets that is considered high utility in $window_{cur}$ but not in $window_{ref}$ and $AHI_{ref} - AHI_{cur}$ as a set of itemsets that is considered high utility in $window_{ref}$ but not in $window_{cur}$.

Once the scanning is done, we will have two sets of itemset, AHI'_{ref} and AHI'_{cur}, that cover $window_{ref}$ and $window_{cur}$ respectively. A test will be conducted on this two sets to see if there is any discrepancy. A detail discussion about the statistical test is covered in the next section. If there is significant difference, the algorithm will report a change in the current stream. The content in $window_{cur}$ and AHI_{cur} will be transferred to $window_{ref}$ and AHI_{ref} respectively. $window_{cur}$ will be cleared to accommodate the next set of sample. The test cycle is repeated again.

4 Statistical Test

An important aspect of the proposed algorithm is the ability to test whether the frequency counts of itemsets in the reference window are different from the frequency counts of those itemsets in the current window. Statistical tests are very helpful. There are many statistical tests [33]; choosing the right one for the given task can be at times challenging as it is in this case. Let us consider the characteristic of the test that we wish to do: (a) it is paired and two-sample – because for the reference and current windows we compare the frequency counts of the same itemsets, (b) it is not continuous, but count data, (c) the underlying distribution is not given, but one may argue in favor of approximate normal distribution considering the fact that the data are frequency counts – typically when it is count data, binomial distribution is used, and when the size is large (as is the case here) one may approximate it to be normal, (d) we are interested in determining whether the differences (irrespective of positive or negative) between the corresponding counts are significant. Considering these criteria, one may choose, paired t-test, non-parameteric tests (Sign test and Wilcoxon Signed-Rank test for paired samples), Chi-square test, test of two proportions, and Fisher's exact test. Let us consider the suitability of these tests for our task in a systematic manner. We devote some space here for describing and analyzing the suitability of these tests in order to give a comprehensive understanding. A simple data is used to compare the suitability among the different methods. Sample 1: Item A - Utility 30, Item B - Utility 70; Sample 2: Item A - Utility 70, Item B - Utility 30. Obviously, these two samples are very different.

4.1 Paired t-test

It is used for two samples that are paired. Two samples can be paired either by self-paring (the same subject has two values corresponding to the two samples) or by matching (there are actually two subjects but they are matched using some common characteristics). In our case the it self-pairing, i.e., the utilities of the same itemset is measured in two samples – reference and current windows. The null hypothesis is that the two population (from which the two samples were drawn) means are the same, or in other words their mean difference is 0. The alternative hypothesis is that the mean difference is not 0. Assuming that the data is approximately normally distributed, t-test is applied. The following equations mathematically depict this. $H_0 : \delta = 0, H_A : \delta \neq 0$ and $t = \frac{\bar{d} - \delta}{\frac{s_d}{\sqrt{n}}}$ where $\bar{d} = \frac{\sum_{i=1}^{n} d_i}{n}$, i.e., the mean difference and d_i is the difference between i^{th} values of reference and current samples, s_d is the standard deviation of the differences d_i, n is the sample size and δ is the population mean difference.

Under H_0 the mean difference is 0. But notice that by taking the mean of the differences, paired t-test is actually homogenizing the differences, both positive and negative. To understand this point, let us run through our example problem described earlier. The two differences are 30-70=-40 and 70-30=40 respectively. The mean difference is 0! Thus the null hypothesis is not rejected which is actually **not true**.

4.2 Nonparametric Tests

These tests are useful particularly when the underlying distributions cannot be assumed to be normal or approximately normal. For paired data two tests are commonly used: sign test and Wilcoxon signed-rank test. In the sign test, just like the paired t-test, the differences between the corresponding values between the two samples are computed and their signs (+ or -) are noted down. The null hypothesis states that the median difference is equal to 0. Note that median (not mean) is used because only the sign but not the magnitude is used. Under the null hypothesis one can expect the number of +ve and -ve signs to be equal. Thus the number of +ve and -ve signs are computed. Notice that as the outcome of each difference is a sign which takes only two possible values with a probability of 0.5, each of these is considered as a Bernoulli trial. Thus the mean number of +ve signs is $n/2$ and standard deviation is $\sqrt{\frac{n}{4}}$. If n is large (≥ 20 one can use a standard normal variable as given below.

$$z_+ = \frac{D - n/2}{\sqrt{\frac{n}{4}}} \tag{4}$$

If $n < 20$, binomial distribution is used. The p-value is determined by computing the twice the probability of obtaining D positive difference – or some number more extreme – given that the null hypothesis is true. In our example problem with only two itemsets, binomial distribution is used. $P(D \geq 1) = P(D = 1) + P(D = 2) = 0.75$ As the p-value is much larger than the assumed level of significance 0.05, the null hypothesis is accepted, i.e., the median difference is 0, although clearly it is not true.

The sign-rank test is a modification of the sign test where in addition to the signs of the differences, their ranks are also considered. The differences are ranked using their absolute values, and then the ranks of the positive difference are separated from the ranks of the negative differences. The sum of these two groups of ranks are taken. Under the null hypothesis which states that the median difference is 0, it is expected that the two sums of ranks (+ve ranks and -ve ranks) will be equal. Then based the on size of the sample (n) either a standard normal distribution (when $n > 12$) or a special distribution (when $n \leq 12$) is used to compute the p-value. In the example data $n = 2$, so we use the special distribution and p-value > 0.05, thus the nul hypothesis is accepted, which is not correct. The reason why both the nonparametric methods failed is because they also grouped the differences **without considering their absolute values**.

4.3 Chi-square Test

Finally we discuss the Chi-square test and some of its variations which are found to be the most suitable tests. Chi-square test is applicable for count data. Data is typically arranged in a contingency table format and the observed frequencies is compared with expected frequencies. See the Table 3 showing the contingency table for the example problem. H_0 : There is no association between the two variables ("samples" and "Itemsets"),or in other words the proportions of the itemsets in the reference sample are identical. H_A : The two proportions are not identical.

Table 3. 2 X 2 Contingency table for the example problem. The values inside bracket () are the expected counts.

	Itemsets		
Sample	A	B	Total
Reference	30(50)	70(50)	100
Current	70(50)	30(50)	100
Total	100	100	200

To perform the test for the counts in a contingency table with r rows and c columns, we calculate the sum

$$X^2 = \sum_{i=1}^{rc} \frac{(O_i - E_i)^2}{E_i} \tag{5}$$

where O_i is the observed count, and E_i is the expected count. Expected count for cell i is equal to the row total multiplied the column total divided by the table total. They are given inside brackets in Table 3. Using our example problem the p-value is computed to be 0 with 1 $(r\text{-}1)^*(c\text{-}1)$ degrees of freedom. The null hypothesis is **rejected**. Thus, it is concluded that the samples are associated with the itemsets, or in other words the proportions of the itemsets in the reference sample are identical.

We can generalize this test to $c > 2$ to suit our need for comparing the reference sample with the current sample. All other computations are similar. Note that one may argue that we could have used "two proportions test". But the limitation of this test is that only **two** proportions (or in our case, items) can be compared. But our need is to compare multiple (more than two) items. One can also use **Fisher's exact test**. Fisher's exact test produces very similar results as Chi-square test. It is more suitable when the counts in the contingency table are very small (< 5) for which Chi-square test is not very suitable. But typically data mining applications have very large size of the data, thus Chi-square test is the best method among all. We did not use any continuity correction (e.g., Yates correction) because the number of degree of freedom $(r\text{-}1)^*(c\text{-}1)$ is very large. This correction is useful when the number of degree of freedom is very small, such as 1.

5   Experimental Evaluation

We conducted extensive experiments to evaluate the performance of *ACD* using different statistical tests (paired *t*-test, Wilcoxon signed-rank test and Chi-square test). All experiments were performed on a dedicated Dell PowerEdge 2600 computer with a processor speed of 3066MHz. We used a synthetic database in our experiments and it was generated using the code from the IBM QUEST project [1]. For the experiments, we will use three different data sets (*StreamA*, *StreamB*, and *StreamC*). The parameter settings for the synthetic data generations are as follows:

StreamA: average size of transaction= 20, average size of maximal pattern= 6.
StreamB: average size of transaction= 15, average size of maximal pattern= 7.
StreamC: average size of transaction= 12, average size of maximal pattern= 5.

These data sets only contain quantity of 0 or 1 for each item in a transaction. To suit our need, we randomly generate the quantity of each item in each transaction, ranging from 1 to 5. The external utility table are also synthetically generated by assighning a utility value to each item randomly, ranging from 0.01 to 10.

Table 4. Number of false alarms generated. There are 100 test points for each data set.

	StreamA	StreamB	StreamC
	Fault	Fault	Fault
t-test	28	35	29
Wilcoxon	2	1	1
Chi-square	5	3	4

5.1   Test for False Alarm

Before running any experiment, we examine how many false alarms each of the statistical test will generate. Without loss of generality, we equate the time duration to a fix number of transactions. For the experiment, we set a minimum utility threshold of 1%. We let *ACD* scan the three data sets using different statistical tests. Each data set contains about 50 million transactions. We set our window size to be 40k. Thus whenever the current window scan passes 500k transactions, *ACD* will conduct a test. There will be about 100 tests in total for each run. The result is presented in Table 4 where a "Fault" represents a false alarm. It shows that when there is no change in the data distribution, the t-test has the highest probability to trigger a false alarm.

5.2   Test for Changes

In the experiment, we investigate the ability of *ACD* to detect changes in the distribution of *AHI*. To do this, we combined the three data sets into one. *ACD* is made to scan pass 50 million transactions. However, after every 1 million transactions, there will be a transition from one data set to another entirely different data set, e.g., *StreamA-StreamB-StreamC-StreamB-StreamA*.... The number of test point remains the same at 100 but out of these 100 tests, 50 are real transitions.

The experiment is reported in Table 5. Here, a "Hit" refers to a success in triggering the alarm whenever there is a real transition. From the table, it is obvious that t-test scores the worst. Although it can score 28 hits, its false alarm rate is too high. Therefore, we cannot conclude if the alarms are triggered by t-test or by random. On the contrary, Wilcoxon test receives the lowest false alarm rate. Unfortunately, it does not respond well to changes. Out of the 50 transitions, it managed to detect only 10. As for Chi-square test, it misses 3 hits and produces 5 false alarms. Clearly, among the three statistical test, Chi-square test is the most sensitive to changes. We let the reference window to be fixed.

Table 6 shows the results of the experiment when the minimum utility threshold is reduced to 0.1%. By lowering the threshold, t-test generates even more false alarms. Only slight changes occur in the other two tests. Based on the overall scores, we are

Table 5. Experiment to detect change with minimum utility threshold set at 1%. There are 100 test points. Half of them are real transitions and the other half are false alarms.

	Hit	Fault
t-test	28	27
Wilcoxon	10	2
Chi-square	47	5

Table 6. Experiment to detect change with minimum utility threshold set at 0.1%. There are 100 test points. Half of them are real transitions and the other half are false alarms.

	Hit	Fault
t-test	35	38
Wilcoxon	8	2
Chi-square	46	6

comfortable to say that when detecting transitional change in data streams, *ACD* with Chi-square test works reasonably well.

5.3   Test for Sensitive

In the previous experiments, we assume that whenever there is a change, the current window is completely filled up with a new data set. However, during the early transition state, it is possible for the reservoir sample to contains a mixture of two different data sets. It will be interesting to examine how well the Chi-square test react to the

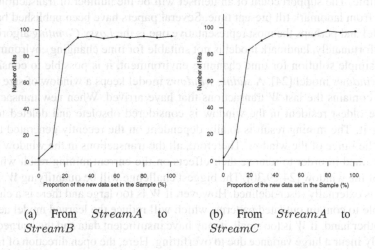

(a) From *StreamA* to   (b) From *StreamA* to
     StreamB   *StreamC*

Fig. 3. Experiment to examine how fast the Chi-square test react when *StreamA* is slowly replaced by new data set

current window if it is partially mixed with new data set. For the experiment, we let the reference window to contain the sample obtained from $StreamA$. As for the current window, we fill it with a sample having a mixture of a new data set and $StreamA$. In figure 3, the X-axis in the graph represents the proportion of the new data set in the sample. We slowly increase the proportion until the current window is completely occupied by the new data set. For every 10 percent increment, we conduct a Chi-square test. This process is repeated for 100 times. The Y-axis of the graph represents the number of hits the test is able to achieve. In figure 3a and 3b, both the graphs converge when the proportion reaches 40%. Beyond that point, there is little increase in the number of hits. Note that even when it is at 20%, Chi-square test can still achieve more than 80 hits. This shows that Chi-square test can be very efficient in early detection of change.

6   Related Work

In this paper, there are two aspects of related work; utility mining and data stream mining. In the field of utility mining, there are several definitions of useful itemset depending on the user's objective. Researches that assign different weight to items have been introduced in [5,6]. The share measure was proposed by [7] to overcome the shortcomings of support. It reflects the impact of the quantity sold on cost or profit of an itemset. [8] proposes a new pruning strategy based on utilities that allow pruning of low utility itemsets to be done by means of a weaker but anti-monotonic condition. Our research follows closely on the work presented in [2,3,4]. [4] is the earliest work on mining temporal high utility itemsets from data streams.

For data stream mining, there are three types of data stream mining models, *landmark*, *sliding windows* and *damped*. The *landmark* model mines all frequent itemsets over the entire history of stream data from a specific time point called landmark to the present time. The support count of an itemset will be the number of transactions it has scanned from landmark till present time. Several papers have been published based on this model and perhaps the most representative one is the *Lossy Counting* algorithm by [13]. Unfortunately, landmark model is not suitable for time changing environment.

As a simple solution for time changing environment, it is possible to consider the *sliding windows* model [24]. A *sliding windows* model keeps a window of size W. The window contains the last W transactions that have arrived. When new transaction arrives, the oldest resident in the window is considered obsolete and deleted to make room for it. The mining result is totally dependent on the recently generated transactions in the range of the window. Therefore, all the transactions in the window need to be maintained in order to remove their effects on the current mining result when they are out of the window [24,29,30]. The biggest challenge will be quantifying W. In most cases, it is externally (user-)defined. However, if W is too large and there is a change, it is possible to contain outdated patterns, which will reduce the learned model accuracy. On the other hand, if W is too small, it may have insufficient data and the learned model will likely incur a large variance due to overfitting. Here, the open direction of research is to design a window that can dynamically decide its width based on the rate of the underlying change phenomenon.

The *damped* model (also called *time-fading* model) assigns different weights to transactions such that new transaction have higher weights than old ones [27,28]. At every moment, based on a fixed decay rate [25] or forgetting factor [26], a transaction processed d time steps ago is assigned a weight ω^d, where $\omega < 1$. As time passes by, the weight decreases. However, *damped* model suffers from similar problems as in *sliding windows* model. The challenge of determining a suitable W is now translated to that of determining a suitable ω. Lastly, Kifer et al.[12] gives a good study for detecting general changes in the data distribution of the data stream. They conduct several experiments in many scenarios and conclude that no statistical test works best in all situations. Our research is different in another salient manner; we focus our work on change detection purely on utility mining.

7 Conclusions

A change detector, *ACD*, is presented in this paper. *ACD* incorporates a statistical tool and is used to detect significant changes in a data stream. Experiments have been conducted to evaluate the performance of *ACD* using different statistical tests and our study demonstrated that Chi-square test is the most suitable for UM. In our work, we have identified two issues. First, as mentioned earlier, we have crudely estimated the sample size based on the Hoeffding bound. Obviously this approach is not good enough as we assume that the amount of an itemset in a transaction to be in binary form. We wish to explore this further. We are currently working in this area and the preliminary results are recorded in [20]. Next, it should be noted that when we use a statistical test to compare two distributions, we are comparing all the itemsets within the histogram in general. There may be some small group of itemsets which behavior changes from time to time. If the majority of the itemsets remain stable, the effect caused by this group can be diluted. Therefore, the test may miss out this group. We consider this group as the outlier. We view outlier as an important issue even though we have not covered in this paper. We leave a more comprehensive investigation on these areas for future research.

Acknowledgement

We would like to express our special appreciation to Zifen Teo for her assistance in conducting the various experiments recorded in this paper.

References

1. Agrawal, R., Imielinski, T., Swami, A.: Fast alogorithms for association rules. In: Int. Conf. Very large Data Bases, pp. 487–499 (1994)
2. Yao, H., Hamilton, H.J., Butz, C.J.: A Foundational Approach to Mining Itemsets Utilities from Databases. In: Proc.of the 4th SIAM Int. Conf. on Data Mining, Florida, USA (2004)
3. Liu, Y., Liao, W.K., Choudhary, A.: A two phase algorithm for fast discovery of high utility itemsets. In: Ho, T.-B., Cheung, D., Liu, H. (eds.) PAKDD 2005. LNCS (LNAI), vol. 3518, pp. 689–695. Springer, Heidelberg (2005)

4. Tseng, V.S., Chu, C.J., Liang, T.: Efficient Mining of Temporal High Utility Itemsets from Data streams. In: UBDM 2006 (2006)
5. Cai, C.H., Fu, A.W., Cheng, C.H., Kwong, W.W.: Mining Association Rules with Weighted Items. In: Proceedings of International Database Engineering and Applications Symposium, IDEAS 1998 (1998)
6. Tao, F., Murtagh, F., Farid, M.: Weighted Association Rule Mining using weighted support and significance framework. In: KDD 2003, pp. 661–666 (2003)
7. Barber, B., Hamilton, H.J.: Extracting Share Frequent Itemsets with Infrequent Subsets. Data Min. Knowl. Discov. 7(2), 153–185 (2003)
8. Chan, R., Yang, Q., Shen, Y.D.: Mining High Utility Itemsets. In: IEEE ICDM 2003, pp. 19–26 (2003)
9. Babcock, B., Babu, S., Datar, M., Motwani, R., Widom, J.: Models and issues in data stream systems. In: PODS (2002)
10. Jiang, N., Gruenwald, L.: Research Issues in Data Stream Association Rule Mining. SIGMOD Record 35(1), 14–19 (2006)
11. Domingos, P., Hulten, G.: A general Framework for mining massive data streams. Journal of Computational and Graphical Statistics 12 (2003)
12. Kifer, D., Ben-David, S., Gehrke, J.: Detecting change in data streams. In: VLDB, pp. 180–191 (2004)
13. Manku, G.S., Motwani, R.: Approximate frequency counts over data streams. In: Bressan, S., Chaudhri, A.B., Li Lee, M., Yu, J.X., Lacroix, Z. (eds.) CAiSE 2002 and VLDB 2002. LNCS, vol. 2590, pp. 346–357. Springer, Heidelberg (2003)
14. Johnson, T., Muthukrishnan, S., Rozenbaum, I.: Sampling algorithms in a stream operator. In: SIGMOD Conference (2005)
15. Olken, F., Rotem, D.: Random Sampling from database files - a survey. In: 5th Intl.Conf. Statistical and Scientific Database Management (April 1990)
16. Cochran, W.G.: Sampling techniques. John Wiley & Sons, Chichester (1977)
17. Toivonen, H.: Sampling large databases for association rules. In: VLDB, pp. 134–145 (1996)
18. Parthasarathy, S.: Efficient Progressive Sampling for Association Rules. In: IEEE ICDM, pp. 354–361 (2002)
19. Zaki, M., Parthasarathy, S., Li, W., Ogihara, M.: Evaluation of sampling for data mining of association rules. In: 7th International Workshop on Research Issues in Data Engineering (1996)
20. Dash, M., Ng, W.: Efficient Reservoir Sampling for Transactional Data Streams. In: IEEE ICDM workshop on Mining Evolving and Streaming Data, Hong Kong (2006)
21. Manku, G.S., Motwani, R.: Approximate frequency counts over data streams. In: Proc. of 28th Int'l Conf. on Very large Databases, Hong Kong (August 2002)
22. Domingos, C., Gavaldà, R., Watanabe, O.: Practical algorithms for on-line sampling. Discovery Science, 150–161 (1998)
23. Yu, X., Chong, Z., Lu, H., Zhou, A.: False positive of false negative: Mining frequent itemsets from high speed transactional data streams. In: Int. Conf on VLDB (2004)
24. Babcock, B., Datar, M., Motwani, R.: Sampling from a moving window over streaming data. In: Proc. SODA (2002)
25. Cohen, E., Strauss, M.: Maintaining time-decaying stream aggregates. In: PODS, pp. 223–233 (2003)
26. Yi, B.K., Sidiropoulos, N., Johnoson, T., Jagadish, H.V., Faloutsos, C., Biliris, A.: Online mining for co-evolving time sequences. In: ICDE, pp. 13–22 (2000)
27. Giannella, C., Han, J., Pei, J., Yan, X., Yu, P.S.: Mining Frequent Patterns in Data Streams at Multiple Time Granularities. In: Kargupta, H., Joshi, A., Sivakumar, K., Yesha, Y. (eds.) Next Generation Data Mining, pp. 191–212. AAAI/MIT (2003)

28. Chen, L., Lee, W.: Finding recent frequent itemsets adaptively over online data streams. In: Proc. of ACM SIGKDD Cof., pp. 487–492 (2003)
29. Chi, Y., Wang, H., Yu, P.S., Muntz, R.R.: Moment: Maintaining closed frequent itemsets over a stream sliding window. In: 4th IEEE ICDM (2004)
30. Lin, C., Chiu, D., Wu, Y., Chen, A.L.P.: Mining frequent itemsets from data streams with a time-sensitive sliding window. Siam Data Mining (2005)
31. Lee, S.D., Cheung, D.W.: Maintenance of Discovered Association Rules: When to update? In: Proc. 1997 ACM-SIGMOD Workshop on Data Mining and Knowledge Discovery (DMKD 1997), Tucson, Arizona (May 1997)
32. Vitter, J.: Random sampling with a reservoir. ACM Transactions on Mathematical Software 11(1), 37–57 (1985)
33. Pagano, M., Gauvreau, K.: Principles of biostatistics, Duxbury, Thomsom Learning,USA (2000)

Teddies: Trained Eddies for Reactive Stream Processing

Kajal Claypool[1,*] and Mark Claypool[2]

[1] MIT Lincoln Laboratories
claypool@ll.mit.edu
[2] Worcester Polytechnic Institute, Worcester, MA
claypool@cs.wpi.edu

Abstract. In this paper, we present an adaptive stream query processor, Teddies, that combines the key advantages of the Eddies system with the scalability of the more traditional dataflow model. In particular, we introduce the notion of adaptive packetization of tuples to overcome the large memory requirements of the Eddies system. The Teddies optimizer groups tuples with the same history into data packets which are then scheduled on a per packet basis through the query tree. Corresponding to the introduction of this second dimension – the packet granularity – we propose an adaptive scheduler that can react to not only the varying statistics of the input streams and the selectivity of the operators, but also to the fluctuations in the internal packet sizes. The scheduler degrades to the Eddies scheduler in the worst case scenario. We present experimental results that compare both the reaction time as well as the scalability of the Teddies system with the Eddies and the data flow systems, and classify the conditions under which the Teddies' simple packet optimizer strategy outperforms the per-tuple Eddies optimizer strategy.

Keywords: Stream Database Systems, Adaptive Query Processing, Adaptive Scheduler.

1   Introduction

The proliferation of the Internet, the Web, and sensor networks have fueled the development of applications that treat data as a continuous stream rather than as a fixed set. Telephone call records, stock and sports tickers, streaming data from medical instruments, and data feeds from sensors are examples of streaming data. As opposed to the traditional database view where data is fixed (passive) and the queries considered to be active, in these new applications data is considered to be the active component, and the queries are long-standing or continuous. In response, a number of systems have been developed [BW01,MWA+03,BBD+02,CCC+02,CDTW00] to address this paradigm shift from traditional database systems to now meet the needs of query processing over streaming data.

---

* This work was done while the author was at University of Massachusetts, Lowell.

J.R. Haritsa, R. Kotagiri, and V. Pudi (Eds.): DASFAA 2008, LNCS 4947, pp. 220–234, 2008.
© Springer-Verlag Berlin Heidelberg 2008

Stream query optimizers represent a new class of optimization problems. The most egregious problem faced by stream query optimizers is the need for query plan adaptation (in some cases continuous adaptation) in reaction to the turbulence exhibited by most data streams. The term turbulence here refers to the fluctuations in fundamental data statistics such as the selectivity, as well as to the variability in the network, where bandwidth, quality-of-service, and latency can modulate widely with time. The turbulence in data streams thus makes a static optimizer, which chooses a query plan once-and-for-all at query set-up time, inadequate. For example, in traditional systems the optimizer makes decisions on scheduling and resource allocation based on the fundamental statistics of the data such as its selectivity. However, in the dynamic case statistics that lead to optimal query plan selection at one point in time may no longer be prudent at a later time. A stream query optimizer must, thus, be able to adapt to changing conditions to deal with this dynamicity. To address this problem, two broad strategies for a dynamic stream query optimizer have been proposed: the *data flow* [CCC+02,MWA+03] and the *Eddies* [AH00,MSHR02] models.

A data flow optimizer typically selects a query plan topology based on past (or expected) stream statistics, and arriving data is processed *en masse* using this query plan topology. While the data flow optimizer is adaptive in theory, in practice it is often difficult to change the overall topology of the query plan without introducing significant delays into the system. The novel Eddies architecture [AH00,MSHR02,Des04,DH04] was proposed to introduce a higher degree of adaptability into the query optimizer. The Eddies model functions at the granularity of individual tuples, electing to find the most efficient evaluation path on a per tuple basis with respect to the conditions that prevail at the time. Thus, unlike the data flow and the traditional models, the Eddies model need not compute apriori the optimal query plan topology for the entire data set, rather it can calculate the optimal path based on current tuple statistics as it processes each tuple. The fast reaction time, the adaptability offered by the Eddies model, comes at a price. As its scheduler operates on a per tuple basis, the Eddies processor cannot take advantage of bulk processing of similar data. Moreover, the metadata required for each tuple to keep track of its progress through the logical query plan makes the Eddies system prohibitively expensive in terms of the memory requirements, thereby limiting its scalability.

In this paper, we propose Teddies, **T**rained **eddies**, – a hybrid system that incorporates the adaptability of the Eddies system with the scalability of the more traditional data flow model. The fundamental building block of the Teddies system is the *adaptive packetization* of tuples to create a tight inner processing loop that gains efficiency by bulk-processing a "train" of similar tuples simultaneously. Corresponding to this new dimension – the packet granularity – we introduce an *adaptive scheduler* that reacts to not only the time varying statistics of the input streams and the selectivity of the operators, but also to the fluctuations in the internal packet sizes. The scheduler, thus, schedules only those packets that are sufficiently filled as determined by a threshold value for each packet type. Additionally, the threshold value for each packet type is adjusted dynamically, growing if the system is creating many tuples of a particular type, and shrinking otherwise. This ensures that there is no starvation – packets containing tuples of all types are eventually scheduled.

The Teddies system uses packets in a manner similar to data flow systems, while at the same time applying an adaptive scheduling mechanism that does not tie down the overall processing to a single query tree topology. The cost of the Teddies system is the introduction of latency into the system since some tuples may have to remain in their respective queues for longer periods of time waiting for tuples with similar metadata to fill the queue to the threshold. We ran a series of experiments to measure the overhead of our adaptive packetization algorithm, as well as a series of experiments to isolate the performance benefits of packetization with respect to the varying cost of a single routing decision. The results show the promise of the Teddies approach over both the data flow and Eddies approaches.

The rest of the paper is organized as follows: Section 2 presents an overview of the data flow and the Eddies system to set the context for the rest of the paper; Section 3 gives an overview of the Teddies system and details its main components. Section 4 outlines our experimental methodology and presents our preliminary experimental results; Section 5 briefly describes the related work; and Section 6 summarizes our conclusions and provides possible future work.

2   Background: Data Flow and Eddies Architecture

2.1   Data Flow Query Processor

In the *data flow* framework, stream operators have dedicated buffers at their inputs, and the query plan chosen by the optimizer is instantiated by connecting the output of some operators to the input of another operator. Figure 1 gives a pictorial depiction of a multi-way join query topology in the data flow model. The operators are scheduled to consume tuples in their input queues in such a way so as to optimize the throughput of tuples in the system. Various strategies and algorithms are used to schedule the operators. However, the topology of the query plan is difficult to change once it has been set up, as the amount of computational state that resides in the operator input queues at any point of time cannot easily be disentangled. Hence, a change to a potentially more advantageous query topology, involving reconnection of some of the operator queues, cannot be done except at special points of time. One possible strategy to address this is to: (1) block all upstream tuples from entering that portion of the query tree that is to be modified; (2) drain all the remaining tuples in the intermediate queues; (3) reconnect the queues in a new plan; and then (4) unblock the upstream tuples to resume execution. If there is a large amount of intermediate state, this blocking strategy is both costly and inefficient. The latency in the reaction time of the data flow architecture is the main motivation for the introduction of the Eddies framework. The Eddies model explicitly keeps track of the intermediate state of tuples in the input queues to enable query

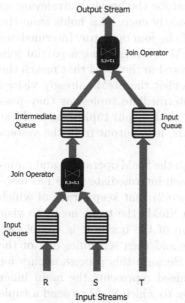

Fig. 1. The Dataflow Query Processing Architecture

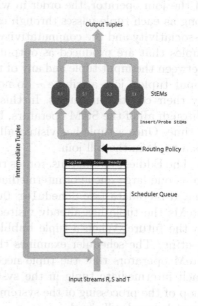

Fig. 2. The Eddies System Architecture

topology rearrangement on the fly without blocking any processing and hence introducing any latency in the system.

2.2 The Eddies Query Processor

The Eddies stream query processor is a highly adaptive architecture for computing stream queries. In Eddies, tuples – input tuples or intermediate tuples – are routed individually through the query plan. Each tuple has associated with it a set of metadata that tracks the progress of the tuple through the query plan. This metadata is unique to each individual query and its structure is determined when the query is first introduced into the system. A query plan consists of the operators that compose the query, such as versions of SELECT, PROJECT, and JOIN operators.

The Eddies system relaxes the requirement to choose a particular query topology by decomposing the join operators, in this case, into sub-operators – called SteMs – and then routing the tuples through the SteMs [Ram01,MSHR02]. There is one SteM associated with each stream attribute that participates in the join. For example, consider the join $R\bowtie_{R.1=S.1}S\bowtie_{S.2=T.1}T$ shown in Figure 2. Here the join attributes $R.1$, $S.1$, $S.2$, and $T.1$ each have an associated SteM. The SteM keeps track of the window of attribute values that have arrived on the stream and allow subsequent tuples from the other streams to probe these stored attribute values to search for a match. The SteM may be implemented, for example, as a hash table on the attribute values to allow for efficient probing. In the case of the join operator, the order in which tuples probe the SteMs is irrelevant as long as each tuple passes through each SteM exactly once. This holds from the associativity and the commutativity property of the join operator. Intermediate tuples that are produced as output from a SteM – representing a partial join between the input tuple and any of the tuples stored in the SteM that match the input tuple's join attribute – do not need to revisit the SteMs already visited by their constituent tuples. In this way the intermediate tuples, as they pass through all of the SteM operators, build up the final output tuples one SteM at a time. Once a tuple has visited all of the SteMs, it is output from the system as a tuple in the full join.

The Eddies system, thus, routes tuples through the SteM operators, and maintains one large queue of intermediate tuples. Each intermediate tuple has associated metadata (see Scheduler Queue in Figure 2) that keeps track of which SteMs the tuple has already visited and which SteMs the tuple needs to visit in the future. When a tuple bubbles to the top of the queue it is eligible for routing. The scheduler examines the metadata and then schedules one of the SteM operators that the tuple needs to visit. Because this process occurs for each intermediate tuple in the system, and indeed represents the main inner loop of the processing of the system, the decision to which SteM to send a tuple must be made efficiently.

The Eddies system uses an *adaptive scheduler*, one version of which keeps track of the number of tuples produced minus the number of tuples consumed for each query operator. In this version of the Eddies adaptive scheduler, a lottery based scheme is used whereby operators gain lottery tickets by consuming input and lose lottery tickets by producing output. For each tuple to be scheduled, the scheduler holds a lottery, with the winning operator selected to process the tuple [MSHR02,DH04]. By scheduling the operator that produces the fewest while consuming the most, the system can react to changing conditions by making a local decision for each tuple. Since each tuple need not follow the same path through the SteM operators, it is possible for the system to find the most efficient path for a given tuple with respect to the prevailing conditions.

3   Teddies Adaptive Query Processor

The Eddies system, while highly reactive, is limited in its scalability by the fine granularity of its routing decisions - that is by its tuple routing. To overcome the drawbacks of the Eddies system, we propose the Teddies system.

Teddies is an adaptive query processing system that leverages the fine-grained adaptivity of pure Eddies with the efficiencies gained from the bulk processing of query operators. A fundamental concept of the Teddies system is a *packet* wherein tuples with the same history (metadata) are grouped together. The Teddies scheduler thus works at the granularity of packets, making routing decisions per packet as opposed to the per tuple routing decisions made by the Eddies system. In a heavily loaded system, this packetization of tuples enables the scheduler to make a reduced number (by a factor of the packet size) of routing decisions per tuple, achieving a corresponding increase in the system efficiency. The potential gains of this amortization is harnessed by the scheduler to enhance the packet routing policies that are employed. In addition, when a particular operator is scheduled by the packet router, the run-time cost of the operator per-tuple is amortized across the number of tuples in a packet. However, the trade-off of grouping tuples into packets is that of a reduced adaptivity, as potentially a finer-grained scheduling policy may respond with minimal delay to the changing characteristics of the input data, and of increased latency for some tuples, as some tuples may now have to wait longer in queues before being scheduled.

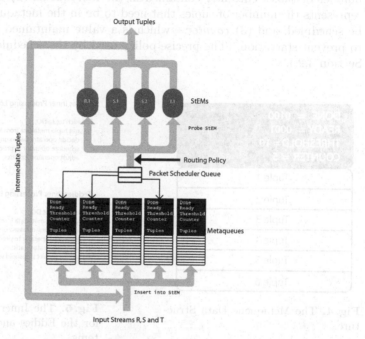

Fig. 3. Teddies System Architecture Illustrating a Three-Way Join Between Three Streams R, S, and T - $R\bowtie_{R.1=S.1}S\bowtie_{S.2=T.1}T$

The overall Teddies design consists of two aspects: the set of data structures used to schedule packets and the algorithm that implements the adaptive packetization scheme.

3.1 The Metaqueue and Scheduler Queue Data Structures

Figure 3 gives an overview of the flow of tuples through the Teddies system for an equi-join $R\bowtie_{R.1=S.1}S\bowtie_{S.2=T.1}T$ defined on three input relations.

The Teddies scheduler design separates the single Eddies scheduler queue into multiple data structures: the *scheduler* queue and the set of *metaqueues*. The scheduler queue maintains references to those metaqueues which are currently eligible for scheduling. The metaqueues, which are the fundamental units that are scheduled, group together tuples with the same metadata history. Thus, in the Teddies system the metaqueues are the specific mechanism used to packetize the data. Figure 4 shows an example metaqueue, where each metaqueue is a set of tuples along with some associated metadata. The associated metadata consists of the READY and DONE bitmaps, which have the same semantics as in the Eddies system. However, in Teddies this metadata is now shared by *all the tuples* in the metaqueue. Additional metadata is used to keep track of the state of the metaqueue, including (1) *occupancy* – which counts the total number of tuples currently contained in the metaqueue; (2) *threshold* – which represents the number of tuples that need to be in the metaqueue before it can be scheduled; and (3) *counter* – which is a value maintained by the scheduler to prevent starvation. (The precise policy used by the scheduler is described in Section 3.2.).

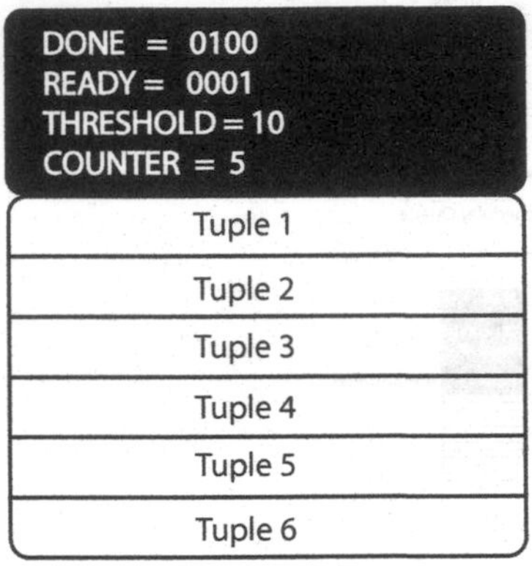

```
DONE    = 0100
READY = 0001
THRESHOLD = 10
COUNTER = 5
```

| Tuple 1 |
| Tuple 2 |
| Tuple 3 |
| Tuple 4 |
| Tuple 5 |
| Tuple 6 |

Fig. 4. The Metaqueue Data Structure

Eddies Inner Processing Loop

For each tuple DO:
    get tuple metadata from top of scheduler queue
    decide operator based on metadata
    schedule operator to process tuple
    adjust operator statistics

Teddies Inner Processing Loop

For each packet DO:
    get packet metadata from the top of scheduler queue
    decide operator based on metadata
    schedule operator to process packet
    adjust operator statistics
    adjust packet threshold values

Fig. 5. The Inner Processing Loops for the Eddies and the Teddies Systems

In Teddies, new input tuples are inserted into the SteM operators *before* they are inserted into the proper metaqueues. This ensures that the insert/probe invariant is maintained: if the input tuples are already ordered by timestamp, then whenever a tuple probes a SteM, all tuples with an earlier timestamp are guaranteed to be present in the SteM, no matter in what order the metaqueues are actually scheduled. This invariant must be maintained to ensure that no potential output tuples are lost. This is in contrast to the policy employed by the Eddies system. In the Eddies system, an input tuple is inserted into the SteMs at the same time that the tuple first probes a SteM: in this manner the insert/probe invariant for the Eddies is maintained.

3.2 The Adaptive Packetization Algorithm

The second major aspect of the Teddies query processor is the adaptive scheduler that provides two functionalities: (1) dynamically adjusting the level of packetization of each individual metaqueue; and (2) making all packet routing decisions – that is deciding which query operator to schedule next for a given packet. This is in contrast to the Eddies adaptive scheduler that focuses only on the routing of individual tuples.

The adaptive packetization algorithm utilizes the metaqueue and scheduler queue statistics to dynamically adjust the number of tuples allowable in each metaqueue. The scheduler queue contains references to unique metaqueues, where every metaqueue is either on the scheduler queue or is currently ineligible to be scheduled. For each round of the scheduler, the top metaqueue of the scheduler queue is processed. The READY bit of the top metaqueue is examined and the router policy is applied. Based on the outcome of the router policy, the entire train of tuples in the metaqueue are processed *en masse* by the selected query plan operator. All tuples output from this operator are either output from the system or placed into a metaqueue based on the READY and DONE bits of the input metaqueue. After all of its tuples are thus processed, the top metaqueue is removed from the scheduler queue, its occupancy is decremented by the number of tuples processed, its counter is reset to zero, and the counter value of any currently unscheduled metaqueue is incremented by one. At this point, as a result of this adjustment, another metaqueue may become eligible to be scheduled; if so, it is placed at the end of the scheduler queue where it awaits its turn to be processed.

The metaqueue selection, i.e., the criterion used to decide the eligibility of a given metaqueue to be scheduled, is based on the three values introduced in Section 3.1: the occupancy, the threshold, and the counter. At the end of each round of scheduler execution, if the occupancy value is greater than or equal to its threshold value, the metaqueue is moved from the unscheduled state and placed at the end of the scheduler queue. The threshold value is a dynamic quantity that is adjusted as follows: if the occupancy of a metaqueue is at least twice its current threshold when the metaqueue is finally processed from the head of the scheduler queue, then the threshold is doubled. On the other hand, if a metaqueue's counter value, after being incremented during a round of the scheduler, is greater than a preset limit (which is a preset system parameter), then the threshold value of this metaqueue is halved. In this way, those scheduled metaqueues tend to accumulate many tuples per round of the scheduler increase their threshold for more efficient processing, and, on the other hand, those unscheduled metaqueues that tend to accumulate few tuples per round lower their threshold to the point that they become eligible to be scheduled, thus avoiding starvation. The dynamics of the threshold value are thus determined by the rate at which a metaqueue gains occupants: high-rate metaqueues accumulate relatively more tuples before being scheduled, allowing for efficient train-processing, while low-rate metaqueues do not starve as their threshold values are lowered in response.

The second function of the Teddies scheduler is implemented by the algorithm that selects the operator to process the metaqueue on the top of the scheduler queue. This part of the scheduler is identical in essence to the Eddies scheduler with one distinction. Consider the inner loop processing of the Eddies and the Teddies system shown in Figure 5. Both schedulers schedule based on the READY bits of the entity on the top of the scheduler queue. However, the Teddies inner loop has an additional step – ("*adjust packet threshold values*") – not shared by the Eddies inner processing loop. This step invokes the adaptive packetization algorithm. Moreover, it should be noted that although all other steps are common to both inner loops, the Teddies inner loop runs fewer times if the data tuples are effectively grouped into packets. This allows the adaptive scheduler to potentially gain efficiencies by trading off the cost of performing the extra packetization step with the reduction in the total number of times the scheduler itself is invoked.

4   Experimental Evaluation

4.1   Experimental Setup and Methodology

To evaluate the Teddies design, we implemented both the Teddies and the Eddies architectures and ran sample queries to compare their relative performance. To ensure an even comparison, the implementations of Teddies and Eddies used as much of the same codebase and data structures as possible. As described in Section 3, the main difference between the two implementations is that Teddies uses a set of metaqueues to store the intermediate tuples while the Eddies system uses a single queue for its intermediate tuples. The tuple routing mechanism was also shared between the two systems, by suitably modifying the Eddies code to handle metaqueues instead of tuples. In addition, for the Teddies system, we modified the query operators (the SteM operators) to allow processing of a train of tuples at a single invocation.

For all experiments, we use a windowed, three-way equality join defined on three input streams (i.e., three relations ordered by timestamp), and varied the join selectivities by changing the size of the defining window. The input streams are multiple integer-valued tuples randomly generated from a uniform distribution. Consequently, each of the possible paths through the three-way join query tree have the same cost. Since all possible query plans have the same cost, the actual scheduling policy used by both the Teddies and the Eddies implementations was not exercised. In addition, since we are interested in measuring the performance difference between single tuple routing and packet tuple routing, we ensured that all routing decisions have equal cost by using a simple random routing policy. The cost of packet routing is the sum of the cost of maintaining the metaqueue threshold data and the cost of routing each packet. Hence, by minimizing the routing component cost, we can isolate the cost/benefits of the packet scheme.

Each experiment consists of measuring the total time required to compute the join query $R\bowtie_{R.1=S.1}S\bowtie_{S.2=T.1}T$ for a given batch size and window size. Batch processing of tuples in this manner gives a measure of the maximum data rate sustainable by the system, as the system never has to spin idly waiting for the arrival of new input data. However, this batch processing introduces the complication of how to introduce new data into the system. The mechanisms used to introduce new tuples into both systems are similar – whenever the respective schedulers cannot proceed because the scheduler queue is empty of intermediate tuples (in the Eddies case) or because the scheduler queue contains no metaqueues above their thresholds (in the Teddies case), a new batch of tuples is added to the scheduler queue (Eddies) or to appropriate metaqueues (Teddies). Thus both systems use a priority scheme whereby new tuples are introduced only when there are no more intermediate tuples available to be scheduled. The number in the new batch of tuples is adjustable and represents the "back pressure" effect that the input has on the system.

4.2   Results and Analysis

The focus of our experiments was to compare the cost of the Teddies system with that of the Eddies system. To this end, the experiments measured the total time required to process a given batch of tuples from randomly generated data sets under varying conditions.

The first set of experiments compared the total runtime cost of the Teddies and the Eddies systems for evaluating the three way join $R \underset{R.1=S.1}{\bowtie} S \underset{S.2=T.1}{\bowtie} T$ with varying join selectivity. We controlled the join selectivity by adjusting the SteM window size – we start with a window size that yields an average selectivity of 1.0, and in each subsequent experiment we double the selectivity by doubling the window size. For this experiment, we used a simple random router policy to consign tuples to operators. The selection of a low cost router was made primarily to highlight the overhead cost of the adaptive packetization mechanism proposed for Teddies.

Figure 6 shows the results of our experiments. The x-axis depicts the join selectivity and the y-axis the runtime ratio. The runtime ratios are *normalized* by dividing the Teddies runtime cost by the total Eddies runtimes (so that Eddies always has ratio 1.0). As can be seen from Figure 6, the total runtime cost of the Eddies system is better than the total runtime cost of the Teddies system for lower join selectivities (between 1.0 and 16.0). However, the Teddies system outperforms the Eddies system at higher selectivities (greater than 32.0). At lower selectivities the opportunities for packetization are fewer, that is, each SteM operator produces few tuples that can be grouped together to form packets. The overhead of the Teddies adaptive packetization is significant in this case. However, higher selectivities result in the formation of larger packets that trade-off the overhead cost of the adaptive processing with the benefits of batch processing, resulting in overall performance improvements.

The second set of experiments were targeted towards measuring the adaptivity of the packetization scheme that has been implemented as a core feature of the Teddies system, and for affirming the results shown in Figure 6. Here, we varied the join selectivities in a manner identical to the join selectivity variation for

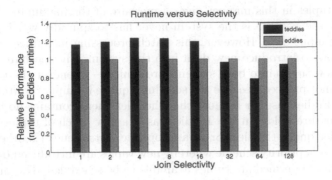

Fig. 6. A Comparison of the Runtimes of Teddies with the Eddies Baseline

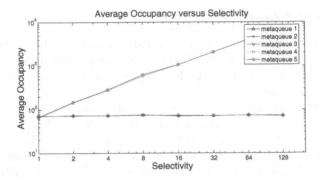

Fig. 7. The Average Occupancy of the Metaqueues for the Join Query

the first set of experiments (Figure 6) and recorded the average occupancies of the metaqueues. To obtain this measure, we instrumented the code as follows – every time a metaqueue is scheduled to be processed by a SteM operator, its occupancy value is recorded and the resulting running average is later computed.

Figure 7 depicts the average occupancy for each of the five metaqueues involved in the evaluation of the three way join $R \bowtie_{R.1=S.1} S \bowtie_{S.2=T.1} T$. Here, the x-axis depicts the join selectivity, while the y-axis shows the average occupancy of the metaqueues in terms of the number of tuples. It can be seen that the metaqueue behavior divides into two primary groups. The first three metaqueues handle only input tuples, and hence have an average occupancy that is independent of the join selectivity. However, the last two metaqueues handle tuples output from the first set of SteM operators. These two metaqueues have an average occupancy that exhibits a relationship linear to the selectivity of the SteMs.

The last set of experiments was geared towards measuring the effect of the cost of the routing decision, i.e., the scheduling cost, on the total runtime cost of the Teddies and the Eddies systems. Both the Teddies and the Eddies systems were used to evaluate the three-way join $R \bowtie_{R.1=S.1} S \bowtie_{S.2=T.1} T$. However, for this

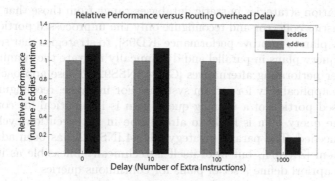

Fig. 8. The Effect of Scheduler Cost

set of experiments, we fixed the join selectivity at 1.0. We simulated the extra scheduling cost by augmenting the random tuple router with an increasing number of dummy instructions every time a routing decision needed to be made. Each dummy instruction simply increments a counter by one. In this manner, we simulate the cost of a more sophisticated tuple router that needs to adapt its decision making to the changing environment.

Figure 8 shows the results of our experiments. Here, the x-axis represents the delay in terms of the total number of extra instructions added per routing decision. The y-axis represents the normalized runtime ratio obtained by dividing the runtime costs by the Eddies runtime cost. Figure 8 shows that as the cost-per-tuple of the Eddies scheduler increases, even in the low selectivity regime (the selectivity here was fixed at 1), the Teddies scheduler achieves significant performance benefits.

Collectively the above graphs (Figure 6-8) present an overall picture of how the Teddies system compares with the Eddies system. Figures 6 and 7, show that the Teddies system performs better when the metaqueues are large – that is when the packet sizes are large. Operating in the high selectivity regime allows Teddies to amortize the cost of maintaining the metaqueues across many tuples. On the other hand, operating in the low selectivity regime when the occupancies of the metaqueues are relatively small, the Teddies adaptive packetization scheme has significant overhead compared to Eddies. However, we point out in Figure 8 that with any reasonable scheduler – a more complex scheduler that presumably makes smarter tuple routing choices – the adaptive packetization of Teddies can provide significant performance benefits even in the lower selectivity regions.

5   Related Work

Adaptive query processing deals primarily with on-the-fly re-optimization of query plans and has been studied in the context of both static databases [Ant96,GC94,INSS92,KD98] and with renewed vigor for continuous queries in stream systems [ZRH04,CCC+02,MWA+03].

Re-optimization strategies in static databases range from those that utilize a run-time statistics collector and reconfigure only the unprocessed portion of the running query plan to improve performance [KD98], to strategies that select and run multiple query plans in parallel and dynamically migrate the running query plans to better performing alternatives [GC94,INSS92]. These approaches have limited direct applicability for stream systems. For instance, reconfiguration of the unprocessed portion of a running query plan is impractical for continuous queries, as the query plan is likely to already be in its execution cycle before migration is needed. The parallel strategy [GC94,INSS92] has been adapted to a stream system [ZRH04], but in of itself is technically infeasible as it is near impossible to apriori define a set of plans for continuous queries.

Migrations strategies in stream systems range from a pause-drain-resume strategy [CCC+02,CCR+03] to adaptation of a parallel strategy [ZRH04] to a tuple-based routing strategy [AH00,MSHR02,DH04]. The pause-drain-resume strategy [CCR+03] has the ability to shutdown processing along a segment of the plan in order to reorder operators within this portion. On completion of the reordering, tuples can resume flow through the previously shutdown segments. A drawback of this strategy is the difficulty of handling stateful operators such as the join operator. Zhu et al. [ZRH04] have developed a moving state and a parallel track strategy to explicitly handle migration of stateful query plans.

Madden et al. [AH00,MSHR02] have developed a novel tuple-based routing called Eddies that shifts away from the traditional re-optimization strategies and makes an optimized decision for each individual tuple. The Eddies system [AH00,MSHR02,Des04,DH04] was designed to allow fine-grained adaptivity at the tuple level. Each tuple in the Eddies model is routed through the query plan independently from other tuples. Tuples are tagged with metadata that allows the scheduler to determine the query operators for a given tuple. The fundamental design choices for an Eddies system implementation include history maintenance for each tuple as it moves through the system, implementation of "stateful" operators such as the join operators, and routing policy decisions for each tuple.

The current Eddies implementation uses bitmaps associated with each tuple to store the metadata. Two types of stateful join operators have been proposed for the Eddies systems: the SteM [Ram01,MSHR02] operator, which implements a version of the symmetric hash join but stores only input tuples in its state hash tables; and the STAIR [DH04] operator that now also stores intermediate tuples in the state hash tables. Storing intermediate tuples in the state enhances efficiency due to reduction of repeated probing. However, it complicates the book-keeping needed to ensure that duplicate tuples are not produced.

An efficient and quickly computed routing policy is crucial for the Eddies system, as a routing decision must be made on every tuple inside the inner loop of the system's operation. One such policy [MSHR02] is the lottery based policy whereby an operator either gains or loses tickets based on how many tuples it consumes or produces. Each time through the inner scheduling loop, a lottery must be held to determine which operator wins the next tuple. In addition, statistics of each operator's behavior must be kept during execution in order to compute the number of tickets each operator holds at any point in time. Other routing policies based on the distribution of attribute values and an estimate of each operator's selectivity have also been proposed in the literature [DH04]. An approach similar to ours that provides virtual batching of tuples has also been proposed [Des04]. Here the scheduler updates the routing decision for each possible tuple history periodically. The period is a fixed size value and each tuple is still routed individually, but the lottery algorithm (or other routing decision algorithm) need not be run for each individual tuple.

6   Conclusions and Future Work

In this paper we have introduced a new type of adaptive query processor. The Teddies adaptive query processor is a hybrid architecture combining the adaptivity of the Eddies model with the efficient bulk processing of the data flow model. We have introduced an adaptive packetization algorithm that fits in the Eddies design scheme but allows for the accumulation of similar tuples into metaqueues, which are then scheduled *en masse* by the query plan router. Our experiments have shown that the packetization algorithm is able to produce large packets in the case of high operator selectivity. Our experiments also show that the bare overhead of packetization is beneficial in the high selectivity regime. Finally, our experiments have shown that the benefits of scheduling tuples in aggregate is beneficial when the cost of making a single routing decision is high, even in the low selectivity regime.

In the future, we plan on investigating the adaptivity of the Teddies architecture. The packetization process gives an adaptive query processor an extra dimension with which to find the most efficient query processing strategy. The above experiments show how this extra dimension given to the adaptive query processor may prove beneficial in the case of high input data rates.

Acknowledgements. The authors would like to thank Henry Kostowski of University of Massachusetts, Lowell, for his contributions to this paper.

References

AH00.   Avnur, R., Hellerstein, J.: Eddies: Continuously Adaptive Query Processing. In: SIGMOD, pp. 261–272 (2000)

Ant96.   Antoshenkov, G.: Dynamic Optimization of Index Scans Restricted by Booleans. In: International Conference on Data Engineering, pp. 430–440 (1996)

BBD+02.   Babcock, B., Babu, S., Datar, M., Motwani, R., Widom, J.: Models and Issues in Data Stream Systems. In: Principles of Database Systems (PODS) (2002)

BW01.   Babu, S., Widom, J.: Continuous Queries over Data Streams. In: Sigmod Record (2001)

CCC+02.   Carney, D., Cetintemel, U., Cherniack, M., Convey, C., Lee, S., Seidman, G., Stonebraker, M., Tatbul, N., Zdonik, S.B.: Monitoring Streams - A New Class of Data Management Applications. In: Int. Conference on Very Large Data Bases, pp. 215–226 (2002)

CCR+03.   Carney, D., Cetintemel, U., Rasin, A., Zdonik, S., Cherniack, M., Stonebraker, M.: Operator Scheduling in a Data Stream Manager. In: Int. Conference on Very Large Data Bases (2003)

CDTW00.   Chen, J., DeWitt, D., Tian, F., Wang, Y.: NiagaraCQ: A Scalable Continuous Query System for Internet Databases. In: SIGMOD, pp. 379–390 (2000)

Des04.   Deshpande, A.: An initial study of overheads of eddies. ACM SIGMOD Record 33(1), 44–49 (2004)

DH04.   Deshpande, A., Hellerstein, J.M.: Lifting the Burden of History from Adaptive Query Processing. In: Int. Conference on Very Large Data Bases (2004)

GC94.   Graefe, G., Cole, R.: Optimization of Dynamic Query Evaluation Plans. In: International Conference on Management of Data (SIGMOD), pp. 150–160 (1994)

INSS92.   Ioannidis, Y., Ng, R.T., Shim, K., Sellis, T.: Parameteric Query Optimization. In: International Conference on Very Large Databases (VLDB), pp. 103–114 (1992)

KD98.   Kabra, N., De Witt, D.J.: Efficient Mid-Query Re-Optimization of Suboptimal Query Execution Plans. In: International Conference on Management of Data (SIGMOD), pp. 106–117 (1998)

MSHR02.   Madden, S., Shah, M., Hellerstein, J.M., Raman, V.: Continuously Adaptive Continuous Queries over Streams. In: SIGMOD (2002)

MWA+03.   Motwani, R., Widom, J., Arasu, A., Babcock, B., Babu, S., Datar, M., Manku, G., Olston, C., Rosenstein, J., Varma, R.: Query Processing, Resource Management, and Approximation in a Data Stream Management System. In: Conference on Innovative Data Systems Research (2003)

Ram01.   Raman, V.: Interactive Query Processing. PhD thesis, UC Berkeley (2001)

ZRH04.   Zhu, Y., Rundensteiner, E., Heineman, G.: Dynamic plan migration for continuous queries over data streams. In: SIGMOD (2004)

Flood Little, Cache More:
Effective Result-Reuse in P2P IR Systems

Christian Zimmer, Srikanta Bedathur, and Gerhard Weikum

Department for Databases and Information Systems
Max-Planck-Institute for Informatics, 66123 Saarbrcken, Germany
{czimmer,bedathur,weikum}@mpi-inf.mpg.de

Abstract. State-of-the-art Peer-to-Peer Information Retrieval (P2P IR) systems
suffer from their lack of response time guarantee especially with scale. To address this issue, a number of techniques for caching of multi-term inverted list
intersections and query results have been proposed recently. Although these enable speedy query evaluations with low network overheads, they fail to consider
the potential impact of caching on result quality improvements. In this paper, we
propose the use of a cache-aware query routing scheme, that not only reduces the
response delays for a query, but also presents an opportunity to improve the result
quality while keeping the network usage low. In this regard, we make three-fold
contributions in this paper. First of all, we develop a cache-aware, multi-round
query routing strategy that balances between query efficiency and result-quality.
Next, we propose to aggressively reuse the cached results of even subsets of a
query towards an approximate caching technique that can drastically reduce the
bandwidth overheads, and study the conditions under which such a scheme can
retain good result-quality. Finally, we empirically evaluate these techniques over
a fully functional P2P IR system, using a large-scale Wikipedia benchmark, and
using both synthetic and real-world query workloads. Our results show that our
proposal to combine result caching with multi-round, cache-aware query routing
can reduce network traffic by more than half while doubling the result quality.

1 Introduction

The Peer-to-Peer (P2P) paradigm is increasingly receiving attention not only in the
context of file sharing, but also in the challenging context of large-scale distributed
information retrieval. The last few years have seen the development of a number of
prototypes of P2P-based information retrieval (P2P IR) systems [1,2,3] which are evaluated extensively under various metrics. These systems hold promise for their almost
effortless scalability to store the entire Web and support search & ranking.

Unfortunately, this promise comes with problems such as their lack of response time
guarantees for queries due to their reliance on unreliable (or slow) networks, and due to
the extra burden they place on the underlying network infrastructure. A few studies have
pointed out these limitations of some classes of P2P IR systems, either through theo-
retic analysis of their architectures [4], or large-scale empirical evaluation [5]. Some
recent research prototypes such as Pier [6], Odissea [3], Alvis Peers [7], and our own
Minerva [1], have made significant algorithmic and architectural advances to overcome
many of these issues.

J.R. Haritsa, R. Kotagiri, and V. Pudi (Eds.): DASFAA 2008, LNCS 4947, pp. 235–250, 2008.
© Springer-Verlag Berlin Heidelberg 2008

In spite of their advances, these P2P IR systems typically operate only within small-scale networks, thus fueling the continuing criticism against their scalability in larger wide-area networks. In particular, the need to contact many peers in the network to achieve good result-quality, where the slowest peer determines the query response time, is considered a key bottleneck in their scalability [8]. Further, some of the peers selected via the initial query routing step may no longer be active during actually query processing. This necessitates additional rounds of peer selection process which further compounds the response time delays. A naive solution is to select many more peers in the initial query routing step itself. But this solution clearly results in a non-optimal use of peers incurring additional network bandwidth usage, and may not be worthwhile in most cases since the typical number of results required is quite small (say, 10 to 20).

1.1   Our Contributions

The idea of caching the results of previous query executions to help the future queries is a well-known paradigm to improve the scalability of search systems [9,10]. However, in the case of P2P IR systems, designing a caching scheme that is both effective, in terms of providing high hit ratio, and at the same time overcomes the network bottlenecks has been considered as an important challenge towards building web-scale solutions [8].

In this paper, we view result caching in P2P IR systems not only as a means of achieving higher performance, but also as an opportunity for *improving result-quality*. We treat result caching as an integrated feature of query routing and processing steps, thus setting up an opportunity for not only boosting the performance but also for improving the result-quality. More specifically, we propose a framework consisting of a cache-aware multi-round query routing strategy interspersed with query processing, with continuous enrichment of cached results. This framework tries to answer a repeated query or its expansions by only using the cache contents, and then, if needed, proceeds to further improve the result-quality via an additional round of query routing (or peer selection) and processing. We show that this seemingly simple idea works amazingly well in keeping network costs low while achieving greater result-quality.

Within this general framework, we first propose and evaluate the standard *Exact Caching* (EC) strategy. EC can be seen as the P2P counterpart of traditional (centralized) result caching where results are used only if a new query matches the earlier query exactly. Although the EC approach is crisp and has an intuitive appeal, it is of limited use if the queries are not repeated exactly with sufficient frequency in the network. This is a clear possibility in Web search environments where users tend to refine their queries either via expansion or reduction of the initial query keyword set.

In order to overcome this limitation, we propose the *Approximate Caching* (AC) approach, where we aggressively reuse only the cached results of subsets of the query, thus producing an *approximate* result with highly controlled query dissemination. Obviously, the approximations produced via AC may be too far off from the actual results. We explore conditions under which this approximation can be effective, i.e., has acceptable quality of results.

Finally, we evaluate these methods over Minerva [1], a fully functional P2P search engine framework, that consists of autonomous peers with local datasets in a

DHT-based network. A given query is routed through the network based on its constituent keywords or query terms, using the distributed directory in the P2P network [3,11]. Our evaluation uses a recently proposed Wikipedia-based P2P IR benchmark that takes into account the real-world data distributions observed over P2P IR [12].

1.2 Outline

The rest of the paper is organized as follows: Section 2 presents an overview of the relevant research. The underlying Minerva P2P system architecture is described in Section 3. Then, Section 4 presents details of our EC and AC strategies that form the key contributions of this paper. Results of our experimental evaluation are shown and analyzed in Section 5, before concluding in Section 6.

2 Related Work

In this section, we briefly review related work on incorporating result caching in the context of P2P IR.These proposals differ mainly in terms of their notion of *cacheable entity* – typically either the final query results or frequently computed inverted list intersections. A uniform index caching mechanism (UIC) as suggested in [13] caches query results in all peers along the inverse query path such that the same query requested by other peers can be replied from their nearby cached results. The DiCAS (Distributed Caching and Adaptive Search) approach [14] extends this by distributing the cached results among neighboring peers and queries are forwarded to peers with high probability of providing the desired cache results. Both UIC and DiCAS cache the final results of a query and reuse them only if a query exactly matches the query that produced the results. It should be noted that UIC and DiCAS are developed in the context of unstructured P2P networks and are not directly applicable in DHT-based structured networks. In [15], the authors present a way to efficiently maintain a distributed version of inverted list intersections that were computed frequently to answer queries. Here, the cacheable entity is a materialized intersection of inverted lists which reduces the bandwidth consumption due to frequent shipping of inverted lists. Similar ideas are also proposed recently in [16]. Although these techniques closely resemble our Approximate Caching (AC) approach, in this scenario we propose to aggressively reuse the top results of subsets of a query as an approximation for the results of the complete query. This idea is useful in many situations where users might need quick, approximate answers so that they can rephrase/expand the query. Our strategy is to maintain result caches at directory peers, thus piggybacking on the standard query routing. In contrast, previous approaches maintain caches at individual peers and build additional cache directory structure to manage these distributed caches [15]. Finally, the main goal of all the caching proposals so far has been to improve only the response time for a query in a P2P search system. To the best of our knowledge, none of these methods have explored the possibility of achieving superior query result-quality by exploiting the cached results.

3   Minerva System Architecture

Before we present the details of our caching framework, we describe the architecture of the underlying P2P IR system, Minerva. The Minerva system [1] is a fully operational distributed search engine. It consists of autonomous peers where each of them has a local document collection, e.g. from its own (possibly thematically focused) Web crawls. The local collection is indexed by inverted lists, one for each term (e.g., word stems) containing document identifiers like URLs and relevance scores based on term frequency statistics. Minerva uses a conceptually global but physically distributed directory layered on top of a Chord-style [17] distributed hash table (DHT) to manage aggregated information about the peers local knowledge in a compact form. Unlike [4], the system uses the Chord DHT to partition the *term space*, such that every peer is responsible for the metadata of a randomized subset of terms within the directory. The system does *not* distribute the actual index lists or even documents across the directory. The Chord DHT offers an efficient *lookup* method to determine a peer responsible for a particular term. This way, the DHT allows scalable access to the global metadata for each term.

In the publishing process, each peer distributes per-term summaries (so-called *posts*) of its local index to the global directory. The DHT determines a peer currently responsible for the term and this peer maintains a *peer-list* of all posts for this term. Each post includes the peer's contact information (IP address and port number) together with statistics to calculate IR-style measures for a term (e.g., the size of the inverted list for the term, and other statistical measures). Posts are not limited to using simple terms as keys. For example, we can also use the directory to store posts about correlated term sets [18], *semantic* information about semistructured documents, QoS agreements, etc. We use a time-to-live (ttl) technique that invalidates posts that have not been updated (or reconfirmed) for a tunable period of time.

3.1   Query Execution

A query with multiple terms is processed as follows. In the preliminary step, the query is executed locally using the query originating peer's local index. If the user is not satisfied, the query is executed over the network. The query execution phase over the network involves two communication steps, namely, the *metadata retrieval* and the *local result retrieval*, which are illustrated in Figure 1.

To retrieve metadata, the peer issues a *peer-list request* to the directory for looking up promising remote peers, for each query term separately. The querying peer uses the DHT to determine the peers responsible for each query term (shown in figure 1(a)).From the retrieved peer-lists, a certain number of most promising peers for the complete query is computed. This step is referred to as *peer selection* or *query routing*. For scalability, the query originator typically decides to ship the query to only a small number of peers based on an expected benefit/cost ratio. The prior literature on query routing has mostly focused on IR-style statistics about the document corpora, most notably, CORI [19], the decision-theoretic framework by [20], and methods based on statistical language models [21]. These techniques have been shown to work well on disjoint data collections, but are insufficient to cope with a large number of autonomous peers that crawl

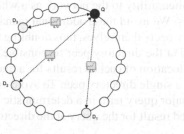   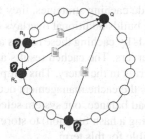

(a) Metadata Retrieval   (b) Local Result Retrieval

Fig. 1. Query Execution in Minerva: (a) Q uses the DHT to send PeerList requests to directory peers D_1, D_2, and D_3; (b) Q sends the query to selected peers R_1, R_2, and R_3, and obtains their local query results

the Web independently of each other, resulting in a certain degree of overlap as popular information may be indexed by many peers. To overcome this, Minerva uses a *overlap-aware* query routing strategy that utilizes compact statistical synopses of peer data collections [22]. To evaluate our caching strategies, we used the standard query routing approach that considers for each peer and query term the number of documents in the peer's local collection that contain the term.

Once this peer selection decision is made, the query is executed completely on each of the remote peers (shown in figure 1(b)), using a local scoring/ranking method and a top-k search algorithm. Each peer returns its top-k results (typically with k equal to 10 to 50), and the results are merged into a global ranking, using either the query originator's statistical model or global statistics derived from directory information (or using some heuristic result merging method [23]).

4 Result Caching at the Directory

In a P2P IR system, the response time for a query is highly dependent on the total number of communication rounds that are necessary to obtain the final result as well as the total number of peers involved in each round. In the standard peer selection and query execution process, the querying peer has to contact the directory in a first step, and then, it has to ask the selected peers for their local results, resulting in two rounds of communication overall. While it is important to request only a small number of peers for the reasons of efficiency, the result-quality can be improved by increasing the number of peers receiving the query. Thus, it is important to balance these two conflicting requirements for effective query routing.

4.1 Caching Framework

As described in Section 3.1 earlier, each round of query processing generates merged peer-lists for the query keywords as well as actual results themselves. These two resources are generated and used by the query initiator. If cached locally – like in standard client-side caching strategies, they are of limited utility to the network as a whole, unless one builds additional overlays or indexes. We avoid these additional management overheads by caching these results at directory peers themselves, thus doubling them as cache servers. The cached results are located at the directory peer responsible for one of the terms in the query. This simplifies the location of cached results for a query, and localizes the cache management decisions to a single directory peer. To avoid concerns about load balance, our system selects one major query term in a deterministic manner (e.g., using a hash function) to store the cached result for the query at the directory peer responsible for this term.

A cached result for a query q consists of two parts. The first part is a set of entries, each of which contains a compact representation of (i) result documents (possibly their URL), and (ii) statistical information about all the query terms involved as well as the source collection. The second part of the cached result is the set of source peer(s) that were involved in its generation. This set is compactly represented using a Bloom-filter [24] data structure, that supports efficient set membership queries.

The query routing and processing steps of Minerva are extended as follows. As before the query originator first sends the peer-list request for all terms in q, and, in addition, sends the *complete query* to all the directory peers to check if a cached result for q is already available in the directory. If there is a cached result that can be used for answering q, then it will be immediately returned by the appropriate directory peer. Otherwise, when the querying peer has computed the final result of the query by merging the local results from participating peers, it sends the full result-set to the directory peer responsible for the cached result of q. The directory peer takes appropriate action to incorporate these results in its cache for use at a later stage. In our experimental evaluation we will also investigate the influence of various cache replacement strategies (e.g., FIFO, LFU, LRU, etc.) using a bounded cache size. Based on this general framework of result caching, we develop the following two strategies for utilizing the cached results: (1) Exact Caching (EC), and (2) Approximate Caching (AC). We describe each of them in detail next.

4.2   Exact Caching (EC)

The first caching approach considers a match for the requested query q, only if the stored result was generated by exactly the same set of query keywords as in q. As shown in Figure 2(a), a cached result for a query q is stored after query execution in the directory. The query initiator selects one of the directory peers responsible for the query terms in a deterministic manner (e.g., using a hash function) and sends the cached result to this directory peer. When requesting all the *peer-lists* for the (same) query terms of q, the directory peer responsible for the cached result returns it to the querying peer. This extension of the metadata retrieval is illustrated in Figure 2(b).

Next, the query initiator checks whether the cached result is sufficient (in terms of its quality, number of results, their freshness, etc.). If satisfied, the query execution stops – thus avoiding the computational and network overheads in rest of the query execution. The overall query execution with EC is much faster, when there is a cache hit for the query, with only one round of network communication and the result-quality is as good as of the previous query execution.

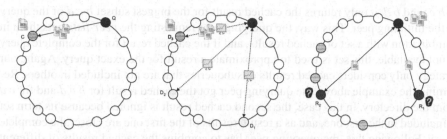

(a) Storing Cached Results   (b) Retrieving Cached Results   (c) Asking Additional Peers

Fig. 2. Exact Caching Approach

Note that in the above step, the querying peer gets, in addition to the cached result, peer-lists for all the query-terms as in the standard query routing. This suggests a way to improve the result-quality as illustrated in Figure 2(c), by sending the complete query additionally to peers that were not involved in the generation of the cached result. In Figure 2(c), represent the peers which are thus ignored, and the filled nodes are the peers involved in the additional round of query processing. This procedure can be easily done by performing the (standard) peer selection, after *excluding* the peers that are already found in the peer-set associated with the cached result. After obtaining local query results from the selected extra peers, they are merged with the current cached result. Clearly, the merged result covers more peers in the network than earlier query execution, while using the network resources required for involving only the extra peers. This quality improvement step can be run repeatedly for the query. Naturally, the improvements diminish as more and more peers are involved in the query execution – but converge to the global result ranking. Note that it is possible to terminate these improvement steps, by specifying a terminating condition – e.g., when a certain percentage of the peers in the network are already involved in the result generation.

4.3 Approximate Caching (AC)

The EC approach outlined above is useful only in a limited scenario wherein a query is repeated exactly by many peers in the network. To overcome this limitation, we introduce our Approximate Caching (AC) strategy by aggressively retrieving and combining the cached results of *subsets* of the query as an *approximation* to the current query. The AC strategy starts with the requesting of *peer-lists* for all query terms. Now, the directory peer responsible for the cached result of the complete query does not store it. Alternatively, the query processing tries to approximate the query result by merging cached results of *subqueries* that are stored in the directory. For this reason, the directory peers requested for the *peer-lists* also return existing cached results for subsets of the query term set. Remember that the directory peers are responsible by design for all possible subsets.

Each directory peer returns the cached results for all the maximal subqueries available in its cached result store. For example, if the requested query is *a b c d* and the directory peer responsible for the term *b* contains the cached results for the queries *b c*

$d, b c$, and $b d$, it only returns the cached result for the biggest subset $b c d$ of the query to the initiating peer. This way, the querying peer requesting the *peer-lists* gets them in combination with a set of cached results, and if the cached result for the complete query is not available, this set is used to approximate a result for the exact query. Again, our strategy only considers cached results of subqueries that are not included in others. Resuming the example above, the querying peer got the cached result for $b c d$ and for $c d$ from the directory. In this case, the second cached result is ignored because its term set is included in the first one, and as a result, results of the first one are closer the complete query. Following this, the querying peer has to combine the cached results of different subqueries to approximate the complete query.

To merge the cached results, we use the following approach: Given a network with peers P and a querying peer $p_q \in P$ requesting an initial multi-term query $q_i = \{t_1, t_2, ..., t_l\}$. Let P_q be the set of directory peers responsible for all query terms. Let $p_i \in P_q$ be responsible for query term t_i and the cached results cr_i. The querying peer p_q requests the *peer-lists* of all query terms by sending t_i and the complete query q to p_i.

The directory peer returns the requested *peer-list* and a set of cached results cr_i' of the queries Q_i' such that for all $q' \in Q_i'$ the following holds: $Q_i' \subseteq Q_i$ and there is no $q^* \in Q_i'$ with $q^* \supset q'$. The querying peer p_q receives $cr' = \bigcup_{i=1}^{l} cr_i'$, the union of all delivered cached result sets for the queries $Q' = \bigcup_{i=1}^{l} Q_i'$. Note that $|Q'|$ is upper bounded by $|q_i|^2$ due to the requirement that the chosen subqueries be maximal. Then, p_q selects $CR'' \subseteq cr'$ containing the cached results for the query set Q'' such that the following holds: $Q'' \subseteq Q'$ and for all $q'' \in Q''$ there is no $q^{**} \in Q''$ with $q^{**} \supset q''$.

Next, D is the set of all documents contained in the cached results of cr'. To get the approximate result for the initial query q, p_q computes the scores for all documents $d \in D$. We extract the document scores $score_{d,p,q}$ out of the cached results and we know that a document d is a local result of peer p for the previous query q such that the final score of this document to get an approximate result ranking is computed as follows:

$$score_d = \max_{p,q}(|q| \cdot score_{d,p,q}) \tag{1}$$

This score computation uses the document scores form the cached results but takes the different query sizes into account. The intuition is that results satisfying longer queries are more selective than others and so these documents get a higher score. We have to consider that more than one cached result can include a document d because it can represent a result for different queries and so, there are multiple scores for the same document. For this reason, the final document score $score_d$ is the maximum of all scores for this document.

5 Experiments

In this section, we present results of our experimental evaluation of both EC and AC, implemented using the Minerva simulation framework. We also provide a detailed cost analysis of our methods using the outcome of our initial experiments. Finally, we present the results using a real-world query-log that provides a conservative view of our methods.

5.1 Experimental Setup

The techniques described in this paper were implemented in the Minerva simulation framework, using Java JDK 1.5. Unless otherwise mentioned we use a network of 1,000 peers, with no disruptions, and a steady network. During query evaluation, we ask every peer involved in the current round of query processing for its top-25 local results, and merge these results to obtain top-50 results for the query.

Benchmarks designed for the evaluation of centralized retrieval systems, are unsuitable for their direct use in P2P IR systems. In view of this difficulty, we use a recently proposed benchmark designed specifically for use in P2P system evaluation [12]. Briefly, the benchmark consists of more than 800,000 documents drawn from Wikipedia corpus, and 99 queries derived from Google Zeitgeist, and an algorithm to distribute the documents among peers with controlled overlap which mimics the real-world behavior. We use the following metrics in our evaluation:

(i) **Relative recall:** This is the fraction of relevant documents returned in the top-k results of the distributed systems. We consider the top-k results of a centralized system that uses the same ranking function as the set of relevant documents in the complete network.

(ii) **Network resource consumption:** This measure tries to capture the system's ability to handle queries efficiently. We are interested in the total network traffic incurred during query processing, number of messages transferred across the network, and also the number of communication rounds.

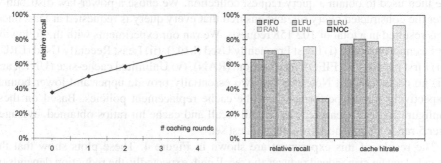

Fig. 3. EC-MS: Relative Recall Gains   **Fig. 4.** EC-MS: Cache Management

5.2 Improving Recall with Exact Caching

The first set of experiments is focused towards the evaluation of the EC method. Since, at the end of one query processing cycle, we obtain the top-50 query results after result merging. These 50 results are cached in the directory node responsible for the query. In the first round of query processing, we disseminate the query to the 50 most promising peers (5% of the network). After this initial round, we involve *additional* 5% of the peers in the network, and merge their local results with the cached results to get the improved top-50 documents. We ran this process until we included 20% of the total network, or there are no more peers which can answer the query.

Figure 3 summarizes the results of this experiment over all the 99 Zeitgeist queries in the benchmark. The x-axis shows the number of query processing rounds (with each round including addition 5% of the network), and the y-axis shows the relative recall concerning the top-50 results obtained so far. As these plots show, the recall steadily increases from a low value of 37%, obtained when only 5% of peers in the network are involved, to about 75% at the end of 4 additional rounds. As we show later in Section 5.4, our method needs much less network resources to obtain practically the same recall as the state-of-the-art single-step query routing to larger number of peers.

5.3   Cache Management Strategies

So far, we assumed that the cached results at the end of a query processing round remain in the cache until the subsequent round. However, the storage space at our disposal for result caching on directory peers are typically bounded and a cache management policy could replace/evict the cached results before the next round of the query. This next set of experiments we describe in this section, evaluates various cache replacement policies operating at each directory peer in terms of their impact to the recall for EC with multi-step query routing.

We assume a common cache-size for all peers in the network, and set it to hold results of just 3 queries (this roughly translates to about 6 Kbytes of cache). We used the set of 99 queries in the benchmark to construct a large synthetic query workload as follows: first, we extracted 135 unique terms in benchmark query-set and then generated all possible one- and two-term queries using these terms. This gave us 9180 queries, which are then used to obtain a query request collection. We chose a power-law distribution over the constructed query set, and ensured that every query is requested at least once. This resulted in a total of 102,158 requests. We ran our experiments with the following replacement policies: (i) Least Frequently Used (LFU), (ii) Least Recently Used (LRU), (iii) First in First out (FIFO), (iv) Random (RAN), (v) Unlimited cache-size (UNL), and (vi) no cache (NOC). Note that last two essentially provide upper and lower bounds respectively for the performance of the cache replacement policies. Based on these configurations we measured the overall recall and cache hit ratios obtained, averaged over 5 random sequences from the request set.

The results of this experiment are shown in Figure 4. These plots show that the bounded cache-size indeed reduces the recall and, expectedly, the reduction depends on the cache replacement policy used. The best performance is obtained using LFU that provides just 12% reduced recall over the best possible in the case of an unbounded cache. Further, the behavaior of recall improvements are commesurate with the hit rates of respective replacement policies. Based on these results, we use LFU in the rest of our experiments to manage each cache.

5.4   Cost Analysis

Now we present a detailed analysis of the network costs – in terms of per query network traffic incurred and the number of messages for every query during its processing of the following scenarios:

(i) *No Caching* (NC) – equivalent to the standard query processing, where we request at once a given number of peers and combine local top-25 documents returned by each of them to obtain a global top-50 results.

(ii) *EC, Single-step* (EC-SS) – EC technique without the multi-step query routing to improve recall. We assume that we query 5% of the peers in the network.

(iii) *EC, Multi-step* (EC-MS) – EC technique along with multi-step query routing that improves recall, where each step of the query routing involves 5% additional peers in the network, until 50% of the network is involved in the query evaluation.

For the sake of simplicity, we assume that the average query size is two, and that each peer returns its top-25 local documents when queried while top-50 documents are expected in the end. In our implementation, a cached result consists of 50 document-entries where each entry has an average size of 34 bytes (30 bytes on average for the URL, and 4 bytes for the real-valued score). The set of peers involved in generating a cached result is represented using a Bloom-filter of a fixed size of 100 bytes. Recall that the Bloom-filter allows set-membership checks with good accuracy. When we combine these with the complete query, the overall size of a cached query result is about 2 Kbytes. Table 1 summarizes the results of our analysis, along with relative recall results of previous sections, for each of the three scenarios we consider.

Table 1. Summary of cost analysis for different caching scenarios

	NC	EC-SS	EC-MS
Relative Recall	0.32	0.32	0.71 (+122%)
Network Traffic (per query)	55.3 Kbytes	23.1 Kbytes (-58.2%)	41.0 Kbytes (-25.9%)
Messages (per query)	106	25.7 (-75.8%)	61.4 (-42.1%)
Response Time (rounds)	2	1.19 (-40.3%)	1.60 (-20.0%)

Our analysis considers costs incurred during each of the following three phases of the query processing in our system: (i) metadata retrieval; (ii) local result retrieval; and (iii) cached result placement.

Metadata Retrieval: For each query, the initiator first needs to request the per-term peer-list that contains the contact information for the candidate peers. A peer-list entry consists of IP number and port (4+ 2 bytes), and the number of documents (2 bytes) as an indicator for the quality of the peer. We found that a random peer-list has about 800 entries on average, resulting in an average peer-list size of 6.4 Kbytes. For two-term queries that we consider in this analysis, this results in a total traffic of 12.8 Kbytes, and can be accomplished using 2 messages for each term. In the case of a cache hit, additional 2 Kbytes (cached query results) are shipped to the initiator. Note that this does not require additional messages since we piggy-back query results along with the shipped peer-list.

Local Result Retrieval: A query is sent to the 50 most promising peers (= 5% of the network) and each peer returns top-25 local documents resulting a total network traffic of ≈42.5 Kbytes per query. This holds for NC, every cache-miss in EC-SS & EC-MS, as well as for each additional step of EC-MS. A total of 100 messages are needed for each of communication step – query request and reply for every peer involved in the current step. In case of EC-SS and EC-MS, if there is a cache hit in the initial step, no costs are incurred in this phase of query processing.

Cached Result Placement: This network traffic occurs only when a new cached query result (or an improved cached query result) needs to be stored in the directory. In our implementation, a cached object is stored at only one directory peer such that additional network traffic of 2 Kbytes are needed, and can be accomplished using 1 message.

Equipped with these costs for individual phases of the query processing, we can now compute the overall network traffic and messages required for each of the settings. We begin with NC, where we first retrieve metadata for 102,158 requests requiring 1,308 Mbytes ($= 102,158 \times 12.8$Kbytes), and perform standard query execution requiring about 4,341 Mbytes ($= 102,158 \times 42.5$Kbytes) of network traffic and more than 10 *million* messages!

When we use EC-SS, during metadata retrieval we cause an additional 165 Mbytes traffic due to the cost for shipping cached results for every cache hit (82,275 cache hits and 2 Kbytes of cached results). This overhead is offset since we save on the traffic during the subsequent query execution, where only a cache miss triggers the need to contact peers for their local top-25 results. It is easy to see that this translates to just 845 Mbytes – less than 20% of the traffic with NC. Following similar analysis, it can be been that in EC-SS, the average traffic per query is more than halved, and the average number of messages required per query is reduced by more than 75%. On the whole, EC-SS conserves the network resources significantly over NC while producing same quality of results.

The EC-MS traffic during metadata retrieval is the same as in EC-SS. However, we incur additional traffic when a cache hit occurs, since we include more peers until half the network is included in query processing. In the workload we used, this situation occurs for 41,138 requests, which generates an additional 2,593 MBytes of traffic during query execution and 122 MBytes for placing results back into the cache. Despite this, the EC-MS setting still saves about 25% of the total network traffic over NC, but, most importantly, can more than *double* the average result-quality over time. Note that these numbers only report on the traffic caused by workload of the queries. Traffic caused by background processes like DHT stabilization or directory maintenance are not considered.

Finally, moving on to quantifying the response times for a query, we argue that the dominating factor is the number of communication rounds between peers. A network traffic overhead of 2 Kbytes for transferring a cached query result does not increase the response time, since we just piggy-back this extra payload with the same message that returns the peer-lists. Further, the extra communication during result placement into its cache in EC-SS and EC-MS scenarios will not affect the response times perceived at the query initiator – thus we ignore them as contributing towards query response time delay.

5.5   Approximate Caching Scenarios

Next, we turn our attention towards evaluating the Approximate Caching (AC) approach that aggressively reuses the cached results for subsets of the query when EC can not be

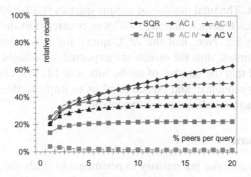

Fig. 5. Performance of AC Strategies

applied. Clearly, being overly aggressive in reusing query subsets can result in highly degraded query results – thus offseting the benefits of network savings. We consider the following 5 situations where AC can be applied on a n-term query q:

1. AC-I – All n subqueries of length $n - 1$ are available in the cache.
2. AC-II – All subqueries of length 2, totaling $n \cdot (n - 1)$ are in the cache.
3. AC-III – Only one $(n - 1)$-length subquery is cached.
4. AC-IV – All single-term subqueries are available in the network cache.
5. AC-V – At least one of $(n - 1)$-length subquery and at least one 2-term query are cached, and, in addition, all query terms are covered using all the cached subqueries.

We generated a set of 4000 random 3- and 4-term queries from the benchmark query set, and compared each of these approaches against the standard query routing (SQR) of Minerva. We evaluate the effectiveness of these AC approaches, in terms of the relative recall obtained as a function of number of peers that have contributed to the cached results used. For e.g., for AC-IV we measure relative recall of answering the query q when 1-20% of of peers in the network are involved in the generation of all single-term queries cached in the network.

Results of these experiments are plotted in Figure 5. Notice that SQR consistently improves in its relative recall as more and more peers are involved in the query evaluation. If only a very small fraction of peers are involved, the performance of AC-I and AC-II are comparable to SQR and even better sometimes. However, the AC-III and AC-IV perform very poorly, and, thus, can not be used. Based on these results, we use AC to reduce message costs, only when at least one biggest sub-query is available and the cached results cover all terms in the original query.

5.6 Experiments with Real-World Query-Log

We ran experiments using a real-world query-log derived from the AOL query-log that was made available for public download early this year. We stripped the log of all the user identification and session information, and considered the queries in their time-order from the first file of the log. This gave us a sequence of $57,344$ requests with

25, 208 unique terms. The total number of unique queries is 39, 640. We used both EC-MS and AC strategies together, although AC was restricted to the conditions derived from experiments in 5.5. Note that the AOL query log is not directly applicable on the Wikipedia benchmark, thus the results are expected to be highly conservative. Our experiments showed that the number of cache hits was 14, 134, about 25% hit rate, at the end of the query sequence, and this enabled us to improve the relative recall from 0.45 to 0.52 on average for all the queries involved.

5.7 Impact of Churn

In the end, we also ran some preliminary experiments to study the impact of churn on the benefits of EC-MS. We simulated churn by randomly removing varying fraction of peers in the network after every 1, 000 queries executed in the network. The results are summarized in Figure 6 under different rates of churn applied. With low rates of churn, the performance of EC-MS is as expected remains fairly good. When the churn-rate increases to 30%, although the performance deteriorates, we still observe high hit-rates and recall improvements when compared to the scenario without caching.

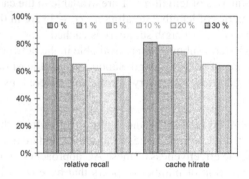

Fig. 6. Impact of Churn on EC-MS

6   Conclusion and Future Work

In this paper, we presented various strategies to take advantage of the previous work of other peers in a P2P search engine. Caching is used to improve the query result-quality, or to avoid additional communication cost such that the query processing gets accelerated. The *Exact Caching* (EC) approach stores cached results in the directory in a smart manner such that caching is integrated in the standard query execution process of our P2P search engine. The querying peer can increase the query result-quality by asking additional peers not involved in the previous result. Our large-scale experiments used a Wikipedia benchmark and a real-world query-log. We investigated various cache replacement strategies and considered churn in the P2P system. If the P2P directory can not deliver information about the exact query but cached results for subqueries, the *Approximate Caching* (AC) approach returns satisfying outcomes. Nevertheless, to reach a good approximation, we make demands on the existing cached results. In a real world experiment we used a large query-log with the combination of our EC and AC approach yielding to satisfying results in terms of relative recall and message costs.

Our caching strategies only work if a query (or a similar query) was executed before. Proactive components could anticipate future queries and could precompute and store cached results to accelerate the query execution. Of course, the precomputation of all possible term combination is impossible such that suitable mechanisms to forecast interesting queries are necessary. One important question for future work is the freshness of cached results: when do we need to update the directory knowledge concerning a query, or when do we have to ignore the cached result because new and better results are available. Another direction takes replication of actual data or metadata into consideration.

References

1. Bender, M., Michel, S., Triantafillou, P., Weikum, G., Zimmer, C.: Minerva: Collaborative P2P Search. In: VLDB (2005)
2. Podnar, I., Rajman, M., Luu, T., Klemm, F., Aberer, K.: Scalable Peer-to-Peer Web Retrieval with Highly Discriminative Keys. In: ICDE (2007)
3. Suel, T., Mathur, C., Wen Wu, J., Zhang, J., Delis, A., Kharrazi, M., Long, X., Shanmuga-sundaram, K.: In: WebDB (2003)
4. Li, J., Loo, B.T., Hellerstein, J.M., Kaashoek, M.F., Karger, D.R., Morris, R.: On the Feasibility of Peer-to-Peer Web Indexing and Search. In: Kaashoek, M.F., Stoica, I. (eds.) IPTPS 2003. LNCS, vol. 2735, Springer, Heidelberg (2003)
5. Zhang, J., Suel, T.: Efficient Query Evaluation on Large Textual Collections in a Peer-to-Peer Environment. In: Peer-to-Peer Computing (2005)
6. Huebsch, R., Hellerstein, J.M., Lanham, N., Loo, B.T., Shenker, S., Stoica, I.: Querying the Internet with PIER. In: Aberer, K., Koubarakis, M., Kalogeraki, V. (eds.) VLDB 2003. LNCS, vol. 2944, Springer, Heidelberg (2004)
7. Luu, T., Klemm, F., Podnar, I., Aberer, M.R.K.: ALVIS Peers: A Scalable Full-text Peer-to-Peer Retrieval Engine. In: P2PIR (2005)
8. Baeza-Yates, R., Castillo, C., Junqueira, F., Plachouras, V., Silvestri, F.: Challenges in Distributed Information Retrieval. In: ICDE (2007)
9. Lempel, R., Moran, S.: Competitive Caching of Query Results in Search Engines. Theoretical Computer Science (2004)
10. Long, X., Suel, T.: Three-level Caching for efficient Query Processing in large Web Search Engines. In: WWW (2005)
11. Crespo, A., Garcia-Molina, H.: Routing Indices For Peer-to-Peer Systems. In: ICDCS (2002)
12. Neumann, T., Bender, M., Michel, S., Weikum, G.: A Reproducible Benchmark for P2P Retrieval. In: ExpDB (2006)
13. Sripanidkulchai, K.: The Popularity of Gnutella Queries and its Implications on Scalability
14. Wang, C., Xiao, L., Liu, Y., Zheng, P.: Distributed Caching and Adaptive Search in Multilayer P2P Networks. In: ICDCS (2004)
15. Bhattacharjee, B., Chawathe, S.S., Gopalakrishnan, V., Keleher, P.J., Silaghi, B.D.: Efficient Peer-To-Peer Searches Using Result-Caching. In: Kaashoek, M.F., Stoica, I. (eds.) IPTPS 2003. LNCS, vol. 2735, Springer, Heidelberg (2003)
16. Skobeltsyn, G., Aberer, K.: Distributed Cache Table: Efficient Query-Driven Processing of Multi-term Queries in P2P Networks. In: P2PIR (2006)
17. Stoica, I., Morris, R., Karger, D.R., Kaashoek, M.F., Balakrishnan, H.: Chord: A Scalable Peer-to-Peer Lookup Service for Internet Applications. In: SIGCOMM (2001)

18. Michel, S., Bender, M., Ntarmos, N., Triantafillou, P., Weikum, G., Zimmer, C.: Discovering and Exploiting Keyword and Attribute-Value Co-occurrences to Improve P2P Routing Indices. In: CIKM (2006)
19. Callan, J.P., Lu, Z., Croft, W.B.: Searching Distributed Collections with Inference Networks. In: SIGIR (1995)
20. Nottelmann, H., Fuhr, N.: A Decision-Theoretic Model for Decentralized Query Routing in Hierarchical Peer-to-Peer Networks. In: Amati, G., Carpineto, C., Romano, G. (eds.) ECiR 2007. LNCS, vol. 4425, Springer, Heidelberg (2007)
21. Si, L., Jin, R., Callan, J.P., Ogilvie, P.: A Language Modeling Framework for Resource Selection and Results Merging.. In: CIKM (2002)
22. Bender, M., Michel, S., Triantafillou, P., Weikum, G., Zimmer, C.: Improving Collection Selection with Overlap-Awareness. In: SIGIR (2005)
23. Meng, W., Yu, C.T., Liu, K.-L.: Building efficient and effective Metasearch Engines. ACM Comput. Surv. (2002)
24. Bloom, B.H.: Space/Time Trade-offs in Hash Coding with Allowable Errors. Commun. ACM (1970)

Load Balancing for Moving Object Management in a P2P Network

Mohammed Eunus Ali[1,2], Egemen Tanin[1,2], Rui Zhang[2], and Lars Kulik[1,2]

[1] National ICT Australia
[2] Department of Computer Science and Software Engineering
University of Melbourne, Victoria, 3010, Australia
{eunus,egemen,rui,lars}@csse.unimelb.edu.au

Abstract. Online games and location-based services now form the potential application domains for the P2P paradigm. In P2P systems, balancing the workload is essential for overall performance. However, existing load balancing techniques for P2P systems were designed for stationary data. They can produce undesirable workload allocations for moving objects that is continuously updated. In this paper, we propose a novel load balancing technique for moving object management using a P2P network. Our technique considers the mobility of moving objects and uses an accurate cost model to optimize the performance in the management network, in particular for handling location updates in tandem with query processing. In a comprehensive set of experiments, we show that our load balancing technique gives constantly better update and query performance results than existing load balancing techniques.

Keywords: P2P Data Management, Spatial Data, Load Balancing.

1 Introduction

Decentralized distributed systems, in particular peer-to-peer (P2P) systems, are an increasingly popular approach for managing large amounts of data. These systems do not have a single point of failure and can easily scale by adding further computing resources. They are seen as economical as well as practical solutions in distributed computing. For example, a police department could deploy a P2P system of hundreds of generic, cheap low-end processing units (nodes) for a traffic monitoring system. With similar constraints and needs, massively multi-player online games as well as new location-based services now form the potential application domains for the P2P paradigm. With recent research in P2P data management, managing complex data and queries is now a reality and such new P2P applications that go beyond file sharing are about to emerge [1].

In this paper, we consider a large set of moving objects maintained by a P2P network of processors. A moving object data management system typically assumes that objects either periodically report their locations or their location changes [2]. The workload of a moving object data management system mainly consists of two parts: handling location updates and processing queries. As it is the case for any distributed system, in a P2P system for moving objects, balancing the workload is essential to optimize the overall performance. However, existing P2P load balancing techniques were designed for stationary data and can produce undesirable workload allocations for moving objects.

J.R. Haritsa, R. Kotagiri, and V. Pudi (Eds.): DASFAA 2008, LNCS 4947, pp. 251–266, 2008.

We show two examples for processing updates and queries that highlight different workload allocations. Fig. 1 shows two different approaches to partition the load from updates for two processing nodes, P_1 and P_2. Fig. 1 (a) shows the initial state with only one node, P_1, managing the entire load (20 updates). The thick lines represent roads and the numbers adjacent to them represent the number of location updates from objects, e.g., cars, moving on the roads during a period of 5 minutes.

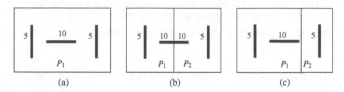

Fig. 1. Examples of workload partitioning for updates

A traditional load balancing scheme might partition the data space as in Fig. 1 (b) for the nodes P_1 and P_2, which results in a perfectly balanced load for P_1 and P_2. The cars on the vertical roads require 5 updates per node. The cars on the horizontal road move from one partition to the other partition and cause two updates: a new update message for the node entered and another update message to delete the object from the node left. As a result, we get a total of $5 + 10 + 5 + 10 = 30$ updates for both nodes, where each node handles 15 updates. However, if the data space is partitioned as in Fig. 1 (c), although the load is not balanced among the two nodes, we only get a total load of 20 updates instead of 30 updates. This interesting example shows that a traditional load balancing scheme can result in higher total workload than unbalanced load partitioning that optimizes the total load with respect to the movement of objects.

Unbalanced partitioning can also improve query processing. Assume the same data distribution as in Fig. 1 (a) and suppose the rectangles shown in Fig. 2 (a) represent two-dimensional range (window) queries. A traditional load balancing scheme might partition the space similar to the previous example leading to the partitioning as in Fig. 2 (b). In this case, node P_1 has to process 8 queries because the data space completely includes 3 query windows and overlaps with 5 query windows; correspondingly node P_2 has to process 7 queries, which leads to a load of 15 queries in total. If we partition the data space as Fig. 2 (c), P_1 has to process 6 queries and P_2 4 queries, which leads to a total load of 10 queries, significantly less than the more balanced partitioning.

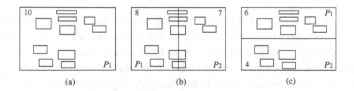

Fig. 2. Examples of workload partitioning for range queries

These examples show that a traditional load balancing scheme can be inefficient for moving objects if their movement is not taken into account. This inefficiency applies to both updates and queries in dynamic spatial settings that use multiple processing nodes. Motivated by these observations, we propose a novel load balancing technique for moving object management in a P2P network. Our technique considers the mobility of moving objects to model and minimize the average cost of handling updates and processing queries. To optimize the overall system performance, we make a trade-off between balancing the workload and minimizing the extra workload overheads for crossing updates and overlapping queries. We propose to model nodes and their workloads using an undirected weighted graph, which allows us to use a graph partitioning algorithm for load partitioning. We then develop an accurate cost model that estimates the cost of handling updates and the processing of queries in order to optimize the performance of the management network. Through an extensive experimental study, we show that our spatial approach to load balancing gives constantly better update and query performance results than existing P2P load balancing techniques.

2   Related Work

Moving object data management in centralized systems has been extensively studied in [3,4,5,6,7]. Recently, for static spatial data, decentralized systems have been developed [1,8,9,10]. In addition, recent research also focused on moving objects in distributed settings [11,12]. None of these existing systems address load balancing issues for moving object data management in P2P systems.

Distributed algorithms and data structures for P2P systems have become the main research topic for large scale distributed data management since early 2000s (e.g., [13,14]). These systems rely on distributed hash tables (DHTs). DHTs maintain logical neighbor relationships between the nodes of a P2P system and each node maintains only a small set of logical neighbors for routing messages closer to the destination nodes. Among these DHTs, the Content Addressable Network (CAN) [13] uses a space partitioning mechanism that can be easily adapted for spatial data. We use CAN as a base approach for large scale moving object data management.

In CAN, with a two-dimensional setting, the data space is mapped onto a $[0, 1] \times [0, 1]$ virtual coordinate space and is divided among the nodes in a P2P system. Thus, the virtual coordinate space is used as an intermediate space to map the data space onto node addresses. Data is stored as (key, value) pairs, in which the key is deterministically mapped onto a point p in the coordinate space and the corresponding pair is stored at the processing node that owns the subregion containing p. The same mapping is used for data retrieval. For two-dimensional spatial data the mapping onto the virtual coordinate space is straightforward. For routing, each node in CAN maintains a routing table containing the IP addresses of the nodes and the extent of the sub-regions adjacent to its own subregion. A node routes a message towards its destination by forwarding it to the neighbor with coordinates which is closest to the destination coordinates. A CAN-based system is built incrementally, where a new node can randomly select an existing node to join in the system by taking over the half of the region from the existing node. Since CAN does not perform any explicit load balancing, a dedicated load balancing strategy is necessary to improve the performance.

Godfrey et al. in [15] propose a load balancing strategy for P2P systems using the concept of virtual servers. The storage and routing occurs at virtual nodes (virtual servers) rather than real nodes (processing nodes) and each real node can host one or more virtual servers. Each virtual server maintains a sub-region. Every virtual server has its own logical address defined by its sub-region and data such as the neighbor table. A virtual server uses a similar greedy algorithm as CAN. An overloaded node transfers some of its virtual servers to an underloaded node. In [15] some nodes in the system act as a load balancer. All nodes send their load information to a randomly chosen load balancing server. Based on the workload of each node, the balancers distribute the virtual servers among the participating nodes to achieve a load balanced system. It is assumed that the overheads for transferring virtual servers among the participating nodes is commonly accepted as negligible in comparison with the benefits that the system can get from load balancing. We use the virtual server based approach for moving object data as a starting point.

Recently, several proposals aim to address the load balancing issues for range partitioned data sets and to preserve the locality of the data. Aspens et al. [16] adopt a pairing strategy in which heavily-loaded machines are placed next to lightly-loaded machines in the data structure to simplify data migration. Similarly, Karger and Ruhl [17] achieve load balancing by periodically moving underloaded nodes next to overloaded nodes. Ganesan et al.'s [18] online load balancing algorithm achieves load balancing by adjusting partition boundaries of range partitioned data and moving data among participating nodes. These techniques are designed for static non-spatial data.

It is important to note that some earlier works [19,20] in spatial database focus on load balancing using parallel systems or a cluster of workstations. Based on the access patterns of the data these systems distribute an index to balance the workload for optimizing the query response time. However, none of these techniques consider the movement patterns of objects and thus are unable to handle extra workload overheads from moving objects.

3   Mobility-Aware Load Balancing

In this section, we propose our load balancing technique that considers the movement of objects. We named our method as Mobility-Aware Load Balancing (MALB). Our technique makes a tradeoff between balancing the workload and minimizing the workload overheads from moving objects to minimize the cost function of the system. In this section, we first give a brief overview of modeling the workload. Then we define a cost function for the system. Finally, we will describe our load balancing technique.

We use CAN as the underlying distributed P2P system. Fig. 3 (a) shows the assignment of four subregions to four processing nodes (or nodes) P_1, P_2, P_3, and P_4 in CAN. Furthermore, we adopt the concept of virtual servers (Section 2) for load balancing, i.e., each node maintains one or more non-contiguous subregions resulting from CAN subdivision. Each subregion is called a virtual server or virtual node. Fig. 3 (b) shows the assignment of a set of virtual nodes $\{vs_4, vs_5\}$, $\{vs_1, vs_2, vs_3, vs_6\}$, $\{vs_7, vs_{10}, vs_{11}\}$, and $\{vs_8, vs_9\}$ representing different subregions to the nodes P_1, P_2, P_3, and P_4, respectively. Note that in the virtual server approach, the load balancer does not consider the movement of objects and queries among subregions while distributing loads. For

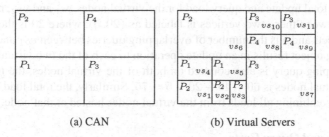

P_2		P_4	
P_1		P_3	

(a) CAN

P_2		P_3 vs_{10}	P_3 vs_{11}
	vs_6	P_4 vs_8	P_4 vs_9
P_1 vs_4	P_1 vs_5	P_3	
P_2 vs_1	P_2 vs_2 P_2 vs_3		vs_7

(b) Virtual Servers

Fig. 3. Region assignment

example, a scenario where a large number of objects cross between vs_4 in P_1 and vs_6 in P_2, the above possibly highly balanced assignment shown in Fig. 3 (b) can result in poor performance due to communication overheads. MALB reduces these communication overheads by only allowing the assignment of virtual nodes from one node to the other if the desired transition optimizes the cost given in Section 3.1 (see equation (1)).

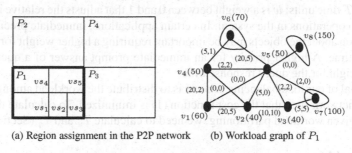

(a) Region assignment in the P2P network   (b) Workload graph of P_1

Fig. 4. Workload representation with a graph

Fig. 4 (a) shows a set of virtual nodes hosted at the node P_1. Each virtual node keeps track of its own load, its neighbors' loads, the number of objects leaving or entering each neighboring virtual nodes, and the overlapping queries with its neighbors. We model the virtual nodes, their relations within a node, and the load relations with neighboring processing nodes as a weighted graph, $G = (V, E)$. Each vertex $v \in V$ represents a virtual node or a neighboring processing node. The weight of a vertex representing a virtual node is the total load of updates and queries of that virtual node and corresponds to the load of its subregion. If a vertex represents a neighboring processing node, its weight reflects the total load obtained from all maintained virtual nodes. An edge $e \in E$ between two vertices in V indicates that the regions represented by these two vertices share a border. Each edge e has a pair of weights (e_{oc}, e_{qc}): e_{oc} represents the number of crossing objects and e_{qc} the number of overlapping queries.

Fig. 4 (a) shows initial assignments of four subregions to four nodes P_1, P_2, P_3, and P_4. Five virtual nodes $\{vs_1, vs_2, vs_3, vs_4, vs_5\}$ hosted at the node P_1 are shown inside thick border lines. The vertices $\{v_1, v_2, v_3, v_4, v_5\}$ represent five virtual nodes of P_1 and the vertices v_6, v_7, and v_8 represent three neighboring nodes P_2, P_3, and P_4, respectively, as shown in Fig. 4 (b). The vertices v_1 and v_2 having weights 60 and 40 represent the total update and query load for the virtual nodes vs_1 and vs_2, respectively. The edge between these two vertices is labeled as $(20, 5)$, where 20 is the number of crossing objects and 5 is the number of overlapping queries between vs_1 and vs_2. Since each crossing object results in an update operation to each of the two virtual nodes, and each overlapping query is also counted on both of the virtual nodes, the total load of these two virtual nodes is $60 + 40 - 20 - 5 = 75$. Similarly, the total load of a node is obtained by combining all loads from the virtual nodes hosted at that node.

3.1 Update and Query Costs

The performance of a moving object management system is determined by its two major tasks: handling updates and processing queries. The performance can be measured by the average response time for handling an update T_u and for processing a query T_q. The system performance can be measured by the following weighted cost function:

$$cost = w \times T_u \times N_u + (1 - w) \times T_q \times N_q, \tag{1}$$

where N_u and N_q are the total number of update requests and queries, respectively, for a period of T time units; w is a weight between 0 and 1 that adjusts the relative importance of the two operations in the system. In certain applications, immediate precise location information about the objects may be important requiring a higher weight for the update response time. A higher priority for an immediate prompt answer of a query, needs a higher weight for the query response time.

The goal of our load balancing scheme is to distribute the workload among the nodes in a P2P network such that the cost function (1) is minimized. To calculate the value of (1) for a given workload partitioning, we need to calculate T_u and T_q (Section 4).

3.2 Algorithm

In this section, we describe an algorithm, named *RegionAdjustment* (Algorithm 1), that every node runs to trade its load by transferring some of its virtual nodes to a neighboring node. If there are multiple neighbors of a node, the node selects the neighbor that result in the highest performance improvement using the cost function. Note that we do not use any explicit load balancing servers. Instead, every node acts as its own load balancing server using only local information.

A node constructs and updates its load interaction graph $G = (V, E)$ periodically. Let V_1 and V_2 be the two initial partitions of V. V_1 is a set of vertices representing a set of virtual nodes of that node and V_2 is a set with a single vertex for the selected neighboring node. Algorithm 1 refines the initial partitions and returns two new partitions that minimizes the cost in equation (1). The cost function is minimal for V_1 and V_2 when their load is equal and they have no crossing objects and overlapping queries. The algorithm determines the vertices to be migrated from V_1 to V_2.

Algorithm 1 pair-wise adjusts the load of virtual nodes between two neighboring nodes by estimating a local approximation of the global cost function equation (1). We show an example run of the algorithm in the workload scenario given in Fig. 4. Suppose node P_1 runs the algorithm to adjust its regions with a neighbor P_2. Initially, the algorithm creates two partitions V_1 and V_2, where $V_1 = \{v_1, v_2, v_3, v_4, v_5\}$ represents the

---

Algorithm 1. RegionAdjustment

1.1 Let X be the set of border vertices between V_1 and V_2;
1.2 Let Y be an empty set of vertices;
1.3 improving $= true$;
1.4 Let initcost be the value of cost using equation (1) for the initial partitions;
1.5 mincost $=$ initcost;
1.6 **while** improving **do**
1.7   improving $= false$;
1.8   **while** X *is not empty* **do**
1.9   **for** *each border vertex $x \in X$* **do**
1.10   Let cost$[x]$ be the value from equation (1) for the partitions assuming x is transferred to the other partition;
1.11   Let v be a vertex in X, such that cost$[v] = \min_{x \in X}$ cost$[x]$;
1.12   Transfer vertex v from the current partition to the other partition;
1.13   **if** cost$[v] <$ mincost **then**
1.14   improving $= true$;
1.15   mincost $=$ cost$[v]$;
1.16   Confirm the vertex migration from the current partition to the other;
1.17   Clear set Y;
1.18   **else**
1.19   Put vertex v in a temporary set Y;
1.20   Remove v from X and set visited flag for v;
1.21   Update X by adding non-visited new border vertices to X and by removing non-border vertices from X;
1.22   **for** *each vertex $y \in Y$* **do**
1.23   Transfer the vertex y from the current partition to the other;
1.24   Clear visited flag for all the vertices;
1.25   Populate X with border vertices excluding the vertices in Y for the next iteration;

---

set of virtual nodes of P_1 and the set $V_2 = \{v_6\}$ corresponds to the neighboring node P_2. In this scenario, X contains the border vertices $\{v_4, v_5\}$, i.e., the set of vertices that shares a border with a vertex in the other partition. Initially, the workload of nodes P_1 and P_2 are 160 and 70, respectively, and the crossing objects and the overlapping queries between these two nodes are $(20 + 5)$ and $(5 + 1)$. The workload of these two nodes can be calculated from the two sets of vertices V_1 and V_2. The response time for each of the 25 crossing updates is determined by the sum of the required update time in the two participating nodes. The response time for each of the 6 overlapping queries is determined by the maximum response time of both nodes. Again, the response time of all other non-crossing updates and non-overlapping queries only depends upon the service time of the corresponding node. Using these workload conditions, the algorithm determines the cost using equation (1) as initcost and determines the initial mincost.

The algorithm calculates the cost for each of the vertices in X in Line 1.9 and 1.10. The cost of a border vertex is the cost in equation (1) for the modified partitions if the vertex migrates to the other partition. If v_5 migrates from V_1 to V_2, the load of V_1 and V_2 would be 123 and 95, respectively, with 9 crossing updates and 4 overlapping queries. Assume the cost for v_5, $cost[v_5]$, is the minimal cost for all candidate vertices.

If $cost[v_5]$ is less than the previous value mincost, then we expect that the migration of v_5 from V_1 to V_2 reduces the overall weighted cost function. Therefore, v_5 migrates from V_1 to V_2 and we update the cost variables in lines 1.14–1.17. This process continues as long as it minimizes the cost function. This algorithm aims to balance the load while minimizing the overhead for crossing updates and overlapping queries.

In summary, the outer loop refines the two initial partitions (lines 1.6–1.25). The inner loop (lines 1.8–1.21) checks for each border vertex if its migration minimizes the cost function. To avoid local minima in Algorithm 1, vertices can temporarily move from one partition to the other (lines 1.18–1.19) even if this move does not immediately optimize the cost. If the algorithm does not find any minima, these vertices are put back (lines 1.22–1.23). Then, the variables are re-set for the next iteration of the outer loop (lines 1.24–1.25). The algorithm stops if no vertex is found whose migration further minimizes the cost.

Our balancing scheme considers both periodic and emergency load balancing measures. Each node periodically wakes up after a time period t_p and runs Algorithm 1 to balance its load with its neighbors. A carefully chosen value of t_p can balance two extreme conditions: a small value can lead to oscillations among participating nodes, and a high value may result in an imbalanced system. To avoid emergencies such as a sudden burst of traffic load, each node ensures that its load does not become higher than some threshold k_n in comparison with the load of its neighbors. A node locks all of its neighbors before running Algorithm 1 to avoid inconsistencies that may arise from concurrent load adjustment procedures among neighbors.

The sole use of local load balancing measures cannot guarantee an optimal load balance among all nodes, specially at dynamic load conditions (e.g., large traffic load for a city center in the morning hours). However, in a P2P system, an all-to-all communication based global load balancing scheme may be problematic in itself. Thus, we adopt a scheme from [18] where the system maintains a skip-graph data structure built on load conditions of the nodes, and can find the nodes with a maximum or minimum load in $O(\log n)$ time. Then, an overloaded node can share the load with the minimum loaded node if the load of the overloaded node is k_g (a threshold value) times higher than the minimum load. Similarly, an underloaded node can share the load with the maximum loaded node. If a new node joins the system or an existing node changes its position to split an overloaded node, the serving node for this request splits its virtual nodes into two sets of virtual nodes. It then retains a set of virtual nodes and gives the other set to the requester. The *Split* procedure is very similar to Algorithm 1. In this paper we do not consider the costs for transferring virtual nodes among the participating processing nodes because this cost is negligible in comparison to total workload for handling large updates and processing queries for a period of time.

4   Cost Model

In algorithm RegionAdjustment, we need to calculate the cost function (1) given a workload partitioning. To calculate the value of (1), in this section, we derive a model to estimate T_u and T_q based on the system knowledge of updates, queries, and the service rate of the processing nodes for a period of T time units.

When update or query requests arrive at a high rate, these requests are queued up in different nodes in the form of messages. In our system, a message is an operation that can be served in one node, while a request may consist of (or create) several messages. A request (update or query) may need to travel several nodes for the desired objective in a P2P system. For example, a range query request that spans over four nodes generates four messages (one message per node). Similarly, an update request may generate two messages (a delete message to the leaving node and an insert message to the entering node) when an object crosses the border of two nodes. A new update or a query message in a node has to wait until all the pending messages are served by the node. We assume that the objects in a node are maintained in a in-memory structure (disk-based structures are not suitable for a very high update rate) so that an update or query message can be processed in a very short amount of time, which is much faster than the queuing time of the message. Therefore, the response time of a message is actually the average queuing time; and the workload is proportional to the number of messages. Given this assumption, we can derive the average response time of a message in a node i as follows.

Let λ_{u_i} and λ_{q_i} be the update and query message arrival rates, respectively, at node i. The total arrival rate at node i is $\lambda_i = \lambda_{u_i} + \lambda_{q_i}$. Let the service rate of node i be μ_i, that is, node i can serve messages at the rate μ_i and assume $\mu_i > \lambda_i$. A node can be seen as a $M/M/1$ queue [21] (where M stands for "Markovian", implying exponential distribution for service times or inter-arrival times). According to Little's law [22], the average queue length (Q) and Average Queueing Time (AQT) of node i are given by the following equations:

$$Q_i = \frac{\frac{\lambda_i}{\mu_i}}{1 - \frac{\lambda_i}{\mu_i}} \tag{2}$$

$$AQT_i = \frac{Q_i}{\lambda_i} \tag{3}$$

According to the analysis given in the previous paragraph, the average response time of a message in node i is also given by (3).

During a period of T time units, $\lambda_{u_i} \times T$ update messages (including update messages created due to objects crossing nodes) arrive and the average response time of a message is AQT_i, so the total time to process the update messages is $\lambda_{u_i} \times T \times AQT_i$. If there are n nodes in the P2P system and the update request rate of the whole system is γ (the actual update requests from data sources), then there are $\gamma \times T$ update requests. So the average service time for an update request T_{us} is given by:

$$T_{us} = \frac{\sum_{k=1}^{n}(\lambda_{u_i} \times T \times AQT_i)}{\gamma \times T} = \frac{\sum_{k=1}^{n}(\lambda_{u_i} \times AQT_i)}{\gamma} \tag{4}$$

However, objects that cross the borders of nodes cause communication overheads from extra update messages. Assume an object crosses the border of a node with a probability p (i.e., the number of objects that crosses the border / the total number of objects) and the average communication time between two nodes is T_c. Then average communication overhead of an update request T_{uc} is $T_{uc} = p \times T_c$.

By adding the service time and the communication time of an update request, we get the average response time of an update request as follows:

$$T_u = T_{us} + T_{uc} \tag{5}$$

Similarly, a query may create several messages for all the intersected nodes by the query. Every intersected node processes its query message and return the answer to the query originating node. The query originating node combines the answers from all the intersected nodes to obtain the answer for the query. Therefore, the response time of the query is the maximum of the response times of all the messages created by the request.

In our system, the total geographic space is normalized into a $[0, 1] \times [0, 1]$ coordinate space and is partitioned among n participating nodes. A window query q of size $w \times h$ whose top-left corner is at (x, y), may intersect 1 to n nodes depending on the location and the size of the query. Let function $f(q_{x,y})$ return the number of nodes k that q intersects. Thereby q creates k query messages and the response time of these query messages are $RT_1, RT_2, ..., RT_k$, respectively, from k participating nodes. Then the response time of q is $\max\{RT_1, RT_2, ..., RT_k\}$. The other cost involved for q is the communication overhead $T_{c_{j \to i}}$ from a source node j (query originating node) to a destination node i. Thus, the response time to the query q, that intersects $f(q_{x,y})$ many nodes can be defined as:

$$g(x, y, q) = \max_{1 \leq i \leq f(q_{x,y})} [T_{c_{j \to i}} + RT_i(Q_i, q_{x,y})] \qquad (6)$$

Given a query size $w \times h$, if we sum up the response time of queries of all possible query locations in the data space and divide the sum by the total number of queries, we find the average response time of the queries for query size $w \times h$ as follows:

$$T_q = \frac{1}{(1 - w) \times (1 - h)} \int_0^{1-w} \int_0^{1-h} g(x, y, q) dx dy \qquad (7)$$

We have derived both T_u and T_q. Substitute T_u and T_q in function 1 by 5 and 7, respectively. We can calculate the cost for a given workload partitioning given the information on crossing objects, update rate, query rate, service rate, etc. However, as in a P2P network each node only has information about its own load and the load of its neighbors, we can only locally calculate the cost and optimize the performance.

5   Experimental Study

We compare the performance of our load balancing algorithm MALB with the virtual server (VS) load balancing technique from [15] on an experimental setup for location-based services.

5.1   Experimental Setup

We use both synthetic and real road networks as shown in Fig. 5. The synthetic road networks are constructed by connecting a large number of small, grid-like, road networks, modeling a set of suburbs connected to each other with freeways. A larger embedded grid is also used as a metropolitan city center. The real road network is from the city of Stockton in San Joaquin County, CA. Again, the density of the network is more at the city center in comparison to the other areas. The movements of the objects within the road networks are generated by the Network-based Generator of Moving

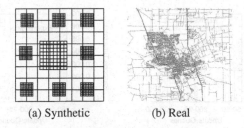

| (a) Synthetic | (b) Real |

Fig. 5. Road networks

Objects [23]. We have used J-Sim [24] to develop our simulation environment, and also used BRITE [25] topology generation tool to create a communication network topology. Each of our nodes are connected with its neighbors through high bandwidth lines (2Gbps). We build the network in an incremental fashion, where each node contacts an existing node to join the system. A summary of experiment parameters (default parameters are shown in bold) are given in Tab. 1. We run each simulation multiple times and give the averages in our results. As we shall see, simple CAN cannot scale well for a skewed load distribution, we present the performance of our technique in comparison to the VS load balancing technique.

Table 1. Summary of parameters for the experiments

Parameter	Value
Road network data	Synthetic, Real
No. of nodes	100
Update arrival rate (per sec.) (λ_u)	4K, 6K, **8K**, 10K, 15K
Query arrival rate (per sec.) (λ_q)	200, **400**, 800, 1600
Service rate in a node (per sec.) (μ)	400
No. of crossing updates in % of all updates	5, 15, **25**, 35, 45
Query size in % of whole data space	1, **5**, 10, 20
Average no. of virtual nodes per node	8
Periodic load balance timeout (sec.)	10
Emergency load balance parameters	$k_n = 2, k_g = 4$

5.2 Evaluation of the Cost Model

We have measured the accuracy of our cost model given in equation (1) using a two node setting. We have chosen a two node setting as our algorithm works on individual nodes and only uses the local neighborhood information available to that node to calculate the cost. For various arrival and service rates, we first calculate the value of the cost function (1), F_{est}. Then we run the experiments to find the experimental values, the actual cost F_{exp}. Finally, we obtain the accuracy of the cost model by using $\frac{|F_{est}-F_{exp}|}{F_{est}}$, that is, the relative error, as the metric. The accuracy of the cost function depends on how well we can estimate the average update response time (AURT) in equation (5) and

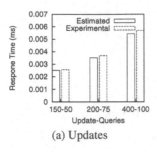 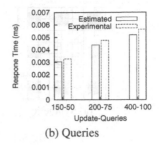

(a) Updates   (b) Queries

Fig. 6. Accuracy of the cost model

the average query response time (AQRT) in equation (7). We plot these two values as functions of different arrival rates in Fig. 6 (a) and 6 (b), respectively. We found that the deviation of the cost function from experimental values is always less than 10%.

5.3 Scalability

Fig. 7 shows that the AURT increases with an increase in update arrival rate and CAN cannot sustain its performance with the increasing load. The graph also shows that the improvements from our approach over the VS remains constant up to an arrival rate of 6000. The improvement increases with the increase in the update arrival rate (up to 28% from the VS approach). The reason for this is, more updates would mean that there are more crossings of objects in the system. In this case, the VS load balancing needs to handle more load due to overhead crossings. Our technique continues to scale better by reducing the overhead crossings from updates.

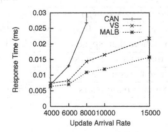

Fig. 7. Effect of update rate

5.4 Effect of w

We measure the value of the cost function defined in (1) by varying the weight, w. As w increases, the value of the cost function increases as we put more weight on updates and the number of updates is much more than the number of queries in the system (Fig. 8). We get a reduced average response time for location updates if $w = 1.0$ and, interestingly, also a good query performance. As we reduce the updates between regions this also reduces the total load in the system leading to good query performances.

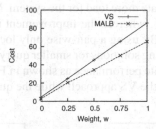

Fig. 8. Effect of weight, w

5.5   Effect of the Number of Region Crossing Object Updates

In this experiment we study the relative performance of the various techniques as the number of crossings among neighboring regions is varied from 5% to 45%. The AURT and AQRT with varying crossing updates are shown in Fig. 9. In Fig. 9 (a), we set $w = 1.0$ because we want to optimize the update performance in the system. Here the x-axis represents the percentage of crossing updates with respect to the total number of updates and the y-axis shows the average response time for an update. The results show that as the number of crossings is increased from 5 to 45 percent, our approach outperforms VS load balancing technique by a large margin, up to 40%. Since the extra workload overheads increase with the increase of crossing updates in the system, and traditional load balancing techniques (e.g., VS) are not aware of these overheads, the performance of the system degrades sharply with larger crossings. On the contrary, MALB reduces crossing overheads while balancing the workload, and thus perform much better than VS. Similarly, Fig. 9 (b) shows the comparison of two load balancing techniques over queries. This figure shows that even for a small number of crossings (i.e., 5%), the improvement is 19%, and as we reach to 45% the improvement becomes equal to 36%.

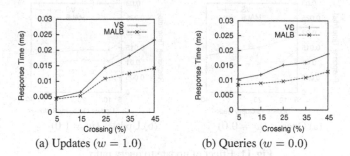

(a) Updates ($w = 1.0$)   (b) Queries ($w = 0.0$)

Fig. 9. Effect of crossing objects

5.6   Effect of Query Size

We have also run experiments by varying the query sizes. Fig. 10 (a) shows that the AQRT increases with an increase in the query size because larger queries overlap with more nodes and thereby generate more load for the system. MALB performs 35% better than the VS approach for 5%, whereas the improvement is only approximately 20% when the query size is 20%. By using a pair-wise only local decision making method we can find better load balancing solutions for smaller query sizes (i.e., 5% -10%). Also, we see that in the case of update performance as shown in Fig. 10 (b), the improvement remains almost constant over the VS approach while the query size varies.

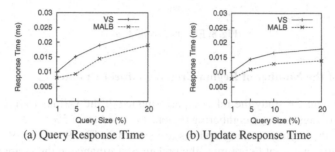

(a) Query Response Time   (b) Update Response Time

Fig. 10. Effect of varying the query size

5.7 Effect of Update to Query Ratio

We vary the update to query ratio from 5 to 40. We keep the update rate constant (8000) while varying the query rate. Fig. 11 (a) shows that our system achieves high performance gains for queries when the update to query ratio is small. As the number of queries increases, MALB can optimize more on the query performance. Fig. 11 (b) shows that the performance gain for updates does not vary as much with the increasing number of queries in the system. Therefore, our technique can achieve high query performance while still being very efficient for updates.

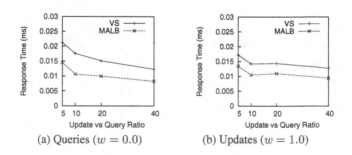

(a) Queries ($w = 0.0$)   (b) Updates ($w = 1.0$)

Fig. 11. Effect of update to query ratio

5.8 Experiments on a Real Road Network

We have run our experiments on a real city road network Stockton in San Joaquin County, CA. We use 32 nodes to share the load for this small setting where the update and query arrival rates are 2000 and 200, respectively. The rest of the parameters are same as in the previous experiments. This setting presents a challenge for our local load balancing algorithm as the city has a dense center with many small roads, only a few highways, and no suburbs (Fig. 5 (b)). Thus, local decisions on the center have a smaller impact on the global load balance. However, we still see in Fig. 12 that MALB outperforms the VS approach and the gains can be up to 20%.

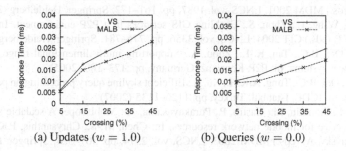

(a) Updates ($w = 1.0$)   (b) Queries ($w = 0.0$)

Fig. 12. Effect of crossing objects in a real road network

6 Conclusions and Future Work

In this paper, we proposed a novel mobility-aware load balancing (MALB) technique for moving object management in a P2P system. MALB considers the movement patterns of objects and achieves better performance than traditional load balancing schemes, which were designed for stationary data. In addition, we optimized the performance of handling updates in tandem with the processing of queries. Through experiments, we show that our load balancing scheme results in constantly better update and query performance results than existing load balancing techniques and the improvement is up to 40%. We show that we can find better load partitions that reduce communication and processing overheads by reducing object updates and queries that span multiple processors. However, accessing information available only at the neighbors can lead to suboptimal results in comparison to a global optimization strategy. Yet, it is not practical to devise a trivially centralized load balancer for a large-scale P2P system. As a direction for future work, we plan to build on our existing pair-wise load balancing scheme to include clustering-based techniques.

References

1. Tanin, E., Harwood, A., Samet, H.: Using a distributed quadtree index in peer-to-peer networks. VLDB Journal 16(2), 165–178 (2007)
2. Wolfson, O., Xu, B., Chamberlain, S., Jiang, L.: Moving objects databases: Issues and solutions. In: SSDBM, Capri, Italy, pp. 111–122 (1998)
3. Cheng, R., Xia, Y., Prabhakar, S., Shah, R.: Change tolerant indexing for constantly evolving data. In: ICDE, Tokyo, Japan, pp. 391–402 (2005)
4. Guting, R.H., Schneider, M.: Moving Objects Databases. Morgan Kaufmann, San Francisco (2005)

5. Lee, M., Hsu, W., Jensen, C., Cui, B., Teo, K.: Supporting frequent updates in R-trees: A bottom-up approach. In: Aberer, K., Koubarakis, M., Kalogeraki, V. (eds.) VLDB 2003. LNCS, vol. 2944, pp. 608–619. Springer, Heidelberg (2004)
6. Tao, Y., Papadias, D., Sun, J.: The TPR*-tree: An optimized spatio-temporal access method for predictive queries. In: Aberer, K., Koubarakis, M., Kalogeraki, V. (eds.) VLDB 2003. LNCS, vol. 2944, pp. 790–801. Springer, Heidelberg (2004)
7. Song, Z., Roussopoulos, N.: Hashing moving objects. In: Tan, K.-L., Franklin, M.J., Lui, J.C.-S. (eds.) MDM 2001. LNCS, vol. 1987, pp. 161–172. Springer, Heidelberg (2000)
8. Guan, J., Wang, L., Zhou, S.: Enabling GIS services in a P2P environment. In: Das, G., Gulati, V.P. (eds.) CIT 2004. LNCS, vol. 3356, pp. 776–781. Springer, Heidelberg (2004)
9. Shu, Y., Ooi, B.C., Tan, K.-L., Zhou, A.: Supporting multi-dimensional range queries in peer-to-peer systems. In: P2P, Konstanz, Germany, pp. 173–180 (2005)
10. Wang, S., Ooi, B.C., Tung, A.K.H., Xu, L.: Efficient skyline query processing on peer-to-peer networks. In: ICDE, Istanbul, Turkey, pp. 1126–1135 (2007)
11. Denny, M., Franklin, M., Castro, P., Purakayasatha, A.: Mobiscope: A scalable spatial discovery service for mobile network resources. In: Chen, M.-S., Chrysanthis, P.K., Sloman, M., Zaslavsky, A. (eds.) MDM 2003. LNCS, vol. 2574, pp. 307–324. Springer, Heidelberg (2003)
12. Gedik, M.-B., Liu, S.M.-L.: Mobieyes: A distributed location monitoring service using moving location queries. IEEE Transactions on Mobile Computing 5(10), 1384–1402 (2006)
13. Ratnasamy, S., Francis, P., Handley, M., Karp, R., Shenker, S.: A scalable content-addressable network. In: SIGCOMM, San Diego, CA, pp. 161–172 (2001)
14. Stoica, I., Morris, R., Karger, D., Kaashoek, M.F., Balakrishnan, H.: Chord: A scalable peer-to-peer lookup service for Internet applications. In: SIGCOMM, San Diego, CA, pp. 161–172 (2001)
15. Godfrey, B., Lakshminarayanan, K., Surana, S., Karp, R., Stoica, I.: Load balancing in dynamic structured P2P systems. In: INFOCOM, Hong Kong, pp. 2253–2262 (2004)
16. Aspnes, J., Kirsch, J., Krishnamurthy, A.: Load balancing and locality in range queriable data structures. In: PODC, Newfoundland, Canada, pp. 25–28 (2004)
17. Karger, D., Ruhl, M.: Simple efficient load-balancing algorithms for peer-to-peer systems. In: Voelker, G.M., Shenker, S. (eds.) IPTPS 2004. LNCS, vol. 3279, pp. 131–140. Springer, Heidelberg (2005)
18. Ganesan, P., Bawa, M., Garcia-Molina, H.: Online balancing of range-partitioned data with applications to peer-to-peer systems. In: VLDB, Toronto, Canada, pp. 444–455 (2004)
19. Kriakov, V., Delis, A., Kollios, G.: Management of highly dynamic multidimensional data in a cluster of workstations. In: Bertino, E., Christodoulakis, S., Plexousakis, D., Christophides, V., Koubarakis, M., Böhm, K., Ferrari, E. (eds.) EDBT 2004. LNCS, vol. 2992, pp. 748–764. Springer, Heidelberg (2004)
20. Lee, M.L., Kitsuregawa, M., Ooi, B.C., Tan, K., Mondal, A.: Towards selftuning data placement in parallel database systems. In: SIGMOD, Dallas, TX, pp. 225–236 (2000)
21. Penttinen, A.: Kendall's notation for queuing models. In: Introduction to Teletraffic Theory. Lecture Notes: S-38, p. 145. Helsinki University of Technology (1999)
22. Little, J.D.C.: A Proof of the Queueing Formula $l = \lambda$ w. Operations Research 9, 383–387 (1961)
23. Brinkhoff, T.: A framework for generating network-based moving objects. GeoInformatica 6(2), 153–180 (2002)
24. J-SIM: http://www.j-sim.org/
25. BRITE: http://www.cs.bu.edu/brite/

Serializable Executions with Snapshot Isolation: Modifying Application Code or Mixing Isolation Levels?

Mohammad Alomari, Michael Cahill, Alan Fekete, and Uwe Röhm

The University of Sydney
School of Information Technologies
Sydney NSW 2006, Australia

Abstract. Snapshot Isolation concurrency control (SI) allows substantial performance gains compared to holding commit-duration readlocks, while still avoiding many anomalies such as lost updates or inconsistent reads. However, for some sets of application programs, SI can result in non-serializable execution, in which database integrity constraints can be violated. The literature reports two different approaches to ensuring all executions are serializable while still using SI for concurrency control. In one approach, the application programs are modified (without changing their stand-alone semantics) by introducing extra conflicts. In the other approach, the application programs are not changed, but a small subset of them must be run using standard two-phase locking rather than SI. We compare the performance impact of these two approaches. Our conclusion is that the convenience of preserving the application code (and adjusting only the isolation level for some transactions) leads to a very substantial performance penalty against the best that can be done through application modification.

1 Introduction

Since the data stored in a database is of value to the organization, it is important to prevent corruption of that data. There are well-known dangers of corruption from the interleaving of concurrent activities; for example, lost update occurs when one transaction T overwrites the value produced by a concurrent transaction U, based on an earlier value which T had read. The inconsistent read anomaly happens when transaction T reads two items, and a concurrent transaction U modifies both items, and T sees the effect of U in one item but not the other. The concept of serializability captures exactly when an interleaved execution is acceptable: when it is equivalent in effect to a serial execution. The essential property is that all integrity constraints are valid at the end of a serializable execution (so long as each transaction separately is written to maintain the constraints).

The standard database textbooks all explain how a database engine can ensure that executions are serializable, by obtaining read and write locks on data and index records before accessing the information, and keeping these locks till

J.R. Haritsa, R. Kotagiri, and V. Pudi (Eds.): DASFAA 2008, LNCS 4947, pp. 267–281, 2008.
© Springer-Verlag Berlin Heidelberg 2008

the transaction commits (or aborts). This concurrency control mechanism is called strict two-phase locking (abbreviated here as 2PL). There are a number of optimizations possible especially in locking index entries, and many vendors implement concurrency control based on some form of 2PL. However, the data integrity guaranteed by 2PL comes at a considerable cost in performance, as a read operation is delayed until the commit of a concurrent transaction which wrote the same item, and as a write operation is delayed until there is no active transaction that has read the item.

For developers who want better performance than 2PL allows, all vendors provide the capability to set a lower isolation level for a transaction. For example, it is usual to offer a READ COMMITTED isolation level in which write locks are held till transaction commit (as in 2PL), but read locks are released as soon as the data has been read. This gives much better performance than 2PL (often throughput is 3 times greater) because updates are not blocked for as long; however it is vulnerable to lost update, inconsistent read, and other anomalies that can violate integrity constraints.

A concurrency control mechanism was introduced by [1] as Snapshot Isolation (abbreviated SI). This uses multiple versions of each data item (so an update copies the old value to a side location, for other transactions to read). The performance of SI is often similar to that of Read Committed, but SI does not suffer from Lost Update or Inconsistent Read. Because SI avoids all the well-known anomalies, some vendors offer it *instead of* 2PL, when users ask for SERIALIZABLE isolation. However, this is risky; while SI does not allow lost update or inconsistent read, there are non-serializable executions possible, and data can be corrupted so that (undeclared) integrity constraints no longer hold. In particular, [1] show an anomaly called Write Skew that is possible with SI. In the meanwhile, there exist DBMS platforms where users can choose SI as Isolation Level SNAPSHOT, as well as being able to have 2PL by setting Isolation Level SERIALIZABLE.

In general, SI allows Write Skew, and therefore it can give non-serializable executions and violation of data integrity. However, some applications have specific patterns of data access, so that for these particular sets of programs, all executions are serializable even if SI is the concurrency control mechanism. The TPC-C benchmark has this property, as was discovered when Oracle recorded (very good) benchmark performance by using SI. Fekete et al [2] introduced a theory which allows one to prove this situation, when it occurs. The theory is based on having the DBA find conflicts between the programs, and represent these conflicts in a graph called Static Dependency Graph (SDG). A simple property of SDG is that there is not a cycle with two consecutive edges of a particular sort (called vulnerable edges). [2] proves that this property guarantees that every execution under SI will be serializable.

While TPC-C works correctly with SI for concurrency control, this is not true for other sets of application programs, What should we do, if we have a set of programs that might execute non-serializably under SI, and we care enough about data integrity that we wish to guarantee serializable execution? On some platforms, we have several courses of action available to us. The authors of [2] propose that one could modify the application programs so that the modified programs do not have non-serializable executions. There are several choices of suitable modifications, such as introducing identity write operations that set an item to its current value, or introducing new items that do not represent any real-world information but can be updated to provide conflicts between transactions. A different approach is proposed in [3], in which some of the transactions run with SI but others use 2PL for concurrency control. Clearly, if all transactions used 2PL, the executions would be serializable; however [3] has identified a minimal set of transactions called *pivots*; the pivots need to use 2PL, but the other transactions can run with SI or 2PL, while guaranteeing that every execution will be serializable.

For the developer, the approach of [3] is much more convenient that of [2]. Allocating each transaction with the appropriate isolation level does not require any changes to the application code, or recompilation. It can be done at run-time, entirely in the client. In contrast, most installations insist on extensive testing before approving any changes to the application code (even ones as simple as Update x=x). In many cases, each application is a stored procedure in the database, so modification requires substantial permissions on the server; but changing isolation level happens in the client without any authorization.

In this paper we address the issue of how much the DBA or developer must give up in performance, for the undoubted convenience of leaving application code unmodified and just changing isolation levels for particular transactions. To answer this, we measure the throughput of the different approaches when applied to the SmallBank benchmark (which we introduced in [4]).

The remainder of this paper is organised as follows: In the next section, we review the related work, and give details of the definitions and results we use from the literature. In Section 3, we present the particular mix of transaction programs on which we do our experiments. Our main contribution is a comparison of the performance impacts of the known different approaches each of which can guarantee serializable executions on a platform, where both strict two-phase locking and SI are available isolation levels. This is in Section 4 of the paper. Section 5 concludes.

2 Background

We first cite important previous research publications that are connected with the topic of this paper. After that, we summarize the insights of those papers that are essential for understanding the new work which we are reporting here.

2.1 Related Work

In 1995, Berenson et al defined a concurrency control mechanism called Snapshot isolation (SI), and they showed that it allowed non-serializable executions [1]. Oracle 7 used a variant of this mechanism to control execution when clients requested that a transaction run with "isolation level serializable" [5], and other DBMS vendors have also used these techniques instead of, or as well as, traditional two-phase locking (2PL).

With the widespread use of SI in common platforms, there has been much research on building replicated data management on top of these platforms, in a variety of system designs [6,7,8,9,10].

Despite the result of [1], that SI allows non-serializable executions, experts were unable to find any non-serializable executions for the TPC-C benchmark [11]. This led to a research agenda exploring conditions in which SI doesn't cause concurrency problems for an application. Bernstein et al [12] gave a proof technique to show that application code executing under SI preserves integrity constraints; the proof must be tailored to the integrity constraint in question. Fekete et al [2] gave a proof technique to show that application code executing under SI has all its executions serializable (and in that case, an arbitrary integrity constraint will be preserved). The proof technique applies to the TPC-C benchmark. This paper also indicates how the application code can be modified without affecting the functionality of each program, modifications which ensure that all executions of the modified code are serializable. Jorwekar et al [13] introduced techniques that allow the analysis of [2] to be done automatically (though with some loss of precision), in a tool. In our earlier work [4], we investigated the performance of different methods proposed in [2], for modifying a set of programs so that they have the property that all executions are serializable. We found that some modifications keep essentially the same performance as SI, while others have a substantial performance impact.

In SQL Server 2005, Microsoft provided a platform which offers clients a choice between 2PL and SI; each transaction can separately indicate which concurrency control mechanism will be applied to its operations. Oracle Berkeley DB also gives this choice of concurrency control techniques, in a record store rather than in a system with SQL queries. The serializability theory of such systems was investigated by Fekete [3].

2.2   Concurrency Control with SI and 2PL

In this paper we consider a platform which offers clients a choice of concurrency control mechanism. Each transaction that is submitted can be run with serializable isolation, using strict two phase locking, or with snapshot isolation. In the following discussion, we explain the key rules that must be enforced by the concurrency control. Each platform may well use different optimizations, storage structures, etc, in implementing these rules.

To support SI, the platform must keep multiple versions of each data item. Every time any transaction modifies a data item, another version is produced, and the various versions might all be accessible. Conceptually, a version is kept forever, except that it is thrown away if the creating transaction aborts; in practice, systems will garbage collect versions from time to time.

When a transaction using 2PL reads the data item, the value returned is the current value at the time when the read occurs (that is, the most recently produced value among those whose creator didn't abort). In contrast, when a transaction using SI reads, it gets the value produced by the most recent transaction, among those that committed before the reading transaction started[1]. This means that a transaction T using SI acts as if it were working on a private copy of the database, called T's snapshot, which captures exactly the effects of transactions which had committed before T began. It is the contents of the snapshot that also determines the calculation of "where clauses" in SQL statements within T.

In such a system, there are locks; these are kept on data items, not on versions. The usual lock modes and conflict rules apply: there can't be simultaneously different transactions holding exclusive locks on the same item, nor can one transaction hold an exclusive lock while others hold shared locks on that same item. A transaction using 2PL must hold some lock (either shared or exclusive) on an item at the time it reads the item's current version; and it must hold an exclusive lock on the item when it creates a new version of the item. A transaction using SI does not need a lock to read a version, nor to produce a new version; however, each version must be "installed" at some time between version creation and the transaction's commit, and the transaction must hold an exclusive lock on the item at the time of the install. The locks obtained by 2PL transactions are held till the transaction completes, and the exclusive locks gained by an SI transaction are also held for the duration of the transaction.

There is one further rule, called "First-Committer-Wins". This says that a transaction T using SI is not allowed to commit if there is any other transaction[2] U such that both the following conditions are true: U committed between the start of T and the commit of T, and there is an item that was modified by both T and U.

We again stress that these descriptions show the functionality required of the transaction management subsystem, but different implementations can do things differently for the sake of efficiency, so long as operations are ordered, and return results are computed, compatibly with these rules. For example, we are not aware of any system which actually takes a snapshot copy of the database at the start of a transaction which is using SI; instead the appropriate version is found when the transaction requests to read the item. In some platforms, the versions are kept around, and a version-aware index scheme finds the appropriate one; in other system designs, only the current version is kept, and the appropriate old version is reconstructed at the time of the read request, from the current version and deltas that were stored to support rollback. Similarly, when a transaction using SI writes an item (and so creates a new version), some systems install this version immediately (which requires obtaining an exclusive lock on the item) while other systems get the lock and install the version only as part of the transaction's commit, so in such a design the lock is released immediately.

---

[1] There is one exception to this rule: if the reader had itself modified the item already, the value returned is the version that the reading transaction had produced.

[2] This applies no matter what concurrency control mechanism U is using.

2.3 Ensuring Serializable Executions

In general, there may be non-serializable executions of a DBMS platform that runs some transactions with SI and others with 2PL. However, research has found a theory which provides a condition on the conflicts between the application programs; if this condition holds, every execution will be serializable (and therefore any integrity constraint will hold at the end of the execution, so long as each application program was written to maintain this constraint. To use this theory, one needs to compute the *Static Dependency Graph (SDG)*. Each transaction program is represented as a node in SDG. There is an edge from program P to program Q exactly when P can give rise to a transaction T, and Q can give rise to a transaction U, with T and U having a conflict (for example, T reads item x and U writes item x). The edge from P to Q is called *vulnerable* if P can give rise to transaction T, and Q can give rise to U, and T and U can execute concurrently[3] with a read-write conflict (also called an anti-dependency); that is, where T reads a version of item x which is earlier than the version of x which is produced by U. Vulnerable edges are indicated in graphs by a dashed line.

Two consecutive vulnerable edges that are within a cycle are called a *dangerous structure* in the SDG. The transaction program that is the common node among the two edges of the dangerous structure is called a *pivot*. If each pivot is using 2PL as its concurrency control mechanism, then every execution of the programs is serializable[4]. (on a DBMS using a mixture of SI and 2PL for concurrency control).

Given a set of application programs, this theory tells us that we can modify or configure the application, without changing the meaning of each program, so that, after the changes, the application will have all executions serializable. The essential of this approach is to take each pivot, and either run it with 2PL, or else make it no longer be a pivot (by changing one edge in the dangerous structure to be non-vulnerable).

If we want to modify the application code so a particular program using SI is not a pivot, we must choose one of the two edges that make up the dangerous structure, and change one or both of the transactions at the end of that edge so that they both write some data item (whenever their parameter values lead them to conflict at all). Provided they write a common item, the First-Committer-Wins rule will prevent the situation where they execute concurrently with a

---

[3] Transactions T and U are concurrent if there is an intersection between the interval from T's start to T's commit, and the interval from U's start to its commit.

[4] To be precise, the theory of [3] was done for a simple model of transactions as sequences of read and write operations. In that case, a pivot must also have the property that there is no chord across the cycle in which the dangerous structure lies, and the rule for allocating isolation levels exactly characterizes when executions are guaranteed to be serializable. However, it is straightforward to extend the theory in the obvious way to application programs with control flow and parameters, just as in [2]. In the extended theory, the pivot is as we define it here, not necessarily in a chord-free cycle. Also the property that pivots run with 2PL is sufficient, but not always necessary, for ensuring serializable executions.

read-write conflict. In the strategy called "Promotion" we take the item that is read by one and written by the other, and include an identity write (which produces a new version with the same value) to accompany the read[5]. Another strategy called "Materialization" introduces write operations into the transactions at both ends of the chosen edge. In our earlier work [4], we showed that the choice of which edges to make non-vulnerable had a large impact on throughput, while the particular technique of modification had less impact. Our later experiments confirm this. Thus in this paper we concentrate on measuring the performance impact of promotion for different edges, and compare this to the impact of running the pivot transactions with 2PL.

We give the name Pivot-2PL to the strategy where every pivot is allocated to use strict 2PL for its concurrency control (by setting isolation level serializable whenever the pivot programs are invoked), and every other application program sets isolation level snapshot. This does not require any rewriting of the application program. Indeed, this strategy can be followed entirely in the client to guarantee serializable executions, without having any extra permissions in the database server. Another feature of this strategy is that once the SDG has been calculated, no choices are made by the DBA; the graph determines the pivots, and thus the configuration of isolation levels. By comparison, in the common situation where each transaction runs a stored procedure, the promotion approaches require rights to alter database objects, and they also require choosing which edge to make non-vulnerable, from the two edges in each dangerous structure.

A characteristic of all these techniques, whether they modify the application programs, or vary the allocated isolation level, is that they rely on knowledge of the full set of application programs. That is, they are not appropriate if there are ad-hoc transactions.

3   The SmallBank Benchmark

The material in this section is condensed from our earlier paper [4], where this benchmark was introduced.

3.1   SmallBank Schema

Our benchmark is a small banking database consist of three main tables: Account (Name, CustomerID), Saving(CustomerID, Balance), Checking(CustomerID, Balance). The Account table represents the customers; its primary key is Name and we declared a DBMS-enforced non-null uniqueness constraint for its CustomerID attribute. Similarly CustomerID is a primary key for both Saving and

---

[5] Some SQL dialects also allow the statement Select ... For Update (SFU). As discussed in [2], on Oracle SFU is treated for concurrency control like an Update, and so promotion can be done by changing the read into SFU; however on our database platform used for the experiments, we have found that using SFU does not always prevent an update in a concurrent transaction, and so SFU can't be used to make an edge non-vulnerable.

`Checking` tables. Checking.Balance and Savings.Balance are numeric valued, each representing the balance in the corresponding account for one customer.

3.2   Transaction Mix

The SmallBank benchmark runs instances of five transaction programs. *Balance*, or Bal(N), is a parameterized transaction that represents calculating the total balance for a customer. It looks up Account to get the CustomerID value for N, and then returns the sum of savings and checking balances for that CustomerID.

DepositChecking, or DC(N,V), is a parameterized transaction that represents making a deposit on the checking account of a customer. Its operation is to look up the Account table to get CustomerID corresponding to the name N and increase the checking balance by V for that CustomerID.

TransactSaving, or TS(N, V), represents making a deposit or withdrawal on the savings account. It increases the savings balance by V for that customer. If the name N is not found in the table or if the transaction would result in a negative savings balance for the customer, the transaction will rollback.

Amalgamate, or Amg(N1, N2), represents moving all the funds from one customer to another. It reads the balances for both accounts of customer N1, then sets both to zero, and finally increases the checking balance for N2 by the sum of N1's previous balances.

WriteCheck, or WC(N,V), represents writing a check against an account. Its operation is to look up Account to get the CustomerID value for N, evaluate the sum of savings and checking balances for that CustomerID. If the sum is less than V, it decreases the checking balance by V+1 (reflecting a penalty of 1 for overdrawing), otherwise it decreases the checking balance by V.

3.3   The SDG for SmallBank

Figure 1 shows the SDG for our benchmark. We use dashed edges to indicate vulnerability, and we shade the nodes representing update transactions; the pivot transaction is shown as a diamond. A read-write conflict exists from Bal to the remaining transactions because Bal reads what the other transactions modify. Similarly there is a read-write conflict from WC toTS. These edges are vulnerable. In contrast, if there is a read-write conflict from WC to Amg on a row of the Saving table, then there is also a write-write conflict on Checking (and so this can't happen between concurrently executing transactions). That is, the edge from WC to Amg is not vulnerable.

We see that the only dangerous structure in Figure 1 is from Balance (Bal) to WriteCheck (WC) to TransactSaving (TS), so the only pivot is WC (shown in Figure 1 as a diamond). The other vulnerable edges run from Bal to programs which are not in turn the source of any vulnerable edge. The non-serializable executions possible are like the one in [14], in which Bal sees a total balance value which implies that an overdraw penalty would not be charged, but the final state shows such a penalty because WC and TS executed concurrently on the same snapshot.

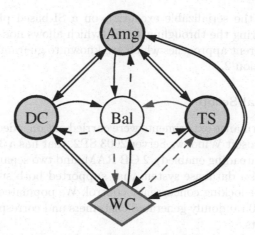

Fig. 1. SDG for SmallBank benchmark

3.4 Ways to Ensure Serializable Executions

As a basis for comparison, we consider the technique where SI is not used at all, and all transactions run with isolation level serializable (we call this Option 2PL-ALL). Our focus, however, is on using SI as much as possible. If one is unable or unwilling to modify the code of the application, one might consider the strategy we call Option Pivot-2PL, which uses 2PL for a minimal set of transactions (in SmallBank, for the pivot WC only, with all other transactions running under SI).

We contrast Pivot-2PL with mechanisms where application code is modified so that the dangerous structure is no longer present. For SmallBank, there are two edges in the vulnerable cycle, and eliminating either gives a minimal set of modifications that still ensure serializable execution.

In Option PromoteWT, we eliminate the dangerous structure by making the edge from WriteCheck to TransactSaving not vulnerable. This is done by promotion: adding an identity update in WriteCheck. To be precise, PromoteWT includes the following extra statement in the code of WC.

```
UPDATE Saving
    SET Balance = Balance
    WHERE CustomerId=:x
```

Finally, in Option PromoteBW, we make sure that the edge from Bal to WC is not vulnerable. This strategy introduces an identity update on the table Checking in the Bal transaction. Note that this makes Bal no longer be a read-only transaction.

4 Evaluation

We have conducted a performance evaluation with our SmallBank benchmark to explore the impact of the different ways of modifying application programs in order to ensure the serializable execution on a SI-based platform. We ran experiments comparing the throughput of SI (which allows non-serializable executions) to the different approaches which are known to guarantee serializability as discussed in Section 2.

4.1   Experimental Setup

The following performance experiments were carried out on a dedicated database server running Microsoft Windows Server 2003 SP2, that has a 3 GHz Pentium 4 CPU (with hyper-threading enabled), 2 GB RAM, and two separate log and data IDE disks. We used a database system that supported both snapshot isolation and strict two-phase locking concurrency control. We populated our SmallBank database with 19000 randomly generated customers and corresponding checking and savings accounts.

The actual test driver is running on a separate client machine that connects to the database server through Fast Ethernet. The client machine is running Windows XP SP2 and is equipped with 1 gigabyte of RAM and a 2.5 GHz Pentium CPU. The test driver is written in Java 1.5.0 and connects via JDBC to the database server. It emulates a varying number of concurrent clients (the multiprogramming level (MPL)) using multithreading.

The test driver runs the five possible transactions, on each invocation the client choses one transaction type randomly. In most of our experiments we chose the transaction type with a distribution of 60% Bal transactions (which are read-only in the unmodified code), and 10% of each of the other transactions. The actual transaction code is executed as stored procedures on the database server, which the client threads invoke with randomly chosen parameters. A fixed portion of the table is a hotspot, and 90% of all transactions deal with a customer which is chosen uniformly in the hotspot. In most of our experiments the hotspot has 100 (of the 19000) customers, but when we consider high contention, we make the hotspot have 10 customers only. The other 10% of transactions access uniformly from the customers outside the hotspot. Our system is a closed system; each thread runs the selected transaction and waits for the reply, after which it immediately (with no think time) initiates another transaction.

Each experiment is conducted with a ramp-up period of 30 seconds followed by a one minute measurement interval. Each thread tracks how many transactions commit, how many abort (and for what reasons), and also the average response time. We repeated each experiment ten times; the throughput figures show the average values plus a 95% confidence interval as error bar. In each experiment we measure some of the approaches that guarantee serializable execution; we also measure the performance when running the unmodified application under SI, and we display graphs to show how much throughput each serializable approach has as a percentage of that possible with SI itself. In almost every situation, the best throughput is that for unmodified applications under SI, but we stress that this choice can have non-serializable executions, unlike all the other options we measure. Thus a DBA who wants to have executions which are serializable must chose another option (2PL-ALL, Pivot-2PL, PromoteBW or PromoteWT), and

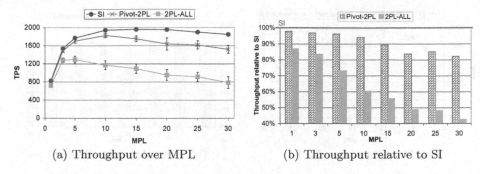

(a) Throughput over MPL   (b) Throughput relative to SI

Fig. 2. Costs for 2PL and SI-serializability when mixing isolation levels

our experiments will show how much, if any, is lost in performance to obtain the guarantee of correct execution.

4.2 Mixing Isolation Levels

We first consider approaches which remove all vulnerable edges from the SDG by adjusting isolation levels (running some or all transactions under 2PL). Figure 2 shows the results: the left graph shows the throughput in transactions per second (TPS) as a function of MPL, and on the right we show the relative performance as compared to the throughput with SI (shown as a horizontal line). As we see, such simple approaches induce hefty performance costs. Throughput for SI increases at first with MPL and reaches a peak around MPL 15 at about 1960 TPS, before it slowly decreases again. Running all transactions using strict two phase locking (2PL-ALL) results in a lot of locking conflicts which are getting worse with increasing MPL: It is between 13% and 57% slower than snapshot isolation with its peak already at 1290 TPS around MPL 5. This is the well-known advantage of SI that it scales better and does not block readers. However note that 2PL, in contrast to SI, does guarantee full serialisability.

We also evaluated the strategy of mixing isolation levels by using snapshot isolation for most transactions, but only execute the pivot transaction using 2PL (Pivot-2PL). This approach gives us a much better performance than running all transactions under 2PL. The costs are between 2% and 18% reduced throughput as compared to SI; peak performance is about 1830 TPS at MPL 10 (thus peak performance suffers by under 7% compared to SI).

4.3 Modifying Transaction Code vs. Mixing Isolation Levels

Next, we compare the different strategies that modify the transaction code to eliminate vulnerable edges from the SDG against the approach of adjusting isolation levels for the pivot. PromoteBW eliminates the vulnerability of the edge from Balance to WriteCheck, PromoteWT eliminates the vulnerability of the

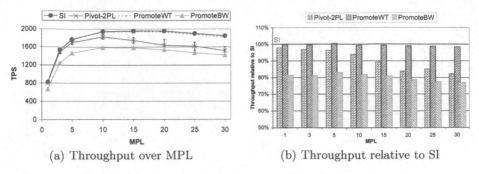

(a) Throughput over MPL   (b) Throughput relative to SI

Fig. 3. Costs of edge-promotion vs. mixing isolation levels

edge from WriteCheck to TransactSavings, and Pivot-2PL runs the WriteCheck transaction in 2PL. The results are shown in Figure 3. We observe that

- Throughput for PromoteWT is basically indistinguisable from that for SI.
- Throughput for PromoteBW is around 20% lower of that of SI.
- The throughput when running the pivot transaction under 2PL is between the two promotion approaches. As we saw above, it starts being 2% lower and ends at 18% less throughput for MPL 30, with a peak about 94% of the peak for SI.

Our results clearly show that with the right strategy, one can make snapshot isolation serializable for very low costs. Eliminating the vulnerability on the WT edge results in better throughput than eliminating the vulnerable BW edge, because the latter modifies a read-only transaction into an update transaction. Mixing isolation levels by executing the pivot transaction under 2PL shows a performance in between. But especially at higher MPL, we see the impact from the extra lock contention with 2PL which reduces throughput drastically. We are exploring this contention effect of Pivot-2PL in more detail in the following subsection.

4.4   High Data Contention

We have also run the experiments where we have reduced the size of the hotspot region from 100 customers (out of 19000 in the table) to only 10 customers. While such high data contention reduces the overall peak performance for each strategy (SI peaks at 1980 TPS with hotspot 100, and at 1780 TPS with hotspot 10), it does not change the overall conclusion that serializability can be obtained without reducing performance compared to SI (see Figure 4). We still find PromoteWT with performance indistinguishable from that of SI. Both Option BW and Pivot-2PL do much worse. Eliminating the vulnerability on BW gives performance which is between 18% and 52% lower than SI. The throughput when mixing isolation levels by running the pivot transaction under 2PL is initially slightly above PromoteBW, but then it starts decreasing with high MPLs. It

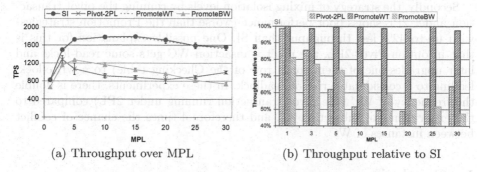

(a) Throughput over MPL   (b) Throughput relative to SI

Fig. 4. Costs of edge-promotion vs. mixing isolation levels - High Contention

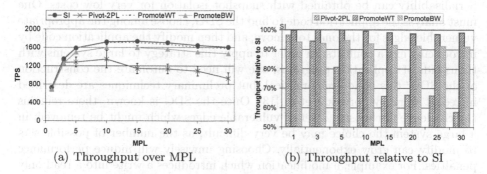

(a) Throughput over MPL   (b) Throughput relative to SI

Fig. 5. Costs of edge-promotion vs. mixing isolation levels - High Update Rates

results in up-to 51% lower throughput than SI, and between MPLs 5 and 20 it is even less successful than PromoteBW. The reason for the performance degradation in Pivot-2PL is the high lock contention for the pivot transaction because of the tiny hotspot and the high MPL. This illustrates again that a SDG analysis of the transaction mix is essential to achieve reasonable performance and serializability under SI at the same time.

4.5   High Update Rates

Finally, we ran the experiments where we were choosing all transaction types uniformly on a hotspot of size 100. This results in a workload comprised of only 20% read-only and 80% update transactions (for unmodified code) as compared to the previous scenarios with 60% readers and 40% updaters. The results in Figure 5 show two notable changes as compared to Figure 3: Firstly, *both* promotion strategies perform quite well. PromoteWT is again indistinguishable from SI, and PromoteBW shows less than 10% reduction in throughput compared to SI. PromoteBW does well in this workload because the additional writes (which turn the Bal transaction into an updater) have less performance impact since Bal itself is much rarer.

Secondly, the strategy of mixing isolation levels by running the pivot transaction under 2PL shows poor performance (worse than the Promotion strategies) with up-to 42% less throughput than SI. One possible explanation for this is that in option Pivot-2PL, the pivot transaction WC gets some read-locks and later upgrades one of them (on a row of the Checking table) to a write-lock, leading to a considerable risk of deadlock. In these experiments, there is double the frequency of WC (the pivot transaction running under 2PL) compared to the experiments of subsection 4.3, and therefore 4 times the chance of conflict between instances of WC.

5   Conclusions

Serialisability can be obtained with snapshot isolation for very low costs. One must analyse the application code to find the SDG, choose among the appropriate vulnerable edges for the ones to remove, and then modify the application code by introducing extra conflicts. In order to apply this strategy to large systems with thousands of application programs, one will need to automate the construction of the SDG; this is not yet practical, but preliminary techniques are described in recent work by Jorwekar et al [13]. Once the SDG is known, there remains the challenge of choosing a set of vulnerable edges which might be removed; in a large system the choice may be very difficult, as the number of possible sets to modify can grow exponentially. Choosing unwisely will induce performance penalties. For example, a modification which introduces a write into a read-only transaction can reduce throughput considerably.

If the platform supports it, the approach Pivot-2PL of mixing isolation levels is convenient. It can be done in the client without rewriting code. There are no choices once the SDG is known: the pivots are determined in polynomial time from the SDG. However convenience comes with a great cost: we found that Pivot-2PL reduced throughput by at least 30% for high MPL with either high contention or high update rate. The bottom line from our experiments is that running the pivot under 2PL only works well if there are few update transactions with few conflicts.

Our experiments justify the inconvenience of modifying the application code, and choosing among the sets of vulnerable edges. The optimal strategy is to remove vulnerable edges between two update transactions by introducing additional updates into the existing code. The resulting performance is basically the same as with standard SI, but it guarantees serialisable executions.

References

1. Berenson, H., Bernstein, P., Gray, J., Melton, J., O'Neil, E., O'Neil, P.: A critique of ANSI SQL isolation levels. In: SIGMOD 1995: Proceedings of the 1995 ACM SIGMOD International Conference on Management of Data, pp. 1–10 (1995)
2. Fekete, A., Liarokapis, D., O'Neil, E., O'Neil, P., Shasha, D.: Making snapshot isolation serializable. ACM Trans. Database Syst. 30(2), 492–528 (2005)

3. Fekete, A.: Allocating isolation levels to transactions. In: Proceedings of the 24th ACM SIGMOD symposium on Principles of Database Systems (PODS 2005), pp. 206–215. ACM Press, New York (2005)

4. Alomari, M., Cahill, M., Fekete, A., Roehm, U.: The cost of serializability on platforms that use snapshot isolation. In: Proceedings of IEEE International Conference on Data Engineering (ICDE 2008) (2008)

5. Jacobs, K.: Concurrency control: Transaction isolation and serializability in SQL92 and Oracle7. Technical Report A33745 (White Paper), Oracle Corporation (1995)

6. Wu, S., Kemme, B.: Postgres-R(SI): Combining replica control with concurrency control based on snapshot isolation. In: Proceedings of the 21st IEEE International Conference on Data Engineering (ICDE 2005), pp. 422–433 (2005)

7. Plattner, C., Alonso, G.: Ganymed: scalable replication for transactional web applications. In: Jacobsen, H.-A. (ed.) Middleware 2004. LNCS, vol. 3231, pp. 155–174. Springer, Heidelberg (2004)

8. Lin, Y., Kemme, B., Patiño-Martínez, M., Jiménez-Peris, R.: Middleware based data replication providing snapshot isolation. In: Proceedings of the 2005 ACM SIGMOD International Conference on Management of Data, pp. 419–430 (2005)

9. Elnikety, S., Pedone, F., Zwaenepoel, W.: Database replication using generalized snapshot isolation. In: 24th IEEE Symposium on Reliable Distributed Systems (SRDS 2005), pp. 73–84 (2005)

10. Daudjee, K., Salem, K.: Lazy database replication with snapshot isolation. In: Proceedings of the 32nd International Conference on Very Large Data Bases (VLDB 2006), VLDB Endowment, pp. 715–726 (2006)

11. Transaction Processing Performance Council: TPC Benchmark C Standard Specification, Revision 5.0 (2001), http://www.tpc.org/tpcc/

12. Bernstein, A.J., Lewis, P.M., Lu, S.: Semantic conditions for correctness at different isolation levels. In: Proceedings of IEEE International Conference on Data Engineering (ICDE 2000), pp. 57–66 (2000)

13. Jorwekar, S., Fekete, A., Ramamritham, K., Sudarshan, S.: Automating the detection of snapshot isolation anomalies. In: Proceedings of the 33rd international conference on Very Large Data Bases (VLDB 2007), pp. 1263–1274 (2007)

14. Fekete, A., O'Neil, E., O'Neil, P.: A read-only transaction anomaly under snapshot isolation. SIGMOD Rec. 33(3), 12–14 (2004)

SemanticTwig: A Semantic Approach to Optimize XML Query Processing

Zhifeng Bao[1], Tok Wang Ling[1], Jiaheng Lu[2], and Bo Chen[1]

[1] School of Computing, National University of Singapore
{baozhife,lingtw,chenbo}@comp.nus.edu.sg
[2] University of California, Irvine
jiahengl@uci.edu

Abstract. Twig pattern matching (TPM) is the core operation of XML query processing. Existing approaches rely on either efficient data structures or novel labeling/indexing schemes to reduce the intermediate result size, but none of them takes into account the rich semantic information resided in XML document and the query issued. Moreover, in order to fulfill the semantics of the XPath/XQuery query, most of them require costly post processing to eliminate redundant matches and group matching results. In this paper, we propose an innovative semantics-aware query optimization approach to overcome these limitations. In particular, we exploit the functional dependency derived from the given semantic information to stop query processing early; we distinguish the output and predicate nodes of a query, then propose a query breakup technique and build a query plan, such that for each distinct query output, we avoid finding the redundant matches having the same results as the first match in most cases. Both I/O and structural join cost are saved, and much less intermediate results are produced. Experiments show the effectiveness of our optimization.

1 Introduction

XML is emerging as a standard for information exchange and representation. Efficient processing of XML queries attracts wide research attentions recently [2,4,10,11,12,3]. The structure of an XML query [7,13] is generally modeled as a twig pattern (i.e. a small tree), while the values of XML elements or attributes are used in selection predicates. Twig pattern edges are either parent-child (P-C) or ancestor-descendant (A-D) relationships, denoted by "/" and "//". E.g. an XML query *book[title= "XML"]/author* returns the authors of all books titled "XML". *author* is the output, and *title= "XML"* is a value-based selection predicate.

So far, many twig pattern matching (TPM) methods have been proposed. Bruno et al. proposed a holistic *TwigStack* join algorithm to avoid producing large intermediate results [2]. Several following approaches [11,12,10,4] suggest different ways to optimize *TwigStack*. However, there are two important issues that the existing TPMs haven't addressed. Firstly, existing TPMs do not distinguish the output nodes and predicate nodes of a query and they return the

J.R. Haritsa, R. Kotagiri, and V. Pudi (Eds.): DASFAA 2008, LNCS 4947, pp. 282–298, 2008.
© Springer-Verlag Berlin Heidelberg 2008

entire twig matches. However, many of them contribute to the same output result, and the entire twig matches are unnecessary for most XPath or XQuery queries. After all twig matches are found, they have to apply a post-processing, which includes eliminating redundant matches having the same query result (and grouping part of results into a set for the variables in *let* clause of XQuery). This post-processing is very costly, as shown in our experiments in section 5. Secondly, none of them makes use of the rich semantic information resided in the XML document and its schema to further optimize query processing.

Motivated by the above two observations, we propose an innovative semantics-aware query optimization approach. There are three directions to exploit the existing information to optimize query processing: (1) optimizations based on the schema information of XML documents; (2) optimizations based on the query structure; (3) optimizations involving data storage and statistics (e.g. indexing). Our work includes the first two: the first optimization (section 4.2) is based on the ORA-SS model [14] described later; the second optimization (section 4.3) is independent of ORA-SS model. To the best of our knowledge, this is the first work that systematically exploit the known semantics in both XML data and XML query to optimize query processing. As a result, for each distinct query output t, we only find the first twig match in XML document that contains t, and the cost of finding the remaining matches that return the same output as the first match is avoided in most cases. The query processing can stop even earlier if we exploit the semantics of XML document. As a result, both the I/O and CPU cost is significantly reduced.

Another challenge is how to efficiently represent the semantic information in XML documents. Although XML is the standard for publishing and exchanging data on the Web, most business data is managed and will remain to be managed by relational database management system (RDBMS) because of their powerful data management services. There is an increasing need to accurately publish relational data as XML documents for Internet-based applications as well as preserve the semantics captured in RDBMS. In the transformation from relational schema to XML schema, some semantics can be captured by DTD and XMLSchema, such as the *identifier constraint* (denoted by "unique" and "key" constraint in XMLSchema), and *participation constraint* of child element on its parent denoted by signs " * ", "?" and " + ". However, richer semantics captured in RDBMS cannot be captured by DTD/XMLSchema, e.g. the *participation constraints* of a parent element on its child, etc. Fortunately, instead of writing down those semantics as comments, notes, or in XML data creator's own mind, we can capture them in ORA-SS schema [14], which is a useful data model and tool for semistructured data. The main advantage of ORA-SS is its ability to distinguish between object classes, relationship types and attributes, and express the degree of an n-ary relationship type. Therefore, in this paper, in order to fully leverage the semantic information in XML data to speedup XML twig query processing, we adopt ORA-SS to model XML data, especially those that are transformed from relational data.

The main contributions of this paper are summarized as follows:

- We exploit important semantics of XML document captured by ORA-SS schema (such as object class identifier, participation constraints, etc.), to derive functional dependencies useful for query optimization. This is the first kind of optimization which depends on the ORA-SS model.
- We propose a *Query Breakup technique* to distinguish the processing of predicate part and output part of a twig query, so for each distinct query result, we only find the first twig match in XML document that contains it, and avoid finding all the remaining redundant matches. This is the second kind of optimization, which is independent of the ORA-SS model.
- We propose *SemanticTwig*, a semantics-aware query processing algorithm which employs the above two kinds of optimizations. Both the I/O and CPU cost reduce significantly compared to existing TPMs. Moreover, our optimization is orthogonal to any existing structural-join based TPMs.
- We perform comprehensive experiments to demonstrate the efficiency and scalability of our approach over existing approaches.

The rest of the paper proceeds as follows. Section 2 reviews the related work. Section 3 gives an overview of ORA-SS. Section 4 propose two query optimization approaches. Section 5 reports experimental results and we conclude in section 6.

2   Related Work

Extensive research efforts have been put into efficient twig query processing with label based structural join. For binary structural join, Zhang et al. [15] proposed a multi-predicate merge join (MPMGJN) algorithm based on region labelling of XML elements. The later work by Al-Khalifa et al. [1] gave a stack-based binary structural join algorithm, called Stack-Tree-Desc/Anc. However, both generate many useless intermediate results for twig query. To solve this problem, Bruno et al. [2] firstly proposed a holistic twig join algorithm called *TwigStack* to solve the problem of useless intermediate results. However, *TwigStack* is only optimal for twig query with only A-D edges in terms of intermediate results. Many subsequent works try to optimize *TwigStack* in terms of I/O. In particular, Lu et al. in [11] introduced a List structure called TwigStackList for a wider range of optimality. Jiang et al. in [10] proposed an XML Region Tree (XR-tree) index structure and a TSGeneric+ algorithm to effectively skip both ancestors and descendants that do not participate in a join. Chen et al. [4] exploits different data partition methods to boost the holism. Lu et al. [12] used Extended Dewey labeling scheme, and proposed a TJFast algorithm to access labels of leaf nodes only. Recently, Chen et al. in [3] proposed a $Twig^2Stack$ algorithm which uses *hierarchical-stacks* to enumeration the twig matches, but it has to maintain a large amount of intermediate results in memory.

Existing semantic related works[5,6] only rely on the integrity constraints captured from the real world knowledge, rather than the schema of XML document. It prepares a set of XML rewriting rules with semantic preserving property from the integrity of the database, and a query is transformed into an efficient and optimized form using these rules. It should be highlighted that sine relational-based ones like XPath accelerator [9] and PPFS+ [8] also focus only on output nodes and handle predicates using the exist clause of SQL; however they cannot handle twig query well.

3 Background on ORA-SS Model

The ORA-SS (Object-Relationship-Attribute model for SemiStructured data) data model has three basic concepts: *object class*, object classrelationship type and relationship typeattribute. An *object class* is similar to an entity type in an ER diagram. A *relationship type* describes a relationship among object classes. *Attributes* are properties belonging to an object class or a relationship type. A full description of the data model can be found in [14].

As Figure 1(b) shows, an ORA-SS *schema* represents an object class as a labeled rectangle, an attribute as a labeled circle. The relationship type between object classes is assumed on any edge between two objects, and described by a label in form of "*name(object_class_list), n, p, c*". Here, *name* denotes the name of relationship type; optional *object_class_list* is the list of participating objects; n indicates the degree of the relationship type; p and c are the participation constraints of parent and child object classes respectively, defined using the min:max notation. All attributes are assumed to be mandatory and single valued, unless the circle contains a "?" indicating it is optional and single valued, "+" indicating it is mandatory and multi-valued, and "*" indicating it is optional and multi-valued. Identifier of an object class is a filled circle. The attribute of a relationship type has the name of the relationship type to which it belongs on its incoming edge, while the attribute of an object class has no edge label.

Figure 1(a) shows the DTD and Figure 1(b) shows an ORA-SS schema diagram for an XML document describing the organization of a university. The rectangles labeled *faculty, department, course, student* and *tutor* are five object classes, and attributes *fname, dname, code, stuNo* and *staffNo* are the identifiers

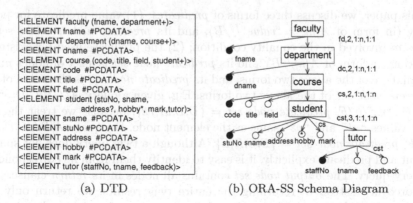

   (a) DTD   (b) ORA-SS Schema Diagram

Fig. 1. Two Schema Representations of XML Document

of *faculty, department, course, student* and *tutor* respectively. For each *student*, *address* is an optional single valued attribute, and *hobby* is an optional multi-valued attribute. There are three binary relationship types, namely *fd, dc* and *cs*. *cs* is a relationship type between *course* and *student*. *dc* represents a one-to-many relationship type, where a *department* can have one or more (1:n) *courses*, and a *course* belongs to exactly one (1:1) *department*. The label *cs* on the edge between *student* and *mark* indicates that *mark* is a single valued attribute of the relationship type *cs*, and a functional dependency (FD) $\{course, student\} \rightarrow mark$ holds. A ternary relationship type *cst* involves *course, student* and *tutor*. The label *cst* on the edge between *tutor* and *feedback* indicates *feedback* is an attribute of relationship *cst*, and an FD $\{course, student, tutor\} \rightarrow feedback$ holds.

4   Semantic Query Optimization

This section investigates how our semantics-aware query optimization works. Our primary objective is to avoid unnecessary computation on finding matches contributing to the same query result and stop query processing as early as possible. The optimization is carried out in two steps: we first try to exploit the *FDs* explicitly given or derived from the semantics of the *XML document* to answer a query through FD's ability of capturing redundancies(section 4.2); if no FD can be exploited, then we make use of the semantics of the *XML query*, i.e. distinguish the output part and predicate part of a query, in which the predicate is only used for existence checking. Therefore, we only need to find the matches that have distinct output results, rather than the matches of the entire query's twig pattern (section 4.3). In addition, our optimization is orthogonal to most existing structural join based TPMs. The optimization approach in section 4.2 relies on the semantics exclusively captured by ORA-SS schema, while section 4.3's approach works without dependence on ORA-SS. Note that, *FD* could be given explicitly instead of using ORA-SS model here.

4.1   Terminology

In this paper, we discuss three forms of *predicates*: (1) value predicate for point query (in form of $A[m = "value"]//B$), and its *predicate node set* is a set of nodes m involved in the equality condition; (2) the predicate used for existence checking (in form of $A[n]//B$), and its *predicate node set* is a set of nodes n; (3) the mixture of the above two forms, and its *predicate node set* is the union of the *predicate node set* of the above two forms. E.g. given a query $A[//B[C[D= "v1"$ *and* $E= "v2"]]]/F$, *predicate node set* $= \{D, E, v1, v2\}$, because we treat the element values "*v1*" and "*v2*" same as the element node. For query $A[B//C[D$ *and* $E]]/F$, *predicate node set* $= \{B, C, D, E\}$. Although a twig query does not specify output and predicate explicitly, it is easy to identify them from its corresponding XQuery query. The *output node set* contains all nodes in its *return* clause.

Moreover, instead of returning the entire twig results, we return only the output part of the query. The result is defined as a set of tuples, and each tuple contains the element labels of output nodes only. If some output node n is a multi-valued attribute and required to return as an ordered set, its matching labels are grouped into a set under their common ancestor before returned. Our approach outperforms the existing TPMs by avoiding the cost on finding as many *redundant matches* as possible, which is defined below.

Definition 1 {Redundant Match}. *For a twig query Q, a match $M1$ is said to be a redundant match, if there exists another match $M2$ found before $M1$, such that $M1$ and $M2$ contribute to the same query output node values.*

4.2   Optimization Via Functional Dependency

Functional dependency (*FD*) is used to model real world constraints and capture redundancies. The given semantics in ORA-SS schema such as the identifier of object class, n-ary relationship and the participation constraint can be used to derive useful *FD* for efficient query processing. Given a query Q, if an FD in form of *predicate node set→output node set* can be derived, then Q is answered by finding the first match of Q appearing in XML document. The I/O cost is saved since it avoids loading remaining labels from each query node's stream; the CPU cost is saved since most XML documents contain redundancies. But it is only true when Q involves the *value predicates*. All Lemmas hold based on this assumption.

FD within an object class. The identifier of an object o can uniquely determine the value of any single-valued attribute of o, and uniquely determine the whole set of values of any multi-valued attribute of o.

Lemma 1. *Given a query Q with predicate node set P and output node set T.*

Case 1: If $\forall t_i \in T$, t_i is a single-valued attribute of some object class O_i, and $\exists p \in P$, such that p is the identifier of O_i; then an FD $P \to T$ holds, and the query result is the first match F of Q over XML data in document order.

Case 2: If some output node $t'_j \in T$ is a multi-valued attribute of some object class O_j, and other output nodes in T are same as Case 1; then Q is answered by finding its first match F (which finds the first value of each t'_j), then retrieving and grouping the remaining values of each t'_j within F into a set.

In Case 1, the overall processing cost is reduced with the utilization of such FD in two ways. Firstly, an object may have more than one occurrence in document, and it's safe to stop query processing after the first match of Q is found. Secondly, there is no need to check the remaining labels in label streams of each query node, which saves I/O cost. In Case 2, we only need to find one match; however, existing approaches have to enumerate all path matches and merge-join them, and finally eliminate redundant match. One important character that distinguishes our approach from the existing TPMs is that we treat the set/non-set elements(i.e. multi/single-valued attributes) separately. The retrieval and grouping of the remaining occurrences of t' are easy and efficient to implement, since their labels are stored sequentially and compactly.

Example 1. Refer to ORA-SS schema in Figure 1(b)[1], the following XQuery query Q retrieves the name and hobby of a student with *stuNo* equal to "u12".
for $s in //student[stuNo="u12"], let $h := $s/hobby
return <stu>{$s/sname}{$h}</stu>

The above query Q's twig pattern is shown in Figure 2. Its *predicate node set* P = {*stuNo, u12*}, and *output node set* T = {*sname, hobby*}. *stuNo* is the identifier of *student* and *hobby* is a multi-valued output, so by Case 2 of Lemma 1, it is enough to find the first match F of Q in XML data, then retrieve and group all labels of remaining *hobbies* within F. A tuple containing *sname* and a set of *hobbies* of this student is returned.

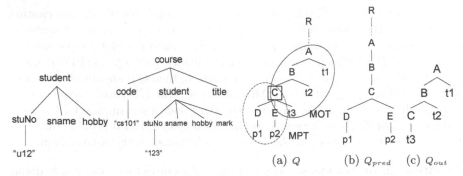

Fig. 2. Example 1   **Fig. 3.** Example 2   **Fig. 4.** Break query (Example 4)

The advantage of our approach is illustrated by an example as below. Assume student *"u12"* has 8 hobbies. Existing TPMs find 8 matches of path *//student/hobby* and join the 8 matches with the match(es) of paths *//stduent/stuNo/u12* and *//student/sname*, while our approach needs only one structural join. Moreover, after finding 8 matches of the entire twig, TPMs group all hobbies, which is a costly post-processing operation; while we handle the grouping in the middle of twig pattern matching. TPMs even need to do redundancy elimination if this student object has more than one occurrences in XML document.

FD in an n-ary relationship type. From ORA-SS schema diagram, the FD in an n-ary relationship is in the form that, identifiers of participating object classes functionally determine the single-valued attributes of the relationship.

Lemma 2. *Given a twig query Q with predicate node set P and output node set T. For each output node $t \in T$,*

Case 1: t is a single-valued attribute of a n-ary relationship R involving n object classes $O_1,...,O_n$, and $\exists p_i \in P$ for each $i \in [1,n]$, s.t. p_i is the identifier of O_i;

Case 2: t is a single-valued attribute of some object class O_k, $\exists p_k \in P$ s.t. p_k is the identifier of O_k. Then in both cases, FD $P \rightarrow T$ holds, and Q can be answered by finding the first match F of Q;

---

[1] All queries throughout this paper refer to the schema diagram shown in Figure 1(b).

Case 3: if some $t \in T$ is a multi-valued attribute of R or any participating object class, and each remaining node $t' \in T$ belongs to either Case 1 or Case 2, then we can adopt the approach in Case 2 *of Lemma 1 to efficiently answer Q.*

Example 2. Find the title of course with code "cs101", and the mark, name and hobby of student taking it, whose stuNo is "123".
*for c in //course, for s in $c[code= "cs101"]/student[stuNo= "123"]$
let $h := $s/hobby$
return <stu> {$c/title$} {$s/sname$} {$s/mark$} {$s/hobby$} </stu>*

Figure 3 shows the twig pattern of the above query Q. The predicate set $P = \{code, stuNo\}$, the output node set T $= \{title, sname, mark, hobby\}$, in which *mark* is a single-valued attribute of relationship type *cs*. From Figure 1(b), it is easy to identify the FD $\{code, stuNo\} \rightarrow mark$, which means a student has exactly one mark for each course taken. $stuNo \rightarrow sname$ and $code \rightarrow title$ also hold. So we can infer $\{code, stuNo\} \rightarrow \{sname, title, mark\}$. So by Lemma 2 Case 3, Q can be answered by finding its first twig match F in XML document, followed by retrieving and grouping the values of remaining *hobbies*.

Participation constraint in n-ary relationship. The 1:1 participation constraint intuitively infers an FD between the participating object classes. A typical example is the *one-to-many relationship type dc* shown in Figure 1(b), in which the 1:1 participation of *course* on relationship type *dc* specifies a course is offered by exactly one department, i.e. *course→department*. The 1:1 participation between some object classes can simplify the functional dependency of an n-ary relationship R. E.g. as Figure 1(b) shows, single-valued attribute *feedback* of ternary relationship *cst* is determined by the 3 participating object classes, i.e. $\{course, student, tutor\} \rightarrow feedback$. The 1:1 participation of *tutor* on *cst* specifies $\{course, student\} \rightarrow tutor$, which means a student has exactly one tutor for each course he takes. Therefore, a new FD $\{course, student\} \rightarrow feedback$ is derived.

Example 3. Find the name of faculty and department offering course "cs101".
*for f in //faculty, d in $f/department, c in $d/course[code= "cs101"]$
return <fac>$f/fname<dept>$d/dname</dept></fac>*

In this query, two binary relationships *fd* and *dc* exist, and the participation of parent node on child node are both 1:1, so two FDs *course→department* and *department→faculty* hold. Then a new FD *course→ {faculty, department}* is inferred. Since the predicate node *code* is the identifier of *course*, the query has a unique answer and we can stop query processing after the first match is found.

4.3   Optimization Via Query Breakup

Motivation. According to the semantics of an XPath/XQuery query, only the output results with no duplicates are expected to return. However, existing approaches are not aware of this distinction of output and predicate nodes in a query, and assume all nodes in a query tree need to be output. They answer a query by applying pattern matching on its entire twig, followed by a costly post-processing which includes the tasks of redundant matches elimination and results grouping. In this section, we aim to fulfill the post-processing task during the pattern matching procedure with low extra cost. In fact we are able to avoid finding as many redundant matches as possible by distinguishing the output nodes and predicate nodes of a query.

Another motivation to propose the query breakup technique is: for each distinct output result of the query Q, there are many redundant matches of the predicate part in XML document. If we can break Q into two sub-twigs corresponding to Q's predicate and output part respectively, and avoid finding those redundant matches, then both I/O and CPU cost are reduced. Because we skip reading the elements of redundant matches into memory, and need less number of structural joins, and process a twig query of smaller size.

Find the breakpoint. The choice of breakpoint is not unique, but it depends on the definition of *predicate node* of a query (defined in section 4.1), and its choice determines the breakup method used. Therefore, the *predicate node*, breakpoint, the breakup method and query optimization based on breakup are defined in a consistent and cooperative way to meet three properties: (1) guarantee the completeness and correctness of query result; (2) read as few elements as possible into memory; (3) less structural join cost. Before introducing the definition of breakpoint, we have the following two definitions.

Definition 2 {*Minimal Predicate Tree (MPT)*}. *The Minimal Predicate Tree of a query Q is the minimal sub-tree of Q's twig tree that covers all nodes in the predicate node set. If there is only one predicate node p, the MPT is a path connecting p and its parent. The root node of MPT is called R_{MPT}.*

Definition 3 {*Minimal Output Tree (MOT)*}. *The Minimal Output Tree of a query Q is the minimal sub-tree of Q's twig tree that covers all nodes in the output node set. If there is only one output node t, the MOT is a path connecting p and its parent. The root node of MOT is called R_{MOT}.*

Definition 4 {Breakpoint}. *The breakpoint BP of the query Q with the predicate node set P and the output node set T is:*

*Case 1: the lowest common ancestor **LCA** of R_{MPT} and R_{MOT} in Q, if there is no A-D relationship between R_{MPT} and R_{MOT}; (line 7 of Algorithm 1)*

Case 2: the node $t \in MOT$, such that t is the ancestor of an output node of Q, and t has the lowest hierarchy along the path downward from R_{MOT} to R_{MPT}, if R_{MOT} is the (self) ancestor of R_{MPT}; (line 3-4 of Algorithm 1)

Case 3: the node $p \in MPT$, such that p is the ancestor of a predicate node of Q, and p has the lowest hierarchy along the path downward from R_{MPT} to R_{MOT}, if R_{MPT} is the (self) ancestor of R_{MOT}. (line 5-6 of Algorithm 1)

Break up the query. Based on the breakpoint BP found, a query Q is broken into two sub-twigs namely Q_{pred} and Q_{out}. Besides the three properties introduced in last subsection, the breakup method should guarantee the union of

---

Algorithm 1. findBreakPoint(Q, MPT, MOT)

---

1 $r1$ = root(MOT); $r2$ = root(MPT)
2 P_1 = path($r1,r2$); P_2 = path($r1,r2$)
3 **if** ($r1.isAncestorof(r2)$) **then**
4      $brkpoint$ = t | t∈MOT ∩ t∈P_1 ∩ ∀t'∈T t'≠t ⇒ level(t)>level(t')
5 **else if** ($r2.isAncestor(r1)$) **then**
6      $brkpoint$ = t | t∈MPT ∩ t∈P_{22} ∩ ∀t'∈P t'≠t ⇒ level(t)>level(t')
7 **else** $brkpoint$ = findLowestComAncestor($r1,r2,Q$)
8 **return** $brkpoint$

Q_{pred} and Q_{out} entirely constitute Q. Since some nodes in Q are not covered by either MOT or MPT, the main problem is how to distribute them into Q_{pred} and Q_{out} to meet the above four properties. Since the definition of *Breakpoint* has three cases, the query breakup is presented to handle the query in each case. In Algorithm 2, lines 1-3 handle the query in Case 1 of Algorithm 1; lines 4-10 handle the query in Case 2; optimization is impossible in Case 3, because no redundant match exists (line 11).

Example 4. We show how *findBreakPoint* and *breakup* work for the query type in *Case 2* of Definition 4. In Figure 4, the query Q has two predicate nodes $p1$ and $p2$ and three output nodes $t1$, $t2$ and $t3$. The Minimal Predicate Tree (MPT) and Minimal Output Tree (MOT) of Q are highlighted by dotted circles in Figure 4(a). A and C are the root of MPT and MOT respectively, and A has a higher hierarchy than C. C is chosen as the breakpoint because it satisfies the three conditions in line 4 of *Algorithm 1*: C is a node in both the path P_1 and MOT, and among the three nodes A, B and C that have an output node as descendant, C has the lowest hierarchy. Next, we follow *Algorithm 2* to break Q. We first find the minimal tree Q_{temp} covering both MPT and C (line 5). Figure 4(b) shows Q_{pred}, which is the minimal tree covering both Q_{temp} and Q's root R(line 6). Q_{rem} resulted from removing Q_{pred} from Q except breakpoint C. Since Q_{rem} is a sub-structure of MOT, Figure 4(c) shows the Q_{out} is MOT(line 7-9).

Algorithm 2. breakup(Q, MPT, MOT, BP)

---

1 **if** ($!isAD(R_{MPT},R_{MOT})$ ∩ $!isAD(R_{MOT},R_{MPT})$) **then**
2   Q_{pred} = findMinTree(MPT, root(Q))
3   Q_{out} = (Q - Q_{pred}) ∪ BP
4 **if** ($isAD(R_{MOT},R_{MPT})$) **then**
5   Q_{temp} = findMinTree(MPT, BP)
6   Q_{pred} = findMinTree(Q_{temp}, root(Q))
7   Q_{rem} = (Q - Q_{pred}) ∪ BP
8   **if** ($Q_{rem}.isSubTree(MOT)$) **then**
9   Q_{out} = MOT
10   **else** Q_{out} = Q_{rem}
11 **else** no optimization is possible

Example 5. The query Q below finds the name of faculty and department, in which a student descendant is called "Bob" and one of his hobby is "tennis".
for $f in //faculty, $d in $f/department, $s in $d//student
where $s/sname= "Bob" and $s/hobby= "tennis"
return <fac>$f/fname<dept>$d/dname</dept></fac>

The *predicate node set* P={*sname,hobby,Bob,tennis*}, and the *output node set* T= {*fname,dname*}. Thus, R_{MPT} is node *student*, and R_{MOT} is node *faculty*. Since *department* is the lowest hierarchy node along the path from *faculty* to *student* and is the ancestor of output node *dname*, the breakpoint is *department* by case 2 of Definition 4. So Q is broken into two sub-twigs Q_{pred} and Q_{out} shown in figure 5(b) and 5(c). If *dname* is removed from the *return* clause of the above query, then the breakpoint is node *faculty* rather than *department*.

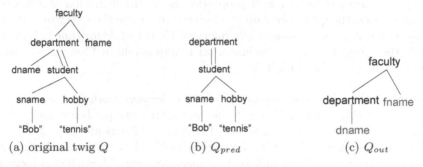

Fig. 5. Breakup a twig query

Optimization based on query breakup. After Q is broken into two sub-queries Q_{pred} and Q_{out}, we build a 4-step query plan to achieve optimization.

Step 1: Initialize the set S_{BP} to empty, which stores the labels of breakpoint BP, each of which contributes to a distinct query output.

Step 2: Once a match F of sub-twig Q_{pred} is found, add the label A of current BP-element into S_{BP}.

Step 3: Keep moving the cursor of BP, i.e. C_{BP} forward, until the current BP-element pointed by C_{BP} is not the descendant of A found in F. Go back to *Step 2* to locate the next match of Q_{pred} until C_{BP} reaches the end.

Step 4: Replace the label stream of BP by S_{BP}, and evaluate the sub-twig query Q_{out}. Note that no join between Q_{pred} and Q_{out} is needed.

- Case 1: If every output node $t \in Q$ is a single-valued attribute, then Q is answered by finding all matches of Q_{out} over the XML document.

- Case 2: If some output node $t' \in Q$ is a multi-valued attribute and declared to return as an ordered set (i.e. t' is declared in the *let* clause of XQuery), then Q_{out} is replaced by $Q_{out'}$ by removing all those t' nodes; and once each match FM' of $Q_{out'}$ is located, all labels of such t' nodes within FM' are retrieved one by one and grouped together.

Example 6. This example illustrates the superiority of our approach. The query is same as *Example 5* except that the output node is *fname*. Suppose the university has 8 faculties, 5 departments per faculty, 100 students per department; in each department *dept*, 3 students are called "Bob" and have a hobby "tennis". *TwigStack* [2] is used as a representative of existing TPMs.

TwigStack decomposes Q into 3 root-to-leaf paths, i.e. *//faculty/fname*, *//faculty/department//student/sname/Bob* and *//faculty/department//student / hobby/tennis*. The number of matches for each path is 8, 8*5*3 and 8*5*3 respectively. Then it joins these path matches to get the entire twig match of Q, which is 8*5*6*0.5 = 120 matches. Finally, it eliminates redundant matches. In the optimal case, total number of labels scanned is 8+8+8*5*3+8*5*3*4 = 716.

In contrast, our approach finds the breakpoint *department*, and breaks Q into Q_{pred} and Q_{out} (Figure 5). Number of matches of Q_{pred} and Q_{out} are both 5*8, so in total 80 matches are found. In the optimal case, 8+8+8*5 = 56 labels are scanned. 8 structural joins are needed in processing Q_{pred}; no structural join is needed in processing Q_{out}. Compared to *TwigStack*, our approach scans smaller number of labels, needs less structural joins and avoid many redundant matches.

4.4   SemanticTwig Optimization Algorithm

Algorithm 3 presents the two kinds of optimizations introduced in section 4.2 and 4.3. If an *FD* in form of {*predicate part→output part*} can be derived from the semantic information in ORA-SS schema, then the query is answered by only finding the first occurrence of its twig pattern in XML document (lines 1-3). Otherwise, we execute the second optimization, i.e. *query breakup*. Firstly, the *Minimal Predicate Tree* and *Minimal Output Tree* are found (line 4-5). The *breakpoint* n is identified by calling Algorithm 1(lines 6), and used to break Q into Q_{pred} and Q_{out}(line 7). Secondly, the labels of n are collected into S_n, each label contributes to a distinct query output(lines 8-13). The cursor C_n of node n keeps moving forward until the current n is not the descendant of the n-element found in last match of Q_{pred}(lines 12-13). Thirdly, based on the shortened label stream of node n(line 14), all matches of sub-twig Q_{out} are located(lines 15-21). *TwigStack* algorithm is the backbone of $findMatch()$ and $find1stMatch()$.

Theorem 1. *Given a twig query Q with specified predicate node set P and output node set T, and an XML database D. Algorithm SemanticTwig correctly returns all the answers for Q on D.*

PROOF: [Sketch] The main difference of *SemanticTwig* and *TwigStack* is the movement of cursors. In *SemanticTwig*, the cursor of breakpoint n skips all elements that are descendants of the n-element A in previous match of Q_{pred}, but the output results R' associated with these descendants are not skipped. Because each time the first match of Q_{pred} is found, all the matches MS of Q_{out} associated with A are located, and R' is in fact a subset of the query results in MS. Therefore, it is safe for C_n to directly jump to the first element which is not the descendant of the A in previous match of Q_{pred}.   □

Algorithm 3. semanticTwig(Q, $predSet$, $outSet$)

---

1 Identify FDs F_1, F_2,..., F_m from ORA-SS schema of XML data
2 **if** FD $Predicate \rightarrow output$ is $derived$ **then**
3   resultSet += find1stMatch(Q)
4 $MPT = $ findMinPredicateTree(Q, $predSet$)
5 $MOT = $ findMinOutputTree(Q, $predSet$)
6 Node $n = findBreakPoint(Q, MPT, MOT)$
7 $\{Q_{pred}, Q_{out}\} = breakup(Q, n, MPT, MOT)$
8 let C_n be the cursor of node n, let S_n be a label set of n
9 **while** $(!end(n))$ **do**
10    $predMatch = find1stMatch(Q_{pred})$
11   S_n += C_n; prevC = C_n; $C_n = C_n$.advance()
12   **while** $(C_n.isDescendantof(prevC))$ **do**
13       $C_n = C_n$.advance()
14 $T' = t' \mid t'$ is multi-valued output $\cap$ t' in let clause of Q
15 Replace the label stream of node n by S_n
16 **while** $(!end(root(Q_{out})))$ **do**
17   fm = findMatch(Q_{out})
18   Let t1,t2,... be labels of each single-valued node of Q_{out}
19   **foreach** $t' \in T'$ **do**
20       $Set_{t'}$ += retrieveLabels(t', fm)
21    $ResultSet$ += <t1,t2,...,$Set_{t'}$,...>
22 return resultSet

Time and Space Complexity

Theorem 2. *Consider an XML database D, a query twig pattern Q consisting of m nodes and ancestor-descendant(A-D) edges only, with specified predicate node set P and output node set T. Algorithm SemanticTwig has the worst-case I/O and CPU complexities linear in the sum of sizes of the m input lists and the output list of Q's sub-twig Q_{out} (which is retrieved via query breakup). The upper bound is $O(m * |R| + |X|)$, where $|X|$ is the size of the m input lists, and $|R|$ is the number of matches of Q_{out}.*

PROOF: *SemanticTwig* first finds the matches of sub-twig Q_{pred} where each match contributes a distinct output result, and it costs $O(m_1 * |R|)$; then it locates all the matches of sub-twig Q_{out} based on the shrinked label stream S_n of node n, and it costs $O(m_2 * |R|)$. Since m_1 and m_2 is number of query nodes in Q_{pred} and Q_{out} respectively, we have $m_1 + m_2 = m + 1$. The cost of reading input streams of each query node is $|X|$. Thus, the total cost is $O(m * |R| + |X|)$.   □

Since *SemanticTwig* calls *twigstack* to find matches of the sub-twigs of Q, and the worst-case size of any stack in TwigStack is proportional to the maximal length of a root-to-leaf path XML database, we have the following results about the space complexity of SemanticTwig.

Theorem 3. *Consider a query twig pattern Q with m nodes and an XML database D. The worst case space complexity of Algorithm SemanticTwig is the minimum*

*of (i) the sum of sizes of the m input lists, and (ii) m times the maximum length of
a root-to-leaf path in D.*

Theorem 4. *The two optimization methods in SemanticTwig are both orthogonal
to all existing structural join based TPMs. The only difference is the implementation of method find1stMatch() and findMatch().*   □

Moreover, *SemanticTwig* is optimal for twig queries with A-D edges only, but suboptimal for queries with P-C edges. But this sub-optimality is due to the suboptimality of *TwigStack* on which our optimization method applies.

5   Experimental Study

We implement *TwigStack* and *SemanticTwig* in JDK 1.4, and run experiments
on a 3.0 GHz Pentium 4 processor PC with 1GB RAM running on windows XP
system. We compare them in terms of intermediate path solutions, I/O cost (i.e.
number of elements read into memory) and total execution time.

In order to evaluate the performance of a particular operation exactly, we
choose DBLP as the real dataset; and generate the synthetic dataset based on
the university's schema diagram shown in Figure 1(b), by manually specifying semantic constraints such as the uniqueness of certain values, the frequency of some
node value in document etc. Three synthetic datasets are used, details summarized in Table 1. *Doc*1 is a non-recursive document corresponding to the schema
in Figure 1(b); *Doc*2 is adapted from *Doc*1, in which *course* becomes a recursive
element, s.t. a course is the prerequisite of other courses. *Doc*3 is adapted from
*Doc*2 by increasing the frequency of some leaf nodes' values. DBLP's summarization is shown in the last row of Table 1.

Table 1. XML Data Sets

Data	Size(MB)	Nodes	Depth
Doc1	10.4	882854	7/4.1
Doc2	11.7	889453	13/5.2
Doc3	18.3	1522218	7/4.5
DBLP	130	3736406	6/2.9

Table 2. Queries over data sets

Q1	//course[code="cs101"]/student[stuNo="u1"]/mark
Q2	//department[.//course/field="www"]/dname
Q3	//student[hobby]/sname
Q4	//faculty[.//student/hobby="tennis"]/fname
Q5	//dblp/article[author]//year
Q6	let $t:=//inproceedings[.//title]/author return {$t}

5.1   SemanticTwig VS TwigStack

Queries Q1-Q4 are chosen for synthetic datasets Doc1-Doc3, and Q5-Q6 are selected for DBLP dataset, as shown in Table 2. Q1 is used to test the effect of
the first optimization method exploiting the *FD* {*course, student*}→*mark*, and Q2-Q6 are used to test the effect of the pure second optimization or the deployment
of both optimizations. In particular, Q3 can exploit the role of multi-valued attribute *hobby* as existence check; Q2 and Q4-Q6 test the effects of query breakup
technique.

SemanticTwig outperforms *TwigStack* on both synthetic and real datasets, as shown in Figure 6 and 7. We also compared it to *TJFast*, and find *SemanticTwig* outperforms *TJFast* in a similar fashion, due to its orthogonality to all structural-join based TPMs. The comparison is further analyzed in terms of the *cost of disk access*, *size of intermediate results* and *query running time*.

Cost of disk access. As shown in Figure 6, *SemanticTwig* reads much fewer elements than *TwigStack* (at least two orders of magnitude, the y-axis's statistics is log-scaled). This huge gap results from the fact that *TwigStack* scans elements for all the query nodes, while *SemanticTwig* scans only the elements of matches contributing to distinct query results, so the elements of redundant matches are skipped. As Figure 6(b) and 6(c) show, *TwigStack* performs even worse for recursive document, while *SemanticTwig*'s I/O cost has small change. E.g. in evaluating Q2, within a certain *department*, once the first *course*-element C whose field is *"www"* is found, we can skip all descendants of C which have the same node type as C, i.e. all *courses* which are the pre-requisite of C can be skipped, since they contribute to the same output value as C; however *TwigStack* still loads them into memory. Figure 6(d) shows our approach reads less elements than *TwigStack* for DBLP.

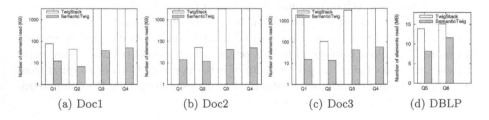

(a) Doc1   (b) Doc2   (c) Doc3   (d) DBLP

Fig. 6. TwigStack vs SemanticTwig on Number of Elements Read (log-scale)

Size of intermediate results. Table 3(a) shows the number of intermediate path solutions generated by *TwigStack* and *SemanticTwig* for non-recursive *Doc1*. The 4th column is the minimal number of merge-joinable path contributing to distinct query answers. *TwigStack* outputs more path solutions than *SemanticTwig* for all queries except Q1 which has the same number of path solutions, because in Q1 a student has only one mark for each course he takes. TwigStack even generates much more partial solutions for a query on recursive XML document, shown in Table 3(b). The reduction of intermediate path solutions on Doc3 is similar to the result on Doc1, we do not show the table due to space limitation. The reduction is 32.8% for $Q5$ and 13.7% for $Q6$ on DBLP. Since each article often has no more than three authors, the reduction is not significant. This is because *SemanticTwig* only generates matches with distinct output result here, while multiple matches contributing to the same result are generated by *TwigStack*.

Table 3. Number of intermediate path solutions

(a) Doc1

Q	Twig	SemTwig	Useful	Reduction
Q1	3	3	3	0%
Q2	18	8	8	55.6%
Q3	213008	53252	53252	75%
Q4	76828	16	16	99.9%

(b) Doc2

Q	Twig	SemTwig	Useful	Reduction
Q1	66	3	3	95.5%
Q2	30	8	8	73.3%
Q3	213008	53252	53252	75.2%
Q4	76828	16	16	99.9%

(c) DBLP

Q	Twig	SemTwig	Useful	Reduction
Q5	331997	223218	223218	32.8%
Q6	703819	607680	697680	13.7%

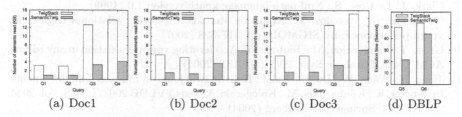

(a) Doc1   (b) Doc2   (c) Doc3   (d) DBLP

Fig. 7. TwigStack vs SemanticTwig on Execution Time

Query running time. Figure 7 reports the total execution time on both synthetic and real datasets. In order to be fair, we do not include the time spent on post-eliminating redundant matches into the *TwigStack*'s total execution time. *SemanticTwig* is at least 4 times faster than *TwigStack*. The improvement is not significant on DBLP, due to the fact that DBLP is a shallow and wide document, and less redundant matches exist for Q5 and Q6. The improvement is much more significant if post-processing time is counted.

6 Conclusion and Future Work

In this paper, we aim to avoid finding redundant matches that return the same results by making use of the semantic information resided in XML document and the issued query. On one hand we utilize the semantics resided in XML document to derive functional dependencies; on the other hand we explore the semantics of an XML query, distinguish its predicate and output nodes and initiate a query breakup technique. These two techniques can be deployed simultaneously in the same query. As a result we propose *SemanticTwig*, which is a novel semantics-aware query optimization algorithm. As part of future work, we want to investigate more semantics useful for efficient XML query processing.

References

1. Al-Khalifa, S., Jagadish, H.V., Patel, J., Wu, Y., Koudas, N., Srivastava, D.: Structural joins: A primitive for efficient Xml query pattern matching. In: ICDE, pp. 141–152 (2002)
2. Bruno, N., Koudas, N., Srivastava, D.: Holistic twig joins: optimal Xml pattern matching. In: SIGMOD, pp. 310–321 (2002)
3. Chen, S., Li, H., Tatemura, J., Hsiung, W., Agrawal, D., Candan, K.S.: $Twig^2$ stack: Bottom-up processing of generalized-tree-pattern queries over Xml documents. In: VLDB, pp. 283–294 (2006)
4. Chen, T., Lu, J., Ling, T.W.: On boosting holism in Xml twig pattern matching using structural indexing techniques. In: SIGMOD, pp. 455–466 (2005)
5. Chippimolchai, P., Wuwongse, V., Anutariya, C.: Semantic query formulation and evaluation for Xml databases. In: WISE, pp. 205–214 (2002)
6. Chippimolchai, P., Wuwongse, V., Anutariya, C.: Towards semantic query optimization for Xml databases. In: ICDE Workshops (2005)
7. Clark, J., De Rose, S.: Xml path language xpath version 1.0 (1999)
8. Georgiadis, H., Vassalos, V.: Xpath on steroids: exploiting relational engines for xpath performance. In: SIGMOD, pp. 317–328 (2007)
9. Grust, T., Van Keulen, M., Teubner, J.: Accelerating xpath evaluation in any rdbms. ACM Trans. Database Syst. 29(1), 91–131 (2004)
10. Jiang, H., Wang, W., Lu, H., Yu, J.: Holistic twig joins on indexed Xml documents. In: Aberer, K., Koubarakis, M., Kalogeraki, V. (eds.) VLDB 2003. LNCS, vol. 2944, pp. 273–284. Springer, Heidelberg (2004)
11. Lu, J., Chen, T., Ling, T.: Efficient processing of xml twig patterns with parent child edges: A look-ahead approach (2004)
12. Lu, J., Ling, T., Chan, C., Chen, T.: From region encoding to extended dewey: On efficient processing of twig pattern matching. In: VLDB, pp. 193–204 (2005)
13. Florescu, D., Boag, S., Chamberlin, D., Robie, J.: Xquery 1.0: An Xml query language (2007)
14. Wu, X., Ling, T.W., Lee, M.-L., Dobbie, G.: Designing semistructured databases using ora-ss model. In: WISE, pp. 171–180 (2001)
15. Zhang, C., Naughton, J., DeWitt, J., Lohman, Q.L.a.: On supporting containment queries in relational database management systems. In: SIGMOD, pp. 425–436 (2001)

Approximate XML Query Answers in DHT-Based P2P Networks

Weimin He and Leonidas Fegaras

University of Texas at Arlington, CSE
Arlington, TX 76019-0015
{weiminhe,fegaras}@cse.uta.edu

Abstract. Due to the increasing number of independent data providers on the web, there is a growing number of web applications that require locating data sources distributed over the internet. Most of the current proposals in the literature focus on developing effective routing data synopses to answer simple XPath queries in structured or unstructured P2P networks. In this paper, we present an effective framework to support XPath queries extended with full-text search predicates over schemaless XML data distributed in a DHT-based P2P network. We construct two concise routing data synopses, termed structural summary and peer-document synopsis, to route the user query to most relevant peers that own documents that can satisfy the query. To evaluate the structural components in the query, a general query footprint derivation algorithm is developed to extract the query footprint from the query and match it with structural summaries. To improve the search performance, we adopt a lazy query evaluation strategy for evaluating the full-text search predicates in the query. Finally, we develop effective strategies to balance the data load distribution in the system. We conduct extensive experiments to show the scalability of our system, validate the efficiency and accuracy of our routing data synopses, and demonstrate the effectiveness of our load balancing schemes.

1 Introduction

Due to the increasing number of independent data providers on the web, there is a growing number of web applications that require locating data sources distributed over the internet. A new computing model, called peer-to-peer(P2P) computing model, has emerged and evolved rapidly to meet the individual users' requirements for sharing and querying data from each other in a pure distributed fashion. In a P2P network, a large number of nodes, called peers, are pooled together to share resources from each other on an equal basis. A peer acts as both a client and server such that both data and query load are distributed among all the peers in the system. A peer may join or leave the system at anytime, which makes P2P systems more flexible, scalable and easier to deploy compared with client-server architectures. More importantly, the network resources can be fully utilized and shared in a P2P system.

J.R. Haritsa, R. Kotagiri, and V. Pudi (Eds.): DASFAA 2008, LNCS 4947, pp. 299–313, 2008.
© Springer-Verlag Berlin Heidelberg 2008

Since XML is rapidly gaining in popularity as a universal data format for data exchange and integration on the web, querying distributed XML data in P2P settings has attracted significant interests in the DB community. Due to the inherent hierarchical structure of XML data, the semantic search over distributed XML data poses more challenges than simple keyword-based search over plain text documents. The key issue to be addressed is how to design efficient and effective routing data synopses to summarize the original XML data, route the user query to the peers that are most likely to own documents that can satisfy the query, and finally return XML fragments as query answers to the user. Most of the current proposals in the literature [3,5,8,13] focus on developing effective routing data synopses to answer simple XPath [15] queries in structured or unstructured P2P networks. In our earlier work [4], we proposed a framework to support XML queries with full-text extension in structured P2P networks. We introduced two types of data synopses, termed content synopses and positional filters, to answer a full-text XPath query in a more precise way when compared with Bloom filter-based approaches. However, the proposed approach is not very efficient because the whole intermediate query results have to be routed from peer to peer along the way during the query evaluation, which might consume much network bandwidth and degrade the search performance.

In this paper, we extend our earlier work [4] and present a more efficient framework for indexing and querying schema-less XML data in DHT-based P2P networks. The work presented in this paper is different from our earlier work in the following aspects. First, instead of extracting and indexing content synopses and positional filters [4], we propose a more concise routing data synopsis, termed peer-document-synopsis(PDS), to summarize peers that own documents containing a path-term pair. PDS can achieve more efficient evaluation of search predicates in the query. Secondly, we have developed a formal query footprint derivation algorithm for structural summary matching. Thirdly, we adopt a lazy query evaluation strategy for efficient full-text search predicates evaluation. Finally, we develop effective schemes to improve the load balancing in the system.

In summary, we make the following contributions in this paper:

- We introduce a concise routing data synopsis, termed peer-document synopsis for the efficient evaluation of full-text search predicates in the user query.
- We develop a general query footprint derivation algorithm for full-text XPath evaluation.
- We adopt a lazy query evaluation algorithm for efficient search predicates evaluation.
- We propose simple schemes to achieve better data load distribution.
- We experimentally validate the system scalability and the efficiency and accuracy of our routing data synopses, and measure the effectiveness of our load balancing schemes.

2   Preliminaries

2.1   Data Placement in DHT-Based P2P Networks

In a DHT-based P2P system, the location of data is determined by some global scheme. More specifically, as a data item I is published by some node, a global hash function is used to map the key K of I to an IP address of a node N in the P2P network and I is placed on node N. One popular mapping approach is called consistent hashing [12], in which both a key and node Id are mapped to the same identifier space. Identifiers are ordered in an identifier circle modulo $|2^m$, where m is the length of an identifier. The data item I associated with the key K is assigned to its successor node, which is the first node whose identifier is equal to or follows the identifier of K in the identifier space. In our framework, we employ Pastry [11] as our P2P back end, in which the key K is assigned to the node whose Id is closest to the Id of the key. The Id of K is derived from a base hash function such as SHA-1 [12].

2.2   Specification of User Queries

Our query language is XPath extended with full-text search. We extend the XPath syntax with a full-text search predicate $e \sim S$, where e is an XPath expression. This predicate returns true if at least one element from the sequence returned by e matches the *search specification*, S. A search specification is a simple IR-style boolean keyword search that takes the form

$$\text{“term”} \quad | \quad S_1 \underline{\text{and}} S_2 \quad | \quad (\, S\,)$$

where S, S_1, and S_2 are search specifications. A term is an indexed term that must be present in the text of an element returned by the expression e.

As a running example used throughout the paper, the following query, Q:

```
//article[//keywords ~ "XML" and "P2P"]
   [//author/lastname ~ "Galanis"]/title
```

searches for the titles of all the articles authored by Galanis that contain the terms "XML" and "P2P" in their keywords.

3   Routing Data Synopses

In our framework, two types of data synopses are constructed to route a user query to the peers who are most likely to own documents that can satisfy the query.

3.1   Structural Summary

To match the structural components in the query during the query evaluation, we construct a type of data synopses, called **Structural Summary**(SS) [7], which

302   W. He and L. Fegaras

is a structural markup that captures all the unique paths in the document. Figure 1(a) gives an example XML document that represents the excerpted bibliography information from DBLP. The structural summary of the example XML document is shown in Figure 1(b). Each node in an SS has a tagname and a unique Id. One SS node may correspond to more than one elements in the original document. For example, the node **article** in Figure 1(b) corresponds to two **article** elements in Figure 1(a). As an XML document is published by a peer, the SS is extracted from the document and each distinct tagname tag is used as the DHT key to route the SS to the peer, denoted by P_{tag}, whose node Id is closet to the Id of tag(The Id of tag is generated from the lowest 128 bits from the 160-bit SHA-1 hash code of tag).

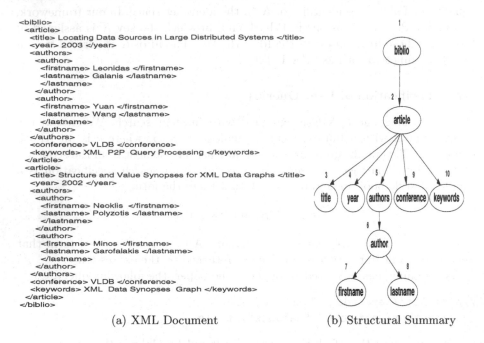

(a) XML Document   (b) Structural Summary

Fig. 1. XML Document & Structural Summary Examples

3.2   Peer-Document Synopses

To evaluate the search predicates in the full-text search query, we construct another type of data synopses, called **Peer-Document Synopsis**(PDS), to summarize the peers who own XML documents that contain a path-term pair. Each PDS is associated with a path-term pair (p, t), where p is a full label path in an SS that match the corresponding path p_q in the query, and t is the term associated with p_q in the query. More specifically, a PDS is a matrix with peer Id as one dimension and document Id as another. To avoid the storage overhead of $PDSs$, instead of using (p, t) as the local indexing key, we employ $(p, hc(t))$ as the indexing key for a PDS, where hc is a hash function that hashes the term

t to some bucket such that all the (p, t) pairs are partitioned into groups. That way, the number of indexed $PDSs$ can be greatly reduced, thus saving the space for storing $PDSs$. Suppose the peer P owns a document D containing n pairs of (p, t), we hash the Id of P to some bucket i along the peer Id dimension and hash the Id of D to some bucket j along the document Id dimension. The cell $pds_{[i,j]}$ in the PDS is set to n, which represents the number of occurrences of (p, t) in document D owned by peer P. An example PDS is shown in Figure 2. The local indexing key for the PDS is $(/biblio/article/keywords, hc(``XML"))$. The cell $pds_{[12,2]} = 4$ indicates the peer 12 owns a document 2 that contains the pair $(/biblio/article/keywords, hc(``XML"))$ 4 times. Note that in the case different peers(documents) are hashed to the same bucket, the value in the cell is the average path-term pair count. As an XML document D is published by some peer P, for each text label path p(the path that leads to the text in the document)in the SS of D, we construct the tuple $< p, peerId, docId, V_t >$ and route it to the peer P_p using the path p as the DHT routing key. Here, V_t is a term-count vector that summarizes the counts of all the terms that are associated with p. The count value in each bucket in V_t is increased by one as a term t under the path p is hashed to that bucket. V_t is used to update the corresponding $PDSs$ on the peer P_p.

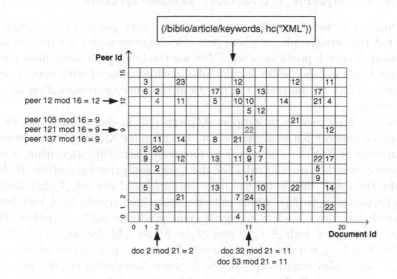

Fig. 2. Peer-Document Synopsis Example

4   Query Processing

In our framework, as the client peer submits a query, the query footprint(QF) is extracted from the query and an arbitrary tagname tag is chosen from the query and serves as DHT key to route QF to the peer P_{tag}, on which we match the footprint QF against the local indexes, through which all the matching structural summaries and distinct label paths corresponding to the query search predicates are retrieved. Based on the retrieved label paths, peer-document synopses associated with those label paths are found on the corresponding peers and qualified documents that may contain all path-term pairs in the query are filtered out and a filtering peer-document synopsis is routed back to the peer P_{tag}. Based on the filtering peer-document synopsis, the original query is sent to the peers that are most likely to own documents that can satisfy the query. Finally, the top-k answers are returned to the client peer. We present the details of our query processing algorithms in the next subsections.

4.1 Query Footprint Derivation

In our framework, The first step in evaluating a full-text XPath query is deriving a query footprint(QF) from the query. A query footprint captures the essential structural components and all the *entry points* associated with the search predicates. The query footprint will be used to find all structural summaries that match the query as well as locating the text label paths that participate in the search specifications of the query. For our running example, the query footprint of Q is:

`//article[//keywords:1][//author/lastname:2]/title`

The numbers 1, and 2 are the numbers of the entry points in the query footprint that indicate the places where peer-document synopses are needed for the evaluation of search predicates(one PDS for the label path associated with entry point 1 and another PDS for the label path associated with entry point 2). The entry points are basically the points in the query corresponding to search specifications.

Since our final goal is to support XQuery [16] in our framework, we have developed a general footprint derivation algorithm for XQuery, whose semantics is a superset of XPath. The recursive footprint derivation algorithm is shown in Figure 3, which gives the rules for the query footprint extraction. It derives two sets: the set $\mathcal{F}$ that holds the pairs $(n, path)$ and the set $\mathcal{J}$ that holds the tuples $(n, path_1, m, path_2)$, which indicate that there should be a join between the two documents n and m using the join predicate $path_1 = path_2$. For an XQuery e, we start with $\mathcal{F} := \emptyset$ and $\mathcal{J} := \emptyset$ and add the set $\mathcal{T}(\llbracket e \rrbracket, \sigma)$ to $\mathcal{F}$. Function $\mathcal{T}(\llbracket e \rrbracket, \sigma)$, not only returns a set of $(n, path)$ pairs to be added to $\mathcal{F}$ at the end, but it also adds more pairs to $\mathcal{F}$ when these pairs are not part of the output and updates $\mathcal{J}$. The notation $\{ e \parallel x \in X, \dots \}$ is a set former notation, similar to tuple relational calculus, while the notation $\mathcal{F} += X$ is equivalent to $\mathcal{F} := \mathcal{F} \cup X$. The environment σ in $\mathcal{T}(\llbracket e \rrbracket, \sigma)$ binds XQuery FLWR variables (ie, variables that appear in a for-loop or a let-binding [16]) to sets of $(n, path)$ pairs. The notation $\sigma[v]$ extracts the binding of the variable v in σ while the notation $\sigma[v/s]$ extends σ with a binding from variable v to a set s.

After the QF is extracted from the query, the client peer chooses a tagname uniformly at random from QF and routes the QF to the peer P_{tag}, on which the structural summaries matching the QF are retrieved.

Translation of XQuery expressions $e, e_1, \ldots, e_n$:

$$\mathcal{T}([\![\$v]\!], \sigma) = \sigma[v]$$
$$\mathcal{T}([\![document()]\!], \sigma) = \{(n, ``")\} \qquad \text{where } n \text{ is a new number}$$
$$\mathcal{T}([\![path]\!], \sigma) = \mathcal{P}([\![path]\!], \sigma)$$
$$\mathcal{T}([\![element(A, L, e)]\!], \sigma) = \mathcal{T}([\![e]\!], \sigma)$$
$$\mathcal{T}([\![e_1, e_2]\!], \sigma) = \mathcal{T}([\![e_1]\!], \sigma) \cup \mathcal{T}([\![e_2]\!], \sigma)$$
$$\mathcal{T}([\![e \sim S]\!], \sigma) = \{(n, x = \text{new}(n)) \mid (n, x) \in \mathcal{T}([\![e]\!], \sigma)\}$$
$$\mathcal{T}([\![e_1 = e_2]\!], \sigma) = \{(n, x = \text{new}(n)) \mid (n, x) \in \mathcal{T}([\![e_1]\!], \sigma)\}$$
$$\cup \{(m, y = \text{new}(m)) \mid (m, y) \in \mathcal{T}([\![e_2]\!], \sigma)\}$$
$$\mathcal{J} \mathrel{+}= \{(n, x, m, y) \mid (n, x) \in \mathcal{T}([\![e_1]\!], \sigma), (m, y) \in \mathcal{T}([\![e_2]\!], \sigma), n \neq m\}$$
$$\mathcal{T}([\![e_1 \text{ and } e_2]\!], \sigma) = \mathcal{T}([\![e_1]\!], \sigma) \cup \mathcal{T}([\![e_2]\!], \sigma)$$
$$\mathcal{T}([\![f(e_1, \ldots, e_n)]\!], \sigma) = \mathcal{T}([\![body]\!], \sigma[x_1/\mathcal{T}([\![e_1]\!], \sigma), \ldots, x_n/\mathcal{T}([\![e_n]\!], \sigma)])$$
$$\text{for defined function } f (\$x_1, \ldots, \$x_n) \{ body \}$$
$$\mathcal{T}([\![\text{for } \$v \text{ in } e_1 \text{ where } e_2 \text{ return } e_3]\!], \sigma) = \mathcal{T}([\![e_3]\!], \sigma[v/\mathcal{T}([\![e_1]\!], \sigma)])$$
$$\mathcal{F} \mathrel{+}= \mathcal{T}([\![e_1]\!], \sigma) \cup \mathcal{T}([\![e_2]\!], \sigma[v/\mathcal{T}([\![e_1]\!], \sigma)]) \cup \mathcal{T}([\![e_3]\!], \sigma[v/\mathcal{T}([\![e_1]\!], \sigma)])$$
$$\mathcal{T}([\![\text{let } \$v := e_1 \text{ where } e_2 \text{ return } e_3]\!], \sigma) = \text{same as } \mathcal{T}([\![\text{for } \$v \text{ in } e_1 \text{ where } e_2 \text{ return } e_3]\!], \sigma)$$
$$\mathcal{T}([\![\text{some}/\text{every } \$v \text{ in } e_1 \text{ satisfies } e_2]\!], \sigma) = \text{same as } \mathcal{T}([\![\text{for } \$v \text{ in } e_1 \text{ return } e_2]\!], \sigma)$$
$$\mathcal{T}([\![e]\!], \sigma) = \emptyset \qquad \text{otherwise}$$

Translation of the path expression, *path*:

$$\mathcal{P}([\![path/A]\!], \sigma) = \{(n, x/A) \mid (n, x) \in \mathcal{P}([\![path]\!], \sigma)\}$$
$$\mathcal{P}([\![path//A]\!], \sigma) = \{(n, x//A) \mid (n, x) \in \mathcal{P}([\![path]\!], \sigma)\}$$
$$\mathcal{P}([\![path/*]\!], \sigma) = \{(n, x/*) \mid (n, x) \in \mathcal{P}([\![path]\!], \sigma)\}$$
$$\mathcal{P}([\![path/@A]\!], \sigma) = \{(n, x/@A) \mid (n, x) \in \mathcal{P}([\![path]\!], \sigma)\}$$
$$\mathcal{P}([\![path[e]]\!], \sigma) = \{(n, x[p]) \mid (n, x) \in \mathcal{P}([\![path]\!], \sigma), (n, p) \in \mathcal{T}([\![e]\!], \sigma[\text{self}/\mathcal{P}([\![path]\!], \sigma)])\}$$
$$\mathcal{F} \mathrel{+}= \mathcal{P}([\![path]\!], \sigma) \cup \mathcal{T}([\![e]\!], \sigma[\text{self}/\mathcal{P}([\![path]\!], \sigma)])$$
$$\mathcal{P}([\![A]\!], \sigma) = \{(n, x/A) \mid (n, x) \in \sigma[\text{self}]\}$$
$$\mathcal{P}([\![e]\!], \sigma) = \mathcal{T}([\![e]\!], \sigma)$$

Fig. 3. Query Footprint Derivation Algorithm

4.2 Structural Summary Matching

As the peer P_{tag} receives the QF from the client peer, it first retrieves its local indexes to find the matching structural summaries against QF. Borrowing ideas from structural joins in XML query processing [17], when an SS is indexed, we employ a similar numbering scheme to encode each node k in a structural summary S by the triple (b, e, l), where b/e is the begin/end numbering of k and l is the level of k in S. In our framework, a peer maintains the following mapping to store structural summaries: $\mathcal{M}_{ss} : tag \rightarrow \{(S, k, b, e, l)\}$, where tag is the tagname of the node k in S, and b, e, l are as defined above. Thus, the key operation in the structural summary matching is a structural join between two tuple streams corresponding to two consecutive location steps in the query footprint. Our index is designed in such a way that the tuples are delivered in major order S and minor order b, so that the structural join can be evaluated in a merge-join fashion. To accelerate query processing and avoid materializing intermediate tuples, we leverage the iterator model in relational databases to form a pipeline of iterators derived from the query footprint (one iterator for each XPath location step). That way, the pipeline processes the structural summary nodes derived from the indexes one-tuple-at-a-time and the intermediate results are never materialized.

Let QF be the structural footprint of a query. Then the structural summary matching is accomplished by the function $\mathcal{SP}[\![QF]\!]$ that returns a set of tuples

(ρ, S, k, b, e, l), where (S, k, b, e, l) is similar to that of $\mathcal{M}_{ss}$ and ρ is a vector of node numbers, such that $\rho[i]$ gives the node number in S corresponding to the ith entry point in QF. The function $\mathcal{SP}[\![QF]\!]$ is defined recursively based on the syntax of QF, generating structural joins for each XPath step. Here, we give two rules as an illustration:

$$\mathcal{SP}[\![QF/tag]\!]$$
$$= \{ (\rho, S, k_2, b_2, e_2, l_2) \mid (\rho, S, k_1, b_1, e_1, l_1) \in \mathcal{SP}[\![QF]\!]$$
$$\wedge (S, k_2, b_2, e_2, l_2) \in \mathcal{M}_{ss}(tag)$$
$$\wedge b_2 > b_1 \wedge e_2 < e_1$$
$$\wedge l_2 = l_1 + 1 \}$$
$$\mathcal{SP}[\![QF : i]\!]$$
$$= \{ (\rho[i] := k, S, k, b, e, l) \mid (\rho, S, k, b, e, l) \in \mathcal{SP}[\![QF]\!] \}$$

The first rule shows how to evaluate a *Child* location step using the index $\mathcal{M}_{ss}$ to retrieve all structural summaries S that contain tag. The second rule assigns the node number k to $\rho[i]$, where i is an entry point in the footprint. That is, it finds the node numbers in the structural summary that correspond to the entry points in the query footprint. From the node numbers in ρ, we can derive the label paths associated with the search predicates in the query from the structural summary. In our running example, the label paths are /biblio/article/keywords and /biblio/article/authors/author/lastname. From these label paths and their associated terms in the query, a list of path-term pairs can be derived to evaluate the search predicates in the query.

4.3   Lazy Query Evaluation

In order to make our query processing more efficient, we have developed a lazy evaluation algorithm for the search predicates evaluation. Our lazy query evaluation algorithm is shown in Algorithm 1. As mentioned above, the result of structural summary matching is a list of path-term pairs. Let the list be $L_{pt} = (p_1, t_1), (p_2, t_2), ..., (p_n, t_n)$, then first p_1 is used as the DHT routing key to route (p_1, t_1) to the peer P_{p_1}, on which $(p_1, hc(t_1))$ is used as the key to search local indexes to find the corresponding peer-document synopsis PDS_1, which is routed to the peer P_{p_2}, on which another peer-document synopsis PDS_2 is retrieved using the key $(p_2, hc(t_2))$. Then PDS_1 and PDS_2 are "added" together and the resulting PDS is routed to the next peer. Finally, a filtering PDS, denoted by PDS_f, is routed back to the peer P_{tag}. The value in each cell of PDS_f is the summation of the corresponding values in each PDS_i. On the next, we rank each non-zero row vector V_j in PDS_f by the descending order of the summation of all the cell values in V_j. Suppose the ranked row vectors are $V_1, V_2, ..., V_m$, we first derive peer Ids and document Ids from V_1 and route the original query Q to those peers and evaluate Q over those documents. If a document d can satisfy the query, the document location of d is returned and put in the top k list. If the top-k threshold value for the user query is satisfied, the query evaluation terminates and the top-k document locations are returned to the client peer. If the threshold value is not satisfied, we choose the next row

---

Algorithm 1. Lazy Evaluation of Search Predicates

Input: K   /* the top-k threshold value for the user query*/
   (p_1, t_1), (p_2, t_2), ... , (p_n, t_n) /* n path-term pairs derived from SS matching */
Output: L_{topK} /* the list of top k qualified XML elements */
1: $L_{topK} := emptyList$;
2: /* Obtain the filtering peer-document synopsis PDS_f */
3: $PDS_f := \Phi$;
4: **for** i = 1 to n-1 **do**
5:    Route (p_i, t_i) and PDS_f to the peer P_{p_i};
6:    Use $(p_i, hc(t_i))$ as the key to retrieve PDS_i from the local indexes of P_{p_i};
7:    $PDS_f := PDS_f \oplus PDS_i$;
8: **end for;**
9: /* Lazy query evaluation guided by the filtering PDS_f */
10: Rank non-zero row vector V_j in PDS_f by descending order of the summation of all cell values in V_j;
11: Let V_1, V_2, ..., V_m be the ranked sequence of vectors;
12: **for** k = 1 to m **do**
13:    Derive the list of peer IDs and document IDs from V_k;
14:    Send original query Q to the peers associated with those peer IDs in the list;
15:    Evaluate Q over the documents associated with those document IDs in the list;
16:    **if** a document d satisfies the query
17:       Route the document location of d to the peer P_{tag} and put it in L_{topK};
18:          **if** the size of L_{topK} > K   **do**
19:             **return** L_{topK} to the user;
20:          **end if;**
21:    **end if;**
22: **end for;**
23: **return** L_{topK} to the user;

---

vector with the highest score and continue the query evaluation until the top-k threshold value is satisfied.

5   Load Balancing

Since our structural summary distribution is purely based on single tagnames, it is likely that peers associated with popular tagnames, such as "name" and "price", would burden a higher load of structural summaries than others, thus need to handle more messages at both data publishing and query processing stages. In order to better balance the data distribution load and query load in our P2P system, we propose a tag-pair based routing scheme for structural summary distribution. Instead of using a single tagname as the DHT routing key, we use a sequence of two tagnames (A,B) as the routing key, where B is a descendant of A in the structural summary. That is, a structural summary is published using all possible tag pairs (A,B) for a descendant B of A, which requires $O(n^2)$ messages for n tagnames, rather than n messages. Then given a query footprint with at least two tagnames, two tagnames A and B are selected uniformly at random from the query footprint and used as the routing key, where B is a descendant of A in the query. As we will demonstrate in our experiments, this tag-pair based routing scheme can achieve better load balancing on the data distribution for structural summaries. Note that since our peer-document synopses are distributed based on label paths, which are tagname sequences, they are naturally better distributed than structural summaries in the P2P network.

6 Experimental Evaluation

6.1 Experimental Setup

We have fully implemented our framework using Java (J2SE 6.0).Our DHT-based P2P back end is FreePastry 2.0 [11], which is responsible for routing various messages associated with data publishing and query evaluation to specific peers. On each peer, Berkeley DB Java Edition 3.2.13 [2] is employed as the local indexing DB. Our experiments were carried out on a WindowsXP machine with 2.8GHz CPU and 512M memory. The maximum heap size for Java Virtual Machine was set to 384MB. We conducted several sets of experiments to extensively measure the efficiency and accuracy of our routing data synopses, and the effectiveness of our load balancing scheme. Our experimental data are synthetically generated from XMark benchmark [14], from which routing data synopses are extracted and populated over the DHT network. For all experiments except those measuring system scalability, the size of the data set is 55.8MB and the number of documents is 11500. In our experiments we create 1150 Pastry node and let each node publishes 10 XML documents over DHT network. Since scalability experiments require large data volumes, we multiply the above data collection by copying when larger volumes are needed. Each peer owns its database to index the meta-data. We designed 10 queries over XMark data to conduct query-related experiments and the queries are shown in Table 1. Query 1 through 5 are simple XPath queries with single search predicate, whereas query 6 through 10 are complex XPath queries with multiple search predicates.

Table 1. Example XMark Queries

Query	Query Expression
Q_1	//item[payment ~ "Creditcard"]/name
Q_2	//item[location ~ "United"]/description
Q_3	//item[description//text ~ "gold"]/payment
Q_4	//open_auction[annotation//text~ "country"]/bidder
Q_5	//open_auction[annotation//text~ "heat"]/bidder
Q_6	//item[payment ~ "Creditcard"][location ~ "United"]/name
Q_7	//item[payment ~ "Creditcard" and "Check"][location ~ "United"]/name
Q_8	//item[payment ~ "Creditcard" and "Cash"] [location ~ "United"][description//text ~"gold"]/name
Q_9	//open_auction [annotation//text ~ "heat"][type ~ "Regular"]/bidder
Q_{10}	//open_auction[annotation//text ~ "heat"][type ~ "Regular"][privacy ~ "No"]/bidder

6.2 Scalability Measurement

In this experiment, we measure the scalability of our system. We first varied the data size from 56MB to 560MB to measure the total data publishing time. As we can see from 4(a), as the data size increases, the data publishing time scales almost linearly in terms of the size of published data. We then varied the number of peers in our system to measure the total data publishing time. As we can see from 4(b), as the network size increases, the total data publishing time increases smoothly with the increasing number of peers. The above results indicate that our system scales gracefully in terms of both data size and network size.

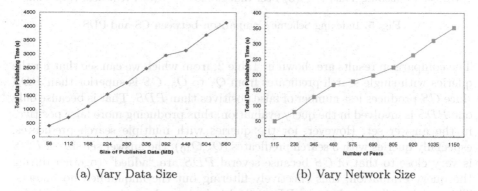

(a) Vary Data Size   (b) Vary Network Size

Fig. 4. Measurement of System Scalability

6.3 Efficiency of PDS-Based Indexing Scheme

In this experiment, we measure the efficiency of our PDS-based meta-data indexing scheme. More specifically, we compared the efficiency of our PDS-based indexing scheme with that of CS-based indexing scheme [4]. The results are shown in Figure 5. As we can see, the average peer publishing time and index size of PDS-based scheme are about 73% and 71% respectively of that of the CS-based scheme. We also compared the search predicates evaluation time between two schemes for all the queries in Table 1. From Figure 5(c), we can see that for most queries, the PDS-based scheme can achieve shorter predicates evaluation time than the CS-based scheme. Especially for those queries with multiple search predicates, such as queries $Q6$, $Q7$, and $Q8$, the PDS-based scheme performs much better than the CS-based scheme because it can avoid routing potentially large intermediate results from peer to peer, thus reducing the query processing overhead.

6.4 Accuracy of Data Synopses

To measure the accuracy of PDS, we compared the query precision of PDS with CS in [4]. We exploited an existing XQuery engine Qizx/open [10] to evaluate each query over the dataset to obtain the accurate relevant set for the query.

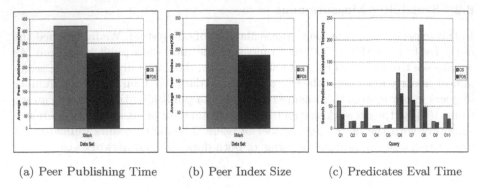

(a) Peer Publishing Time   (b) Peer Index Size   (c) Predicates Eval Time

Fig. 5. Indexing Scheme Comparison between CS and PDS

The comparison results are shown in Table 2, from which we can see that for the queries with single search predicates from Q_1 to Q_5, CS is superior than PDS since CS produces less number of false positives than PDS. That is because only one PDS is involved in the query evaluation, thus producing more false positives in the answer set. However, for the queries with multiple search predicates, especially queries with 3 search predicates (Q_8,Q_{10}), the answer set size of PDS is very close to that of CS because several PDS are "added" together during the query evaluation, thus effectively filtering out the unqualified documents. This result indicates that our PDS-based indexing scheme is more suitable for complex XPath queries with multiple search predicates. Notice that since XMark data are homogeneous, the number of false positives for some queries are high. For example, the number of false positives for Q_3 is 1001 for CS, which is much higher than the size of relevant set. That is because among 11500 documents, although only 138 documents exactly match the query (satisfy both structural and search predicates), a lot more documents match only the structural part in the query, thus the answer set size is much larger than the relevant set size. However, under the heterogeneous environment in a real world application, the answer set size would be much smaller.

Table 2. Accuracy Comparison between CS and PDS

Query	Relevant Set Size	Answer Set Size		# of False Positives	
		CS	PDS	CS	PDS
Q_1	3142	3142	4620	0	1478
Q_2	3574	3586	5380	12	1806
Q_3	138	1139	2040	1001	1902
Q_4	73	601	1120	528	1047
Q_5	105	630	1560	525	1455
Q_6	2717	2790	3360	73	643
Q_7	1662	2314	2820	652	1158
Q_8	54	521	580	467	526
Q_9	63	337	420	274	357
Q_{10}	14	269	340	255	326

6.5   Effectiveness of Load Balancing Scheme

In order to measure the effectiveness of our tag-pair based load balancing scheme, we fixed the number of peers to 1150 and varied the number of XML documents published in P2P network to compare the number of involved peers associated with structural summary distribution between the single-tag based scheme and tag-pair based scheme. The results are shown in Figure 6. As we can see from Figure 6, when the tag-pair based scheme is employed to publish structural summaries, the number of involved peers becomes larger than that when only a singe tagname is used to construct the DHT key for the structural summary publication. Furthermore, as the number of published documents increases, the tag-pair based scheme can make more and more peers involved in the data distribution than the single-tag based scheme, which indicates that our load balancing scheme is more effective for large scale P2P networks. Note that since only a single DTD exists for XMark benchmark and thus structural summaries are homogeneous for different documents, the percentage of involved peers is relatively low for both schemes. In a real heterogeneous P2P environment, our tag-pair based load balancing scheme is expected to achieve better data load distribution outcomes.

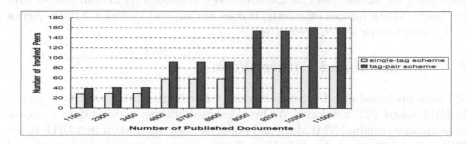

Fig. 6. Comparison between Load Balancing Schemes

7   Related Work

Over the past few years, the popularity of P2P applications and XML has brought many researchers on investigating the problem of indexing and querying distributed XML data sources in P2P networks [5,3,8,13,4]. Some researchers are interested in querying XML data in unstructured P2P networks. In [8], the authors proposed two multi-level Bloom filters, termed Breath Bloom Filter and Depth Bloom Filter, to summarize the structure of an XML document for efficient path query routing in unstructured P2P networks. They also advocated building a hierarchical organization of nodes by clustering together nodes with similar content. Although their approach is effective for simple linear XPath queries, it can not handle the descendant axis in the query effectively and precisely. Kd-synopsis [13] is another graph-structured routing synopsis based on length-constrained FBsimulation relationship, which allows the balancing of the precision and size of the synopsis according to different space constraints on peers with heterogeneous capacity. Although kd-synopsis is more precise than the approach in [8], it can deal with only simple Branching Path Queries(BPQ) [6] without full-text search predicates.

Due to the inherent better scalability of structured P2P systems, much work has also been done on query processing of XML data in DHT-based P2P systems. Galanis et al. [5] proposed a meta-data indexing scheme and query evaluation algorithm over distributed XML data in structured P2P networks. In their framework, a distributed catalog service is distributed across the data sources themselves and is responsible for routing user queries to the relevant peers. They also proposed a structure-based key splitting scheme for load balance. However, their framework only supports general XPath queries without full-text search predicates and thus is more suitable for processing data-centric XML data instead of document-centric XML data. In [3], the authors proposed XP2P, which indexes XML data fragments in a structured P2P system based on their concrete paths that unambiguously identify the fragments in the document(by using positional filters in the paths). The search key used for fragment indexing is the hash value of its path. Thus, XP2P can answer simple, but complete XPath queries (without predicates or descendant-of steps) very quickly, in one peer hop, using the actual query as the search key. Although XP2P can answer simple linear XPath queries efficiently, it does not support complex XPath queries with conditions or search predicates.

8   Conclusion and Future Work

We have presented an effective framework for answering full-text XPath queries in DHT-based P2P networks. We first designed efficient routing data synopses to summarize original XML data and populate these meta-data in a DHT-based P2P system. Then we developed an effective query processing framework and adopted a lazy search predicate evaluation strategy to optimize the query processing. In addition, we proposed an effective load balancing scheme for the data distribution. Finally, we conducted comprehensive experiments to measure the system scalability, the efficiency and accuracy of our data synopses and the effectiveness of our load balancing schemes. As our future work, we plan to adopt classical top-k algorithms in DB to further reduce the query processing time in our framework.

Acknowledgments. This work is supported in part by the National Science Foundation under the grant IIS-0307460.

References

1. Abiteboul, S., Manolescu, I., Preda, N.: Constructing and Querying Peer-to-Peer Warehouses of XML Resources. In: Proc. of ICDE, Tokyo, Japan, pp. 1122–1123 (2005)
2. Berkeley DB. http://www.oracle.com/database/berkeley-db

3. Bonifati, A., Matrangolo, U., Cuzzocrea, A., Jain, M.: XPath Lookup Queries in P2P Networks. In: Proc. of ACM WIDM, Washington DC, USA, pp. 48–55 (2004)
4. Fegaras, L., He, W., Das, G., Levine, D.: XML Query Routing in Structured P2P Systems. In: Moro, G., Bergamaschi, S., Joseph, S., Morin, J.-H., Ouksel, A.M. (eds.) DBISP2P 2005 and DBISP2P 2006. LNCS, vol. 4125, pp. 13–24. Springer, Heidelberg (2007)
5. Galanis, L., Wang, Y., Jeffery, S.R., DeWitt, D.J.: Locating Data Sources in Large Distributed Systems. In: Aberer, K., Koubarakis, M., Kalogeraki, V. (eds.) VLDB 2003. LNCS, vol. 2944, pp. 874–885. Springer, Heidelberg (2004)
6. Kaushik, R., Bohannon, P., Naughton, J.F., Korth, H.: Covering indexes for branching path queries. In: Proc. of SIGMOD, Madison, USA, pp. 133–144 (2002)
7. Kaushik, R., Bohannon, P., Naughton, J.F., Shenoy, P.: Updates for Structure Indexes. In: Bressan, S., Chaudhri, A.B., Li Lee, M., Yu, J.X., Lacroix, Z. (eds.) CAiSE 2002 and VLDB 2002. LNCS, vol. 2590, pp. 239–250. Springer, Heidelberg (2003)
8. Koloniari, G., Pitoura, E.: Content-Based Routing of Path Queries in Peer-to-Peer Systems. In: Bertino, E., Christodoulakis, S., Plexousakis, D., Christophides, V., Koubarakis, M., Böhm, K., Ferrari, E. (eds.) EDBT 2004. LNCS, vol. 2992, pp. 29–47. Springer, Heidelberg (2004)
9. Koloniari, G., Pitoura, E.: Peer-to-peer management of XML data: issues and research challenges. SIGMOD Record 34, 6–17 (2005)
10. Qizx/open, http://www.axyana.com/qizxopen/
11. Pastry, http://freepastry.rice.edu
12. Stoica, I., et al.: Chord: A Scalable Peer-to-Peer Lookup Protocol for Internet Applications. IEEE/ACM Trans. on Networking 11, 17–32 (2003)
13. Wang, Q., Jha, A.K., Ozsu, M.T.: An XML Routing Synopsis for Unstructured P2P Networks. In: Proc. of the 7th Int. Conference on Web Age Information Management Workshop(WAIMW), p. 23, Hongkong, China (2006)
14. XMark, http://www.xml-benchmark.org/
15. XML Path Language (XPath) 2.0, http://www.w3.org/TR/xpath20/
16. XQuery 1.0: An XML Query Language, http://www.w3.org/TR/xquery/
17. Zhang, C., et al.: On Supporting Containment Queries in Relational Database Management Systems. In: Proc. of SIGMOD, Santa Barbara, USA, pp. 425–436 (2001)

Efficient Top-k Search Across Heterogeneous XML Data Sources

Jianxin Li[1], Chengfei Liu[1], Jeffrey Xu Yu[2], and Rui Zhou[1]

[1] Swinburne University of Technology, Melbourne, Australia
{jili,cliu,rzhou}@ict.swin.edu.au
[2] Chinese University of Hong Kong, China
yu@se.cuhk.edu.hk

Abstract. An important issue arising from XML query relaxation is how to efficiently search the top-k best answers from a large number of XML data sources, while minimizing the searching cost, i.e., finding the k matches with the highest computed scores by only traversing part of the documents. This paper resolves this issue by proposing a bound-threshold based scheduling strategy. It can answer a top-k XML query as early as possible by dynamically scheduling the query over XML documents. In this work, the total amount of documents that need to be visited can be greatly reduced by skipping those documents that will not produce the desired results with the bound-threshold strategy. Furthermore, most of the candidates in each visited document can also be pruned based on the intermediate results. Most importantly, the partial results can be output immediately during the query execution, rather than waiting for the end of all results to be determined. Our experimental results show that our query scheduling and processing strategies are both practical and efficient.

1 Introduction

Over decades, processing top-k query has been extensively studied in different research areas, such as relational databases [1,2,3], multimedia databases [4,5,6,7,8,9] and keyword search [10,11]. Recently, Efficiently computing top-k answers to XML queries is gaining importance due to the increasing number of XML data sources and the heterogeneous nature of XML data. Top-k queries on approximate answers are appropriate on structurally heterogeneous data. On the one hand, it is difficult for users to formulate their queries exactly and search the exact answers. On the other hand, an XML query may have a large number of answers, and returning all answers to the user may not be desirable. The top-k approach can limit the cardinality of answers by returning k answers with the highest scores.

The problem of finding top-k answers within a large XML repository has been studied in [12], where an adaptive strategy is proposed for filtering out some unqualified candidates. However, its searching overhead is expensive due to frequent adaptivity among possible candidates and dynamic sort of partial matches. Furthermore, this work only considers query evaluation in a single XML document.

J.R. Haritsa, R. Kotagiri, and V. Pudi (Eds.): DASFAA 2008, LNCS 4947, pp. 314–329, 2008.
© Springer-Verlag Berlin Heidelberg 2008

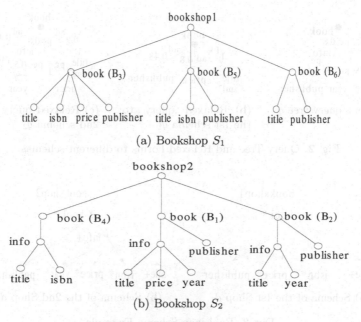

(a) Bookshop S_1

(b) Bookshop S_2

Fig. 1. Bookshop Example

For many applications, it is more meaningful to find top-k results from multiple heterogeneous XML data sources. In this paper, we target this problem by proposing a bound-threshold based scheduling strategy (in short BT strategy) with the help of schema information of each XML data source. We are not required to evaluate a top-k query over all data sources. Instead, we schedule the query to the most relevant ones by leveraging the schema information, which can produce top k results as early as possible and output each result immediately after it is generated.

Example 1. Consider two bookshop XML data sources in Figure 1 that maintain the partial or full information of each book: *title, isbn, price, publisher* and *year*. To search for two books (top-k=2) that contain "XML" in their titles and also include other specific information: expected price, published time and publisher, we can represent it as a tree pattern query q in Figure 2(a) where nodes are labeled by element tags, leaf nodes are labeled by tags and values, and edges are XPath axes (e.g. *pc* for parent-child, *ad* for ancestor-descendant). The root of the tree (shown in a solid circle) represents the distinguished node.

A naive solution to the above top-2 query is to retrieve the two most relevant books from each source and then select the more relevant ones by comparing their scores. However, this approach is not desirable for a large number of data sources due to amount of unnecessary processing cost. To solve this problem, we deploy XML schema information because a schema embodies to some extent the maximal structural information in the corresponding data sources that conform to the schema. For example, in Example 1, the two bookshop sources conform

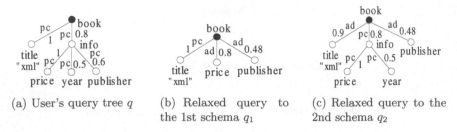

(a) User's query tree q (b) Relaxed query to the 1st schema q_1 (c) Relaxed query to the 2nd schema q_2

Fig. 2. Query Tree and Relaxed Forms to different schemas

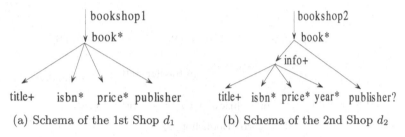

(a) Schema of the 1st Shop d_1 (b) Schema of the 2nd Shop d_2

Fig. 3. Bookshop Schema Example

to schema d_1 in Figure 3(a) and d_2 in Figure 3(b), respectively. Apparently, we can see that d_2 is more similar to the query q than d_1 in Figure 3. Consequently, we may expect to get more relevant results from S_2 than S_1 at the first instance. To achieve this, query relaxation can be used[13,14]. The top-2 query q against d_1 and d_2 can be relaxed into q_1 and q_2, which are shown in Figure 2(b) and Figure 2(c), respectively. In other words, q_2 is more similar to q than q_1, therefore, we may schedule to evaluate q_2 on S_2 first. If we are able to get enough qualified books from S_2, we do not need to evaluate q_1 on S_1 at all. By a qualified book, we mean that the book contains more required information than any book in all other sources. In the example, two approximate results B_1 and B_2 in S_2 are qualified because both of them contain more information than any book in S_1 with regards to the original query q. As such, these two books may be returned as the results for the top-2 query.

However, not every approximate result returned from q_2 is qualified. For example, if we are evaluating a top-3 query, B_4 in S_2 may not be qualified because it contains less information than B_3 in S_1. This is because that a result (XML fragment) conforming to a schema may not necessarily contain all structural information of the schema. For a schema represented in DTD, we are allowed to specify disjunctive semantics (denoted as "|") and optional semantics (denoted as "*" or "?"). In other words, XML documents conforming to the same schema may vary in their structures. If we fail to find any or enough qualified results in one data source, we may try the next most relevant data source, say S_1 in the example to work out the results or rest of the results, say B_3 in the example for the top-3 query.

From the motivating example, it is not hard to find that for a large number of data sources, the processing time can be reduced significantly by adaptively scheduling user's query on the most relevant data source at the time and progressively evaluating it according to schema information. Bearing this in mind, we design our upper/lower bounds and threshold for the BT-based strategy. This paper makes the following contributions:

- It proposes a BT-based scheduling strategy for efficiently searching top k results where we can skip most of the XML data sources according to schema information and also prune most candidates in each visited data source.
- It guarantees that results generated can be output immediately without waiting for the end of query evaluation.
- It provides stress tests and large-scale performance experiments that demonstrate the scalability and significant benefits of the proposed scheduling strategies.

The rest of the paper is organized as follows. In Section 2, we first introduce preliminary knowledge for query relaxation and then formalize top-k problem over a large number of XML data sources. Section 3 presents an overview of the BT-based strategy based on schema information and any scoring function that is used to measure the relevance of XML fragments with regards to a relaxed query. The detailed discussions of the BT-based scheduling strategy and its corresponding algorithms are given in Section 4. The experimental results are reported in Section 5. Finally, we discuss the related work and conclude the study in Section 6 and Section 7, respectively.

2   Preliminary and Problem Statement

XML Query Relaxation: In this paper, we represent XML queries as Tree Patterns [15] and allow users to add a weight on each edge to show their preferences on different path steps. We relax an XML query to different data sources based on the structural information provided in their corresponding DTDs while compute the changing weight of edges in query.

To guarantee that a relaxed query can be evaluated on a source, we need to collect information such as whether a node in the query appears in a DTD, the relationships between two query nodes, and the cardinality of a query node in a DTD, etc. Given a query q and a DTD d, the target of query relaxation is to find the relaxed query q' that is best suited to be evaluated on sources that conform to d by calling a set of relaxation operations: *deleting* a node, *generalizing* a pc-edge into a ad-edge and *promoting* a node [13,14]. The minimal requirement for the relaxed query q' is that the root node r is kept in q'. For example, the query q in Figure 2(a) does not match exactly with the schema d_1 of the 1st shop in Figure 3(a). Therefore, in order to provide considerate and reliable service for users, relaxing the query against DTDs for the conformed documents is strongly in demand. According to the structural information in the schema d_1, we firstly delete the nodes *info* and *year*. And then the nodes *price* and *publisher* are promoted under the distinguished node *book* where they are connected with ancestor-descendant (ad) edges. The relaxed query is shown in Figure 2(b).

Problem Statement: Consider a weighted top-k query q and a large number of data sources $\{S_1, S_2, ..., S_n\}$ that conform to different DTDs $\{d_1, d_2, ..., d_n\}$ respectively. Let $\{q_1, q_2, ..., q_n\}$ be the set of weighted relaxed queries of q with regards to the set of DTDs. Our aim in this paper is to efficiently search top k results by scheduling the evaluation of $\{q_1, q_2, ..., q_n\}$ over the set of data sources.

3   Overview

As stated in the problem statement, a set of relaxed queries $\{q_1, q_2, ..., q_n\}$ can be generated from the original query q based on the conforming DTDs $\{d_1, d_2, ..., d_n\}$, respectively. To start with, we require to rank the similarity between each q_i and q. During the query evaluation on a data source, we also check if a returned result is qualified or not. In this regard, we need a scoring function.

In a tree pattern query q, a user may specify, on an edge $e(v_1, v_2)$, how close v_1 and v_2 are associated with each other. To compute the weight of a query q, a naive function is to combine the weights of all edges in the query together where the edges are assumed to be independent from each other [14]. However, according to common understandings about XML queries, we find (1) the more path steps there are between two nodes, the less related the two nodes may be; (2) nodes lying on different paths are not related with each other, i.e., nodes are only related with their ancestors and descendants. Keeping these two features in mind, we introduce the concept of extended edge weight between a pair of nodes with *ad* relationship, $ad(v_i, v_j)$. The extended edge weight can be derived by multiplying weights along the path from v_i to v_j. Extended edge weights can be computed on the fly when required, or be calculated out beforehand, and then maintained dynamically.

Definition 1. *Extended Edge Weight:* *Let a WTPQ $q = (V, E, r, w)$, for two nodes $v_i, v_j \in V$ ($i \neq j$), the extended edge weight between v_i and v_j, denoted as $w_e(v_i, v_j)$, is defined as follows: if $ad(v_i, v_j)$ or $ad(v_j, v_i)$ holds, let $w_1, w_2, ..., w_n$ be weights on edges along the path from v_i to v_j, $w_e(v_i, v_j) = \prod_{t=1}^{n} w_t$; otherwise, $w_e(v_i, v_j) = 0$. Note that extended edge weight is a symmetric function on v_i and v_j, i.e., $w_e(v_i, v_j) = w_e(v_j, v_i)$.*

Based on Definition 1, we can score the weight of a tree pattern query q by summing all extended edge weights of q, i.e., $score(q) = \sum_{\forall v_i, v_j \in V, ap(v_i, v_j)} w_e(v_i, v_j)$ where $ap(v_i, v_j)$ means either $ad(v_i, v_j)$ or $pc(v_i, v_j)$. Similarly, we can measure the similarity of a potential result rooted at any node v in source S with q by summing the weights of those extended edges that match q. We denote this as $score(v, q)$. For q_1 and q_2 in Figure 2, we have $score(q_1) = w_e(book, title) + w_e(book, price) + w_e(book, publisher) = 1 + 0.8 + 0.48 = 2.28$ and $score(q_2) = w_e(book, title) + w_e(book, info) + w_e(book, publisher) + w_e(book, price) + w_e(book, year) + w_e(info, price) + w_e(info, year) = 0.9 + 0.8 + 0.48 + 0.8 \times$

$1 + 0.8 \times 0.5 + 1 + 0.5 = 4.88$. For the potential results, we have $score(B_1, q_2)$ = $score(q_2)$ because B_1 covers all edges of q_2; $score(B_2, q_2)$ = $score(q_2)$ − $w_e(book, price)$ − $w_e(info, price)$ = 3.08; Similarly, we have $score(B_3, q_1)$ = $score(q_1)$ = 2.28 and $score(B_4, q_2)$ = 1.7, and scores for B_5 and B_6 are less than that of B_4.

Come back to our problem, we are now able to compute the scores of the relaxed queries $\{q_1, q_2, ..., q_n\}$ as $score(q_1), score(q_2), ..., score(q_n)$ respectively. The BT-based strategy we propose is based on the concepts of upper/lower bounds [16] and threshold. We initialize the upper bound $U(i)$ and lower bound $L(i)$ of each source S_i as $score(q_i)$ and zero, respectively. To start our adaptive scheduling, we first choose the data source to be evaluated as S_{k1} if $U(k_1)$ = $max\{U(i)|1 \leq i \leq n\}$ (i.e., the highest upper bound) and the threshold σ = $U(k_2)$ = $max\{U(i)|1 \leq i \leq n \wedge i \neq k_1\}$ (i.e., the next highest upper bound). Then we start to evaluate q_{k1} on S_{k1} by probing an edge $e(v_1, v_2)$ of q_{k1} at a time. If $e(v_1, v_2)$ cannot be found in S_{k1}, $U(k_1)$ will be decreased; otherwise, $L(k_1)$ will be increased. The probing continues for next edge of q_{k1} until either $L(k_1) \geq \sigma$ or $U(k_1) < \sigma$. If $L(k_1) \geq \sigma$, all the candidates may become possible results depending on the value of k required in top-k. If the number of candidates equals to k, all the candidates can be returned as qualified results and the process stops; if the number of candidates is less than k, all the candidates can also be returned as qualified results (with the adjustment of k) but the probing process continues on S_{k1}; otherwise, more probing is required to refine the qualified results. If $U(k_1) < \sigma$, we will continue the process. The next data source to be evaluated will be S_{k2} and the threshold will be chosen based on the updated list of the upper bounds. The process stops until k results are returned.

In our example, S_2 is chosen as the the data source to be evaluated first because $U(2)$ = $score(q_2) > U(1)$ = $score(q_1)$ at the beginning. If we have a top-2 query, B_1 and B_2 in S_2 will be returned as qualified results because both $score(B_1, q_2)$ and $score(B_2, q_2)$ are no less than the threshold $U(1)$ = $score(q_1)$. If we have a top-3 query, we will first have B_1 and B_2 in S_2 as qualified results but the probing in S_2 continues until B_4 is met. At this time, $U(2)$ is decreased to $score(B_4, q_2)$ = 1.7, which is less than the threshold $U(1)$ = $score(q_1)$ = 2.28. So the next source to be evaluated is switched to S_1, and its threshold is $score(B_4, q_2)$ = 1.7. Since $score(B_3, q_1)$ (=2.28) is greater than the new threshold, B_3 becomes the third qualified result.

If the number of the data sources is large, we can avoid to evaluate most of the data sources based on the BT scheduling strategy. In addition, the qualified results can be returned immediately without waiting for all results to be determined.

4   Scheduling Strategy for Top-k Queries

In this section, the BT-based scheduling strategy for evaluating top-k queries will be discussed in detail. Specifically, in Section 4.1 we introduce data source and result determination properties that can be applied to schedule query evaluation over different data sources. Then, in Section 4.2 static/dynamic strategies are proposed to evaluate the edges, which can reduce unnecessary computational cost. Finally, in Section 4.3 we design a set of algorithms for the BT-based scheduling.

4.1 Data Source and Result Determination Properties

From the overview of the BT-based scheduling strategy in Section 3, we can get the following two properties.

Property 1. **Data Source Determination and Switching:** At any time of query evaluation, we always evaluate the data source S_{k1} that has the highest upper bound $U(k_1) = max\{U(i)|1 \leq i \leq n\}$. When an edge $e(v_1, v_2)$ in q_{k1} is evaluated on the data source S_{k1}, if it turns out that $e(v_1, v_2)$ cannot be successfully evaluated on the fragments rooted from all of the distinguished nodes of S_{k1}, then the upper bound $U(k_1)$ will be decreased by $U(k_1) = U(k_1) - score(v_2, q_{k1}) - w_e(v_1, v_2)$. Suppose that the threshold $\sigma = U(k_2)$, then we have:

- If the updated upper bound $U(k_1)$ is still larger than or equal to the threshold σ, then we need to continuously evaluate other edges in the query over the current data source S_{k1}.
- If the updated upper bound $U(k_1)$ becomes lower than the threshold σ, then the current data source S_{k1} needs to be suspended and query evaluation will be switched to the data source S_{k2}.

Property 2. **Result Determination:** When an edge $e(v_1, v_2)$ in q_{k1} is evaluated on the data source S_{k1}, if it turns out that $e(v_1, v_2)$ can be successfully evaluated on the fragments rooted from some of the distinguished nodes of S_{k1}, then the lower bound $L(k_1)$ will be increased by $L(k_1) = L(k_1) + w_e(v_1, v_2)$. Suppose that the threshold $\sigma = U(k_2)$ and the updated lower bound becomes larger than σ. Then we can affirm that some candidates generated so far in S_{k1} must be qualified as top-k results. We divide the set of candidates in S_{k1} into two groups G_1 that satisfies $e(v_1, v_2)$ and G_2 that does not, then the two groups will have different upper/lower bounds. Suppose that $\sigma \geq U(k_1)(G_2)$, then we have:

- If $|G_1| = k$, all the candidates in group G_1 can be returned as the qualified results and searching task would be terminated.
- If $|G_1| < k$, all the candidates in group G_1 can be returned as the qualified results and the k value will be decreased by $k = k - |G_1|$. Then the group G_2 should be evaluated if it is not suspended. If all the other groups in the data sources have been suspended, then we should switch to the next data source based on Property 1.
- If $|G_1| > k$, we will evaluate other edges in the query q_{k1} on G_1 to find the top k results.

Consider the top-2 query in our example again. We first evaluate q_2 on S_2 because $U(2)$ is larger than $U(1)$ (Property 1). Then we will choose some edges in q_2 to be evaluated, such as $(book, title)$, and $(book, info)$. All the edges can be found in the candidates of S_2. After that, the lower bound of the data source will increase to 1.7 (i.e., $L(2) = 0.9 + 0.8$). Then suppose we continue to evaluate $(info, year)$, at this point, we have two groups. The group G_1 of B_1 and B_2 satisfies $(info, year)$ while the group G_2 of B_4 does not. The lower bound of G_1 is increased to 2.6 $(L(2)(G_1) = 0.9 + 0.8 + 0.8 \times 0.5 + 0.5)$ while the upper bound of G_2 is decreased $(U(2)(G_1) = 4.48 - 0.9 = 3.58)$. When $(info, price)$ is evaluated, the upper bound of G_2 is further dropped to 1.78 $(3.58 - 0.8 \times 1 - 1)$. To this point, the 2 candidates in G_1 can be output as qualified results because $L(2)(G_1) > \sigma$ and $U(2)(G_2) < \sigma$. The process stops here.

4.2   Edge Selection and Unqualified Edge Reduction

According to the above properties, the relationships among $U(k_1)$, $L(k_1)$ and σ need to be checked during query evaluation. Obviously, changing the value of the three variables will produce different query evaluation sequences over the large number of data sources. But the changing is likely to be influenced to some extent by the next edge that will be evaluated. Therefore, the selection of next edge can also affect the performance of query evaluation. In this section, we first introduce three ways to determinate the next edge. Then we discuss how to filter unqualified edges during query evaluation.

Intuitively, there are three processing strategies for determining next edge: *random* i.e., the next edge can be evaluated at random; *min_weight* i.e., the edge with the minimal weight can be evaluated first and *max_weight* i.e., the edge with the maximal weight can be evaluated first. For the first two strategies, the possibility that some data sources would be visited frequently is likely to be increased to some extent, which may lead to unnecessary costs. For the third one, at every time the edge with the maximal weight is selected to be evaluated, so that it has the higher possibility to increase the score of $L(k_1)$ if the edge can be found, otherwise, the score of $U(k_1)$ would be decreased at most. Both of the trends are likely to locate the data sources as early as possible that can return the answers. Therefore, the last one would yield a better performance.

Besides next edge selection, the determination of selection range is another important factor to improve the performance of query evaluation. A simple method is to consider all edges together at the beginning and rank them based on the weights of their corresponding subtrees, denoted as *static* style. Although it makes next edge selection very easy in real application, some edges that should be filtered out based on the intermediate feedbacks have to be still evaluated. Therefore, an optimized approach is proposed to incrementally expand the selection range, denoted as *dynamic* style. The reason that the dynamic approach can do better than the static one depends on the disjunctive and optional semantics in DTD. For example, if an edge (e.g. x/y) in a query is specified as optional in a DTD and does not exist in a data source conforming to the DTD, then all the edges coming from the element y are not required to be evaluated because they cannot exist in the current fragments. Therefore, if we expand the selection range in a dynamic style, some edges can be filtered beforehand based the intermediate results.

Algorithm 1. BT-based Scheduling Strategy

---

input: a set of weighted relaxed queries $\{q_1, q_2, ..., q_n\}$ rooted at $\{r_1, r_2, ..., r_n\}$ and a set of data sources $\{S_1, S_2, ..., S_n\}$
output: top k results

1: call for the function *computingScore()* in Algorithm 4 to compute query weight
   as upper bound for each data source and denote the two highest upper bound as
   $U(k_1)$ and $\sigma = U(k_2)$ where $U(k_1) \geq U(k_2)$, $L(k_1) = 0$;
2: //$\{S_{k1}$ will be first evaluated$\}$
3: put all candidates in S_{k1} into group G;
4: **if** $ch(r_{k1}) \neq \phi$ **then**
5:     list l = sortAllChildNodes($ch(r_{k1})$);
6:     ScheEval(l, q_{k1}, G, $U(k_1)(G)$, $L(k_1)(G)$, σ) in Algorithm 2;

4.3 BT-Based Scheduling Strategy

We use Algorithm 1 to initialize query evaluation over the data source S_{k1}. Algorithm 4 is used to compute the weight of each subtree in the query q_{k1} and $score(q_{k1})$ is taken as the initial value of the upper bound $U(k_1)$. Based on the BT scheduling strategy, we always evaluate the query q_{k1} on the data source S_{k1} with the highest upper bound $U(k_1)$ at any point. Then all the candidates in the data source S_{k1} can be clustered initially into one group G by using index or other technologies. After that, we will evaluate the edges in the query q_{k1} in a similar *breadth-first search* (BFS). To this end, three functions are deployed during query evaluation: *sortAllChildNodes()* sorts a list of nodes based on the weight of the subtrees rooted at these nodes where any traditional sorting algorithm can be applied (e.g., Insert Sort in [17]); *mergesort()* merges two sorted lists like Merge Sort in [17], which can improve the sorting efficiency because the previous list has been sorted before; *getFirstNode()* gets the first node from the sorted list l. At last, we will call for Function *ScheEval()* to probe a data source. Based on the evaluated results, we determine how to proceed at next step. The detailed procedure is described in Algorithm 2.

In Algorithm 2, we first get a node v with the function *getFirstNode()* and evaluate the edge $e(v'parent, v)$ over the group of candidate nodes G. There are three possibilities. **(1)** Line 2 - 2: If no candidates in G satisfy the evaluated edge e, then the upper bound $U(k_1)(G)$ for the group will get a penalty $score(v)$, i.e., subtracting the score of the subtree rooted at v from the current upper bound. After that, we will compare the updated $U(k_1)(G)$ with the threshold σ. If $U(k_1)(G)$ is lower than σ, the current group will be suspended. And then previous groups or next data source will be evaluated depending on the conditions $\sigma = U(k_1)(G_x)$ or $\sigma = U(k_2)$, respectively. **(2)** Line 2 - 2: If all candidates in G satisfy the evaluated edge e, then the lower bound $L(k_1)(G)$ for the group will be increased by summing the extended weight $w_e(v'parent, v)$ of the edge. If $L(k_1)(G)$ is higher than σ, it means that the current group contains part or all results that can be determined by Function *determineCandidates()* in Algorithm 3. Otherwise, we open the child nodes of the node v to expand the current range of edges because the group of candidates can not be determined based on

---

Algorithm 2. ScheEval(a list l, query q, group G, $U(k_1)(G)$, $L(k_1)(G)$, σ)

1: **while** $l \neq \phi$ **do**
2:   node v = getFirstNode(l) and delete the node v from the list l;
3:   evaluate the edge $e(v'parent, v)$ in query q over the candidates G;
4:   **if** No candidates in G satisfy the edge e **then**
5:     $U(k_1)(G) = U(k_1)(G) - score(v, q) - w_e(e)$;
6:     **if** $U(k_1)(G) < \sigma$ **then**
7:       suspend the current group G;
8:       **if** $\sigma == U(k_1)(G_x)$ **then**
9:         Switching to probe the group G_x in the current data source S_{k1};
10:        ScheEval($l, q, G_x, U(k_1)(G_x), L(k_1)(G_x)$);
11:      **else**
12:        Switching to the next data source S_{k2} due to $\sigma = U(k_2)$;
13: **else if** All candidates in G satisfy the edge e **then**
14:    $L(k_1)(G) = L(k_1)(G) + w_e(v'parent, v)$;
15:    **if** $L(k_1)(G) \geq \sigma$ **then**
16:      determineCandidates();
17:    **else**
18:      list l' = sortAllChildNodes($ch(v)$) and list l = mergeSort(l, l');
19: **else**
20:    //{Partial candidates in G satisfy the edge e}
21:    divideGroup(e, q, G) into two groups G_1 that satisfies the edge and G_2 that doesn't and putActiveGroup(G_2);
22:    $U(k_1)(G_2) = U(k_1)(G_2) - score(v, q) - w_e(v'parent, v)$;
23:    $L(k_1)(G_1) = L(k_1)(G_1) + w_e(v'parent, v)$;
24:    **if** $U(k_1)(G_2) > \sigma$ **then**
25:      $\sigma = U(k_1)(G_2)$;
26:      //{The group G_2 in k_1 data source would be evaluated at next step.}
27:    **if** $L(k_1)(G_1) \geq \sigma$ **then**
28:      determineCandidates();
29:    **else**
30:      list l' = sortAllChildNodes($ch(v)$) and list l = mergeSort(l, l');

---

the current edge e so far. **(3)** Line 2 - 2: Most of the time, only part candidates in G satisfy the edge e, e.g., a subgroup G_1 of candidates satisfy while another subgroup G_2 of candidates do not. We use the function $divideGroup(e, q, G)$ to divide the group G of candidates into G_1 and G_2. Then we compute the upper bound, lower bound for each group. For G_1, its upper bound $U(k_1)(G_1)$ does not change, but its lower bound $L(k_1)(G_1)$ will increase. For G_2, its upper bound $U(k_1)(G_2)$ will decrease, however its lower bound $L(k_1)(G_2)$ keeps unchanged. Obviously, we have $U(k_1)(G_1) > U(k_1)(G_2)$. Therefore, we prefer searching in group G_1 to G_2 while cache group G_2 with Function $putActive$-$Group()$. If $\sigma < U(k_1)(G_2)$, we should take $U(k_1)(G_2)$ as the new threshold for the current group G_1. And if $L(k_1)(G_1)$ is greater than or equal to the updated threshold σ, we will call for Function $determineCandidates()$ in Algorithm 3. Otherwise, a new edge need to be evaluated on the current group of candidates.

Algorithm 3. Function: determineCandidates()

1: **if** $|G| = k$ **then**
2:     return k results while **Stop** searching;
3: **else if** $|G| < k$ **then**
4:     return λ results and k = k - $|G|$;
5:     **if** $\sigma == U(k_1)(G_x)$ **then**
6:         Switching to probe the group G_x in the current data source S_{k1};
7:         ScheEval(l, q, G_x, $U(k_1)(G_x)$, $L(k_1)(G_x)$);
8:     **else**
9:         Switching to the next data source S_{k2} due to $\sigma = U(k_2)$;
10: **else**
11:     list $l' = $ sortAllChildNodes($ch(v)$) and list $l = $ mergeSort(l, l');
12:     ScheEval(l, q, G, $U(k_1)(G)$, $L(k_1)(G)$);

Algorithm 3 can be designed to determine the correct ones from the group if we find that a group of candidates in a data source would contain the correct answers for top-k query. There are three ways to process the candidates in G. (1) If $|G| = k$, the group of candidates are correct answers for top-k query and searching is terminated; (2) If $|G| < k$, the group of candidates are part of the correct answers and the value of k will be decreased by $k = k - |G|$. At next step, we would probe the previous groups G_x in the current data source S_{k1} if we have $\sigma = U(k_1)(G_x)$ or switch to the next data source S_{k2} if we have $\sigma = U(k_2)$ (3) Otherwise, we will expand the edges and continuously evaluate them over the group G for determining the k best ones.

Algorithm 4. ComputingScore()

---

input: a weighted query rooted at r
output: a query that every subtree is marked with scores

1: push(the root r, a stack S);
2: **while** the stack is not empty $S \neq \phi$ **do**
3:     $v = $ getStackTop(S);
4:     $existEdgeScore = $ getEdgeScore(v);
5:     **if** $ch(v) \neq \phi$ **then**
6:         **for all** $v_c \in ch(v)$ **do**
7:             $newEdgeScore = $ getEdgeScore(v_c);
8:             $currentEdgeScore = existEdgeScore \times newEdgeScore$;
9:             updateEdgeScore(v_c, $currentEdgeScore$);
10:            push v_c into the stack S;
11:            **ComputingScore**(v_c);
12:    **else**
13:        pop(a node, a stack S);
14:        $v_x = $ getStackTop(S);
15:        $xEdgeScore = $ getEdgeScore(v_x);
16:        updateEdgeScore(v_x, $xEdgeScore + existEdgeScore$);

Algorithm 4 is used to mark the weight for each subtree in *deepth-first search* style. For each internal node v (i.e., $ch(v) \neq \phi$), we should push it into the stack S while update its score by computing the extended edge weight between the node v and its ancestor. For each leaf node or internal node that its child nodes have been processed, we will pop the node from the stack S while update its parent's score by propagating its score to its parent. Two important functions *getEdgeScore()* and *updateEdgeScore()* are used to retrieve and update the score of each node, respectively.

5   Experiments

The presented algorithms for the BT strategy are implemented in a Java prototype using JDK 1.4. B+-tree indexes are used to access the nodes in each data source. Wutka DTDparser[1] is used to analyze the source DTDs and extract their structural information. We run our experiments on an Intel P4 3GHz PC with 512M memory.

Table 1. Designed Queries

q_1:	//item [./description /parlist]
q_2:	//item [./description /parlist /mailbox /mail [./text]]
q_3:	//item [./mailbox /mail /text [./keyword and ./xxx] and ./name and ./xxx]

Dataset and Queries: We use XMark XML data generator[2] to generate a number of data sets, of varying sizes and other data characteristics, such as the fanout (MaxRepeats) and the maximum depth, using the *auction.dtd* and changed versions by deleting some nodes. We also use the XMach-1[3] and XMark[4] benchmarks, and some real XML data. The results obtained are very similar in all cases, and in the interest of space we present results only for the largest auction data set that we generated. We evaluate the presented algorithms using the set of queries shown in Table 1 where the symbol "xxx" is added as noise node that do not appear in the DTD. In our query set, we consider the structural difference between the query and the DTD, such as the edge "parlist/mailbox" does not exist in the DTD. It will be adjusted by calling for previous query relaxation. We also take into account two semantics in DTD, such as the edge "description/parlist" satisfies disjunctive semantics and the nodes "mail" and "text" satisfy optional semantics.

Test Results: In our experiments, we also implement the Naive strategy that first retrieves top k matches from each data source and then selects the k most

---

[1] Wutka DTD parser. http://www.wutka.com/dtdparser.html

[2] Xmark XML data generator. http://monetdb.cwi.nl/xml/index.html

[3] XMach-1. http://dbs.unileipzig. de/en/projekte/XML/XmlBenchmarking.html

[4] The XML benchmark project. http://www.xml-benchmark.org

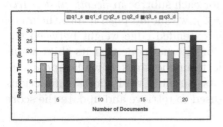

Fig. 4. Static Sort vs. Dynamic Sort

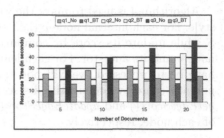

Fig. 5. No Schedule vs. BT Schedule

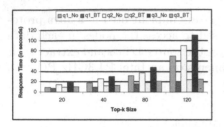

Fig. 6. Varying Top-k Size

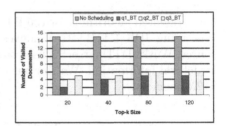

Fig. 7. Varying Top-k Size

relevant ones from the intermediate results. Our test results show that the BT scheduling strategy is faster than the Naive strategy to search top-k matches over multiple data sources. Especially, when the number of data sources or the value of top-k are large, more benefits would be gained.

Figure 4 shows that dynamic sort-based BT scheduling strategy can improve the performance more than static sort-based BT scheduling strategy, in terms of evaluation of unqualified edges for some documents where the three queries are evaluated over 5, 10, 15 and 20 number of XML documents respectively and top-k is set as 80. In the following paragraph, we mainly choose the experiments about dynamic sort-based BT scheduling strategy in different conditions. Figure 5 shows BT scheduling strategy outperforms Naive strategy greatly. Two appealing features can be obtained: one the one hand, the larger the number of XML documents to be searched, the more benefits the BT scheduling strategy can gain; on the other hand, the BT scheduling strategy has good scalability, i.e., the increasing trends will become slow after the number of XML documents is relatively large. For example, the trends evaluating the three queries over 10, 15, and 20 documents are much slower than the trend between 5 and 10 documents. This is because a larger number of documents have higher possibility to contain noise documents. Figure 6 and Figure 7 illustrate the performance when we vary the size of top-k value across 20, 40, 80 and 120 where all the three queries are evaluated over 15 documents. From Figure 6, the both strategies can gain similar time cost when the top-k value is small. But the gap between the BT scheduling strategy and the Naive strategy will become much larger when top-k is 120. In addition, Figure 7 shows the number of documents that need to be visited in order to answer the three queries. Obviously, for the Naive strategy, all the documents require to be checked. However, for the BT scheduling strategy, only part of the documents are visited during query evaluation. Furthermore, the same number of documents are traversed for q_2 and q_3 when top-k is 80 or 120. This is because some elements in query like *mailbox* are distributed in the same documents when we design our data sets.

6   Related Work

Top-k query processing has been extensively studied in the literature. In relational databases, existing work has focused on extending the evaluation of SQL queries for top-k processing. None of these works follows an adaptive query evaluation strategy. Carey and Kossmann [1] optimize top-k queries when the scoring is done through a traditional SQL order by clause, by limiting the cardinality of intermediate results. Other works [2,3] use statistical information to map top-k queries into selection predicates which may require restarting query evaluation when the number of answers is less than k. Over multiple repositories in a mediator setting, Fagin et al. propose a family of algorithms [4,5,6], which can evaluate top-k queries that involve several independent subsystems, each producing scores that are combined using arbitrary monotonic aggregation functions. These algorithms are sequential in that they completely process one tuple before moving to the next tuple. The Upper [7], MPro [8] and TPUT [9] algorithms show that interleaving probes on tuples results in substantial savings in execution time. In addition, Upper [7] uses an adaptive per-tuple probe scheduling strategy, which results in additional savings in execution time when probing time dominates query execution time.

Recently in [10,11], top-k keyword queries for XML have been studied via proposals extending the work of Fagin et al., [5,18] to deal with a bag of single path queries. Adaptivity and approximation of XML queries are not addressed in their work. Marian et. al. in [12] explore an adaptive top-k query processing strategy in XML, which permits different query plans for different partial matches and maximizes the best scores. Based on the intermediate results, the irrelevant answers for the top-k query may be pruned as early as possible. But they do not discuss top-k query evaluation over a larger number of different data sources. Furthermore, the correct results can not be determined until all candidates are evaluated.

Different from previous work, we study top-k search over a large number of XML data sources and focus on issues such as exploring the BT-based scheduling strategy that not only skips many data sources without probing for top-k evaluation, but also prunes the unqualified distinguished nodes of each visited data source. Additionally, we also deploy an adaptive query relaxation strategy to filter out some unqualified edges in the query for some data sources based on schema information, which can further improve query evaluation efficiency.

7 Conclusions

The primary contribution of this paper lies in the BT-based scheduling strategy that we proposed. Based on the strategy, we are able to avoid the evaluation of big number of data sources, and prune unqualified results in the visited data sources. Besides, the strategy also satisfies monotonic feature for returning qualified results. The experimental results demonstrated the BT scheduling strategy can gain more benefits when the value of top-k and the number of data sources are large. Additionally, the results also shown that the BT scheduling strategy can skip most of data sources during query evaluation. Therefore, it is appropriate and practical for the BT scheduling strategy to be applied to XML searching system.

Acknowledgments. This work was supported by grants from the Australian Research Council Discovery Project (DP0559202) and the Research Grant Council of the Hong Kong Special Administrative Region, China (CUHK418205).

References

1. Carey, M.J., Kossmann, D.: On saying enough already! in sql. In: SIGMOD Conference, pp. 219–230 (1997)
2. Bruno, N., Chaudhuri, S., Gravano, L.: Top-k selection queries over relational databases: Mapping strategies and performance evaluation. ACM Trans. Database Syst. 27(2), 153–187 (2002)
3. Chen, C.-M., Ling, Y.: A sampling-based estimator for top-k query. In: ICDE, pp. 617–627 (2002)
4. Fagin, R.: Fuzzy queries in multimedia database systems. In: PODS, pp. 1–10. ACM Press, New York (1998)
5. Fagin, R., Lotem, A., Naor, M.: Optimal aggregation algorithms for middleware. In: PODS, pp. 102–113. ACM Press, New York (2001)
6. Fagin, R.: Combining fuzzy information: an overview. SIGMOD Record 31(2), 109–118 (2002)
7. Marian, A., Bruno, N., Gravano, L.: Evaluating top-k queries over web-accessible databases. ACM Trans. Database Syst. 29(2), 319–362 (2004)
8. Chang, K.C.-C., won Hwang, S.: Minimal probing: supporting expensive predicates for top-k queries. In: SIGMOD Conference, pp. 346–357 (2002)
9. Cao, P., Wang, Z.: Efficient top-k query calculation in distributed networks. In: PODC, pp. 206–215 (2004)
10. Theobald, M., Schenkel, R., Weikum, G.: An efficient and versatile query engine for topx search. In: VLDB, pp. 625–636 (2005)
11. Kaushik, R., Krishnamurthy, R., Naughton, J.F., Ramakrishnan, R.: On the integration of structure indexes and inverted lists. In: SIGMOD Conference, pp. 779–790 (2004)
12. Marian, A., Amer-Yahia, S., Koudas, N., Srivastava, D.: Adaptive processing of top-k queries in Xml. In: ICDE, pp. 162–173 (2005)
13. Schlieder, T.: Schema-driven evaluation of approximate tree-pattern queries. In: Jensen, C.S., Jeffery, K.G., Pokorný, J., Šaltenis, S., Bertino, E., Böhm, K., Jarke, M. (eds.) EDBT 2002. LNCS, vol. 2287, pp. 514–532. Springer, Heidelberg (2002)

14. Amer-Yahia, S., Cho, S., Srivastava, D.: Tree pattern relaxation. In: Jensen, C.S., Jeffery, K.G., Pokorný, J., Šaltenis, S., Bertino, E., Böhm, K., Jarke, M. (eds.) EDBT 2002. LNCS, vol. 2287, pp. 496–513. Springer, Heidelberg (2002)
15. Lakshmanan, L.V.S., Ramesh, G., Wang, H., Zhao, Z(J.): On testing satisfiability of tree pattern queries. In: VLDB, pp. 120–131 (2004)
16. Bruno, N., Gravano, L., Marian, A.: Evaluating top-k queries over web-accessible databases. In: ICDE, p. 369 (2002)
17. Cormen, T.H., Leiserson, C.E., Rivest, R.L., Stein, C.: Introduction to Algorithms, 2nd edn. MIT Press, Cambridge (2001)
18. Fagin, R.: Combining fuzzy information from multiple systems. In: PODS, pp. 216–226 (1996)

Example-Based Robust DB-Outlier Detection
for High Dimensional Data

Yuan Li[1] and Hiroyuki Kitagawa[2]

[1] Graduate School of Systems and Information Engineering
[2] Center for Computational Sciences
University of Tsukuba, Tennoudai 1–1–1, Tsukuba, Ibaraki, 305–8573 Japan
ly0007@kde.cs.tsukuba.ac.jp, kitagawa@cs.tsukuba.ac.jp
http://www.kde.cs.tsukuba.ac.jp

Abstract. This paper presents a method of outlier detection to identify exceptional objects that match user intentions in high dimensional datasets. Outlier detection is a crucial element of many applications like financial analysis and fraud detection. Scholars have made numerous investigations, but the results show that current methods fail to directly discover outliers from high dimensional datasets due to the curse of dimensionality. Beyond that, many algorithms require several decisive parameters to be predefined. Such vital parameters are considerably difficult to determine without identifying datasets beforehand. To address these problems, we take an Example-Based approach and examine behaviors of projections of the outlier examples in a dataset. An example-based approach is promising, since users are probably able to provide a few outlier examples to suggest what they want to detect. An important point is that the method should be robust, even if user-provided examples include noises or inconsistencies. Our proposed method is based on the notion of DB- (Distance-Based) Outliers. Experiments demonstrate that our proposed method is effective and efficient on both synthetic and real datasets and can tolerate noise examples.

Keywords: Outlier, DB-Outlier, High-dimensional Data, Example.

1 Introduction

Outlier detection aims to discover abnormal objects in datasets. Unusual objects often involve useful information that may be more interesting than common objects. Techniques of outlier detection can be used in many popular applications such as credit card fraud, money laundering and the analysis of statistic data of professional athletes. Such applications generally process high dimensional datasets. Most traditional research attempts to define algorithms based on distance [3] or density [7]. However, it has been indicated that in high dimensional space, data distribution is sparse, which means that every data point becomes a good outlier candidate based on the definition of distance or density. Consequently, traditional algorithms become impractical when processing high dimensional datasets.

J.R. Haritsa, R. Kotagiri, and V. Pudi (Eds.): DASFAA 2008, LNCS 4947, pp. 330–347, 2008.

It has been certified that meaningful outliers are likely to be defined by examining the behavior of the data in low dimensional projections [5]. In other words, we can simplify high dimensional data problems to low dimensional problems by investigating the behaviors of projections, namely subspaces of a dataset. Therefore, we employ the Subspace-Based method [5,9] to address the curse of dimensionality.

On the other hand, in most cases, a few decisive predetermined parameters are necessary for outlier detection algorithms. Such parameters are usually key factors in detecting outliers, so correspondingly they are quite difficult to determine beforehand. Compared with predetermining vital parameters, users can more easily provide outlier examples containing their intentions. The Example-Based method is shown to be promising in discovering hidden user views of outliers [1]. Hence, outlier examples are good useable inputs as a substitute for vital parameters. Further, we try to use outlier examples to mitigate the curse of dimensionality.

We propose a method that first finds an optimal subspace where outlier examples are outstanding more significantly than in any other subspaces. Then it reports objects having similar characteristics to examples in the optimal subspace as outliers. Actually, outlier detection is operated on the optimal subspace with lower dimensionality.

In this paper, we apply the concept of Distance-Based outliers to identify DB-Outliers corresponding with users' intentions. The Distance-Based method is a simple, classic and commonly utilized algorithm. In addition, distance is also a frequently used standard in many technical and knowledge areas.

Our previous work [1] proposed an approach to detecting outliers with examples, but did not consider high dimensional datasets. Another work [2] targeted high dimensional datasets based on a Grid-Based approach inspired by [5,9]. The Grid-Based approach is efficient, but cannot assure the quality of outcomes. The comparison between the method [2] and the one proposed here is shown in Section 4.3. Our primary idea of Example-Based DB-Outlier detection in high dimensional datasets and preliminary experiments are demonstrated in [10]. Preliminary experiments demonstrated the simple primary approach worked well in discovering 2-D (two dimensional) optimal subspace. However, if noise is interfused in outlier examples, the method will fail to pick out the optimal subspace.

Here we present an improved robust approach that can abide noise outlier examples. We test our advanced proposed method intensively on both synthetic and real datasets, and conduct experiments comparing our method with our previous work [2] in terms of time and quality. Our intensive experiments have brought comprehensive results that reveal our method to be effective and efficient in detecting outliers based on user viewpoints in high dimensional datasets.

2   Distance-Based Outlier

Since our goal is to detect DB-Outliers, we must give a general introduction on DB-Outlier at the beginning. In this section, we mainly present the definition and two detection algorithms of DB-Outliers.

2.1 Fundamental Concept

The notion of DB-Outlier studied here is the same as Knorr and Ng's work [3]:

An object O in a dataset T is a DB(p, D)-Outlier if at least fraction p of the objects in T lie greater than distance D from O.

The parameter p is the minimum fraction of objects in a dataset that must reside outside an outlier's D-neighborhood. For convenient explanation and calculation, we employ another parameter M, which denotes the maximum portion of objects within an outlier's D-neighborhood as follows:

$$M = N(1 - p) \qquad N : datasize \tag{1}$$

The detection of DB-Outliers can be elucidated as detecting those objects that have no more than M neighbors in their D-neighborhoods.

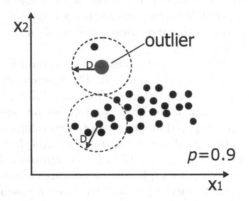

Fig. 1. DB-Outlier

Here is an example illustrated in Figure 1 to clarify the notion of DB-Outlier. This example describes a 2-D space. There are 30 points (viz. data size N=30), D equals to the length of leader lines, and parameter $p=0.9$. Here, $M = N(1-p) = 30 * (1 - 0.9) = 3$. According to the DB-Outlier definition, if there are no more than 3 points in the dotted circles (D-neighborhoods), the central points can be regarded as outliers. There are only two points in the circle above, so in this example, the big point is an outlier.

2.2 Detecting Algorithms

Two algorithms are used to mine DB-Outliers in our proposed method. One is the simple algorithm; the other is the Cell-Based algorithm [3].

– Simple Algorithm

The simple algorithm executes based on the definition of DB-Outlier. It examines the D-neighborhood centered at each object. As soon as M data points are found in an object's D-neighborhood, algorithm marks this object as a non-outlier and changes to examine another object. If, after checking all the mutual distances of an object, it finds no more than M D-neighbors, this object is recognized as a DB-Outlier. This algorithm is easy. However, it takes much time to calculate mutual distances.

– Cell-Based Algorithm

The Cell-Based algorithm (see [3] for detailed direction) uses a cell structure's properties to rule out non-outliers quickly. It quantizes all data objects into a space that has been partitioned into cells. Such cells have a particular size correlating with parameter D. The majority of non-outliers can be eliminated from outlier candidates by counting the number of objects in these cells. Hence, a Cell-Based algorithm can save a lot of processing time; nevertheless, the Cell-Based algorithm shows worse performance than a simple algorithm when the dimensionality of a dataset is high (more than 4 [3]). It implies that if the optimal projection's dimensionality is lower than 5, we prefer the Cell-Based algorithm to identify DB-Outliers; otherwise, we choose the simple algorithm.

3 Proposed Method

We demonstrated our primary idea of Example-Based DB-Outlier detection in high dimensional datasets in [10]. However it can only work with good outlier examples. Thus, we have amended our proposal to strengthen our proposed method. Our target in this paper is to propose a robust approach to detecting DB-Outliers in high dimensional datasets. In this section, we present our improved method, which can tolerate noise interfusing in outlier examples.

The main inputs of our method are outlier examples. First, we look for the most suitable subspace where outlier examples are isolated more significantly than in any other subspace. Such optimal subspace generally has low dimensionality because outlier examples have only a portion of the exceptional attributes in the real world. If noise is interfused in outlier examples, we prefer that the subspace be the optimal subspace, where as many outlier examples perform abnormal appearances as possible. After discovering the optimal subspace, we seek objects that also reside in sparse areas, just as with outlier examples in this subspace. Such objects having similar characteristics to outlier examples will be reported as outliers.

There is a problem in looking for the most suitable subspace: before examining all subspace candidates whose dimensionalities vary from 1 to the total dimensions, we cannot confirm which subspace is the one required. Let the full dimensionality of a dataset be d, and the dimensionality of the optimal subspace be k, $\binom{d}{k}$ combination of candidates should be examined. Because of the uncertain value of k, there are altogether $\sum_{k=1}^{d} \binom{d}{k} = 2^d - 1$ possible combinations that should be checked in seeking out the most appropriate combination. If d is big, it becomes exhausting or even impossible to discover such desired subspace by examining all candidate combinations. Therefore, the brute force method, which checks all candidate combinations, is infeasible when dealing with high dimensional datasets. For this reason, we exploit a Genetic Algorithm (GA) [8] to select the optimal subspace in less time.

3.1   Summarization of GAs

Genetic algorithms simulate the processes of survival of the fittest, as in Charles Darwin's theory of evolution. This variety of algorithms creates competition among a population of solutions and follows the principles of survival of the fittest. There are three phases in GAs named selection, crossover, and mutation. GAs go over the three steps and try to identify the best solution to a certain problem. Each solution (in this paper a solution stands for a subspace) can be evaluated by a Fitness Value function and solutions that have bigger Fitness Value should generate more copies. First, some solutions are selected as parents, and then crossed over under rules to bring forth better children whose Fitness Values are superior to parents. Mutation happens with extremely low probability. However, it may create unexpected solutions that have much bigger Fitness Values than their parents. GAs repeat the three steps until the convergence criterion is reached. In our proposed method, we use a GA instead of brute force to seek the optimal subspace.

3.2   Procedure of Proposed Method

This section overviews the procedures of our proposed method. There are three main steps in our proposed method:

1. **Detecting Most Suitable Subspace with a GA**
   Except noises, outlier examples should be recognized as outliers in the optimal subspace. Therefore, at first, we detect a subspace in which most outlier examples reside lonelier than in any other subspaces with a Genetic Algorithm.
2. **Parameter Selection**
   After discovering the most suitable subspace, some pairs of parameters p and D are selected automatically to detect DB-Outliers in this subspace. Such outliers have similar characteristics to outlier examples.
3. **Outlier Report**
   With different parameters (p, D), we may receive different results in the same subspace. Hence, in the parameter section step, we produce several pairs of parameters (p, D) in order to get precise results. In the last step, we not only report outliers detected in the most suitable subspace, but also report the goutlier-nessh degree of each outlier.

3.3   Detecting Most Suitable Subspace with a GA

This section introduces how to find the optimal subspace. This task is accomplished by a GA, so next we present how the GA works.

(1) Fitness Value Function

A Fitness Value function is used in evaluating solutions. It is relevant to screening out preferable solutions. Defining a proper Fitness Value function is a crucial component of the GA. Here, we demonstrate the calculation of Fitness Value.

Let Nm denote the number of D-neighbors. When the value of parameter D grows longer from 0, the number of D-neighbors, namely Nm of an isolated point, has poor growth for a certain time. In contrast, the value of Nm of an ordinary object mushrooms as the D grows from the word go, since the density of a normal object's D-neighborhood is high. According to this property, we build a Distance-Nm chart for each solution subspace. A D-Nm (Distance-Nm) line describes the relationship between distance D and Nm of an object. Isolated objects have different types of D-Nm lines from the normal type. In the optimal subspace, outlier examples are isolated, so their D-Nm lines stay away from those of normal objects. Therefore, we make good use of such a property to define the Fitness Value function.

Figure 2 describes a Distance-Nm space. In this figure, two curved lines stand for D-Nm lines of an outlier example and an ordinary object, respectively. The more the outlier examples are isolated, the farther the two lines keep away from each other. Let A_{NO} be the area encircled by D-Nm lines of outlier examples and normal objects. We use the value of A_{NO} to measure the goodness of outlier examples. That means A_{NO} can also be used to examine subspaces.

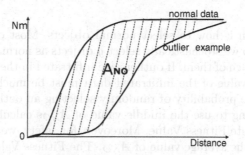

Fig. 2. Distance-Nm Lines

If noise is interfused in outlier examples, we cannot find such a subspace in which all the outlier examples exhibit abnormal behavior. The A_{NO} of a noise is extremely small, so it differs significantly from that of isolated examples. Taking noise into account, the optimal subspace should contain as many isolated outlier examples as possible. Further, it should have a big A_{NO}. Also, users usually wish to find a subspace with lower dimensionality. Consequently, we define the Fitness Value function as follows:

$$f = \frac{A_{NO} * C_o}{k * C_e} \tag{2}$$

f: Fitness Value.

A_{NO}: *the area encircled by D-Nm lines of outlier examples and normal objects.*

C_o: *number of outlier examples isolated in the subspace.*
C_e: *number of all outlier examples.*
k: *dimensionality of a dataset.*

Each outlier example has a D-Nm line. If the D-Nm line of an outlier example is intermixed with normal objects' D-Nm lines, such an example will be recognized as noise. Optimal subspace should have the least noise. For those isolated outlier examples, we choose only the maximum values of D-Nm lines to calculate A_{NO}. The explanation graphic is shown in Figure 3.

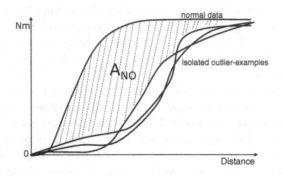

Fig. 3. Distance-Nm Lines of Examples

Another problem is how to select normal objects. Most objects are normal data in datasets, so we stochastically select g objects as normal objects, and calculate A_{NO} with each of them. If outliers are infiltrated in the selected "normal" objects, the A_{NO} value of the infiltrated data must be much smaller than the others. Besides, the probability of randomly selecting an outlier is actually low. Thus, it is promising to use the middle value of A_{NO}s calculated with selected g objects to compute Fitness Value. Moreover, for safety, we do the same work q times to obtain the average value of A_{NO}. The Fitness Value function can be reformed as follows:

$$f = \frac{C_o * \overline{A_{NO}}}{C_e * k} = \frac{C_o * \sum_{n=1}^{q} Mid(A_{NO})(n)}{C_e * q * k} \tag{3}$$

$Mid(A_{NO})(n)$: *The nth middle value of A_{NO}s.*

It is meaningless to calculate the real irregular area encircled by D-Nm lines. We uniformly select a certain number of points $(D_1, D_2, ..., D_m)$ along the Distance line in each Distance-Nm space, and then add up all the gaps between D-Nm lines of isolated outlier examples and normal objects at each distance point. We regard the summation as the value of A_{NO}.

In a dataset, different attributes have different value scopes. To treat every attribute fairly, we normalize each attribute to a certain range such as 0~1 beforehand. If the dimensionality of a subspace is k, the mutual distance of two data points is by far $\sqrt{k}$, as we use Euclidean metric distance function ($distance = \sqrt{(x_1 - y_1)^2 + (x_2 - y_2)^2 + ... + (x_n - y_n)^2}$). Figure 4 illustrates the calculation of A_{NO}. There are m points selected on the distance line. In this illustration, the normalized distance line is no longer than $\sqrt{k}$. Only the double arrowed lines need to be added to calculate A_{NO} (viz. $A_{NO} = gap_{D1} + gap_{D2} + gap_{D3} + gap_{D4} + ... + gap_{Dm-2} + gap_{Dm-1} + gap_{Dm} = 0 + gap_{D2} + gap_{D3} + gap_{D4} + .. + gap_{Dm-2} + 0 + 0$).

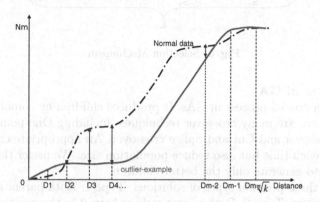

Fig. 4. Calculation of A_{NO}

(2) Selection of GA

A solution denoting a subspace can be represented by a binary code string. Each position of the code string stands for a dimension. Therefore, a subspace is composed of those dimensions whose position values are 1. For instance, string 010100 designates a two dimensional subspace constructed by the second and the fourth dimensions of a 6-D dataset.

Let the population size be Z. We stochastically select Z solutions as parents for the next generation. A rank selection mechanism is used to choose parents. All the solutions are ranked in descending order of their Fitness Values. We can then draw a line whose length equals to the sum of all solutions' Fitness Values. Each solution occupies a section of the Fitness Value line. In other words, the length of a section of a solution is the same as the solution's Fitness Value. Solutions with bigger Fitness Values obviously occupy longer sections. If L denotes the length of the Fitness Value line, L can be calculated by function: $L = f_1 + f_2 + ... + f_Z$. Except for the first step, the selection mechanism moves along the Fitness Value line with equivalent steps. The selection mechanism stops at a section after each step and picks the solution representing this section as a parent. The standard step size is calculated by the function below:

$$step = \frac{L}{Z+1} = \frac{f_1 + f_2 + ... + f_Z}{Z+1} \tag{4}$$

The first step size is a random number that is less than the standard step size. This selection mechanism selects solutions with bigger Fitness Values easily. Figure 5 shows the operation of the selection mechanism.

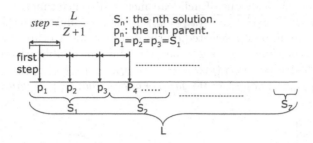

Fig. 5. Selection Mechanism

(3) Crossover of GA

Crossover is a crucial process in GAs. It produces children by combining parent solutions. There are many crossover techniques, including One-point crossover, Uniform Crossover and Cut and Splice crossover. An appropriate crossover not only can retrench time but also reduce population size. We prefer the optimized crossover [2] to generate only the better children.

There are three position types of solutions. Type 1: Both parent strings have 1 in the positions. Type 2: Both parent strings have 0 in the positions. Type 3: Otherwise. For the crossover rule, children must keep the same position values as the parent in Type 1 and Type 2. Therefore, in the crossover section, children are produced from the possible combinations of positions in Type 3. As the Fitness Value decreases with dimensionality, we begin to examine possible combinations from low dimensionality. We cite the pseudo code of the optimized crossover [2] to demonstrate the process in Figure 6.

(4) Mutation of GA

Mutation happens with very low probability and creates unexpected children, whereby it may produce some surprising solutions that are far superior to their parents. In our mutation step, we randomly reverse some position values to

```
Input: Two parent strings: p1, p2
Output: A child string: child
Algorithm:
begin
        T1:= Set of positions where p1=p2=1;
        T2:= Set of positions where p1≠p2;
        s:= Solution(T1);
        While s is worse than p1 and p2 do begin
                for each Vᵢ∈T2
                        Qᵢ:= Solution(Vᵢ∪T1);
                s:= Best solution in all the Qᵢ;
                T2:= T2-(1_positions(s)-T1);
                T1:= 1_positions(s);
                end;
        child:= s;
        return child;
end
```

```
Algorithm: Solution(T)
begin
        return a solution whose values of
        positions in T are 1 and 0 for the rest.
end
Algorithm: 1_positions(s)
begin
        return a set of positions where the
        values are 1 in s;
end
```

Fig. 6. Optimized Crossover Algorithm [2]

make new solutions. Every position of a solution string has a probability of being inversed. Since the probability is low (we set the probability 0.001 in our experiments), in most cases, the positions keep the original value.

(5) Convergence Criterion

The GA repeats selection, crossover, and mutation until it reaches the convergence criterion. The convergence criterion is generally defined that a certain percent $y\%$ solutions in the population have the same Fitness Values. The solution with the biggest Fitness Value in the convergent population is the optimal subspace.

3.4 Parameter Selection

As soon as the most suitable subspace is found, several pairs of parameters (p, D) are selected automatically to detect DB-Outliers in this subspace. This section explains how our proposed method automatically selects parameters (p, D). We still interpret this procedure with a Distance-Nm chart.

Given parameters (p, D), pair of $(M=N(1-p), D)$ decides a point in the Distance-Nm space. If the point is the demarcation point that distinguishes between outliers and ordinary data points. Namely, objects whose D-Nm lines go under this point are recognized as outliers; otherwise, the objects are normal data.

We only care about the D-values where isolated outlier examples' D-Nm lines stay far away from normal objects, because, with such distances, we can detect other outliers having similar characteristics to examples. Here, we introduce the process of parameter selection with a graphic illustrated in Figure 7.

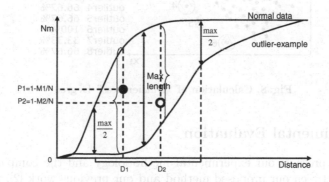

Fig. 7. Selection of Parameters (p,D)

First, we randomly select points on the distance line as parameter Ds between $\frac{max}{2}$ lines (in Figure 7, there are only two Ds). $\frac{max}{2}$ lines mark the position where the gaps between outlier examples' and normal objects' D-Nm lines are half of the maximum gap. We then trace out vertical lines with these selected distance points. Next, we also randomly choose one point on each vertical line segment surrounded by D-Nm lines of outlier examples and normal objects. Each point on the vertical line decides a pair of (M, D). Such points can separate outliers having similar characteristics to outlier examples from normal data. Parameter p can be easily computed by function $p = 1 - M/N$. in Figure 7; two points produce two different pairs of (p, D).

3.5 Outlier Report

For the same dataset, different pairs (p, D) bring different results. To make our results more accurate, we detect DB-Outliers with several pairs of (p, D) produced in the "Parameter Selection," and not only report outliers detected in the optimal subspace, but also the "outlier-ness" degree. The calculation of "outlier-ness" degree is shown in Figure 8.

In Figure 8, eight outliers were detected with three different pairs of parameters (p, D). Outliers detected with $(p1, D1)$, $(p2, D2)$ and $(p3, D3)$ are separately marked by rhombuses, triangles and circles. Outlier1 is selected three times with the three pairs of parameters, thus the "outlier-ness" degree of outlier1 is 100%. Outlier2 is only found with one pair of (p, D), so its "outlier-ness" degree is one third, namely 33.33%. Conspicuously, the higher the "outlier-ness" degree, the more similar the outlier is to outlier examples.

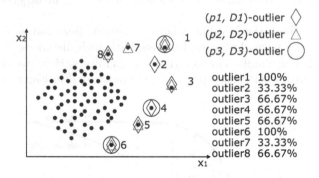

Fig. 8. Calculation of "Outlier-ness" Degree

4   Experimental Evaluation

This section presents our experimental methodology and the comparison performances between our proposed method and our previous work [2]. All of our experiments were run on a Microsoft Windows XP machine having 1GB of main memory.

4.1 Datasets

We test our proposed method on one synthetic, two real, and one commingled dataset (see Table 1 for descriptions). The comparison experiments are operated on two synthetic datasets and one real dataset (see Table 2 for details).

Table 1. Datasets for Verification (Dim denotes the dimensionality of a dataset.)

Dataset	Dim	Description
Synthetic Data	30	10,000 data points, which are normally distributed in 30 dimensions.
Abalone Data	8	Abalone data, obtained from the UCI machine learning repository, 4177 examinations of abalones with 8 attributes.
NBA Data	20	Statistic NBA players' data extracted from the official NBA website, 425 instances of players with 20 attributes.
Exchange Rate Data	30	Money exchange rate data, 1,000 data points, which are holding two attributes that record the recent 4 years' highest and lowest exchange rate between the New Zealand dollar and Japanese Yen, and distributed uniformly in the remaining 28 dimensions.

Table 2. Datasets for Comparison (Dim denotes the dimensionality of a dataset.)

Dataset	Dim	Description
Synthetic Data I	30	10,000 data points, which are normally distributed in 30 dimensions.
Synthetic Data II	30	10,000 data points, which are uniformly distributed in 28 dimensions and holding a group of outliers in the remaining oblique line distributed 2-D subspaces.
Abalone Data	8	Abalone data, obtained from the UCI machine learning repository, 4177 examinations of abalones with 8 attributes.

4.2 Verification Experiments and Results

Here, we mainly demonstrate experiments to test our method over the synthetic and real datasets listed in Table 1. In our experiments, we select 5 objects as normal data (g=5) for calculating A_{NO}, and do the calculation 5 times (q=5) to produce a credible average A_{NO}. If a majority of outlier examples perform ordinary behaviors in a subspace, we do not compute the Fitness Value of such subspace and let it be 0, even though some A_{NO} values of isolated outlier examples may be extraordinarily high. In our experiments, 10 pairs of (p, D) are created in the optimal subspace for detecting outliers.

Figure 9 shows the distributions of two subspaces (a) (b) of the synthetic datasets. We input two groups of outlier examples including noise. The two groups of outlier examples are isolated in subspace (a) (except noise) but performing ordinary appearances in subspace (b). The first group of examples reside farther from the great major points. Our method is to discover subspace (a) as the optimal projection. The results of the two groups of inputs are shown in Figure 10. Figure 10 (1) shows outliers detected with the first group examples, and (2) shows the result for the second group. The detected outliers in both results have more than 60% "outlier-ness" degrees. The result demonstrates that outcomes have a close relationship with input examples. We also input some

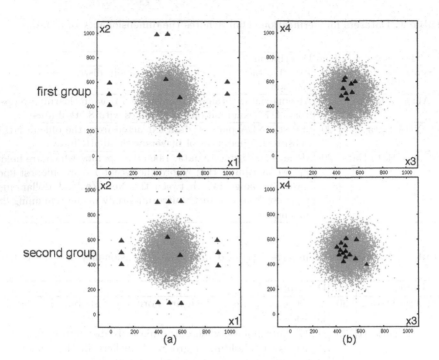

Fig. 9. Outlier Examples in Synthetic Dataset

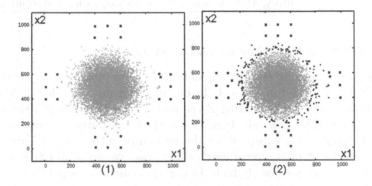

Fig. 10. Detected Outliers in Synthetic Dataset

outlier examples (containing noises) isolated in 3-D and 4-D subspaces, and the optimal subspaces are detected precisely with our approach. Due to the difficulty of ploting out 3-D and 4-D graphics, we only list the summary of these experimental results in Table 3.

For the other datasets, we provide only one group of outlier examples (Each group contains 30% of noise examples). Figure 11 shows the results of the abalone dataset. Since most outlier examples are isolated in the Diameter-Whole Weight subspace, our proposed method marked this subspace as the optimal subspace

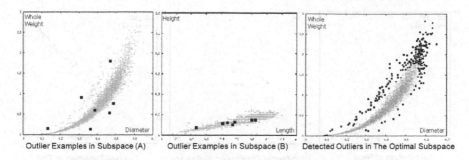

Fig. 11. Abalone Dataset

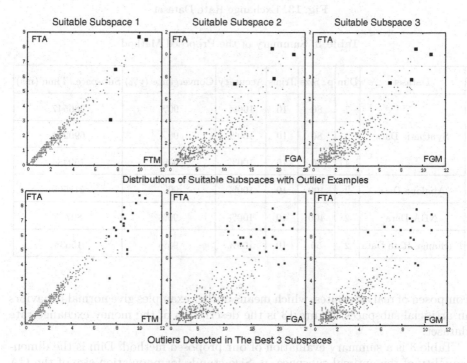

Fig. 12. NBA Dataset

and reported DB-Outliers detected in this subspace. Generally, the attributes of real datasets are not independent, so outlier examples are sometimes isolated in not only one subspace. In addition to the most suitable subspace having the biggest Fitness Value, the second optimal subspace may also be promising. Figure 12 exhibits the distributions of the best three subspaces of the NBA dataset with outlier examples. FTA, FTM, FGA and FGM are abbreviations of Free Throw Attempts, Free Throw Made, Field Goals Attempts, and Field Goals Made, respectively. Outliers picked up in the best three subspaces are some special players. The optimal subspace of the money exchange rate dataset is

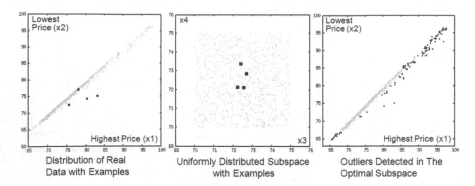

Fig. 13. Exchange Rate Dataset

Table 3. Summary of the Proposed Method

Dataset	Dim	p_Size	Trials	Accuracy	Convergence (y%)	Subspace_Time (ms)
	2	80	10	100%	90%	286547
Synthetic Data	3	80	10	100%	90%	690281
	4	80	10	100%	90%	759434
Abalone Data	2	100	10	100%	90%	218599
NBA Data	2	100	10	100%	90%	80755
Exchange Rate Data	2	50	10	100%	90%	15635

composed of real attributes, which means outlier examples give normal behaviors in artificial subspaces. Figure 13 is the description of the money exchange rate dataset.

Table 3 is a summary evaluation of our proposed method. Dim is the dimensionality of the optimal subspace, p_Size stands for population size of the GA and Accuracy describes the probability of discovering the optimal subspaces during 10 trials. Since there are three suitable subspaces of the NBA data based on user examples, the Accuracy of the NBA dataset denotes the number of times the best three subspaces are found. Subspace_Time records the average time to discover the optimal subspace.

4.3   Comparison Experiments and Results

The objective of this series of experiments is to contrast the results between our idea and the previous work demonstrated in paper [2]. We intend to compare the processing time and quality of the two methods. Such experiments were made

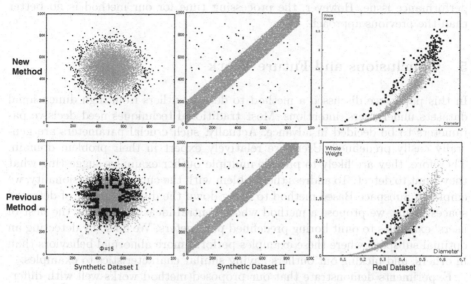

Fig. 14. Comparison Results

Table 4. Summary of Comparison Results

Type of Method	p_Size	Subspace_Time (ms)	Outlier_Time (ms)	Convergence
New (Simple Algorithm)	80	95586	317925	90%
New (Cell-Based Algorithm)	80	-	147156	90%
Previous (ϕ=15)	80	11881	79	90%
Previous (ϕ=22)	80	18625	125	90%

on two synthetic and one real dataset (see Table 2 for descriptions). Noise is not interfused in the experiments here, since the method [2] cannot tolerate noise examples.

The outcomes of our proposed method (new method) and the previous approach are illustrated in Figure 14, and the comparative evaluation is summarized in Table 4. ϕ is a virtual parameter that had to be predefined in the previous approach. Outlier_Time denotes the time to detect outliers in the optimal subspace. The experimental data in Table 4 are gathered from the results of synthetic data I.

Figure 14 gives strong evidence that our proposed method has better quality and can process more kinds of datasets. Beyond that, our method does not need solid, predefined inputs. The unpredictable parameter ϕ plays an important role in detecting outliers. It not only affects the quality issue, but also the performance issue. However, the processing time for our method is no better than the previous approach.

5   Conclusions and Future Work

In this paper, we discussed a method to detect outliers from high dimensional datasets under users' intentions. Most traditional techniques need decisive parameters to be decided in advance. Actually, such crucial parameters are generally easily predefined. Users are relatively expert in their problem domain. Therefore, they are likely to provide multiple outlier examples suggesting what they want to detect. To address the problem with the curse of dimensionality, we employ a Subspace-Based method to bring down the dimensionality of detected spaces. Thus, we propose a method whose central ideas are making the best of users' examples to omit boring predefined parameters. We do so by detecting an optimal subspace where these examples perform more abnormal behaviors than in others, and picking out outliers having similar characteristics to examples.

Experiments demonstrate that our proposed method works well with different kinds of datasets, including synthetic and real datasets. It can cope with arbitrary dimensional subspaces. In addition, a few dummy outlier examples (noise) cannot influence the outcome. Our new method can manage more types of datasets than our previous work. The previous work still needs a predefined decisive parameter, despite utilization of the outlier examples it made. However, the previous method's time performance is better than ours.

Improvement to addressing the problem of processing time is our future research issue. In the future, we plan to do research on producing hybrid methods to improve processing time.

Acknowledgements

This research has been supported in part by the Grant-in-Aid for Scientific Research from JSPS(#18200005) and MEXT(#19024006).

References

1. Zhu, C., Kitagawa, H., Papadimitriou, S., Faloutsos, C.: OBE: Outlier By Example. In: Dai, H., Srikant, R., Zhang, C. (eds.) PAKDD 2004. LNCS (LNAI), vol. 3056, pp. 222–234. Springer, Heidelberg (2004)
2. Zhu, C., Kitagawa, H., Faloutsos, C.: Example-Based Outlier Detection for High Dimensional Datasets. IPSJ Transactions on Databases 46(SIG5), 120–129 (2005)
3. Knorr, E.M., Ng, R.T.: Algorithms for Mining Distance-Based Outliers in Large Datasets. In: Proc. VLDB, pp. 392–403 (1988)
4. http://www.ics.uci.edu/~mlearn/MLRepository.html
5. Aggarwal, C.C., Yu, P.S.: Outlier Detection for High Dimensional Data. In: Proc. SIGMOD Conf., pp. 37–46 (2001)

6. Beyer, K., Goldstein, J., Ramakrishnan, R., Shaft, U.: When is Nearest Neighbors Meaningful? In: Proc. Int. Conf. Database Theory, pp. 217–235 (1999)
7. Breuning, M.M., Kriegel, H.P., Ng, R.T., Sander, J.: LOF: Identifying Density-Based Local Outliers. In: Proc. SIGMOD Conf., pp. 93–104 (2000)
8. Goldberg, D.E.: Genetic Algorithms in Search, Optimization and Machine Learning. Addison Wesley, Reading (1989)
9. Aggarwal, C.C., Yu, P.S.: An Effective and Efficient Algorithm for High-dimensional Outlier Detection. The VLDB Journal 14(2), 211–221 (2005)
10. Li, Y., Kitagawa, H.: DB-Outlier Detection by Example in High Dimensional Datasets. In: Proc. Proc. 3rd IEEE International Workshop on Databases for Next-Generation Researchers (SWOD) (2007)

A Novel Fingerprint Matching Method by Excluding Elastic Distortion

Keming Mao, Guoren Wang, and Ge Yu

Northeastern University, Shenyang 110004, China
wanggr@mail.neu.edu.cn

Abstract. Fingerprint matching is a key issue in fingerprint recognition systems. Although there already exist many researches about fingerprint matching algorithm, it is still a challenging problem for reliable person authentication because of the complex distortions involved in the fingerprint. The non-linear deformation can lead to the change of both position and direction of minutiae and hence decrease the reliability of the minutiae. In this paper, we propose a novel minutiae-based fingerprint matching algorithm. First a new structure Adjacent Feature Union(AFU) is defined, which is invariant to rotation and translation. AFU represents the local feature of a fingerprint and it is used to align the fingerprint. Moreover, the information of the ridge frequency and block orientation is utilized to construct a distortion-tolerant model. Using this model the position and direction of the minutiae can be readjusted and thus enhance the performance of the matching. The proposed method can minimize the effect of the elastic distortion in fingerprint. Experiment results demonstrate the effectiveness of the matching method.

1 Introduction

Biometric is the automatic identifications of an individual that is based on physiological or behavioral characteristics. Among all the biometrics(ie., face, fingerprints, hand geometry, iris, retina, signature, voice print, facial thermogram, hand evin, gait, ear, odor, etc), fingerprint is the most widely used one. A fingerprint is a pattern of ridges and valleys on the surface of the finger. The uniqueness of a fingerprint can be determined by the overall pattern of ridges and valleys as well as the local characteristics [1]. Fingerprint is currently the most common and trusted biometric for personal identifications in virtue of its uniqueness and immutability [2,3]. The key issue of the fingerprint recognition is the matching algorithm. Given two fingerprint images, the matching task is to judge whether they are identical or not.

This paper introduces a fingerprint matching method, which can properly handle the elastic distortion in fingerprint. Our method consists of two stages: fingerprint alignment and global matching. In the fingerprint alignment stage, we define a new structure AFU(Adjacent Feature Union), which is used to characterize the local features in fingerprint. In this way, a fingerprint can be composed of some AFUs. Because the distortion in local area is relatively small, the template and the query fingerprint can be aligned by comparing the corresponding

J.R. Haritsa, R. Kotagiri, and V. Pudi (Eds.): DASFAA 2008, LNCS 4947, pp. 348–363, 2008.

AFUs. The local features used in [4,6,7] only include the relative information between minutiae like distance and angle. AFU also incorporate the textural feature, and this can make AFU more complete and precise. Then the reference minutiae pair is obtained by comparing the corresponding AFUs in template and query fingerprint. The minutiae of template and query fingerprint are converted into the polar coordinate system with respect to the reference minutiae pair in a similar way employed in [2,4,8]. In the global matching stage, the aligned minutiae are used to determined whether two fingerprints are identical. In [4,8], they directly use the radial distance and angle in the polar coordinate system to compare the minutiae. Here a novel method is devised to deal with the distortion in fingerprint. The ridge frequency and block orientation are utilized to readjust the radial distance and angle of the minutiae of the template and query fingerprint under the polar coordinate system respectively. An adaptive threshold according to the radial distance of the minutiae is also applied for matching. The proposed method is applicable to large distortion instance and can improve the effectiveness of the matching process.

Our fingerprint matching method is described in detail in the following sections. Section 2 mainly presents the problem statement and the existing challenging. In Section 3 we propose the fingerprint alignment using AFU and global matching using the distortion-tolerant model. Experimental results are described in Section 4. The related work is given in Section 5. Section 6 concludes this paper.

2   Problem Statement

A fingerprint is composed of ridges and valleys and there are also local ridge discontinuities, known as minutiae as shown in Fig 1 (a). The location and type of minutiae that exist in a fingerprint make the fingerprint unique. There are two basic types of minutiae: ridge ending and ridge bifurcation. As shown in Fig 1 (b), ridge endings are the locations where ridges terminate and ridge bifurcations are the locations where a ridge splits into two separate ridges. The arrowhead of dashed line is the direction of the minutiae.

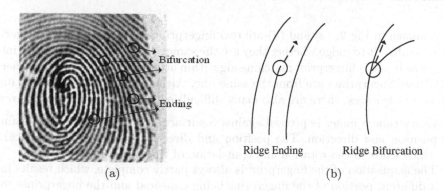

(a)   (b)

Fig. 1. Fingerprint Minutiae

A typical fingerprint contains up to 80 minutiae, but the combination of the minutiae makes the fingerprint unique and most fingerprint identification systems are based on matching minutiae [3]. A minutia m_i can be represented as equation (1) :

$$m_i = \{x_i, y_i, \theta_i, t_i\} \tag{1}$$

where x_i and y_i are the coordinate of m_i; $\theta_i(\theta_i \in [0, 2\pi))$ is the direction of m_i; t_i is the type of the m_i(Ending or Bifurcation). Therefore a fingerprint can be expressed as equation (2):

$$F = \{m_1, m_2, \ldots, m_N\} \tag{2}$$

where N is the number of minutiae in fingerprint. Given a template and a query fingerprint, let $F^T = \{m_1^T, m_2^T, \ldots, m_N^T\}$ and $F^Q = \{m_1^Q, m_2^Q, \ldots, m_M^Q\}$ denote minutiae lists of the template and query fingerprint respectively. So the fingerprint matching can be regarded as the problem of point matching and we can determine if they are identical by matching the two minutiae lists. However, with the presence of noisy and elastic distortion as well as the matching practicability, the fingerprint matching problem is still far beyond solved.

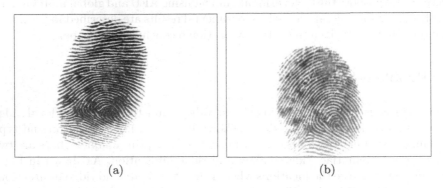

(a)   (b)

Fig. 2. (a) and (b) are from the same finger. In order to know the correspondence between the minutiae of these two fingerprint images , all the minutiae must be precisely localized and the deformation must be recovered.

As shown in Fig 2, (a) and (b) are two fingerprints from the identical finger, but it is difficult to judge whether they are the same or not. Given two fingerprint images as input, a fingerprint matching algorithm attempts to determine whether or not two fingerprints are from the same one. Although the fingerprint has its uniqueness features, there are also many difficulties in the matching procedure:

- Every time a finger is pressed against a surface, it is applied with a certain position and direction. The position and direction could not be identical, which leads to the rotation and translation of the fingerprint.
- The acquisition of the fingerprint is always partly complete, which results in a different portion of the fingerprint being captured and the fingerprints to be matched could have only a small area of overlap.

- If the skin is dry, sweaty, diseased or injured, some parts of the ridges may lost and some noises may appear. Even under ideal conditions, noise will be present in all fingerprint images to some extent.
- The fingerprint is obtained by a mapping from a 3D data to a 2D data. Moreover, different part of the fingerprint may be got with unequal pressure. This generates the non-linear deformation due to the elastic distortion of the fingerprint, which makes the location and the orientation of the ridges altered.

3   Fingerprint Matching

In this paper, a minutiae-based fingerprint alignment and matching method is proposed. The method comprises two stages: (1) Alignment stage. In this stage, the reference minutiae pair is obtained from the template and query minutiae lists by comparing the corresponding AFUs, then both the template and query minutiae lists are converted into the polar coordinate system according to the reference minutiae pair; and (2) Global matching stage. Minutiae lists obtained above are readjusted on the basis of textural information of the fingerprint, then the matching algorithm is implemented.

3.1   AFU and Fingerprint Alignment

In [4,6,15], the methods they used employ the topology of the minutiae to construct the local features and obtain the reference minutiae pairs. For example, k-Nearest Neighbor in [4,15], constructing triangle in [6]. A new structure is proposed in this paper to characterize the local feature information in a fingerprint.

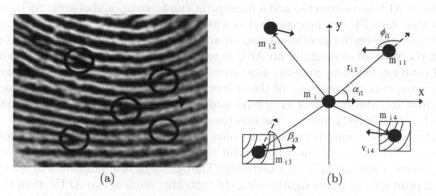

Fig. 3. The Structure of AFU

Fig 3 (a) shows the local structure of a fingerprint, we define a new representation of local information called Adjacent Feature Union(AFU) as shown in Fig 3 (b). An AFU can be defined as follows:

1. For a given minutiae m_i, regarding the direction of m_i as the X-axis direction, choose the nearest neighboring minutia such that the nearest neighbor are

chosen from each of the four quadrants anti-clockwise, which are named $m_{i1}, m_{i2}, m_{i3}, m_{i4}$ respectively.

2. As shown in table 1, a group of 6 elements$(r_{ik}, \alpha_{ik}, \phi_{ik}, v_{ik}, \beta_{ik}, t_{ik})(1 \leq k \leq 4)$ is used to represent the k_{th} neighbor's relation to the central minutia m_i.

Table 1. Elements description in AFU

Symbol	Descriptions
r_{ik}	The Euclidean distance between m_i and m_{ik};
α_{ik}	The angle difference between θ_{m_i} and the angle of the line connected m_i and m_{ik};
ϕ_{ik}	The angle difference between $\theta_{m_{ik}}$ and the angle of the line connected m_i and m_{ik};
v_{ik}	The gray variance of the block which m_{ik} belongs to;
β_{ik}	The angle difference between θ_{m_i} and the orientation of the block which m_{ik} belongs to;
t_{ik}	The type of m_{ik}.

where the value of gray variance of the block and the block orientation are obtained from the fingerprint image preprocessing stage (The fingerprint image is of gray scale and is divided into some nonoverlapping blocks in preprocessing stage. If m_i and m_j are corresponding minutiae pair then the value of v_i resembles v_j. The block orientation is the ridge orientation belongs to this block. Note that the block orientation is between 0 and 2π, so β_{ik} needs to be adjusted to the range of $[0, \pi/2)$. Then β_{ik} is invariant to rotation and translation).

So an AFU is constructed and a fingerprint can be composed of some AFUs in this way. An AFU can be expressed as a vector $(r_{ik}, \alpha_{ik}, \phi_{ik}, v_{ik}, \beta_{ik}, t_{ik})(1 \leq k \leq 4)$. Under the conditions of small distortion appears in a local area. It can be seen that the value of elements in an AFU is relative and is invariant to translation and rotation. Different from the local structure proposed in [4,6,8], the AFU not only comprises the topology of the central minutia and its neighbor minutiae, but also includes the block gray level and block orientation of the fingerprint, which can effectively improve the exactness of the alignment.

The reference minutiae pairs must have the similar local feature information when the template and query fingerprint are identical, a reference minutiae pair must have a similar AFU to a certainty. The AFUs of the template and query fingerprint are compared rightly using the matching score of two AFUs, then the reference minutiae pair is obtained. The matching score of two AFUs $AFUScore$ is calculated by equation (3):

$$AFUScore(AFU_i, AFU_j) =$$

$$\sum_{k=1}^{4} (w_1 \times |r_{ik} - r_{jk}| + w_2 \times |\alpha_{ik} - \alpha_{jk}| + w_3 \times |\phi_{ik} - \phi_{jk}|$$

$$+ w_4 \times |v_{ik} - v_{jk}| + w_5 \times |\beta_{ik} - \beta_{jk}| + w_6 \times |t_{ik} - t_{jk}|) \qquad (3)$$

where $w_1, \ldots, w_6$ are weights of each element in AFU. Here, the higher $AFUScore$ two AFUs have, the more similar they are.

The AFU pair having the highest $AFUScore$ provides the reference minutiae pair for aligning the global structure of the fingerprint. Given the template and query fingerprint, the reference minutiae pair is obtained by comparing their AFUs and then two fingerprints are aligned according to the reference minutiae pair. The alignment is carried out by transforming the minutiae lists of the template and query fingerprint into polar coordinate system. Let $T = \{(x_1^T, y_1^T, \theta_1^T), \ldots, (x_M^T, y_M^T, \theta_M^T)\}$ and $Q = \{(x_1^Q, y_1^Q, \theta_1^Q), \ldots, (x_N^Q, y_N^Q, \theta_N^Q)\}$ denote the sets of minutiae in the template and query fingerprint respectively. After deciding the reference minutiae pair, the minutiae in the template and query fingerprints are all transformed into polar coordinate system using equation (4):

$$\begin{pmatrix} r_i \\ e_i \\ \theta_i \end{pmatrix} = \begin{pmatrix} \sqrt{(x_i^* - x_r)^2 + (y_i^* - y_r)^2} \\ \arctan \frac{(y_i^* - y_r)}{(x_i^* - x_r)} - \theta_r \\ \theta_i^* - \theta_r \end{pmatrix} \tag{4}$$

where $(x_i^*, y_i^*, \theta_i^*)$ are the coordinate and direction of a minutia m_i; (x_r, y_r, θ_r) are the coordinate and direction of the reference minutia m_r; (r_i, e_i, θ_i) is the representation of m_i in the polar coordinate system(r_i denotes the radial distance, e_i denotes the radial angle, and θ_i is the angle difference between θ_i^* and θ_r). Then two list p^T and p^Q are obtained, which represent the template and query fingerprint after transformation respectively. Usually, the list is sorted according to the radial distance.

The aligned global structure is used to determined whether the two fingerprints are generated from the same finger. Two or three reference minutiae pairs that have higher $AFUScore$ can also be regarded as the candidate to take precautions against that reference minutiae pair with highest $AFUScore$ may be the fake one.

3.2   Global Matching

If the two fingerprints are from the same finger and the right reference minutiae pair is chosen, each pair of corresponding minutiae is completely coincident in ideal instance. In this way, the matching problem is simple and straightforward. However, owning to the intrinsic elastic distortion, it is difficult to locate the position and direction of minutiae in fingerprint exactly. Therefore, the matching algorithm must be flexible to stand the distortion due to non-linear deformation in fingerprint. Generally, such an elastic matching can be achieved by placing a bounding box around each template minutia, which can restrict the corresponding minutia in the input image to be within this box. In [4], the bounding box is fixed, and in [2] the size of bounding box is determined by the radial distance of the minutiae automatically. These methods could dispose the problem brought by deformation to some extent. But the deformation in the big area could be very large and complicated [5]. So a novel method is proposed, which utilizes the textual information(the ridge frequency and block orientation) between minutiae to readjust the radial distance and angle of minutiae under the polar coordinate system. In virtue of getting rid of the deformation in the fingerprint, so that we can make certain the position and direction of minutiae to enhance the effectiveness of the matching process.

The deformation in fingerprint can be ignored within a relative small area, but the accumulation of the distortion in a big area could give a strong impact on the position and the direction of the minutiae as shown in Fig 2, and this severely influences the matching. Here the radial distance and angle of the minutiae is readjusted by employing the frequency and orientation information of the block, which can be gained at the fingerprint image preprocessing and the minutiae extracting stage. As shown in Fig 4 (a), m_r^T and m_i^T are the minutiae in the template fingerprint, and m_r^Q, m_i^Q are corresponding minutiae in query fingerprint. m_r^T and m_r^Q are reference minutiae pair. The elastic distortion results in the radial distance and angle changed. r and r_1 are corresponding radial distances. A_1 and A_2 are changes of radial angle and minutiae direction. The information of ridge frequency and block orientation between minutiae can be used to restore the radial distance and direction of the minutiae.

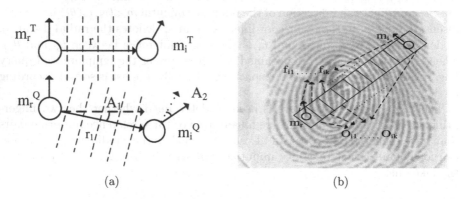

(a)   (b)

Fig. 4. Elastic Distortion in Fingerprint

As shown in Fig 4 (a), when the elastic distortion presence in the fingerprint, the ridge frequency between minutiae changes according to the degree of the distortion, hence the distance between minutiae also changed. If the ridge frequency decreases, the distance between minutiae will be magnified; if the ridge frequency increases, the distance between minutiae will be shrinked. So we can use the ridge frequency to resume the distance between minutiae. $(f_{i1},\ldots,f_{ik})$ in Fig 4 (b) denote the frequencies of the blocks which connect the minutiae m_i with the reference minutia point, where $(f_{i1},\ldots,f_{ik})$ can be obtained by tracing the path from m_r to m_i. Let f_{r_i} denote the average frequency between m_r and m_i, which can be calculated by equation (5):

$$f_{r_i} = \frac{1}{k} \sum_{j=1}^{k} f_{ij} \qquad (5)$$

Let f_{avg} denotes the average ridge frequency of the whole region of the template fingerprint image. f_{avg} is calculated by equation (6), N is the number of the total blocks of the template fingerprint and f_k is the ridge frequency of corresponding block.

$$f_{avg} = \frac{1}{N} \sum_{k=1}^{N} f_k \tag{6}$$

The ratio of f_{avg} and f_{ri} is used to characterized the distortion of radial distance. In this way, the value of r_i can be readjusted. r'_i is the readjusted radial distance according to equation (7):

$$r'_i = \sqrt{(x_i^* - x_r)^2 + (y_i^* - y_r)^2} \times \frac{f_{avg}}{f_{ri}} \tag{7}$$

The normalization process is employed to the radial distance of the minutiae because if the query fingerprint and template fingerprint are from the same finger then they must have the same f_{avg}. Moreover, all the block frequencies between m_r and m_i are considered for computing.

The regulation to the angle of minutiae is applied in succession. It is explicitly shown in Fig 4 (a) that the distortion in fingerprint can affect the angle of the minutiae. As the block orientation changes, the radial angle and the direction of the minutiae change, so A_1 and A_2 appear. If the block orientation change clockwise, then the radial angle and the direction of the minutiae minish; if the block orientation change anticlockwise, then the radial angle and the direction of the minutiae augment. So the changes of orientation between two minutiae can be used to readjust the radial angle and direction of the minutiae. As illustrated in Fig (4) (b), $(O_{i,1}, \ldots, O_{i,k})$ denote the orientation of blocks from m_r to m_i. The angle of the minutiae is readjusted according to $(O_{i1}, \ldots, O_{ik})$. Let $C_{O_{ik}}$ denotes the change of orientation between kth and $(k+1)th$ block following the equation (8). And C_{O_i} denotes the accumulation of orientation changes from m_r to m_i on the basis of equation (9):

$$C_{O_{ik}} = \begin{cases} (O_{i,k} - O_{i,k+1}) + \pi & \text{if } O_{i,k} \leq \xi \text{ and } |O_{i,k+1} - \pi| \leq \xi, \\ (O_{i,k} - O_{i,k+1}) - \pi & \text{if } O_{i,k+1} \leq \xi \text{ and } |O_{i,k} - \pi| \leq \xi, \\ (O_{i,k} - O_{i,k+1}) & \text{otherwise} \end{cases} \tag{8}$$

$$C_{O_i} = \sum_{j=1}^{k-1} C_{O_{ij}} \tag{9}$$

It is known that if m_i and m_p are corresponding minutiae pair and if there is no elastic distortion in the fingerprint, C_{O_i} and C_{O_p} are identical. The difference between C_{O_i} and C_{O_p} compensates for the affection of the angle of the minutiae brought by the deformation. Therefore, e_i and θ_i are readjusted using the equations (10) and (11):

$$e'_i = \arctan \frac{(y_i^* - y_r)}{(x_i^* - x_r)} - \theta_r - C_{O_i} \tag{10}$$

$$\theta_i' = \theta_i^* - \theta_r - C_{O_i} \tag{11}$$

In this way, the radial distance and radial angle have been readjusted, and the global matching stage is put up using these readjusted values. In such a case, a point pattern matching can be achieved by counting the number of matched minutiae pairs. With a proper bounding box, whether two minutiae are matched and whether two fingerprint are identical can be determined by the number of matched minutiae pairs.

Let m_k^T and m_l^Q are two minutiae to be compared from template and query fingerprint respectively. r_{dif}, e_{dif} and θ_{dif} are the differences between r, e and θ of m_k^T and m_l^Q in the polar coordinate system respectively. They are calculated by the following equations (12), (13) and (14):

$$r_{dif} = |r_k - r_l| \tag{12}$$

$$e_{dif} = |e_k - e_l| \tag{13}$$

$$\theta_{dif} = |\theta_k - \theta_l| \tag{14}$$

An adaptive bounding box $(r_{th}, e_{th}, \theta_{th})$ is used to judge whether two minutiae m_k^T and m_l^Q are matched, and the size of the bounding box is changeable in conformity to the normalized radial distance r_k. The value of $(r_{th}, e_{th}, \theta_{th})$ are determined according to the following equations:

$$r_{th} = \begin{cases} r_L & \text{if } r_k \leq D_L, \\ r_H & \text{if } r_k > D_H, \\ r_L + \frac{r_k - D_L}{D_H - D_L} \times (r_H - r_L) & \text{otherwise} \end{cases} \tag{15}$$

$$e_{th} = \begin{cases} e_L & \text{if } r_k \leq D_L, \\ e_H & \text{if } r_k > D_H, \\ e_L + \frac{r_k - D_L}{D_H - D_L} \times (e_H - e_L) & \text{otherwise} \end{cases} \tag{16}$$

$$\theta_{th} = \begin{cases} \theta_L & \text{if } r_k \leq D_L, \\ \theta_H & \text{if } r_k > D_H, \\ \theta_L + \frac{r_k - D_L}{D_H - D_L} \times (\theta_H - \theta_L) & \text{otherwise} \end{cases} \tag{17}$$

where D_L and D_H are the boundaries of r_k. Note that if r_{dif}, e_{dif} and θ_{dif} between two minutiae are within the bounding box $(r_{th}, e_{th}, \theta_{th})$, these two minutiae are considered matched.

3.3   Matching Algorithm

Algorithm 1 describes the procedure of the fingerprint matching. After the minutiae extraction, minutiae transformed into polar coordinate system and minutiae readjustment, the number of matched minutiae pairs is determined by comparing elements in the template minutiae list with the elements in the query minutiae

---

Algorithm 1. Fingerprint Matching

Input: F^T, F^Q : the template and query fingerprint image.
Output: *Count* : the number of matched minutiae between two fingerprints.
 1: Gain two minutiae lists M^T and M^Q from F^T and F^Q respectively.
 2: Find the best matched AFUs as the reference minutiae pair.
 3: Convert the minutiae to polar coordinate system according the reference minutiae pair, gain two list P^T and P^Q.
 4: Readjust the minutiae in P^T and P^Q, result in $P^{T\prime}$ and $P^{Q\prime}$.
 5: **for** each element M_i^T in $P^{T\prime}$ **do**
 6:    **for** each element M_j^Q in $P^{Q\prime}$ **do**
 7:      **if** $(abs(r_j^Q - r_i^T) > r_{th})$ **then**
 8:        break;
 9:      **end if**
10:      **if** $((r_{dif}, e_{dif}, \theta_{dif})$ between M_i^T and M_j^Q are within the bounding box) **then**
11:        $Count + +$;
12:        remove M_j^Q from $P^{Q\prime}$;
13:        break;
14:      **end if**
15:    **end for**
16: **end for**
17: **return** *Count*;

---

list. Here the template and query minutiae lists are sorted by the radial distance of the minutiae, and the iterate can terminate in time while the r_{dif} between query fingerprint minutiae and template fingerprint minutiae exceeds r_{th}. If two minutiae are within the bounding box, they are treated as matched. Once they are confirmed, the minutiae in query minutiae list is removed in the interest of time consuming .

Then the matching score S is computed according to *Count* by equation (18):

$$S = \sqrt{\frac{n^2}{N_T N_Q}} \times 100 \tag{18}$$

where n is the number of matched minutiae between two fingerprints; N_T and N_Q are the numbers of minutiae in the template and query fingerprint respectively.

4 Performance Evaluation

In the fingerprint preprocessing stage, fingerprint features are extracted from a fingerprint image. The algorithm proposed by [16] is employed to solve this problem. This method involves five major steps. The first step is the segmentation and normalization of the fingerprint image, which distinguishes the useful and unuseful blocks of the fingerprint. Then orientation estimation, ridge frequency estimation and Gabor filtering are applied to the image obtained in the previous stage, which wipe off the noises and make the image more smooth.

The next step is binarisation and thinning of the image. Finally, the minutiae are extracted from the thinning fingerprint image. The location, direction, and the type are stored for each minutia. The ridge frequency, block orientation and gray variance of the blocks each minutia point resides in are also stored, which are important information in construction of the AFU, alignment of the two fingerprints and the global matching stages.

In order to evaluate the performance, the matching algorithm is tested on FVC2000 DB1 [17] and FVC2002 DB1 [18]. Both two databases consist of 800 images (100 distinct fingers, 8 instances each). The proposed method is implemented using Java programming language and tested on a Celeron-M 1500MHZ CPU 512 RAM computer.

4.1   Parameter Estimation

The parameters are estimated according to the performance characters FRR (False Rejection Rate) and FAR(False Acceptance Rate). Here, the FRR is the number of identical image pairs which have not been matched divided by the total number of corresponding image pairs; the FAR is the number of nonidentical image pairs which are matched divided by the number of total nonidentical image pairs.

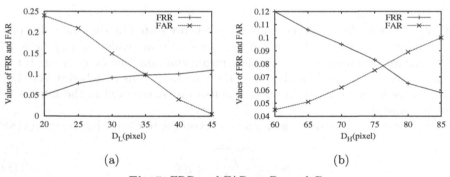

(a)   (b)

Fig. 5. FRR and FAR vs D_L and D_H

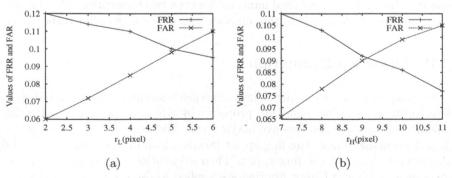

(a)   (b)

Fig. 6. FRR and FAR vs r_L and r_H

Fig 5 shows that with the increase of D_L, the FRR augments and the FAR falls; with the increase of D_H, the FRR falls and the FAR augments. The parameter D_L is the lower boundary of the radial distance. If the radial distance cannot exceed the D_L, the r_{th} keeps a small value, which makes the bounding box stricter. On the contrary, if parameter D_H increases, the r_{th} becomes larger as a result. In this way, the bounding box is relaxed.

Fig 6 demonstrates the scalability of the FRR and FAR with respect to the parameters r_L and r_H. It is shown that the FRR is in inverse proportion to r_L and r_H, FAR is in proportion to r_L and r_H. The reason is that with the increase of r_L and r_H, the size of the bounding box is also magnified. This can lead to the results that the increase of the FAR and decrease of the FRR.

The parameters (e_L, e_H) and (θ_L, θ_H) are also experimentally evaluated. It can be found that (e_L, e_H) and (θ_L, θ_H) have a similar effect on the performance character FRR and FAR with parameters (r_L, r_H). It is shown that the proposed method can provide a low EER with proper parameters.

4.2   Test for Distribution of Matching Score

The distributions of correct and incorrect matching score are shown in Fig 7. It can be seen from this figure that the matching scores are distributed mainly in two regions. One corresponds to the incorrect matching scores, which values are lower than 25, while the other is associated with the correct matching scores with a value around 50. This shows that the proposed method is capable of distinguishing fingerprints by setting an appropriate threshold.

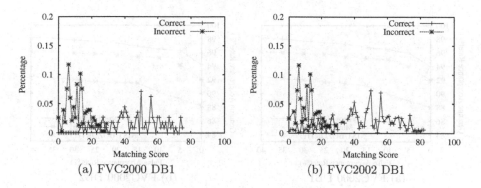

(a) FVC2000 DB1   (b) FVC2002 DB1

Fig. 7. Distributions of correct and incorrect matching score

4.3   Test for Matching Time

Fig 8 summarizes the performance of matching time. The average time for the matching method is 0.02 seconds. It is need to explain that the matching time depends on the number of the minutiae, so the quality of the fingerprint image and the minutiae extraction is vital to the matching stage.

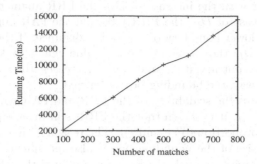

Fig. 8. Runtime vs. Number of matches

4.4 Test for ROC Curve

Another performance parameter of a biometric system can be shown as a Receiver Operating Characteristic(ROC) curve that plots the Genuine Accept Rate against the FAR at different thresholds on matching score. The ROC curves are calculated under the optimized parameters obtained above. As shown in Fig 9, the proposed method is compared with the minutiae-based [19] and popular NIST/BOZORTH3 matching method. It can be seen that our method exceeds the performance of the other two matching methods. It shows that the AFU structure is more exact, and the readjustment according to the textural information can decrease the FAR and enhance the Genuine Acceptance Rate helpfully.

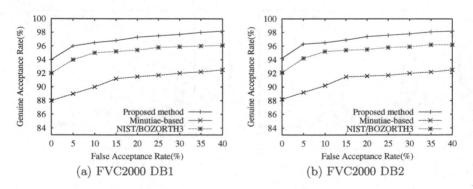

Fig. 9. ROC Curves on two Data Set

5   Related Work

Fingerprint matching has been widely investigated recently and many matching algorithms have been proposed. These algorithms can be mainly classified into the following two categories: Texture-based method [10,11]. In [10], each fingerprint image is filtered in a number of directions and a fixed-length feature vector

FingerCode is extracted in the central region of the fingerprint. The matching stage computes the Euclidean distance between the FingerCode of the template and query fingerprint. [11] uses 8 Gabor filters at various orientations to extract the ridge feature maps to represent, align and match fingerprint image. These methods obviate the need for extracting the minutiae point and the fingerprint is represented by a fixed-length vector. Texture-based methods need to contain more texture information but could not describe the local features properly, especially the minutiae information.

Minutiae-based method [2,7,9,12,13,14] determine the matching result of two fingerprint mainly by the number of matched minutiae pairs. [12] estimates the rotation and translation parameters using the generalized Hough transform. The matching score is then calculated by tabulating the number of corresponding minutiae. For each minutia in [13], the number and type of other minutiae within a given radius is recorded. This information can be used to find potential matches in another minutiae set. The space of all possible transformations is discretized into a finite set of values, and for each pair of potentially matching minutiae the translation and rotation necessary to align them is calculated. After testing all possible matching minutiae pairs, the parameter space is used to select the most likely translation and rotation parameters. In [2], when extracting the minutiae, the shape and location of its associated ridge is also recorded. After the alignment of the corresponding minutiae, a string representation of both minutiae sets is constructed and each character in the string corresponds to a single minutia. With the restriction of a bounding box, a dynamic-programming string matching algorithm is implemented to define an 'edit distance' between the strings corresponding to the template and query minutiae set. The method proposed in [7] use the Delaunay triangulation of minutiae, then apply the RBF to estimate the transformation parameters. [9] samples the ridges of corresponding minutiae, and records the distance and direction of the sampled points, which are employed to describe the local feature of fingerrpint. In [14], four neighbors of a minutia with a fixed angle and distance are used to construct the AOV, and then estimated whether or not the AOVs from two fingerprint are matched. The number of matched AOV pairs is then used to determine the final fingerprint matching score. Comparing to the texture-based method, these methods are mainly incorporating both local and global features for dealing with the minutiae matching problem.

6   Conclusions and Future Work

In this paper, a new minutiae-based fingerprint matching algorithm is presented, which uses the textural information to readjust the position and the direction of the minutiae. By comparing the AFUs of the template and query fingerprint, the reference minutiae pair is obtained. And the minutiae are converted into the polar coordinate system according to the reference minutiae pair. After applying the readjustment, the matching stage are carried out based on the readjusted minutiae. The experiment results show that the performance and efficiency are significant.

Our future work will focus on three aspects. First, it is useful to add some new features for our fingerprint matching technique. Second, we will concentrate on constructing the elastic deformation model more exact by using other techniques. Finally, the matching performance are seriously affected by the fingerprint image preprocessing stage and other image enhancement algorithms will be investigated to extract features more accurately.

Acknowledgments. This research was partially supported by the National Natural Science Foundation of China (Grant No. 60573089 and 60773219) and National Basic Research Program of China (Grant No. 2006CB303103).

References

1. Pankanti, S., Prabhakar, S., Jain, A.K.: On the individuality of fingerprint. Jounal of IEEE Trans. on Pattern Analys. Machine Intellig 24(8), 1010–1025 (2002)
2. Anil, K., Jain, A.K., Hong, L., Bolle, R.M.: On-Line Fingerprint Verification. Jounal of IEEE Trans. on Pattern Analys. Machine Intellig 19(4), 302–314 (1997)
3. Yager, N., Amin, A.: Fingerprint verification based on minutiae features: a review. Jounal of Pattern Anal. Appl. 7(1), 94–113 (2004)
4. Jiang, X., Yau, W.-Y.: Fingerprint Minutiae Matching Based on the Local and Global Structures. In: Proc. of ICPR, pp. 6038–6041 (2000)
5. Miklos, Z., Vajna, K.: A fingerprint verification system based on triangular matching and dynamic time warping. Jounal of IEEE Trans. Pattern Anal. Mach. Intell 22(11), 1266–1276 (2000)
6. Germain, R.S., Califano, A., Colville, S.: Fingerprint Matching Using Transformation Parameter Clustering. Jounal of IEEE Computational Science and Engineering 4(4), 42–49 (1997)
7. Jiang, X., Yau, W.-Y.: An efficient algorithm for fingerprint matching. In: Proc. of ICPR, pp. 1034–1037 (2006)
8. Pradeep, S.N., Jain, M.D., Balasubramanian, R., Bhargava, R.: Local and Global Tree Graph Structures for Fingerprint Verification. In: Proc. of SPPRA, pp. 287–293 (2006)
9. He, Y., Tian, J., Li, L., Chen, H., Yang, X.: Fingerprint Matching Based on Global Comprehensive Similarity. Jounal of IEEE Trans. Pattern Anal. Mach. Intell 28(6), 850–862 (2006)
10. Jain, A.K., Prabhakar, S., Hong, L., Pankanti, S.: Filterbank-based fingerprint matching. Jounal of IEEE Transactions on Image Processing 9(5), 846–859 (2000)
11. Ross, A., Reisman, J., Jain, A.K.: Fingerprint Matching Using Feature Space Correlation. In: Proc. of Biometric Authentication, pp. 48–57 (2002)
12. Ratha, N.K., Karu, K., Chen, S., Jain, A.K.: A Real-Time Matching System for Large Fingerprint Databases. Jounal of IEEE Trans. Pattern Anal. Mach. Intell. 18(8), 799–813 (1996)
13. Hrechak, A., McJigj, J.: Automated fingerprint recognition using structural matching. Jounal of Patt. Recog. 23(8), 893–904 (1990)
14. Ng, G.S., Tong, X., Tang, X., Shi, D.: Adjacent Orientation Vector Based Fingerprint Minutiae Matching System. In: Proc. of ICPR, pp. 528–531 (2004)
15. Wahab, A., Chin, S.H., Tan, E.C.: Novel Approach to Automated Fingerprint Recognition. Jounal of IEEE Proc. Vision, Image and Signal Processing 145(3), 160–166 (1998)

16. Hong, L., Wan, Y., Jain, A.K.: Fingerprint Image Enhancement: Algorithm and Performance Evaluation. Jounal of IEEE Trans. Pattern Anal. Mach. Intell. 20(8), 777–789 (1998)
17. Maio, D., Maltoni, D., Cappelli, R., Wayman, J.L., Jain, A.K.: FVC2000: Fingerprint Verification Competition. Jounal of IEEE Trans. Pattern Anal. Mach. Intell. 24(3), 402–412 (2002)
18. Maio, D., Maltoni, D., Cappelli, R., Wayman, J.L., Jain, A.K.: FVC2002: Second Fingerprint Verification Competition. In: Proc. of 16th Int'l Conf.Pattern Recognition, pp. 811–814 (2002)
19. Jain, A.K., Hong, L., Pankanti, S., et al.: An identity-authentication system using fingerprints. Proc. of the IEEE 85(9), 1365–1388 (1997)

Approximate Clustering of Time Series Using Compact Model-Based Descriptions

Hans-Peter Kriegel, Peer Kröger, Alexey Pryakhin,
Matthias Renz, and Andrew Zherdin

Institute for Computer Science
Ludwig-Maximilians-University of Munich
Oettingenstr. 67, 80538 Munich, Germany
{kriegel,kroegerp,pryakhin,renz,zherdin}@dbs.ifi.lmu.de

Abstract. Clustering time series is usually limited by the fact that the length of the time series has a significantly negative influence on the runtime. On the other hand, approximative clustering applied to existing compressed representations of time series (e.g. obtained through dimensionality reduction) usually suffers from low accuracy. We propose a method for the compression of time series based on mathematical models that explore dependencies between different time series. In particular, each time series is represented by a combination of a set of specific reference time series. The cost of this representation depend only on the number of reference time series rather than on the length of the time series. We show that using only a small number of reference time series yields a rather accurate representation while reducing the storage cost and runtime of clustering algorithms significantly. Our experiments illustrate that these representations can be used to produce an approximate clustering with high accuracy and considerably reduced runtime.

1 Introduction

Clustering time series data is a very important data mining task for a wide variety of application fields including stock marketing, astronomy, environmental analysis, molecular biology, and medical analysis. In such application areas the time series have usually an enormous length which has a significantly negative influence on the runtime of the clustering process. As a consequence, a lot of research work has focused on efficient methods for similarity search in and clustering of time series in the past years.

Time series are sequences of discrete quantitative data assigned to specific moments in time, i.e. a time series X is a sequence of values $X = \langle x_1, \ldots, x_N \rangle$, where x_i is the value at time slot i. This sequence is often also taken as a N-dimensional feature vector, i.e. $X \in \mathbb{R}^N$.

The performance of clustering algorithms for time series data is mainly limited by the cost required to compare pairs of time series (i.e. the processing cost of the used distance function). As indicated above, time series are usually very large containing several thousands of values per sequence. Consequently, the

J.R. Haritsa, R. Kotagiri, and V. Pudi (Eds.): DASFAA 2008, LNCS 4947, pp. 364–379, 2008.

comparison of two time series can be very expensive, particularly when considering the entire sequence of values of the compared objects. The most prominent approaches to measure the similarity of time series are the Euclidean distance and Dynamic Time Warping (DTW). The choice of the distance function mainly depends on the application. In some applications, the Euclidean distance produce better results whereas in other applications, DTW is superior. The big limitation of DTW is its high computational cost of $O(N^2)$ while the Euclidean distance between two time series can be computed in $O(N)$. Since we consider large databases and long time series (i.e. large values of N) in this paper, we focus on the Euclidean distance as similarity function in the following.

In general, if we apply the Euclidean distance to the entire sequences, this is also only adequate for short time series. In case of long time series, we face two problems: The distance computation requires rather high runtimes and, if the time series are indexed by a standard spatial indexing method such as the R-Tree [1] or one of its variants, this index will perform rather bad due to the well-known curse of dimensionality. Thus, the common way is to create adequate but considerably shorter approximations of the data retaining essential features of interest. According to this schema there exist a lot of approaches for dimensionality reduction resulting in suitable time series representations that allow efficient similarity distance computations. However, since the distance computations performed on the approximations do not reflect the exact similarity, they can either be used as a filter step of the data mining task or the preliminary results can be directly taken to approximately solve the problem if the results satisfactorily agree with the exact query response. In the first case, the approximations should fulfill the lower bounding property to guarantee complete results. The advantage of the second solution over the first one is that the approximations do not need to fulfill this lower bounding property which makes it easier to find a proper approximation technique. Furthermore, the second method will yield considerably lower response times because no refinements are required. However, the challenge of the second solution is that the distances on the approximations should accurately estimate the distances on the exact time series in order to achieve satisfying results (i.e. approximate results of high accuracy).

The question at issue is which approximation we should use. Adequate time series approximations can be built by means of mathematical models. Most approaches use models which are based on approximations in time, i.e. models that describe how a time series depends on the time attribute (cf. Section 2). The common characteristics of these techniques are that the approximation quality decreases with increasing length of the time series assuming a constant approximation size. In this paper, we propose a method for the approximation of time series based on mathematical models that explore dependencies between different time series. We represent each time series by an adequate combination of a set of specific reference time series (usually these reference time series can easily be determined e.g. by a domain expert). The resulting representation consists of some low-dimensional feature vector that can easily be indexed by means of any Euclidean index structure. The similarity distance used for the clustering is

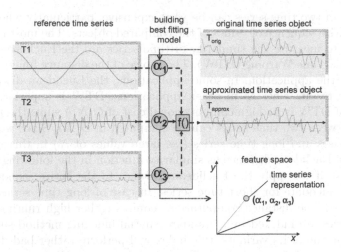

Fig. 1. Model-based time series representation

computed by applying the parameters that specify the combination. Consequently, the cost of the clustering process depend only on the number of reference time series rather than on the length of the time series. As we will see in our experiments, the number of reference time series can usually be very small in order to achieve rather accurate results.

Figure 1 illustrates our approach. A set of reference time series (marked as "T1", "T2", and "T3" on the left hand side of Figure 1) is used to approximate an original time series T_{orig} (shown on the upper right hand side of Figure 1) by an arbitrary complex combination T_{approx}. In case of Figure 1 this is a combination of the coefficients $\alpha_1, \ldots, \alpha_3$ representing the three input time series using a function f. The resulting approximated time series (marked as "output" in the middle of the right hand side of Figure 1) is similar to the original time series. For clustering, the approximation is represented by a feature vector of the coefficients of the combination (cf. lower right hand side of Figure 1).

The rest of the paper is organized as follows. In Section 2, we survey related work. In Section 3, we introduce the notion of mathematical models and describe our powerful method for the calculation of compact representation for time series based on the idea of mathematical models. Section 4 presents results of versatile experimental evaluation. Finally, we conclude our paper in Section 5 with a short summary and show directions for further research.

2   Related Work

In general, a time series of length d can be viewed as a feature vector in a d-dimensional space. As discussed above, we focus on similarity in time, i.e. we assume that the similarity of time series is represented by the Euclidean distance of the corresponding feature vector. Since for long time series d is usually large, the efficiency and the effectiveness of data analysis methods is rather limited due to the *curse of dimensionality*. Thus, several more suitable representations of time series data, e.g. by reducing the dimensionality, have been proposed. Most of them are based on the GEMINI indexing approach [2]: extract a few key *features* for each time series and map each time sequence X to a point $f(X)$ in a lower dimensional feature space, such that the distance between X and any other time series Y is always lower-bounded by the Euclidean distance between the two points $f(X)$ and $f(Y)$. For an efficient access, any well known spatial access method can be used to index the feature space. The proposed methods mainly differ in the representation of the time series which can be classified into *non data adaptive methods*, including DFT [3] and extensions [4], DWT [5], PAA [6], and Chebyshev Polynomials [7], as well as *data adaptive* methods, including SVD [8,9], APCA [10], and cubic splines [11].

Contrary to our approach, all these approximation techniques represent time series by a set of attributes describing how the time series depend on time. As a consequence, the approximation quality of these methods decreases with increasing length of the time series assuming a constant number of approximation attributes.

In [12] the authors use a clipped time series representation rather than applying a dimensionality reduction technique. Each time series is represented by a bit string indicating the intervals where the value of the time series is above the mean value of the time series. This representation can be used to compute an approximate clustering of the time series. The bit level representations are compressed using standard compression algorithms in order to reduce the I/O cost and to speed-up the clustering task. Unfortunately, the authors did not propose any index structure for the approximation data. Each similarity search task results in a full scan over the approximated data.

For clustering time series data, most of the various clustering methods proposed in the past decades have been successfully applied. A general overview over clustering methods is given in [13].

In this paper, we claim the following contribution. We propose a novel compact approximation method for time series data that is independent of the length of the time series. The resulting representation can be indexed using any Euclidean index structure and is rather accurate for an approximate clustering of the database.

3   Mathematical Models for Time Series Data

Mathematical modeling is a powerful method for the description of real-word processes by a compact mathematical representation (e.g., mathematical models of physical or chemical processes). In this section, we introduce a formal definition of mathematical models. Additionally, we describe our method for the description of large time series data by using a compact representation based on the idea of mathematical models.

3.1   Mathematical Model

We start with an informal discussion of the notion of a mathematical model. A mathematical model is an approximate description of a class of certain objects and their relationships. This approximate description is given by mathematical formulas. In context of time series data, a mathematical model describes dependencies between recorded time series data called *inputs* or *exploratory variables* and time series data called *outputs* or *dependant variables* of an observed process. For instance, we can model the relationship between the air pressure in an enclosed container w.r.t. the temperature of the surrounding environment. The observation of both pressure values and temperature values are given in the form of time series. The values of pressure are used as values of the dependant variable. The values of temperature are used as values of the exploratory variable. More formally, a mathematical model can be defined as follows.

Definition 1 (Mathematical Model)
A mathematical model $\mu = (\boldsymbol{X}, \boldsymbol{\alpha}, f)$ for a dependent variable Y (output) consists of a set of exploratory variables $X_1, \ldots, X_k$ called inputs and a mathematical function $f(\boldsymbol{X}, \boldsymbol{\alpha})$ that is used to describe the dependency between the variable Y and the variables $X_1, \ldots, X_k$, where $\boldsymbol{\alpha}$ denotes the model parameters also called coefficients *of the model. The general form of the model is given by $Y = f(\boldsymbol{X}, \boldsymbol{\alpha}) + \varepsilon$, where ε denotes the random error.*

In this definition, the exploratory variables $X_1, \ldots, X_k$ are inputs of the model. The model parameters $\boldsymbol{\alpha}$ are the quantities that are estimated during the modeling process. The value ε represents the random error that makes the relationship between the dependant variable and the exploratory variables a "statistical" one rather than a perfect deterministic one. This statistical character is justified by the fact that the functional relationship holds only in average (i.e., not for each data point).

In general, for building a mathematical model for a time series Y of measured values as a dependant variable we need a mathematical function f and a set $\rho = \{\rho_1, \ldots, \rho_k\}$ of input time series also called *reference time series*. Usually, f and ρ can be given by a domain expert or can be choosen by examining a small sample of the time series in the database. The goal is to find the "best fitting" model. Obviously, in order to find this "best fitting", the random error ε should be minimized. This minimization can be achieved by calculating suitable model parameters $\boldsymbol{\alpha}$. In the last decades, several methods were proposed that allow us to fit the model to the real time series data (i.e. to calculate the model parameters $\boldsymbol{\alpha}$ so that the random error ε is minimized). The most popular method is Least-Squares Estimation which we will use in the following.

Let us consider some examples of mathematical functions that are typically used in mathematical modeling. For a time series Y that fits a straight line with an unknown intercept and slope, there are two parameters $\boldsymbol{\alpha} = (\alpha_1, \alpha_2)$, and one exploratory variable X such that $f(\boldsymbol{X}, \boldsymbol{\alpha}) = \alpha_2 \cdot X + \alpha_1$.

Figure 2 illustrates an example for the approximation of a more complex time series $Y = DV$ by a mathematical model using four reference time series

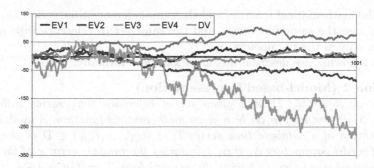

Fig. 2. An example for relationship between a dependant variable (DV) and four exploratory variables (EV1-4)

$\rho_1 = EV1$, $\rho_2 = EV2$, $\rho_3 = EV3$, and $\rho_4 = EV4$ that are combined as given by $DV = EV1 + 2 \cdot EV2 - 4 \cdot EV3 - EV4$.

Thus, the mathematical model describing $Y = DV$ consists of the set of reference time series $\rho = \{\rho_1, \rho_2, \rho_3, \rho_4\}$ and the function $f(\rho, \boldsymbol{\alpha}) = \rho_1 + 2 \cdot \rho_2 - 4 \cdot \rho_3 - \rho_4$ and $\alpha_1 = 1$, $\alpha_2 = 2$, $\alpha_3 = -4$, and $\alpha_4 = -1$.

To summarize, a mathematical model provides an elegant method of describing the relationship between a dependent variable (output time series) and a set of exploratory variables (reference time series). In general, it can use any complex mathematical function such as the combination of quadratical and logarithmical functions to approximate this relationship. In order to express the relationship formally, parameters of a given mathematical function need to be fitted.

3.2 Representation of Time Series Based on Mathematical Models

In this section, we introduce the intuition behind our compact representation of a time series and present a novel technique that transforms a very long time series into a compact representation.

Let us consider a given set of reference time series ρ and a given mathematical model $\mu = (\rho, \boldsymbol{\alpha}, f)$. Each time series $T_i \in \mathcal{D}$ in the database can now be considered as a dependant variable Y_i. Values of the dependant variable Y_i can be approximated by values of the mathematical model $\mu_i = (\rho, \boldsymbol{\alpha}_i, f)$ that contains the model parameters $\boldsymbol{\alpha}_i$ that are fitted in order to approximate the values of the dependant variable Y_i as exactly as possible. Thus, the given mathematical model μ_i describes relationships between the reference time series ρ and the approximated time series T_i (i.e., it expresses how strong Y_i depends on each of the reference time series) by means of the model parameters $\boldsymbol{\alpha}_i$. Obviously, dependant variables Y_i and Y_j with similar dependencies should have very similar mathematical models μ_i and μ_j, i.e. the parameters $\boldsymbol{\alpha}_i$ and $\boldsymbol{\alpha}_j$ will be rather similar. In other words, if the underlying physical processes represented by measured values in the database have similar character, their mathematical models look very similar. This relation between original processes and mathematical models is justified by the fact that we consider dependencies based on the same form of the mathematical function and the same reference time series, i.e. all the models μ_i use the same function f and the same set of reference time series ρ but differ only in the parameters α_i.

In the following, we describe this intuition more formally:

Definition 2 (Model-based Representation)
Let $\rho = \rho_1, \ldots, \rho_k \subseteq \mathcal{D}$ be a given set of reference time series with $\rho_j = \langle \rho_{j,1}, \ldots, \rho_{j,N} \rangle$ and let $f(\rho, \alpha)$ be a given mathematical function. A model-based representation of a database time series $T_i = \langle t_{i,1}, \ldots, t_{i,N} \rangle \in \mathcal{D}$ is given by a vector of model parameters α_i if α_i minimizes the random error ε of the mathematical model $\mu = (\rho, \alpha, f)$ having the general form $T_i = f(\rho, \alpha_i) + \varepsilon$.

In the example shown in Figure 2, the model-based representation of the time series DV w.r.t. the reference time series $\rho = \{EV1, EV2, EV3, EV4\}$ is given by a vector $\alpha = (1, 2, -4, -1)$. Let us not that, in this case, we describe a time series of length 1,000 by a short model-based representation with four coefficients.

To summarize, we describe each time series by a small set of model parameters of a mathematical model the shape of which is identical for all time series in the database. The size of our model-based representation is independent on the length of the time series in the underlying database but depends only on the number of reference time series. In particular, the approximation exactness of our model-based representation only depends on the applied model function and the reference time series. Therefore, we can achieve an arbitrary level of exactness of the approximation by choosing a model function and a set of reference time series that are most appropriate for the given application area.

3.3   Model-Based Similarity of Time Series

The bottom line of clustering time series is the distance (or similarity) measure used to decide about the similarity of time series. As discussed above, we focus on similarity in time. Thus, for our approach we use the most prominent time-based distance measure for time series, the Euclidean distance. The Euclidean distance is commonly used for the dimensionality reduction techniques mentioned in Section 2.

For long time series, the computation of the Euclidean distance is very expensive. Furthermore the well-known curse of dimensionality limits the efficiency of indexing methods to speed-up similarity queries. For this reason we propose to compute the similarity using the representations based on mathematical models consisting of only a few coefficients (model parameters). We can show that our model-based similarity distance based on the model parameters accurately approximates the Euclidean distance between the original time series. The approximation accuracy mainly depends on how good the model fits to the original time series, i.e. how accurate the model approximates the original time series.

When defining the similarity of time series based on the model parameters, we need to consider that the pairwise similarities between our reference time series need not be identical. An illustrative example is shown in Figure 3. The depicted three time series T_1, T_2, and T_3 are represented by a model μ that is based on

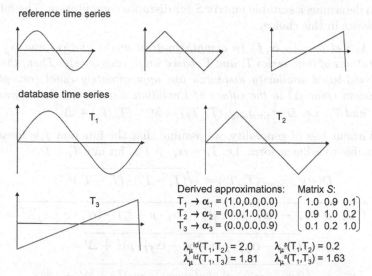

Fig. 3. Motivation for the use of the Mahalanobis-distance

the three reference time series presented at the top of Figure 3. Since T_1 is equal to the first reference time series, the coefficients of the model-based representation of T_1 are given by $\alpha_1 = (1.0, 0.0, 0.0)$. Accordingly, the coefficients of the model-based representation of T_2 which is equal to the second reference time series are given by $\alpha_2 = (0.0, 1.0, 0.0)$. The coefficients of the model-based representation of T_3 which is nearly equal to the third reference time series are given by $\alpha_2 = (0.0, 0.0, 0.9)$. If we compare the Euclidean distance between T_1 and T_2 (denoted by $\lambda_\mu^{Id}(T_1, T_2)$ in the Figure) with the Euclidean distance between T_1 and T_3 (denoted by $\lambda_\mu^{Id}(T_1, T_3)$) we see that $\lambda_\mu^{Id}(T_1, T_2) > \lambda_\mu^{Id}(T_1, T_3)$. This is rather unintuitive because the original time series T_1 is much more similar to the original time series T_2 than to T_3. This is because we do not consider that the first reference time series is more similar to the second than to the third.

Thus, we need to consider these different pair-wise similarities of our reference time series when computing the similarity between the model parameters α_i and α_j of two time series $T_i, T_j \in \mathcal{D}$. This can be done using the well-known *Mahalanobis-distance* between the vectors α_i and α_j, formally

Definition 3 (Model-based Similarity Distance)
Let $T_i, T_j \in \mathcal{D}$ be two time series and let $\mu = (\rho, \alpha, f)$ be a mathematical model where α_i and α_j are the representations of T_i and T_j based on μ, respectively. The model-based similarity distance $\lambda_\mu^S(T_i, T_j)$ *between T_i and T_j is defined by*

$$\lambda_\mu^S(T_i, T_j) = \sqrt{(\alpha_i - \alpha_j) \cdot S \cdot (\alpha_i - \alpha_j)^T}.$$

The key part of the Mahalanobis-distance is the matrix S that is used to rank the pair-wise combinations of the reference time series. Thus, an important issue is to determine a suitable matrix S for distance computation. The following lemma assists in this choice.

Lemma 1. *Let $\mu = (\rho, \alpha, f)$ be a mathematical model and α_i and α_j be the representations of time series T_i and T_j based on μ, respectively. Then, the values of the model-based similarity distances are approximately equal (except for a small random error Δ) to the values of Euclidian distance on the original time series T_i and T_j, i.e. $Dist_{Euclidian}(T_i, T_j) = \lambda_\mu^{\rho \cdot \rho^T}(T_i, T_j) + \Delta$.*

Proof. Without loss of generality, we assume that the function f is linear or is transformable to a linear form, i.e. $T_i = \alpha_i \cdot \rho + \varepsilon_i$ for any $T_i \in \mathcal{D}$.

$$Dist_{Euclidian}(T_i, T_j) = \sqrt{(T_i - T_j) \cdot (T_i - T_j)^T} =$$

$$\sqrt{((\alpha_i \cdot \rho + \varepsilon_i) - (\alpha_j \cdot \rho + \varepsilon_j)) \cdot ((\alpha_i \cdot \rho + \varepsilon_i) - (\alpha_j \cdot \rho + \varepsilon_j))^T} =$$

$$\sqrt{((\alpha_i - \alpha_j) \cdot \rho) \cdot ((\alpha_i - \alpha_j) \cdot \rho)^T + \Delta'} =$$

$$\sqrt{((\alpha_i - \alpha_j) \cdot \rho) \cdot (\rho^T \cdot (\alpha_i - \alpha_j)^T) + \Delta'} =$$

$$\sqrt{(\alpha_i - \alpha_j) \cdot (\rho \cdot \rho^T) \cdot (\alpha_i - \alpha_j)^T + \Delta'} =$$

$$\lambda_\mu^{\rho \cdot \rho^T}(T_i, T_j) + \Delta$$

The lemma states that the model-based similarity distance approximates the Euclidian distance on the original time series[1] by an error of Δ, if $S = \rho \cdot \rho^T$ where ρ is a matrix consisting of the reference time series. Obviously, Δ depends on the random errors ε_i and ε_j of the model-based approximation of T_i and T_j. Thus, if the errors of the approximation are small (which is a design goal of the approximation and is realized by the Least-Squared Error method), then Δ will also be small. As a consequence, if we set $S = \rho \cdot \rho^T$, the model-based similarity distance will be a rather accurate approximation.

Let us note that if S is the unity matrix, i.e. $S = \rho \cdot \rho^T = Id$, the Mahalanobis-distance is identical to the Euclidean distance. In Figure 3, the model-based similarity distance between T_1 and T_2 as well as between T_1 and T_3 using $S = \rho \cdot \rho^T$ denoted by $\lambda_\mu^S(T_1, T_2)$ and $\lambda_\mu^S(T_1, T_3)$, respectively, are compared with the corresponding Euclidean distance on the model-based representations. As it can be seen, the model-based similarity distance using the Mahalanobis-distance more accurately reflects the intuitive similarity of the original time series than the Euclidean distance on the model-based representations.

Furthermore, let us note that our method may have a slight increase of the CPU cost because we use the Mahalanobis-distance rather than the Euclidean distance used by the existing methods. However, since the size of our approximations is independent of the length of the original data items this marginal CPU performance loss leads to a great benefit in terms of I/O-cost especially when dealing with long time series.

---

[1] Please recall that we focus on similarity in time rather than similarity in shape, and, thus use the Euclidean distance as the baseline.

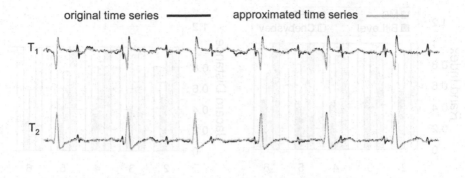

time series length = 1024,   # model coefficients = 44

Fig. 4. Approximations for sample time series

3.4 Choosing the Reference Time Series

Obviously, an important aspect of our model-based representation of time series is the choice of the reference time series. As already sketched, this choice can usually be done by a domain expert. However, if no such domain expert is at hand, we need a procedure for this choice.

In general, the reference time series should have a high correlation to a subset of the remaining time series in the database. Inspired by this intuition, we propose to use the following procedure to derive a set of reference time series. Let us assume that we want to choose k reference time series. We simply cluster the time series using a k-medoid clustering algorithm, e.g. PAM [13]. This yields a set of k cluster medoids (time series), each representing its corresponding cluster. All time series of a cluster are strongly correlated to the corresponding cluster medoid. Obviously, taking these medoids for the derivation of the reference time series should be a very good choice. In addition, in order to avoid wasting to much computational costs, we propose to perform the PAM clustering only on a small sample of the database. In practice, a sample rate of about 1% has shown a sufficient high clustering accuracy.

3.5 Efficient Approximative Clustering

Based on the previously defined similarity distance measure, we can apply any analysis task to time series data. Our main goal is to yield an efficient clustering of the database $\mathcal{D}$ of time series using the approximative representations while generating clusters with sufficient quality. In other words, approximative clustering implements the idea that a user may want very fast response times while accepting a considerable decrease of accuracy. This is a reasonable setting in many application domains. For our experiments, we used the most prominent clustering method k-means. However, our approximate representation can be integrated into any other clustering algorithm the user is most accustomed to. The key issue for approximate clustering is of course to generate accurate results, i.e.

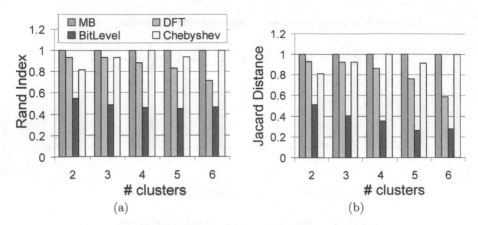

Fig. 5. Cluster quality for varying number of clusters (DS1)

the used approximations should describe the original time series considerably well. We will show in our experiments that a very small set of parameters for the time series approximations is sufficient to achieve high quality clustering results even if the original time series that should be clustered are very long. The example shown in Figure 4 indicates the potentials of our approximation. Two rather complex time series T_1 and T_2 of length 1,024 are compared with some sample approximations using only 44 coefficients. As it can be seen, the compressions approximate the real time series rather accurate.

4   Evaluation

We implemented our method and comparison partners in Java 5. All experiments were performed on a workstation featuring two 3 GHz Xeon CPUs and 32GB RAM. We used four datasets for our experiments, three artificial datasets (DS1, DS2 and DS4) and one real world dataset (DS3) as depicted in Table 4.

name	type	length N
DS1	artificial	2,560
DS2	artificial	6,000
DS3	real world	1,024
DS4	artificial	2,000-14,000

The reference time series of the first two artificial datasets DS1 and DS2 are generated by random walk. The corresponding datasets are built by a linear combination of the reference time series compounded by the identity, square, cube and first and second derivatives.

In order to demonstrate that our approach can handle versatile data, we composed the dataset DS3 in the following way. It consists of real-world time series from the following application areas: (1) wing flutter[2], (2) cutaneous potential recordings of a pregnant woman[3], (3) data from a test

---

[2] http://homes.esat.kuleuven.be/ smc/daisy/daisydata.html
[3] http://www.tsi.enst.fr/icacentral/base_single.html

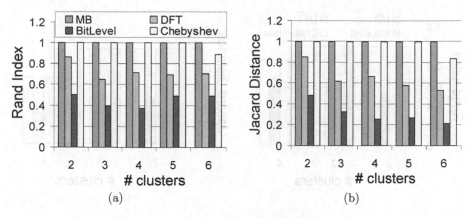

Fig. 6. Cluster quality for varying number of clusters (DS2)

setup of an industrial winding process[4], (4) continuous stirred tank reactor[5]. In this dataset (DS3), the reference time series were given by domain experts. The forth dataset DS4 is generated using the Cylinder-Bell-Funnelmethod[6]. It is an artificial dataset that covers the complete spectrum of stationary/ non-stationary, noisy/ smooth, cyclical/ non-cyclical, symmetric/ asymmetric etc. data characteristics. We used a PAM clustering ($k = 4$) of a random sample in order to derive the reference time series as described above. The k parameter was determined by standard methods.

We compare our mathematical model based time series approximation (MB) with the following competing approximation techniques: *Bit Level* using clipped time series representations as proposed in [12], Discrete Fourier Transformation (DFT) [3] and representations by means of Chebyshev polynomials (Chebyshev) [7]. The competing techniques are evaluated by the approximation quality of k-means clusterings.

Model description of the test datasets. For the mathematical function of the model we used a linear combination of the original set of reference time series, the quadrature and cubature of the reference time series, and the first and second derivation of the reference time series in time. In fact, using n model parameters we only required $n/5$ reference time series.

The overall number of model parameters used for the experiments are justified to the datasets. We used 101 parameters for DS1 and DS2, 51 parameters for DS3 and 20 for DS4. In order to be comparable to the competitors, we used the same number of coefficients for DFT and Chebyshev based approximations.

Measuring clustering quality. For the experimental evaluation of our approach, we built reference clusterings based on the Euclidean distance between the

---

[4] http://homes.esat.kuleuven.be/ smc/daisy/daisydata.html
[5] http://www.fceia.unr.edu.ar/isis/cstr.txt
[6] http://waleed.web.cse.unsw.edu.au/phd/html/node119.html

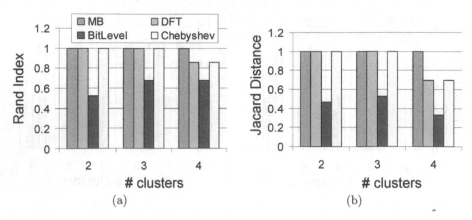

Fig. 7. Cluster quality for varying number of clusters (DS3)

original time series and measured the clustering quality w.r.t. this reference clustering. For the clustering quality measure, we used the two most prominent clustering evaluation methods, the *Rand Index* and the *Jacard Distance* [14].

Experiments on clustering quality. In the first experiment, we examine the quality of our approximation method for a varying number of clusters based on the three datasets DS1 (cf. Figure 5), DS2 (cf. Figure 6) and DS3 (cf. Figure 7). Over all competitors, the *Bit Level* approach yields the lowest clustering quality for all experiments and experimental settings. In comparison we achieve a quality which is at least two times higher than that of the *Bit level* approach. In our experiments, our approach outperforms the method based on DFT coefficients and is even better than the approach using the Chebyshev polynomials when increasing the number of searched clusters.

Against the competitors, our approach achieves optimal clustering quality, even on the real world dataset. This can be justified by the fact that our model-based similarity distance reflects the Euclidean distance on the original time series very accurately.

Dependency on time series length. In the next experiment, we examine how the cluster quality depends on the size of the time series. Figure 8(a) and Figure 8(b) depict the results. Obviously, the characteristic of both dimensionality reduction approaches DFT and Chebyshev is that the clustering quality decreases drastically with increasing time series length. In contrast, we achieve high quality over all investigated time series lengths applying our model-based approximations. Similar to our approach, the Bit Level approach keeps nearly constant quality even for long time series, but yields rather low performance.

Runtime comparison. Last but not least, we compared the speed-up of our method in comparison to the original Euclidean distance in terms of CPU time. For that purpose, we varied the length of the time series of DS4. The results

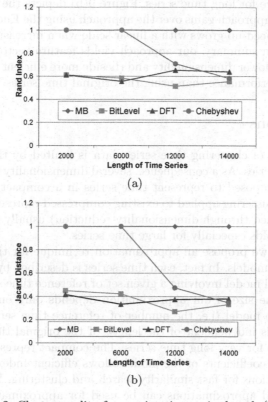

(a)

(b)

Fig. 8. Cluster quality for varying time series length (DS4)

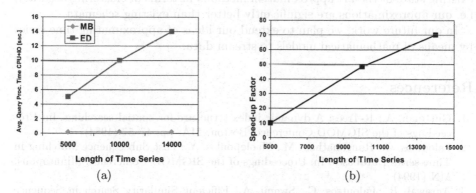

(a)

(b)

Fig. 9. Performance of our model-based approach vs. Euclidean distance

are illustrated in Figure 9(a). As expected, our model-based approach (marked with "MB" in the figure) scales constant, while the Euclidean distance (marked with "ED" in the figure) scales linear wr.t. the length of the time series. It can be further observed that our model-based approach clearly outperforms the

Euclidean distance for long time series. Figure 9(b) depicts the speed-up factor our model-based approach gains over the approach using the Euclidean distance. Obviously, this speed-up grows with a linear scale when increasing the length of the time series. In summary, our approach yields feature vectors of a constant and considerably lower dimensionality and (beside more efficient indexing) yields to better CPU performance than using the original time series.

5   Conclusions

The performance of clustering time series data is limited by the length of the considered time series. As a consequence, several dimensionality reduction methods have been proposed to represent time series in a compact approximation. Approximative clustering applied to existing compressed representations of time series (e.g. obtained through dimensionality reduction) usually suffers from low accuracy. This holds especially for large time series.

In this paper, we propose an approximation technique for time series based on mathematical models. In fact, each time series is described by the coefficients of a mathematical model involving a given set of reference time series. The great benefit is that the size of our approximation depends only on the number of coefficients of the model (i.e. the number of reference time series). In particular, our method is independent of the length of the original time series and is thus suitable also for very long time series. The compact representation using a feature vector of coefficients of the model allows efficient indexing of the time series approximations for fast similarity search and clustering. We further show how our proposed approximations can be used for approximate clustering. In our experimental evaluation, we illustrate that our novel method outperforms existing state-of-the-art approximation methods in terms of clustering accuracy, i.e. our approximations are significantly better than existing schemata.

In our future work, we plan to extend our ideas of approximating time series by means of mathematical models to stream data.

References

1. Guttman, A.: R-Trees: A dynamic index structure for spatial searching. In: Proceedings of the SIGMOD Conference, Boston, MA, pp. 47–57 (1984)
2. Faloutsos, C., Ranganathan, M., Maolopoulos, Y.: Fast Subsequence Matching in Time-series Databases. In: Proceedings of the SIGMOD Conference, Minneapolis, MN (1994)
3. Agrawal, R., Faloutsos, C., Swami, A.: Efficient Similarity Search in Sequence Databases. In: Proc. 4th Conf. on Foundations of Data Organization and Algorithms (1993)
4. Wichert, S., Fokianos, K., Strimmer, K.: Identifying Periodically Expressed Transcripts in Microarray Time Series Data. Bioinformatics 20(1), 5–20 (2004)
5. Chan, K., Fu, W.: Efficient Time Series Matching by Wavelets. In: Proceedings of the 15th International Conference on Data Engineering (ICDE), Sydney, Australia (1999)

6. Yi, B.K., Faloutsos, C.: Fast Time Sequence Indexing for Arbitrary Lp Norms. In: Proceedings of the 26th International Conference on Very Large Data Bases (VLDB), Cairo, Egypt (2000)
7. Cai, Y., Ng, R.: Index Spatio-Temporal Trajectories with Chebyshev Polynomials. In: Proceedings of the SIGMOD Conference (2004)
8. Korn, F., Jagadish, H., Faloutsos, C.: Efficiently Supporting Ad Hoc Queries in Large Datasets of Time Sequences. In: Proceedings of the SIGMOD Conference, Tucson, AZ (1997)
9. Alter, O., Brown, P., Botstein, D.: Generalized Singular Value Decomposition for Comparative Analysis of Genome-Scale Expression Data Sets of two Different Organisms. Proc. Natl. Aca. Sci. USA 100, 3351–3356 (2003)
10. Keogh, E., Chakrabati, K., Mehrotra, S., Pazzani, M.: Locally Adaptive Dimensionality Reduction for Indexing Large Time Series Databases. In: Proceedings of the SIGMOD Conference, Santa Barbara, CA (2001)
11. Bar-Joseph, Z., Gerber, G., Jaakkola, T., Gifford, D., Simon, I.: Continuous Representations of Time Series Gene Expression Data. J. Comput. Biol. 3-4, 341–356 (2003)
12. Ratanamahatana, C.A., Keogh, E., Bagnall, A.J., Lonardi, S.: A Novel Bit Level Time Series Representation with Implication for Similarity Search and Clustering. In: Ho, T.-B., Cheung, D., Liu, H. (eds.) PAKDD 2005. LNCS (LNAI), vol. 3518, Springer, Heidelberg (2005)
13. Han, J., Kamber, M.: Data Mining: Concepts and Techniques. Academic Press, London (2001)
14. Halkidi, M., Batistakis, Y., Vazirgiannis, M.: On Clustering Validation Techniques. Intelligent Information Systems Journal (2001)

An Approach for Extracting
Bilingual Terminology from Wikipedia

Maike Erdmann, Kotaro Nakayama, Takahiro Hara, and Shojiro Nishio

Dept. of Multimedia Engineering,
Graduate School of Information Science and Technology,
Osaka University
1-5 Yamadaoka, Suita, Osaka 565-0871, Japan
{erdmann.maike,nakayama.kotaro,hara,nishio}@ist.osaka-u.ac.jp
http://www-nishio.ist.osaka-u.ac.jp

Abstract. With the demand of bilingual dictionaries covering domain-specific terminology, research in the field of automatic dictionary extraction has become popular. However, accuracy and coverage of dictionaries created based on bilingual text corpora are often not sufficient for domain-specific terms. Therefore, we present an approach to extracting bilingual dictionaries from the link structure of Wikipedia, a huge scale encyclopedia that contains a vast amount of links between articles in different languages. Our methods analyze not only these interlanguage links but extract even more translation candidates from redirect page and link text information. In an experiment, we proved the advantages of our methods compared to a traditional approach of extracting bilingual terminology from parallel corpora.

1 Introduction

Bilingual dictionaries are required in many research areas, for instance to enhance existing dictionaries with technical terms [1], as seed dictionaries to improve machine translation results, in cross-language information retrieval [2] or for second language teaching and learning. Unfortunately, the manual creation of bilingual dictionaries is not efficient since linguistic knowledge is expensive and new or highly specialized domain specific words are difficult to cover.

In recent years, a lot of research has been conducted on the automatic extraction of bilingual dictionaries. Dictionary extraction from large amounts of bilingual text corpora is an emerging research area. However, that approach faces several issues. Particularly, for very different languages or for domains where sufficiently large text corpora are not available, accuracy and coverage of translation dictionaries are rather low.

Therefore, in order to provide a high accuracy and high coverage dictionary, we propose the extraction of bilingual terminology from multilingual encyclopedias like Wikipedia. Wikipedia is a very promising resource as the continuously growing encyclopedia already contains more than 5 million articles in several hundred languages and a broad variety of topics. We already proved that Wikipedia can

J.R. Haritsa, R. Kotagiri, and V. Pudi (Eds.): DASFAA 2008, LNCS 4947, pp. 380–392, 2008.

be used to create an accurate association thesaurus [3][4] since it has a very dense link structure.

In addition, Wikipedia has a lot of links between articles in different languages. If we regard the titles of Wikipedia articles as terminology, it is easy to extract translation relations by analyzing the interlanguage links, assuming that two articles connected by an interlanguage link are likely to have the same content and thus an equivalent title.

On the other hand, an article in the source language has usually at most one interlanguage link to an article in the target language. Thus, creating a dictionary from interlanguage links only leads to a low coverage for cases where several correct translations for a term exist.

Therefore, we propose new methods to improve the coverage while maintaining a high accuracy. Our methods use redirect pages as well as link texts to extend the number of translations for a given term. In order to evaluate our methods, we extracted Japanese translations for 200 English sample terms and compared the accuracy and coverage of these translations to the translations extracted from a parallel corpus.

The remainder of this paper is organized as follows. We give an overview on manual dictionary construction and on the state of art in automatic dictionary construction from bilingual texts in section 2 and present our approach in section 3. In section 4, we describe the experiment we conducted to evaluate our methods and discuss its results. Finally, we conclude the paper in section 5.

2   Related Work

For bilingual dictionary construction, we can distinguish two approaches: manual and automatic dictionary creation. We discuss both approaches in the following subsections.

2.1   Manual Dictionary Construction

The traditional way of creating bilingual dictionaries is the manual compilation by human effort. Nowadays, paper-based dictionaries are being more and more replaced by machine readable dictionaries. Besides, those dictionaries are often not created by linguists but voluntarily by a large community of second language learners and other users.

For translations from English to Japanese, one of the most commonly used dictionaries is the freely available online dictionary EDICT. The JMdict/EDICT project [5] was started in 1991 by Jim Breen and the dictionary file has been extended by a large amount of people since then. It comprises more than 99,300 terms as of 2004 including even an impressive large amount of entries for domain-specific terms.

However, even with the aid of a large community, the manual creation of a dictionary is a time-consuming process. In the case of EDICT, it took over 10 years and the effort of numerous people to achieve the current dictionary size. Even though it now covers an impressively high number of terms, latest terms and domain-specific terms are not covered exhaustively. In addition, the correctness of dictionary entries is not guaranteed when e.g. language learners participate, thus the refinement of dictionary entries is time-consuming as well.

2.2 Automatic Dictionary Construction

Nowadays, a lot of machine readable documents in multiple languages are being created every day and often published on the Internet for everyone to access. That has lead to the idea of automatically creating bilingual dictionaries using these resources, thus reducing the burden of manual dictionary compilation. In today's research, mainly two approaches can be distinguished. The first approach uses parallel corpora, bilingual text collections consisting of texts in one language and their translations into another language. For Japanese-English dictionary extraction, e.g. corpora of paper abstracts [6] or software documentations [7] have been exploited.

However, while for high frequency terms usually good results can be achieved, the accuracy decreases drastically when the term to be translated is not present in the corpus in a large quantity. This is often the case for domain-specific terms.

Furthermore, the accuracy of these dictionaries is rather low for language pairs from very different language families like Japanese and English, since the construction relies on natural language processing. For instance, in Asian languages sentence boundaries tend to be in different places than in sentences of European languages. Besides, Fung and McKeown [7] stated that a parallel corpus often does not contain exact translations. For grammatical reasons, or just in order to add supplementary information not generally known by the readers of one language version, some text can be added. Respectively, text can be omitted or presented in a different way in one language.

Another problem is that not all domains and languages have sufficiently large parallel corpora available, thus the coverage of the dictionary remains insufficient. Also the collection, e.g. due to copyright restrictions, preparation and analysis of large parallel corpora can be troublesome.

For these reasons, for language pairs such as Japanese and English, the use of comparable corpora is also interesting. A comparable corpus contains not exact translations but texts from the same domain. Thus we can assume that similar terminology is covered. Among others, research using a corpus of Japanese patent abstracts with non-verbatim English translations [1] or research using newspaper articles [2][8] have been conducted. Although it is much easier to collect a comparable corpus than a parallel corpus, it is even more difficult to obtain a sufficient accuracy.

Altogether, the usage of parallel or comparable corpora for automatic dictionary construction is a very interesting approach. However, achieving sufficient accuracy and coverage is still difficult for less frequent terms as well as for certain language pairs and text domains.

English Article: Japanese Article:

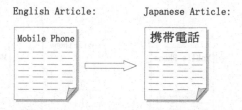

Fig. 1. Interlanguage Link Example

3 Proposed Methods

Our idea is to use a multilingual Web-based encyclopedia such as Wikipedia for extracting bilingual terminology. Wikipedia currently contains more than 5 million articles. It covers general topics, domain specific topics as well as proper nouns, containing even latest terminology since Wikipedia is being updated all the time. Moreover, Wikipedia contains a lot of links among its articles, not only within the articles of one language but also between articles of different languages. As opposed to the plain text in bilingual corpora, Wikipedia links contain to some extent semantic information. For instance, an interlanguage link usually indicates that one page title is the translation of the other. This can decrease difficulties of dictionary creation caused by natural language processing issues.

Wikipedia is being created manually by a large number of contributors. However, we can reuse the contributions for the creation and maintenance of the bilingual dictionary, thus no additional human effort is needed. Apart from that, when using a pivot language such as English, we can translate words from and to minor languages even if there is no direct translation.

3.1 Wikipedia Link Structure

In order to create a high accuracy and high coverage dictionary, we analyzed several kinds of link information. Prior to describing our methods, we illustrate the link structure information usage in the following clauses.

Interlanguage Links. An interlanguage link in Wikipedia is a link between two articles in different languages as shown in Figure 1. We assume that in most cases, the titles of two articles connected by an interlanguage link are translations of each other.

Redirect Pages. Redirect pages in Wikipedia, shown in Figure 2 (left), are pages containing no content but a link to another article (target page) in order to facilitate the access to Wikipedia content.

When a user accesses a redirect page, he will automatically be redirected to the target page. Redirect pages are usually strongly related to the concept of the target page. They often indicate synonym terms, but can also be abbreviations, more scientific or more common terms, frequent misspellings or alternative spellings etc.

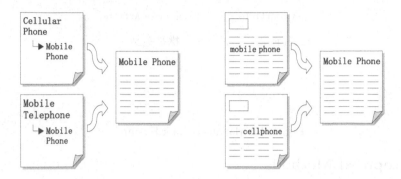

Fig. 2. Redirect Page (left) and Link Text (right) Examples

Link Texts. A link text, shown in Figure 2 (right), is the text part of a link, i.e. the text that is presented to the user in the browser where he clicks to reach the target page.

In Wikipedia, the title of the target article is displayed as the link text by default. However, link texts can be changed freely by creating so called piped links.

We extract the link text information by analyzing all internal links, i.e. links within one language version of Wikipedia. We already realized that link texts are usually strongly related to the target page title [3][4]. In many cases, they differ only in capitalization, but sometimes they are changed in other ways to fit in the sentence structure of the linking article. Therefore, they can help to overcome NLP problems such as finding a translation for a term in plural form when there is only a dictionary entry for the singular form. In some cases however, link text contains unprofitable terms, such as style information in form of HTML tags.

Forward/Backward Links. For all the above mentioned kinds of links, we distinguish the link direction. As shown in Figure 3, a forward link is an outgoing hyperlink and a backward link is an incoming hyperlink of a Web page. Both forward and backward links of Wikipedia articles are useful information for extracting translation candidates. Furthermore, the number of backward links is a valuable factor for estimating the quality of a translation candidate as we describe in the following subsections.

3.2   Extraction of Translation Candidates

In the following clauses, we describe how we extract a baseline dictionary from interlanguage links and present three methods for enhancing that dictionary; the redirect page method (RP method), the link text method (LT method), and the combination of both methods (RP ∪ LT method). Some of the variables defined in the following clauses are visualized in Figure 4.

3.3   Extraction of Interlanguage Links

At first, we create a baseline dictionary from Wikipedia by extracting all translation candidates from interlanguage links. The flow is described as follows.

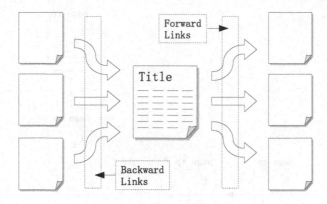

Fig. 3. Forward and Backward Links

For a term to be translated, a Wikipedia source page sp is extracted if its title s is equivalent to that term. In cases where the term is equivalent to the title of a redirect page, the corresponding target page is used as sp and its title as s.

In the second step, in case sp has an interlanguage link to a page tp in the target language, the title t of tp will be chosen as the translation, thus the set of translation candidates TC is defined as:

$$TC(s) = \{t\} . \tag{1}$$

Enhancement by Redirect Pages. The idea of the RP method is to enhance the dictionary with the set of redirect page titles R of all redirect pages of page tp. The list of translation candidates TC is hence defined as:

$$TC(s) = \{t\} \cup R(tp) . \tag{2}$$

As mentioned before, not all redirect pages are suitable translations. Therefore, we want to assign a score to all extracted translation candidates and filter doubtful terms through a threshold.

We found out experimentally that the number of backward links of a page can be used to estimate the accuracy of a translation candidate, because redirect pages where the title is wrong or semantically not related to the title of the target page usually have a small number of backward links. This approach has already proved effective in creating the Wikipedia Thesaurus [3][4].

We calculate the score of a redirect page title r of a redirect page rp by comparing the number of backward links of rp to the sum of backward links of tp and all its redirect pages.

The score s_{rp} is hence defined by the formula:

$$s_{rp} = \frac{|\text{Backward links of } rp|}{|\text{Backward links of } tp \text{ and of all redirect pages of } tp|} . \tag{3}$$

We can calculate the score of the target page title t in an analogous manner. Usually, redirect pages have much less backward links than target pages.

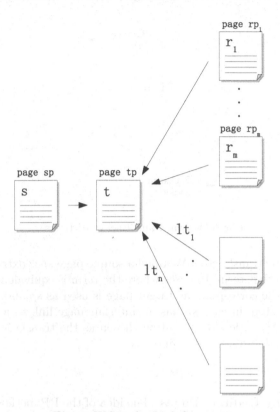

Fig. 4. Wikipedia Link Structure

However, redirect pages with more backward links than the corresponding target page also exist, indicating that the redirect page title is a good translation candidate, potentially even better than the target page title.

Enhancement by Link Texts. The LT method enhances the dictionary created from interlanguage links with the set of link texts LT of all inner language backward links of tp. The list of translation candidates TC is thus defined as:

$$TC(s) = \{t\} \cup LT(tp) . \tag{4}$$

Like for the RP method, we filter unsuitable translations extracted by the LT method by setting a threshold. We calculate the score s_{lt} of a link text lt by comparing the number of backward links of tp containing the link text lt to the total number of backward links of tp:

$$s_{lt} = \frac{|\text{Backward links of } tp \text{ with link text } lt|}{|\text{Backward links of } tp|} . \tag{5}$$

3.4   Enhancement by Redirect Pages and Link Texts

At last, we combine the RP method and the LT method, thus the list of translation candidates TC can be enhanced as follows:

$$TC(s) = \{t\} \cup R(tp) \cup LT(tp) .\tag{6}$$

The overall score s of a translation candidate $c \in TC$ can now be calculated by the following formulas.

If c is the target page title or a redirect page title and at the same time a link text $(c \in (\{t\} \cup R) \wedge c \in LT)$, the score is the weighted sum of s_{rp} and s_{lt}:

$$s = (w_{rp} \cdot s_{rp}) + (w_{lt} \cdot s_{lt}) .\tag{7}$$

If c is the target page title or a redirect page title but not a link text $(c \in (\{t\} \cup R) \wedge c \notin LT)$, the score is calculated by only s_{rp}:

$$s = (w_{rp} \cdot s_{rp}) .\tag{8}$$

If c is a link text but neither the target page title nor a redirect page title $(c \in LT \wedge c \notin (\{t\} \cup R))$, the score is calculated by only s_{lt}:

$$s = (w_{lt} \cdot s_{lt}) .\tag{9}$$

The variables w_{rp} and w_{lt} represent weight factors to normalize the score. For our experiment, we chose $w_{rp} = w_{lt} = 0.5$. We also tested other values but could not detect significant influence on the result.

4 Evaluation

We conducted an experiment in which we compared the translations of 200 terms extracted by our methods to the translations extracted from a parallel corpus. In the following, we describe the experiment and discuss its results.

4.1 Extraction from Wikipedia

We downloaded the English and Japanese Wikipedia database dump data from November/December 2006 [9] containing 3,068,118 English and 455,524 Japanese pages (including redirect pages). From that data, we extracted all interlanguage links, link texts and redirect pages as well as the number of backward links for each page. In total, we extracted 103,374 interlanguage links from English to Japanese, 108,086 interlanguage links from Japanese to English, 1,345,318 English and 91,898 Japanese redirect pages, 7,215,301 different English and 2,019,874 different Japanese link texts. In order to improve the accuracy, we applied several thresholds to filter terms with a low score.

4.2 Extraction from a Parallel Corpus

We compared the translations extracted by our approach to a dictionary extracted from the parallel corpus JENAAD [10]. With 150,000 one-to-one sentence alignments in each language, that corpus consisting of Japanese and English versions of Yomiuri newspaper articles is one of the largest, compared with other Japanese-English parallel corpora. The corpus has the advantage of being already sentence-aligned (each sentence in one language is paired with exactly one sentence in the other language) and the Japanese text is split into chunks, a procedure that is indispensable to isolate terms since the Japanese language does not use word boundaries.

We used the IBM Models 1-5 [11] in combination with the Hidden Markov Model [12], both standard models in word alignment research, to train the parallel corpus.

The training was accomplished using the open source training tool GIZA++ [13] and the translation candidates were then extracted from the inverse probability table created by GIZA++. Each line of the table consists of a word in the source language, a translation and a score (in the following referred to as s'). In total, we extracted 1,033,086 translation pairs. The coverage of the dictionary, however, is much smaller than expected from the number of translation pairs, since it contains a lot of noise, i.e wrong translations with very low scores. In order to improve the accuracy, it was therefore crucial to define thresholds to filter terms with low values of s'.

4.3   Term Selection

The experiment was conducted on 200 English terms. Since the parallel corpus which we used for the experiment contains newspaper articles, we chose only test terms from the newspaper domain in order to ensure a fair evaluation. The terms consist exclusively of nouns since the titles of Wikipedia articles usually are nouns. Apart from that, only terms consisting of one word were selected because the dictionary created by GIZA++ does not translate word compounds. The terms were divided into two categories.

High Frequency Terms: 100 terms were high frequency terms which we selected automatically using the most frequent nouns in the parallel corpus.

Low Frequency Terms: 100 terms were low frequency terms. These terms were chosen by native speakers and people fluent in English. These persons were asked to list up technical terms found in English newspapers. We call these terms low frequency terms since they appear in the parallel corpus much less frequently than the terms in the first category, even though the term selectors were not instructed to choose low frequency words. We further split the low frequency terms into two categories with 50 terms that can be found in the dictionary EDICT and 50 terms that cannot be found in EDICT.

Most of the selected terms are common nouns such as "agreement", "franchise" or "presidency". Only in the category of low frequency terms not found in EDICT, half of the terms are proper nouns such as "Balkan" and "HIV", since it proved extremely difficult to find common nouns from the newspaper domain that are not covered in EDICT. However, proper nouns in Japanese are usually not written in the Latin alphabet, thus it still makes sense to cover these terms in a dictionary.

4.4   Comparison Criteria

We used the two standard criteria precision and recall to compare accuracy and coverage of our methods and the parallel corpus approach.

The precision measures the accuracy by calculating how many of the extracted translation candidates are correct:

$$precision = \frac{|\text{Extracted correct translations}|}{|\text{All extracted translations}|} . \tag{10}$$

The recall measures the coverage by calculating how many correct translations have been extracted by a method compared to the total number of correct translations. It is not trivial to estimate the total number of correct translations, since it cannot be calculated automatically. In our experiment, we estimated the value by using the manually counted number of correct translations from EDICT, since it contains a large amount of translations not only for general but also for domain-specific terms:

$$recall = \frac{|\text{Extracted correct translations}|}{|\text{Correct translations in EDICT}|} . \tag{11}$$

The term evaluation as well as the counting of correct EDICT translations were conducted by totally 12 judges, mostly native speakers of Japanese with a sufficient English proficiency.

4.5   Experiment Results

In the following, we discuss the results of our experiment based on precision, recall and the absolute number of extracted correct translations (in parentheses) shown in Table 1. For the parallel corpus approach, the results without using a threshold are not included, because the number of translation candidates would have been too high for manual evaluation.

High Frequency Terms. For high frequency terms, our methods achieve better results than the parallel corpus approach. Comparing our methods to using interlanguage links only (ILL method), we can see that for the threshold 0.5, the results are almost identical with the ILL method. For lower thresholds, the recall increases but this is at the expense of the precision. Thus, only for applications in which a high recall is more important than a high precision (i.e. cross-language information retrieval), our methods perform better than the ILL method.

Besides, the recall of our methods is rather low compared to the number of translations in EDICT. We believe that this is because high frequency terms are often well-known general terms and thus there is no need to cover them in Wikipedia articles. For instance, for the terms "work" and "situation", no translation candidates could be extracted from Wikipedia.

Another reason for the low recall is that if the term to be translated is ambiguous such as "party" or "diet", we often get only translations for one meaning of the word. Moreover, in some cases we do not get any translation at all, since disambiguation pages in Wikipedia often do not contain interlanguage links.

Table 1. Experiment Results

method	threshold	High Freq. Terms		Low Freq. Terms in EDICT		Low Freq. Terms not in EDICT	
		precision	recall	precision	recall	precision	recall
ILL	–	0.835	0.122 (66)	0.788	0.234 (26)	0.677	– (21)
	–	0.639	0.182 (99)	0.475	0.342 (38)	0.536	– (37)
RP	$s_{rp} > 0.01$	0.742	0.164 (89)	0.648	0.315 (35)	0.623	– (33)
	$s_{rp} > 0.1$	0.792	0.140 (76)	0.763	0.261 (29)	0.683	– (28)
	$s_{rp} > 0.5$	0.833	0.120 (65)	0.781	0.225 (25)	0.680	– (17)
	–	0.393	0.309 (168)	0.366	0.577 (64)	0.464	– (70)
LT	$s_{lt} > 0.01$	0.727	0.171 (93)	0.608	0.405 (45)	0.545	– (48)
	$s_{lt} > 0.1$	0.824	0.129 (70)	0.744	0.261 (29)	0.733	– (33)
	$s_{lt} > 0.5$	0.833	0.120 (65)	0.774	0.216 (24)	0.700	– (21)
	–	0.389	0.341 (185)	0.333	0.613 (68)	0.443	– (77)
RP ∪ LT	$s > 0.01$	0.752	0.195 (106)	0.629	0.396 (44)	0.623	– (48)
	$s > 0.1$	0.821	0.144 (78)	0.744	0.261 (29)	0.739	– (34)
	$s > 0.5$	0.831	0.118 (64)	0.781	0.225 (25)	0.667	– (18)
	$s' > 0.01$	0.305	0.420 (228)	0.128	0.252 (28)	0.115	– (7)
Parallel Corp.	$s' > 0.1$	0.637	0.214 (116)	0.324	0.207 (23)	0.280	– (7)
	$s' > 0.5$	0.811	0.079 (43)	0.500	0.135 (15)	0.545	– (6)

Low Frequency Terms in EDICT. For low frequency terms which can be found in EDICT, our methods achieve much better results than the parallel corpus approach. However, like for the high frequency terms, our methods do not perform better than the ILL method except when a lower precision can be accepted in return for a high recall.

The recall of our methods is better than for high frequency terms. That is because low frequency terms are often domain-specific terms such as "entrepreneurship" or "communism." Therefore, their coverage in Wikipedia is high compared to high frequency terms. Apart from that, the ambiguous term problem is less relevant in this category, since domain-specific terms are usually well defined.

For the parallel corpus, on the other hand, good translation results can only be achieved when a term is contained in the corpus in high quantity. For that reason, using a multilingual encyclopedia such as Wikipedia is more efficient for low frequency term translation.

Low Frequency Terms not in EDICT. Since we cannot calculate the recall value for this category, we only use the absolute number of correctly extracted terms to measure the coverage. We can see that also for low frequency terms which cannot be found in EDICT, the Wikipedia approach achieves much better results than the parallel corpus approach. Besides, unlike the other categories, our methods can for the threshold 0.1 achieve at the same time higher precisions and higher recalls than the ILL method, thus our methods perform better than using only interlanguage links.

The results in this term category show that our methods can not only achieve better results than methods of automatic dictionary construction from bilingual text corpora, but also improve our baseline dictionary. Moreover, since all terms in this category are not included in EDICT, we can see that our methods are also valuable to enhance manually constructed dictionaries. Wikipedia contains very specialized domain-specific terms not covered in EDICT and we thus can extract translations for even terms such as "jihadist" or "al-Quaeda."

5   Conclusion and Future Work

In this paper, we presented our approach of bilingual dictionary extraction from Wikipedia, a multilingual encyclopedia. We proposed three methods for extracting terminology which are using not only interlanguage links but also redirect page and link text information.

Our conviction that Wikipedia is an invaluable resource for bilingual dictionary extraction and that redirect pages and link texts are helpful to enhance a dictionary constructed from interlanguage links has been confirmed in our experiment. Our methods are very useful for specialized domain-specific terms (low frequency terms not in EDICT), because accuracy and coverage are much better than that of the bilingual text corpus approach and also better than the baseline dictionary created from interlanguage links only. Besides, we evaluated the methods only for single words. Domain-specific words are often word compounds and for those both accuracy and coverage are probably significantly better than for the bilingual text corpus approaches. Our dictionary can be accessed freely under the following URL.

http://wikipedia-lab.org:8080/WikipediaBilingualDictionary

We believe that Wikipedia will become even much more comprehensive in near future which will also result in a better coverage. For general terms, especially for word groups other than nouns, it is also promising to integrate manually constructed dictionaries like EDICT. For context-sensitive translations (e.g. machine translation), we can benefit from combining our approach with the parallel corpus approach.

We are planning to further enhance the accuracy and coverage of our dictionary by analyzing the redirect pages and link texts not only in the target language but also in the source language. It is also promising to analyze the text content of Wikipedia articles. By doing that, we are hoping to find translation candidates even when the interlanguage link is missing. Likewise, two Wikipedia articles connected by an interlanguage link but having utterly different content might advise against including them in the dictionary.

Acknowledgment

This research was supported in part by Grant-in-Aid on Priority Areas (18049050), and by the Microsoft Research IJARC CORE project.

References

1. Shimohata, S.: Finding translation candidates from patent corpus. In: Proceedings of the Machine Translation Summit, September 12-16, 2005, pp. 50–54 (2005)
2. Sadat, F., Yoshikawa, M., et al.: Bilingual terminology acquisition from comparable corpora and phrasal translation to cross-language information retrieval. In: The Companion Volume to the Proceedings of Annual Meeting of the Association for Computational Linguistics, July 2003, pp. 141–144 (2003)
3. Nakayama, K., Hara, T., Nishio, S.: A thesaurus construction method from large scale web dictionaries. In: IEEE International Conference on Advanced Information Networking and Applications (AINA 2007), pp. 932–939 (2007)
4. Nakayama, K., Hara, T., Nishio, S.: Wikipedia mining for an association web thesaurus construction. In: Benatallah, B., Casati, F., Georgakopoulos, D., Bartolini, C., Sadiq, W., Godart, C. (eds.) WISE 2007. LNCS, vol. 4831, Springer, Heidelberg (2007)
5. Breen, J.W.: Jmdict: a japanese-multilingual dictionary. In: COLING Multilingual Linguistic Resources Workshop (August 2004)
6. Tsuji, K., Kageura, K.: Automatic generation of japanese-english bilingual thesauri based on bilingual corpora. Journal of the American Society for Information Science and Technology 57(7), 891–906 (2006)
7. Fung, P., McKeown, K.: A technical word- and term-translation aid using noisy parallel corpora across language groups. Machine Translation 12(1-2), 53–87 (1997)
8. Kaji, H.: Adapted seed lexicon and combined bidirectional similarity measures for translation equivalent extraction from comparable corpora. In: Proceedings of the Conference on Theoretical and Methodological Issues in Machine Translation, October 4-6, 2004, pp. 115–124 (2004)
9. Wikimedia Foundation: Wikimedia downloads, http://download.wikimedia.org/
10. Utiyama, M., Isahara, H.: Reliable measures for aligning japanese-english news articles and sentences. In: Proceedings of the Annual Meeting of Association for Computational Linguistics, pp. 72–79 (2003)
11. Brown, P.F., Pietra, V.J.D., Pietra, S.A.D., Mercer, R.L.: The mathematics of statistical machine translation: parameter estimation. In: Proceedings of the International Conference on Computational Linguistics, vol. 19(2), pp. 263–311 (1993)
12. Vogel, S., Ney, H., Tillmann, C.: Hmm-based word alignment in statistical translation. In: Proceedings of the Conference on Computational Linguistics, pp. 836–841 (1996)
13. Och, F.J., Ney, H.: Improved statistical alignment models. In: Proceedings of the Annual Meeting of the Association for Computational Linguistics, October 2000, pp. 440–447 (2000)

Cost-Effective Web Search in Bootstrapping for Named Entity Recognition

Hideki Kawai[1], Hironori Mizuguchi[2], and Masaaki Tsuchida[2]

[1] NEC C&C Innovation Research Laboratories,
8916-47, Takayama-cho, Ikoma, Nara, Japan
h-kawai@ab.jp.nec.com
[2] NEC Service Platforms Research Laboratories,
8916-47, Takayama-cho, Ikoma, Nara, Japan
{hironori@ab,m-tsuchida@cq}jp.nec.com

Abstract. In this paper, we propose a cost-effective search strategy framework to extract keywords in the same semantic class from the Web. Constructing a dictionary based on the bootstrapping technique is one promising approach to harnessing knowledge scattered around the Web. Open web application programming interfaces (APIs) are powerful tools for the knowledge-gathering process. However, we have to consider the cost of API calls because too many queries can overload the search engines, and they also limit the number of API calls. Our goal is to optimize a search strategy that can collect as many new words as possible with the least API calls. Our results show that the optimized search strategy can extract 64,642 words in five different domains with a precision of 0.94 with only 1,000 search API calls.

Keywords: Dictionary Construction, Bootstrapping, Web API, Search Strategy.

1 Introduction

From the advent of the Internet, a vast amount of knowledge has been accumulated on the Web in various forms. Information Extraction (IE) can be one of the important technologies for harnessing distributed and heterogeneous knowledge on the Web. In an IE system, Named Entity Recognition (NER) is a key task that identifies linguistic expressions that refer to a specific entity, such as the names of people, organizations, or locations [1-3]. The extraction patterns may be manually hard-coded or may be generated automatically using machine learning technique with a large amount of annotated training corpora [4-6].

Bootstrapping is an unsupervised approach to constructing dictionaries for the NER task [7-9]. Starting from a small set of keywords (*seed words*) in the same semantic class, the algorithm can grow the seed words by iterating pattern generation and instance extraction processes. The pattern generation process finds all occurrences of seed words in a corpus, and it detects globally common expressions around seed words as patterns. At the same time, it also calculates the reliability of patterns because a low quality pattern can yield many false instances. The instance

J.R. Haritsa, R. Kotagiri, and V. Pudi (Eds.): DASFAA 2008, LNCS 4947, pp. 393–407, 2008.

extraction process extracts new instances by highly reliable patterns, and assigns a score for each instance based on the reliability of extraction patterns. Then, only high-scored instances can be added to seed words, because false instances can generate false extraction patterns during later iterations.

Because the document scanning processes to enumerate global patterns and extract new instances are expensive and time consuming, some previous work often relied on relatively small-sized clean text collections containing typical expressions such as news corpora [10-11]. However, we can exploit a huge amount of web pages as a text corpus with web search engines. With the emerging trend of Web 2.0 [12], web search engines such as Yahoo! [13], Google [14] and MSN [15] have opened their Application Programming Interfaces (APIs). KnowItAll [8] is one of the search-oriented bootstrapping systems.

However, most search-oriented systems have ignored the cost of search API calls. Although the search APIs are available at no charge so far, it would be inappropriate to overload the search engines. So we should regard the search API calls as a cost like a disk access in memory management. Actually, the cost is not only response time, but is also associated with some limitations for the usage of search APIs. For example, even if more than one million pages match a query, search APIs return only a small subset (*result set*) of URLs at once. Usually, the size of the result set is only 10 to 100 depending on the search engine provider. Additionally, the maximum number of pages for a query and the number of API calls per day are also limited.

In this paper, we propose a Cost-Effective Search Strategy (CESS) framework for an automatic dictionary construction. Our goal is to provide an algorithm that can rapidly grow the seed words with fewer search API calls. The key ideas underlying our solution are (1) multi-seed combination query, and (2) single-use local pattern. To illustrate our approach, suppose that you randomly pick five actor names and input them as an 'AND' query to the search engine, what kind of pages can be retrieved? Probably, some pages in the search result may be written in a structured manner such as movie listings and actor index. Within a listing page, local context around the five names can be coherent, and you may also find many other instances that appear in the same context. So, you can use the local context as an extraction pattern, instead of a global pattern. However, this pattern can be applied only once because other pages are written in different formats. Our experimental results demonstrate the efficiency and robustness of our framework.

In summary, our main contributions are:

- We describe a brief overview of the CESS framework, and show how we model the bootstrapping process. (Section 3)
- We illustrate the main idea of multi-seed combination query, and propose a search strategy based on the distribution of the seed words (Section 3.1).
- We define a structural coherence of a web page, and show how we detect a single-use local pattern and extract new seeds (Section 3.2).
- We demonstrate the efficiency and robustness of our framework by constructing large dictionaries with limited search API calls (Section 4).

2 Related Work

For information extraction tasks, semantic lexicons that consist of word lists in the same semantic class have been almost always constructed by hand because general purpose dictionaries, such as WordNet [16], do not contain the necessary domain-specific vocabulary. Early approaches for automatic dictionary construction, such as AutoSlog [4], CRYSTAL [5], and RAPIER [6] relied on domain-specific rules or a supervised approach based on machine learning technique.

While bootstrapping approaches adopted by AutoSlog-ST [7], KnowItAll [8] and Pasca et al. [9] rely on an unsupervised manner, input for those systems is basically only un-annotated texts and a handful of sets of seed words or extraction rules However, the document scanning process to find global patterns is tedious and time consuming. Brin [17] estimated that only the single extraction phase would have taken a couple of days to run over 27 million web pages.

Web search engines are powerful tools for obtaining the occurrence of seed words in web pages. KnowItAll [8] uses multiple search engines including Google, Alta Vista, and Fast. In their experiments, they fetched 313,349 web pages over 92 hours using recursive query expansion. Soderland et al. [18] also used search engines to validate extracted instances. According to their validation process, they compute co-occurrence statistics for combinations of new instances and discriminators, such as "Chicago" and "the cities of $<City>$". So if there are m instances and n discriminators, they have to issue $m \times n$ queries to validate all instances. Thus, most traditional works relied on a very greedy search strategy, or they just ignored the cost of API calls.

By contrast, we focus on providing a cost-effective search strategy that can save the number of search API calls during each phase of bootstrapping. The main features of the CESS framework are (1) multi-seed combination query and (2) single-use local pattern. The multi-seed combination query aims to get pages including seed words as a list or table. And instead of enumerating global patterns, we detect a coherent local pattern within a single page, and use it for instance extraction only once within the page. The benefit of these features is not only reducing search cost but also making our framework robust for false instances.

There is another pattern extraction technique that focuses on structured web pages. Wrapper induction systems [19-21] are able to learn extraction patterns in the given information sources. By contrast, the CESS framework has a search strategy to find the information sources. Because the search results contain heterogeneous pages, our framework separates the wheat from the chaff based on the structural coherence of each page, and integrates results from thousands of sites. SEAL [22] exploits a character-level wrapper which is similar to our single-use local pattern, but they do not consider search strategy for the bootstrapping algorithm.

3 CESS Framework

Figure 1 shows the bootstrapping algorithm in the CESS framework. This framework consists mainly of two processes, search and extraction. Here, we describe the detailed procedures:

```
CESS_Bootstrap(S,α){
    apiCalls ← 0
    ITbl ← φ
    while apiCalls < α
        // Search Process
        Q ← QueryBuilder(S)
        D ← Crawler(Q)
        apiCalls = apiCalls + 1
        // Extraction Process
        for each d in D
            P ← PatternGenerator(d,S)
            add InstanceExtractor(d,P) to ITbl
        endfor
        add SeedSifter(ITbl) to S
    endwhile
    return(S);
}
```

Fig. 1. Bootstrap algorithm in CESS framework

The algorithm receives an initial set of seed words $S = \{s_1, s_2, ..., s_n\}$ and a maximum number of search calls α from a user. At the first stage, the algorithm initializes an API call counter *apiCalls* to be zero and an instance table *ITbl* to be empty. In the search process, the procedure *QueryBuilder(S)* picks up ω words from the seed words S and combines them into an 'AND' query Q. Next, the procedure *Crawler(Q)* calls the search API with the query Q and gets a result set D, which consists of web pages returned from the search API. Then, the algorithm increments *apiCalls* by 1. Note that the procedure *Crawler(Q)* calls the search API only once. So the number of pages in the result set D is at most the size of a result set (generally, 10 to 100 depending on the search API). This is because getting only the first result set for one query can grow the seed words faster than getting all result sets for one query. For example, if you can search only three times, which search strategy can grow the seed words faster: getting the first result sets for three queries, such as "s_1 AND s_2 AND s_3", "s_4 AND s_5 AND s_6" and "s_7 AND s_8 AND s_9", or getting three result sets for "s_1 AND s_2 AND s_3"? Intuitively, the former strategy can grow the seed words faster because new instances extracted from the result sets can be more diverse than that of the latter strategy. Also, note that the procedure *QueryBuilder()* has a query history internally to avoid yielding the same word combination as the query issued before. Likewise, the procedure *Crawler()* also has an access history to avoid accessing the same page redundantly.

In the extraction process, for each page d, the procedure *PatternGenerator(d, S)* searches each seed word $s_1, s_2, ..., s_n$, and if any seed words are found in page d, the procedure takes statistics about contexts around them. And if there are any coherent expressions around the seed word, the procedure puts them in the pattern set P. Otherwise, it makes P empty. Then, the procedure *InstanceExtraction(d, P)* scans the page d and extracts instances I matching any pattern p in P. It also evaluates the structural coherence of page d based on the number of extracted instances $|I|$. If more than τ instances are extracted from page d, the procedure regards d as a well-structured page such as a listing page. We call τ the listing likelihood parameter. If page d is a listing page, the procedure makes a tuple $<i, p>$ for each pair of extracted instances i and matching patterns p, and adds all touples to the instance table $ITbl$. Otherwise, it discards the extracted instances I because we regard the quality of instances I from unstructured pages to be low. If the extraction pattern P is empty, the procedure $InstanceExtraction(d, P)$ makes $ITbl$ empty. Then, the procedure $SeedSifter(ITbl)$ scans for each instance i in the instance table and counts unique patterns, which are the pairs of i. We also assumed that the instance with higher pattern variation is more likely to be a good instance, so only if the number of unique patterns around i is bigger than a threshold θ, the procedure adds the instance i to the seed words S. We call θ the seed sifter threshold. The main reason why we use the variation of pattern instead of frequency is because the validation based on frequency can overestimate the instances from database-driven pages. It sounds like a paradox, because our search strategy focuses on retrieving structured pages such as listing pages. But in our pilot study, we found that the good instances are written in a coherent style within a page and also written in various styles in different pages. So our framework can take advantage of the heterogeneity of web pages in $SeedSifter()$. Finally, the algorithm returns the seed words S if $apiCalls$ equals α.

3.1 Multi-seed Combination Query

Here, we describe a search strategy: How to choose the seed words for a query. The simplest way is to just randomly combine ω words from seed words, and input them as an 'AND' query to a search engine. Basically, the ratio of well-structured pages in a result set can be higher for larger ω, but too many ω can yield no results. So the number of words ω can be a parameter for search strategy optimization. We call ω the seed parameter.

Additionally, a search strategy based on uniform random sampling can cause a problem when the seed words grow larger. Supposing that the seed words consist of 10,000 actor names, it is highly possible that a combination query based on uniform random sampling contains some unfamiliar actors' names, which can yield no search results due to the sparseness of seed words. Additionally, low quality seed words can also slip into the query, so we propose a power distribution-based search strategy that can selectively use high-frequency and high-reliability words as a query. We defined the score of a seed word s_j as follows:

$$Score(s_j) = v_j / v_{max} , \tag{1}$$

where v_j is the number of pattern variations of s_j computed in the procedure $SeedSifter()$, and v_{max} is the maximum number of pattern variations in the seed words S. Note that $Score(s_j)$ takes a value from 0 to 1. This score can represent both the frequency and reliability of the word approximately.

Let $rand$ be a function that returns uniform random real numbers in $[0, 1)$, i.e. the probability of $rand < \alpha$ with $0 < \alpha < 1$ is α. And let '$\wedge$' denote a power operator. The search strategy in the procedure $QueryBuilder()$ is shown in Fig. 2. At first, the algorithm initializes sampled seeds SS to be empty. Next, the algorithm generates two uniform random numbers j and k, where $0 \leq j < |S|$, and $0 \leq k < 1$. And if k is less than $\{Score(s_j)\}^\gamma$, the seed word s_j is removed from S and added to SS. We call γ the search strategy parameter. If the size of SS becomes the same as the word parameter ω, the algorithm returns a query combining SS with 'AND' operators. Otherwise, the algorithm generates two random numbers again to pick another seed word.

In this algorithm, the probability that a seed word is chosen as a query depends on the score of the seed word and the search strategy parameter γ. The parameter γ plays an important role because the distribution of scores in seed words S is not uniform. To illustrate, suppose that there are 10,000 seed words in S, and 10 of them are scored 1.0, and 1,000 of them are scored 0.1. In this case, if the parameter γ is 1, the probability $Pr(1.0)$ that a word with score 1.0 is chosen as a query keyword is $(10/10,000) \times 1.0^1 = 0.001$, while the probability $Pr(0.1)$ that a word with score 0.1 is chosen as query keyword is $(1,000/10,000) \times 0.1^1 = 0.01$. Because $Pr(1.0) < Pr(0.1)$, it is still highly possible that a query contains many low-frequency and low-quality words. However, if the parameter γ is 3, $Pr(1.0) = 0.001$, and $Pr(0.1) = 0.0001$. In this case, the query can contain more high-frequency and high-quality words.

More formally, we assumed that the distribution of scores in S obeys the power law approximately. In our model, the number of words $w(t)$ in S with score t can be given as below:

$$w(t) = at^{-b} , \tag{2}$$

where a and b are constants. Here, we can define the probability $Pr(t)$ as follows:

$$Pr(t) = c \times \frac{w(t)}{|S|} \times t^\gamma , \tag{3}$$

where c is a normalizing constant to ensure the probability sums to 1 over all values of t. Now, we describe how a combination query will differ from the search strategy parameter γ. Suppose a search strategy such that the probability $Pr(t)$ is proportional to the score t. In this strategy, if $u = kt$, $Pr(u)$ should be equal to $kPr(t)$. This condition along with using (2) and (3) yields the following equation:

$$c \times \frac{a(kt)^{-b}}{|S|} \times (kt)^\gamma = k \times c \times \frac{at^{-b}}{|S|} \times t^\gamma . \tag{4}$$

By solving this equation, we find:

$$\gamma = b + 1 . \tag{5}$$

If the distribution of scores in seed words S follows Zipf's law, the exponent b is close to unity. So when $\gamma = 2$, the probability that a seed word with the score t can be chosen as a query keyword is proportional to t. And if $\gamma > 2$, the search strategy can be strongly biased to choose a seed word with a higher score. And note that $\gamma = 0$ corresponds to the search strategy based on the uniform random sampling we mentioned above.

```
QueryBuilder(S){
    SS ← φ
    while |SS| < ω
        j ← rand * |S|
        k ← rand
        if k < Score(S[j])^γ then
            remove S[j] from S and add to SS
        endif
    endfor
    Q ← combine SS into an AND query
    return(Q);
}
```

Fig. 2. Search strategy in QueryBuilder

3.2 Single-Use Local Pattern

Our pattern generation is not a tag-based but a character-based algorithm. This is because coherent patterns can be found not only between HTML tags but also in the fixed text format. For example, suppose a movie title extraction from a page where the movie title and the year is listed in the fixed text format, like "- Modern Times (1936)", "- Bride of Frankenstein (1935)", "- Casablanca (1944)". The character-based algorithm can find the left-hand pattern as "- ", and the right-hand pattern as " (19". Of course, " (20" can also be the right-hand pattern, if the page also includes 2000s movies. And HTML tags can be treated as a kind of fixed text format.

The algorithm of the procedure *PatternGenerator*() is shown in Figure 3. This algorithm consists of two main processes, i.e., pattern detection and coherence validation. At first, the algorithm initializes a left-hand pattern set *LP*, a right-hand pattern set *RP*, and a combination pattern set *P* to be empty. And it also initializes a match counter *matchCnt* to be 0.

In the pattern detection process, for each *s* in seed words *S*, the algorithm scans the page *d* and finds every matching point, and counts up *matchCnt*. It also adds 2 to 50-letter suffixes followed by *s* to *LP*, and likewise adds 2 to 50-letter prefixes following s to *RP*. At this point, *LP* and *RP* contain all patterns around the seed words that occurred in page *d*.

In the coherence validation process, the coherence of each pattern is validated by the function defined as follows:

$$coherence(xp) = \frac{occurence(xp)}{matchCnt} ,\qquad (6)$$

where *xp* denotes a left-hand pattern *lp* in *LP* or a right-hand pattern *rp* in *RP*, and *occurrence(xp)* denotes the occurrence of *lp* in *LP* or the occurrence of *rp* in *RP*. A higher *coherence(xp)* indicates that the seed words are written with the fixed pattern *xp* in page *d*. And a lower *coherence(xp)* indicates that many seed words are written in a different pattern from *xp*. So the process discards every pattern *xp* such that *coherence(xp)* is less than a threshold μ. We call μ the pattern coherence parameter.

```
PatternGenerator(d,S){
    LP ← φ
    RP ← φ
    P ← φ
    matchCnt ← 0

    // Pattern Detection
    for each s in S
        for each occurrence of s in d
            matchCnt ← matchCnt + 1
            for k = 2 to 50
                add k-letter suffix followed by s to LP
                add k-letter prefix following s to RP
            endfor
        endfor
    endfor

    // Coherence Validation
    remove all lp s.t. coherence(lp) < μ from LP
    remove all rp s.t. coherence(rp) < μ from RP

    add all LP-RP pairs <lp, rp> to P

    return P
}
```

Fig. 3. Local pattern detection in PatternGenerator

Finally, the procedure *PatternGenerator*(d, S) returns the pattern set P. These patterns strongly depend on the style of the page where the pattern set is generated. Let's get back to the movie title and year example. The left-hand pattern "- " and the right-hand pattern " (19" are based on the style of the page. And it is difficult to find other pages that are written in the same style. Thus, this pattern can be applied only once to the page where the pattern comes from.

3.3 Robustness for False Seeds

Unfortunately, if a false instance slips through the validation process in the procedure *SeedSifter*(), it becomes a false seed. For a bootstrapping algorithm, robustness for false seeds is important because a false seed can propagate exponentially by successive iteration processes. Here, we illustrate the robustness of our framework, derived from the multi-seed combination query and single-use local pattern.

In our framework, false seeds can cause bad effects in the following two cases:

Case 1: Query Contamination -- False seeds are used as a query.
Case 2: Pattern Contamination -- False patterns are generated by false seeds.

To consider query contamination, assume that 10% of your seed words are false seeds, and you use single word query like a traditional approach. The probability that you use a false seed as a query and get a completely bad search result is 0.1. However, what if you use a 3-seed combination query in our framework? The case in which you get completely bad search results is when all three of the seeds in a query are false seeds. The probability of this case is only $0.1^3 = 0.001$. The probability that two out of three are false seeds is also small, i.e., $_3C_2 \times 0.1^2 \times 0.9 = 0.027$. The probability that one out of three is a false seed is $_3C_1 \times 0.1 \times 0.9^2 = 0.247$, but the search results can be fair because over half of the query consists of good seeds. Likewise, the more you use seed words for a query, the less you get bad search results. And even if a few false seeds slip into the query, it can hardly cause a serious problem for the search results.

For the pattern contamination, the pattern validation function *coherence*() in the procedure *PatternGenerator*() plays an important role. If a page is unstructured, no contexts can have enough coherence as extraction patterns. This case is not a problem because the patterns around the false seeds are also ignored. How about a page that has a fixed structure around correct seed words? This is also not a problem because only correct patterns can pass through coherent validation and contexts around false seeds cannot survive. Of course, it is possible the contexts around false seeds can reduce the coherence of correct patterns and yield no extraction patterns. In this case, we may lose some good instances, but it is not so serious because high precision is much more critical than high coverage. The worst case is that a page is structured for only false instances, but the chances are slim. And even if false patterns are generated, their usage is limited to only once within the page. Therefore, the effect of pattern contamination is not likely to propagate through successive iterations.

4 Experimental Evaluation

We implemented a bootstrapping algorithm in the CESS framework using Perl, and conducted extensive experiments on five domains, i.e., Actor, Film, Location, Company and School. To avoid complaints of arbitrariness, we chose existing rankings for seed words. For the Actor domain, we picked 20 names of actors/actresses who won the Academy Award[1] (for Best Actor/Actress, and Best Supporting Actor/Actress) between 2002 and 2006. For the Film domain, we picked 20 movie titles that won the Academy Award for Directing between 1987 and 2006. For the Location domain, we picked the top 20 city names from the world's most expensive cities in 2007 as surveyed by Mercer Human Resource Consulting.[2] For the Company domain, we picked the top 10 companies in the Fortune 500 in 2007.[3] And for the School domain, we picked the top 10 American research universities surveyed by Arizona State University.[4]

We used the Yahoo! Web API [13] as the search engine. Parameters for the API were quite simple. A parameter *result* that specifies the size of the result set is designated as 50, and a parameter *start* that specifies the starting result position was

---

[1] Oscar.com, http://www.oscar.com/
[2] Mercer Human Resource Consulting, http://www.mercerhr.com/
[3] Fortune 500 2007, http://money.cnn.com/magazines/fortune/fortune500/
[4] The Top American Research Universities 2006, http://mup.asu.edu/research2006.pdf

fixed at 0. This means that we got only 50 URLs in the first result set for every query. We also set parameter *format* to 'html' and parameter *language* to 'en' to limit result pages to English HTML pages.

4.1 Extraction Parameters

To compare various search strategies on an equal footing, first we optimized three extraction parameters, i.e., the pattern coherence parameter μ, listing likelihood τ, and seed sifter threshold θ. To decide these parameters, we conducted a pilot experiment in the Actor domain by collecting about 1,000 pages with 3-seed combination queries based on uniform random distribution. In the pilot experiment, changing the value of some parameters, we randomly sampled and checked the correctness of extracted instances by the procedure *InstanceExtractor()* and new seed words updated by the procedure *SeedSifter()*. Finally, we decided the best combination of parameters $(\mu, \tau, \theta) = (0.5, 5, 5)$. In the following experiments, we fixed these parameters.

4.2 Search Strategies

Now, we compare several search strategies using two main criteria, i.e., the size and precision of final seed words. In the following experiment, we fixed the number of search API calls to 200. Although we conducted the same experiments in the Film domain, the results showed quite similar trends, so we only show and discuss the details of the Actor domain results.

(A) Uniform Random Sampling-Based Search Strategy ($\gamma = 0$)
To illustrate the basic characteristic of multi-seed combination query, we discuss the fundamental results shown in Table 1 derived by the uniform random sampling-based search strategy. Remember that this strategy corresponds to the search strategy parameter $\gamma = 0$. In Table 1, ω denotes the seed parameter. For the precision evaluation, we randomly sampled 500 words from the final seed words S', and manually checked whether every sampled name could be found as an actor/actress using IMDb,[5] and computed the number of correct names divided by 500.

According to Table 1, 2-seed combination query ($\omega = 2$) achieved the largest number of final seed words $|S'| = 26,564$. However, the precision 0.81 for $\omega = 2$ was the worst among other queries, while larger seed parameter ω have achieved remarkable precision, between 0.96 to 1.00. Therefore, the seed parameter $\omega = 3$ is the best considering both the number of final seed words and precision. The reason for the high precision for larger seed parameter ω can be explained by the robustness of the CESS framework discussed in Section 3.3. The typical false seeds are typos of actor/actress names, and people in different categories such as musicians and fashion models. We also conducted an experiment for $\omega = 1$, but the single word query could not retrieve any listing pages, so the seed words did not grow at all.

Next, we discuss the reason why $|S'|$ decreased as ω increased in Table 1. The total size of result sets RS in Table 1 indicates that higher ω yielded fewer search results.

---

[5] IMDb, http://www.imdb.com/

The total number of accessed pages AP also showed the same trend. So, do the multi-seed combination queries containing many seed words always yield few results? No. Growth rates of seed words as a function of search API calls are shown in Fig. 4. In Fig. 4(a), for all ω, the seed words grew at the beginning of bootstrapping, but the higher ω was, the faster the number of seed words saturated. This reason can be explained by Fig. 4(b). At the first 10 searches of bootstrapping, the average size of result sets for every ω is almost maximum (= 50). This indicates that even a 6-seed combination query yielded more than 50 matching pages at the beginning. However, as shown in Fig. 4(b), the average size of result sets rapidly decreased just after several dozen of the subsequent iterations because of the sparseness of seed words mentioned in Section 3.1. Note that we omitted plots for $\omega = 4$ and 5 in Fig. 4(b) to improve visualization, but the plots for these parameters fell between the plots for $\omega = 3$ and 6.

We found another effect of seed parameter ω for the search efficiency. In Table 1, we computed the skip ratio based on the total size of result sets RS and the total number of accessed pages AP. As mentioned in Section 3, the procedure $Crawler()$ skips pages registered in the access history. So, the skip ratio indicates how much the result set is duplicated. According to Table 1, the duplication of the result set increased from 0.07 to 0.45 almost linearly to the seed parameter ω.

However, a query with larger seed parameter ω can retrieve more listing pages that contain more new seed words. As shown in Table 1, the ratio of listing pages in the accessed pages increased up to 0.57 as the seed parameter ω increased. And the number of new seeds extracted in a listing page also increased from 26.6 to 30.5 with the increase of ω. However, the benefit is not enough to overcome the negative effect in the size of result set RS and the skip ratio.

Table 1. Fundamental Results in Uniform Random Sampling-Based Search Strategy

ω	# of Final Seed Words $|S'|$	Precision	Total Size of Result Sets RS	Accessed Pages AP	Skip Ratio $(RS-AP) / RS$	Listing Pages LP	Listing Page Ratio LP / AP	New Seeds in a Listing $|S'| / LP$
2	26,564	0.81	4,661	4,338	0.07	1,000	0.23	26.6
3	15,678	0.97	1,581	1,341	0.15	569	0.42	27.6
4	11,218	0.96	982	735	0.25	402	0.55	27.9
5	6,777	1.00	670	396	0.41	224	0.57	30.3
6	5,300	0.98	590	323	0.45	174	0.54	30.5

(B) Power Distribution Based Search Strategy ($1 \leq \gamma \leq 4$)

We summarize the effect of search strategy parameter γ for the size of final seed words $|S'|$ in Fig. 5. Note that the result of $\gamma = 0$ in Fig. 5 corresponds to the result of the uniform distribution-based search strategy shown in Table 1. In Fig. 5, the search strategy with $\omega = 3$, $\gamma = 3$ achieved the largest final seed words $|S'| = 27,344$. The precision of these final seeds was 0.93 and was still fine. So we concluded that for the search strategy with $\omega = 3$, $\gamma = 3$ is the best.

Next, we look more deeply into the effect of the search strategy parameter γ. In Fig. 5, it seems as if there are two different behaviors. The size of final seed words $|S'|$

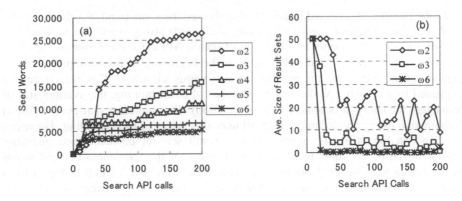

Fig. 4. Growth rate of seed words

for $\omega = 2$ decreased monotonically with the increase of γ, but $|S'|$ for $\omega \geq 3$ increased and had a peak at $\gamma = 3$. However, we found that these two different behaviors stemmed from the same reason, i.e., the trade-offs between the number and the redundancy of the search result shown in Fig. 6.

Figure 6(a) shows that the total size of result set RS increased with the increase of search strategy parameter γ. This indicates that the multi-seed combination query contained high-frequency words and yielded more search results by larger γ. Remember that we called search API 200 times, and the maximum size of the result set per query was 50, so the total size of the result set is at most 10,000. Thus, the queries with $\omega = 2$ and 3 achieved the maximum size of total result sets, and even 6-seed combination queries achieved about 80% of the maximum size. However, according to Fig. 6(b), the redundancy of new seed words also increased with the total size of the result set. The number of new seed words in a listing page decreased almost linearly to γ. This indicates that extracted instances can be highly redundant when γ increases. The reason is because a multi-seed combination query for higher γ tends to contain the same high-frequency word, and the variation of extracted instances may be decreased by these similar queries. We also checked the redundancy of the result set via the skip ratio. It also increased with the increase of γ. The range of the skip ratio at $\gamma = 4$ was from 0.15 ($\omega = 2$) to 0.69 ($\omega = 6$). Thus, for $\omega = 2$, the negative effect of redundancy shown in Fig. 6(b) is always stronger than the positive effect shown in Fig. 6(a), so the net gain shown in Fig. 5 is always decreasing. However, for $\omega \geq 3$, the positive effect is stronger than the negative effect until $\gamma = 3$, and the negative effect becomes dominant at $\gamma = 4$, so the net gain in Fig. 5 showed a peak at $\gamma = 3$.

In this experiment, we tried only integers for the search strategy parameter γ. However, if we used real numbers for γ, we would find a peak between $\gamma = 2$ and 3. In the Film domain, the best performance was achieved at $\omega = 3$, $\gamma = 2$. Remember the score distribution of seed words discussed in Section 3.1. It is interesting that the best performance was achieved by the search strategy such that the sampling probability is close to proportional to the score of seed words.

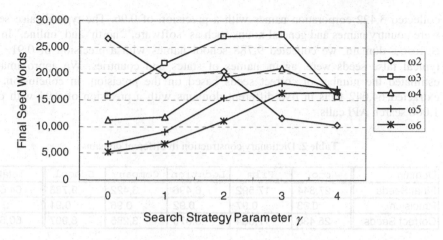

Fig. 5. Effect of search strategy parameter γ

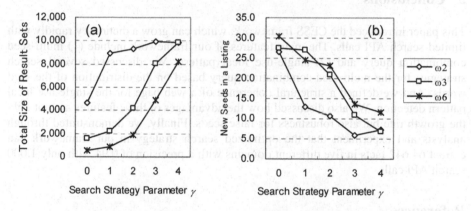

Fig. 6. Trade-offs for search strategy parameter γ

4.3 Results in Different Domains

We show the brief results in different domains in Table 2. In this experiment, we fixed the seed parameter ω and the search strategy parameter γ to $(\omega, \gamma) = (3, 3)$. And we also fixed the number of search API calls to 200. In addition to precision evaluation in the Actor domain, we randomly sampled 500 words from the final seed words and checked manually. For the Film domain, we also used IMDb. For the rest of the domains, we used search engines to find actual entities referred to by the name in the final seed words.

We found 17,592 movie titles in the Film domain with a precision of 0.97. The typical false seeds were only typos. In the Location domain, we found 6,496 city names with a precision of 0.92. The typical false seeds were location names referring to broader areas such as states, countries and continents. If these broader names are allowed as location names, the precision would be 0.97. In the Company domain, we collected 3,422 corporation names with a precision of 0.96. The typical false seeds were country names and general nouns such as 'software,' 'north' and 'online.' In the School domain, we collected 9,788 school names with a precision of 0.91. The typical false seeds were, again, names of states and countries. We approximately estimated the number of correct seeds based on the precision. In conclusion, we extracted 64,642 facts in five different domains with a precision of 0.94 with only 1,000 search API calls.

Table 2. Dictionary construction in different domains

Domain	Actor	Film	Location	Company	School	Total
Final Seeds	27,344	17,592	6,496	3,422	9,788	64,642
Precision	0.93	0.97	0.92	0.96	0.91	0.94
Correct Seeds	25,430	17,064	5,976	3,285	8,907	60,663

5 Conclusions

This paper introduced the CESS framework, which can grow a dictionary rapidly with limited search API calls. The main features of our framework include (1) multi-seed combination query and (2) single-use local pattern. We discussed several search strategies for the multi-seed combination query based on the distribution of the seed words. And we defined a structural coherence of a web page for the single-use local pattern detection. We also discussed how the advantages of these features are not only the growth rate but also robustness for false seeds. Finally, we demonstrated through analysis and experiment that the optimized search strategy in our framework can extract 64,642 facts in five different domains with a precision of 0.94 with only 1,000 search API calls.

References

1. Bikel, D.M., Milier, S., Schwartz, R., Weischedel, R.: Nymble: A high-performance learning name-filter. In: Proc. the Fifth Conference on Applied Natural Language Processing (1997)
2. Seymore, K., McCallum, A., Rosenfeld, R.: Learning hidden Markov model structure for information extraction. In: AAAI Workshop on Machine Learning for Information Extraction (1999)
3. Collins, M., Singer, Y.: Unsupervised models for named entity classification. In: Proceedings of the Joint SIGDAT Conference on Empirical Methods in Natural Language Processing and Very Large Corpora (1999)
4. Riloff, E.: Automatically constructing a dictionary for information extraction tasks. In: Proc. of the Eleventh National Conference on Artificial Intelligence, pp. 811–816. AAAI Press / The MIT Press (1993)
5. Soderland, S., Fisher, D., Aseltine, J.: W. Lehnert, W.: Crystal: Inducing a conceptual dictionary. In: Proc. the Fourteenth International Joint Conference on Artificial Intelligence, pp. 1314–1319 (1995)

6. Califf, M.E., Mooney, R.J.: Relational learning of pattern-match rules for information extraction. In: Working Notes of AAAI Spring Symposium on Applying Machine Learning to Discourse Processing (1998)
7. Riloff, E., Jones, R.: Learning dictionaries for information extraction using multi-level bootstrapping. In: Proc. the Sixteenth National Conference on Artificial Intelligence, pp. 1044–1049. AAAI Press / The MIT Press (1999)
8. Etzioni, O., Cafarella, M., Downey, D., Kok, S., Popescu, A.-M., Shaked, T., Soderland, S., Weld, D.S., Yates, A.: Web-scale information extraction in KnowItAll. In: Proc. the 13th International Conference on World Wide Web (2004)
9. Pasca, M., Lin, D., Bigham, J., Lifchits, A., Jain, A.: Organizing and searching the World Wide Web of facts - step one: the one-million fact extraction challenge. In: Proc. AAAI 2006 (2006)
10. Thelen, M., Riloff, E.: A bootstrapping method for learning semantic lexicons using extraction pattern contexts. In: Proc. Conference on Empirical Methods in Natural Language Processing, pp. 214–222 (2002)
11. Hasegawa, T., Sekine, S., Grishman, R.: Discovering relations among named entities from large corpora. In: Proc. ACL, pp. 415–422 (2004)
12. O'Reilly, T.: What is Web 2.0: Design patterns and business models for the next generation of software, http://www.oreillynet.com/lpt/a/6228
13. Yahoo! Developer Network, http://developer.yahoo.com/
14. Google Code, http://code.google.com/
15. Windows Live Developer Center, http://msdn.microsoft.com/msn/default.aspx
16. WordNet, http://wordnet.princeton.edu/
17. Brin, S.: Extracting patterns and relations from the World Wide Web. In: Proc. the International Workshop on the World Wide Web and Databases, pp. 172–183 (1998)
18. Soderland, S., Etzioni, O., Shaked, T., Weld, D.: The use of Web-based statistics to validate information extraction. In: AAAI workshop on Adaptive Text Extraction and Mining (2004)
19. Kushmerick, N.: Wrapper induction: efficiency and expressiveness. In: AAAI Workshop on AI and Information Integration (1998)
20. Zhai, Y., Liu, B.: Web data extraction based on partial tree alignment. In: Proc. the 14th International Conference on World Wide Web (2005)
21. Chuang, S.-L., Chang, K.C.-C., Zhai, C.: Context-Aware wrapping: synchronized data extraction. In: Proc. the 33rd Very Large Data Bases Conference (VLDB) (2007)
22. Wang, R.C., Cohen, W.W.: Language-independent set expansion of named entities using the Web. In: Proc. IEEE International Conference on Data Mining (2007)

Learning Bayesian Network Structure from Incomplete Data without Any Assumption

Céline Fiot[1], G.A. Putri Saptawati[2],
Anne Laurent[1], and Maguelonne Teisseire[1]

[1] LIRMM - Univ. Montpellier II, CNRS
161 rue Ada, 34392 Montpellier, France
{fiot,laurent,teisseire}@lirmm.fr
[2] Institut Teknologi Bandung
Jl. Ganesha 10, Bandung 40132, Indonesia
putri@informatika.org

Abstract. Since most real-life data contain missing values, reasoning and learning with incomplete data has become crucial in data mining and machine learning. In particular, Bayesian networks are one machine learning technique that allows for reasoning with incomplete data, but training such networks on incomplete data may be a difficult task. Many methods were thus proposed to learn Bayesian network structure from incomplete data, based on multiple structure generation and scoring of their adequacy to the dataset. However, this kind of approaches may be time-consuming. Therefore we propose an efficient dependency analysis approach that uses a redefinition of probability calculation to take incomplete records into account while learning BN structure, without generating multiple possibilities. Some experiments on well-known benchmarks are described to show the validity of our proposal.

1 Introduction

Graphical models [1] are tools combining two different areas: graph theory and probability theory. They are often used to illustrate and work with conditional independencies and probabilistic relationships between variables in a given problem. Among graphical models, Bayesian networks are often involved in tasks requiring to reason under uncertainty. In particular, many approaches allow for incomplete data. But the task of training Bayesian networks with incomplete datasets is more complex.

Actually, there are two classes of methods for building the graphical structure of Bayesian networks from complete datasets. Scoring-based algorithms consist in generating many likely structures and then scoring them using a fitness function measuring how well each possible graph fits the data. Constraint-based algorithms – also called dependency analysis approaches – build the graph directly from the data thanks to probability calculations and conditional independency tests.

Some approaches tackle the problem of missing values either by deleting observations with missing values or using ad-hoc techniques to impute lacking

J.R. Haritsa, R. Kotagiri, and V. Pudi (Eds.): DASFAA 2008, LNCS 4947, pp. 408–423, 2008.

information. Such procedures may however lead to biased results, and when imputing a single value for unassigned attributes, to an overconfidence in the results of the analysis. Some specific algorithms have also been shown to be successful for learning Bayesian network structure from complete data, and learning parameters for a fixed network. Other scoring-based algorithms have finally been developed using an estimation of the missing observations on the basis of available information and data distribution. Though this may be time-consuming and resource-demanding.

Since very few methods are able to use incomplete cases as a base to determine the structure of a Bayesian network by a constraint-based approach, we introduce in this paper an efficient dependency analysis approach to handle incomplete records while learning BN structure. Our proposal consists in a redefinition of probability calculation that allows for information incompleteness without missing value imputation. Then we adapt the efficient Three Phase Dependency Analysis algorithm proposed by [2] to our new probability definitions, while computing conditional independency tests. Some experiments on classical benchmarks are described. They show the validity of our proposal for generating Bayesian network structure underlying incomplete datasets.

This paper is organized as follows: in Section 2, we introduce the definition and principles used in the context of Bayesian network learning; then in Section 3, we detail the basis of our approach and our new definitions for probability calculation. Before concluding in Section 5, some results of experiments are developed by Section 4.

2   Bayesian Network and Incomplete Data

Regarding a data set, a Bayesian network gives both a qualitative and quantitative description of the dependencies existing between data attributes. First, these dependencies are visually described by a *directed acyclic graph* (DAG). In this graph, each vertex, or node, corresponds to an attribute in the database and directed edges between nodes show the dependencies between related attributes. Then, each node is associated with a *conditional probability* table, which gives, for each value of the node attribute, its probability considering the value of the attribute parent nodes.

2.1   Directed Acyclic Graph (DAG)

A *graph*, or *undirected graph*, is defined as a set of *nodes*, also called *vertices*, and a set of *edges*, or *arcs*, each being a pair of nodes. If the vertices linked by each edge are ordered, then the edges have a direction assigned to them and the graph is called a *directed graph*. A *chain* is a series of nodes where each successive node is connected to the previous node by an edge. A *path* is a chain with the further constraint for directed graphs that each connecting edge in the chain has the same direction as the chain. A *cycle* is a path that starts and ends at the same node. A *directed acyclic graph*, or *DAG*, is a directed graph that has no cycles.

The terms *parent* and *child* define the relationship between two vertices connected by a directed edge from the parent to the child. Two vertices are said to be *adjacent* when they are connected by an undirected edge.

2.2   Bayesian Network

A Bayesian network is a specific graphical model as it is a concise representation of the joint probability distribution for a large set of attributes in a database. Each attribute can be considered as a random variable associated with several values. Then, for a set of variables $\mathcal{V}$, a Bayesian network consists of a *directed acyclic graph* that encodes a set of conditional dependency and independency assertions about variables in $\mathcal{V}$, and a set of local probability distributions associated with each variable. Together, these components define the joint probability distribution for $\mathcal{V}$.

Pearl [3] defines a Bayesian network as a triplet $[\mathcal{V}, G, P(V_i|P_a(V_i))]$, where:

- $\mathcal{V} = \{V_1, ..., V_n\}$ is a *set of random discrete variables*;
- G is a *directed acyclic graph* whose nodes represent variables V_i, and whose arcs encode the *conditional dependencies* between the variables;
- $P(V_i|p_a(V_i))$ describes the conditional probability distribution of each variable V_i given its immediate parents $pa(V_i)$ in the graph G.

The edges in the Bayesian network encode a particular factorization of the joint probability distribution. In general, the joint probability function for any Bayesian network representing the set of nodes $\mathcal{V}$ is given by

$$P(\mathcal{V}) = \prod_{i=1}^{n} P(V_i|pa(V_i))$$

meaning that the joint probability of all the variables is the product of the probabilities of each variable given its parents' values. Then, the graph describes these dependencies. For any given edge between variables V_i and V_j linked by a causal relationship, the edge will be directed, from the cause variable to the effect variable. If there is just a correlation between the two variables, the edge will be undirected [4]. Two conditionaly independent variables have no direct impact on each other's values. However, any path through intermediary variables that separates two conditionally independent variables shows how they affect each other.

2.3   Learning Bayesian Networks

As a Bayesian network is constituted by one qualitative component, the DAG, and one quantitative component, the conditional probability distribution, learning Bayesian network comes down to two tasks. First the structure describing the dependencies is designed, then the conditional probabilities of each node are calculated. Two approaches generally aimed at learning the structure. The first one is based on scoring an *a priori* designed structure. The second one uses constraints and conditional independency tests to build the graph.

Scoring-based approaches [5,6,7,8] select the DAG that best fits the data among several ones a priori designed. The purpose of learning is then to evaluate each previously designed model and to determine the most adapted one. These methods require to specify scoring functions in order to evaluate how well each network matches the training data. The approaches based on model selection use some criterion to measure the degree to which a network structure (equivalence class) fits prior knowledge and data. A search algorithm is then used to find an equivalence class that receives a high score by this criterion.

Constraint-based algorithms, also called dependency analysis algorithms, build the DAG structure by identifying the conditional independency relationships among the variables [9,10,11,12]. These methods are based on the causal sufficiency hypothesis: *for every pair of measured variables in the training data, all their common parents are also measured.* Thus, the graph is built solely on the set of data, without external knowledge. Vertices of the DAG are drawned from the variables within the dataset and the edges according to the observed dependencies between variables within the data.

Once the structure has been designed, each node of the DAG is associated with a table of conditional probabilities that gives for each value of each variable the path to follow in the DAG.

Depending on the problem, either the topology or the probability distributions or both may be pre-defined by hand or may be learned from the data.

In this paper, we consider that the structure is unknown but that all the variables can be identified (i. e. there is no *hidden variables* [13]). Within a context where some variables are randomly unassigned, we tackle the problem of learning the structure of a Bayesian network given data, using an approach based on the information theory.

2.4 Handling Missing Values

In data mining and machine learning, missing value handling is a significant problem as most real-life data contain unassigned variables.

One advantage of belief networks is that they allow reasoning with incomplete data [13]. Many inference algorithms can indeed be used to estimate the probability of any variable, that has not been measured, conditionally to the values of measured variables. But complete data are often required for training such networks.

As described in the previous section, there are two different problems related to the presence of missing values while learning Bayesian network. One evaluates the probability parameters despite unassigned variables, the second assesses the dependencies and learns the graph structure in spite of incomplete observations.

Some approaches answer to the problem of missing values either by deleting observations with missing values or by using ad-hoc techniques to impute missing information. Such procedures may however lead to biased results, and in case of imputing a single value for unassigned attributes, to an overconfidence in the results of the analysis.

Therefore specific algorithms have been developed and several methods have been shown to be successful for learning both network structure and parameters from complete data, and learning parameters for a fixed network [14]. But very few methods are capable of using incomplete cases as a basis to determine the structure of a Bayesian network by a constraint-based approach.

Most of techniques that determine the Bayesian network structure of an incomplete dataset are indeed based on model scoring and selection: these proposals use an a priori known structure and compare it to the observed distribution. Well-known methods typically involve the use of the EM algorithm [15] or Markov Chain Monte-Carlo methods, such as Gibbs sampling [16]. The basic strategy underlying these methods is based on the *Missing Information Principle* [17]: fill in the missing observations on the basis of the available information.

Thus [18,19,20] propose different algorithms based on extensions of the expectation-maximization algorithm for model selection problems. [21] and [22] describe approaches based on stochastic search and evolutionary algorithm that approximates a maximum likelihood approach to score the network by evolving samples of incomplete data. Unfortunately, these processes are usually highly resource demanding, their convergence rates may be slow, and their execution time heavily depends on the number of missing values.

Therefore [23] uses an entropy maximization procedure to include information regarding the nature of the missing data mechanism and thus considerably saving computation time when compared to Gibbs sampling.

Other scoring-based approaches are based on estimation of missing data for both parameter and structure learning. [24,25] introduce a deterministic method to estimate the conditional probabilities in a Bayesian network which does not rely on the Missing Information Principle. Then, the *Bound and Collapse* algorithm is proposed for parameter estimation and model selection from incomplete data. However this algorithm also relies on an assumed pattern of missing data that may be either provided by an external source of information or may be estimated from the available information under the assumption that data are missing at random. This approach is extended in [26] to learn the graphical structure of a Bayesian network from a possibly incomplete database, using estimation of missing data.

More recently, [27] describes an imputation-based approach for model learning from incomplete data, where possible completions of the data are scored together with the observed part of the data. [28] and [29] describe an algorithm on the ground of an extended evolutionary programming method. It uses fitness function based on expectation, which converts incomplete data to complete data. [30] introduces an approach for assessing the predictive distribution of missing values that is then combined to any learning algorithm.

2.5 Objectives

The main disadvantage of scoring-based approaches is that they rely on determining among numerous structures the one that best fits the data. So this kind of approaches requires *a priori* expert knowledge to design few structures to be tested or to generate every possible structures from the data. This processes are thus highly resource demanding and in the case of incomplete data, the runtime may become very high.

Therefore in this paper we propose a constraint-based approach to learn Bayesian network from a randomly incomplete dataset without assessing or deleting missing values nor generating several possible structures. We use observed data without requiring any external information or estimation of missing value distribution.

Our method is based on a redefinition of the probability functions taking into account that some variable values are unknown. We developed our algorithm for learning Bayesian network structure adapting the efficient Three Phase Dependency Analysis (TPDA) algorithm proposed by [2] to make it use our own probability definitions. Finally, we ran several experiments to show the feasibility, validity and robustness of our approach.

In the following section, we introduce the principles we based our work on. Then, we detail our new definitions and prove that they satisfy all the conditions required from a probability measure. Last, we describe the overall learning algorithm and run a brief example. In Section 4, we present some results of our experiments.

3   TPDA for Incomplete Databases

Our approach is based on the same principles as the ones used by the RAR algorithm [31] for association rule mining [32] in incomplete databases. The data formalism of association rules is indeed quite similar to the one of Bayesian network: the dataset is a relational table consisting in records in which values are associated with attributes. That corresponds to random variables in the context of Bayesian networks.

Our main idea is based on the RAR method. It consists in disabling incomplete elements, that is within our context, incomplete records. As for the RAR algorithm for association rules mining, we will restrain to complete records to compute the conditional probabilities. In other words, when an incomplete record is scanned, only filled-in attributes are considered for probability calculation. Thus each conditional probability is computed on a partial database, but the whole dataset is used to find the whole set of dependencies.

3.1   Overall Principle

The RAR algorithm (Robust Association Rules), proposed by [31], allows the user to consider incomplete data while association rule mining within incomplete relational databases, thanks to partial and temporary omission of such incomplete records. The main idea consists in taking only filled-in attributes in incomplete records into account. The whole database is not used to discover each rule but the whole set of rules.

This technique is based on the valid database concept, which is a complete dataset for a given itemset, i.e. a set of attributes or variables. The remaining part of the database is temporary ignored. In order to consider this partition of the dataset, definitions of support (percentage of records in database that include the rule items) and confidence (probability for a record to contain the right part of the rule knowing it contains the left part) were reformulated.

For instance, to calculate the support of the rule $X_1 = y \rightarrow X_4 = n$, records R_1 and R_2 are disabled, since the first one is incomplete for X_1 and the second for X_4. Then support and confidence of this rule are computed on the set of records $\{R_3, R_4, R_5\}$. But support and confidence of the rule $X_1 = y \rightarrow X_5 = n$ would be computed on the set $\{R_2, R_3, R_4, R_5\}$.

Table 1. An incomplete dataset

	X_1	X_2	X_3	X_4	X_5
R_1	?	y	y	n	n
R_2	y	n	n	?	y
R_3	y	y	n	y	n
R_4	y	n	n	y	n
R_5	n	?	y	y	y

Learning the structure of a Bayesian network by a constraint-based approach such as TPDA algorithm requires to compute conditional probabilities and probabilistic conditional independency tests. We here apply the formalism of the RAR algorithm to define a new probability measure. This measure will then be used by our implementation of the TPDA algorithm to run the conditional independency tests and thus to build the DAG structure.

So, let us consider a set of random variables V and their realisations v, the set $\mathcal{R}$ of records r in the database DB can be divided into three disjoint subsets (Figure 1). The set of records filled in with the corresponding value v_i for each variable V_i of V is denoted by $\mathcal{R}_V$. The set of records filled in with at least one value different from the set v is denoted by $\mathcal{R}_{\overline{V}}$. And the set of records for which at least one value v is unfilled, i. e. is missing and we do not know if $r(V_i) = v_i$ or not, is denoted by $\mathcal{R}_V^*$.

For each set of variables V, only the subsets $\mathcal{R}_{\overline{V}} \cup \mathcal{R}_V$ are kept to determine the conditional probabilities on V. These subsets represent the *valid database* for V. Incomplete records are *disabled* for V.

Definition 1. *A* valid database *is a database only containing complete records for a given set of random variables, i.e. each value of each record in the data corresponds to an identified values v of $\mathcal{D}om(V)$.*

On Figure 1, the valid database for V is $\mathcal{R}_{\overline{V}} \cup \mathcal{R}_V$.

Definition 2. *A record is* disabled *for an instanciation of a set of variables V if it is incomplete for V (i.e. we cannot decide whether it includes V or not). The set of records disabled for a set V is denoted by $Dis(V)$.*

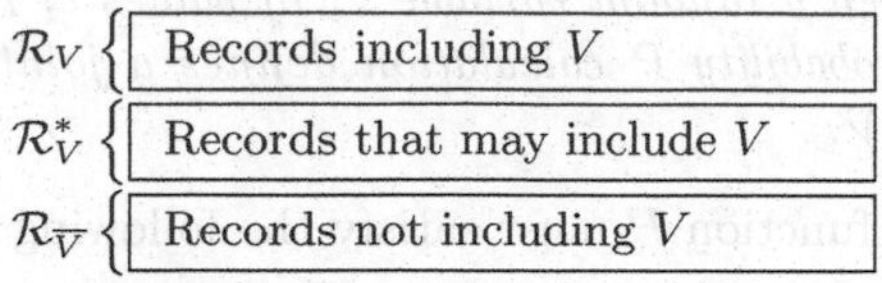

$\mathcal{R}_V$ { | Records including V |

$\mathcal{R}_V^*$ { | Records that may include V |

$\mathcal{R}_{\overline{V}}$ { | Records not including V |

Fig. 1. Partition of the database according to V inclusion

Note that $Dis(V)$ corresponds to the set $\mathcal{R}_V^*$ on Figure 1 and that $Dis(V) = \mathcal{R}\backslash\mathcal{R}_{\overline{V}} \cup \mathcal{R}_V$.

For instance, in the dataset described by Fig. 1, the valid database for X_1 is composed of records R_2 to R_5 and $Dis(X_1) = \{R_1\}$. The valid database for $V = \{X_1, X_4\}$ is $\{R_3, R_4, R_5\}$, and $Dis(V) = \{R_1, R_2\}$.

Building a valid database leans on temporary disabling records that contain missing values for variables in the set of random variables. This implies a redefinition of the probability calculation to consider the database partial deactivation.

3.2   Redefining Calculation of Probabilities

The probability definition is modified in order to take into account the valid database concept, and thus that only one part of the dataset is used for each probability calculation.

Definition 3. *The probability of an event v_i is estimated by the appearance rate of this event among the records that can include it. It is defined as the ratio of the number of records r such that $r(V_i) = v_i$ by the number of records that are filled in for V_i (complete records for V_i). It is given by:*

$$P(V_i = v_i) = \frac{card\left(\{r \in \mathcal{R} | r(V_i) = v_i\}\right)}{card(\mathcal{R}) - card(Dis(V_i))} \tag{1}$$

Considering a set of random variables $V = \{V_1, ..., V_n\} \subseteq \mathcal{V}$ and a joint probability function P defined on $\mathcal{V}$, the probability of a joint event $P(v_1, ..., v_n)$ is computed considering the set of records that are complete for all the variables $V_1, ...V_n$. Then the previous formula can be expressed as follows:

$$P_{V_1, ..., V_n}(v_1, ..., v_n) = \frac{card\left(\bigcap_{i \in [1,n]} \{r \in \mathcal{R} | r(V_i) = v_i\}\right)}{card(\mathcal{R}) - card(Dis(V_1, ..., V_n))} \tag{2}$$

For instance on Fig. 1, to compute $P(X_1 = y)$, we find $Dis(X_1) = \{R_1\}$, then $P(X_1 = y) = card(\{R_2, R_3, R_4\})/5 - card(Dis(X_1)) = 3/(5 - 1) = 0.75$. If we compute $P(X_1 = y, X_2 = n)$, then $Dis(X_1, X_2) = \{R_1, R_5\}$ and $P(X_1 = y, X_2 = n) = card(\{R_2, R_4\})/(5 - card(Dis(X_1, X_2))) = 2/(5 - 2) = 0.67$.

Proposition 1. *Given a random variable V_i of values v_i in domain $\mathcal{D}(V_i)$, the redefinition of the probability P calculation defines a joint probability function over the variable set $\mathcal{V}$.*

Proof. A probability function P must satisfy the following properties:

1. for every event $A \in \mathbf{A}$, $0 \le P(A) \le 1$;
2. for the impossible event $\varnothing$ and the certain event Ω, $P(\varnothing) = 0$ and $P(\Omega) = 1$;
3. if the events $A_i \in \mathbf{A}$ are finite or countably many mutually exclusive events $(A_i A_k = \varnothing$ for $i \ne k)$, then $P(\bigcup_{i=1}^{n} A_i = \Sigma_{i=1}^{n} P(A_i)$

We will denote $r(V_i)$ the value of attribute/variable V_i in the record r and $card(S)$ will denote the cardinality of a subset S of records.

1. We consider the database partitioning, $\{r \in \mathcal{R} | r(V_i) = v\} \subseteq \mathcal{R} \backslash Dis(V_i)$, which implies that $card(\{r \in \mathcal{R} | r(V_i) = v\}) \le card(\mathcal{R}) - card(Dis(V_i))$. Then, as a cardinality is necessarily greater than or equal to zero and assuming that at least one record in the database is complete for V_i, we obtain

$$0 \le \frac{card(\{r \in \mathcal{R} | r(V_i) = v\})}{card(\mathcal{R}) - card(Dis(V_i))} \le 1 \Rightarrow \forall v, 0 \le P_{V_i}(v) \le 1$$

2. A record necessarily contains a value, missing or not, for a variable V_i then $card(\{r \in \mathcal{R} | r(V_i) = \varnothing\}) = 0$ and $P_{V_i}(\varnothing) = 0$. Then we show that $P_{V_i}(\bigcup_{v \in \mathcal{D}(V_i)} V_i = v) = 1$:

$$\bigcup_{v \in \mathcal{D}(V_i)} \{r \in \mathcal{R} | r(V_i) = v\} = \mathcal{R} \backslash Dis(V_i) \Rightarrow \frac{card(\bigcup_{v \in \mathcal{D}(V_i)} \{r \in \mathcal{R} | r(V_i) = v\})}{card(\mathcal{R}) - card(Dis(V_i))} = 1$$
$$\Rightarrow P_{V_i}(\bigcup_{v \in \mathcal{D}(V_i)} V_i = v) = 1$$

3. $\forall A_j \in V_i | \forall j \ne k, A_j \cap A_k = \varnothing, P_{V_i}(\bigcup_j A_j) = \sum_j P_{V_i}(A_j)$.

   Within our context, such an event A_j corresponds to a set of values v for a random variable V_i. In other words, it can be expressed by the formula

   $$\forall A_j, \exists D_j \subseteq \mathcal{D}(V_i) | A_j = \bigcup_{v \in D_j} \{r \in \mathcal{R} | r(V_i) = v\}$$

   We use this formulation to prove that the last condition is satisfied.

   $$P_{V_i}(\bigcup_j A_j) = P_{V_i}(\bigcup_j \bigcup_{v \in D_j} V_i = v) = \frac{card(\bigcup_j \bigcup_{v \in D_j} \{r \in \mathcal{R} | r(V_i) = v\})}{card(\mathcal{R}) - card(Dis(V_i))}$$

   As the events A_j are disjoint, $\quad = \sum_j \frac{card(\bigcup_{v \in D_j} \{r \in \mathcal{R} | r(V_i) = v\})}{card(\mathcal{R}) - card(Dis(V_i))}$

   $$= \sum_j P_{V_i}(A_j)$$

So, for all random variables V_i, the new measure P_i defines a probability measure on each variable. $\qquad \square$

Proposition 2. *Given a set of random variables V, every function defined by $P_{W \subseteq V}(\bigcap_{W_i \in W} W_i = w_i)$, computed by the formula 2, is a joint probability function.*

Proof. We have to prove that for all set W of random variables such that $W \subseteq V$, the function P_W defined by

$$P_W(\cap_{W_i \in W} W_i = w_i) = \frac{card(\cap_{W_i \in W}\{r \in \mathcal{R}|r(W_i) = w_i\})}{card(\mathcal{R}) - card(Dis(W))}$$

is a joint probability function. $Dis(W)$ denotes the set of records disabled for W, i.e. the set of records for which at least one variable in W is unassigned.

First we show that $P_W(\cap_{W_i \in W} W_i = w_i)$ is in $[0, 1]$. As $\{r \in \mathcal{R}|\cap_{W_i \in W} r(W_i) = w_i\} \subseteq \mathcal{R}\backslash Dis(W)$, we can simply show that $P_W(\cap_{W_i \in W} W_i = w_i) \leq 1$, using the same proof as previously. Moreover, as it is defined by set cardinalities, it is necessarily greater than 0.

Then we prove that $\sum\limits_{w_i \in D(W_i)} M_W(\cap_{W_i \in W} W_i = w_i) = 1$.

$\cup_{w_i \in \mathcal{D}(W_i)}\{r \in \mathcal{R}|\cap_{W_i \in W} r(W_i) = w_i\} \quad = \mathcal{R}\backslash Dis(W)$

$\Rightarrow card(\cup_{w_i \in \mathcal{D}(W_i)}\{r \in \mathcal{R}|\cap_{W_i \in W} r(W_i) = w_i\} = card(\mathcal{R}) - card(Dis(W))$

events $\cap_{W_i \in W} W_i = w_i$ being mutually exclusive,

$\Rightarrow \sum\limits_{w_i \in \mathcal{D}(W_i)} card(\{r \in \mathcal{R}|\cap_{W_i \in W} r(W_i) = w_i\}) \quad = card(\mathcal{R}) - card(Dis(W))$

$\Rightarrow \sum\limits_{w_i \in \mathcal{D}(W_i)} \frac{card(\{r \in \mathcal{R}|\cap_{W_i \in W} r(W_i) = w_i\})}{card(\mathcal{R}) - card(Dis(W))} = \sum\limits_{w_i \in \mathcal{D}(W_i)} M_W(\cap_{W_i \in W} W_i = w_i) = 1$

$\square$

3.3   Learning Algorithm

The proposition 2 allows us to apply all the formalisms defined for Bayesian network learning methods with complete data. Our approach is based on the generic principle of constraint-based learning methods. More precisely, we implemented our algorithm from the Three Phase Dependency Analysis algorithm developed by [2], using the probability formulæ introduced in the previous section for computing the conditional independency tests. The overall algorithm lays on three elementary steps:

1. conditional independencies are uncovered from the data using statistical tests,
2. then, these independencies are used to build a partially directed acyclic graph (PDAG) in two steps,
   - edges X–Y of an undirected fully connected graph are deleted for each pair of independent variables (X, Y),
   - the undirected graph then obtained is partially directed using the discovered conditional independencies;

3. last the PDAG is completed applying the following rules:
   - if there is an edge such that $X \to Y$ and Z is adjacent to Y but not to X, then if there is an undirected edge between Y and Z, this edge is directed from Y to Z ($Y \to Z$),
   - if there exists a directed path from X to Y and an undirected edge between X and Y, then this edge should be oriented from X to Y ($X \to Y$) to avoid building cycles.

4   Experiments

Our experiments were done to compare network structures generated by TPDA with complete data with the one obtained running TPDA adapted for handling missing values (TPDAID), using our own definitions for probability calculations, detailed by section 3. The goal was to show the validity of our redefinition of probabilities within the context of training Bayesian network with incomplete data. We also aimed at measuring the robustness of our proposal to various incompleteness rate of the datasets.

4.1   Datasets

The results detailed here were obtained on several standard benchmarks often used by the Bayesian network community. The characteristics of these datasets created from the various belief networks are described by Table 2.

Incomplete datasets were generated from the complete ones. Missing values were randomly inserted in the database, replacing some attribute values. For each complete database, we thus created six incomplete datasets respectively containing 5%, 10%, 20%, 30%, 40% and 50% of missing values.

Table 2. Characteristics of the datasets

Dataset	# of attributes	# of records
Fire Network [33]	6	10,000
Asia / Chest Clinic Network [34]	8	5,000
Alarm Network [35]	37	10,000

4.2   Results

For each complete dataset and then incomplete datasets, we ran TPDA or TP-DAID and so for each dataset we generated the Bayesian network structure. Then we compared the graphs resulting from training with incomplete data to those resulting from training with complete data. To do so we analyzed the number of missing or additional edges and the number of wrong directions.

Figure 3(a) shows the comparison between the graphs obtained for the *Fire* dataset according to the incompleteness rate. The original Bayesian network contains five edges, it is described by Figure 2(a).

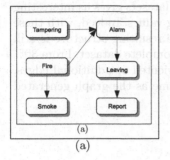
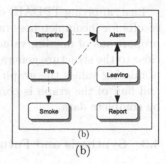

(a)

(b)

Fig. 2. (a): The *Fire* Bayesian network structure (taken from [33]); (b): The *Fire* Bayesian network structure obtained from 20% of missing values dataset

For 5% and 10% of missing values, the graphs resulting of TPDAID are exactly the same as the complete database graph. Then, for 30 to 50% of missing values, the number of additional edges increases to 1. On the *Fire* dataset it seems that the incompleteness rate influences more the number of wrong directions. Indeed, the graphs contain at least one wrong direction from 20% of missing values, as shown by Figure 2(b); the dashed arrows are the same as in the original network, the boldfaced one is the one modified.

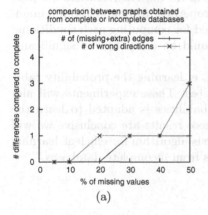
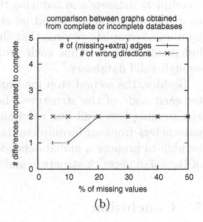

(a)

(b)

Fig. 3. (a): Results for the *Fire* dataset; (b): Results for the *Asia* dataset

Figure 3 describes differences observed between graphs generated by TPDA with complete data and those generated by TPDAID with incomplete data. Figure 3(b) shows the comparison between graphs obtained for the *Asia* dataset according to the incompleteness rate. We can observe that results are not as good as with the *Fire* dataset (Figure 3(a)) for low incompleteness rates. However the quality of the graph remains stable as the percentage of missing values increases.

Last, applying TPDAID on the *Alarm* incomplete datasets gives us interesting results for incompleteness rates below 30% of missing values. Indeed, with these proportions of missing values – from 5 to 20% – the resulting structure is still close to the structure generated by TPDA using the complete dataset. From 30% of missing values the graph contains half of badly-oriented or additional edges, and half of the graph is correctly built, being the same as the graph generated by complete data.

4.3 Synthesis and Future Work

This work requires further validation on larger datasets and also against existing algorithms handling missing values. Especially, a runtime vs. quality comparison would be interesting, as the overall complexity of our proposal is much lower than the one of scoring-based approaches.

However, through our experiments we observed that the data distribution and the incompleteness rate have an influence on the quality of the results returned by our approach. Analysing the different results, we consider that our approach is robust until around 25% to 30% of missing values in the training dataset. But, there are important differences if we compare results obtained on datasets containing many or few variables. The number of records also influences the results.

Therefore, we focus now on improving the quality of the graph trained on incomplete datasets and reducing the influence of the data distribution. We plan to define a parameter, based on statistical properties, to bound the minimum number of complete records that should be used for computing each conditional independency test. Thus each probability should be computed on significant-enough valid databases.

Besides, the second step for learning BN, i. e. learning the probability table for each node of the structure, should be tested. These experiments will aim at assessing how well our redefinition of probabilities is adapted to learn the parameters from an incomplete dataset. If these results are conclusive we will be able to propose a global dependency analysis algorithm for efficient learning of Bayesian network structure and parameters from incomplete databases.

5 Conclusion

In this paper, we introduced a new method for learning Bayesian networks from incomplete data. Unlike existing algorithms that are based on model scoring and selection, or on assessing or imputing missing values, our approach is based on dependency analysis and a redefinition of the probability calculations.

Thus, while other approaches are resource-demanding or time-consuming because of multiple iterations, our algorithm generates the graph computing conditional independency tests with a reformulation of probabilities. This redefinition is based on the principle that incomplete records contain some certain information exactly handled, assigned attributes, and an uncertain part of information that should be ignored, missing values. This hypothesis enables us to compute probabilities without multiple iterations for estimating missing values nor a biasing prior imputation, and without requiring external knowledge.

As shown by the preliminary experimental results, this new approach leads to quite good results for databases containing up to 30% of missing values. However, it can be improved by taking data distribution and statistical results into account to refine the probability calculation. We thus plan to develop a global algorithm that will learn both structure and parameters for Bayesian network from incomplete datasets, based on our redefinition of probabilities that handle uncertainty contained in incomplete data.

Acknowledgement

The authors would like to thank Liva Ralaivola for the fruitfull discussions we had and André Mas, and Cécile Low-Kam for their remarks.

References

1. Whittaker, J.: Graphical models in applied multivariate statistics. John Wiley & Sons, Inc, Chichester (1990)
2. Cheng, J., Bell, D., Liu, W.: Learning belief networks from data: an information theory based approach. In: The 6th ACM International Conference on Information and Knowledge Management, pp. 207–216 (1997)
3. Pearl, J.: Probabilistic Reasoning in Intelligent Systems. Morgan Kaufmann, San Francisco (1988)
4. Cowell, R.G., Dawid, A.P., Lauritzen, S.L., Spiegelhalter, D.J.: Probabilistic networks and expert systems. Statistics for engineering and information science. Springer, Heidelberg (1999)
5. Cooper, G.F., Herskovits, E.: A bayesian method for the induction of probabilistic networks from data. Machine Learning 9(4), 309–347 (1992)
6. Spiegelhalter, D.J., Dawid, A.P., Lauritzen, S.L., Cowell, R.G.: Bayesian analysis in expert systems. Statistical Science 8, 219–282 (1993)
7. Lam, W., Bacchus, F.: Learning bayesian belief networks: An approach based on the mdl principle. Computational Intelligence 10, 269–293 (1994)
8. Heckerman, D., Geiger, D., Chickering, D.M.: Learning bayesian networks: The combination of knowledge and statistical data. Machine Learning 20(3), 197–243 (1995)
9. Chow, C.K., Liu, C.N.: Approximating discrete probability distributions with dependence trees. IEEE Transactions on Information Theory 14, 462–467 (1968)
10. Pearl, J., Verma, T.S.: A theory of inferred causation. In: Principles of Knowledge Representation and Reasoning (KR 1991), pp. 441–452 (1991)
11. Spirtes, P., Glymour, C., Scheines, R.: Causation, Prediction, and Search. Lecture Notes in Statistics. Springer, Heidelberg (1993)
12. Spirtes, P., Meek, C.: Learning bayesian networks with discrete variables from data. In: 1st International Conference on Knowledge Discovery and Data Mining (KDD 1995) (1995)
13. Heckerman, D.: A tutorial on learning with bayesian networks. In: The NATO Advanced Study Institute on Learning in graphical models, pp. 301–354 (1998)

14. Lauritzen, S.L.: The em algorithm for graphical association models with missing data. Computational Statistics and Data Analysis 19, 191–201 (1995)
15. Dempster, A.P., Laid, N.M., Rubin, D.B.: Maximum likelihood from incomplete data via the em algorithm. Journal of the Royal Statistical Society 39(1), 1–38 (1977)
16. Chickering, D.M., Heckerman, D.: Efficient approximations for the marginal likelihood of bayesian networks with hidden variables. Machine Learning 29(2-3), 181–212 (1997)
17. Little, R.J.A., Rubin, D.B.: Statistical analysis with missing data. John Wiley & Sons, Inc., Chichester (1987)
18. Friedman, N.: Learning belief networks in the presence of missing values and hidden variables. In: 14th International Conference on Machine Learning, pp. 125–133 (1997)
19. Friedman, N.: The bayesian structural em algorithm. In: 14th Conference on Uncertainty in Artificial Intelligence, pp. 129–138 (1998)
20. Leray, P., François, O.: Bayesian network structural learning and incomplete data. In: International and Interdisciplinary Conference on Adaptive Knowledge Representation and Reasoning (AKRR 2005), pp. 33–40 (2005)
21. Myers, J.W., Laskey, K.B., Levitt, T.S.: Learning bayesian networks from incomplete data with stochastic search algorithms. In: 15th Conference on Uncertainty in Artificial Intelligence (UAI 1999) (1999)
22. Myers, J.W., Laskey, K.B., Dejong, K.: Learning bayesian networks from incomplete data using evolutionary algorithms. In: Genetic and Evolutionary Computation Conference (GECCO 1999) (1999)
23. Cowell, R.G.: Parameter estimation from incomplete data for bayesian networks. In: International Workshop on Artificial Intelligence and Statistics, pp. 193–196 (1999)
24. Ramoni, M.F., Sebastiani, P.: The use of exogenous knowledge to learn bayesian networks from incomplete databases. In: Liu, X., Cohen, P.R., R. Berthold, M. (eds.) IDA 1997. LNCS, vol. 1280, Springer, Heidelberg (1997)
25. Ramoni, M.F., Sebastiani, P.: Parameter estimation in bayesian networks from incomplete databases. Intelligent Data Analysis 2(1), 139–160 (1998)
26. Ramoni, M.F., Sebastiani, P.: Learning bayesian networks from incomplete databases. In: 13th Conference on Uncertainty in Artificial Intelligence (UAI 1997), pp. 401–408 (1997)
27. Riggelsen, C., Feelders, A.J.: Learning bayesian network models from incomplete data using importance sampling. In: 10th International Workshop on Artificial Intelligence and Statistics, pp. 301–308 (2005)
28. Li, X., He, X., Yuan, S.: Learning bayesian networks structures from incomplete data: An efficient approach based on extended evolutionary programming. In: Ho, T.-B., Cheung, D., Liu, H. (eds.) PAKDD 2005. LNCS (LNAI), vol. 3518, pp. 474–479. Springer, Heidelberg (2005)
29. Li, X., He, X., Yuan, S.: A new method of learning bayesian networks structures from incomplete data. In: Duch, W., Kacprzyk, J., Oja, E., Zadrożny, S. (eds.) ICANN 2005. LNCS, vol. 3697, pp. 261–266. Springer, Heidelberg (2005)
30. Riggelsen, C.: Learning bayesian networks from incomplete data: An efficient method for generating approximate predictive distributions. In: Jonker, W., Petković, M. (eds.) SDM 2006. LNCS, vol. 4165, Springer, Heidelberg (2006)
31. Ragel, A., Cremilleux, B.: Treatment of missing values for association rules. In: Wu, X., Kotagiri, R., Korb, K.B. (eds.) PAKDD 1998. LNCS, vol. 1394, pp. 258–270. Springer, Heidelberg (1998)

32. Agrawal, R., Imielinski, T., Swami, A.N.: Mining Association Rules between Sets of Items in Large Databases. In: The ACM SIGMOD International Conference on Management of Data, pp. 207–216 (1993)
33. Poole, D., Mackworth, A., Goebel, R.: Computational Intelligence. Oxford University Press, Oxford (1998)
34. Lauritzen, S.L., Spiegelhalter, D.J.: Local computations with probabilities on graphical structures and their application to expert systems, 415–448 (1990)
35. Beinlich, I.A., Suermondt, H.J., Chavez, R.M., Cooper, G.F.: The ALARM monitoring system: A case study with two probabilistic inference techniques for belief networks. In: The 2nd European Conference on Artificial Intelligence in Medicine (1989)

Ranking Database Queries with User Feedback: A Neural Network Approach

Ganesh Agarwal, Nevedita Mallick, Srinivasan Turuvekere,
and ChengXiang Zhai

Department of Computer Science,
University of Illinois at Urbana-Champaign, IL, USA

Abstract. Currently, websites on the Internet serving structured data allow users to perform search based on simple equality or range constraints on data attributes. However, to begin with, users may not know what is *desirable* to them precisely, to be able to express it accurately in terms of primitive equality or range constraints. Additionally, in most websites, the results provided to users can be sorted with respect to values of any one particular attribute at a time. For the user, this is like *searching for a needle in a haystack* because the user's notion of *interesting* objects is generally a function of multiple attributes.

In this paper, we develop an approach to (i) support a family of functions involving multiple attributes to rank the tuples, and (ii) improve the ranking of results returned to the user by incorporating user feedback (to learn user's notion of *interestingness*) with the help of a neural network. The user feedback driven approach is effective in modeling a user's intuitive sense of desirability of a tuple, a notion that is otherwise near impossible to quantify mathematically. To prove the effectiveness of our approach, we have built a middleware for an application domain that implements and evaluates these ideas.

1 Introduction

Relational databases provide flexibility to users to query structured databases using various operations. However, they are limited in the type of searches a user can perform. For example, they only support the retrieval of data items which match a particular filtering criteria initially set by the user. But, in real world use-cases, it is very rare that users know the exact ranges for all the attribute inputs for their search. They may usually have some preferences but are willing to accept results which deviate a little from the search criteria they initially submit. Generally, they may want to differentiate between search conditions by giving different weightage to each of them.

Therefore, there is a need for a mechanism by which a user can pose a query and get the results ranked according to his notion of "interestingess", i.e, top-K results in a preference order. The framework should be able to include results which do not exactly match all criteria but match some of the desired criteria more heavily compared to other tuples. A mechanism for user feedback can also

J.R. Haritsa, R. Kotagiri, and V. Pudi (Eds.): DASFAA 2008, LNCS 4947, pp. 424–431, 2008.

be added, so that after users view the initial top-K results, they can indicate which tuples were of more interest to them. Depending on the feedback provided, the system can estimate the weights of different attributes and rank the result set to reflect the user's notion of interesting results.

2   Related Work

There has been recent interest in ranking the results of database query according to some function that models user preference and return the top-K tuples. Research in this area has spanned addition of new relational algebra constructs to support tuple ranking, query optimization for efficient execution of top-K queries, and middleware-based strategies that use traditional *SQL* constructs and try to minimize the number of database tuples retrieved.

Li et al. [1] introduced a systematic and principled framework, by extending relational algebra and query optimizers, to support ranking as a first-class construct in relational database systems. In this work, the ranking function is specified in the query. In contrast, our approach learns the ranking function through user feedback. There has been extensive study [2,3,4,5] of strategies for efficient execution of top-k queries. Under the top-K queries model used in [2], users specify target values for the attributes of a relation, and expect in return the tuples that best match these values. In [3] and [4], the authors discuss the principle of handling trade-off between different preferences while combining them. A recent work [6] studies how to find "best-k" results based on fuzzy matching and ranking of tuples. Agrawal et al. [7] also support fuzzy querying. They try to optimize the queries by exploiting existing indexes and database heuristics on cardinality and selectivity of relevant attributes. Also, they try to automatically infer the similarity between attribute values using the attribute distributions; which may not be aligned with users' notion of similarity.

3   Design Details

In this paper, we have proposed an approach for preference-based user searches. Our work tries to fill the gap between [2] and [7]. We explain our approach using our framework implemented over Realtor.com, a popular real estate search engine portal. However, the same concept is applicable to any other domain of user search. Our approach is to take a minimum amount of initial input from the user, in the form of constraints and relative weights on attributes. We relax these constraints to some extent to obtain a working set of tuples from the database, and incorporate a machine learning based framework to learn the user's preferences (iteratively) and present the top-K tuples in ranked order.

3.1   The Neural Network Approach

We are using a Neural Network based approach to build a framework, which will initially rank the tuples according to a weighted average of the user's provided

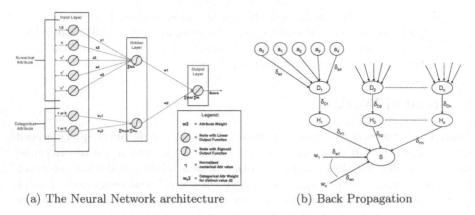

(a) The Neural Network architecture   (b) Back Propagation

Fig. 1. Neural Network and Back Propagation

input. We use a scoring function to evaluate score for each tuple and then rank the tuples according to their score values. As shown in Fig. 1(a), the Neural Network has **three layers**. The input nodes correspond to the attribute inputs.

For numerical attributes, the user will give a range of acceptable values. For categorical attributes the user will select a set of desired attributes. One hidden node corresponds to the normalized desirability value for an attribute. The output nodes correspond to the scoring function and hence will give the final score for each tuple. The initial network is built on the basis of the user provided ranges/values and weights for the attributes.

3.2 Numerical Attributes

Let A_i denote a numerical attribute for which the user provides a desirable range $[l_i, u_i]$. We relax this range (based on the weights between attributes and the interval length) and fetch a suitable initial working set of tuples in the new range $[l_i^*, u_i^*]$ for attribute A_i.

The increase in range interval is $\Delta_i = (1 - w_i) * (u_i - l_i)$, where w_i is weight of attribute A_i relative to other attributes. Thus, attributes with higher weights are relaxed to lesser extents. The new interval is given by $l_i^* = l_i - \Delta i$ and $u_i^* = u_i + \Delta i$.

We define a function $P(t(A_i) = x_i | C_i)$ to denote a *desirability function*, which gives the desirability value for a tuple t, where attribute A_i has value x_i and satisfies condition C_i. It is a degree four polynomial: $D(x) = a_0 + a_1 x + a_2 x^2 + a_3 x^3 + a_4 x^4$ The desirability distribution graph (with normalized values) in Fig. 2(a) shows the initial distribution. The curve will have high desirability values in the range provided by the user. This is evident from the steep slope which the curve takes initially and then almost flattens around the middle of the range provided by the user. The coefficients of the desirability function correspond to the weights of the connections between the input nodes and the hidden node of the Neural Network. We use a *Sigmoid function* to normalize the desirability function values at each hidden node. For a hidden node corresponding to attribute A_j having value x, the function is defined as: $H_j(x) = \frac{1}{1+e^{-a(D_j(x)-h)}}$.

3.3 Categorical Attributes

Since there is no ordering of categorical values, we cannot fit any desirability function. However, there may still be some similarity/desirability value associated with each value that is obtained initially from user. These desirability values are stored as the weights of the connections between the respective first level input nodes and the corresponding hidden node. The number of input nodes for categorical values can be either (i) the number of distinct values specified by the user, or (ii) the maximum number of distinct values based on the categorical attribute itself. In the latter case, the weights of the connections connecting distinct values not specified by the user are initialized to 0. For each tuple, therefore, the input is really a unit vector of n dimensions. We used a weighted sum as the aggregation function and a Sigmoid function as the node output function.

3.4 Tuple Score

The overall scoring function yielding the output of the Neural Network is a linear function of the hidden nodes' output normalized w.r.t. to the weights w_j of the attributes. For a tuple t, it can be defined as $S(t) = \sum_{j=1}^{n} w_j H_j(t(A_j)) / \sum_{j'=1}^{n} w_j'$.

3.5 Learning Using User Feedback

Suppose the user has provided feedback for m tuples $t_1, t_2, \ldots, t_m$. The user will give a score on a scale of 0 to 1 for these tuples. Suppose there are n attributes $A_1, A_2, \ldots, A_n$. Let $t_i(A_j)$ denote the value of the attribute A_j of tuple t_i.

The *mean squared error MSE function* on the set of m tuples to which the user has provided feedback can be defined as:

$$E(\boldsymbol{a}, \boldsymbol{w}) = \frac{1}{m} \sum_{i=1}^{m} |S(t_i) - US_i|^2 \tag{1}$$

Here, US_i is the user's provided score for tuple i. Since the target function and the number of feedback tuples are fixed, E is only a function of the weights a_{kj} and w_j of the Neural Network. The aim of our learning algorithm will be to adjust the weights using a *gradient descent method* so as to minimize the mean squared error E.

The coefficients and weights of each attribute will be adjusted as follows:

$$a_{kj}^{new} = a_{kj} - \epsilon \frac{\partial E}{\partial a_{kj}} \; ; \; w_j^{new} = w_j - \epsilon \frac{\partial E}{\partial w_j} \tag{2}$$

where $\frac{\partial E}{\partial a_{kj}}$ and $\frac{\partial E}{\partial w_j}$ denote the partial differentiation of MSE with respect to coefficients and weights of the attributes respectively. ϵ represents the contribution of learning in a single batch of feedback (rating of m tuples) by the user.

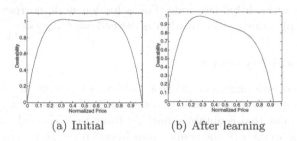

(a) Initial   (b) After learning

Fig. 2. Desirability function

We have adopted *Back propagation* strategy because it utilizes local computation at each node of the Neural Network and hence simplifies the computation
involved in the gradient descent method.

For each tuple t_i, where $i = 1 \ldots m$, the partial derivatives will be propagated
back through the nodes as shown in Fig. 1(b). The partial derivatives of MSE
are computed as follows:

$$\delta_S = 2\left(S(t) - US\right) \; ; \quad \delta_{H_j} = \delta_S \frac{w_j}{\sum_{j'=1}^{n} w_{j'}} \tag{3}$$

$$\delta_{w_j} = \delta_S \frac{\left(\sum_{j'=1}^{n} w_{j'}\right) H_j - \left(\sum_{j'=1}^{n} w_{j'} H_{j'}\right)}{\left(\sum_{j'=1}^{n} w_{j'}\right)^2} \tag{4}$$

$$\delta_{D_j} = \delta_{H_j} H_j\left(t(A_j)\right)\left(1 - H_j\left(t(A_j)\right)\right) \; ; \quad \delta_{a_{jk}} = \delta_{D_j}\left(t(A_j)\right)^k \tag{5}$$

The overall adjustment of weights will evaluate the above measures for each
of the m feedback tuples $t_1, t_2, \ldots, t_m$ using m back propagation computations.
The coefficients of the desirability function for each attribute and the weights of
the attributes will be updated as follows:

$$a_{kj}^{new} = a_{kj} - \epsilon \frac{1}{m} \sum_{i=1}^{m} \delta_{a_{kj}}(t_i) \; ; \quad w_j^{new} = w_j - \epsilon \frac{1}{m} \sum_{i=1}^{m} \delta_{w_j}(t_i) \tag{6}$$

4   Experiments

We tested our system extensively for various user scenarios. We would like to
present 3 user cases here.

Firstly, we tested how our system works for the case (Case 1) of simple ranking
based on one particular attribute. The system was queried for houses within a
particular range of price. The initial results obtained are ranked as per the
initial desirability curve for the attribute price as shown in Fig. 2(a). However,
when the user gives positive feedback for 2 tuples with lower price and negative
feedback for 2 tuples with higher price, the network is able to learn the user's

Table 1. Detailed results of Case 2

GT RANK	GROUND TRUTH (GT) HOUSE NAME	AREA	AGE	REALTOR REALTOR RANK	OUR ALGORITHM INITIAL RANK	OUR ALGORITHM RANK AFTER FEEDBACK
1	525 N HICKORY	1,371	1	35	45	4
2	2506 STRICKER LANE	2,345	2	43	8	1
3	3009 WEEPING CHERRY	2,335	5	36	5	2
4	1401 MITTENDORF	1,215	5	15	70	7
5	2109E PENNSYLVANIA	1,244	7	19	12	6
6	1308 W EADS	1,284	10	1	20	5
7	309 S NEW	1,666	16	18	14	3
8	1317 E HARDING DRIVE	1,715	19	17	46	8
9	1905 #2 CHRISTOPHER	970	22	41	77	31
10	2103 MONONA COURT	1,060	24	6	56	35
11	508 E MECHENRY	2,572	25	64	11	9
12	1738 WESTHAVEN	2,810	26	28	17	10
13	114 E PARK	2,500	26	55	18	11
14	801 BURKWOOD	2,390	26	62	9	17
15	1902 S GEORGE HUFF	2,372	26	58	80	21
16	715 W VERMONT	2,120	26	42	19	15
17	1806 GOLFVIEW DRIVE	2,044	26	48	15	12
18	2115 RANSOM PLACE	2,026	26	60	38	13
19	2507 LYNDHURST	2,014	26	34	13	14
20	2508 S LYNN	1,992	26	40	10	16
PRECISION AT 10				6	5	8
PRECISION AT 20				6	13	17
MEAN AVERAGE PRECISON (MAP)				0.320	0.421	0.952

interest in lower values of price and reflects it in the new modified desirability curve for the attribute price shown in Fig. 2(b). The system now gives higher scores to lower priced houses. In the figure, the normalized values of 0, 0.2, 0.8 and 1 represents the extended minimum value, actual minimum value, actual maximum value and the extended maximum value of price respectively. After learning, we can see that the desirability of houses decreases from normalized value of price from 0.2 to 0.8 . However, it should be noted that if the user provides feedback indicating preference for houses with prices outside the original range, then the desirability curve would be modified accordingly by the network to reflect higher desirability for prices outside the original range specified in the query. Importantly, the system uses this knowledge for the user's future searches too. However, the main strength of our system is when the user's interest in houses is based on their values for multiple attributes. Let us examine two such cases.

Let us take a scenario (Case 2) of a couple looking to buy a house. They query for houses giving two zip-codes as it is acceptable for them to live in any of the two adjacent neighboring cities. The husband queries for houses giving higher weight to area ranging from 1,000 to 2,500 sq. ft. and lower weight to age ranging from 1 to 50. However, on seeing the initial set of results, the wife demands for newer homes and provides positive feedback for 1 new home and and negative feedback for 2 older homes. The system incorporates the knowledge in the feedback and presents houses to the users with the new ranking. Next, we form the "Ground Truth" for this query. We initially extend the range of attributes and obtain a list of houses satisfying this range. We then manually rank these houses such that lower aged houses are ranked higher and if multiple houses have the same age, the houses with greater area are ranked higher. The top 20 houses in this list is taken as "Ground Truth" for this query as shown

Table 2. Detailed results of Case 3

GROUND TRUTH (GT)				REALTOR	OUR ALGORITHM	
GT RANK	HOUSE NAME	PRICE	AREA	REALTOR RANK	INITIAL RANK	RANK AFTER FEEDBACK
1	1118 W PARK	$145,000	2,160	31	16	5
2	603 W CLARK	$149,800	1,827	18	9	2
3	611 W CLARK	$164,900	1,920	7	2	1
4	2002 KAREN CT	$154,900	1,725	14	8	16
5	712 DEVONSHIRE	$169,900	1,932	6	10	6
6	309 S NEW	$154,900	1,666	15	5	3
7	808 S FOLEY	$177,000	2,160	32	28	15
8	1615 GLENN PARK	$129,900	1,476	20	14	8
9	914 S LYNN	$149,900	1,431	17	6	7
10	505 S GARFIELD	$175,000	1,771	4	25	10
PRECISION AT 10				3	6	8
MEAN AVERAGE PRECISON (MAP)				0.337	0.477	0.839

in Table 1.[1] Considering these 20 houses as the set of relevant results and all other houses as non-relevant, we measure precision at 10, precision at 20 and Mean Average Precision(MAP) for all three approaches. It can be seen that the performance using feedback in our system tends to outperform the rest by a large margin.

Next, we simulate the case (Case 3) of a user who wants to buy a house with large area but can only afford about $150,000. He issues a query for houses in range of $125,000 to $175,000 with area ranging from of 1,000 to 2,500 sq. ft. Once again, on receiving initial set of results, he provides very minimum feedback of 2 positive and 2 negative examples. Here, the query is highly subjective and it is not possible to get a standard ranking based on sorting them in order of an atttribute. Hence, we form the ground truth by taking the average of the ranking provided by multiple users simulating the same user query. The results can be seen in Table 2.[2] We also evaluated our system based on 11-point

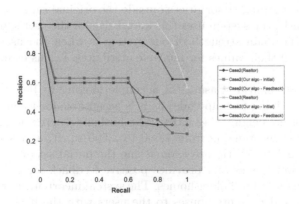

Fig. 3. 11-point Interpolated Average Precision-Recall Curve for Case 2 and 3

---

[1] As Realtor.com does not support querying for 2 zip-codes simultaneously, the overall ranking is taken by merging the two separate ranked lists in a round-robin fashion.

[2] In this case, only one zip-code was taken.

interpolated average precision-recall graph as shown in Fig. 3. This graph also shows the superiority of our algorithm which uses feedback over the other two alternatives.

5   Future Work and Conclusion

We conclude with some directions for future work. Presently, the system considers the attributes to be independent. However, the network can be modeled to learn different correlation and dependencies between attributes. One could also develop better learning strategy for categorical attributes in cases when the attributes are related to each other or have some ordering between them, which would help rank the results even better. We have chosen a degree 4 curve to represent the initial curve for all attributes. The system could be extensively tested for other different types of curves. Also, the architecture of the existing framework can be optimized for parallel calculations on multi processor machines, which enables faster processing of huge amounts of data.

The chief contributions of our work include a novel ranking mechanism which takes a weighted measure of multiple attributes rather than ranking it based on values of a single attribute; and the use of feedback in the search mechanism, which helps in presenting results to user which are based on their perception of "interestingness". Also, the middleware developed is very scalable as it does not assume any initial knowledge of the attributes of the given domain and hence can be easily adapted to work in any other domain of interest.

References

1. Li, C., Chang, K.C.-C., Ilyas, I., Song, S.: RankSQL: Query Algebra and Optimization for Relational Top-k Queries. In: Proc. of the ACM SIGMOD Conference, Baltimore, Maryland, USA (2005)
2. Bruno, N., Chaudhuri, S., Gravano, L.: Top-k selection queries over relational databases: Mapping strategies and performance evaluation. ACM Transactions on Database Systems(TODS) 27(2) (2002)
3. Fagin, R.: Combining fuzzy information from multiple systems. In: Proc. of the 15th ACM SIGACT-SIGMOD-SIGART Symposium on Principles of Database Systems (PODS), Montreal, Canada (1996)
4. Agrawal, R., Wimmers, E.L.: A framework for expressing and combining preferences. In: Proc. of the ACM SIGMOD Conference, Dallas, Texas, USA (2000)
5. Re, C., Dalvi, N., Suciu, D.: Efficient Top-k Query Evaluation on Probabilistic Data. In: Proc. of the 23rd International Conference on Data Engineering (ICDE) (2007)
6. Tao, T., Zhai, C.: Best-k Queries on Database Systems. In: Proc. of the 15th ACM Conference on Information and Knowledge Management (CIKM) (2006)
7. Agrawal, S., Chaudhuri, S., Das, G., Gionis, A.: Automated Ranking of Database Query Results. In: Proc. of First Biennial Conference on Innovative Data Systems Research (CIDR), Asilomar, California, USA (2003)
8. Haykin, S.: Neural Networks: A Comprehensive Foundation, 2nd edn. Prentice-Hall, Englewood Cliffs (1998)

Supporting Keyword Queries on Structured Databases with Limited Search Interfaces

Nurcan Yuruk[1,*], Xiaowei Xu[1], Chen Li[2], and Jeffrey Xu Yu[3]

[1] University of Arkansas at Little Rock, USA
[2] University of California, Irvine, USA
[3] Chinese University of Hong Kong, Hong Kong
{nxyuruk,xwxu}ualr.edu, chenli@ics.uci.edu, yu@se.cuhk.edu.hk

Abstract. Many Web sources provide forms to allow users to query their hidden data. For instance, online stores such as Amazon.com have search interfaces, using which users can query information about books by providing conditions on attributes of title, author, and publisher. We propose a novel system framework that supports keyword queries on structured data behind such limited search forms. It provides user-friendly query interfaces for users to type in IR-style keyword queries to find relevant records. We study research challenges in the framework and conduct extensive experiments on real datasets to show the practicality of our framework and evaluate different algorithms.

1 Introduction

Recently the Web has been deepened due to more online databases that hide their data behind search interfaces. These data sources provide forms to allow users to query their data. For instance, online bookstores such as Amazon.com have search forms to accept queries on book information. Using such a form users can provide conditions on attributes such as title, author, and publisher. A typical query is "Finding books whose title contains words database and system, and whose author contains the word ullman." Given such a query, the source returns all the books satisfying these conditions. Since these data sources provide a tremendous amount of valuable information, it is becoming increasingly important to support queries on their data to meet the demand of various business applications. Recently there have been intensive studies on how to support queries over Web sources with such limited querying interfaces. (i.e., [1], [2], [3], [4]). In the paper we propose a domain-independent system framework, called SKUA[1], which supports Information-Retrieval (IR)-style keyword queries on data sources through their limited, structured search interfaces. Given such a data source and a user query that consists of a list of keywords, our goal is to find the records in the hidden database that contains all these keywords. One main advantage of having such a system is to allow other applications to access the information at the data sources through a simple IR-style query interface. This feature is especially important in search engines built on top of these data sources.

---

[*] Stands for "Supporting Keyword qUeries on structured datA".

J.R. Haritsa, R. Kotagiri, and V. Pudi (Eds.): DASFAA 2008, LNCS 4947, pp. 432–439, 2008.
© Springer-Verlag Berlin Heidelberg 2008

One challenge arises naturally: given a set of keywords in a query and a structured search interface at a source, how to fill out the form in order to retrieve the records satisfying the query conditions? The mapping problem becomes more challenging when we want to build a domain-independent framework and domain-specific knowledge might not be available.

Our solution to the problem is to build a dictionary of word frequencies in different attributes at the source. The dictionary has information about a subset of words in the hidden database, and the frequency of a word on an attribute, and/or the frequency of a word in those sampled records. When a user query is posted, we utilize this dictionary to generate a plan to access the data source to retrieve relevant information.

2 Related Work

Recently, there have been increased interests to support IR queries from hidden web sources. Successful systems are developed for relevant resource discovery and answering IR queries from hidden web sources [2], [3], [4]. In comparison, SKUA focuses on structured data behind structured search forms.

Supporting IR-style keyword queries over relational DBMS is an important functionality as allowing simple IR-style keyword queries significantly facilitates finding information for casual users. It provides valuable information to the user without requiring any knowledge of the database schema or any need to write SQL queries. DBXplorer [17], BANKS [18] are those systems proposed to address the problem. However, all these systems either depend on the knowledge of the database schema or require a direct access to the database. Our SKUA framework supports keyword queries on a single structured relation based only on the word-frequency information.

SKUA needs to map keywords in a user query to the attributes in the search interface provided by the hidden source. The main goal of mapping the keywords to the attributes of a structured database can be defined as word sense disambiguation in a broader sense, since it needs the interpretation of keywords and attributes before being able to map them to each other. Although word sense disambiguation ([14], [15]) and semantic tagging [16] are well studied, existing methods cannot be directly applied here, since the proposed approaches of SKUA do not rely upon any lexical or syntactic information.

3 Problem Formulation and Complexity

Keyword Query: We are given a set of keywords $S = \{w_1, w_2, \ldots, w_m\}$ and a hidden database DB with a searchable interface of $I(A_1, A_2, \ldots, A_k)$, which relies on relation $R(A_1, A_2, \ldots A_k)$. Our goal is to retrieve all the records containing all the keywords, more specifically, an answer set Ans(S), such that

$$Ans(S) = \{r \mid r \in R \land S \subseteq \bigcup_{i=1}^{k} r.A_i\} \quad (1)$$

where r.A_i is the set of words in the A_i value of record r. To retrieve the records in the answer set, we need to map the keywords to the searchable attributes. Each mapping is a query to the hidden source using the interface I. The number of possible source queries (called query cost) is exponential in terms of the number of keywords.

Proposition 1: Given a searchable interface $I(A_1,A_2,\ldots A_k)$, the number of all possible source queries for a set of keywords $S = \{w_1,w_2,\ldots,w_m\}$ is

$$p = \sum_{i=1}^{k} C(m,i).k^i \tag{2}$$

Each source query will return a subset of relevant records together with some irrelevant records. An answer to a user keyword query $S = \{w_1,w_2,\ldots,w_m\}$ is a record r that contains all the keywords $w_1,w_2,\ldots,w_m$ in its attribute values, i.e., $r \in Ans(S)$ based on Equation (1). Let $Ans(Q)$ denote the set of answers returned by a source query Q. keywords.

Proposition 2: Let $Q_1,Q_2,\ldots,Q_p$ be all possible source queries for a keyword query S.

$$Ans(S) = \bigcup_{i=1}^{p} Ans(Qi) \tag{3}$$

Definition 1 (Keyword Query Problem): Given a keyword query S and an interface I, find a plan, which is a subset of all possible source queries, such that all answers are covered by the queries in the plan and the total cost is minimal. More formally, a plan is a subset C of all possible source queries $\{Q_1,Q_2,\ldots,Q_p\}$. The plan C has two properties. First, it returns all answers to the keyword query:

$$Ans(S) = \bigcup_{q \in C} Ans(q) = \bigcup_{i=1}^{p} Ans(Qi) \tag{4}$$

Second, the plan cost, which is measured by the total cost of the source queries in C, is minimized:

$$C = \underset{D \subseteq \{Q_1,Q_2,\ldots,Q_p\}}{\arg\min} \left(\sum_{q \in D} c(q) \right) \tag{5}$$

where c(q) is the processing cost for source query q. Even we know the answer set $Ans(Q_i)$ for source query Q_i, the problem is NP-complete. The problem is even more challenging due to the lack of prior knowledge of $Ans(Q_i)$. Each source query Q_i has to be executed in order to obtain such information.

4 Search Algorithms

The naïve exhaustive search algorithm is computationally prohibitive. In this section we propose efficient algorithms to solve this problem. All the algorithms first try to reduce the search space then aim to achieve a high recall of the answers. We first introduce the following proposition.

Proposition 3: The number of returned records is anti-monotone to the number of used keywords in the source queries. Assume that the used keywords in source queries Q_1 and Q_2 are s_1 and s_2 respectively; and $s_1 \subset s_2$. Then the set of returned records for Q_1 is a superset of that for Q_2. More specifically, $Rt(Q_1) \supset Rt(Q_2)$, where $Rt(Q_i)$ is the set of returned records for Q_i (i=1,2).

4.1 Machine Learning (ML) Algorithms

Some source queries could return an empty answer. If we can predict whether the answer to a source query is empty, we can reduce the cost by only issuing those source queries that return a non-empty set of answers. One possible approach is to use machine learning algorithms, as presented below.

Training Dataset: To construct a training dataset from the source, we first generate a set of keyword queries. For each such keyword query, we consider all the possible source queries that use all these keywords. If a source query returns a non-empty answer, it is labeled as class "1", and class "0" otherwise.

Predictive Models: Once a training dataset is built, we can train a predictive model for source queries by using supervised machine learning algorithms. We use the trained model to predict if this source query will return a non-empty answer set. All these source queries are ranked descending order based on the predicted likelihood. The source queries are then issued in the sorted order. If the prediction is accurate we can avoid issuing source queries that return an empty set of answers, thus reduce the query cost.

4.2 Greedy Search Algorithms

Proposition 3 implies that using more keywords in a source query reduces the number of the returned records and thus the number of interactions to download the records. On the other hand, the number of possible source queries could increase as the number of keywords increases. Therefore, we want to limit the number of source queries using a greedy strategy. We start with source queries using a single keyword and augment a source query with additional keywords if its number of returned records is too large. We rank the keywords based on the number of returned records containing the keywords, which is also called *selectivity* (or *frequency*.

Definition 2: Let w be a keyword. The selectivity of w is $s(w) = |\{r \mid w \in r \wedge r \in D\}|$, where D is the source.

We rank the query keywords based on a decreasing order of their selectivity. The greedy search algorithm starts by choosing the most selective keyword to execute source queries. This keyword is filled into all the attributes to generate source queries. If a source query is expected to return too many records, we need to augment it by adding the next most selective keyword, and so on. The algorithm only requires the knowledge of the keyword selectivities, which are stored in the dictionary. In the case where a selectivity of a keyword (in the database or on an attribute) is missing in the dictionary, we assume this frequency is low. Since the dictionary used by the search algorithm is incomplete, there are two variations of the Greedy search algorithm.

Definition 3: Probabilistic Greedy (P-Greedy for short): as the name implies, it is the probabilistic version of the Greedy search algorithm, where missing frequencies in the dictionary are treated as a zero frequency even though the actual values may not be zero. Thus, there might be some false dismissals.

Definition 4: Cautious Greedy (C-Greedy for short): A hundred percent recall is guaranteed by executing every source query regardless of its corresponding frequency.

5 Dictionary Construction

The task of dictionary construction is to extract a set of words from the database and calculate their estimated frequency information from the returned records for the database and different attributes. A query-based sampling approach has been proven to be successful for resource selection ([2], [5]) and information extraction [8]. We can use this query-based sampling approach to construct a dictionary D by iteratively querying database DB, retrieving the answer, and calculating the frequencies of the words in the returned answers.

6 Experiments

We evaluated SKUA on the following data sets. The first one was the DBLP Computer Science Bibliography Database [7]. The second one was the movie dataset provided by Professor Gio Wiederhold from Stanford University [6]. We also used a book database collected by Cai-Nicolas Ziegler in a 4-week crawl from the Book-Crossing community [9].

Quality Metric: We consider the following two measures to compare the algorithms:

1. Recall: the percentage of the number of records satisfying the conditions in a user query that can be retrieved by the plan generated by an algorithm.

2. Query cost: the number of source queries sent to the underlying database. The search algorithms were evaluated on several generated workloads.

6.1 ML Algorithms on Sampled Dictionary

Three different machine learning (ML) algorithms, namely Support Vector Machines (SVM)[10], [11], Decision Tree (DT) c4.5 [12], Naïve Bayesian (NB) [13], were evaluated. We conducted experiments for all three methods using complete dictionary. As SVM performed best among others (and due to space constraints) only SVM results will be presented for sampled dictionary. Figure 1 shows the SVM results of recall using a dictionary of the DBLP dataset constructed based on sampling. When the user query had only one keyword (QW1), recall rate was quick to reach hundred percent (after just sending two source queries).Though, for same number of source queries recall rates are 0.85, 0.74, 0.62 for QW2, QW3, and QW4 respectively.

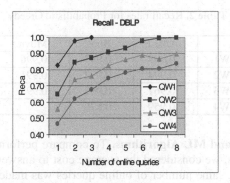

Fig. 1. Recall vs online query cost for SVM; DBLP dataset

6.2 Greedy Algorithms on Sampled Dictionary

We considered four scenarios for greedy algorithms, namely Greedy-DB, Greedy-Rand, P-Greedy-DB', C-Greedy-DB'. In the Greedy-DB case, we used a complete dictionary that was derived from the original database, containing all the keywords and their accurate frequency information. Although having complete dictionary information may not be realistic, these results are shown for comparison purposes. Yet another extreme case is that the dictionary does not have any information at all, and the search algorithm randomly selects a keyword in a user query to grow the query plan tree. This scenario is called Greedy-Rand. In the third scenario, called P-Greedy-DB', we used the Probabilistic-Greedy algorithm on a dictionary constructed by query-based sampling. In the last case, called C-Greedy-DB' we used the Cautious-Greedy algorithm for the same Sampled Dictionary. (The suffix –DB' means that the dictionary was constructed based on sampling.)

We present the query-cost results for different scenarios for DBLP in Table 1. As expected, Greedy-DB has best and Greedy-Rand has worst performance. The more interesting observation, which demonstrates the practicality of the proposed framework, is that P-Greedy-DB' having very close results to ideal case Greedy-DB. In average, query cost is decreased by 2 as compared to its C-Greedy counterpart.

Among all these cases, P-Greedy-DB' was the only one that cannot guarantee a 100% recall. We show its recall results in Table 2. If we take DBLP as an example, all the recalls are greater than 95%. Both considering the improvement on the performance and high recall rates, we can conclude that P-Greedy outperforms C-Greedy.

Table 1. Average query cost for DBLP

	Greedy-DB	P-Greedy-DB'	C-Greedy-DB'	Greedy-Rand
1QW	1.765	1.815	3	3
2QW	2.295	3.045	4.959	5.997
3QW	1.901	2.787	4.621	6.441
4QW	1.568	2.671	4.224	6.36

Table 2. Recall rates for Probabilistic Greedy

	DBLP	Book	Movie
QW1	0.998	0.997	0.996
QW2	0.994	0.992	0.991
QW3	0.961	0.962	0.99
QW4	0.959	0.876	-

Comparing Greedy and ML Algorithms. To compare performance of ML models and greedy algorithms, we considered the average cost to answer a user query for the greedy algorithm. The same number of online queries was made to the source by the ML algorithms. Then we compared their achieved recalls. Table 3 shows the results of SVM and P-Greedy-DB' assuming an average query cost of 3. It shows that the P-Greedy algorithm achieved a higher recall for the same number of source queries than SVM. Also note that the greedy algorithm does not require an expensive training phase in comparison with the ML algorithms.

Table 3. Recall rates for P-Greedy and SVM; DBLP dataset

	P-Greedy	SVM
QW1	0.998	1
QW2	0.994	0.875
QW3	0.961	0.762
QW4	0.959	0.678

7 Conclusions

As the web is becoming deepened, searching and extracting information from hidden sources becomes increasingly important. In this paper, we proposed a framework, called SKUA, to support IR-style keyword queries using a limited interface from a hidden Web source. We have presented the system framework, and the corresponding optimization problem to answer a keyword query efficiently. Empirical evaluations using real datasets have shown that the proposed greedy search algorithms can retrieve all the answers to a user query from the underlying database with a small number of source queries.

References

1. Zhang, Z., He, B., Chang, K.C.C.: Understanding web query interfaces: Best-effort parsing with hidden syntax. In: ACM SIGMOD Conference, Paris, France, pp. 107–118 (2004)
2. Ipeirotis, P., Gravano, L.: Distributed search over the hidden web: Hierarchical database sampling and selection. In: VLDB Conference, Hong Kong, China, pp. 394–405 (2002)
3. Li, C., Chang, E.: Query planning with limited source capabilities. In: ICDE Conference, San Diego, CA, pp. 400–412 (2000)

4. Liu, Z., Luo, C., Cho, J., Chu, W.W.: A probabilistic approach to metasearching with adaptive probing. In: ICDE Conference, Boston, USA, pp. 547–559 (2004)
5. Callan, J., Connell, M.: Query-based sampling of text databases. ACM Transactions on Information Systems 19(2), 97–130 (2001)
6. UCI KDD Arhive Movies Dataset, http://kdd.ics.uci.edu/databases/movies/movies.html
7. Digital Bibliography & Library Project, http://www.informatik.uni-trier.de/~ley/db/
8. Agichtein, E., Gravano, L.: Querying text databases for efficient information extraction. In: ICDE Conference, Bangalore, India, pp. 113–124 (2003)
9. Ziegler, C.N., McNee, S.M., Konstan, J.A., Lause, G.: Improving Recommendation Lists Through Topic Diversification. In: 14th International World Wide Web Conference, Chiba, Japan, pp. 22–32 (2005)
10. Joachims, T.: Making large scale support vector machine learning practical. Advances in Kernel Methods: Support Vector Machines. MIT Press, Cambridge (1998)
11. A library for support vector machines, http://www.csie.ntu.edu.tw/~cjlin/libsvm/
12. Quinlan, J.R.: C4.5: Programs for Machine Learning. Morgan Kaufman, San Francisco (1993)
13. Mitchell, T.: Machine Learning. McGraw-Hill, New York (1997)
14. Zhang, G., Chu, W.W., Meng, F., Kong, G.: Query Formulation from High-Level Concepts for Relational Databases. In: UIDIS, pp. 64–75 (1999)
15. Peh, L.S., Ng, H.T.: Domain Specific Semantic Class Disambiguation using Wordnet. In: Fifth Workshop on Very Large Corpora, Hong Kong, China, pp. 56–64 (1997)
16. Boufaden, N.: An Ontology-based Semantic Tagger for IE system. In: 41st Annual Meeting on Association for Computational Linguistics, Sapporo, Japan, pp. 7–14 (2003)
17. Agrawal, S., Chaudhuri, S., Das, G.: DBXplorer: A System for Keyword-Based Search over Relational Databases. In: ICDE Conference, San Jose, CA, p. 5 (2002)
18. Aditya, B., Bhalotia, G., Chakrabarti, S., Hulgeri, A., Nakhe, C., Sudarshan, P.S.: BANKS: Browsing and Keyword Searching in Relational Databases. In: Bressan, S., Chaudhri, A.B., Li Lee, M., Yu, J.X., Lacroix, Z. (eds.) CAiSE 2002 and VLDB 2002. LNCS, vol. 2590, pp. 1083–1086. Springer, Heidelberg (2003)

Automated Data Discovery in Similarity Score Queries

Fatih Altiparmak[1], Ali Saman Tosun[1,2],
Hakan Ferhatosmanoglu[1], and Ahmet Sacan[1,3]

[1] The Ohio State University, Dept. of Computer Sci. & Eng., Columbus, OH
{altiparm,sacan,hakan}@cse.ohio-state.edu
[2] The University of Texas at San Antonio, Dept of Computer Science
tosun@cs.utsa.edu
[3] Middle East Technical University, Dept. of Computer Eng., Ankara, Turkey

Abstract. A vast amount of information is being stored in scientific databases on the web. The dynamic nature of the scientific data, the cost of providing an up-to-date snapshot of the whole database, and proprietary considerations compel the database owners to hide the original data behind search interfaces. The information is often provided to researchers through similarity-search query interfaces, which limits a proper and focused analysis of the data. In this study, we present systematic methods of data discovery through similarity-score queries in such "uncooperative" databases. The methods are generalized to multi-dimensional data, and to L-p norm distance functions. The accuracy and performance of our methods are demonstrated on synthetic and real-life datasets. The methods developed in this study enable the scientists to obtain the data within the range of their research interests, overcoming the limitations of the similarity-search interface. The results of this study also present implications in data privacy and security areas, where the discovery of the original data is not desired.

1 Introduction

An ever growing amount of information is being served on the Web. Some of this information is in the form of inter-linked HTML pages, which are crawled, indexed, and made accessible by the search engines such as Google (google.com), Yahoo (yahoo.com), and MSN (msn.com). The portion of the Web that is accessible via search engines is termed the *surface Web*. A far greater amount of information is believed to be hidden behind databases whose content is not accessible through static URL links. It was estimated that this *deep Web* contained 7,500 terabytes of data – 500 times larger than the surface Web [5].

The information in the hidden Web is available only as a response to dynamically issued queries to the search interface of the databases. Recent efforts have focused on categorizing these databases at the absence of content summaries [13], or on building meta-search engines that provide a unified search interface to these databases [20]. However, most of these efforts were limited to *text databases* and ignored the *numeric databases* prevalent in scientific repositories.

J.R. Haritsa, R. Kotagiri, and V. Pudi (Eds.): DASFAA 2008, LNCS 4947, pp. 440–451, 2008.
© Springer-Verlag Berlin Heidelberg 2008

In scientific databases, the data is usually represented as multi-dimensional feature vectors. Due to the nature of the data and the large quantity of information, similarity search has emerged to be the de facto form of query in scientific applications such as high-energy physics [24], geographic information systems (GIS) [6], financial time series databases [14], medical imaging [19, 17], and bioinformatics [15].

The degree of similarity between objects in a database is often quantified by a distance measure, e.g., Euclidean distance, operating on the multi-dimensional data objects or the feature vectors extracted from the data objects. For example, a user may pose a query over a medical database asking for X-rays that are similar to a given X-ray in terms of Euclidean distance of multi-dimensional texture feature vectors [18, 16]. 3D Shape histograms of proteins are used to identify their similarities [1]. Similarity query is usually implemented by finding the closest feature vector(s) to the feature vector of the query data. This type of query is known as nearest neighbor (NN) query [21] and it has been extensively studied in the past [11, 12, 23, 2, 3, 9, 4, 7]. A closely related query is the ϵ-range query where all feature vectors that are within ϵ neighborhood of the query point q are retrieved.

Bandwidth and resource constraints and the continuous nature of the data acquisition itself, whether it be from manual contributions from researchers or automated sensor input, make the maintenance of an up-to-date, downloadable snapshot of the whole database unfeasible. Therefore the database providers limit the data access to the similarity query interface they provide. Still other providers practice this limitation due to privacy or proprietary concerns.

Even though similarity query over the database is one of the first steps of gaining valuable information about the entity under consideration, it is insufficient for further scientific investigation. The scientists often seek to acquire a portion of the database relevant to their research question. In this study, we overcome the limitation imposed by the similarity query interface, and show that the data of interest can be automatically retrieved while minimizing the burden on the resource constraints of the database owners.

The results of this study also have critical implications in database security, where the discovery of the data is not desired by the providers. Specifically, we show that the whole database can be discovered through similarity queries. We give recommendations for preventive measures where privacy or proprietary concerns lead the query interface limitation.

We have previously identified two main models of similarity search queries [22]:

- *Reply Model.* Client queries vector x and database responds with the closest k vectors y_i $(i = 1 \ldots k)$.
- *Score Model.* Client queries vector x and database responds with similarity score $\| x - y \|$, where y is the closest vector in the database to the x.

In this paper, we focus on the Score Model, where a rigorous analysis was missing. The contributions of this paper can be summarized as follows:

- Data discovery through similarity score queries is proven under a general probing strategy.
- A strategy using query histories is developed to improve the efficiency of the data discovery
- The methods are generalized to multi-dimensional case, and to $L_p - norm$ distance measures

2   Methods

In the score model, upon receiving a similarity query x from user, the database responds with the similarity score $\| x - y \|$ [8], where y is the closest point in the database to x. Assume that l_1, l_2, and u_1, u_2 are the coordinates of the two corners on the same diagonal of the minimum rectangle bounding the region of interest in 2-dimensional space. Further assume that the closest distance between any two points in the database is given as c.

A basic data discovery approach where the database consists of a *single n-*dimensional vector y was given in [22]. We reiterate this basic approach and its proof here in Algorithm 1 for completeness. The function $createVector(d, i, v)$ creates a d dimensional vector which has 0 in all dimensions but v in the ith dimension. The n-dimensional vector y can be discovered using $n + 1$ queries. So, for 2 dimensions, 3 queries would be needed.

Algorithm 1. Algorithm to discover y

1: n = dimensionality of vectors x and y
2: $q_1 = sim\_search([0, 0, ...0])$
3: **for** i= 1 to n **do**
4:     $q_2 = sim\_search(createVector(n, i, 1))$
5:     $y_i = \frac{q_2^2 - q_1^2 - 1}{-2}$
6: **end for**

Lemma 1. *Algorithm 1 discovers y.*

Proof. Consider i^{th} iteration of the loop. We have $q_1 = (\sum_{k=1}^{n}(y_i)^2)^{1/2}$ and $q_2 = ((y_i - 1)^2 + \sum_{k=1, k \neq i}^{n}(y_i)^2)^{1/2}$. By simple algebra we get $q_2^2 - q_1^2 = -2y_i + 1$. We can solve this equation to find y_i.

Now let us consider a database with large number of tuples. In this case, the basic approach in Algorithm 1 can not be used since queries q_1 and q_2 can return scores based on different vectors in database. The example shown in Figure 1 returns a score of 1 for similarity search $(0, 0)$ and returns a score of $1/2$ for similarity search $(0, 1)$. These are not comparable since their distances are to different nodes.

In the following sections, we first explain our proposed discovery strategy for 2 dimensions using l_2 (Euclidian) distance and then generalize it to n dimensions using l_p norm as the distance metric. For notational clarity, we first consider the

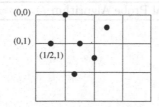

Fig. 1. Discovery of multiple points. Algorithm 1 is insufficient in correctly discovering the data.

problem of discovering the whole database; i.e., all the data points lie within the region of interest. The extension to discovery of a region of the database is trivial and discussed later.

2.1   Discovery of Multiple Data Points in 2 Dimensions

In order to discover every point in the database, closely located queries need to be sent to ensure that all the queries return a score to the same data point. Assume that the whole space is divided into equally spaced squares and that the length of an edge of the square, which is also the distance between two consecutive probes, is $c\prime$. In the best case, $n + 1{=}3$ probes having the coordinates x,y,x,y+$c\prime$,x+$c\prime$,y are required to return scores to the same data point to guarantee discovering it.

Recall that we have the probes and the associated scores, but do not have any information identifying the data point for the returned score. Therefore, to discover data points when the databases use the score model, we must take advantage of all the information at hand: the returned scores and the distance between probes, $c\prime$. The task becomes finding a probe distance $c\prime$ such that putting a condition on the returned scores for the nearby probe points guarantees that the score returned by each of them is to the same data point. In the following lemma we put a condition, maximum score, on the returned distance for a probe to guarantee that each of its $c\prime$ distanced neighbors returns a score to the same data point.

Lemma 2. *Let the score for a probe $p = (x, y)$ be δ where $\delta \leq \frac{c}{4}$ and the closest point to probe p be $q = (s, t)$. Then the closest point to probes $p_2 = (x + \frac{c}{4}, y)$, $p_3 = (x - \frac{c}{4}, y)$, $p_4 = (x, y + \frac{c}{4})$, and $p_5 = (x, y - \frac{c}{4})$ is $q = (s, t)$.*

Proof. Consider the distance $d(p_2, q)$. By triangle inequality we have $d(p_2, q) \leq d(p_2, p) + d(p, q) \leq \delta + \frac{c}{4}$. Therefore, $d(p_2, q) \leq \frac{c}{4} + \frac{c}{4} \leq \frac{c}{2}$. Since c is the smallest distance between pairs, closest point to probe $p_2 = (x + \frac{c}{4}, y)$ is $q = (s, t)$. Proofs for p_3, p_4, p_5 are similar.

The required distance, $c/4$, and $c\prime$, $\frac{c}{4}$, are selected such that sum of them is $\leq c/2$ and at least one of the corners of the square a data point lies in returns a score $\leq$ the required distance, $c/4$ to the point. Algorithm shown in Figure 2 utilizes lemma 2 to discover all points in a two dimensional space.

Algorithm 2. The General Probe Algorithm

1: $\alpha_1 = \lceil \frac{u_1 - l_1}{c/4} \rceil + 1$

2: $\alpha_2 = \lceil \frac{u_2 - l_2}{c/4} \rceil + 1$

3: **for** $i = 0$ to $\alpha_1 - 1$ **do**

4:     **for** $j = 0$ to $\alpha_2 - 1$ **do**

5:         $probe[1] = l_1 + i\frac{c}{4}$

6:         $probe[2] = l_2 + j\frac{c}{4}$

7:         $dist_{i,j} = \text{dist\_search}(probe)$

8:     **end for**

9: **end for**

10: **for** $i = 0$ to $\alpha_1 - 2$ **do**

11:     **for** $j = 0$ to $\alpha_2 - 2$ **do**

12:         **if** $dist_{i,j} \leq c/4$ **then**

13:             $y_1 = \dfrac{\frac{dist_{i+1,j}^2 - dist_{i,j}^2}{c/4} - 2(l_1 + i\frac{c}{4}) - c/4}{-2}$

14:             $y_2 = \dfrac{\frac{dist_{i,j+1}^2 - dist_{i,j}^2}{c/4} - 2(l_2 + j\frac{c}{4}) - c/4}{-2}$

15:             **if** y not in database **then**

16:                 save y

17:             **end if**

18:         **end if**

19:     **end for**

20: **end for**

Theorem 1. *The General Probe Algorithm in Algorithm 2 discovers the whole database.*

Proof. To make the proof complete we need to show for each data point that

(1) at least one probe exists such that distance between
    the probe and the data point is $\leq \frac{c}{4}$ and
(2) algorithm guarantees to find the data point.

The proof for each item will be made separately.

(1) Since voronoi regions cover the whole region, each data point lies in a square having edge of $\frac{c}{4}$. Hence, the longest distance between a data point and the nearest corner of the square it lies into can be $c\sqrt{2}/8$ which is less than $c/4$. Thus, at least one of these probes will return a score $\leq \frac{c}{4}$ to the point.

(2) As stated by Lemma 2, each of the probes with indices $\{i, j\}, \{i+1, j\}$ and $\{i, j+1\}$ returns the distance to the the same data point if the returned distance for $\{i, j\} \leq \frac{c}{4}$. Let us call this data point y. We have $dist_{i,j} = ((l_1 + i\frac{c}{4} - y_1)^2 + (l_2 + j\frac{c}{4} - y_2)^2))^{1/2}$ and $dist_{i+1,j} = ((l_1 + (i+1)\frac{c}{4} - y_1)^2 + (l_2 + j\frac{c}{4} - y_2)^2))^{1/2}$. By simple algebra we get $dist_{i+1,j}^2 - dist_{i,j}^2 = \frac{c}{4}(-2y_1 + 2(l_1 + i\frac{c}{4}) + \frac{c}{4})$. We can solve this equation to find y_1. We can find y_2 similarly by considering probe with indices $\{i, j+1\}$ instead of $\{i+1, j\}$. Since we have shown that such a probe exist for each data point, the algorithm will discover the whole database.

The General Probe Algorithm divides the space into a grid and queries the database with the corners of this grid. The distance between two consecutive probes is $c/4$ to guarantee that for each data point, at least one of the probes belonging to the corners of the square that the data point lies in returns a score less than or equal to the required distance, $c/4$. Therefore, there is at least one probe within $c/4$ distance to each point and by using Lemma 2 the point can be discovered.

2.2 A Progressive Probing Algorithm

A very small value for c, the minimum distance between data points, can cause the General Probe Algorithm to generate an infeasible number of probes. As suggested in [22], a progressive probing scheme can be utilized to hierarchically sample the database, and in turn, to discover the data in finer detail as the number of probes increases.

The progressive scheme is expected to be better at discovering more of the database with earlier queries since it spreads the probe points. The total of the indices is used and the level of a probe is calculated by the modula operation. For example, for two dimensions modula of the sum of row number and column number is used for calculating the level of a probe. As an example, *levelno* for a two dimensional space divided into 5 columns and 5 rows is 4 ($\lceil$max row index $+ 1 +$ max column index $+ 1\rceil$). The probe pattern of progressive discovery for this space is given in Figure 2.

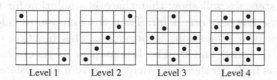

Level 1   Level 2   Level 3   Level 4

Fig. 2. Levels of Progressive Probing in 2 dimensions

2.3 Exploiting the Query History

The General Probe Algorithm is based on querying the database at each of the interval points. However, some of these probes may be redundant, and it is possible to eliminate such probes based on the information obtained from previous queries. To discover every point in the database, the space can be divided in a way that there exist $(n+1)$ probes for each data point that return score$\leq c/2$ to the point. The returned scores for most of the queries will be larger than $c/2$, which would guarantee that the nearby probes return scores to the same data point. It is possible to eliminate unnecessary probes based on the information obtained from previous queries.

In Figure 3, the database returns distance value s for a query probe p whose nearest neighbor is the data point y. The $c\prime$ is taken as $c/4$ as found above. Lemma 2 states all probes having distance $c\prime$ to a probe p will return a score to the same data point if the returned score for p is $\leq c/4$. In the proof of this Lemma we showed that all these probes will have at most a score of $c/2$. Since

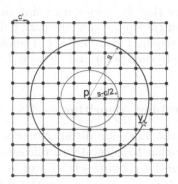

Fig. 3. Elimination of Redundant Probes

there is no other data element within the outer circle, any probe in the inner circle cannot have a score $\leq c/2$. Hence, the probes in the inner circle except p are redundant and can be eliminated.

2.4   Extension to Higher Dimensions and to l_p Norm

We will extend the solution presented for 2 dimensions to n dimensions in three steps. As a first step, we will show that there is a solution for the simple case where the database has only one point. Then the function to calculate the total number of probes in n dimensions is provided. This function depends on the distance metric (l_p norm) and n. The last step is to generalize Lemma 2, on which solution was built, to the case where we have n dimensions and use the l_p norm as the metric.

Definition 1. l_p *norm between* n *dimensional vectors* x *and* y *is defined as* $l_p(x, y) = (\sum_{i=1}^{n} |x_i - y_i|^p)^{1/p}$.

The solution shown in Algorithm 3 is an extended version of the solution in Algorithm 1. There are two differences between these solutions. The first difference is that we send a value of -1 instead of a value of 1 in the appropriate dimension when we are gathering information about a dimension. The other difference is that while for 2 dimensions using l_2 norm we can find an exact solution for y_i, we can only show that a unique real solution exists for n dimensions using l_p norm.

Algorithm 3. Algorithm to discover y for n dimensions

1: $q_1 = sim\_search([0, 0, ...0])$
2: **for** $i = 1$ to n **do**
3:     $q_2 = sim\_search(createVector(n, i, -1))$
4:     Compute y_i
5: **end for**

Theorem 2. *Algorithm in Algorithm 3 discovers y for every l_p.*

Proof. Consider i^{th} iteration of the loop. Using the definition of l_p norm, we have

$$q_1 = (\sum_{j=1}^{n} |y_j|^p)^{1/p}$$

and

$$q_2 = (|y_i + 1|^p + \sum_{j=1,j\neq i}^{n} |y_j|^p)^{1/p}$$

By using above equation we have

$$q_2^p - q_1^p = |y_i + 1|^p - |y_i|^p$$

Left hand side is known since we know q_1, q_2 and p. As q_1 and q_2 are returned for the same data point, we know that this equation has at least one real solution, y_i. Therefore, we need to prove the uniqueness of the solution to discover y_i. We need to show that this is an increasing function in each of the possible 3 intervals, ≥ 0, $[-1, 0)$, and < -1, to prove the uniqueness of the solution.

◊ $y_i \geq 0$: The expression becomes $(y_i + 1)^p - (y_i)^p$. If we take the derivative with respect to y_i we get

$$p(y_i + 1)^{p-1} - p(y_i)^{p-1}$$

Since $p \geq 1$ above expression is always positive. Therefore, in this interval the function is increasing.

◊ $-1 \leq y_i < 0$: The expression becomes $(y_i + 1)^p - (-y_i)^p$. If we take the derivative with respect to y_i we get

$$p(y_i + 1)^{p-1} + p(-y_i)^{p-1}$$

Above expression is always positive therefore, in this interval the function is increasing.

◊ $y_i < -1$: The expression becomes $(-(y_i + 1))^p - (-y_i)^p$. If we take the derivative with respect to y_i we get

$$-p(-(y_i + 1))^{p-1} + p(-y_i)^{p-1}$$

If we reorganize the expression we get

$$p((-y_i)^{p-1} - (-(y_i + 1))^{p-1})$$

Since the first term in the parenthesis is always greater than the second one, the derivative is always positive in this interval. Hence, in this interval the function is increasing.

The function is increasing in all of the intervals, thus it is an increasing function. As a result, independent from p we have a unique real solution for y_i and it can be computed by Newton's method.

The total number of probes for n dimensions is found by multiplying the number of probes needed for each dimension separately. The following theorem considers this fact and the gives the method to find the common distance between closest probes, $c\prime$.

Theorem 3. *For the score model, general scheme in n dimensions utilizing l_p norm as the distance metric requires $\prod_{i=1}^{n} \lceil \frac{u_i - l_i}{c\prime} \rceil + 1$ probes to discover all elements of an n dimensional database where $c\prime$ is equal to the minimum of $c/4$ and $\frac{c}{2(n)^{1/p}}$.*

Proof. The general scheme divides the whole space into hypercubes having an edge of $c\prime$ which is selected to guarantee that the returned score for at least one of the probes belonging to the corners of the region that a data point lies in is $\leq$ the required distance, $c/4$. To make this guarantee, $c/4$ should be the longest distance between a point and its closest probe. Hence, the diagonal of the hypercube should be $c/2$ and an edge of it, $c\prime$ should be $\frac{c}{2(n)^{1/p}}$. However, we should also consider the requirement that $c\prime$ should be $\leq c/4$. So, $c\prime$ is the minimum of $c/4$ and $\frac{c}{2(n)^{1/p}}$ for n dimensions using l_p norm.

Lemma 2 is at the heart of the solution for 2 dimensions using l_2 norm shown in Figure 2. The extended version of this Lemma should be used for the solution for n dimensions using l_p norm. We will not make the proof, instead we will verify that the sum of the required distance, $c/4$ and the $c\prime$ is $\leq c/2$. The value of $c\prime$ for n dimensions using l_p norm is given in Theorem 3 as the minimum of $\frac{c}{2(n)^{1/p}}$ and $c/4$. So, the sum will be less than or equal to $c/2$. As a result, if the distance returned for a probe is less than or equal to $c/4$, all probes which are in the $c\prime$ distance to this probe will return a score to the same data point. In total, there are $2n$ such probes.

The proof outline for proving the existence of a unique real solution when the database contains a large number of data points, and the probe that has a score $\leq c/4$ is considered with its $c\prime$ neighbors is to follow the same path as Theorem 2.

3   Experiments and Results

The methods developed above are applied on four datasets. Three of the datasets are 2-dimensional each of which represents a different type of distribution. The first dataset is latitude and longitude of road crossings in Maryland. This data set is a good example for uniformly distributed data. The second dataset has points mostly clustered in one region. The third skewed dataset is a correlated data typically seen in time series data such as stock price movements [10]. For each of these three data sets, if the data set contained more than 1000 points, we used a subsampled version with 1000 points. The fourth data set is a clinical data obtained from a pharmaceutical company[1]. The data contains measurements of

---

[1] We wish to thank Pfizer, Inc. for kindly providing the patient dataset.

4 blood ingredients for 244 patients. Half of these patients were suffering from arthritis, while the other half were from a healthy control group.

The General and Progressive methods were applied with query history to eliminate unnecessary probes. Since we do not utilize any distance $> c/2$ while discovering a data point, we used already sent queries to eliminate probes which have scores more than $> c/2$. Results are summarized in Figure 4.

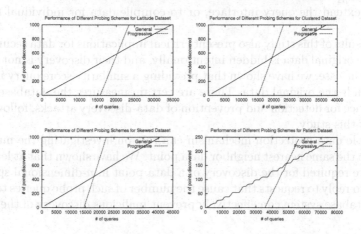

Fig. 4. Performance of General and Progressive probing using history

Because of the leveling strategy used in the progressive model, queries that are next to each other are not sent until the last level. Therefore, this strategy does not find any points until it starts to send the probes on this last level. On the other hand, the distribution of probes resulting from the progressive scheme eliminates more probes than the general scheme. This phenomenon is shown in the graphs in Figure 4. The progressive scheme does not begin to discover points until the general scheme has already discovered about half the database, but it completely discovers the database before the general scheme does.

4 Discussion

Web-based search engines and many biomedical and clinical databases utilize similarity search as their major type of query. In this paper we showed how the data in numeric databases can be discovered through similarity score queries. The methods we have developed were extended to multi-dimensional data, and to l_p-norm distance measures. Using a progressive scheme and exploiting the results of previous queries, we were able to improve the performance over the general probing strategy significantly.

Using the methods we have developed, it is now possible to discover data within a range, or the whole database using results of similarity score queries. This effectively removes the limitation imposed by the query interfaces and lets the researchers extract the data of their interest through the query interface channel they have been provided with. Instead of downloading an outdated snapshot of the whole database, the researchers can obtain the up-to-date information for the portion of the database that they are interested in.

We believe that the data discovery methods provided here would also relieve the database providers of the burden of providing customized data to each request coming from the researchers. The database providers will not have to spend effort to extend the query interface, or to compile data for individual research interests.

The results of this study also present critical implications for data security [22], where the original data is hidden intentionally, and their discovery is not desired. If this is the case, we have shown that providing a similarity score query interface can not hide the original data. There are certain measures the database owners can practice for detection and prevention of data-discovery attacks, following the results of this study.

A simple data protection mechanism can rely on investigating the number of queries to the same nearest neighbor data point. We have shown that at least $n+1$ probes are required for the discovery of a data point in n-dimensional space. By refusing to reply to requests that cause the number of such probe queries to exceed n, the database system can effectively prevent malicious discovery of the data.

References

[1] Ankerst, G., Kastenmüller, M., Kriegel, H., Seidl, T.: Nearest neighbor classification in 3d protein databases. In: Proc. 7th Int. Conf. on Intelligent Systems for Molecular Biology (ISMB 1999) (1999)

[2] Arya, S., Mount, D.M., Netanyahu, N.S., Silverman, R., Wu, A.Y.: An optimal algorithm for approximate nearest neighbor searching. In: 5th Ann. ACM-SIAM Symposium on Discrete Algorithms, pp. 573–582 (1994)

[3] Berchtold, S., Bohm, C., Keim, D., Kriegel, H.: A cost model for nearest neighbor search in high-dimensional data space. In: Proc. ACM Symp. on Principles of Database Systems, Tuscon, Arizona, June 1997, pp. 78–86 (1997)

[4] Beyer, K., Goldstein, J., Ramakrishnan, R., Shaft, U.: When is nearest neighbor meaningful. In: Int. Conf. on Database Theory, Jerusalem, Israel, January 1999, pp. 217–225 (1999)

[5] BrightPlanet.com. The deep web: Surfacing hidden value (2000) Accessible at, http://brightplanet.com

[6] Cheng, X., Dolin, R., Neary, M., Prabhakar, S., Kanth, K.V.R., Wu, D., Agrawal, D., Abbadi, A.E., Freeston, M., Singh, A.K., Smith, T.R., Su, J.: Scalable access within the context of digital libraries. In: Advances in Digital Libraries, pp. 70–81 (1997)

[7] Ciaccia, P., Patella, M.: PAC nearest neighbor queries: Approximate and controlled search in high-dimensional and metric spaces. In: Proc. Int. Conf. Data Engineering, San Diego, California, March 2000, pp. 244–255 (2000)

[8] Du, W., Atallah, M.: Protocols for secure remote database access with approximate matching. In: 7th ACM Conference of Computer and Communications Security (ACMCSS 2000), The First Workshop on Security and Privacy in E-commerce (2000)

[9] Ferhatosmanoglu, H., Stanoi, I., Agrawal, D., Abbadi, A.E.: Constrained nearest neighbor queries. In: Jensen, C.S., Schneider, M., Seeger, B., Tsotras, V.J. (eds.) SSTD 2001. LNCS, vol. 2121, Springer, Heidelberg (2001)

[10] Ferhatosmanoglu, H., Tuncel, E., Agrawal, D., El Abbadi, A.: Vector approximation based indexing for non-uniform high dimensional data sets. In: Proceedings of the 9th ACM Int. Conf. on Information and Knowledge Management, McLean, Virginia, November 2000, pp. 202–209 (2000)

[11] Ferhatosmanoglu, H., Tuncel, E., Agrawal, D., El Abbadi, A.: Approximate nearest neighbor searching in multimedia databases. In: Proc of 17th IEEE Int. Conf. on Data Engineering (ICDE), Heidelberg, Germany, April 2001, pp. 503–511 (2001)

[12] Indyk, P., Motwani, R.: Approximate nearest neighbors: Towards removing the curse of dimensionality. In: 30th ACM Symposium on Theory of Computing, Dallas, Texas, May 1998, pp. 604–613 (1998)

[13] Ipeirotis, P.G., Gravano, L., Sahami, M.: Probe, count, and classify: Categorizing hidden web databases. In: SIGMOD Conference (2001)

[14] Jacob, K.J., Shasha, D.: Fintime – a financial time series benchmark (March 2000), http://cs.nyu.edu/cs/faculty/shasha/fintime.html

[15] Kahveci, T., Singh, A.K.: Efficient index structures for string databases. The VLDB Journal, 351–360 (2001)

[16] Korn, F., Sidiropoulos, N., Faloutsos, C., Siegel, E., Protopapas, Z.: Fast and efficient retrieval of medical tumor shapes. IEEE Transactions on Data Engineering (TKDE 1998) (1998)

[17] Korn, F., Sidiropoulos, N., Faloutsos, C., Siegel, E., Protopapas, Z.: Fast nearest neighbor search in medical image databases. The VLDB Journal, 215–226 (1996)

[18] Korn, F., Sidiropoulos, N., Faloutsos, C., Siegel, E., Protopapas, Z.: Fast nearest neighbor search in medical image databases. In: Proceedings of the Int. Conf. on Very Large Data Bases, Mumbai, India, pp. 215–226 (1996)

[19] Korn, F., Sidiropoulos, N., Faloutsos, C., Siegel, E., Protopapas, Z.: Fast and effective retrieval of medical tumor shapes. IEEE Trans. Knowl. Data Eng. 10(6), 889–904 (1998)

[20] Meng, W., Yu, C.T., Liu, K.-L.: Building efficient and effective metasearch engines. ACM Computing Surveys 34(1), 48–89 (2002)

[21] Roussopoulos, N., Kelly, S., Vincent, F.: Nearest neighbor queries. In: Proc. ACM SIGMOD Int. Conf. on Management of Data, San Jose, California, May 1995, pp. 71–79 (1995)

[22] Tosun, A.S., Ferhatosmanoglu, H.: Vulnerabilities in similarity search based systems. In: CIKM, pp. 110–117. ACM, New York (2002)

[23] Weber, R., Bohm, K.: Trading quality for time with nearest-neighbor search. In: Proc. Int. Conf. on Extending Database Technology, Konstanz, Germany, March 2000, pp. 21–35 (2000)

[24] Whalley, M.R.: The Durham-RAL high energy physics database - HEPDATA. Computer Physics Communications 57(1-3), 536–537 (1990)

Efficient Algorithms for Node Disjoint Subgraph Homeomorphism Determination*

Yanghua Xiao, Wentao Wu, Wei Wang, and Zhenying He

Department of Computing and Information Technology
FuDan University, ShangHai, China
{Shawyanghua,wentaowu1984}@gmail.com,{weiwang1,zhenying}@fudan.edu.cn

Abstract. Recently, great efforts have been dedicated to researches on the management of large-scale graph-based data, where node disjoint subgraph homeomorphism relation between graphs has been shown to be more suitable than (sub)graph isomorphism in many cases, especially in those cases where node skipping and node mismatching are desired. However, no efficient algorithm for node disjoint subgraph homeomorphism determination (ndSHD) has been available. In this paper, we propose two computationally efficient ndSHD algorithms based on state spaces searching with backtracking, which employ many heuristics to prune the search spaces. Experimental results on synthetic data sets show that the proposed algorithms are efficient, require relatively little time in most of cases, can scale to large or dense graphs, and can accommodate to more complex fuzzy matching cases.

1 Introduction

Graph-based pattern matching is one of the key issues underlying large-scale graph-based data management, which recently has attracted more and more research interests, due to the broad applications of graph-based data. Existing graph pattern matchings based upon *subgraph isomorphism* cannot represent the fuzzy matching in some cases where *node skipping or node mismatching is allowed.* For example, as shown in Figure 1, although G_2 is not a subgraph of G_1, G_2 still can be regarded as matched to G_1 if node skipping or node mismatching is allowed.

Such kind of fuzzy matching is desired in various real applications. For example, the discovery of frequent conserved subgraph patterns from protein interaction networks [1,2] is an important and challenging work in evolutionary and comparative biology, where 'conserved' just means the inexact graph pattern matching allowing node mismatch and node skipping. Similarly, in social network analysis, the direct connection between nodes usually is not the focus; instead, the high-level topological structure with independent paths contracted is of great interest.

---

* The work was supported by the National Natural Science Foundation of China under Grant No.60303008, No.60673133, No.60703093; the National Grand Fundamental Research 973 Program of China under Grant No.2005CB321905.

J.R. Haritsa, R. Kotagiri, and V. Pudi (Eds.): DASFAA 2008, LNCS 4947, pp. 452–460, 2008.

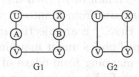

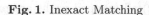

G1   G2

Fig. 1. Inexact Matching

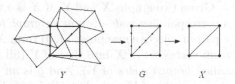

Y   G   X

Fig. 2. Topological Minor

Using *Graph Minor* theory [4], the abstract topological structure in many real applications can be described as *topological minor*, and the relation between abstract topological structure and its detailed original graph can be described as *node/vertex disjoint subgraph homeomorphism*. However, to determine whether a pattern graph P is a topological minor of data graph G is non-trivial, and has been proved to be NP-complete when P and G are not fixed [3]. Although Robertson and Seymour [4] have proposed a framework to solve *minor containment* problem that is a generalization of topology containment problem and [5] has implemented the framework, no practically efficient algorithm has been available to solve *ndSHD*, to the best of our knowledge. Here, we propose two algorithms that are based upon state space searching with backtracking. To improve the efficiency, many heuristics have been integrated into the searching procedures to prune the search spaces.

2   Preliminaries

We begin this section with some basic notations. Let $G = (V, E, l)$ be a *vertex labeled graph*, where V is the set of vertices, E is the set of edges and $E \subseteq V \times V$, and l is a label function $l : V \to L$, giving every vertex a label. The vertex set of G is referred to as $V(G)$, and the edge set is referred to as $E(G)$. A *path* P in a graph is a sequence of vertices $v_1, v_2, ..., v_k$, where $v_i \in V$ and $v_i v_{i+1} \in E$ for each i. The vertices v_1 and v_k are linked by P and are called its *ends*. The number of edges of a path is its *length*, and the path of length k is denoted as P^k. A path is *simple* if its vertices are all distinct. Particularly, a group of paths are *independent* if none of the paths has an inner vertex on another path.

As described in [7], a *topological minor* of a graph is obtained by contracting the independent paths of one of its subgraphs into edges. For example, in Figure 2, X is a topological minor of Y, since X can be obtained by contracting the independent paths of G that is a subgraph of Y.

Formally, as shown in Figure 2, if we replace all the edges of X with independent paths between their ends, so that these paths are *pairwise node independent*, i.e. none of these paths has an inner vertex on another path, then G is a *subdivision* of X. Furthermore, if G is a subgraph of Y, then X is a *topological minor* of Y. As a subdivision of X and a subgraph of Y, if G is obtained by replacing all the edges of X with independent paths with length from l to h, then G is an *(l, h)-subdivision* of X and X is an *(l, h)-topological minor* of Y.

Given two graphs X and Y, if X is a topological minor of Y, then there exists a corresponding *node disjoint subgraph homeomorphism* from X into Y, which is a pair of injective mappings (f, g) from X into Y. Here, f is an injective mapping from vertex set of X into that of Y (all the mapped nodes under mapping f are called *branch nodes* of Y). And g is an injective mapping from edges of X into simple paths of Y such that (1) for each $e(v_1, v_2) \in E(X)$, $g(e)$ is a simple path in Y with $f(v_1)$ and $f(v_2)$ as two ends; and (2) all mapped paths are pairwise independent.

3   Algorithm Framework

3.1   A Rudimentary Algorithm

To determine whether G_1 is an (l, h)-*topological minor* of G_2 is equivalent to find a pair of mappings (f, g) between these two graphs. The final solution of the determination can be described as $\mathcal{M} = (NM, EPM)$, where $NM \subseteq V_1 \times V_2$ is the node match set and $EPM \subseteq E_1 \times (P^l \cup ... \cup P^h)$ is the edge-path match set. All the mapped nodes of G_2 can be denoted by $NM^{(2)}$, and all the mapped paths of G_2 can be denoted by $EPM^{(2)}$.

The process of finding the homeomorphism mapping can be suitably described by means of *State Space Representation* [9]. Each state s of the matching process can be associated with a partial mapping solution $\mathcal{M}_s = (NM_s, EPM_s)$, where NM_s and EPM_s are the node match set and edge-path match set at state s, respectively. Obviously, $\mathcal{M}_s$ contains all the matches we have found so far and will probably be a subset of some final match set $\mathcal{M}$. The algorithm framework based on state space searching is shown as follows.

Algorithm ndSHD1(G_1, G_2, l, h)
Input: G_1, G_2:vertex labeled graphs; l:minimal path length; h:maximal path length.
Output: If G_1 is an (l, h)-topological minor of G_2 return *true* and return *the **first** found node disjoint subgraph homeomorphism* (f, g), otherwise return *false*.

1. Initial(M, R);/*Initialize SHD, generate necessary path information, initialize the basic data structures M and R, which will be described in the following section.*/
2. $s \leftarrow \emptyset$; /*Initialize state as empty state.*/
3. **while** NodeMappingSearch(s,M,R) /*Search in node mapping space. */
4.   **if** EdgePathMappingSearch(s,M,R) /*Search in edge-path mapping space.*/
5.   **return true**;
6. **return false**;

For example, given two graphs shown in Figure 3, let (l, h) be $(2, 2)$, which means the edges in G_1 can only be mapped to the paths in G_2 with length 2. The running procedure under above parameters is shown in Figure 4. The result of the determination is true, and the node mappings are $NM = \{1 - 2, 2 - 8, 3 - 6, 4 - 4\}$ and the five edge-path mappings are $EPM = \{12 - 218, 13 - 296, 14 - 234, 23 - 876, 34 - 654\}$.

Please note that the answer to the problem is sensitive to the given parameter (l, h). If (l, h) is $(3, 3)$, G_1 is not a topological minor of G_2. The influence of

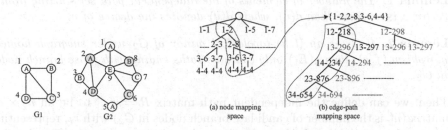

Fig. 3. Running Example   **Fig. 4.** Two-Level State Space Searching

parameter (l, h) on topology containment determination has been discussed in [8] in detail.

3.2 Basic Data Structures

As described above, we need two basic data structures, one is used to represent the node mapping information; the other is used to represent (l, h) independent path information of G_2. For the former, we use node compatible matrix; the latter, we use independent path matrix as well as a path index structure. Both of them are changing with the transition of the matching state.

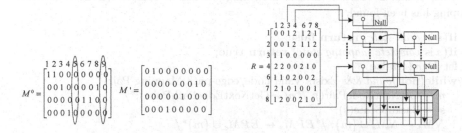

Fig. 5. M^0 and M' **Fig. 6.** R and its associated path index structure

We define node compatible matrix $M = [m_{ij}]$ to be an n_1 (rows)×n_2 (columns) matrix whose elements are 1's or 0's, where n_1 and n_2 are the number of nodes in G_1 and G_2, respectively. At the final success state, we can get a final mapping matrix $M' = [m'_{ij}]$ whose elements are 1's or 0's, such that each row contains exactly one 1 and each column contains no more than one 1. The final mapping matrix represents a valid one-to-one mapping between nodes of G_1 and G_2, while the initial compatible matrix M^0 represents the probable mappings between nodes of G_1 and G_2. Obviously, for each element m'_{ij} of M', $(m'_{ij} = 1) \rightarrow (m^0_{ij} = 1)$.

Clearly, reducing the number of 1's in M is the key to reduce the node mapping space, which is the basic idea of Lemma 1. When constructing independent path matrix and its associated path index structure, we only need to enumerate all the (l, h)-paths between all *candidate branch node* pairs, which is based on Lemma 2.

Lemma 1. *The number of elements of the independent path set starting from a vertex v is no more than $d(v)$, where $d(v)$ denotes the degree of v.*

Lemma 2. *If G_1 is an (l, h)-topological minor of G_2 under subgraph homeomorphism (f, g), then $g(E_1)$ only contains paths ending with those branch nodes in G_2.*

Then, we can define the independent path matrix $R = [r_{ij}]$ to be an $n_2' \times n_2'$ matrix (n_2' is the number of candidate branch nodes in G_2) with r_{ij} representing the number of (l, h) paths between the node pair (v_i, v_j) in G_2. The detailed path information is stored in an array of lists $RLists$, where each list contains path addresses that point to the physical storage of the paths for some r_{ij}.

Figure 5 and Figure 6 show these two basic data structures used in the running case shown in Figure 3.

3.3   State Space Searching

The procedure of node mapping space searching and edge-path mapping space searching are similar to each other. These two procedures are shown as follows.

Algorithm Node/EdgePathMappingSearch1 (s,M,R)
Input: s:the current matching state; M:the current node compatible matrix; R: the current independent path matrix.
Output: *found*: a boolean variable indicating whether a complete node/edge-path mapping has been found.

1. **if**(s is *dead state*) **return false**;
2. **if**(s is *complete mapping state*) **return true**;
3. let *found*←**false**
4. **while**(**not** *found* && Exists Valid node/edge-path Mapping Pair)
5. $m \leftarrow$ GetNextNodePair(); /*$m \leftarrow$ GetNextEdgePathPair();*/
6. $s' \leftarrow$ BackupState(s);
7. $NM_s \leftarrow NM_s \cup \{m\}$; /*$EPM_s \leftarrow EPM_s \cup \{m\}$*/
8.   Refine(M,R);
9.   *found*←Node/EdgePathMappingSearch(s, M, R);
10.   **if**(*found*) **return true**;
11. $s \leftarrow$ RecoverState(s');
12. **return false**;

From lines 1-2, we can see that when a new state s arrives, s can be a *dead state* or *success state*(complete mapping state). The state space search arrives at a *success state* if all the node mappings or edge-path mappings have been found, which means $|NM_s| = |V_1|$ or $|EPM_s| = |E_1|$. The node mapping state space search arrives at a *dead state* if there is a row with all 0's in node compatible matrix M of the current state, i.e. $\exists i, |NM_s| \leq i \leq n_1$,s.t. $\sum_{1 \leq j \leq n_2} m_{ij} = 0$. And the edge-path mapping state space search arrives at a *dead state* if there is no path between some pair of branch nodes, i.e., $\exists i, |EPM_s| \leq i \leq n_2'$,s.t. $\prod_{(f^{-1}(node(i)), f^{-1}(node(j))) \in E_1} r_{ij} = 0$, where $node(i)$ gets the vertex corresponding to the i-th column in matrix R.

3.4   Refinement Procedure

To traverse all possible mapping branches is time-consuming, so space pruning is essential for ndSHD. For this purpose, we devise two refinement procedures on R and M, respectively. Lemma 3 and 4 show the correctness of refinement on R, and Lemma 5 shows the correctness of refinement on M.

Lemma 3. *In the matching process, let s be the current state, if $v \in NM_s^{(2)}$ and M_s will be a partial solution of some final solution M, then any path with v as inner vertex will not $\in EPM^{(2)}$.*

Lemma 4. *In the matching process, let s be the current state, if $p \in EPM_s^{(2)}$ and M_s will be a partial solution of some final solution M, then any path passing trough the inner vertex of p will not $\in EPM^{(2)}$.*

Lemma 5. *In the matching process, let s be the current state, if $(v_i, v_j) \in NM(v_i \in V_1, v_j \in V_2)$ and M_s will be a partial solution of some final solution M, then the following statements hold true:*

1. $\prod r_{j'k} > 0$, *where* $j' = Index(v_j), k \in Index(V)$ *and* $V = \{v_2 | v_1 \in Adjacent (v_i) \land (v_1, v_2) \in NM_s\}$.
2. $\forall v' \in V', \exists v \in V_2$ *such that* $l_2(v) = l_1(v')$ *and* $r_{j'k} > 0$, *where* $j' = Index(v_j)$, $k = Index(v)$ *and* $V' = \{v' | v' \in Adjacent(v_i) \cap (V_1 - NM_s^{(1)})\}$.
3. *The path set consisting of the paths to which all mentioned $r_{j'k}$'s in (1) and (2) indicate is independent.*

In the above statements, the function $Index(v)$ gets an index in R for a node v in G_2; and $Adjacent(v)$ obtains the adjacent vertex set of v.

3.5   More Efficient Searching Strategy

A basic observation of the above refinement procedures is that the constraint resulting from an edge-path match will be more restrictive than that resulting from a node match. Hence, a better strategy is to try edge-path match as early as possible, rather than performing edge-path match until complete node match has been found. We denote these two strategy as s_1 (old strategy) and s_2(new strategy), respectively; and algorithms employing two strategies are denoted as *ndSHD1* and *ndSHD2*, respectively. Intuitively, in *ndSHD2* the searching procedure will meet with the dead state very early if the current searching path will not lead to a successful mapping solution, thus the searching procedure will fast backtrack to try another mapping solution.

The framework of algorithms *ndSHD2* and *Node/EdgePathMappingSearch2* are similar to that of *ndSHD1* and *Node/EdgePathMappingSearch1*, and thus the details are omitted here due to the space limitation.

4   Experimental Evaluation

To test the efficiency of the algorithms, we generate the synthetic data sets according to the random graph [10] model that links each node pair by probability p. All generated graphs are vertex labeled undirected connected graphs. We also randomly label vertices so that the vertex labels are uniformly distributed. We implement the algorithm in C++, and carry out our experiments on a Windows 2003 server machine with Intel 2GHz CPU and 1G main memory.

The efficiency of the algorithms is influenced by the following factors: N_1: node size of G_1, N_2: node size of G_2, M_1: average degree of G_1, M_2: average degree of G_2, (L, H): the minimal and maximal path length. The efficiency also can be influenced by the number of vertex labels.

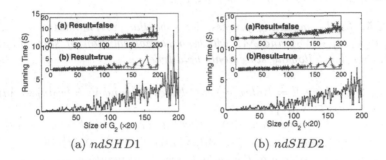

(a) $ndSHD1$ (b) $ndSHD2$

Fig. 7. Efficiency and scalability with respect to the growth of size of data graph (G_2). The inset of (a), (b) show runtime of all running cases where the determination result is *true*, *false* respectively.

First we will demonstrate the scalability with respect to the growth of the size of nodes of data graph G_2. We use a complete graph with 4 uniquely labeled nodes as a minor graph; we generate overall 200 data graphs G_2 with node size varying from 20 to 4000 in increment of 20. The average degree of each data graph is fixed as 4 and nodes of each graph are randomly labeled as one of overall 20 labels. L and H are fixed as 1 and 3, respectively, meaning that the path length is in the range of $[1, 3]$. Thus the parameters can be denoted as $N_1 4 M_1 3 M_2 4 L 1 H 3$. From Figure 7, we can see that $ndSHD1$ and $ndSHD2$ both are approximately linearly scalable with respect to the number of nodes in G_2, irrespective of the result of the determination.

Next we will show the scalability of $ndSHD1$ and $ndSHD2$ with respect to the size of G_1. We fix some parameters as $M_1 4 N_2 4 k L 1 H 3$ and vary the size of G_1 from 6 to 82 in increment of 4 to generate 20 minor graphs. Each minor graph is uniquely labeled. Two data graphs are used, one has average degree M_2 as 8 and the other as 20. These two data graphs are randomly labeled as one of 200 labels. Figure 8(a)(I) ($M_2 = 8$) and (II) ($M_2 = 20$) show the results. As can be seen, $ndSHD1$ and $ndSHD2$ both are approximately linearly scalable with respect to the number of nodes in G_1. In (II), running time of $ndSHD1$ is not

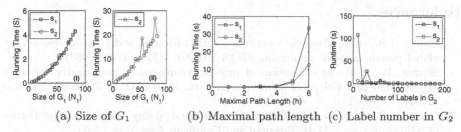

(a) Size of G_1 (b) Maximal path length   (c) Label number in G_2

Fig. 8. Scalability of two algorithms

available when $M_2 = 20$, meaning that all running cases need time larger than one hour.

Figure 8(b) shows the runtime of the algorithm with respect to (l, h). Parameters of this experiment are set as $N_1 4 M_1 3 N_2 1 k M_2 8 L 1$. The minor graph is uniquely labeled; data graphs are randomly labeled as one of 20 labels. As can been seen, the broader the range is, the longer the running time is; and the runtime of *ndSHD1* and *ndSHD2* both increase dramatically with the growth of upper bound of path length. However, the increasing speed of *ndSHD2* is slower than that of *ndSHD1*, which implies that *ndSHD2* is more efficient than *ndSHD1* with respect to larger h. The super linearly growth of the runtime with the increase of upper bound of the path length can be partly attributed to the exponential growth of the number of potentially mapped paths. However, in various real applications, larger upper bound of path length is non-meaningful when performing fuzzy matching on graph data, and usually upper bounds less than 3 are desired.

To examine the impact of the number of vertex labels on the performance of *ndSHD1* and *ndSHD2*, we use a uniquely labeled graph with 6 nodes and 15 edges as minor graph, a graph with 1000 nodes and 4000 edges as data graph. We randomly labeled the data graph from 10 labels to 200 labels in increment of 10 to generate 20 different labeled data graphs. L and H are set as 1 and 3, respectively. The result of this experiment is shown in Figure 8(c). Clearly, runtime of *ndSHD1* and *ndSHD2* substantially decrease with the growth of number of labels of G_2, which confirms to what we have expected, since larger number of labels in G_2 can reduce the node mapping space between minor graph and data graph. We also can see that *ndSHD2* outperforms *ndSHD1* to a great extent when the number of labels is small.

5   Conclusions

In this paper, we investigated the problem known as node disjoint subgraph homeomorphism determination; and proposed two practical algorithms to address this problem, where many efficient heuristics have been exploited to prune the futile searching space. The experimental results on synthetic data sets show that our algorithms are scalable and efficient. To the best of our knowledge, no practical algorithm is available to solve node disjoint subgraph homeomorphism determination.

References

1. Kelley, R.B., et al.: Conserved pathways within bacteria and yeast as revealed by global protein network alignment. PNAS 100(20), 11394–11399 (2003)
2. Sharan, R., et al.: Identification of protein complexes by comparative analysis of yeast and bacterial protein interaction data. In: RECOMB 2004, pp. 282–289 (2004)
3. Garey, M.R., Johnson, D.S.: Computers and Intractability. A Guide to the Theory of NP-completeness. W.H. Freeman and Company, New York (2003)
4. Robertson, N., Seymour, P.D.: Graph minors. XIII: The disjoint paths problem. Journal of Combinatorial Theory 63, 65–110 (1995)
5. IIIya, V.: Hicks:Branch Decompositions and Minor Containment. Networks 43(1), 1–9 (2004)
6. Ullmann, J.R.: An Algorithm for Subgraph Isomorphism. Journal of the ACM 23, 31–42 (1976)
7. Diestel, R.: Graph Theory. Springer, Heidelberg (2000)
8. Jin, R., Wang, C., Polshakov, D., Parthasarathy, S., Agrawal, G.: Discovering frequent topological structures from graph datasets. In: KDD 2005, Chicago,USA, pp. 606–611 (2005)
9. Nilsson, N.J.: Principles of Artificial Intelligence. Springer, Heidelberg (1982)
10. Erdös, P., Rényi, A.: On random graphs. Publicationes Mathematicae, 290–297 (1959)

The Chronon Based Model for Temporal Databases

D. Anurag and Anup K. Sen

Indian Institute of Management Calcutta, India
anurag@email.iimcal.ac.in, sen@iimcal.ac.in

Abstract. Among various models like TSQL2, XML and SQL:2003 for
temporal databases, TSQL2 is perhaps the most comprehensive one and
introduces new constructs for temporal query support. However, these
constructs mask the user from the underlying conceptual model and essentially
provides a "black box" flavour. Further, the temporal attributes are not directly
accessible. In this paper, we propose a "chronon" based conceptual (relational)
model, which works on a tuple timestamping paradigm. Here, the time attribute
is treated in similar fashion as any other attribute and is available to the user for
inclusion in SQL statements, making the model easy to understand and use. A
corresponding efficient physical model based on the attribute versioning format
akin to XML/SQL:2003 has also been proposed and the mapping between the
proposed conceptual and physical models has been described. We evaluate the
model by comparing with TSQL2 and SQL:2003.

Keywords: Temporal Database, Chronon Model, TSQL2.

1 Introduction

The need for a temporal database has been felt for over two decades now [2]. A
temporal database attempts to capture time varying information in an efficient
manner. However, till date there exists no universally accepted standard. Among
various models proposed, the TSQL2 standard is perhaps the widely studied one.
TSQL2 incorporates new keywords and aims to shield the user from the underlying
complexity of the conceptual model. However, this hiding of the temporal component
through the keywords costs the model flexibility and the naturalness of the relational
model is lost. Thus the motivation for a new model is to develop a simplistic
conceptual model for temporal databases and maintain the flavour and transparence of
the relational model. The novel *chronon*[1] model extends the relational model through
a time attribute. This attribute is allowed to be used in ways similar to any other
attribute. Through examples we show the simplicity and flexibility achieved. Further,
we show the advantage of the chronon model over TSQL2 where only a single
additional construct to the existing SQL standard is needed. We evaluate the model
through a comparison with TSQL2 and SQL:2003. Finally, for the chronon model to
be successfully implemented, we need an efficient physical model and this is

---

[1] A chronon is the smallest indivisible time unit [14].

J.R. Haritsa, R. Kotagiri, and V. Pudi (Eds.): DASFAA 2008, LNCS 4947, pp. 461–469, 2008.
© Springer-Verlag Berlin Heidelberg 2008

borrowed from the concepts developed in SQL:2003 and XML. Thus the objectives of this paper can be summarized as follows:

- To propose a new chronon based conceptual model for temporal databases; the model uses tuple timestamping paradigm which enables the user to access the time attribute as any other attribute in SQL; only a single additional construct, MERGE BY, is required to solve the coalescing issues in temporal projection;
- To provide a corresponding physical model for efficient implementation and develop the mapping from the conceptual to the physical.

The conceptual modeling of temporal schema using ER diagramming received attention early and a complete review of the various constructs can be found in [7]. Many conceptual models have been proposed, however, no current standard exists [16]. The TSQL2 model, proposed by Richard Snodgrass, was adapted [14] and pushed to be included in the SQL3 standard but was cancelled in 2001 [11]. Subsequently, SQL:2003 has introduced features that provide indirect temporal support [4]. Substantial research has also looked at the viability of XML as a temporal database [3, 4, 5]. The commercial products based on temporal models proposed are very limited in number [16]. Without going into the aspects of each model, we introduce the new chronon model.

2 The Chronon Based Temporal Model

Temporal support provided by TSQL2, XML or SQL:2003 is based on either tuple versioning or attribute versioning (Tables 1,2). Introducing temporal support in either way brings about, five key issues [15]: (a) Key Constraint, (b) Referential Constraint, (c) Temporal Projection, (d) Joins and (e) Updates, which we discuss below. TSQL2 follows the tuple versioning approach while SQL:2003 and XML follow attribute versioning. TSQL2 has single mindedly focused on simplifying the SQL statements by introducing new constructs. However, we content the problem lies not in just the lack of features in SQL, but in the conceptualization of the tuple versioning approach. A user writes SQL queries based on the conceptual model. For example, for a non temporal relational model, the conceptualization of a table of values is straight forward. However, the inclusion of valid and transaction time introduces complexities. TSQL2, by masking the user from the underlying time attributes, has moved the model away from the natural transparence of the relational model. This makes the model difficult to comprehend and has inhibited its flexibility. The temporal attributes, for example, valid times, are not directly available to the user. Thus the motivation for a new model is to develop a simpler conceptualization to ease the writing of SQL queries and maintain the transparence of the relational model.

Table 1. Tuple Versioning

emp_ID	salary	dept	valid_start	valid_end
01	10000	10	1-Jan-06	31-Dec-06
01	20000	10	1-Jan-07	now

Table 2. Attribute Versioning

emp_ID	Salary		dept	
01	10000		10	
	valid_start	valid_end	valid_start	valid_end
	1-Jan-06	31-Dec-06	1-Jan-06	now
	20000			
	valid_start	valid_end		
	1-Jan-07	now		

Inspired by the representation of time in [11], the chronon conceptual model is made to comprise of a single time attribute (say valid time) for an entire relation. i.e., a relation $R(A_1,A_2,A_3)$ is written as $R(A_1,A_2,A_3,T)$. Here T holds the time instant the relation was *saved*. Conceptually, a relation is saved every instant.

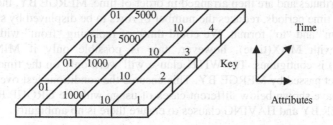

Fig. 1. The chronon based conceptual (relational) model

Fig. 1 shows the state of the tuple for every time instant. Here to define instant, we borrow the definition of chronon from [14]. *A chronon is the smallest indivisible time unit.* Thus, T2 – T1 = 1 chronon. The keys of the relation are assumed to be time invariant (for time-variant keys see [6]). We have also limited our discussion to valid time. Transaction time can be modeled similarly. Consider a relation, emp(*empID,salary,deptID,time*). The first tuple was added on 1-Jan-07 (Table 3).

Table 3. emp table on 1-Jan-07

empID	salary	deptID	time
01	1000	10	1-Jan-07

Table 4. emp table on 2-Jan-07

empID	salary	deptID	Time
01	1000	10	1-Jan-07
01	1000	10	2-Jan-07

For simplicity, the granularity of time is assumed to be a day. On 2-Jan-07, the table would look as in Table 4. There is no change in the attributes except time. The second tuple implies empID *01* continues to have a salary of *1000* and belongs to department *10* on *2-Jan-07*. *This second tuple is not inserted explicitly by the user, but is (conceptually) saved by the DBMS*. This automatic (virtual) save will continue until the tuple is deleted. The user *believes* the daily save occurs. The physical model instead stores data in an efficient manner. Now, to appreciate the simplicity of the model, consider the different problems associated with temporal databases. To start with, the user creates a table by mentioning the time attribute explicitly: *CREATE TABLE emp(empID, salary, deptID, time)*

Key Constraint: The user can conveniently state time as part of the key and thus ensures the key is not overlapping in time. No new construct is needed.

Referential Constraint: The user simply includes the time attributes in the syntax. This is the standard SQL construct without any need for enhancements: *FOREIGN KEY(depID,time) REFERENCES mgr(depID,time)*

Temporal Projection: Here, a new construct is needed which will coalesce the values along time. Assuming today is 6-Jan-07, a *SELECT * FROM emp* would result in Fig. 2(a). Thus we propose a construct MERGE BY, with syntax as shown in Fig. 2(b). The MERGE BY clause will work only on the time attribute and can include a condition on this time attribute using a WITH clause. Its operation can be thought of as being similar to ORDER BY; the rows of the result set are ordered based on values of other attributes and are then arranged in order of time. MERGE BY, then, for every contiguous time periods, reduces the number of rows to be displayed by showing time in the "from" and "to" format. We could think of replacing "from" with MIN(time) and "to" with MAX(time), however, this is possible only if MIN(time) and MAX(time) is contiguous. The WITH clause will work only on the time attribute of the result set passed by MERGE BY. This ensures the conditions test over contiguous time. We have shown below different cases of usage with the GROUP BY, MERGE BY, ORDER BY and HAVING clauses to ensure there is no ambiguity.

empID	salary	deptID	time
01	1000	10	1-Jan-07
01	1000	10	2-Jan-07
01	1000	10	3-Jan-07
01	2000	10	4-Jan-07
02	1000	20	4-Jan-07
01	2000	10	5-Jan-07
02	1000	20	5-Jan-07
01	2000	10	6-Jan-07
02	1000	10	6-Jan-07

Fig. 2(a). Query Output

```
SELECT attribute-and-function-list
FROM table-list
[ WHERE row-condition ]
[ GROUP BY grouping-attribute(s) ]
[ HAVING group-condition ]
[ MERGE BY time-column [WITH
time-condition] ]
[ ORDER BY attribute-list ]
```

Fig. 2(b). MERGE BY Syntax

```
SELECT * FROM
emp
MERGE BY time
```

Fig. 3(a). Query

empID	salary	deptID	Time
01	1000	10	1-Jan-07 to 3-Jan-07
01	2000	10	4-Jan-07 to now
02	1000	20	4-Jan-07 to 5-Jan-07
02	1000	10	6-Jan-07 to now

Fig. 3(b). Output of Query 3(a)

The query of Fig.3(a), with MERGE BY, will result in the output as shown in Fig.3(b) after coalescing the tuples along time. Note that now represents today's date.

CASE 1: MERGE BY and WITH: The result of MERGE BY can be tested for conditions on time using the WITH clause as seen below:

SELECT empID, time FROM emp MERGE BY time WITH time >= '4-Jan-07'

empID	Time
02	4-Jan-07 to now

Fig. 4(a). Query   **Fig. 4(b).** Query Output

CASE 2: MERGE BY and GROUP BY: The GROUP BY clause with the standard definition can work on the time attribute in similar ways as any other attribute. The SELECT clause can include the aggregate operators linked with GROUP BY. The inclusion of MERGE BY will now effect in arranging the rows with the result of the aggregate operator included.

SELECT depID, COUNT(*), time FROM emp GROUP BY depID, time MERGE BY time

deptID	COUNT	Time
10	1	1-Jan-07 to 5-Jan-07
20	1	4-Jan-07 to 5-Jan-07
10	2	6-Jan-07 to now

Fig. 5(a). Query   **Fig. 5(b).** Query Output

CASE 3: GROUP BY and HAVING: The HAVING clause can include the columns specified in the GROUP BY clause as in the standard case. Note that this means, the inclusion of the time attribute is legal. See example below.

SELECT empID, time FROM emp GROUP BY empID,time HAVING (time >= '4-Jan-07') //notice we could have achieved the same // using only a WHERE clause

empID	Time
01	4-Jan-07
02	4-Jan-07
01	5-Jan-07
02	5-Jan-07
01	6-Jan-07
02	6-Jan-07

Fig. 6(a). Query   **Fig. 6(b).** Query Output

SELECT empID, time FROM emp MERGE BY time ORDER BY time DESC

empID	Time
02	4-Jan-07 to now
01	1-Jan-07 to now

Fig. 7(a). Query   **Fig. 7(b).** Query Output

CASE 4: MERGE BY and ORDER BY: The ORDER BY clause works in the standard way and can include the time attribute. Notice that arranging according to time is equivalent to arranging along the "from" time.

To summarise, *no default action of standard SQL statements have been modified.* A single new clause MERGE BY is introduced. It can have a condition to be tested on the time attribute through a WITH clause.

Updates: The user simply updates the tuple whose time value is in the range. For example, *UPDATE emp SET salary = 20000 WHERE time >= '1-Jan-07'.* The update is now a single logical statement. Multiple operations are avoided.

Joins: The treatment of the time attribute is in the same manner as any other attribute and thus the joins follow the standard SQL syntax.

The chronon model, thus, *requires only one construct to answer all the difficulties of temporal support in databases.* In contrast TSQL2 requires five new constructs.

Comparison with TSQL2 and SQL:2003: We now run through the examples provided by TSQL2 in [14] and bring out the conceptual simplicity of the chronon model. The SQL:2003 statements have been tested on ORACLE 10g EXPRESS EDITION. The queries assume two tables employee(eno,name,street,city) and salary(eno,amount)

QUERY 1: List the number of employees with salary > 5000 in each city with the time periods.	
TSQL2	VALIDTIME SELECT city,count(*) FROM employee e, salary s WHERE e.eno=s.eno AND salary > 5000 GROUP BY city
SQL:2003	SELECT c.city, count(*) FROM employee e, TABLE (city) c, salary s, TABLE(amount) a WHERE e.eno=s.eno AND a.amount > 5000 AND a.valid_start <= c.valid_start AND a.valid_end > c.valid_end GROUP BY c.city;
CHRONON MODEL	SELECT city,count(*),time FROM employee e, salary s WHERE e.eno = s.eno AND e.time = s.time AND amount > 5000 GROUP BY (city,time) MERGE BY time
NOTES	• The MERGE BY in the chronon model includes the COUNT(*) with the city while displaying the time in the "from" and "to" format • In SQL:2003, although the answer would be correct, it misses out if finer granularity than that of the city time period exists in amount.
QUERY 2: Update salary of "Therese" to 6000 for year '06.	
TSQL2	VALIDTIME PERIOD '[1-Jan-06 - 31-Dec-06]' UPDATE salary SET amount=6000 WHERE eno IN (SELECT eno FROM salary WHERE name ='Therese')
SQL:2003	// we couldn't think of any other way than an embedded program.
CHRONON MODEL	UPDATE salary SET amount=6000 WHERE eno IN (SELECT eno FROM salary WHERE name='Therese' AND time >= '1-Jan-06' AND time <= '31-Dec-06') AND time >= '1-Jan-06' AND time <= '31-Dec-06'
NOTES	• TSQL2 makes the assumption of consistent time across the two sub-queries. • The explicit mention of "time" in the chronon model makes the query much more readable and flexible

Fig. 8. Queries comparing TSQL2, SQL:2003 and chronon model

The chronon model although using only an additional construct, is much more expressive and flexible - count the number of days using COUNT(time), or determine the first or the last day using MIN(time) or MAX(time) respectively.

Mapping of the chronon model to the physical model: The physical model exploits the attribute versioning scheme for optimal disk usage. The mapping is provided below.

CREATE TABLE: The standard constructs to be followed. A new parameter "chronon" introduced to indicate the valid time temporal support. Internally, attribute versioning is made.

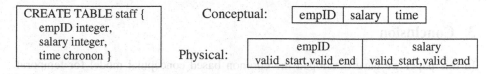

Fig. 9. Conceptual to Physical mapping for table creation

INSERT/DELETE/UPDATE: Fig. 10 captures the relationship between the value of time supplied by the user and today's date, indicated by "now". Case 2 is the most common scenario and forms the default behaviour. See Fig. 11.

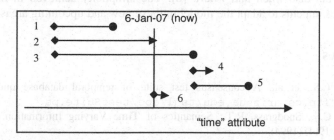

Fig. 10. Plot of INSERT cases. The diamonds at the start of each timeline indicates "valid_start". The dots/arrows indicate "valid_end". An arrow indicates the value is "now".

Conceptual:

empID	salary	time
10	1000	4-Jan-07
10	1000	5-Jan-07
10	1000	6-Jan-07

INSERT INTO staff VALUES (10,1000,'4-Jan-07')

Physical:

empID valid_start,valid_end	salary valid_start,valid_end
10 valid_start=4-Jan-07, valid_end=now	1000 valid_start=4-Jan-07, valid_end=now

Fig. 11. The mapping of the conceptual to the physical model with today's date being 6-Jan-07

DELETE/UPDATE follow standard SQL syntax (DELETE FROM staff WHERE time > 5-Jan-07 AND time < 10-Jan-07; UPDATE staff SET salary=2000 WHERE time=5-Jan-07). The physical model is updated with the appropriate change in the value of the attributes and "valid_start" and/or "valid_end". Cases 1 and 2 (Fig.10) occur when retroactive data is to be deleted/updated. Cases 4, 5 and 6 occur when data valid in the future has been inserted and now there is a need to alter them. A DELETE/UPDATE can also potentially split a contiguous block into two. This transpires to two tuples in the physical model. Thus the logical DELETE/UPDATE in the chronon model remains a single operation while the multiple steps needed in the physical model are caught by the mapping.

3 Conclusion

In this paper we have developed the chronon based conceptual model for temporal databases and showed the simplicity and flexibility achieved using only a single additional construct, MERGE BY. We have evaluated the model against TSQL2 and shown its advantages. There are a few issues to be addressed in the future. We need to analyse the complexities introduced due to the inclusion of transaction time support in the chronon model. We are also particularly interested in testing the model in a Spatio-Temporal setting where the need is to store not only *what* is important to an application but also *when* and *where* [6]. The simplicity achieved in the temporal domain gives impetus to adapt the model in these new and upcoming areas.

References

1. Jensen, C.S., et al.: A consensus test suite of temporal database queries (1993), ftp://ftp.cs.arizona.edu/tsql/doc/testSuite.ps
2. Jensen, C.S., Snodgrass, R.T.: Semantics of Time Varying Information. Information Systems 19(4) (1994)
3. Wang, F., Zaniolo, C.: XBiT: An XML-based Bitemporal Data Model. In: Atzeni, P., Chu, W., Lu, H., Zhou, S., Ling, T.-W. (eds.) ER 2004. LNCS, vol. 3288, Springer, Heidelberg (2004)
4. Wang, F., Zaniolo, C., Zhou, X.: Temporal XML? SQL Strikes Back! In: 12th International Symposium on Temporal Representation and Reasoning (TIME) (June 2005)
5. Wang, F., Zaniolo, C.: An XML-Based Approach to Publishing and Querying the History of Databases. Internet and Web Information Systems 8(3) (September 2005)
6. Allen, G., Bajaj, A., Khatri, V., Ram, S., Siau, K.: Advances in Data Modeling Reasearch. Communications of the Association of Information Systems 17 (2006)
7. Gregerson, H., Jensen, C.S.: Temporal Entity-Relationship Models – A Survey. IEEE transactions on knowledge and Data Engineering 11(3) (1999)
8. Clifford, J., Dyreson, C., Isakowitz, T., Jensen, C.S., Snodgrass, R.T.: On the Semantics of now in Databases. ACM transactions on database systems 22(2) (June 1997)
9. Bohlen, M.H., Snodgrass, R.T., Soo, M.T.: Coalescing in Temporal Databases. In: VLDB (1996)
10. Elmasri, R., Navathe, S.B.: Fundamentals of Database Systems. Pearson Education, London (2003)

11. Snodgrass, R.T., Ahn, I.: A Taxonomy of Time in Databases. In: ACM SIGMOD International Conference on Management of Data (1985)
12. Snodgrass, R.T.: The TSQL2 Temporal Query Language. Springer publication, Heidelberg (1995)
13. Snodgrass, R.T.: Managing temporal data – a five part series., Database programming and design, TimeCenter technical report (1998)
14. Snodgrass, R.T., Bohlen, M.H., Jensen, C.S., Steiner, A.: Transitioning temporal support in TSQL2 to SQL3. Temporal Databases: Research and Practice, 150–194 (1998)
15. Snodgrass, R.T.: Developing Time Oriented Database Applications in SQL. Morgan Kaufmann Publishers, San Fransisco, California (2000)
16. Snodgrass, R.T.: Official website in University of Arizona, http://www.cs.arizona.edu/~rts/sql3.html

Main Memory Commit Processing: The Impact of Priorities

Heine Kolltveit and Svein-Olaf Hvasshovd

Department of Computer and Information Science
Norwegian University of Science and Technology

Abstract. Distributed transaction systems require an atomic commitment protocol to preserve ACID properties. The overhead of commit processing is a significant part of the load on a distributed database. Here, we propose approaches where the overhead is reduced by prioritizing urgent messages and operations. This is done in the context of main memory primary-backup systems, and the proposed approaches is found to significantly reduce the response time as seen by the client. Also, by piggybacking messages on each other over the network, the throughput is increased. Simulation results show that performance can be significantly improved using this approach, especially for utilizations above 50%.

1 Introduction

The increasing complexity of modern computer systems put increasing demands on the performance of all components of systems. Databases are commonly an important part of these systems. Thus, it is crucial that they are fast and reliable. The ratio between storage capacity and cost of main memory (MM) have been increasing exponentially, which has contributed to the advent of MM databases and applications. Since the data in MM databases is always in MM, slow disk accesses are replaced with faster MM accesses to data [1,2], improving the potential performance.

MM databases can be classified by how *logging* and *persistence* are handled. The first determines whether the log is saved to disk or not. The latter dictates whether the log is persistently stored or not. If the log is saved to disk, persistence is achieved if it is on disk before the transaction commits. If it is only stored in MM, it must be replicated to a process with a different failure mode [3] to make it persistent. Here we discuss only *pure* main memory systems, where neither data nor log resides on disk. Both data and the log are replicated to achieve persistence.

Atomic commitment protocols are designed to ensure that all participants of a transaction reach the same decision about a transaction. As soon as the decision is reached and has been persistently stored, an *Early Answer* [4] can be given to the client. This has been shown to significantly reduce transaction response time [5].

This paper is motivated by the observation that operations (or tasks) and messages before the early answer are more urgent than those after. Also, larger messages take less time to send and receive per byte than smaller messages. Thus, by *piggybacking* messages from a single source to a common destination, the transmission of messages is more effective.

J.R. Haritsa, R. Kotagiri, and V. Pudi (Eds.): DASFAA 2008, LNCS 4947, pp. 470–477, 2008.

The main contributions of this paper are the usage of priorities and piggybacking in a main memory distributed database setting and simulation results showing the performance benefits gained by using this approach. The results of such an evaluation can favorably be used by system developers to improve the performance of MM applications, e.g., in a shared-nothing fault-tolerant DBMS like ClustRa [4]. In addition, the simulation framework developed in [5] has been improved to facilitate piggybacking and priorities.

This paper only simulates failure-free execution where all transactions commits. Since this is the dominant execution path, it is where the potential for performance gains is greatest.

The rest of the paper is organized as follows. Section 2 gives an overview of related work. Section 3 presents the priority and piggybacking schemes. Section 4 outlines the simulation model and parameters used for the simulations in Section 5. The simulations are compared to a statistical analysis in Section 6, while Section 7 gives some concluding remarks.

2   Related Work

Several 2PC-based [6,7] modifications where performance issues are handled exist [8]. Presumed commit and presumed abort [7] both avoid one flushed disk write, by assuming that a non-existent log record means that the transaction has committed or aborted, respectively. Transfer-of-commit, lazy commit and read-only commit [9], sharing the log [7,10] and group commit [11,12] are other optimizations. An optimization of the presumed commit protocol [13] reduces the number of messages, but requires the same number of forced disk writes.

One-phased commit protocols have also been proposed [10,14,15,16,17,4]. These are based on the early prepare or unsolicited vote method by Stonebraker [18] where the prepare message is piggybacked on the last operation sent to a participant. In this way, the voting phase is eliminated. However, these approaches may inflict strong assumptions and restrictions on the transactional system [15].

1-2PC [19,20] is an approach which dynamically chooses between one-phased execution and two-phased. Thus, it provides the performance advantages of 1PC when applicable, while also supporting the wide-spread 2PC.

There exists some pure main memory commit protocols that ensures persistence. R2PC and R1PC use synchronous messages to backups to persistently store the log, while C2PC and C1PC executes the commit processing in a circular fashion [21,5]. The ClustRa commit protocol [4], CCP, is another protocol. It is a highly parallel one-phased protocol. R2PC, R1PC, C2PC, C1PC and CCP are the protocols used in this paper. See [22] for a more thorough presentation of them. A dynamic coordinator algorithm called DPC-P [5] together with the Early Answer optimization [4] is also used in this paper.

3   Priorities and Piggybacking

The early answer optimizations allows us to divide messages and tasks into two separate groups: The ones before the reply and the ones after. Intuitively, to achieve a short response time, the ones before the reply should be prioritized. This section first presents a scheme for prioritizing operations, and then three schemes for prioritizing and piggy-backing messages. Pseudo-code for the algorithms are presented fully in [22].

Prioritizing messages before the reply leads to longer locking of data at the participants. However, we assume a large number of records and minimal data contention. Therefore, the extra time spent before releasing the locks does not degrade the system.

3.1 Priority of Tasks by the Transaction Manager

All nodes are assumed to have a transaction manager, which manages the transactional requests. In front of it there is a queue of tasks. An `urgent` flag is set for the task, if it is an operation that happens before the reply to the client. When the transaction manager is ready to perform a task, it starts at the front of the queue and checks if the first task is urgent. If it is, it is immediately processed. If not, the task's `waited` field is checked. If it has waited longer than a predetermined maximum waiting period, it is processed, else the algorithm searches for the first urgent task in the queue. If none is found, the first in the queue is chosen. We call this priority by the transaction manager *PRI*. The default way without using PRI is called NOPRI.

3.2 Priority and Piggybacking Messages over the Network

The messages sent over the network can be prioritized similarly to the tasks at the transaction manager. In addition, messages can be grouped together using piggyback-ing. *Piggybacking* is a technique where messages going to the same destination are appended on each other. This reduces the total load on the system since there is a quite large initiation cost for messages, and the total cost per bit sent decreases as the number of bits increases.

The rest of this section presents four alternate ways to chose which messages to piggyback.

PB1: No Piggybacking And No Priority. In this first setup, the queue of outgoing messages are served on a first-come first-served basis. No messages are given priority and no piggybacking occurs. This is the default way to process outgoing messages.

PB2: Piggybacking, But No Priority For PB2, the queue is served on a first-come first-served basis. However, all the messages in the queue which are going to the same destination are piggybacked on the message. Thus, the number of required messages is reduced. Intuitively, this increase the throughput and decrease the response time.

PB3: Piggyback From Non-Urgent Queue In PB3, the non-urgent messages are moved to a separate queue. When an urgent message is found, the non-urgent queue is checked for messages with the same destination. These are piggybacked on the ur-gent message. To avoid starvation of non-urgent messages a timeout is used. After the timeout expires for the first message in the non-urgent queue, it is sent, and the rest of the non-urgent queue is checked for messages to piggyback.

Table 1. The input parameters of the simulation

Parameter	Value	Parameter	Value	
Simulation Model	Open	Send Message	$(26 + 0.055 * B)$ μs	if B < 1500,
Context Switch	3.5 μs		$(73 + 0.030 * B)$ μs	else
Timeslice	5 ms	Receive	$(52 + 0.110 * B)$ μs	if B < 1500,
Simulation Time	200 s	Message	$(146 + 0.060 * B)$ μs	else
Capture Time	160 s [40 - 200]	Long	700	DoWork,
Simulation runs	10	Operations	μs	DoWorkAndPrepare
Subtransactions	3	Medium	150	BeginTxn,Prepare,
Transactions	Single tuple update	Operations	μs	Abort, Commit
# of simulated nodes	20	Short	50	WorkDone, Vote,
CPU Utilization	Variable	Operations	μs	Aborted,Committed

PB4: Piggyback From Urgent and Non-Urgent Queue PB4 is similar to PB3, but extends it by piggybacking messages from both outgoing queues. Thus, more messages are potentially piggybacked on the same message. As for PB3, a timeout is used to avoid starvation of non-urgent messages.

4 The Simulation Model

To evaluate the performance of the commit protocols a discrete-event simulator was developed using Desmo-J [23]. The input parameters of the system are given in Table 1. The model and the parameters have been more extensively explained in [22]. However, the cost of sending and receiving messages has changed. In line with update transactions, the messages sent over the network are assumed to be small, i.e. 500 bytes for the first message to reach each participant for each transaction, and 50 bytes for the others. The formulas for the response time are found by taking the average of 100.000 round-trip times for UDP datagrams with 10 bytes intervals. Measurements of CPU usages indicate that it takes twice as long to receive, as to send a message.

5 The Simulation Results

The simulations were performed using the system model presented in Section 4. The results from executing the main memory protocols and the chosen optimizations are presented in this section.

5.1 Average Response Time

Simulations were performed for all combinations of protocols, priority and piggybacking algorithms. The resulting average response times versus throughputs are plotted in Figure 1. The DPC-P optimizations have been chosen since it gives the best performance [5]. PB1, PB2, PB3 and PB4 is plotted as circles, squares, triangles pointing up and triangles pointing down, respectively. NOPRI is shown in white, and PRI in black. The results for each protocol have been plotted for throughputs up to 97.5% utilization of the original protocol combined with the optimization not using any piggybacking (PB1) and NOPRI.

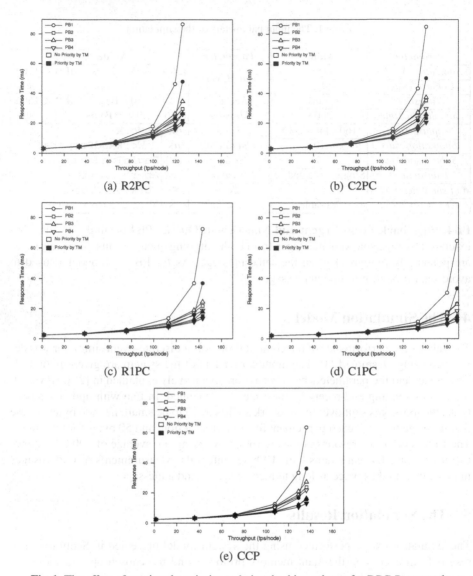

(a) R2PC

(b) C2PC

(c) R1PC

(d) C1PC

(e) CCP

Fig. 1. The effect of varying the priority and piggybacking scheme for DPC-P-protocols

Figure 1 clearly show that PB1 with NOPRI is the worst performer. The PRI approach results in faster responses to the client, at the costs of longer holding of the locks at the participants, since the commit messages might be delayed. At low utilizations, PB3 performs better than PB2 since the first prioritizes messages to send, while the latter only uses piggybacking. However, as the utilization increases and the queues grows, PB2 outperforms PB3, because PB3 will only send one prioritized messages each time, while PB2 might send more. PB4 does not have this problem since both the urgent and the non-urgent queues are checked for messages to piggyback.

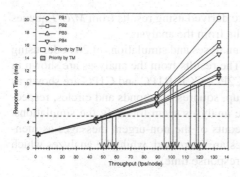

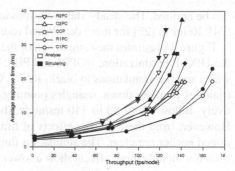

Fig. 2. The maximum response time for the 95% quickest transactions while executing C1PC - DPC-P

Fig. 3. Comparing analysis and simulations for all protocols using DPC-P, NOPRI and PB3

The combination of PRI and PB4 yields the best performance for all protocols, optimizations and utilizations. This is caused by the fact that it always piggybacks all messages in the outgoing queues going in the same direction and always prioritizes the most urgent tasks to be performed by the TM. Priority of tasks by TM has a greater impact on the performance than piggybacking for utilizations up to approximately 90%. From there on, piggybacking using PB4 has more effect.

5.2 Response Time Demands

Real-time databases typically require that most (i.e. 90%, 95%, 97%, 98% or 99%) transactions reply within a given time limit [4]. Figure 2 plots the maximum of the 95% shortest response times using the best performing protocol, C1PC with the DPC-P optimization. The 5 and 10 millisecond limits are shown in dashed and dotted lines, respectively. The downward pointing arrows mark the throughput at the intersections between the simulations and the dashed and dotted lines.

The simulation results for C1PC - DPC-P are presented in Figure 2. At the 5 millisecond limit, PB4 tolerates 6% more transactions than PB1, both executed with NOPRI. For PRI, the same difference is 5%. The difference between executing with PRI and PB4, and with NOPRI and PB1 is 20%. For the 10 millisecond limit and NOPRI, PB2, PB3 and PB4 tolerate approximately 5% − 10% more transactions than PB1. With PRI, it is 6% − 9%. Comparing PB1 with and without PRI, a 20% improvement is found. The best combination, PB4 with PRI, tolerates 30% more transactions than the NOPRI and PB1.

As for the average response time in Section 5.1, priorities give greater impact on performance of the protocols than piggybacking. Also, the simulations using PB4 and PRI shows the best performance.

6 Analysis

The simulation results are verified by using the analytical model used in [5]. The model uses an multiple class open queuing network with exponentially distributed inter-arrival times. It assumes uniformity for all nodes, such that single-server queueing theory [24]

can be applied. The steady-state solutions are derived using results from $M/G/1$ queues [24]. Refer to [22] for more details and results from the analysis.

Figure 3 illustrates the comparison of the analyses and simulations of 20 nodes using the DPC-P optimization, NOPRI and PB3. The results from the analyses are shown in white and the simulations in black. R2PC, C2PC, CCP, R1PC and C1PC are shown as triangles pointing down, triangles pointing up, squares, diamonds and circles, respectively. Below about 80 to 110 tps/node, the analyses seem to verify the simulations. However, from there on, the effects of timeouts of the non-urgent messages as mentioned earlier are clear. The simulation times are increased, while the analyzes which does not include these effects show a lower response time.

7 Evaluation and Conclusion

This paper has presented new approaches to prioritize tasks and piggyback messages in the context of main memory commit processing. The approaches have been evaluated by simulation using existing main memory commit protocols and optimizations. The results show that the new approaches can improve the performance significantly for these protocols, especially for utilizations above 50%.

The best piggybacking algorithm approximately halves the average response time at 90% utilization compared to not using any piggybacking. The improvement is less for lower utilizations. Prioritizing messages has even more impact. At 90% utilization the improvement in the average response time is around 56%. Combining those two, the total improvement in the average response time is more than two-thirds.

Using the response time requirement that 95% of the transactions respond within a time limit of 10 ms, a 23% and 30% improvement was found using piggybacking and priorities compared to not using these techniques for C2PC and C1PC, using DPC-P.

For further work, the protocols, optimizations and piggybacking and priority approach should be implemented in a existing main-memory system. This would give results for real world applications. Also, the effect of longer lock holding and the resulting increase in data contention caused by the priority approach should be investigated.

References

1. De Witt, D.J., Katz, R.H., Olken, F., Shapiro, L.D., Stonebraker, M.R., Wood, D.: Implementation techniques for main memory database systems. In: Proc. of SIGMOD (1984)
2. Garcia-Molina, H., Salem, K.: Main memory database systems: An overview. IEEE Transactions on Knowledge and Data Engineering 04 (1992)
3. Cristian, F.: Understanding fault-tolerant distributed systems. Communications of the ACM 34 (1991)
4. Hvasshovd, S.O., Torbjørnsen, Ø., Bratsberg, S.E., Holager, P.: The ClustRa telecom database: High availability, high throughput, and real-time response. In: Proc. of VLDB (1995)
5. Kolltveit, H., Hvasshovd, S.O.: Performance of Main Memory Commit Protocols. Technical Report 06/2007, NTNU, IDI (2007)
6. Gray, J.: Notes on data base operating systems. In: Flynn, M.J., Jones, A.K., Opderbeck, H., Randell, B., Wiehle, H.R., Gray, J.N., Lagally, K., Popek, G.J., Saltzer, J.H. (eds.) Operating Systems. LNCS, vol. 60, Springer, Heidelberg (1978)

7. Mohan, C., Lindsay, B., Obermarck, R.: Transaction management in the R* distributed database management system. ACM Trans. Database Syst. 11 (1986)
8. Samaras, G., Britton, K., Citron, A., Mohan, C.: Two-phase commit optimizations and trade-offs in the commercial environment. In: Proc. of ICDE (1993)
9. Gray, J., Reuter, A.: Transaction Processing: Concepts and Techniques. Morgan Kaufmann, San Francisco (1993)
10. Stamos, J.W., Cristian, F.: A low-cost atomic commit protocol. In: Proc. of SRDS (1990)
11. Gawlick, D., Kinkade, D.: Varieties of concurrency control in IMS/VS Fast Path. IEEE Database Eng. Bull. 8 (1985)
12. Park, T., Yeom, H.Y.: A consistent group commit protocol for distributed database systems. In: Proc. of PDCS (1999)
13. Lampson, B., Lomet, D.: A new presumed commit optimization for two phase commit. In: Proc. of VLDB (1993)
14. Abdallah, M., Pucheral, P.: A single-phase non-blocking atomic commitment protocol. In: Quirchmayr, G., Bench-Capon, T.J.M., Schweighofer, E. (eds.) DEXA 1998. LNCS, vol. 1460, Springer, Heidelberg (1998)
15. Abdallah, M., Guerraoui, R., Pucheral, P.: One-phase commit: Does it make sense? In: Proc. of ICPADS (1998)
16. Lee, I., Yeom, H.Y.: A single phase distributed commit protocol for main memory database systems. In: Proc. of the Int. parallel and distributed processing symposium (2002)
17. Stamos, J.W., Cristian, F.: Coordinator log transaction execution protocol. Distributed and Parallel Databases 1 (1993)
18. Stonebraker, M.: Concurrency control and consistency of multiple copies of data in distributed ingres. IEEE Trans. Software Eng. 5 (1979)
19. Al-Houmaily, Y.J., Chrysanthis, P.K.: 1-2PC: The one-two phase atomic commit protocol. In: Proc. of SAC, pp. 684–691. ACM Press, New York (2004)
20. Yousef, J., Al-Houmaily, P.K.C.: ML-1-2PC: an adaptive multi-level atomic commit protocol. In: Benczúr, A.A., Demetrovics, J., Gottlob, G. (eds.) ADBIS 2004. LNCS, vol. 3255, pp. 275–290. Springer, Heidelberg (2004)
21. Kolltveit, H., Hvasshovd, S.O.: The Circular Two-Phase Commit Protocol. In: Proc. of Int. Conf. of Database Systems for Advanced Applications (2007)
22. Kolltveit, H., Hvasshovd, S.O.: Main Memory Commit Processing: The Impact of Priorities - Extended Version. Technical Report 11/2007, NTNU, IDI (2007)
23. Page, B., Kreutzer, W.: The Java Simulation Handbook. Simulating Discrete Event Systems with UML and Java. Shaker Verlag (2005)
24. Jain, R.: The Art of Computer Systems Performance Analysis. Wiley & sons, Chichester (1991)

Association Rules Induced by Item and Quantity Purchased

Animesh Adhikari[1] and P.R. Rao[2]

[1] Department of Computer Science, S P Chowgule College, Margao, Goa - 403 602, India
animeshadhikari@yahoo.com
[2] Department of Computer Science and Technology, Goa University, Goa - 403 206, India
pralhaad@rediffmail.com

Abstract. Most of the real market basket data are non-binary in the sense that an item could be purchased multiple times in the same transaction. In this case, there are two types of occurrences of an itemset in a database: the number of transactions in the database containing the itemset, and the number of occurrences of the itemset in the database. Traditional support-confidence framework might not be adequate for extracting association rules in such a database. In this paper, we introduce three categories of association rules. We introduce a framework based on traditional support-confidence framework for mining each category of association rules. We present experimental results based on two databases.

Keywords: Confidence; Support; Transaction containing an item multiple times.

1 Introduction

Pattern recognition [1], [15] and interestingness measures [9] are two important as well as interesting topics at the heart of many data mining problems. Association analysis using association rules [1], [3] has been studied well on binary data. An association rules in a database are expressed in the form of a forward implication, $X \rightarrow Y$, between two itemsets X and Y in a database such that $X \cap Y = \phi$. Association rule mining in a binary database is based on support-confidence framework established by Agrawal et al. [1]. Most of the real life databases are non-binary, in sense that an items could be purchased multiple times in a transaction. It is necessary to study the applicability of traditional support-confidence framework for mining association rules in these databases.

A positive association rule in a binary database BD expresses positive association between itemsets X and Y, called the *antecedent* and *consequent*, of the association rule respectively. Each itemset in BD is associated with a statistical measure, called *support* [1]. Support of itemset X in BD is the fraction of transactions in BD containing X, denoted by $supp(X, BD)$. The interestingness of an association rule is expressed by its support and confidence measures. The support and confidence of an association rule $r: X \rightarrow Y$ in a binary database BD are defined as follows: $supp(r, BD) = supp(X \cap Y, BD)$, and $conf(r, BD) = supp(X \cap Y, BD) / supp(X, BD)$. An association

J.R. Haritsa, R. Kotagiri, and V. Pudi (Eds.): DASFAA 2008, LNCS 4947, pp. 478–485, 2008.

rule *r* in *BD* is *interesting* if *supp*(*r*, *BD*) ≥ *minimum support*, and *conf*(*r*, *BD*) ≥ *minimum confidence*. The parameters, minimum support and minimum confidence, are user input given to an association rule mining algorithm.

Association rules in a binary database have limited usage, since in a real transaction an item might be present multiple times. Let *TIMT* be the type of a database such that a Transaction in the database might contain an Item Multiple Times. In this paper, a database refers to a *TIMT* type database, if the type of the database is unspecified. Then, the question comes to our mind whether the traditional support-confidence framework still works for mining association rules in a *TIMT* type database. Before answering to this question, first we take an example of a *TIMT* type database *DB* as follows.

Example 1. Let *DB* = {{A(300), B(500), C(1)}, {A(2), B(3), E(2)}, {A(3), B(2), E(1)}, {A(2), E(1)}, {B(3), C(2)}}, where $x(\eta)$ denotes item x purchased η numbers at a time in the corresponding transaction. The number of occurrences of itemset {A, B} in the first transaction is equal to minimum {300, 500}, i.e., 300. Thus, the total number of occurrences of {A, B} in *DB* is 304. Also, {A, B} has occurred in 3 out of 5 transactions in *DB*. Thus, the following attributes of itemset X are important consideration for making association analysis of items in X: number of occurrences of X in *DB*, and number of transactions in *DB* containing X.   ●

Rest of the paper is organized as follows. In Section 2, we study association rules in a *TIMT* type database and introduce three categories of association rules. In Section 3, we introduce a framework based on traditional support-confidence framework for mining each category of association rules. We study the properties of proposed interestingness measures in Section 4. In Section 5, we discuss a method for mining association rules in a *TIMT* type database. Experimental results are provided in Section 6. We discuss related work in Section 7.

2 Association Rules in a *TIMT* Type Database

We are given a *TIMT* type database *DB*. A transaction in *DB* containing p items could be stored as follows: {$i_1(n_1)$, $i_2(n_2)$, ..., $i_p(n_p)$}, where item i_k is purchased n_k numbers at a time in the transaction, for i = 1, 2, ..., p. Each itemset X in a transaction is associated with the following two attributes: transaction-itemset frequency (*TIF*), and transaction-itemset status (*TIS*). These two attributes are defined as follows: $TIF(X, \tau, DB) = m$, if X occurs m times in transaction τ in *DB*. $TIS(X, \tau, DB) = 1$, for $X \in \tau$, $\tau \in DB$, and $TIS(X, \tau, DB) = 0$, otherwise. Also, each itemset X in *DB* is associated with the following two attributes: transaction frequency (*TF*), and database frequency (*DF*). These two attributes are defined as follows: $TF(X, DB) = \sum_{\tau \in DB} TIS(X, \tau, DB)$, and $DF(X, DB) = \sum_{\tau \in DB} TIF(X, \tau, DB)$

When an item is purchased multiple times in a transaction then the existing generalized frameworks [7], [8] might not be adequate for mining association rules, since they are based on a binary database. The following example shows why the traditional support-confidence framework is not adequate for mining association rules in a *TIMT* type database.

Example 2. Let there are three *TIMT* type databases DB_1, DB_2, and DB_3 containing five transactions each. $DB_1 = \{\{A(1000), B(2000), C(1)\}, \{A(5), C(2)\}, \{B(4), E(2)\},$ $\{E92), F(1)\}, \{F(2)\}\}$; $DB_2 = \{\{A(1), B(1), C(2)\}, \{A(1), B(1), E(2)\}, \{A(1), B(1),$ $F(1)\}, \{A(1), B(1), G(2)\}, \{H(3)\}\}$; $DB_3 = \{\{A(500), B(600)\}, \{A(700), B(400),$ $E(1)\}, \{A(400), B(600), E(3)\}, \{G(3)\}, \{A(200), B(500), H(1)\}\}$. The numbers of occurrences of itemset $\{A, B\}$ in transactions of different databases are given below.

Table 1. Distributions of itemset $\{A, B\}$ in transactions of different databases

Database	Trans #1	Trans #2	Trans #3	Trans #4	Trans #5
DB_1	1000	0	0	0	0
DB_2	1	1	1	1	0
DB_3	500	400	400	0	200

In Table 1, we observe the following points regarding itemset $\{A, B\}$: (i) It has high database frequency, but low transaction frequency in DB_1, (ii) In DB_2, it has high transaction frequency, but relatively low database frequency, and (iii) It has high transaction frequency and high database frequency in DB_3.   •

Based on the above observations, it might be required to consider database frequencies and transaction frequencies of $\{A\}$, $\{B\}$ and $\{A, B\}$ to study association between items A and B. Thus, we could have the following categories of association rules in a *TIMT* type database: (*I*) Association rules induced by transaction frequency of an itemset, (*II*) Association rules induced by database frequency of an itemset, and (*III*) Association rules induced by both transaction frequency and database frequency of an itemset. The goal of this paper is to provide frameworks for mining association rules under different categories in a *TIMT* type database.

3   Frameworks for Mining Association Rules

Each framework is based on traditional support-confidence framework for mining association rules in a binary database.

3.1   Framework for Mining Association Rules under Category *I*

Based on the number of transactions containing an itemset, we define *transaction-support* (*tsupp*) of the itemset in a database as follows.

Definition 1. Let X be an itemset in *TIMT* type database DB. Transaction-support of X in DB is given as follows: $tsupp\ (X, DB) = TF(X, DB) / |DB|$.   •

Let X and Y be two itemsets in DB. An itemset X is *transaction-frequent* in DB if $tsupp\ (X, DB) \geq \alpha$, where α is user-defined *minimum transaction-support* level. We define transaction-support of association rule $r: X \rightarrow Y$ in DB as follows: $tsupp\ (r, DB) = tsupp\ (X \cap Y, DB)$. We define *transaction-confidence* (*tconf*) of association rule r in DB as follows: $tconf(r, DB) = tsupp(X \cap Y, DB) / tsupp(X, DB)$. An association rule r in DB is *interesting* with respect to transaction frequency of an itemset if $tsupp(r, DB) \geq minimum\ transaction\text{-}support$, and $tconf(r, DB) \geq \beta$, where β is user-defined *minimum transaction-confidence* level.

3.2 Framework for Mining Association Rules under Category *II*

Let $X = \{x_1, x_2, ..., x_k\}$ be an itemset in database DB. Also, let τ be a transaction in DB. Let item x_i be purchased η_i numbers at a time in τ, for $i = 1, 2, ..., k$. Then, $TIF(X, \tau, DB) = $ minimum $\{\eta_1, \eta_2, ..., \eta_k\}$. Based on the frequency of an itemset in a database, we define *database-support* (*dsupp*) of an itemset as follows.

Definition 2. Let X be an itemset in *TIMT* type database DB. Database-support of X in DB is given as follows: $dsupp\ (X, DB) = DF\ (X, DB)\ /\ |DB|$.   •

An item in a transaction could occur more than once. Thus, $dsupp\ (X, DB)$ could be termed as the *multiplicity* of itemset X in DB. An important characteristic of a database is the average multiplicity of an item (*AMI*) in the database. Let m be the number of distinct items in DB. We define *AMI* in a *TIMT* type database DB as follows: $AMI(DB) = \sum_{i=1}^{m} dsupp(x_i, DB)/m$, where x_i is the i-th item in DB, for $i = 1$, $2, ..., m$. An itemset X is *database-frequent* in DB if $dsupp(X, DB) \geq \gamma$, where γ is user-defined *minimum database-support* level. Let Y be another itemset in DB. We define *database-support* of association rule $r: X \rightarrow Y$ in DB as follows: $dsupp(r, DB)$ $= dsupp(X \cap Y, DB)$. Also, we define *database-confidence* (*dconf*) of association rule $r: X \rightarrow Y$ in DB as follows: $dconf(r, DB) = dsupp(X \cap Y, DB)\ /\ dsupp(X, DB)$. An association rule $r: X \rightarrow Y$ is *interesting* with respect to database frequency of an itemset if $dsupp(r, DB) \geq minimum\ database\text{-}support$ and $dconf(r, DB) \geq \delta$, where δ is user defined *minimum database-confidence* level.

3.3 Framework for Mining Association Rules under Category *III*

An association rule r in *TIMT* type database DB is *interesting* with respect to both transaction frequency and database frequency of an itemset if $tsupp(r, DB) \geq \alpha$, $tconf(r, DB) \geq \beta$, $dsupp(r, DB) \geq \gamma$, and $dconf(r, DB) \geq \delta$. The parameters α, β, γ, and δ are defined in Sections 3.1 and 3.2. The parameters are user defined inputs given to a category *III* association rule mining algorithm. Based on the framework, we extract association rules in a database as follows.

Example 3. Let $DB = \{\{A(1), B(1)\}, \{A(2), B(3), C(2)\}, \{A(1), B(4), E(1)\}, \{A(3), E(1)\}, \{C(2), F(2)\}\}$. Let $\alpha = 0.4$, $\beta = 0.6$, $\gamma = 0.5$, and $\delta = 0.5$. Transaction-frequent and database-frequent itemsets are given in Tables 2 and 3, respectively. Interesting association rules under category *III* are given in Table 4.

Table 2. Transaction-frequent itemsets in DB

itemset	A	B	C	E	AB	AE
tsupp	0.8	0.6	0.4	0.4	0.6	0.4

Table 3. Database-frequent itemsets in *DB*

itemset	A	B	C	E	F	AB	AC	AE	BC	CF	ABC
dsupp	1.4	1.6	0.8	0.4	0.4	0.8	0.4	0.4	0.4	0.4	0.4

Table 4. Interesting association rules in *DB* under category *III*

$r: X \rightarrow Y$	$tsupp\,(r, DB)$	$tconf\,(r, DB)$	$dsupp\,(r, DB)$	$dconf\,(r, DB)$
A→ B	0.6	0.75	0.8	0.57143
B→ A	0.6	1.0	0.8	0.5

3.4 Dealing with Items Measured in Continuous Scale

We discuss here the issue of handling items that are measured in continuous scale. Consider the item milk in a departmental store. Let there are four types of milk packets: 0.5 kilolitre, 1 kilolitre, 1.5 kilolitres, and 2 kilolitres. The minimum packaging unit could be considered as 1 unit. Thus, 3.5 kilolitres of milk could be considered as 7 units of milk.

4 Properties of Different Interestingness Measures

Transaction-support and transaction-confidence measures are the same as the traditional support and confidence measures of an itemset in a database, respectively. Thus, they satisfy all the properties that are satisfied by traditional measures.

Property 1. $0 \leq tsupp\,(Y, DB) \leq tsupp\,(X, DB) \leq 1$, for itemsets X *and* $Y\,(\supseteq X)$ in *DB*.   •

Transaction-support measure satisfies anti-monotone property [11] of traditional support measure.

Property 2. $tconf\,(r, DB)$ lies in $[tsupp(r, DB), 1]$, for an association rule r in *DB*.   •

If an itemset X is present in transaction τ in *DB* then $TIF(X, \tau, DB) \geq 1$. Thus, we have the following property.

Property 3. $tsupp\,(X, DB) \leq dsupp\,(X, DB) < \infty$, for itemset X in *DB*.   •

In this case, database-confidence of an association rule r might not lie in $[dsupp(r, DB), 1.0]$, since database-support of an itemset in *DB* might be greater than 1 (please see Table 3). But, the database confidence of an association rule r in *DB* satisfies the following property.

Property 4. $dconf\,(r, DB)$ lies in $[0, 1]$, for an association rule r in *DB*.   •

5 Mining Association Rules

Many interesting algorithms have been proposed for mining positive association rules in a binary database [2], [6]. Thus, there are several implementations [4] of mining positive association rules in a binary database. In the context of mining association rules in a *TIMT* type database, we shall implement apriori algorithm [2], since it is simple and easy to implement. For mining association rules in a *TIMT* type database, we could apply apriori algorithm. For mining association rules under category *III*, the pruning step of interesting itemset generation requires testing on two conditions: minimum transaction-support and minimum database-support. The interesting association rules under category *III* satisfy the following two additional conditions: minimum transaction-confidence and minimum database-confidence.

6 Experiments

We present the experimental results using real databases *retail* [5] and *ecoli*. The database *ecoli* is a subset of *ecoli database* [10]. All the experiments have been implemented on a 2.8 GHz Pentium D dual core processor with 512 MB of memory using visual C++ (version 6.0) software. Due to unavailability of *TIMT* type database, we have applied a preprocessing technique. If an item is present in a transaction then the number of occurrences of the item is generated randomly between 1 and 5. Thus, a binary transactional database gets converted into a *TIMT* type database. Top 4 association rules under category *I* (sorted on transaction-support) and category *II* (sorted on database-support) are given in Tables 5 and 6, respectively.

Table 5. Association rules in different databases under category *I*

database	(α, β)	(antecedent, consequent, *tsupp*, *tconf*)			
retail	(0.05, 0.2)	(48, 39,0.3306, 0.6916)	(39, 48, 0.3306, 0.5751)	(41, 39, 0.1295, 0.7637)	(39,41, 0.1295, 0.2252)
ecoli	(0.1, 0.3)	(48,50, 0.9583, 0.9817)	(50, 48, 0.9583, 0.9583)	(44, 48, 0.1399, 0.4029)	(40, 50, 0.1369 0.4104)

Table 6. Association rules in different databases under category *II*

database	(γ, δ)	(antecedent, consequent, *dsupp*, *dconf*)			
retail	(0.07, 0.4)	(48,39, 0.7255, 0.5073)	(39, 48, 0.7255, 0.4208)	(41, 39, 0.2844, 0.5584)	(38, 39, 0.2574, 0.4848)
ecoli	(0.1, 0.2)	(48,50, 2.7583, 0.9231)	(50, 48, 2.7583, 0.9176)	(44, 48, 0.3005, 1.0000)	(40, 50, 0.2828, 1.0000)

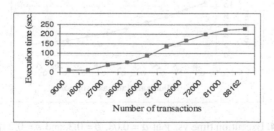

Fig. 1. Execution time vs. size of database at $\alpha = 0.05$, $\gamma = 0.07$, $\beta = 0.2$, and $\delta = 0.4$ (*retail*)

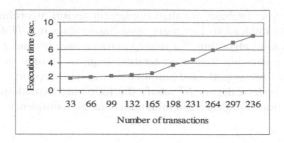

Fig. 2. Execution time vs. size of database at $\alpha = 0.1$, $\gamma = 0.1$, $\beta = 0.3$, and $\delta = 0.3$ (*ecoli*)

Also, we have obtained execution times for extracting association rules at different database sizes. As the size of a database increases, the execution time also increases. We have observed such phenomenon in Figures 1 and 2.

Also, we have obtained execution times for extracting association rules at different minimum database-supports. As the value of minimum database-support increases, the number of interesting itemsets decreases. So, the time required to extract category *II* association rules also decreases. We have observed such phenomenon in Figures 3 and 4.

The graph of execution time versus α at a given tuple (γ, β, δ) is similar to the graph of execution time versus γ at a given tuple (α, β, δ). As the value of minimum transaction-support increases, the number of interesting itemsets decreases. So, the time required to extract category *I* association rules also decreases.

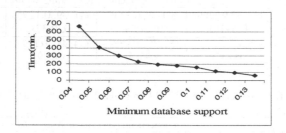

Fig. 3. Execution time vs. γ at $\alpha = 0.04$, $\beta = 0.2$, and $\delta = 0.4$ (*retail*)

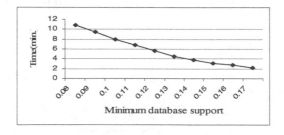

Fig. 4. Execution time vs. γ at $\alpha = 0.08$, $\beta = 0.3$, and $\delta = 0.2$ (*ecoli*)

7 Related Work

Agrawal and Srikant [2] have proposed apriori algorithm that uses breadth-first search strategy to count the supports of itemsets. The algorithm uses an improved candidate generation function, which exploits the downward closure property of support and makes it more efficient than earlier algorithm [1]. Han et al. [6] have proposed data mining method *FP*-growth (frequent pattern growth) which uses an extended prefix-tree (*FP*-tree) structure to store the database in a compressed form. In the context of interestingness measures, Tan et al. [9] have presented an overview of twenty one interestingness measures proposed in statistics, machine learning and data mining literature.

8 Conclusion

The traditional support-confidence framework has limited usage in association analysis of items, since a real life transaction might contain an item multiple times. The traditional support-confidence framework is based on the frequency of an itemset in a binary database. In the case of a *TIMT* type database, there are two types of frequency of an itemset: transaction frequency, and database frequency. Due to these reasons, we get three categories of association rules in a TIMT type database. We have introduced a framework for mining each category of association rules. The proposed frameworks are effective for studying association among items in real life market basket data.

References

1. Agrawal, R., Imielinski, T., Swami, A.: Mining association rules between sets of items in large databases. In: Proceedings of SIGMOD Conference, pp. 207–216 (1993)
2. Agrawal, R., Srikant, R.: Fast algorithms for mining association rules. In: Proceedings of the International Conference on Very Large Data Bases, pp. 487–499 (1994)
3. Antonie, M.-L., Zaïane, O.R.: Mining Positive and Negative Association Rules: An Approach for Confined Rules. In: Boulicaut, J.-F., Esposito, F., Giannotti, F., Pedreschi, D. (eds.) PKDD 2004. LNCS (LNAI), vol. 3202, pp. 27–38. Springer, Heidelberg (2004)
4. FIMI (2004), http://fimi.cs.helsinki.fi/src/
5. Frequent itemset mining dataset repository, http://fimi.cs.helsinki.fi/data/
6. Han, J., Pei, J., Yiwen, Y.: Mining frequent patterns without candidate generation. In: Proceedings of SIGMOD Conf. Management of Data, pp. 1–12 (2000)
7. Steinbach, M., Kumar, V.: Generalizing the notion of confidence. In: Perner, P. (ed.) ICDM 2004. LNCS (LNAI), vol. 3275, pp. 402–409. Springer, Heidelberg (2004)
8. Steinbach, M., Tan, P.-N., Xiong, H., Kumar, V.: Generalizing the notion of support. In: Proceedings of KDD, pp. 689–694 (2004)
9. Tan, P.-N., Kumar, V., Srivastava, J.: Selecting the right interestingness measure for association patterns. In: Proceedings of SIGKDD Conference, pp. 32–41 (2002)
10. UCI ML repository, http://www.ics.uci.edu/~mlearn/MLSummary.html
11. Zaki, M.J., Ogihara, M.: Theoretical foundations of association rules. In: Proceedings of DMKD, pp. 7.1–7.8 (1998)

An Indexed Trie Approach to Incremental Mining of Closed Frequent Itemsets Based on a Galois Lattice Framework

B. Kalpana[1], R. Nadarajan[2], and J. Senthil Babu[3]

[1] Department of Computer Science, Avinashilingam University for Women, Coimbatore- 43
kalpanabsekar@yahoo.com
[2] Department of Mathematics and Computer Applications, PSG College of Technology,
Coimbatore-4
nadarajan_psg@yahoo.co.in
[3] Final Year MCA, PSG College of Technology, Coimbatore-4
senthilbaboo@yahoo.com

Abstract. Incrementality is a major challenge in the mining of dynamic data-bases. In such databases, the maintenance of association rules can be directly mapped into the problem of maintaining closed frequent itemsets. A number of incremental strategies have been proposed earlier with several limitations. A serious limitation is the need to examine the entire family of closed itemsets, whenever there are insertions or deletions in the database. The proposed strategy relies on an efficient and selective update of the closed itemsets using an indexed trie structure. The framework emphasizes on certain fundamental and structural properties of Galois Lattice theory to overcome the limitations of the earlier approaches. Apart from facilitating a selective update, the indexed structure removes the necessity of working with a wholly memory resident trie.

1 Introduction

Frequent Itemset Mining (FIM) is a demanding task common to several important data mining applications that look for interesting patterns within large databases. The problem can be stated as follows:

Let $I = \{a_1 \dots a_m\}$ be a finite set of items, and D a dataset containing N transactions, where each transaction $t \in D$ is a list of distinct items. Given an itemset, its support is defined as the number of transactions in D that includes I. Mining of frequent itemsets requires us to discover all itemsets which have a support higher than the min-support threshold. This results in the exploration of a large search space given by the powerset(I). The huge size of the output makes the tasks of analysts very complex since they have to extract useful knowledge from a large amount of frequent patterns. Generation of closed itemsets is a solution to the above problem. Closed frequent itemsets are a condensed representation of frequent itemsets and can be magnitudes fewer than the corresponding frequent itemsets representing the same knowledge obtainable from frequent itemsets. In this study we adopt a Formal Concept Analysis (FCA) framework, so that the strategy exploits certain Galois Lattice properties to make an intelligent and selective traversal of the closed itemset lattice. FCA has a specific place for data organization, information engineering, data mining and

J.R. Haritsa, R. Kotagiri, and V. Pudi (Eds.): DASFAA 2008, LNCS 4947, pp. 486–495, 2008.

reasoning. It may be considered as a mathematical tool that unifies data and knowledge for information retrieval. FCA as an analytic method provides a systematic search, which ensures a lossless enumeration of closed itemsets.

2 Related Work

A number of strategies exist for the closed frequent itemset mining problem. Most of the algorithms adopt a strategy based on search space browsing and closure computation. An effective browsing strategy aims at identifying a single itemset for each equivalence class in the lattice. Some algorithms use closure generators [4]. Whatever may be the closure computing technique, the task of checking for duplicates is an expensive operation. Several algorithms like the Charm [12], Closet [9], Closet+[10] perform an explicit subsumption check operation which is expensive in terms of time and space. The DCI-Closed[4] adopts different strategies to extract frequent closed itemsets from dense and sparse datasets, it also uses a memory efficient duplication avoidance technique. Close[7] and A-Close[5] employ a level wise approach and are Apriori inspired strategies. In [1] the authors make use of a sliding window and limit the memory space using a CET. The work in [11] has a strong theoretical background in Formal Concept Analysis. FCA offers a huge potential for excellent representation mechanisms, search space browsing strategies and reduction techniques within the knowledge discovery paradigm. Our work is largely motivated by Galicia-T and Galicia-M [11]. The Galicia-T uses a complete search of the trie whenever there is an increment. This proves to be exponential to the transaction and attribute size. The Galicia-M overcomes the upsurge in memory encountered in Galicia-T by a selective update policy, however it works on a flat set of closed itemsets resulting in redundancy in the storage of closed itemsets. The indexing scheme also proves to be expensive in terms of storage. Our contribution in the indexed trie approach is an improvisation of the Galicia-T and Galicia _M by way of an efficient index over the trie which serves the dual purpose of a selective update and reduces the constraint on the main memory. We also incorporate the necessary logic for handling deletions in the framework.

3 The FCA Framework – Closed Itemset Lattice and Incremental Mining

We present in this section some basic definitions relevant to our study and discuss the properties useful for computing closed itemsets incrementally. Further details on formal concept analysis can be referred from [2,3].

Data Mining Context: A data mining context is a triple $D = (O,I,R)$. O and I are finite sets of objects and database items respectively. $R \subseteq O \times I$ is a binary relation between objects and items. Each couple $(o,i) \in R$ denotes the fact that the object $o \in O$ has the item $i \in I$. The data mining context can be a relation, a class or the result of an SQL / OQL query.

Galois Connection: Let $D = (O,I,R)$ be a data mining context. For $o \subseteq O$ and $i \subseteq I$, we define:

$f(O): P(O) \rightarrow P(I)$

$f(O) = \{i \in I \mid \forall o \in O, (o,i) \in R\}$

$g(I): P(I) \rightarrow P(O)$

$g(I) = \{o \in O \mid \forall i \in I, (o,i) \in R\}$

$f(O)$ associates with O all items common to all objects $o \in O$ and $g(I)$ associates with I all objects containing all items $i \in I$. A couple of applications (f,g) is a Galois connection between the powerset(O) and the powerset(I). The operators gof in I and fog in O are Galois closure operators. Given the Galois connection (f,g), the following properties hold for all $I, I_1, I_2 \subseteq I$ and $O, O_1, O_2 \subseteq O$ [7].

(1) $I_1 \subseteq I_2 \Rightarrow g(I_1) \supseteq g(I_2)$	(1') $O_1 \subseteq O_2 \Rightarrow f(O_1) \supseteq f(O_2)$
(2) $I \subseteq gof(I)$	(2') $O \subseteq fog(O)$
(3) $gof(gof(I)) = gof(I)$	(3') $fog(fog(O)) = fog(O)$
(4) $I_1 \subseteq I_2 \Rightarrow gof(I_1) \subseteq gof(I_2)$	(4') $O_1 \subseteq O_2 \Rightarrow fog(O_1) \subseteq fog(O_2)$
(5) $fog(g(I)) = g(I)$	(5') $gof(f(O)) = f(O)$
(6) $O \subseteq g(I) \Leftrightarrow I \subseteq f(O)$	

Closed Itemset: Let $C \subseteq I$ be a set of items from D. C is a closed itemset iff $gof(C)=C$. The smallest (minimal) closed itemset containing an itemset I is obtained by applying gof to I.

Closed Itemset Lattice: Let L_c be the set of closed itemsets (concepts) derived from D using the Galois connection . The pair $L_c = (C, \leq)$ is a complete lattice called the closed itemset lattice having the following properties:

1. A partial order on the lattice elements such that, for every element $C_1, C_2 \in Lc$, $C_1 \leq C_2$, Iff $C_1 \subseteq C_2$. i.e.
   C_1 is a sub-closed itemset of C_2 and C_2 is a sup-closed itemset of C_1.
2. All subsets of Lc have one upper bound, the join element $\vee$ and one lower bound, the meet element $\wedge$.

For a closed pair of sets (X,Y) where $X \in P(O)$ and $Y \in P(I)$, $X=Y'$ and $Y=X'$. X is called the extent and Y the intent of the concept.

The partial order $Lc = (C, \leq)$ is a complete lattice, called the Galois or concept lattice with joins and meets defined as follows:

$$V^k_{i=1}(X_i, Y_i) = ((\cup^k_{i=1} X_i)'', \cap^k_{i=1} Y_i),$$
$$\wedge^k_{i=1}(X_i, Y_i) = (\cap^k_{i=1} X_i, (\cup^k_{i=1} Y_i)'')$$

The Galois lattice provides a hierarchical organization of all closed pairs which can be exploited to execute a systematic computation and retrieval based on a selective update. The inherent structure facilitates easy insertion and deletions with minimal recomputation.

Upper cover: The upper cover of a concept f denoted $Cov^u(f)$ consist of the concepts that immediately cover f in the lattice. The set $Cov^u(f) = \{f_1 \mid f_1 \in Lc$ and $f \subset f_1$ and there does not exist $f_2 \in Lc \mid f \subset f_2 \subset f_1\}$.

The converse of this applies to lower covers.

Minimal Generator: An itemset $g \subseteq I$ is said to be a minimal generator of a closed itemset f, iff $Y(g) = f$ and there does not exist $g_1 \subset g \mid Y(g_1) = f$.

Given a sample database in Table .1, its associated concept lattice is shown as a line diagram in figure1. We denote the original database as D containing transactions 1-5, D' as the augmented database with the increment transaction 6. Lc as the concept lattice (fig.1) corresponding to D and Lc' as the concept lattice corresponding to the updated database D' (Fig.3).

Table 1. Sample Database

TID	ITEMS
1	A,C,D
2	B,C,E
3	A,B,C,E
4	B,E
5	A,B,C,E
6	A,B,D,E

Concepts #1 and #8 are the top and bottom concepts of the Galois Concept Lattice respectively (Fig.1). For a minimum support of 50 %, concepts #2, #3, #4, #5 are said to be closed and frequent. Maximal concepts lie towards the top of the lattice - a property that is exploited while mining the trie of closed itemsets.

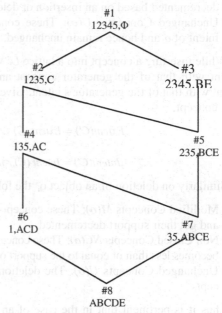

Fig. 1. Galois Concept Lattice Lc for Sample Database(D)

3.1 The Incremental Approach

Incremental approaches exploit the previous mining result and avoid recomputation of the closed itemsets from scratch. While the closed itemset computation is closely mapped to the concept lattice generation problem, there is also a close similarity between the incremental updates of lattice substructures and the incremental mining of closed itemsets. While this study is not involved in exploring the intricacies of the lattice structural updates, we make use of certain fundamental and structural properties involved in lattice updates to perform a selective update of the closed itemset family when insertions and deletions take place in a dynamic database.

Integration of a new object / transaction o_i is mainly aimed at the insertion in to Lc of all concepts whose intent does not correspond to the intent of an existing concept, and is an intersection of $\{o_i\}'$ with the intent of an existing concept. On insertion of an object three sets of concepts arise in the original lattice Lc transforming it to the new lattice Lc'.

(i)   Generator Concepts- $G(o)$:These concepts give rise to new concepts and help in the computation of the new concept's extent and intent. A generator is chosen as the minimal of all concepts that generate a given new concept in the concept lattice.
(ii) Modified Concepts- $M(o)$: These concepts incorporate o_i into their extents, while their intents are stable. The support of the modified concepts get incremented or decremented based on an insertion or deletion operation.
(iii) Unchanged Concepts- $U(o)$: These concepts do not incorporate the extent or intent of o_i and hence remain unchanged.

While inserting a concept into a lattice Lc we perform a union of the new concept's extent with that of the generator concept and an intersection of the new concept's intent with that of the generator's intent. Given C a generator concept in Lc and C' the new concept,

$$Extent(C') = Extent\ (C) \cup Extent\ (C'). \qquad (1)$$

$$Intent(C') = Intent\ (C) \cap Intent\ (C'). \qquad (2)$$

Similarly on deletion of an object o_i, the following sets of concepts arise.

(i)   Modified Concepts $M(o)$: These concepts remove o_i from their respective extents and get their support decremented.
(ii) Non Closed Concepts $NC(o)$: These concepts become non closed since their support becomes less than or equal to the support of their minimal super closed concepts.
(iii) Unchanged Concepts $U(o)$: The deletions do not have any effect on these concepts.

Thus it is pertinent that in the case of an insertion or deletion, our focus is on a substructure of Lc that is the order filter generated by the object concept o_i. This is denoted by $\uparrow(Vo_i)$. Detailed definitions related to order filter can be found in [2,11].

4   Indexed Trie Approach – Design and Implementation

The approach relies on an indexed trie structure which facilitates a selective update of the trie by retrieving branches of the trie corresponding to an increment or decrement. We aim to keep the memory requirements to a minimum by proposing a two level indexing structure on the trie. Further selective update and pruning mechanisms have been incorporated to cut down on the search space and time while considering increments and deletions to the database. Since the trie facilitates common prefixes to be merged and stored, a compact representation is achieved. In this framework the indexed trie provides for a selective update, while closely reflecting the structure of the Galois concept lattice. Given below are definitions of a few terms used in our work:

Maximal Concept: A maximal concept refers to the concept with the highest support among the concepts indexed to an item.

Index Chain: A sequence of concepts to which an item is indexed, starting with the maximal concept . The super closed concept of the current concept becomes the next concept in the index chain.

Super closed concept: Given two concepts *C1* and *C2*. *C2* is a Super closed concept of *C1* if *Intent(C2)⊃Intent(C1)* and *Extent(C2)⊂Extent(C1)*

The trie corresponding to the sample database *D* is shown in Fig.2. Closed itemsets are shaded. Each node in the trie contains item, pointers to the child nodes, pointer to the parent node, support in case the node is closed, a flag and a sibling pointer to enable access to the next node in the index chain. The index chain is indicated by a dotted line for items A and B alone. The sibling pointer which points to the same item in other branches, is used in identifying the next concept in the index chain. The strategy treats each object as an increment and performs an intersection with concepts along the index chain retrieved from the index entries corresponding to the increment / decrement. The first level index comprises of the pair (item, maximal concept) and it maps each item to its maximal concept in the concept lattice. This is indicated in Table 2 ,maximal concept for each item is marked bold. The second level of index connects each concept to its super closed concept in the index chain. It is sufficient to store the first level index alone in main memory and retrieve portions of the trie, corresponding to the index entries. Once the index entries corresponding to the items in the increment/decrement are retrieved, we use the index chain of each indexed item to traverse the trie hopping from one closed itemset to another to determine whether they get modified or act as generators for a new concept. Generators and modified concepts along a chain are marked off in subsequent index chains. For the given example, entries in table 2 which are struck off, are not examined.

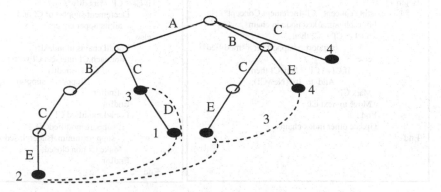

Fig. 2. Trie for the Sample Database

Let *CI* denote the concept being examined and *CI′* the concept corresponding to the increment.

If *CI* ∩ *CI′* = *CI* then the *CI* gets modified. The support is incremented by one. If *CI* ∩ *CI′* = *New CI* then it is added to the trie; its support is one more than the support of its minimal generator. Intersection operations are done by making a backward traversal up a branch of the trie from a closed concept in the index chain. The backward traversal is advantageous in that, we are able to reach the root of the selected branch without traversing unnecessary sub-branches. The overhead of unnecessary intersections encountered in Galicia-T is therefore avoided. Subset pruning is based on the following observations:

Lemma 1. If *C1* and *C2* are two modified concepts then the common lower cover of *C1* and *C2* also gets modified i.e. $C1 \wedge C2$ is modified.

Proof. *C1* and *C2* are modified $\Rightarrow$ *Intent (C1)* $\subseteq$ *Intent of (CI')*
$\Rightarrow$ *Intent (C2)* $\subseteq$ *Intent of (CI')*
$\Rightarrow$ *Intent (C1 $\wedge$ C2)* $\subseteq$ *Intent of (CI')*
$\therefore$ $C1 \wedge C2$ also gets modified.

Lemma 2. A generator concept cannot be a generator again or a modified concept for the same increment.

Proof. Case1: *C1* is a generator $\Rightarrow$ $C1 \not\subset (CI')$
     $\therefore C1$ cannot be modified.
    Case2: *C1* is a generator $\Rightarrow$ *extent (CI')* $\subset$ *of extent (C1)*.

Since we retrieve concepts indexed by the sorted items in *CI'*, we start examining concepts with the maximal support in an index chain. It is obvious that a generator of *CI'* is minimal and therefore can be a generator only once for a given increment.

The pseudocode for handling increments and deletion is given below:

```
Procedure Increment_Trie(transaction t)

For each index chain corresponding to
            maximal concepts retrieved from index[t]
Begin
    /* CI –existing concept, CI'-Increment Concept*/
        For each unmarked node in chain
            if CI ∩ CI' = CI then
                    support = support+1;  /*modified*/
            else
                    If CI ∩ CI' = NewCI then
                        Add_to_trie(NewCI);
            Mark CI;
            Move to next CI;
        End For
        Update other index chains;
End
```

```
Procedure Del_Trie(Transaction t)
    Begin
        CI = ∪(max CI's)
            retrieved from index[t];
        if CI = CI' / *modify*/
            Decrement support of CI and
                all its upper covers
        else
            For all chains in index[t]
            For each CI intersect CI with t
                if CI=t //modify
                    Support = support – 1;
            Endfor
        Endfor
        For all modified CI
            If support(modifiedCI) ≤
                support(minimalsuperclosedset)
            Make CI non closed;
        Endfor
    End
```

As an example consider TID 6 in the sample database as an increment. Index entries with maximal concepts in bold is shown in the Table.2. The entries corresponding to the items in the transaction A,B,D,E are underlined. The concepts which are struck off on entries corresponding to B, D, E are pruned. The new concept lattice after the increment ABDE is shown in fig.3. New concepts #2 , #6, #7, #10 are shown encircled. The generator concepts for the above new concepts are #5, #9, #11, #12 respectively and are underlined. The concepts #1 and #4 are modified. They are marked bold. The uppermost concept gets modified by default. Similarly, all items get indexed to the bottom element of the lattice ABCDE though it does not represent a transaction.

Table 2. Index of CI's for Items

Items	Index Chain
A	AC,ACD,ABCE,ABCDE
B	BE, BCE,~~ABCE,ABCDE~~
C	C,AC,ACD,BCE,ABCE,ABCDE
D	~~ACD, ABCDE~~
E	~~BE,BCE,ABCE,ABCDE~~

Deletion in the concept lattice Lc is simpler. If a concept has the same intent as the concept to be deleted (CI') then the CI and all its upper covers get modified. Their support is decremented. Otherwise the indexed entries in CI' are retrieved and each CI is examined to determine whether it gets modified. Pruning is similar as in the case of increments. After the modification phase the modified CI's are compared with their minimal closed supersets and if the support becomes less than that of the minimal closed supersets, they become non closed.

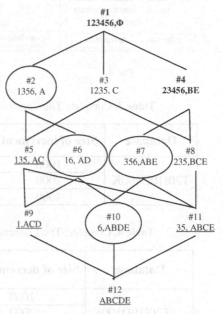

Fig. 3. Galois Concept Lattice for D'

5 Experimental Results

Synthetic databases used in the experiments were generated using the IBM dataset generator .The experiments were conducted on a 1.8 Ghz Pentium IV processor with 1GB RAM running on Windows XP. We compare the performance of our approach with Galicia-M[11]. Statistics for the maximum size of the memory resident database for T25I10D10K, T20I10D100K, T10I10D200K are indicated in fig.4. The number of items was set to 1000. Results indicate that the size of the memory resident database is orders of magnitude smaller than that of Galicia-M. Time for computing the closed itemsets on increments of sizes 1000, 2000,3000, 4000 are indicated for T10I8D100K and T20I10D100K in fig.5 and fig.6 respectively. T20I10D100K is slightly dense than T10I8D100K. In both the cases the time taken by the indexed trie approach is lesser than Galicia-M indicating the efficacy of the indexing and pruning mechanism. The execution times in comparison with increment sizes indicate that the approach is scalable.

Tables 3 and 4 gives a comparison of time required for a complete trie construction Vs a selective update for both increments and deletions on T20I10D100K. Table 5 shows the evolution of the closed itemsets and frequent closed itemsets before and after augmenting the database. The figures corresponding to T10I8D100K, T20I10D100K and its respective execution times in fig.5 and 6 clearly favor the Indexed Trie approach proposed in this study. The tables clearly indicate the efficiency of selective update in the case of both insertions and deletions. For mining the trie based on a minimum support threshold, a BFS traversal was adopted.

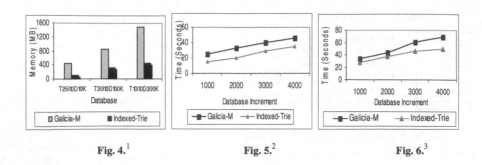

Fig. 4.[1]   Fig. 5.[2]   Fig. 6.[3]

Table 3. Complete Trie Construction Vs Selective Update (Insertions)

Database	Size of increment	Time (Seconds)	
		Complete Trie	Selective Update
T20I10D100K	1000	158.2	26.2
	2000	238	38.23
	4000	316	50.13

Table 4. Complete Trie Construction Vs Selective Update (Deletions)

Database	Size of decrement	Time (Seconds)	
		Complete Trie	Selective Update
T20I10D100K	1000	101.12	18.26
	2000	198.3	31.13
	4000	289.36	47.23

Table 5. Comparison of the effective sizes of the CI and FCI families before and after the increment

Database	Original Database		Augmented Database (+1000)	
	CI	FCI (Supp:0.05%)	CI	FCI (Supp: 0.05%)
T25I10D10K	4,20,144	22,326	4,23,124	23,189
T10I8D100K	98,23,101	17,216	98,32,091	17,612
T20I10D100K	1,28,68,438	3,13,406	1,29,12,413	3,15,103
T10I10D200K	1,86,23,105	11,01,402	1,87,15,110	11,01,703

6 Conclusion

The proposed framework makes use of an efficient selective update policy and pruning using an indexed trie structure. This proves effective in avoiding duplicates. The

---

[1] Indexed trie Vs Galicia-M Memory consumption.
[2] Execution Times for different increment sizes on T10I8D100K.
[3] Execution Times for different increment sizes on T20I10D100K.

necessary logistics for deletion has also been incorporated. The approach clearly wins over the Galicia-M in terms of reduced time and storage.The approach can be adopted for applications where large dynamic databases need to be mined.

References

[1] Chi, Y., Wang, H., Yu, P.S., Muntz., R.R.: Moment Maintaining Closed Frequent Itemsets Over a Stream Sliding Window. In: Perner, P. (ed.) ICDM 2004. LNCS (LNAI), vol. 3275, pp. 59–66. Springer, Heidelberg (2004)

[2] Davey, Priestly.: Introduction to Lattices and Order, 2nd edn. Cambridge University Press, Cambridge (2002)

[3] Ganter, B., Wille., R.: Formal Concept Analysis Mathematical Foundations. Springer, Heidelberg (1999)

[4] Lucchese, C., Orlando, S., Perego., R.: Fast and Memory Efficient Mining of Frequent Closed Itemsets. IEEE Transaction on Knowledge and Data Engineering 18, 21–36 (2005)

[5] Pasquier, N., Bastide, Y., Taouil, R., Lakhal., L.: Discovering Frequent Closed Itemsets For Association Rules. In: Beeri, C., Bruneman, P. (eds.) ICDT 1999. LNCS, vol. 1540, pp. 398–416. Springer, Heidelberg (1998)

[6] Pasquier, N., Bastide, Y., Taouil, R., Lakhal., L.: Closed Set Based Discovery Of Small Covers For Association Rules. In: BDA 1999, pp. 361–381 (1999)

[7] Pasquier, N., Bastide, Y., Taouil, R., Lakhal, L.: Efficient Mining Of Association Rules Using Closed Itemset Lattices. Information Systems 24(1), 25–46 (1999)

[8] Pasquier, N., Bastide, Y., Taouil, R., Stumme, G., Lakhal., L.: Generating A Condensed Representation For Association Rules. Journal of Intelligent Information Systems 24(1), 29–60 (2005)

[9] Pei, J., Han, J., Mao, R.: Closet: An Efficient Algorithm For Mining Frequent Closed Itemsets. In: ACM SIGMOD Workshop on Research Issues in Data Mining and Knowledge Discovery, pp. 21–30 (2000)

[10] Pei, J., Han, J., Wang, J.: Closet+. Searching For Best Strategies For Mining Frequent Closed Itemsets. In: 9th ACM SIGKDD International Conference on Knowledge Discovery and Data Mining (2003)

[11] Valtchev, P., Missaoui, R., Godin, R., Meridji, M.: Generating Frequent Itemsets Incrementally: Two Novel Approaches Based On Galois Lattice Theory. Journal of experimental and theoretical AI 14(2-3), 115–142 (2002)

[12] Zaki, M.J., Hsaio, C.: Efficient Algorithms For Mining Closed Itemsets And Their Lattice Structure. IEEE Transactions on Knowledge and Data Engineering 17(4), 462–478 (2005)

Mining Frequent Patterns in an Arbitrary Sliding Window over Data Streams

Guohui Li, Hui Chen*, Bing Yang, and Gang Chen

School of Computer Science and Technology,
Huazhong University of Science and Technology, Hubei 430074, P.R.C.
Tel.:+86-27-87543104
chen_hui@smail.hust.edu.cn

Abstract. This paper proposes a method for mining the frequent patterns in an arbitrary sliding window of data streams. As streams flow, the contents of which are captured with *SWP-tree* by scanning the stream only once, and the obsolete and infrequent patterns are deleted by periodically pruning the tree. To differentiate the patterns of recently generated transactions from those of historic transactions, a time decaying model is also applied. The experimental results show that the proposed method is efficient and scalable, and it is superior to other analogous algorithms.

1 Introduction

Recently, data streams[1] are found in a variety type of applications including network monitoring, financial monitoring, sensor networks, web logging and large retail store transactions. In the past few years, many methods have been proposed to mine all kinds of knowledge of data streams, among of which, frequent pattern mining [1,2,3,4] attracts extensive interests. C. Giannella etc. [2] proposed an *FP-stream* approach to mine the frequent patterns in data streams at multiple time granularities. In which a titled-time windows technology is used to capture the recent changes of streams at a fine granularity, but long term changes at a coarse granularity. J. H. Chang and W. S. Lee [3] presented an *esDec* method to adaptively mine the recent frequent itemsets over online data streams. in this method, in order to differentiate the information of the recently generated transactions from those of the old transactions, a decaying mechanism [2,3] is used. Furthermore, C. K.-S. Leung and Q. I. Khan [4] proposed a tree structure named *DStree* to exactly mine the frequent sets in a sliding window.

In this paper, an *MSW* method is proposed to mine the frequent patterns in an arbitrary sliding window of online data streams. As data stream flows, a new transaction arrives and enters into the sliding window, and the contents of which is incrementally updated into a compact prefix-tree named *SWP-tree*. At the same time, the oldest transaction in the sliding window is removed. To save the space of the tree, an operation of pruning *SWP-tree* is periodically performed to delete the obsolete information of those historic transactions. Furthermore, a

---

* Corresponding author.

J.R. Haritsa, R. Kotagiri, and V. Pudi (Eds.): DASFAA 2008, LNCS 4947, pp. 496–503, 2008.

time decaying model is used to differentiate the information of recently generated transactions from that of the historic transactions. At any time, an *SWP-tree* always keeps the up-to-date contents of a data stream, and a query for recent frequent patterns could be responded immediately after the request is submitted. Compared with other analogous methods, the *SWP-tree* is more efficient to capture the contents of recent stream data. The *MSW* method is more scalable to precess all kinds of data streams, and it is more compatible to mine the frequent patterns in a sliding window of arbitrary size.

2   Preliminaries

2.1   Definitions

Let $\mathcal{A}=\{a_1, a_2, \ldots, a_m\}$ be a set of data items which are sorted according to a total order $\prec$ (where $\prec$ is one kind of predefined orders), $DS=\{T_1, T_2, \ldots, T_N, \ldots\}$ be an unbounded data stream, where T_i denotes the transaction generated at i^{th} turn in DS. A transaction is a subset of $\mathcal{A}$ and each has an unique identifier *tid*. For any a pattern $\mathcal{P}(\mathcal{P}\subseteq\mathcal{A})$, the *frequency* of $\mathcal{P}$ is the number of the transactions containing $\mathcal{P}$ in DS and denoted by $freq(\mathcal{P})$. A sliding window SW contains the latest N transactions in DS.

Definition 1. For any a transaction $T=\{a_1, a_2, \ldots, a_k\}$, if the items in which are rearranged according to the total order $\prec$, we can get another data sequence $T'=\{a'_1, a'_2, \ldots, a'_k\}$, then T' is called the **projection** of T.

Definition 2. Given the *user support threshold* $\theta(0<\theta<1)$ and *maximum support error* $\epsilon(0<\epsilon<\theta)$, the patterns in a data stream of size N can be divided into three categories: for any a pattern $\mathcal{P}$, (1) $\mathcal{P}$ is **frequent** if $freq(\mathcal{P})\geq\theta N$, (2) $\mathcal{P}$ is **subfrequent** if $\epsilon N\leq freq(\mathcal{P})<\theta N$, and (3) $\mathcal{P}$ is **infrequent** if $freq(\mathcal{P})<\epsilon N$.

Although we are only interested in *frequent* patterns, we have to maintain *subfrequent* patterns since they may become *frequent* later. Besides, we want to discard *infrequent* patterns since the number of *infrequent* patterns is really large, and the error will be no more than ϵ because of the loss of support from *infrequent* patterns[2]. So in mining frequent patterns over an online data streams, it is usual to choose $(\theta-\epsilon)$ as the *minimum support threshold* for no false negative[1].

2.2   Time Decaying Model

If let **decaying factor** $f(0<f<1)$ denote the decaying rate of a frequency for a time-unit, the decayed frequency $freq_d(\mathcal{P}, T_1)$ of pattern $\mathcal{P}$ be r when the first transaction T_1 arrives, then when the second transaction T_2 arrives, the decayed frequency count $freq_d(\mathcal{P}, T_2) = freq_d(\mathcal{P}, T_1) \times f + r$ since the frequency should be decayed by a rate f. Analogically, the decayed frequency $freq_d(\mathcal{P}, T_i)$ when transaction T_i arrives can be gotten by following formula:

$$freq_d(\mathcal{P}, T_i) = \begin{cases} r & if\ i = 1 \\ freq_d(\mathcal{P}, T_{i-1})\times f + r & if\ i \geqslant 2 \end{cases}, r = \begin{cases} 1 & if\ \mathcal{P} \subseteq T_i \\ 0 & otherwise \end{cases}$$

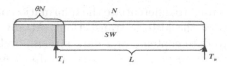

Fig. 1. The critical circumstance when mining a sliding window of size N

To evaluate the accuracy of our method for mining an arbitrary sliding window, a metric, **coverage**[1], is introduced. Let $\mathcal{P_A}$ denote a set of obtained frequent patterns of a data stream, $\mathcal{P_B}$ denote the actual frequent sets in the same stream, the **coverage** of the mining results can be defined as $coverage = \frac{\mathcal{P_A} \cap \mathcal{P_B}}{\mathcal{P_B}}$. In order to guarantee the *coverage* to be 100 percent, all real frequent patterns must be selected with no false negative when mining a data stream.

Now let us consider the critical situation showed in fig.1, the size of the sliding window SW is N, T_N is the current transaction of data stream, T_i is one of the transactions containing $\mathcal{P}$, and the θN transactions which contain $\mathcal{P}$ are highlighted with grey background. It is obvious that the real frequency of $\mathcal{P}$ is θN, so $\mathcal{P}$ is *frequent* and should be selected when mining SW.

However, after applying time decaying mode TDM, the frequency of $\mathcal{P}$ is decayed as time goes by. If the decayed frequency of $\mathcal{P}$ at the time T_N arrives is denoted as $freq_d(\mathcal{P}, T_N)$, then $freq_d(\mathcal{P}, T_N) = f^{N-\theta N} + \ldots + f^{L-1} + \ldots + f^{N-1}$, where f^{L-1} is the affect of transaction T_i on the decayed frequency of $\mathcal{P}$ at the time T_N arrives, T_i is one of the transactions which contain pattern $\mathcal{P}$, L is the size of interval between T_N and T_i, f is the decaying factor.

As analyzed in last subsection, when mining an online data stream, it is feasible to choose $(\theta - \epsilon)$ as the *minimum support threshold* when mining a data strema. So to achieve the *coverage* to be 100%, the decayed frequency of $\mathcal{P}$ when T_N arrives should satisfy the following condition.

$$freq_d(\mathcal{P}, T_N) = f^{N-\theta N} + \ldots + f^{L-1} + \ldots + f^{N-1} \geqslant (\theta - \epsilon)N$$

Based on the above formula, the value range of f can be got as follow.

$$1 > f \geqslant \sqrt[(2N-\theta N-1)]{[(\theta - \epsilon)/\theta]^2}$$

Obviously, the minimum of f is a function on variables: *user support threshold* θ, *maximum support error* ϵ and the size N of sliding window SW. If given θ and ϵ, we could choose an appropriate decaying factoe f for any an sliding window of size N. And if the value of f is bigger than $\sqrt[(2N-\theta N-1)]{[(\theta - \epsilon)/\theta]^2}$ but less than 1, the *coverage* of the mining result will keep 100%.

3   Mining Arbitrary Sliding Windows

3.1   Sliding Window Pattern Tree

An **SWP-tree** is a tree structure based on *FP-tree*[5]. Similarly, an *SWP-tree* also contains a prefix-tree and a frequent-item-header table *f_list*. However, an

FP-tree is designed for mining frequent patterns in a static database, so several improvements are essential to adapt for mining an online data stream. (1) Firstly, except for *item-name, count, node-link, parent-link, sibling-link* and *first-child-link*, another data field named *tid* is added in each node of an *SWP-tree*, and it registers the identifier of the latest transaction which contains the pattern corresponding to the node. (2) Secondly, all nodes of a path in an *SWP-tree* are sorted in the total order ≺ instead of in their frequency descending order. (3) Thirdly, all entries of the *f_list* are sorted in the total order ≺ too.

3.2 Incrementally Updating SWP-Tree

When a new transaction $T = \{a_1, a_2, \ldots, a_k\}$ arrives, the detailed process of incrementally updating the contents of T into an *SWP-tree* can be described as follow. First, get the projection of T and denote it as $T' = \{a'_1, a'_2, \ldots, a'_k\}$. Second, define a pointer *curNode* which points the position in the pattern tree to insert the new item, and initial it to point the *root* of the *SWP-tree*. Third, get the first item from T' and name it as *a*. Four, Search the children nodes of the node *curNode* pointing to find the node which carries the same *item-name* with *a*. If found such a node, suppose it named as *tNode*, then update its frequency count and *tid* field, and following, and next, set *curNode* to point *tNode*. If Found not, then create a new node with its *count* initialized to 1, its *tid* initialized to the identifier of T, its *parent-link* linked to *curNode*, its *sibling-link* linked to the first child of *curNode* and its *node-link* linked to the nodes with the same item-name via the node-link structure, and next, set *curNode* to point the new node just created. Five, if T' is not null, then set T' to be the rest of itself except *a*, and repeated the steps from third to fifth till T' is null.

If let $|Update(a)|$ denote the cost of incrementally updating the item *a* into an *SWP-tree*, the cost of incrementally updating a transaction T into *SWP-tree* could be denoted as $O(|Update(a)| \times |Length(T)|)$, where $|Length(T)|$ is the number of the items in T.

Property 1. *In each branch of an SWP-tree, the occurrence count and timestamp of an ancestor node are no less than those of its descendant nodes.*

Rationale. An *SWP-tree* is a tree based on the *FP-tree* [5]. If two or more transactions share a common prefix, to save the space of the pattern tree, the shared parts are merged using one prefix structure as long as the *count* registered properly. Moreover, the items closer to the *root* of a pattern tree have chance to share more prefix strings. Suppose *tNode* be a tree node in the pattern tree, if a transaction contains *tNode*, then it must contain all the ancestor nodes of *tNode* too. Therefore, in each branch of an *SWP-tree* , the occurrence count and timestamp (the value of *tid* field) of an ancestor node will be no less than those of its descendant nodes.

3.3 Pruning SWP-Tree

When incrementally updating *SWP-tree*, many *infrequent* patterns are updated into the pattern tree. Additionally, many patterns kept in *SWP-tree* become obsolete as time goes by. To save the space of the *SWP-tree*, those *infrequent* itemsets should be deleted too when pruning the tree. However, it is costly to keep on pruning the pattern tree immediately after one transaction is processed. In this paper, every $n = \lceil 1/\epsilon \rceil$ transactions, the operation of pruning tree is performed to delete the nodes corresponding to obsolete or *infrequent* patterns.

Suppose the size of the sliding window be N, and the operation of pruning pattern tree be performed just after transaction T_N has been processed. Then for any a node *tNode* in the *SWP-tree*:

(1) If $|T_N.tid - tNode.tid| \geqslant N$, then *tNode* is a node corresponding to the pattern of obsolete transactions. So *tNode* and all its descendent nodes can be deleted based on property 1. When a node which corresponds an obsolete pattern is pruned, only the information of those transactions which have been removed from the sliding window is deleted. So no false error is brought to the current mining result. However, when *tNode* is pruned, if $tNode.count \times f^{|T_N.tid-tNode.tid|} \geqslant \epsilon N$, then the affection of historic transactions on current mining results is still big, and it should also be eliminated by updating all the ancestor nodes of *tNode* according to the following formula.

$$node_i.count = node_i.count - tNode.count \times f^{|node_i.tid-tNode.tid|}$$

where $node_i$ is one of the ancestor nodes of *tNode*.

(2) Let the decayed frequency of *tNode* be ξ, $\xi = tNode.count \times f^{|T_N.tid-tNode.tid|}$. If $|T_N.tid - tNode.tid| < N$ and ξ satisfies the condition in proposition 1, then *tNode* is thought to be *infrequent*, where $T_N.tid$ is the identifier of T_N, *tNode.tid* the value kept in the *tid* field of *tNode*.

Proposition 1. *For any a pattern maintained in an SWP-tree, if its decayed frequency is ξ, and $\xi \leqslant \frac{1}{1-f}\left[f^{N-\theta N} - \frac{f^{N-1}}{f-N(\theta-\epsilon)(f-f^2)}\right]$, then the pattern could be pruned without bringing false error into the mining results.*

Proof. As showed in figure 1(a), suppose at the time when transaction T_N arrives, there are $x(x < \theta N)$ transactions which contain pattern $\mathcal{P}$ in the sliding window SW of size N. And in the next $\theta N - x$ time unit, another $\theta N - x$ transactions containing $\mathcal{P}$ arrive (see in fig.1(b)). Obviously, at this time, $\mathcal{P}$ is *frequent*, so $\mathcal{P}$ should be selected when mining the sliding window SW.

Suppose the decayed frequency of $\mathcal{P}$ at the time when transaction T_N (see in fig.1(a)) arrives be $\xi = f^{N-\theta N+x-1} + f^{N-\theta N+x-2} + \ldots + f^{N-\theta N}$. If ξ is so small that pattern $\mathcal{P}$ is *infrequent* and deleted when pruning the *SWP-tree* before other

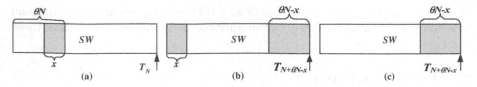

Fig. 2. The three different states of sliding window SW

transaction containing $\mathcal{P}$ arrives, then the state of the sliding window SW at the time when transaction $T_{N+\theta N-x}$ arrives is showed in fig.1(c).

As mentioned above, pattern $\mathcal{P}$ is *frequent* if without pruning the pattern tree. To assure that the *coverage* will remain to be 100%, pattern $\mathcal{P}$ must be selected when mining SW even though some information of $\mathcal{P}$ has been deleted. Thus, the decayed frequency of $\mathcal{P}$ under the situation in figure 3(c) should still satisfy the following condition.

$$f^{\theta N-x-1} + f^{\theta N-x-2} + \ldots + f^0 \geqslant (\theta - \epsilon)N \tag{1}$$

Additionally, based on the analysis in subsection 2.2, the decaying factor f should satisfy the following condition.

$$f^{N-1} + f^{N-2} + \ldots + f^{N-\theta N} \geqslant (\theta - \epsilon)N \tag{2}$$

Based on the above two formulas, it is easy to get the following condition.

$$\xi \leqslant \frac{1}{1-f}\left[f^{N-\theta N} - \frac{f^{N-1}}{f - N(\theta - \epsilon)(f - f^2)} \right] \tag{3}$$

Obviously, event though the *infrequent* patterns are deleted when pruning $SWP\text{-}tree$, the *coverage* of the mining result could remain 100%.   □

Based on the above analysis, all *infrequent* and obsolete nodes can be deleted without bringing any false negative. In our method, the operation of pruning an $SWP\text{-}tree$ is performed every $n = \lceil 1/\epsilon \rceil$ transactions, and the detailed processes can be described as follow. First, fetch an entry from the f_list starting from the head, and name it as e. Second, visit the first node in the $SWP\text{-}tree$ via the $node\text{-}link$ of e, and denote it by $tNode$. Third, if $tNode$ is obsolete or *infrequent*, then delete it, or else pass to the next node via its $node\text{-}link$. Repeat the third step till the $node\text{-}link$ is null. Fourth, set e to be the next entry of itself in f_list, and repeat the steps from second to fourth till all entries in f_list are processed.

3.4   Frequent Pattern Selection

When selecting frequent patterns from an $SWP\text{-}tree$, each data item kept in the tree should be first checked whether it is frequent. If it is, then select all patterns containing it. Or else, pass to the next item. The detailed processes can be described as follow: Firstly, get an entry from f_list starting from the bottom and name it as e. Secondly, calculate the decayed frequency $freq_d(e)$ of e at present, and $freq_d(e) = \sum_{j=0}^{s} node_j.count \times f^{|node_j.tid - T.tid|}$, where s denotes the size of the nodes which carry the same *item-name* with e, $node_j$ is one of the nodes, and T is the latest transaction has be processed before frequent pattern selection starts. Thirdly, if $freq_d(e) \geqslant (\theta - \epsilon)N$, then select all patterns containing e from the $SWP\text{-}tree$ using the $FP\text{-}growth$ method. Or else, set e to be the next entry of itself in f_list, and repeated the steps from second to third till all entries in f_list have be processed.

4   Simulation

All experiments were performed on a PC with a CPU PIV 1.8GHz and 768MB memory. All program were implemented in VC++. The test datasets were generated by IBM synthetic data generator with the total number of items being 10K. In the experiments, the *user support threshold* θ was set to be 0.001 and the *maximum support error* ϵ was fixed to be $0.1 \times \theta$, and we compared our method with *FP-stream*[2], *estDec*[3] and *DStree* [4] methods.

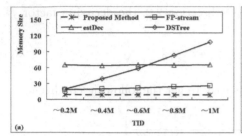 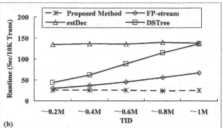

Fig. 3. The performance compare of the four methods

Firstly, we evaluate the performance of the proposed method and the three comparative methods. In those simulations, data set T10I4D1000K is experimented and firstly divided into five intervals each consists of 0.2M transactions. We first compare the maximum memory usages (MMU) of the four methods during each interval and the experiment results are showed in figure 4(a). The MMUs of the three methods except *DStree* remain same during the five intervals, and among of which, *estDec* method has the biggest MMU, *FP-stream* is the next and the proposed method is the smallest. When the *DStree* method running during the data stream, the MMU of which linearly increases.

Secondly, we also compare the average processing time (APT) of the four methods. As showed in figure 4(b), during the five intervals, the APTs of the proposed method and *estDec* method are uninfluenced. However, the APTs of the other two methods are increasing during the stream, and the increasing speed of *DStree* is much faster than that of the *estDec* method.

Thirdly, to evaluate the scalability of the proposed method on different data streams, we compare the four methods by conducting experiments on data sets T15I6D200K, T15I10D200K, T20I10D200K, T20I15D200K and T25I20D200K. In those simulations, the average processing time (APT) is measured. As showed in figure 5(a), with the average transaction length (T) and the average maximum pattern length (I) increasing, the APTs of four methods increase, but compare the increasing speeds, *estDec* is the fastest, the followings are *FP-stream*and *DStree* in turn, and the proposed method is the last.

Fourthly, we also compare the scalability of the four methods on mining a big sliding window, In the experiments, data set T10I4D1000K is used, and the experimental results are showed in figure 5(b). When the sliding window

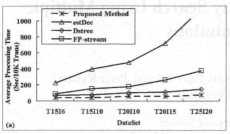 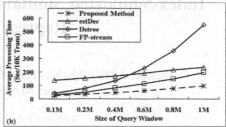

Fig. 4. (a) The scalability compare of the four methods

is small, the average processing times (APTs) of *DStree*, *FP-stream* and the proposed method are near, and they are much smaller than that of *estDec*. With the sliding window enlarging, the APTs of all the four methods increase, but that of *DStree* increased at the rapidest rate and it exceeds that of the *estDesc* when N is about 0.5M. Compared with *DStree*, the increasing speed of APT in *FP-stream* is comparatively slow. The change of APTs in *estDec* and proposed method is the slowest and similar.

5   Conclusion

In this paper, an *SWP-tree* was designed to dynamically maintain the up-to-date contents of an online data stream by scanning it only once, and an *MSW* method was to proposed to mine the frequent patterns in an sliding window of arbitrary size. This method could answer a request with no false negative. Extensive experimental results show that the proposed method is efficient and superior to other existing algorithms.

References

1. Yu, J.X., Chong, Z., Lu, H., Zhang, Z., Zhou, A.: A false negative approach to mining frequent itemsets from high speed transactional data streams. Information Sciences 176, 1986–2015 (2006)
2. Giannella, C., Han, J., Pei, J., Yan, X., Yu, P.S.: Mining frequent patterns in data streams at multiple time granularities. In: Data Mining: Next Generation Challenges and Future Directions, pp. 191–212. AAAI/MIT Press (2004)
3. Chang, J.H., Lee, W.S.: Finding recent frequent itemsets adaptively over online data streams. In: ACM SIGKDD 2003, August 2003, pp. 487–492. ACM Press, New York (2003)
4. Leung, C.K.-S., Khan, Q.I.: DStree: A tree structure for the mining of frequent sets from data streams. In: ICDM 2006, December 2006, pp. 928–932. IEEE press, Los Alamitos (2006)
5. Han, J., Pei, J., Yin, Y.: Mining frequent patterns without candidate generation. In: ACM SIGMOD 2000, May 2000, pp. 1–12. ACM press, New York (2000)

Index-Supported Similarity Search Using Multiple Representations

Johannes Aßfalg, Michael Kats, Hans-Peter Kriegel, Peter Kunath,
and Alexey Pryakhin

Institute for Computer Science
Ludwig-Maximilians-University of Munich
Oettingenstr. 67, 80538 Munich, Germany
{assfalg,kriegel,kunath,pryakhin}@dbs.ifi.lmu.de

Abstract. Similarity search in complex databases is of utmost interest in a wide range of application domains. Often, complex objects are described by several representations. The combination of these different representations usually contains more information compared to only one representation. In our work, we introduce the use of an index structure in combination with a negotiation-theory-based approach for deriving a suitable subset of representations for a given query object. This most promising subset of representations is determined in an unsupervised way at query time. We experimentally show how this approach significantly increases the efficiency of the query processing step. At the same time the effectiveness, i.e. the quality of the search results, is equal or even higher compared to standard combination methods.

1 Introduction

Similarity search is an important issue in a broad range of applications like the retrieval of multimedia, biological, spatial, and CAD objects. In order to handle complex domain-specific objects, a feature extraction is typically applied. The feature extraction aims at transforming characteristic object properties into feature values. The extracted feature-values can be interpreted as a vector in a multidimensional vector space called feature space. The most important characteristic of a meaningful feature space is that whenever two of the objects are similar, the associated feature vectors have a small distance according to an appropriate distance function (e.g., the Euclidean distance). Thus, similarity search on complex objects can be naturally translated into a k-nearest neighbor (kNN) query in a feature space. Objects are usually described by several feature spaces in order to capture various object properties. Thus, one of the most promising approaches for effective similarity search in databases is to exploit the properties of multiple feature spaces or representations. Though the effectiveness can be improved by using multiple representations, the efficiency of the multi-represented similarity search should also be addressed. For answering a kNN query, we have to consider all available representations. This can be accomplished in two ways. We can either perform a kNN query on all representations independently and combine the results, or we can combine all feature spaces into a single feature space and perform a kNN query on this combined feature space. Each of these two approaches has its drawbacks. The first approach

J.R. Haritsa, R. Kotagiri, and V. Pudi (Eds.): DASFAA 2008, LNCS 4947, pp. 504–511, 2008.

yields potentially different kNN ranking results for the different representations and it is not obvious how to derive a combined answer. The seconde approach suffers from the well-known "curse of dimensionality". In this paper, we propose a novel approach for efficient, multi-represented similarity search where each representation uses its own index structure. In the first step, our approach performs a pre-selection in order to reduce the number of available representations to a small subset S of the most accurate representations. This reduction is based on the coalitional game theory. Furthermore, our approach allows to calculate such a most promising subset S dynamically (i.e., the subset S is computed depending on the given query object). In contrast to existing approaches (e.g., entropy-based methods [1]) which apply an effective but supervised technique for similarity search with multiple representations, we propose an unsupervised approach. Furthermore we outline an algorithm for answering kNN queries using separate index structures for each representations. Instead of using similarity distances, we follow the idea of negotiation game theory and apply self confidence and so-called payoff values in order to rank multi-represented objects.

2 Related Work

Similarity search based on multiple representations has attracted considerable attention in several research communities. However, to the best of our knowledge, no existing technique dynamically calculates a suitable coalition of representations and supports efficient multi-represented kNN query processing in an unsupervised way. The existing approaches can be grouped into two categories: indexing of multi-represented objects and combining several similarity measures corresponding to different representations.

In [2], the M^2-tree is proposed that combines information from multiple metric spaces within a single index structure. The main drawback of the M^2-tree is that it combines features spaces statically, i.e. independently of the current query object. Furthermore, the combination function has to be known beforehand. In contrast to that, the approach of [3,4] derives a linear combination of metrics dynamically, i.e. based on a given query. However, all available representations are considered, in contrast to our approach that dynamically selects a small subset of relevant representations.

An overview of combining approaches in information retrieval can be found in [5]. According to [5], the most common way to combine representations is the use of the weighted sum of distances in each representation. To find proper weights for each representation, several approaches were proposed that rely on user feedback. Further approaches to approximate weights employing user feedback are described in [6,7]. In comparison to our approach, these methods employ global weights and do not use dynamic and unsupervised adjusting.

The authors of [8] introduce a technique based on the entropy impurity measure. In comparison to our method, the proposed technique requires a set of labeled objects. An unsupervised technique for the weighted combination of multiple representations was proposed in [9], but this method does not consider efficient data access using index structures. It is furthermore only applicable in combination with summarization which is not a necessary element of general multi-represented similarity search.

3   Adapting Coalitional Game Theory for Similarity Search

3.1   Preliminaries

Comparability of Feature Spaces. Usually, the similarity distance values of different representations do not have a common scale. This problem is called the comparability problem. To overcome this problem, normalization methods are applied. We use the most common Min-Max normalization which calculates the maximum (max) and the minimum (min) of the original distance values. Afterwards, a distance value d is mapped to the normalized distance d_n where $d_n = (d - min)/(max - min)$.

Coalitional Games. We consider the problem of combining similarity information of different representations as a game theory problem, in particular as a so called coalitional game. For a detailed introduction, we refer the reader to [10].

A game is a tuple (N, V), where $N = \{r_1 ... r_n\}$ is the set of players participating in the game. In our approach these players correspond to the available representations, each one trying to suggest its own similarity distance as the best one. V is a function which assigns the so called payoff or gain value to a subset $U \subseteq N$. U is called a coalition.

Each player can choose from a predefined set of strategies, each yielding a certain payoff. In our approach each representation can choose between n possible moves: either not to cooperate or to cooperate with one of the $n - 1$ other representations. A representation is more likely to cooperate with another one, if they are similar to each other. A representation is more likely not to cooperate with another representation if either no similar representation is available for a given query or if the representation is very confident of its own similarity distance (cf. Section 3.2).

A game can be described by a $n \times n$ payoff matrix M listing the payoff values for each possible move. The diagonal holds the gain values for the decision not to cooperate while the entry in the i-th row and the j-th column corresponds to the gain for player i to cooperate with player j. Section 3.3 describes how to use M to determine the winning coalition, i.e. to determine a subset of representations that is used to calculate the similarity of database objects for a given query object. The overall idea of our negotiation-game-theory-based approach (NGT) is depicted in Figure 1(a).

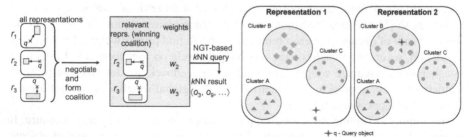

(a) Overview of our negotiation-based approach for efficient similarity search.

(b) Estimating the suitability of different representations for a given query q based on close-by clusters in the feature spaces.

Fig. 1. Overview and general idea

3.2 Calculating Gains for Coalitions of Representations

This section describes the calculation of the gain values for the different strategies. Given a query, several influences have to be taken into account.

The first influence is the position of the query object in the different feature spaces. Intuitively, if the feature vector corresponding to the query object is far away from a cluster in a feature space, this might indicate that the current representation considers the query as an outlier. Assuming there are indeed similar objects in the database this means the current representation is less suitable for describing and answering the current query. This idea is illustrated in Figure 1(b), where the first representation might be less useful for the given query q. In order to obtain clustering information we exploit the clustering properties of tree-like index structures. We propose to use an X-tree [11] because it is a common index structure for high-dimensional data and efficiently supports kNN queries. As the directory of a tree-like index structure is a good approximation of the underlying data distribution, for representation i we calculate the distance d_i between q and the nearest cluster as the MINDIST [12] between q and a directory node.

The second influence on the negotiation behavior of different representations is their similarity to each other. More similar representations should more willingly cooperate with each other. As we normalized all distances as described above, we are able to use the next-cluster values d_i as described before to compare different representations. Let in the following μ be the mean value and σ be the standard deviation of the distances d_i. We define the gain for representation r_i not to cooperate as

$$M_{i,i} = (1 - d_i^2) \cdot (1 - |d_i - \mu|^2)$$

The first factor of the product reflects the quality of representation r_i for a given query as described above. The second factor of the formula compares the quality of r_i to the average quality of all representations. Thus, a representation has the largest motivation not to cooperate with another representation if its next-cluster distance is small and at the same time similar to the average next-cluster distance.

The remaining matrix entries $M_{i,j}$, $i \neq j$, indicate the payoff for representation r_i for cooperating with representation r_j. A cooperation with r_j can either increase or decrease the gain for r_i, i.e. $M_{i,j} = M_{i,i} \cdot changeFactor$. The change factor takes the following considerations into account: The smaller the difference between d_i and d_j compared to σ, the more likely is a cooperation. If the difference between d_i and d_j is larger than σ, r_i and r_j are not allowed to cooperate. In order to form stable coalitions we have to prevent the case where r_i decides to cooperate with r_j while at the same time r_j decides to cooperate with r_i. Only a one-directional cooperation is allowed. We favor the representation whose next-cluster distance is nearer to the average μ. This leads to the following definition for non-diagonal payoff matrix elements:

$$M_{i,j} = M_{i,i} \cdot changeFactor = M_{i,i} \cdot \begin{cases} -1 & \text{if } |d_i - \mu| < |d_j - \mu| \vee |d_i - d_j| > \sigma \\ (1 + (\sigma - |d_i - d_j|)) * \frac{1 + |d_i - \mu|}{1 + |d_j - \mu|} & \text{else} \end{cases}$$

3.3   Determining the Winning Coalition

After having calculated all payoff values we are now able to determine the winning coalition of representations, i.e. the coalition with the highest gain. The gain of a coalition is the sum of the gain values of the decisions that have led to the coalition. In order to identify the strongest coalition we determine the maximal entry of each row, as this entry indicates the best strategy of the associated representation. Then, for a certain column, we sum up all these maximal values that can be found in this column. This yields the payoff for the coalition consisting of the representations whose best gain values have been summed up.

3.4   Deriving Weights

Aside from limiting the number of representations involved in answering a query, we can use the above generated matrix to calculate weights for the remaining representations. We use these weights to perform high-quality kNN queries very efficiently as described in Section 4. Let C be the coalition. The weight w_i of the coalition member r_i is calculated as the ratio between $M_{i,i}$ and the average non-cooperating gains of all participating representations: $w_i = (M_{i,i} \cdot |C|)/(\sum_{r_j \in C} M_{j,j})$. Intuitively, $M_{i,i}$ reflects how confident representation r_i is about its own quality. So, weights larger than 1 indicate a confidence above the average.

4   Efficient kNN Query Processing on Multiple Representations

In the following, we use the above described weights to calculate a weighted linear combination of these representations. As in the previous section, we are using the principles of the negotiation game theory for combining the representations. In our experiments, we observed that using the weighted sum yields the best results for our NGT-based approach.

A ranking is performed on each of the relevant representations in order to answer a kNN query for a given query object. The following technique bases on the assumption of searching in a tree-like index structure. The kNN query algorithm uses the well-known Hjaltason-Samet ranking algorithm [13] and orders all objects in a single priority queue. The priority queue of the ranking is initialized with the roots of all representations of the best coalition. The ranking priority queue is organized in descending order w.r.t. gain-based value as described below. We propose to calculate the priority of the ranked directory pages dp similar to the representation selection in the previous section. The gain formula has to be adapted as follows: $priority = gain_{dp} * w_i$, where $gain_{dp} = (1 - MinDist(dp)^2)$. In each iteration of the algorithm, the first object is removed from the ranking queue. For each entry of a directory page, we calculate the priority according to the above formula and insert it into the ranking priority queue. In case of a data page, we process all objects within this page. Each retrieved object is added to the result priority queue, where the sum of the weighted gains in all relevant representations is used as priority value. In case an object is already in the queue, its priority value is updated. The gain is calculated as the gain for a directory page, instead of the MINDIST value we use the Euclidean distance value. The idea behind using the sum of the gain values as priority is that objects which have been retrieved in a lot of representations are ranked higher. Furthermore, we also test if the currently retrieved or updated object fulfills a stop condition. We terminate the kNN algorithm if there is no change in the first k objects of our result queue. Once this is the case, it is most likely that no further gain value has enough impact to alter the kNN result queue.

5   Experimental Evaluation

We performed our evaluation on four different datasets. Please note that each dataset is described by three or more representations because our NGT approach is only applicable if at least three representations are available. As mentioned in Section 3, all datasets are organized in an X-tree. Unless noted otherwise, we conducted 30 kNN queries with randomly chosen query objects and averaged the results.

The **NTU** dataset is based on a subset of the NTU 3D Model Benchmark [14] and consists of 549 objects in 46 classes. We extracted three representations for this dataset with an average dimensionality of 80. The **Music** dataset contains 516 songs taken from 15 different music genres. We generated 6 different feature representations per song with an average dimensionality of 500. The **Proteins** dataset consists of 2465 objects taken from the SWISS-PROT [15] protein database. We derived 18 feature representations with an average dimensionality of 20. The last dataset is the synthetic CBF **Timeseries** dataset [16] for which we calculated 9 different feature representations with an average dimensionality of 20.

5.1   Efficiency Evaluation

At first, we turned our attention to the efficiency of our proposed approach. We compared our approach with the standard combination rules SUM, PROD, MIN, and MAX. Figure 2(a) depicts the runtime results of a 5NN query for all four datasets. Because our approach executes the kNN query only for a subset of the available representations, it yields a significant speedup on all datasets. Even when only three different representations are available (*NTU*), our NGT approach achieves a speedup of about factor 3.

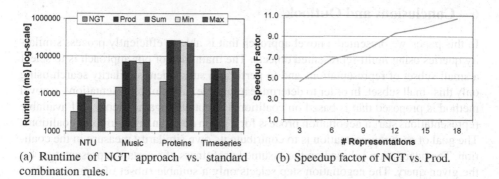

(a) Runtime of NGT approach vs. standard combination rules.

(b) Speedup factor of NGT vs. Prod.

Fig. 2. Efficiency evaluation

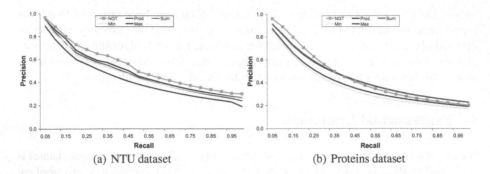

(a) NTU dataset   (b) Proteins dataset

Fig. 3. Quality of NGT approach vs. standard combination rules

For the *Proteins* dataset, we executed another experiment in order to evaluate the relationship between the runtime and the available representations (see Figure 2(b)). We started with three representations and added another three representations in every step. As already demonstrated in the first evaluation, our NGT approach shows an advantage in runtime even for three representations. This advantage increases when more representations are available up to a factor of 10.7 for 18 representations. We observed the NGT approach does not favor a particular subset of representations. Instead, the best coalition is always chosen dynamically from all available representations.

5.2 Effectiveness Evaluation

In this section, we demonstrate the effectiveness of our NGT approach. Figure 3 depicts the quality of our NGT approach and the four standard combiners for the NTU and the Proteins dataset. This figure displays precision-recall plots, which were obtained by performing kNN queries where $k = |dataset|$. For all datasets, the effectiveness of the NGT approach is either comparable or even better than the traditional combination rules. On the *NTU* dataset, our NGT technique is able to outperform all other combiners for all recall values. For the other datasets, the result quality of the NGT method is comparable to the standard combination rules.

6   Conclusions and Outlook

In this paper, we presented a novel approach that is able to efficiently process similarity queries using multi-represented objects. The main idea of our approach is to select a small subset of representations and to perform a subsequent similarity search using only this small subset. In order to determine a suitable subset of representations, a novel method is proposed that is based on coalitional negotiation game theory. All available representations start a negotiation process for a given query in order to form coalitions. The goal of each representation is to contribute its own similarity measure in the coalition. The negotiation is based on the assumed usefulness of a certain representation for the given query. The negotiation step selects only a suitable subset of representations that is used to answer the query. Furthermore we introduced an efficient kNN query algorithm which operates on the selected representations. We demonstrated the efficiency of the proposed approach on four datasets. Furthermore, our novel approach yields a comparable or even higher effectiveness on all considered datasets. As future work we plan to turn our attention to further aspects of the game theory. For example, it seems promising to use a mix of coalitional and behavioristic negotiation theory to determine suitable coalitions.

References

1. Bustos, B., Keim, D., Saupe, D., Schreck, T., Vranic, D.: Automatic selection and combination of descriptors for effective 3d similarity search. In: Proc. ICME (2004)
2. Ciaccia, P., Patella, M.: The M^2-tree: Processing complex multi-feature queries with just one index. In: DELOS Workshop: Information Seeking, Searching and Querying in Digital Libraries (2000)
3. Bustos, B., Keim, D., Schreck, T.: A pivot-based index structure for combination of feature vectors. In: ACM Symposium on Applied Computing (2005)
4. Bustos, B., Skopal, T.: Dynamic similarity search in multi-metric spaces. In: Proc. MIR (2006)
5. Croft, W.B.: Advances in Information Retrieval: Recent Research from the CIIR. Kluwer Academic Publishers, Dordrecht (2000)
6. Chua, T.S., Low, W.C., Chu, C.X.: Relevance feedback techniques for color-based image retrieval. In: Proc. MMM (1998)
7. Rui, Y., Huang, T.S., Mehrotra, S.: Content-based image retrieval with relevance feedback in mars. In: Proc. ICIP (1997)
8. Bustos, B., Keim, D.A., Saupe, D., Schreck, T., Vranic, D.V.: Using entropy impurity for improved 3d object similarity search. In: Proc. ICME (2004)
9. Kriegel, H.P., Kröger, P., Kunath, P., Pryakhin, A.: Effective similarity search in multimedia databases representations. In: Proc. MMM (2006)
10. von Neumann, J., Morgenstern, O.: Theory of games and economic behavior (2004)
11. Berchtold, S., Keim, D.A., Kriegel, H.-P.: The X-Tree: An index structure for high-dimensional data. In: Proc. VLDB (1996)
12. Roussopoulos, N., Kelley, S., Vincent, F.: Nearest neighbor queries. In: Proc. SIGMOD (1995)
13. Hjaltason, G., Samet, H.: Incremental similarity search in multimedia databases
14. Chen, D.-Y., Tian, X.-P., Shen, Y.-T., Ouhyoung, M.: On visual similarity based 3d model retrieval. EUROGRAPHICS (2003)
15. Boeckmann, B., Bairoch, A., Apweiler, R., Blatter, M.-C., Estreicher, A., Gasteiger, E., Martin, M.J., Michoud, K., O'Donovan, C., Phan, I., Pilbout, S., Schneider, M.: The SWISS-PROT Protein Knowledgebase and its Supplement TrEMBL in 2003. Nucleic Acid Research (2003)
16. Saito, N.: Local feature extraction and its application using a library of bases. PhD thesis, Yale University, New Haven, Connecticut (1994)

Distance Based Feature Selection for Clustering Microarray Data

Manoranjan Dash and Vivekanand Gopalkrishnan

Nanyang Technological University,
50 Nanyang Avenue, Singapore
{asmdash,asvivek}@ntu.edu.sg

Abstract. In microarray data, clustering is the fundamental task for separating genes into biologically functional groups or for classifying tissues and phenotypes. Recently, with innovative gene expression microarray data technologies, thousands of expression levels of genes (features) can be measured simultaneously in a single experiment. The large number of genes with a lot of noise causes high complexity for cluster analysis. This challenge has raised the demand for feature selection – an effective dimensionality reduction technique that removes noisy features. In this paper we propose a novel filter method for feature selection. The suggested method, called ClosestFS, is based on a distance measure. For each feature, the distance is evaluated by computing its impact on the histogram for the whole data. Our experimental results show that the quality of clustering results (evaluated by several widely used measures) of K-means algorithm using ClosestFS as the pre-processing step is significantly better than that of the pure K-means.

Keywords: Feature Selection, Clustering, Distance Function, Microarray Data.

1 Introduction

In gene expression analysis, there is an enormous amount of data with informational riches produced by microarray technology. Hence data mining techniques have been applied to extract useful information in this domain. Gene clustering is used to find groups of genes that are highly co-regulated or similar in expression patterns, while sample clustering is applied to discover the distinction between cancerous and normal tissues or to identify phenotypes. With innovative gene expression microarray data technologies, the number of genes in a single sample test have rapidly increased (high dimensionality) [16], but the number of sample tests remain small. This unique nature of the domain leads to difficulties for clustering samples because: a) there is much noise in the data and b) many genes become irrelevant and/or redundant to the underlying clusters. Therefore, feature selection techniques that quickly select a reasonably small subset of genes for clustering has increasingly received focus.

Ideally, feature selection is a preprocessing technique that finds an "optimal" subset of informative features from the original set D such that the underlying structure of D is retained in the subset. In this paper, we introduce an effective feature selection algorithm using a filter technique for clustering samples from microarray

J.R. Haritsa, R. Kotagiri, and V. Pudi (Eds.): DASFAA 2008, LNCS 4947, pp. 512–519, 2008.

data. It selects features to form minimum "distance" - appropriately defined - from the whole data. By this process, a number of features which are necessary for underlying clusters are retained, leading to a significant improvement on the quality of clustering results (partitions) compared to that of clustering without feature selection.

In the rest of paper, we briefly introduce the objectives of feature selection and describe related work in Section 2. Section 3 proposes our dimensionality reduction algorithm to select a subset of features. Section 4 describes evaluation methods that we use for comparing our approach, and presents some experimental results and Section 5 concludes the paper.

2 Background and Related Work

In the last several years a number of methods for feature selection for clustering have been proposed, most of which belong to the *wrapper* approach. A wrapper method uses a clustering algorithm to evaluate the candidate feature subsets. Wrapper methods can be categorized based on whether they select features for the whole data (*global type*) or just for a fraction of the data in a cluster (*local type*). The global type assumes a subset of features to be more important than others for the whole data while the local type assumes each cluster to have a subset of important features.

Examples of global methods are [5],[6],[9] and CLIFF [15]. The method described in [9] uses *K*-means for evaluation of subsets of features. In [6], EM (Expectation–Maximization) and trace measure are used for evaluation. The authors also propose visual aids for the user to decide the optimal number of features. In [5], features are ranked and selected for categorical data. Forward and backward search techniques are used to generate candidate subsets. To evaluate each candidate subset, these methods measure the category utility of the clusters by applying COBWEB [7], a hierarchical clustering algorithm for categorical data. In [14], authors proposed an objective function for choosing the feature subset and finding the optimal number of clusters for a document clustering problem using a Bayesian statistical estimation framework.

Examples of local wrapper methods are [1],[2] and [15]. Projected clustering (ProClus [1]) finds subsets of features defining (or important for) each cluster. ProClus first finds clusters using *K*-medoid considering all features and then finds the most important features for each cluster using Manhattan distance. CLIQUE [2] divides each dimension into user given divisions. It starts with finding dense regions (or clusters) in 1-dimensional data and works upward to find *j*-dimensional dense regions using a candidate generation algorithm. In CLIFF [15], the authors suggest methods that iteratively select a subset of features and generate a reference partition from the subset by a clustering algorithm. Then based on this partition, they select important features, according to their level of relevance and redundancy.

Performance of these wrapper methods depend on the clustering. As any clustering algorithm requires some parameters such as number of clusters or its equivalent, the selected features are very much influenced by the choice of parameter. In this work, we propose a *filter* method with the aim of independently selecting the most relevant and irredundant subset of features.

Feature selection methods typically search through the subsets of features and try to find the best one among the competing 2^m candidate subsets according to some evaluation function, where m is the number of features in the original data set. But this procedure is exhaustive as it tries to find only the best one, and may be too costly and practically prohibitive even for a medium-sized m. Other feasible methods based on heuristic or random search methods attempt to reduce computational complexity by compromising optimality. The goal of this paper is to select the important original features for clustering thus reducing the data size (and the computational time) and at the same time improving knowledge discovery performance and comprehensibility. Any feature selection algorithm for clustering should take into account the following two scenarios: a) where a single feature may define clusters independently of others, or b) where individual features do not define clusters but correlated features do.

3 Feature Selection Algorithm for Clustering

Our task now is to develop a method that can extract features which are correlated and relevant to the underlying clusters. The proposed algorithm is primarily based on the idea of sampling transactions from database according to the distance caused between the selected subset and the whole data. Traditional sampling methods require one or more computationally intensive passes over the entire database and can be prohibitively slow. Use of a simple random sample, however, may lead to unsatisfactory results. The problem is that such a sample may not adequately represent the entire data set due to random fluctuations in the sampling process. This difficulty is particularly apparent at small sizes. Thus, researches for better sampling methods have been in much focus.

The key measure of sampling is the distance, which is calculated based on frequency of items. The problem of sampling for feature selection can be defined as follows: select a set of features such that the "distance" of the data with only the selected features, from the data with all features is as small as possible. In other words, the selected features cause least difference between histograms of itemsets in the whole data and in the subset. Let D be the initial dataset with m features and n tuples; S_0 be the output subset of features with m_0 features produced by the proposed algorithm, and D_{S0} be the initial dataset. We are interested in minimizing $Dist(D, D_{S0})$, where $Dist$ is a distance function based on frequencies of individual tuples.

Data conversion
Since frequency-based sampling is especially designed for categorical count data, numerical data such as microarray data must be converted into transactional format. We first discretize each gene using variations of 0-μ, 1-σ method; where μ and σ are mean and standard deviation, respectively. This discretization method assumes the mean to be at 0; values in the range $[\mu-u^*\sigma, \mu+v^*\sigma]$ are assigned the discrete value 2, values in the range $[-\infty, \mu-u^*\sigma]$ are assigned the discrete value 1, and values in the range $[\mu+v^*\sigma, +\infty]$ are assigned the discrete value 3. Now each sample test T_i is an m-tuple, with values from the set $\{1, 2, 3\}$. We represent each tuple T_i by three m-tuple items $T_{ij} \mid j: 1, 2, 3$, such that the kth value of item $T_{ij} = 1$, when the kth value of sample test $T_i = j$, and 0 otherwise. Now the data consists of transactions (genes) having 3^*n item T_{ij}, where $i = 1, ..., n$ and $j = 1, 2, 3$. Our goal is to find a small sample such that

the discrepancy between the support of the 1-itemsets in the sample and those in the whole data is as small as possible. The discrepancy is computed as the distance of 1-itemset frequencies between the superset D and any subset S_0 as follows:

$$Dist(D,S_0) = \sum_{T_{ij} \in D} (f(T_{ij};D) - f(T_{ij};S_0))^2 \tag{1}$$

where $f(T_{ij}; D)$ or $f(T_{ij}; S_0)$ is frequency of item T_{ij} in D or S_0 respectively.

The ClosestFS algorithm

There are two search strategies for a sampling algorithm: trimming or growing. In the trimming process, sampling algorithms often start with a relatively large simple random sample of transactions and deterministically trim the sample to create a final subsample whose "distance" from the complete database is as small as possible. In the growing process, the final subsample is initially empty and candidate transactions are added to the subsample if the distance between the subsample and original data is the smallest. The stopping criterion is a predetermined number of transactions of interest.

We use the growing strategy in our sampling algorithm since only a small number of features (about 100s) carry necessary information for clustering. Our algorithm, called ClosestFS, is shown in Fig. 1. In each iteration, we remove a feature f_i and measure $Dist(D, D_i)$, where D_i is D without feature f_i. Given k number of features to be selected in each iteration, we add to S_0 the top k features for which the produced distances are higher than others. The idea is *if a feature f_i is a good representative of the whole data, then removing it will result in larger $Dist(D, D_i)$ than keeping it.* By setting $k = 1$, ClosestFS becomes a greedy search and the total number of $Dist$ evaluations required to create the final subsample is $O(mm_0 - m_0^2)$. Here, we simply use $k = m_0$; thus the computational complexity is $O(m)$. ClosestFS satisfies the first scenario described in the previous section by selecting features based on its individual merit measured by distance, and the second by combining the top-ranked features.

Input	D, m_0, k //number of genes selected in each iteration
Output	S_0
1:	$S_0 := \{\}$
2:	Compute $f(T_{ij}; D)$, for each T_{ij} in D
3:	while $(|S_0| < m_0)$ do
4:	for all f_i in D do
5:	$D_i := D - \{f_i\}$
6:	Calculate $Dist(D, D_i)$
7:	$\Sigma := \{ f_i \mid Dist(D, D_i)$ is one of k largest distances$\}$
8:	$S_0 := S_0 \cup \Sigma$
9:	for all T_{ij} in D
10:	Compute $f(T_{ij}; S_0)$
11:	Return S_0

Fig. 1. Algorithm ClosestFS

4 Empirical Study

To demonstrate the usefulness of our *filter* feature selection method for clustering, we compare performance of *K*-means algorithm with and without feature selection in terms of quality of clustering results and execution time. For the case of non feature selection, we simply run the K-means on the original data set to obtain a partition and then compute the validation measures. In the case of feature selection, we first discretize the data set; run ClosestFS to get a subset of 450 genes; then feed the subset to the clustering algorithm (we denote this combination by CFSK-means). We chose 450 genes in order to maximize the values of the quality indexes to include as many relevant features as possible. Please note that the clustering algorithm has a random seed selection and uses Euclidean as distance measure.

Validation Method
We observe that the quality of a clustering algorithm is measured both by the ability to determine the optimal number of clusters and the goodness of the clustering result. Thus, we present three criteria to test our method: (i) influence on finding the optimal number of clusters, (ii) the goodness of the computed partition in itself, and (iii) the goodness of the computed partition in comparison with the pathological/historical labeling of the sample tests based on their gene expression profiles (true partition).

These criteria of optimal number of clusters, intra-cluster quality and inter-cluster quality are validated using popular metrics. To validate the number of clusters, we used *silhouette index* [11] because of the popularity and the correctness which is evaluated through statistical tests. It has been applied for evaluation of clustering validity by deciding the goodness of the number of clusters [10]. To indicate the quality of an individual clustering result, we applied the *separation index* (S_{avg}) [12]. This kind of index refers only to the computed partition, not to the true partition. For verifying whether a generated partition goes well with the true partition which is the result of labeling sample tests in a pathological/historical manner, measures of agreement such as are *Jaccard index* [12] and *Adjusted Rand index* [8] are computed. In all cases, higher values are better. Detailed descriptions of this validation process are presented in [3].

Data Used for This Study
From commonly used microarray data, we conducted experiments over two data sets: leukemia and lung cancer. The data sets are downloaded from *Kent Ridge Biomedical Data Set Repository*[1], and their details can be obtained from [11] and [12].

Results and Analysis
Fig. 2 (a) and (b) shows the silhouette values versus the number of clusters u with $2 \leq u \leq 10$ for Leukemia and Lung cancer data sets. Although both algorithms are able to select exactly the true number of clusters, it can be seen that CFSK-means outperforms K-means for every resultant partition. For leukemia data, the silhouette of K-means is between 0.04 and 0.1, while the corresponding values produced by CFSK-means ranges from 0.06 to 0.2. For Lung cancer data, the range of silhouette

---

[1] http://sdmc.lit.org.sg/GEDatasets/Datasets.html

generated by K-means is [0.05, 0.1], and the range of CFSK-means is [0.06, 0.18]. As a result, the difference of silhouette values of partitions generated by CFSK-means is more significant than that of K-means. In addition, the difference between the highest value and others in the case of CFSK-means is considerably large. Therefore, we argue that our ClosestFS feature selection algorithm helps K-means improve the sensitivity of the silhouette index in determining the optimal number of clusters.

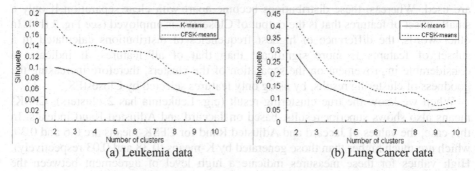

(a) Leukemia data   (b) Lung Cancer data

Fig. 2. Silhouette values for two datasets

Table 1. Quality comparison of K-means and CFSK-means

Datasets	Separation		Jaccard		Adjusted Rand	
	K-means	CFSK-means	K-means	CFSK-means	K-means	CFSK-means
Leukemia	0.05	**0.36**	0.32	**0.45**	0.22	**0.31**
Lung Cancer	0.05	**0.15**	0.37	**0.69**	0.16	**0.51**

Table 1 lists both intra-partition quality and inter-partition quality indexes to evaluate the clustering result in itself and in comparison with the true partition, respectively. These values are averages for all clusterings for each dataset, where the number of clusters is varied from 2 to 10. According to separation index related to only the computed partition, ClosestFS is particularly helpful to separate clusters since for all numbers of clusters, CFSK-means outperforms K-means. The average separation value of CFSK-means over the numbers of clusters is 0.36 for leukemia

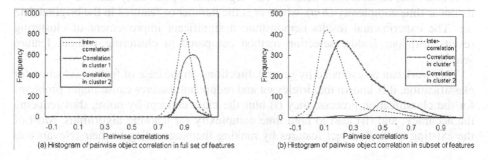

(a) Histogram of pairwise object correlation in full set of features   (b) Histogram of pairwise object correlation in subset of features

Fig. 3. Histograms of pairwise object correlations for Leukemia data

data, which is significantly higher than the corresponding values of K-means (0.05). In Lung cancer data, the corresponding values are respectively 0.15 and 0.05.

Furthermore, it has been pointed out that the good quality of a partition is obtained if the histograms of correlations of pairs of objects in the same clusters (intra-cluster distribution) and in different clusters (inter-cluster distribution) are Gaussian-like distribution and well-separated [12]. But as we see in Fig. 3 (a) for the leukemia data set[2], when the number of clusters is 2, this condition is not guaranteed if all features are used. Whereas, these distributions become apart with clear separation if only a small subset of features that is the output of ClosestFS is employed (see Fig. 3 (b)). In other words, the difference of highest frequencies of distributions calculated in a subset of features is more significant than that of all features. It indicates a considerable improvement on the separation of the clusters, therefore increasing the goodness of clustering results, by using only features selected by ClosestFS.

When we know the true clustering result (e.g. Leukemia has 2 clusters), CFSK-means also shows superior results, based on Jaccard and Adjusted Rand indexes. In this case, the values of Jaccard and Adjusted Rand for CFSK-means are 0.6 and 0.39, which are much higher than those generated by K-means (0.38 and 0.03 respectively). High values for these measures indicate a high level of agreement between the clustering results and the biological categorization. Hence, for these data, it is derived that contrary to K-means with feature selection, K-means running from the full data set produces low-quality partitions. In other words, feature selection improves the goodness of a clustering result by discarding unnecessary genes and noise in the data, leading to make clusters prominent. With this consideration, we claim that our feature selection method is able to remove unnecessary genes while retaining the underlying structure of the whole data, that is, the correlation among features and the relevance to underlying clusters.

5   Conclusions and Future Work

The major contribution of this work is to emphasize the importance of selecting a subset of relevant features for underlying clusters. It can be concluded that feature selection helps clustering algorithms in both identifying the optimal number of clusters and improving the "accuracy" of computed partition with the true partition. Another contribution is that we propose a new distance based filter method for feature selection in an unsupervised manner. This algorithm is practically feasible because of its linear time complexity $O(m)$, where m is the number of features in the original data set. The experimental results demonstrate a significant improvement of clustering results with our feature selection method compared to clustering without feature selection.

This work can be extended by several directions. In the case of feature selection for classification, it is known that irrelevant and redundant features cause major problems for the classification process: they (i) blur the class concept by noise, thus reducing the accuracy; and (ii) lead to high time complexity for learning algorithms. Most of the existing methods select features by ranking them according to their relevance to

---

[2] Due to lack of space, all results cannot be shown here. Readers may refer to [3] for details.

class concept. As a result, they cannot remove redundant features. Furthermore, assessing the degree of redundancy of a feature against a subset is a highly complicated task. Thus, efficient and effective methods to remove both irrelevant and redundant features are required. To that end, we have proposed a new two-phase algorithm for feature selection [4].

References

[1] Aggarwal, C.C., Procopiuc, C., Wolf, J.L., Yu, P.S., Park, J.S.: Fast algorithms for projected clustering. In: Proc. of ACM SIGMOD (1999)

[2] Agrawal, R., Gehrke, J., Gunopulos, D., Raghavan, P.: Automatic subspace clustering of high dimensional data for data mining applications. In: Proc. of ACM SIGMOD (1998)

[3] Dash, M., Gopalkrishnan, V.: Distance Based Feature Selection for Clustering Microarray Data, Technical Report, School of Computer Engineering, Nanyang Technological University, Singapore (March 2007)

[4] Dash, M., Gopalkrishnan, V.: Two Way Focused Classification. In: Song, I.-Y., Eder, J., Nguyen, T.M. (eds.) DaWaK 2007. LNCS, vol. 4654, Springer, Heidelberg (2007)

[5] Devaney, M., Ram, A.: Efficient feature selection in conceptual clustering. In: Proc. of ICML (1997)

[6] Dy, J.G., B.C.E.: Visualization and interactive feature selection for unsupervised data. In: Proc. of ACM SIGKDD (2000)

[7] Fisher, D.H.: Knowledge acquisition via incremental conceptual clustering. Machine Learning 2, 139–172 (1987)

[8] Hubert, L., Arabie, P.: Comparing partitions. Journal of Classification 2, 193–218 (1985)

[9] Kim, Y.S., Street, W.N., Menczer, F.: Feature selection in unsupervised learning via evolutionary search. In: Proc. of ACM SIGKDD (2000)

[10] Luo, F., Khan, L., Bastani, F., Yen, I.-L., Zhou, J.: A dynamically growing self-organizing tree (DGSOT) for hierarchical clustering gene expression profiles. Bioinformatics 20, 2605–2617 (2004)

[11] Rousseeuw, P.J.: Silhouettes: a graphical aid to the interpretation and validation of cluster analysis. J. Comput. Appl. Math 20, 53–65 (1987)

[12] Sharan, R., Shamir, R.: CLICK: A Clustering Algorithm with Applications to Gene Expression Anaysis. In: Proc. of ISMB, pp. 307–316 (2000)

[13] Tamayo, P., Slonim, D., Mesirov, J., Zhu, Q., Kitareewan, S., Dmitrovsky, E., Lander, E., Golub, T.R.: Interpreting patterns of gene expression with self-organizing map: Methods and application to hematopoietic differentiation. Proc. Natl. Acad. Sci. USA 96, 2907–2912 (1999)

[14] Vaithyanathan, S., Dom, B.: Model selection in unsupervised learning with applications to document clustering. In: Proc. of ICML (1999)

[15] Xing, E.P., Karp, R.M.: CLIFF: clustering of high-dimensional microarray data via iterative feature filtering using normalized cuts. Bioinformatics 17, 306–315 (2001)

[16] Yu, L., Liu, H.: Redundancy based feature selection for microarray data. In: Proc. of KDD, pp. 737–742 (2004)

Knowledge Transferring Via Implicit Link Analysis

Xiao Ling, Wenyuan Dai, Gui-Rong Xue, and Yong Yu

Department of Computer Science and Engineering
Shanghai Jiao Tong University
No. 800 Dongchuan Road, Shanghai 200240, China
{shawnling,dwyak,grxue,yyu}@apex.sjtu.edu.cn

Abstract. In this paper, we design a *local* classification algorithm using *implicit link analysis*, considering the situation that the labeled and unlabeled data are drawn from two different albeit related domains. In contrast to many *global* classifiers, e.g. Support Vector Machines, our local classifier only takes into account the neighborhood information around unlabeled data points, and is hardly based on the global distribution in the data set. Thus, the local classifier has good abilities to tackle the non-*i.i.d.* classification problem since its generalization will not degrade by the bias w.r.t. each unlabeled data point. We build a local neighborhood by connecting the similar data points. Based on these *implicit links*, the Relaxation Labeling technique is employed. In this work, we theoretically and empirically analyze our algorithm, and show how our algorithm improves the traditional classifiers. It turned out that our algorithm greatly outperforms the state-of-the-art supervised and semi-supervised algorithms when classifying documents across different domains.

1 Introduction

Supervised classification [1,2] requires a large number of labeled data. However, manual labeling is very expensive and time-consuming. Many investigations focus on the situations that the labeled data are scarce, and the unlabeled data are used to enhance the classification performance [3,4]. Actually, most of these researches ignore the fact that there might be quite a lot of existing labels from the similar domains. For example, the Blog documents and Web pages come from different albeit related domains; there are hardly any labeled Blog documents, while there are plenty of labeled Web pages, e.g. the pages in Open Directory Project (ODP). It is quite wasteful not to use these label information. However as a result of the domain difference, their distributions differ due to the different word usage for the documents in the two domains. The non-*i.i.d.* data violate the basic assumption of the traditional classification techniques, and thus the traditional classifiers cannot cope well with the *cross-domain* classification problem. Note that the *cross-domain learning*, which is one simplified case of *transfer learning* [5,6,7,8], transferring knowledge across tasks and domains.

In this paper, we focus on the problem of classifying documents across domains. Recall the Web page and Blog entry example. The training data (the Web pages) are under domain $\mathcal{D}_{in}$ and the test data (the Blog entries) under domain $\mathcal{D}_{out}$ are available. We call $\mathcal{D}_{in}$ *in-domain*, and $\mathcal{D}_{out}$ *out-of-domain*. In addition, it is assumed that $\mathcal{D}_{out}$ and

J.R. Haritsa, R. Kotagiri, and V. Pudi (Eds.): DASFAA 2008, LNCS 4947, pp. 520–528, 2008.

$\mathcal{D}_{in}$ are related and share some common knowledge, which makes knowledge transferring feasible. Our general goal is to classify the test data from $\mathcal{D}_{out}$ accurately by
transferring knowledge from the labeled data from $\mathcal{D}_{in}$.

We regard this classification problem as a labeling problem in a graph where both
labeled and unlabeled documents are represented as nodes. The edges are built based on
the *similarity* between two nodes. Such connections are so-called *Implicit Links*. Based
on the assumption of Markov Random Fields (MRF) that the label of each node is only
dependent on its immediate neighbors, we adopt the *Relaxation Labeling* [9] technique
to address the labeling problem. Initially, the labels for those unlabeled data are assigned
using a global classifier and then these labels are iteratively updated according to the
local neighborhood information. Since both labeled and unlabeled documents may exist
among the neighbors due to the similarity of domains, the iterative adjustments are
in fact implicitly transferring knowledge to the target domain. This is why our local
classifier is capable of handling the cross-domain problems.

Some prior works use labeled data from in-domain to solve problems under the target domain. Wu & Dietterich [10] investigated how to exploiting in-domain data in *k-
Nearest-Neighbors* and SVM algorithm. Daumé III and Marcu [11] utilized additional
in-domain labeled data to train a statistical classifier under the *Conditional Expectation
Maximum* framework. Those in-domain data play a role as auxiliary data in tackling
the scarcity of out-of-domain training data. In these work, the auxiliary data serve as a
supplement to the ordinary training data. In contrast, our work do not need any training examples in the target domain. Note that, it is possible, because the in-domain and
out-of-domain data share come common knowledge as we assumed, for the in-domain
model to learn from the out-of-domain data.

2   Transferring Knowledge through Relaxation Labeling

2.1   Problem Definition

For conciseness and clarity, we mainly focus on binary classification on the textual data
from different domains. Given two document sets $\mathcal{S}_{in}$ and $\mathcal{S}_{out}$ from in-domain $\mathcal{D}_{in}$ and
out-of-domain $\mathcal{D}_{out}$ respectively, each element $\mathbf{d}_i$ in two sets is represented by a feature
vector. In the binary classification setting, the label set is $\{+1, -1\}$, that is $c(\mathbf{d}_i)$ equals
$+1$ (positive) or -1 (negative) where $c(\mathbf{d}_i)$ is $\mathbf{d}_i$'s true label. As assumed in Section
1, $\mathcal{D}_{in}$ and $\mathcal{D}_{out}$ are different albeit related. The objective is to find the hypothesis h
which satisfies $h(\mathbf{d}_i) = c(\mathbf{d}_i)$ for as many $\mathbf{d}_i \in \mathcal{S}_{out}$ as possible.

2.2   Local Classifier Using Labeled Neighbors

When only the content information is considered, the most probable class label c_i for
each document d_i maximizes $\Pr(c_i|\tau(d_i))$ where $\tau(d_i)$ is the textual information of
d_i. However, the fact that the labeled and unlabeled data come from different domains
curbs the generalization ability since the model will fit the training data, but will not
cope well with the test data.

In order to circumvent this obstacle, the class labels of similar documents are also
worthy considering. We build a graph with nodes representing documents and edges by

implicit links. Hereinafter, the terms "document" and "node" are used interchangeably. Each document is connected to its most similar documents. With these links, we prefer the class label which maximizes $\Pr(c_i|\tau(d_i), N_i)$ where N_i is the immediate neighborhood of d_i. This immediate neighborhood assumption characterizes the first-order *Markov Random Field*. In this subsection, it is assumed that the labels of neighbors are all known, although this assumption does not hold in our problem setting. In the next subsection, the model will be extended to cope with neighbors without labels. Applying the Bayes Rule to $\Pr(c_i|\tau(d_i), N_i)$, it is obtained that

$$\Pr(c_i|\tau(d_i), N_i) = \frac{\Pr(\tau(d_i), N_i|c_i) \cdot \Pr(c_i)}{\Pr(N_i, \tau(d_i))}. \tag{1}$$

Assume the content of the document $\tau(d_i)$ has no direct coupling with its neighbors' labels. And $\Pr(N_i, \tau(d_i))$ is regarded as a constant since the task is to classify d_i. Then (1) is spanned into

$$\Pr(c_i|\tau(d_i), N_i) \propto \Pr(\tau(d_i)|c_i) \cdot \Pr(N_i|c_i) \cdot \Pr(c_i). \tag{2}$$

Assuming that given the class label of a node d_i, all its neighbors are independent with each other,

$$\Pr(N_i|c_i) = \prod_{d_j \in N_i} \Pr(d_j|c_i). \tag{3}$$

Combining (2) and (3), we obtain that

$$
\begin{aligned}
c_i &= \arg\max_{c_i} \Pr(c_i|\tau(d_i), N_i) \\
&= \arg\max_{c_i} \Pr(\tau(d_i)|c_i) \cdot \Pr(c_i) \prod_{d_j \in N_i} \Pr(d_j|c_i).
\end{aligned}
\tag{4}
$$

2.3   Classification with Out-of-Domain Unlabeled Data

As mentioned in the last subsection, the assumption that the labels of all neighbors are known is hardly satisfied. It is to say that all the similar documents of an unlabeled document are labeled, which is rarely possible in the cross-domain setting. To utilize the neighbors without labels, the *Relaxation Labeling* (abbreviated as RL) [12] technique is adopted here. In the RL process, with the initial labels, updates for unlabeled data are carried out iteratively.

Intuitively, the neighborhood of a certain node d is more likely to be given the same label of d. Both the test instances and the training ones are allowed to be the neighbors of the test nodes. The neighbors from the training data partially supervise the labeling while at the same time the test neighbors help not only correctly update labels but also avoid the bias by the constraints of local consistency. The Relaxation Labeling technique here reduces the cross-domain bias because in the iteration it enables the unlabeled data to be classified by themselves. In this view, the Relaxation Labeling updates the labels iteratively and thus gradually transfers knowledge across domains.

With the implicit links in previous subsection, we denote G^K to be all the information known in the graph. In this notation, the most probable label c_i is the one that can maximizes

$$\Pr(c_i|G^K) = \sum_{N_i^U} \Pr(c_i|G^K, N_i^U) \cdot \Pr(N_i^U|G^K) \tag{5}$$

where c_i is the class label corresponding to d_i and N_i^U represents the set of d_i's neighbors still with "unknown" label. The summation is over all possible assignments of N_i^U.

Using the independence assumption of the class label for each d_j among N_i^U,

$$\Pr(N_i^U|G^K) = \prod_{d_j \in N_i^U} \Pr(c_j|G^K). \tag{6}$$

Similarly with previous subsection, the class label of one document is dependent on its local content as well as its similar documents (i.e. its immediate neighbors).

$$\Pr(c_i|G^K, N_i^U) = \Pr(c_i|N_i^K, N_i^U) \tag{7}$$

where N_i^K is the neighborhood with "known" labels. Combining (6) and (7) and manipulating it into an iterative solution, we obtain

$$\Pr(c_i|G^K)^{(r+1)} = \sum_{N_i^U} \left[\prod_{d_j \in N_i^U} \Pr(c_j|G^K)^{(r)} \Pr(c_i|N_i^K, N_i^U)^{(r)} \right] \tag{8}$$

where $\Pr(c_i|N_i^K, N_i^U)$ can be treated as $\Pr(c_i|N_i)$ in the previous subsection where the labels of N_i are all known. The superscript (r) denotes the iteration number.

Since the number of the terms in the summation (8) is exponential to the size of unlabeled documents, the computation is intractable. To reduce the computation expense, we adopted the *hard labeling* method in [13], whose main idea is to use the most probable initial labels of those unknown neighbors to alleviate the consuming computation of summation.

$$\Pr(c_i|G^K) \approx \Pr(c_i, N_i^{U'}|N_i^K) \tag{9}$$

where $N_i^{U'}$ is the neighborhood with the most probable assignment for class labels. This hard labeling is seen as a rough approximation of (8). However, the magnitude of other terms is often small compared to the selected assignment, and therefore the hard labeling method may work well. We also consider the soft version of labeling strategy [13], which selectively takes more terms of the summation (8) into computation. Empirically, it is comparable to the "hard labeling" strategy. The details are omitted due to the space limit. Algorithm 1 gives the outline of our method. After the initializations, the algorithm iterates until the convergence and then it outputs the predicted labels of unlabeled data.

Algorithm 1. Transfer Knowledge by Relaxation Labeling (TKRL)

Input :
labeled and unlabeled data from in-domain and out-of-domain respectively,
the initial labels for unlabeled data via a basic classifier,
parameter k for building the graph.
Output : the final labels for unlabeled data
Initialization:
$oldlabel = null$, $newlabel$ = initial labels,
build the graph with all information including the content of each d_i and each immediate
neighborhood N_i, s.t. $|N_i| = k$ for each i.
Iteration:
while $oldlabel \neq newlabel$ **do**
  $oldlabel = newlabel$.
  estimate the prior probability $\Pr(c = +1)$ and $\Pr(c = -1)$.
  for each d_i **do**
    estimate the conditional probabilities $\Pr(c_i = +1|d_i)$, $\Pr(c_i = -1|d_i)$
  end for
  for d_i in unlabeled data **do**
    update its label in $newlabel$ according to (9)
  end for
end while
return $newlabel$

---

3 Experimentation

3.1 Data Sets

To validate our algorithm, we developed a series of cross-domain data sets based on
20 Newsgroups[1], Reuters-21578[2] and SRAA[3]. The basic idea is to utilize the hierarchy
of the data sets. The task is defined as classifying top categories. Each top category
is split into two disjoint parts with different sub-categories, one for training and the
other for test. Therefore the training and test data come from different domains. Take
SRAA as an example, which is a Simulated/Real/Aviation/Auto UseNet data set for
document classification. For the data set `real vs simulated`, we use the documents in `real-auto` and `sim-auto` as in-domain data, while `real-aviation`
and `sim-aviation` as out-of-domain data. Other tasks were generated in a similar
way. On these textual data, regular preprocessing was done including tokenization into
bag-of-words, converting into low-case words, stop-word removing and stemming. We
also carried out feature selection by thresholding Document Frequency [14]. In our
experiments, Document Frequency threshold is set to 3.

The data from different domains are certainly under different distributions. To verify our data design, we calculated *Kullback-Leibler Divergence* (K-L Divergence) [15]
based on Term Frequency for each data set, which measures distance between

---

[1] http://people.csail.mit.edu/jrennie/20Newsgroups/
[2] http://www.daviddlewis.com/resources/testcollections/
[3] http://www.cs.umass.edu/ mccallum/data/sraa.tar.gz

Table 1. Description of the data sets for cross-domain text classification, and the error rates of each classifier. "$\mathcal{D}_{in}-\mathcal{D}_{out}$" means training on $\mathcal{D}_{in}$ and testing on $\mathcal{D}_{out}$; "$\mathcal{D}_{out}$-CV" means 10-fold cross-validation on $\mathcal{D}_{out}$.

Data Set	Documents			K-L	SVM		TSVM	NBC	TKRL
	$|\mathcal{D}_{in}|$	$|\mathcal{D}_{out}|$	$|\mathcal{W}|$		$\mathcal{D}_{out}$-CV	$\mathcal{D}_{in}-\mathcal{D}_{out}$			
real vs simulated	8,000	8,000	14,433	1.161	0.032	0.266	0.130	0.245	**0.126**
auto vs aviation	8,000	8,000	14,433	1.126	0.033	0.228	0.102	0.136	**0.099**
rec vs talk	3,669	3,561	19,412	1.102	0.003	0.233	0.040	0.269	**0.032**
rec vs sci	3,961	3,965	18,152	1.021	0.007	0.212	0.060	0.153	**0.058**
comp vs talk	4,482	3,652	17,918	0.967	0.005	0.103	0.097	0.025	**0.022**
comp vs sci	3,930	4,900	18,379	0.874	0.012	0.317	0.183	0.206	**0.100**
comp vs rec	4,904	3,949	18,903	0.866	0.008	0.165	0.098	0.216	**0.046**
sci vs talk	3,374	3,828	20,057	0.854	0.009	0.226	0.108	0.258	**0.056**
orgs vs places	1,079	1,080	4,415	0.329	0.085	0.454	0.436	0.375	**0.339**
people vs places	1,239	1,210	4,562	0.307	0.113	0.266	0.231	0.217	**0.188**
orgs vs people	1,016	1,046	4,771	0.303	0.106	0.297	0.297	0.282	**0.272**

distributions. More formally, $\mathrm{KL}(\mathcal{D}_1||\mathcal{D}_2) = \sum_i \mathcal{D}_1(i) \log_2 \frac{\mathcal{D}_1(i)}{\mathcal{D}_2(i)}$ where $\mathcal{D}_1$ and $\mathcal{D}_2$ are two distributions. As listed in the fifth column of Table 3.1, the K-L Divergence values of all the data sets are all far larger than zero which means that they come from different distributions. This observation justifies that our design is reasonable. Also, we calculated the error rates using the SVM classifier across the domains ($\mathcal{D}_{in}-\mathcal{D}_{out}$) and only within the test set (D_{out}-CV). The relative low error rates in D_{out}-CV prove that the test data are out-of-domain.

3.2 Experimental Results

To evaluate the effectiveness of our method, we compare it to two supervised methods: the SVM and the Naive Bayes classifier (NBC) as well as a semi-supervised method: the TSVM (Transductive SVM) classifier [4] by their error rates. The SVM and TSVM classifiers are implemented by SVM^{light} [4]. The Naive Bayes Classifier is implemented using Laplace Smoothing. Each document is then represented by a feature vector with Term Frequency in our algorithm. When applying SVM or TSVM to these data (mentioned in the next subsection), the tf-idf values are used. Through comparing with traditional supervised classifier, it is seen that the different domains the training and test data come from bring classification much difficulty and hence poor performance. Although the semi-supervised classifier fully utilizes the unlabeled data in the classification process, it still works under the identical-domain assumption.

Our method (named TKRL) aims at handling the cross-domain problem, which achieves high performance in this cross-domain data setting. In the implementation, we use the Naive Bayes classifier to give the initial labels and adopt cosine measure for building the graph. In Table 3.1, we see that semi-supervised algorithm TSVM (8th column) always outperforms the supervised algorithm SVM (7th column) and NBC (9th column) almost all the time. It is because taking unlabeled data into account is in some

---

[4] http://svmlight.joachims.org

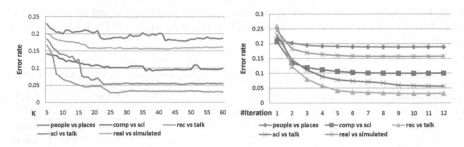

Fig. 1. The error rate curve against parameter k **Fig. 2.** The convergence curve of five tasks

sense partially transferring supervisory knowledge into the target domain. However the transferring is not complete. On the other hand, it is noticed that NBC performs better than SVM. We believe that NBC is less influenced by the domain difference than SVM due to its simple independence assumption. Employing implicit link analysis, our method aims at handling data under different domains and in fact TKRL achieves the lowest error rates through all eleven tasks. However, the performance of certain data sets are still unsatisfactory. It is mainly attributed to the noise in the data. In the three Reuters-21758 tasks, the test error by SVM is not satisfactory yet. It is mainly because of the data noise and thus less common knowledge between domains. Note that our algorithm achieves improvements on the classical classifiers.

Parameter Sensitivity. Only one parameter k exists in our algorithm, which limits the size of immediate neighborhood. We enumerate the value of k ranging from 5 to 60 to evaluate its influence on performance. Figure 1 displays the error rate curve on the five representative tasks. It is observed that our algorithm is not very sensitive to k when k is greater than 30 since the rest of the curves are quite stable. Empirically, we set $k = 30$ to get better performance.

Convergence. In Fig. 2, we plot the error rate along with each iteration step on five tasks. Experimentally, our algorithm converges after several iterations. Generally, the

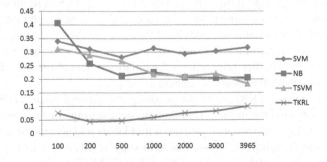

Fig. 3. The error rate curve against the size of training data

iteration process needs around 9 steps on average. From Fig. 2, we observe that the error rate decreased by a large amount in the first several iterations.

Size of Training Examples. We also investigate the influence by the size of training examples. A portion of examples in `comp vs sci` are randomly chosen for training, from 100 examples to all. From Fig. 3, we observed that TKRL reaches the lowest error rate at the size of 200 training examples. It is because if the training data are fewer, the information from labeled data will be too scarce; on the contrary, if the data from in-domain are more than enough, the in-domain knowledge will impact and deteriorate the out-of-domain classification performance.

4 Conclusion and Future Work

In this paper, we design a method for the cross-domain classification problem where only labeled data from in-domain are available for predicting the class labels of unlabeled data from out-of-domain. Our local classifier labeled the test data only considering the neighborhood information. We leverage implicit link analysis for this cross-domain classification. Experimental evaluations reveal that our method is very effective on handling the cross-domain problems. There are several directions for future extensions. We wish to test on another kind of data, such as images. It is also interesting to find an online way of classification, that is the test data are incrementing.

Acknowledgements. All authors are supported by a grant from National Natural Science Foundation of China (NO. 60473122). We thank the anonymous reviewers for their great helpful comments.

References

1. Lewis, D.D.: Representation and learning in information retrieval. PhD thesis, Amherst, MA, USA (1992)
2. Boser, B.E., Guyon, I., Vapnik, V.: A training algorithm for optimal margin classifiers. In: Proceedings of the Fifth Annual Workshop on Computational Learning Theory (1992)
3. Zhu, X.: Semi-supervised learning literature survey. Technical Report 1530, University of Wisconsin–Madison (2006)
4. Joachims, T.: Transductive inference for text classification using support vector machines. In: Proceedings of Sixteenth International Conference on Machine Learning (1999)
5. Schmidhuber, J.: On learning how to learn learning strategies. Technical Report FKI-198-94, Fakultat fur Informatik (1994)
6. Thrun, S., Mitchell, T.: Learning One More Thing. IJCAI, 1217–1223 (1995)
7. Caruana, R.: Multitask Learning. Machine Learning 28(1), 41–75 (1997)
8. Ben-David, S., Schuller, R.: Exploiting task relatedness for multiple task learning. In: Proc. of the Sixteenth Annual Conference on Learning Theory COLT 2003 (2003)
9. Pelkowitz, L.: A continuous relaxation labeling algorithm for markov random fields. IEEE Transactions on Systems, Man and Cybernetics 20(3), 709–715 (1990)
10. Wu, P., Dietterich, T.: Improving SVM accuracy by training on auxiliary data sources. In: Proceedings of the Twenty-First International Conference on Machine Learning, pp. 871–878

11. D.I.H., Marcu, D.: Domain Adaptation for Statistical Classifiers. Journal of Artificial Intelligence Research 1, 1–15 (1993)
12. Chakrabarti, S., Dom, B., Indyk, P.: Enhanced hypertext categorization using hyperlinks. In: SIGMOD, pp. 307–318 (1998)
13. Angelova, R., Weikum, G.: Graph-based text classification: learn from your neighbors. In: SIGIR, pp. 485–492 (2006)
14. Yang, Y., Pedersen, J.: A comparative study on feature selection in text categorization. In: Proceedings of the Fourteenth International Conference on Machine Learning (1997)
15. Kullback, S., Leibler, R.: On Information and Sufficiency. The Annals of Mathematical Statistics 22(1), 79–86 (1951)

Exploiting ID References for Effective Keyword Search in XML Documents

Bo Chen[1], Jiaheng Lu[2], and Tok Wang Ling[1]

[1] School of Computing, National University of Singapore
{chenbo,lingtw}@comp.nus.edu.sg
[2] University of California, Irvine
jiahengl@uci.edu

Abstract. In this paper, we study novel *Tree + IDREF* data model for keyword search in XML. In this model, we propose novel *Lowest Referred Ancestor (LRA) pair*, *Extended LRA (ELRA) pair* and *ELRA group* semantics for effective and efficient keyword search. We develop efficient algorithms to compute the search results based on our semantics. Experimental study shows the superiority of our approach.

1 Introduction

Keyword search is a convenient way to query XML documents since it allows users to easily issue keyword queries without the knowledge of complex query languages and/or the structure of underlying data.

Existing approaches. Majority of the research efforts in XML keyword search focus on keyword proximity search in either *tree model* or *general digraph model*. Both approaches generally assume a smaller sub-structure of the XML document that includes all query keywords indicates a better result.

In tree model, *SLCA (Smallest Lowest Common Ancestor)* ([7,3,8,6]) is a simple and effective semantics for XML keyword proximity search. Each SLCA result of a keyword query is a smallest XML node[1] that 1) covers all keywords in its descendants and 2) has no single proper descendant to cover all query keywords. However, the SLCA semantics based on tree model does not capture ID reference information which is usually present and important in XML databases. As a result, it may return a large tree including irrelevant information. For example, Figure 1 shows an XML data for computer science department in a university. Consider keyword query "Smith Database", which looks for whether Smith teaches some Database course. In this case, the SLCA is the overwhelming root of the whole XML database, which will frustrate the searcher.

On the other hand, XML documents can be modeled as digraphs to take into account ID reference edges. The key concept in digraph model is called *reduced subtrees* ([2,5]), which searches for minimal connected subtrees in graphs. However, the problem of finding all reduced subtrees and enumerating results by

---

[1] When we say an XML node in this paper, we refer to the subtree rooted at the node.

J.R. Haritsa, R. Kotagiri, and V. Pudi (Eds.): DASFAA 2008, LNCS 4947, pp. 529–537, 2008.
© Springer-Verlag Berlin Heidelberg 2008

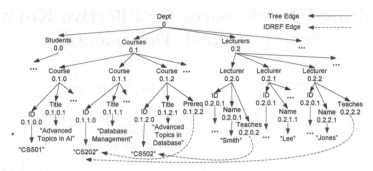

Fig. 1. Example XML document of computer science department (with Dewey IDs)

increasing sizes of reduced subtrees is NP-hard. Thus, most previous algorithms (e.g. [4]) on XML digraph model are intrinsically expensive, heuristics-based.

Our Approach. In this paper, we study a novel special graph, *Tree + IDREF* model, to capture the same ID references in digraph model which are missed in tree model; and meanwhile to achieve better efficiency than general digraph model by distinguishing reference edges from tree edges in XML.

In particular, we first propose novel *LRA pairs (Lowest Referred Ancestor pairs)* semantics. Informally, LRA pairs semantics returns a set of lowest ancestor pairs such that each pair in the set are connected by ID references and the pair together cover all keywords in their subtrees. For example, consider the query "Smith Database" in Figure 1 again. The result of LRA pairs semantics is the pair of nodes Lecturer:0.2.0 and Course:0.1.2 that are connected by ID reference and together cover all keywords. Then, we extend LRA pairs that are directly connected by ID references to node pairs that are connected via intermediate node hops by a chain of ID references; which we call *ELRA pairs (Extended Lowest Referred Ancestor pairs)*. Finally, we further extend ELRA pairs to *ELRA groups* to define the relationships among two or more nodes which together cover all keywords and are connected with ID references.

Contributions and Organization. Our contributions are as follows:

(1) We introduce *Tree + IDREF* data model for keyword proximity search in XML databases. In this model, we propose novel LRA pairs, ELRA pairs and ELRA groups as complements of well-known SLCA semantics (Sections 2 and 3).

(2) We study and analyze efficient polynomial algorithms to evaluate keyword queries based on proposed semantics (Section 4).

(3) We conduct extensive experiments with our keyword search semantics. Results prove the superiority of proposed model and search semantics (Section 5).

2   Background and Data Model

ID references in XML. In many XML databases, ID references are present and play an important role in eliminating redundancies and representing relationships between XML elements. For example, in Figure 1, references indicate important *teach* relationships between Lecturer and Course elements. Without ID references, the relationships has to be expressed in further nested structures (e.g. each lecturer is nested and duplicated in each course she/he teaches or vice versa), potentially introducing harmful redundancies.

Data Model. We model XML as special digraphs, *Tree + IDREF*, $G = (N, E, E_{ref})$, where N is a set of nodes, E is a set of tree edges, and E_{ref} is a set of ID reference edges between two nodes. Each node $n \in N$ corresponds to an XML element, attribute or text value. Each tree edge denotes an parent-child relationship. We denote a reference edge from u to v as $(u,v) \in E_{ref}$. In this way, we distinguish the tree edges from reference edges in XML. The subgraph $T = (N, E)$ of G without ID reference edges, E_{ref}, is a tree. When we talk about parent-child (P-C) and ancestor-descendant (A-D) relationships between two nodes in N, we only consider tree edges in E of T.

3 Search Semantics

3.1 LRA Pairs Semantics

Definition 1. *(reference-connection) Two nodes u, v with no A-D relationship in an XML database have a reference-connection (or are reference-connected) if there is an ID reference between u or u's descendant and v or v's descendant.*

Definition 2. *(LRA pairs) In an XML database, LRA pairs semantics of a list of keywords K returns a set of unordered node pairs $\{(u_1, v_1), (u_2, v_2), ..., (u_m, v_m)\}$ such that for any (u_i, v_i) in the set,*
  (1) u_i and v_i each covers some and together cover all keywords in K; and
  (2) there is a reference-connection between u_i and v_i; and
  (3) there is no proper descendant u' of u_i (or v' of v_i) such that u' forms a pair with v_i (or v' forms a pair with u_i resp.) to satisfy conditions (1) and (2).

For example, there is a reference-connection between nodes Lecturer:0.2.0 and Course:0.1.2 in Figure 1 since there is an ID reference edge between their descendants (Teaches:0.2.0.2 and ID:0.1.2.0). Therefore, Lecturer:0.2.0 and Course:0.1.2 form an LRA pair for query "Smith Advanced Database", whose SLCA is the overwhelming root. Note reference-connected Lecturers:0.2 and Courses:0.1 do not form an LRA pair for "Smith Advanced Database" since their descendants already form a lower connected pair to cover all keywords.

3.2 ELRA Pairs Semantics

Now, we extend the direct reference-connection in LRA pairs to a chain of connections as *n-hop-connection* in *Extended LRA (ELRA) pairs* semantics.

Definition 3. *(n-hop-connection) Two nodes u, v with no A-D relationship in an XML database have an n-hop-connection (or are n-hop-connected) if there are $n - 1$ distinct intermediate nodes $w_1, ...w_{n-1}$ with no A-D pairs in $w_1, ...w_{n-1}$ such that $u, w_1, ..., w_{n-1}, v$ form a chain of connected nodes by reference-connection;*

Definition 4. *(ELRA pairs) In an XML database, ELRA pairs semantics of a list of keywords K returns a set of unordered node pairs $\{(u_1, v_1), (u_2, v_2), ..., (u_m, v_m)\}$ such that for any (u_i, v_i) in the set,*

(1) u_i and v_i each covers some and together cover all keywords in K; and

(2) there is an n-hop-connection between u_i and v_i for an upper limit L of the number of intermediate hops; and

(3) there is no proper descendant u' of u_i (or v' of v_i) such that u' forms a pair with v_i (or v' forms a pair with u_i resp.) to satisfy conditions (1) and (2).

ELRA pairs semantics returns a set of pairs such that two nodes in each pair are lowest n-hop-connected to together cover all keywords. The upper limit (L) of the length of n-hop-connection can be determined at system tuning phase. Then, if users are interested in more results, the system can progressively increase L.

For example, in Figure 1, Lecturer:0.2.0 and Course:0.1.1 are connected by a 2-hop-connection via node Course:0.1.2. Thus, with L set as 2 or more, this pair will form an ELRA pair for query "Smith Database Management", which can be understood as Database Management is a prerequisite of the course that Smith teaches. We can see from this example that ELRA pairs semantics has better chance to find smaller results than LRA pairs semantics since LRA pairs are the lowest pairs with direct reference-connection while ELRA pairs are the lowest pairs with connections up to L intermediate hops including reference-connection.

3.3   ELRA Group Semantics

Finally, we extend ELRA pairs semantics to ELRA groups semantics to define relationships among two or more connected nodes that cover all keywords.

Definition 5. *(ELRA groups) In an XML database, ELRA groups semantics of K keywords returns a set of node groups $\{G_1, G_2, ..., G_m\}$ s.t. $\forall G_i$ $(1 \leq i \leq m)$,*

(1) all nodes in G_i each covers some and together cover all K keywords; and

(2) there is a node h_i to connect all nodes in G_i by n-hop-connection with up to L' as the number of intermediate hops; we call h_i as the hub for G_i; and

(3) there is no proper descendants of any node u in G_i to replace u to cover the same set of keywords as u and are n-hop-connected $(n \leq L')$ to the hub; and

(4) there is no proper descendant d of G_i's hub h_i such that d is the hub of another ELRA group G_d and $(G_d \cup \{h_i\}) \supseteq G_i$.

Similar to ELRA pairs, we can set the value for L' at system tuning phase for ELRA groups semantics and L' can be increased upon users' requests.

Compared to SLCA and ELRA pairs, ELRA groups can potentially find more and smaller nodes in the result. For example, in Figure 1, with L' set as two, for query "Jones Smith Database", the node group Course:0.1.1, Course:0.1.2, Lecturer:0.2.0 and Lecturer:0.2.2 form an ELRA group result. Both node Course:0.1.1 and Course:0.1.2 can be considered as the hubs in this ELRA group result.

4   Algorithms for Proposed Search Semantics

4.1   Data Structures

The data structures that we adopt in this paper are *keyword inverted lists* and *connection table*. Keyword inverted lists are standard structures for keyword search. Each inverted list stores all the Dewey labels of a keyword.

The connection table maintains one connection-list, List(u), for each node u in the XML document such that List(u) contains all the lowest nodes (v) that have direct reference-connection to u in document order. From the Dewey label of v, we can easily get all v's ancestors that are not ancestors of u so that they are also reference-connected to u. Indexes can be built on top of the connection table to facilitate efficient retrieval of the connection list for a given node.

For example, Figure 2 shows the B+ tree indexed connection table for the XML data in Figure 1. In Figure 2, we can see node 0.1.1 has reference connection to 0.1.2.2, thus we can tell that 0.1.1 is also reference-connected to 0.1.2.

4.2   Sequential-Lookup Algorithm

We present Sequential-Lookup algorithm, to compute all search results for the proposed semantics in Algorithm ??. The inputs to the algorithm are k inverted lists of keywords $(I_1,...,I_k)$, the connection table CT and the upper limits for connection chain lengths of ELRA pairs (L) and ELRA groups semantics (L'). The outputs are SLCA, ELRA pair and group results. The main idea is to use three steps to compute SLCAs (line 1), ELRA pairs (line 3) and ELRA groups (line 4) by calling corresponding functions. Note we can get all LRA pairs by setting the limit L of ELRA pairs as one.

The time complexities of each step for Sequential-Lookup algorithm are:

- $O(kd|N_{min}| \log |N_{max}|)$ for SLCA (based on the analytical result of [8]),
- $O(d \sum_{i=1}^{k} |N_i|(|E_L| + kd|Q_L| \log |N_{max}|))$ for ELRA pairs and,
- $O(|Q_{L'}|d^2 \sum_{i=1}^{k} |N_i|(|E_{L'}| + kd|Q_{L'}| \log |N_{max}|))$ for ELRA groups,

where k is the number of keywords; d is the maximum depth of the XML documents; $|N_{min}|$, $|N_{max}|$ and $|N_i|$ are the sizes of shortest, longest and i^{th} inverted lists in the query respectively; E_L and Q_L are the maximum number of edges and nodes reached by depth-limited search with chain length limit L for ELRA pairs; and finally $E_{L'}$ and $Q_{L'}$ are the maximum number of edges and nodes reached by depth-limited search with chain length limit L' for ELRA groups.

4.3   Rarest-Lookup Algorithm

Since every ELRA pair (or ELRA group) must include at least one node (or its ancestor) from the shortest inverted list of query keywords, it is sufficient to only check the shortest inverted list for all results. Therefore, we further propose Rarest-Lookup algorithm to scan nodes in the shortest (rarest) inverted list and their connected nodes to compute ELRA pairs and groups. The only changes for

---

Algorithm 1. Sequential-Lookup

Input: Keyword lists $I_1, I_2, ..., I_k$, connection table CT, the upper limits L, L'
Output: SLCA, ELRA_P, ELRA_G
1 SLCA=computeSLCA($I_1, I_2, ..., I_k$);   // adopt existing algorithms for SLCA
2 let I_{seq} be the sort-merged list of $I_1, I_2, ..., I_k$;
3 ELRA_P=computeELRA_P($I_{seq}, I_1, I_2, ..., I_k, CT, L$) ;
4 ELRA_G=computeELRA_G($I_{seq}, I_1, I_2, ..., I_k, CT, L'$) ;
5 **return and output** SLCA, ELRA_P, ELRA_G ;

Function computeELRA_P ($I_{seq}, I_1, I_2, ..., I_k, CT, L$)
7 initial empty ELRA_P ;   //mapping each u to $\forall v$ s.t. $u \& v \in$ ELRA pair
8 **for** *(each self-or-ancestors u of each node in I_{seq} in top-down order)* **do**
9   get $I_i, ..., I_m$ whose keywords u does not cover;
10   **if** *($u \notin$ ELRA_P **and** u does not cover all keywords)* **then**
11   Q=getConnectedList(u, CT, L);
12   S_u=computeSLCA($Q, I_i, ..., I_m$);
13   remove $\forall v \in S_u$ from S_u s.t. v covers all keywords ;
14   ELRA_P.put(u, S_u);
15   **for** *($\forall a$ s.t. a is ancestor of u **and** $a \in$ ELRA_P)* **do**
16   S_a = ELRA_P.get(a); $S_a = S_a - S_u$;   // set difference
17   ELRA_P.update(a, S_a);
18 **return** ELRA_P;

Function computeELRA_G ($I_{seq}, I_1, I_2, ..., I_k, CT, L'$)
20 initial empty ELRA_G ;   // mapping each u to a group G s.t. u is a hub to
                              // connect $\forall v \in G$ as an ELRA group
21 **for** *(each self-or-ancestors u of each node in I_{seq} in top-down order)* **do**
22   G=getELRAGroup(u) ;
23   **if** *($G \neq null$)* **then**   ELRA_G.put(u, G) ;
24   **for** *($\forall a$ s.t. a is ancestor of u **and** $a \in$ ELRA_G)* **do**
25   **if** *($G \cup \{a\} \supseteq$ ELRA_G.get(a))* **then**   ELRA_G.remove(a) ;
26   Q=getConnectedList(u, CT, L');
27   **for** *(each self-or-ancestors q of each node in Q in top-down order)* **do**
28   G=getELRAGroup(q) ;
29   **if** *($G \neq null$)* **then**   ELRA_G.put(q, G) ;
30   **for** *($\forall a$ s.t. a is ancestor of u **and** $a \in$ ELRA_G)* **do**
31   **if** *($G \cup \{a\} \supseteq$ ELRA_G.get(a))* **then**   ELRA_G.remove(a) ;
32 **return** ELRA_G ;

Function getELRAGroup (h)
34 **if** *(h cover all keywords)* **then**   **return** *null* ;
35 Q = getConnectedList(h, CT, L') ;
36 initial empty set G ;
37 **for** *(each $I_i \in I_1, I_2, ..., I_k$)* **do**
38   Y = getSLCA(I_i, Q) ;
39   remove $\forall y \in Y$ from Y s.t. y covers all keywords ;
40   **if** *(Y is empty **and** h does not cover I_i's keyword)* **then**   **return** *null* ;
41    $G = G \cup Y$;
42 **return** G ;

Function getConnectedList (u, CT, L)
44 return the list of lowest nodes computed by depth-limited search from CT
   that have n-hop-connection ($n \leq L$) to u in document order ;

Data	File size	Keyword inverted lists		Connection table	
		creation time	size	creation time	size
DBLP	362.9MB	321 sec	145.7MB	81 sec	1.62MB
XMark	113.8MB	193 sec	140.3MB	234 sec	13.7MB

Fig. 2. Connection Table of XML tree in Figure 1

Fig. 3. Data size, index size and creation time

the Rarest-Lookup algorithm from Sequential-Lookup are at lines 2–4. Instead of getting the sort-merged list of all query keyword inverted lists, we pass the shortest (rarest) inverted list to the functions for ELRA pairs and groups.

The time complexities of each step for Rarest-Lookup algorithm are:

- $O(kd|N_{min}|\log|N_{max}|)$ for SLCA,
- $O(d|N_{min}|(|E_L| + kd|Q_L|\log|N_{max}|))$ for ELRA pairs and,
- $O(|Q_{L'}|d^2|N_{min}|(|E_{L'}| + kd|Q_{L'}|\log|N_{max}|))$ for ELRA groups,

where the variables are the same as those in Sequential-Lookup's complexity.

5 Experimental Evaluation

5.1 Experimental Settings

Hardware and implementation. We use a normal PC with Pentium 2.6GHz CPU and 1GB memory. All codes are written in java. We set the limits of connection chain length as two for ELRA pairs and one for ELRA groups.

Datasets and indexes creation. We choose both real DBLP and synthetic XMark datasets in our experiment. Two datasets are pre-processed to create the inverted lists and connection tables. They are stored in Disk with Berkeley DB [1] B-trees. The details of file sizes and index creation of the two datasets are shown in Figure 3. Note that the connection table of DBLP is small due to the incomplete citation information of the data.

Queries and performance measures. For each dataset, we generate random queries of 2 to 5 keywords long, with 50 queries for each query size. We use these queries to compare the efficiency of 1) Sequential-lookup with Rarest-Look algorithm and 2) our algorithms with heuristic Bi-directional expansion in general digraph model.

5.2 Comparison of Search Efficiency Based on Random Queries

Sequential-lookup v.s. Rarest-lookup. In Figure 4, we show the efficiency comparisons between Sequential-lookup and Rarest-lookup in computing ELRA pairs (Seq_P and $Rarest_P$) and ELRA groups (Seq_G and $Rarest_G$). It is clear that Rarest-lookup achieves much better efficiency in both datasets. Rarest-lookup is also more scalable to queries of more keywords.

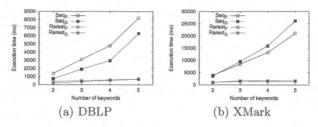

(a) DBLP   (b) XMark

Fig. 4. Time Comparisons between Rarest-lookup and Sequential-lookup

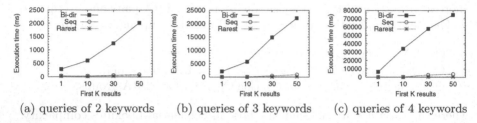

(a) queries of 2 keywords   (b) queries of 3 keywords   (c) queries of 4 keywords

Fig. 5. Time comparisons between Bi-Directional Expansion and proposed algorithms for computing first-k results in XMark

Tree + IDREF v.s. general digraph model. Now, we compare the efficiency of our algorithms with Bi-directional expansion (Bi-dir for short) heuristics ([4]) to find first-k results. Figure 5 shows Bi-directional (Bi-dir) is significantly slower than Sequential-lookup (Seq) and Rarest-lookup (Rarest) for XMark data. For DBLP data with less amount of ID references, the result shows even more significant advantage of our approach over Bi-dir, which is omitted due to space limitation. The reasons for the inefficiency of Bi-dir are: Firstly, at each expansion, Bi-dir needs to find the best node to expand among all expandable nodes. Secondly, Bi-dir involves floating point numbers in computing and comparing the goodness of expandable nodes. Thirdly and most importantly, when Bi-dir computes or updates the goodness of a node, it has to recursively propagate the goodness to all neighbors to improve their goodness until no nodes' goodness can be improved.

6   Conclusion

Motivated by the limitations of existing tree and general digraph model, we propose tree + IDREF model for XML keyword proximity search. In particular, we propose novel LRA pairs, ELRA pairs and groups semantics to exploit information in ID references and meanwhile leverage the efficiency benefit of tree model. Experimental evaluation shows the advantage of our methods. In the future, we would like to study relevance oriented ranking for keyword search in our model.

References

1. Berkeley DB, http://www.sleepycat.com/
2. Cohen, S., Kanza, Y., Kimelfeld, B., Sagiv, Y.: Interconnection semantics for keyword search in Xml. In: Proc. of CIKM Conference, pp. 389–396 (2005)
3. Guo, L., Shao, F., Botev, C., Shanmugasundaram, J.: XRANK: Ranked keyword search over XML documents. In: SIGMOD, pp. 16–27 (2003)
4. Kacholia, V., Pandit, S., Chakrabarti, S., Sudarshan, S., Desai, R., Karambelkar, H.: Bidirectional expansion for keyword search on graph databases. In: Proc. of VLDB Conference, pp. 505–516 (2005)
5. Kimelfeld, B., Sagiv, Y.: Efficiently enumerating results of keyword search. In: Proc. of DBPL Conference, pp. 58–73 (2005)
6. Li, Y., Yu, C., Jagadish, H.V.: Schema-free XQuery. In: VLDB, pp. 72–83 (2004)
7. Schmidt, A., Kersten, M.L., Windhouwer, M.: Querying Xml documents made easy: Nearest concept queries. In: ICDE, pp. 321–329 (2001)
8. Xu, Y., Papakonstantinou, Y.: Efficient keyword search for smallest LCAs in XML databases. In: Proc. of SIGMOD Conference, pp. 537–538 (2005)

Storage Techniques for Multi-versioned XML Documents

Laura Irina Rusu[1], Wenny Rahayu[2], and David Taniar[3]

[1,2] LaTrobe University, Department of Computer Science & Computer Eng, Australia
lirusu@students.latrobe.edu.au, wenny@cs.latrobe.edu.au
[3] Monash University, School of Business Systems, Clayton, VIC 3800, Australia
David.Taniar@infotech.monash.edu.au

Abstract Following the increased popularity of using XML documents for information exchange between applications and representing semi-structured data, the issue of warehousing collections of XML documents has become strongly imperative. The focus of this paper is on the storage of dynamic (multi-versioned) XML documents. In this paper we present four different storage methods for dynamic XML documents, and evaluate each of them using two performance indicators, namely I/O loading and versioning costs.

1 Introduction

The growing use of XML format for the exchange of information between legacy systems as well as for representing semi-structured data in web applications [9] have stimulated the need for the development of XML-specific data warehousing models.

In our previous work [13, 14] we have identified and described: (i) static XML documents and (ii) dynamic XML documents. A *static XML document* does not change its content and/or structure in time and does not possess a temporal element; each document is seen solely as a snapshot of data. A *dynamic XML document* may change its structure and/or content regularly due to certain business processes. It inherently contains the notion of time, whereby at least one element indicates the temporal coverage of the specific version of the document.

Changes to XML documents make them dynamic. Each set of changes applied to an XML document will produce a new XML document, which is considered a new version of the old one. However, versioning for dynamic XML documents has a different nature from versioning in traditional relational databases.

Therefore, the issue of warehousing collection(s) of versions of XML documents is imperative when the access to the historical data is required. In this paper, we present a *thorough analysis of four different storage techniques* for dynamic XML documents, aiming to find the one which is the most flexible and efficient, regardless of the type and timestamp of the requested historic version.

2 Related Work

A substantial amount of work has been carried out during the last few years to find solutions to the issue of warehousing XML documents; several research directions

J.R. Haritsa, R. Kotagiri, and V. Pudi (Eds.): DASFAA 2008, LNCS 4947, pp. 538–545, 2008.

are: conceptual design [8, 9], design based on user requirements and XML schema [16], OLAP and XML data cubes [10], modelling and querying [11], Peer-to-Peer XML warehousing [1, 2] etc.

The majority of work has considered XML documents as static; hence the work in warehousing dynamic XML documents is still in its infancy. One of the first works is [6], which looked at the versions management issue for multi-versioned XML documents in the context of Xyleme project [12]; the discussion was also continued and extended in [4]. The authors introduced the concepts of 'persistent identifier' and 'completed delta' and the storage policy was to keep the most current version and a document showing all changes between versions.

The authors of [16] argue that the abovementioned storage policy does not respond well to complex queries, and they propose a technique which stores successive versions of an XML document in a 'V-document' which increases, incrementally, with each new version. In [13], it is acknowledged that the approach in [16] resolves the issue of redundancy to a certain degree, but there are still issues to be addressed, such as redundancy in storing the temporal attributes, overheads in performing updates, difficulty in determining the actual types of changes, etc.

While analysing the existing approaches in warehousing dynamic XML documents we found that none of the existing works has addressed and highlighted the more complex issues related to the dynamic XML warehousing task, e.g. efficiency of storage and retrieval of historical versions. Hence, we perceived the need for a systematic study of different possibilities of storing data from multi-versioned XML documents. We have identified four different possible storage techniques and this paper proposes to analyse them, determine their strengths and weaknesses, and decide which is the most efficient, based on two cost indicators.

3 Multi-versioned XML Storage Techniques

For the purpose of the techniques presented in the next sections, the term *XML document* (*XMLDoc*) refers to a well-formed and valid XML document, having unique identifiers attached to each node (the method used to assign IDs to nodes is not critical) and an attached tree-representation T.

In the case of dynamic (multi-versioned) XML documents, each new version is a new *XMLDoc* as defined above, obtained by modifying the previous version; this is done by applying a combination of *insert*, *update* and *delete* operations.

Definition 1. Given a dynamic XML document, in two states, D and D', where $D \neq D'$, a generic **XMLdelta** document (noted Δ) records the changes of the document from one state to another. It consists of a set of basic operations O (update, insert, delete), which, upon execution on the document in state D, will return the document in state D'.

Definition 2. Given two successive versions $XMLDoc_1$ and $XMLDoc_2$ of an XML document *XMLDoc*, **XMLDeltaForward** is defined as the *XMLDelta* which records the changes made to $XMLDoc_1$ to produce $XMLDoc_2$ and **XMLDataBackward** as the *XMLDelta* which records the changes made to $XMLDoc_2$ to produce $XMLDoc_1$.

3.1 Storage Technique 1 – SFD (Store the First XML Document and all Forward Deltas)

Definition 3. An **ordered collection of versions of XML** documents is defined as $D = \{D_0, D_1, \ldots D_n\}$, where:

o D_0 represents the initial version of the XML document and D_n represents the last (most current) version of the XML document;
o $D_1, D_2, \ldots D_{n-1}$ are the intermediate versions of the XML document;
o The document versions are ordered in time, that is $D_0 <_T D_i <_T D_j <_T D_n$, where i < j< n. *Note* that '$<_T$' symbol signifies the precedence in time of the left-side item relating to the right-side item ($a <_T b$ means that '*a* happens before *b* happens').

Definition 4. Considering the ordered collection of versions of an XML document D, defined as per Definition 3, an **ordered collection of forward deltas** is defined as $C = \{\Delta_1, \Delta_2, \ldots \Delta_n\}$, where:
o each Δ_i is a forward delta and $\Delta_i = D_i - D_{i-1}$; the *starting forward delta* is $\Delta_1 = D_1 - D_0$; the *ending(last available) forward delta* is $\Delta_n = D_n - D_{n-1}$;
o the deltas in the collection C are *consecutive*, that is, $\Delta_j = D_j - D_i$ where $j = i + 1$; more, the collection C of deltas is *ordered* ($\Delta_1 <_T \Delta_i <_T \Delta_j <_T \Delta_n$) and *linear*.

Definition 5. The **SFD storage method** is defined as the initial version of the XML document plus the set of forward deltas associated, that is $SFD = D_0 \cup \{\Delta_i \mid 1 \leq i \leq n\}$.

The *size of the SFD storage* is given by the sum of the sizes of all forward deltas, plus the size of the initial version:

$$| SFD | = | D_0 | + \sum_{i=1}^{n} |\Delta_i | \tag{1}$$

Definition 6. The **SFD versioning technique** is the recursive process of retrieving an historic D_k version of the XML document, as follows: $\forall i, 1 \leq i \leq k \leq n, D_i = D_{i-1} + \Delta_i$ and $D_1 = D_0 + \Delta_1$; the versioning process is *ordered* (Δ_i can be applied on D_{i-1} only after Δ_{i-1} has been applied on D_{i-2}) and *linear* (only Δ_i and no other forward delta is applied on D_{i-1}).

Considering the above definitions, the *cost of retrieving the kth version* will be the sum of the costs of the intermediary steps, which is the cost of applying the operations from each forward delta Δ_i on the previous constructed version D_{i-1}:

$$Cost_{SFD}^{K} = Cost(\Delta_1) + Cost(\Delta_2) + \ldots + Cost(\Delta_k) = \sum_{i=1}^{k} Cost(\Delta_i) \tag{2}$$

In Eq. (2), each $Cost(\Delta_i)$ is calculated as the sum of the costs assigned for each individual operation (insert, update, delete) from each forward delta Δ_i.

One major weakness of the SFD approach is the inflexible process when a closer-to-date version is required.

3.2 Storage Technique 2 – SLD (Storing Last version of the XML Document and the Backward Deltas)

Definition 7. Considering the ordered collection of versions of an XML document D defined as per Definition 3, an **ordered collection of backward deltas** is defined as $C' = \{\Delta'_i \mid 1 \le i \le n\}$, where:

o each Δ'_i is a backward delta and $\Delta'_i = D_{i-1} - D_i$; the *starting backward delta* is $\Delta'_1 = D_0 - D_1$ and the *ending backward delta* is $\Delta'_n = D_{n-1} - D_n$;

o the backward deltas in C' collection are: *consecutive* ($\Delta'_j = D_i - D_j$ where $j = i+1$), *ordered* ($\Delta'_1 <_T \Delta'_i <_T \Delta'_j <_T \Delta'_n$) and *linear*.

Definition 8. The **SLD storage method** is defined as the last version of the XML document plus the set of backward deltas, that is $SLD = D_n \cup \{\Delta'_i \mid 1 \le i \le n\}$.

The *size of the SLD storage* is given by the sum of the sizes of all backward deltas, plus the size of the last version:

$$| SLD | = | D_n | + \sum_{i=1}^{n} | \Delta'_i | \qquad (3)$$

Definition 9. The **SLD versioning technique** is the recursive process of retrieving an historic D_k version of the XML document as follows: $\forall i, 1 \le i \le n-k, D_{n-i} = D_{n-i+1} + \Delta'_{n-i+1}$. The versioning process is *ordered* (Δ'_{n-i} is applied on D_{n-i} only after Δ'_{n-i+1} has been applied on D_{n-i+1}) and *linear* (Δ'_{n-i} and no other delta is applied on D_{n-i}).

The *cost of retrieving the kth version* when using the SLD storage technique is the sum of the costs of applying the operations from each backward delta Δ'_i on the previously obtained D_i:

$$Cost_{SLD}^{k} = Cost(\Delta'_n) + Cost(\Delta'_{n-1}) + ... + Cost(\Delta'_{k+1}) = \sum_{j=k+1}^{n} Cost(\Delta'_j) \qquad (4)$$

In Eq.(4), each $Cost(\Delta'_i)$ is calculated as the sum of the costs assigned to each individual operation (insert, update, delete) in each backward delta Δ'_i.

In a similar way with the issue identified in the SFD approach, the major weakness of the SLD storage technique is the inflexible process to be followed when an early version is required. In each of the two methods previously presented, the user is limited to using the same method (SFD, respectively SLD) to retrieve the desired historic version, even when it would be easier and / or cheaper to get it using the opposite method. Therefore, a combination of the SFD and SLD techniques is presented next, where the cost method is employed to determine the best direction (forward or backward) of reconstructing the documents during the versioning process.

3.3 Storage Technique 3 - SFLD (Storing First version, Last version, and the Deltas)

This technique uses the concept of "completed delta" which can act as either forward or backward delta, depending on the direction of reconstruction. For two versions of the XML document ($XMLDoc_1$ and $XMLDoc_2$), the operations which are recorded in the completed delta are actually a combination of the operations from the forward delta and backward delta associated.

Definition 10. The **SFLD storage technique** is defined as the last version of the XML document plus the initial version of the XML document plus the set of completed deltas, that is $SFLD = D_0 \cup D_n \cup \{\Delta_i^c \mid 1 \le i \le n\}$, where Δ_i^c is the completed delta between D_{i-1} and D_i .

The *size of the SFLD storage* is given by the sum of all completed deltas' sizes, plus the size of the first and initial versions (D_0, respectively D_n):

$$\left|SFLD\right| = \left|D_0\right| + \left|D_n\right| + \sum_{i=1}^{n} \left|\Delta_i^c\right| \tag{5}$$

Every time when a new version D_i of the document arrives, the completed delta Δ_i^c is built and two separate costs are calculated, that is C_i^f (the forward cost of transforming D_{i-1} into D_i) and C_i^b (the backward cost of transforming D_i into D_{i-1}).

Note that the formulas for calculating C_i^f and C_i^b are the same as those used in the SFD and SLD techniques, proposed in the previous subsections; hence,

$$C_i^f = Cost(\Delta_i) \text{ and } C_i^b = Cost(\Delta_i') \text{ where } 1 \le i \le n \tag{6}$$

Combining equations (2), (4) and (6), we can say that

$$Cost_{SFD}^k = \sum_{i=1}^{k} C_i^f \text{ and } Cost_{SLD}^k = \sum_{j=k+1}^{n} C_j^b \tag{7}$$

Definition 11. Considering the SFLD storage method presented above, the **SFLD versioning technique** is a process of retrieving a historic D_k version of the XML document, which observes the following conditions:

- $\forall k, 1 \le k \le n$, SFLD uses SFD versioning technique if $Cost_{SFD}^k \le Cost_{SLD}^k$
- $\forall k, 1 \le k \le n$, SFLD uses SLD versioning technique if $Cost_{SLD}^k < Cost_{SFD}^k$

In other words, when the user attempts to retrieve the kth version of the XML document using the SFLD storage technique, the cost of recreating it starting from the initial version D_0 by using the forward method ($Cost_{SFD}^k$) is compared with the cost of recreating it starting from the last version D_n by using the backward method ($Cost^k_{SLD}$) and the direction which shows the smaller cost is chosen.

The SFLD proposal attempts to overcome the limitations of the SFD and SLD methods; that is, it does not force a compulsory direction of reconstructing the intermediate documents, and considers the costs involved. The calculation of both C_i^f and C_j^b values for each incoming version of the dynamic document might seem to be an overhead for the process, but it is certainly justified by the great flexibility in choosing the lowest-cost direction of reconstruction (versioning) of the required document.

3.4 Storage Technique 4 - CΔ (the Consolidated Delta)

We have showed that the SFLD method is much more flexible than SFD and SLD, by allowing the selection of optimal direction of versioning. Though, all three techniques raise a number of *issues*, relating especially to the redundancy of data or operations:

 o *Redundancy of data stored in deltas*;
 o *Redundancy in recreating elements in consecutive versions*;
 o *Redundancy in retrieving historical versions*;
 o SFD, SLD and SFLD require a *system of tracking* of the deltas.

A different method of storing the changes between versions of dynamic XML documents was proposed in [13], by using the concept of *consolidated delta*. In summary, this is the initial document plus the changes stored on top of it, with some clear rules to follow while building the consolidated delta.

The main difference between consolidated delta and SFD, SLD and SFLD techniques is that there is no operational cost involved in versioning using consolidated delta. When version D_k is required, we do not need to recreate the entire set of intermediate documents from D_n to D_k (as it would have been done in SLD or SFLD) or the set of intermediate documents from D_1 to D_k (as it would have been done in SFD or SFLD). Instead, we query the consolidated delta for the elements which have the timestamp T_k (please see [13] for more details about the versioning process).

4 Performance Indicators and Experimental Results

In this section, we compare two indicators which are representative for all four techniques detailed in the previous section.

Definition 12. Given a storage technique S, the **Loading cost** involved in retrieving any historic version D_k, noted as $I/OCost^k_S$, is defined as the cost of loading in memory all the documents required in order to reconstruct the requested version. This is calculated by dividing the total size of the required documents to the memory page size (noted here with σ).

Definition 13. Given a storage technique S, the **Cost of versioning** (or the **Operational Cost**) for retrieving any historic version D_k, noted as $OpCost_S^k$, is given by the total cost of the operations necessary to reconstruct the requested version, divided by the total number of nodes in the reconstructed document (N_k).

Table 1 shows how the above indicators are calculated, for the techniques presented earlier in the paper (where N_k is the number of the nodes in the retrieved version D_k).

Table 1. Loading costs and efficiency formulas for the presented techniques

Storage technique	$I/OCost_S^k$ (loading cost)	$OpCost_S^k$ (versioning cost)
SFD	$\left(\|D_0\| + \sum_{k=1}^{n} \|\Delta_i\| \right)/\sigma$	$Cost_{SFD}^k / N_k = \left(\sum_{i=1}^{k} Cost(\Delta_i) \right)/N_k$
SLD	$\left(\|D_n\| + \sum_{j=k+1}^{n} \|\Delta_j'\| \right)/\sigma$	$Cost_{SLD}^k / N_k = \left(\sum_{j=k+1}^{n} Cost(\Delta_j') \right)/N_k$
SFLD	$MIN\left(I/OCost_{SFD}^k, I/OCost_{SLD}^k \right)$	$MIN\left(OpCost_{SFD}^k, OpCost_{SLD}^k \right)$
CΔ	$\|C\Delta\|/\sigma$	$Cost_{C\Delta}^k = 0$ (no operations required) $\Rightarrow Cost_{C\Delta}^k / N_k = 0$ (max efficiency)

We have conducted a number of experiments in order to compare the performance indicators. We have used four dynamic XML documents from [15], of various sizes. For each document, we have applied different percentages of mixed and random changes, in order to obtain multiple successive versions. For each sequence of versions, we have determined the collections of forward deltas, backward deltas and completed deltas; also, we have calculated the C_i^f and C_j^b costs and built the consolidated delta(s). From the results of our set of experiments, we appreciate that the consolidated delta approach is very efficient in most of the versioning circumstances. The I/OCost using consolidated delta is indeed dependent on the number of changes between document's versions, especially for a large document, but there is the option of splitting it into smaller consolidated deltas and using only those required by the versioning process. From the operational cost point of view, though, the results obtained using CΔ are the best, because no operations are involved in versioning.

5 Conclusions

In this paper, we have analysed four different storage techniques and finally we have concluded that the consolidated delta option is the most efficient one for the purpose of warehousing multi-versioned XML documents. We have calculated and compared two performance indicators (loading and versioning costs) for each of the four options.

References

1. Abiteboul, S., Manolescu, I., Preda, N.: Peer-to-peer warehousing of XML resources, 20èmes Journées Bases de Données Avancées (BDA 2004). Montpellier, Actes (Informal Proceedings), pp. 343–346 (2004)
2. Abiteboul, S., Manolescu, I., Preda, N.: Constructing and Querying Peer-to-Peer Warehouses of XML Resources. In: Proceedings of the 21st International Conference on Data Engineering (ICDE 2005), Tokyo, Japan, pp. 1122–1123. IEEE Computer Society, Los Alamitos (2005)
3. Chien, S.Y., Tzotras, V.J., Zaniolo, C., Zhang, D.: Storing and Querying Multiversion XML Documents using Durable Node Numbers. In: Proceedings of the 2nd International Conference on Web Information Systems Engineering (WISE 2001), Kyoto, Japan, pp. 232–241. IEEE Computer Society, Los Alamitos (2001)
4. Cobena, G., Abiteboul, S., Marian, A.: Detecting Changes in XML Documents. In: Proceedings of the 18th International Conference on Data Engineering (ICDE 2002), San Jose, CA, US, pp. 41–45. IEEE Computer Society, Los Alamitos (2002)
5. Elmasri, R., Navathe, S.: Fundamentals of Database Systems, 5th edn. Addison Wesley, Reading (2006)
6. Marian, A., Abiteboul, S., Cobena, G., Mignet, L.: Change-Centric Management of Versions in an XML Warehouse. In: Proceedings of the 27th International Conference on Very Large Data Bases (VLDB 2001), pp. 581–590. Morgan Kaufmann Publishers, San Francisco (2001)
7. Mignet, L., Barbosa, D., Veltri, P.: The XML Web: A first study. In: proceedings of the International WWW Conference (WWW 2003), Budapest, Hungary, pp. 500–510 (2003)
8. Nassis, V., Rajugan, R., Dillon, T., Rahayu, W.: Conceptual Design of XML Document Warehouses. In: Kambayashi, Y., Mohania, M., Wöß, W. (eds.) DaWaK 2004. LNCS, vol. 3181, pp. 1–14. Springer, Heidelberg (2004)
9. Nassis, V., Dillon, T., Rajugan, R., Rahayu, W.: An XML Document Warehouse Model. In: Li Lee, M., Tan, K.-L., Wuwongse, V. (eds.) DASFAA 2006. LNCS, vol. 3882, pp. 513–529. Springer, Heidelberg (2006)
10. Park, B.K., Han, H., Song, I.L.: A Multidimensional Analysis Framework for XML Warehouses. In: Tjoa, A.M., Trujillo, J. (eds.) DaWaK 2005. LNCS, vol. 3589, pp. 32–42. Springer, Heidelberg (2005)
11. Pokorny, J.: XML Data Warehouse: Modelling and Querying. In: Proceedings of the 5th International Conference on Databases and Information Systems II (Baltic DB&IS 2002), Estonia, pp. 67–80. Kluwer Academic Publishers, Dordrecht (2002)
12. Xyleme, L.: A Dynamic warehouse for XML Data of the Web. IEEE Data Engineering Bulletin 24(2), 40–47 (2001)
13. Rusu, L.I., Rahayu, W., Taniar, D.: Maintaining Versions of Dynamic XML Documents. In: Ngu, A.H.H., Kitsuregawa, M., Neuhold, E.J., Chung, J.-Y., Sheng, Q.Z. (eds.) WISE 2005. LNCS, vol. 3806, pp. 536–543. Springer, Heidelberg (2005)
14. Rusu, L.I., Rahayu, W., Taniar, D.: Warehousing Dynamic XML Documents. In: Tjoa, A.M., Trujillo, J. (eds.) DaWaK 2006. LNCS, vol. 4081, pp. 175–184. Springer, Heidelberg (2006)
15. Sigmod XML dataset (2007), available online at: www.cs.washington.edu/datasets
16. Wang, F., Zaniolo, C.: Temporal queries in XML Document Archives and Web Warehouses. In: Proceedings of the 10th International Symposium on Temporal Representation and Reasoning / 4th International Conference on Temporal Logic (TIME-ICTL 2003), pp. 47–55. IEEE Computer Society, Los Alamitos (2003)

Twig'n Join: Progressive Query Processing of Multiple XML Streams

Wee Hyong Tok, Stéphane Bressan, and Mong-Li Lee

School of Computing
National University of Singapore
{tokwh,steph,leeml}@comp.nus.edu.sg

Abstract. We propose a practical approach to the progressive processing of (FWR) XQuery queries on multiple XML streams, called Twig'n Join (or TnJ). The query is decomposed into a query plan combining several twig queries on the individual streams, followed by a multi-way join and a final twig query. The processing is itself accordingly decomposed into three pipelined stages progressively producing streams of XML fragments. Twig'n Join combines the advantages of the recently proposed TwigM algorithm and our previous work on relational result-rate based progressive joins. In addition, we introduce a novel dynamic probing technique, called Result-Oriented Probing (ROP), which determines an optimal probing sequence for the multi-way join. This significantly reduces the amount of redundant probing for results. We comparatively evaluate the performance of Twig'n Join using both synthetic and real-life data from standard XML query processing benchmarks. We show that Twig'n Join is indeed effective and efficient for processing multiple XML streams.

Keywords: XML, Progressive Join.

1 Introduction

The ubiquity of network accessible XML data necessitates the design of XML query processors which can process complex queries over multiple XML data streams. We need to devise XML query processors for XML languages such as XPath or XQuery that supports the processing of structural and predicate constraints as well as join queries [1] over multiple XML data streams. In order to ensure a good user experience, the XML query processors must deliver initial results quickly, and maintain a consistent high result throughput. Main memory is limited and when it is full, data needs to be flushed to disk. As we need to produce results progressively with a high throughput, we need to effectively manage the XML data that is kept in memory and favor data that is most likely to contribute to the result. A key insight is to make use of statistics from either the input (i.e. data) or output (i.e. result) distributions. The problem of effective management of in-memory data was first studied in [2,3] for relational equijoins. However, to our knowledge, no work exists for progressive XML join processing.

J.R. Haritsa, R. Kotagiri, and V. Pudi (Eds.): DASFAA 2008, LNCS 4947, pp. 546–553, 2008.
© Springer-Verlag Berlin Heidelberg 2008

In this work, we propose a practical approach, called Twig'n Join (TnJ), to the progressive processing of XQuery queries on multiple XML streams. A For-Where-Return (FWR) XQuery query is first decomposed into a query plan which consists of twig queries over the input streams, followed by a multi-way join and a final twig query. TnJ processes the query plan on-the-fly and delivers the results progressively. The novelty of this approach comes from the decomposition of the XQuery queries into several independent components. This reduces the complexity for the design of XQuery query processing algorithms. We also introduce a dynamic probing technique, called Result-Oriented Probing (ROP), which is to effectively determine an optimal probing sequence for the multi-way join. This significantly reduces the amount of redundant probing for results. Using both synthetic and real-life data from standard XML query processing benchmarks, we comparatively evaluate the performance of the Twig'n Join variants. We show that Twig'n Join, with the Result-Oriented Probing (ROP) is an effective and efficient technique for processing multi-way XML value joins.

The rest of the paper is organized as follows: In section 2, we discuss related work. In Section 3, we propose Twig'n Join, a progressive join algorithm for processing multiple XML data streams. We conduct the performance evaluation in Section 4. Due to space constraints, the results for other XML benchmarks and real-life datasets are omitted from the paper. Interested readers are referred to the extended version of this paper [4] for more details. We conclude in Section 5.

2   Related Work

Many techniques [5,6,7,8] have been proposed for processing XPath and XQuery queries on single XML document or stream. In [5], the BEA/XQRL processor was proposed to support pipelined execution by using an iterator model over the data stream. [7] proposed transformation techniques to enable XQuery queries to be evaluated in one-pass. In addition, [7] proposed code generation techniques (from the XQuery queries) to handle user-defined aggregates and recursive functions. [8] proposed the *TwigM* machine, an efficient non-blocking method for evaluating twig queries over a single XML data stream. *TwigM* assumes an input sequence of SAX events (i.e. startElement, endElement), and uses a stack-based structure to compactly encode the solutions to the Twig join. The output consists of XML fragments. None of these techniques considered XML query processing over multiple XML data streams.[1] proposed a Massively Multi-Query Join Processing (MMQJP) technique for processing value joins over multiple XML data streams. MMQJP can only deliver results when the XML documents have arrived entirely. In contrast to MMQJP, our proposed technique delivers results progressively as portions of the streamed XML documents arrived.

3   Twig'n Join

In this section, we discuss a practical approach to the progressive processing of (FWR) XQuery queries on multiple XML streams, called Twig'n Join (TnJ).

Given two XML data streams, R and S, where the XML data are delivered tag by tag from remote data sources. Twig pattern (extracted from the XQuery query) T_r and T_s are defined for R and S respectively. XML result fragments F_r and F_s are produced for portions of the XML documents that matches T_r and T_s respectively. The user define a set of join attributes A in which the XML fragments can be joined. A result $<F_r, F_s>$ is reported if F_r and F_s fulfill the join attribute condition defined by A. Our goal is to be able to progressively deliver the result.

A FWR XQuery query can be decomposed into three parts: (1) Structural filtering on the input streams (2) Predicate Processing and (3) Structural filtering on the results. We assume that a XQuery pre-processor will parse the (FWR) XQuery expression and generate a query plan. During predicate processing, we perform value-based filtering as well as process the joins between the input streams. In our work, we focus on join processing. Figure 1 shows a possible query plan generated. We note that further optimization of the query plan is possible. However, we consider query optimization as an orthogonal issue, and do not explore it in this paper.

The query plan consists of several twig queries on the individual XML streams, followed by predicate processing and a final twig query. XML data are continuously streamed from remote sites. The data is then matched using the two twig matching operators $(TM_A$ and $TM_B)$. The output from TM_A and TM_B (XML fragments) are then joined using a join operator (i.e. predicate processing).

We use the TwigM machine [8] to efficiently perform the twig matches on the streaming XML data, and a hash-based join for joining the data. Whenever a new XML fragment, f_d, is produced by the TwigM machine, we first use it to probe the corresponding hash partition from the other data stream. Based on the join predicates defined, we check each of the XML fragments found in the partition. Results are output whenever the join predicates are satisfied. In addition, a counter, *numResults*, keeps track of the results produced by each of the partitions. The counter is updated when all the results have been produced.

After probing, the XML fragment is then inserted into its corresponding hash partition. When intermediate XML fragments are continuously produced by the twig matching operators, the memory might become full. Thus, some of the in-memory XML fragments need to be flushed to disk. We make use of the generic Result Rate-based flushing (RRPJ) technique used in [3] to determine the XML fragments to be flushed to disk.

3.1   Multi-way Join

In this section, we discuss how we can generalize Twig'n Join for processing XQuery queries on multiple XML streams.

Existing multi-way join techniques for relational equi-join, such as MJoin [9], can be used as to handle the multi-way between the XML fragments that are produced. The performance of the multi-way join is dependent on the probing sequence. For example, MJoin sorts the hash partitions based on the join selectivity. The key intuition is that by probing partitions with a lower join selectivity

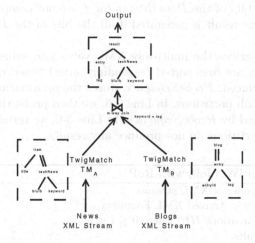

Fig. 1. Query Execution Plan

first, we can filter away tuples that will not generate any result early during the multi-way join. This helps to reduce the number of unnecessary probes to the remaining un-probed hash partitions. When the inputs to the multi-way join consists of intermediate results from a pipelined process, it is difficult to determine the join selectivity of the hash partitions. For example, the XML fragments used in the join are produced by the TwigM machines. Even if the join selectivity of the join attribute for the base XML streams can be accurately determined, it is not straightforward to determine the join selectivity of the intermediate XML fragments produced. In addition, determining the join selectivity apriori might not be useful if the join selectivity changes during the lifetime of the multi-way join.

In order to deal with the problem of determining an effective probing sequence for the multi-way join, we propose a novel technique, called Result-Oriented Probing (RoP). RoP dynamically determines the order of the hash partitions to be probed in the multi-way join. RoP tracks the number of partial results that are produced by each hash partition. In our work, we consider only multi-way join queries with the same join predicates.

Each XML fragment, produced by the TwigM machine is augmented with a bitmap (i.e DoneBitmap). The *DoneBitmap* is used to determine whether the XML fragment has been used to probe a particular partition. *DoneBitmap* consists of n bits. Each bit corresponds to one of the n XML data streams. When a XML fragment is created, its *DoneBitmap* is initialized to 0s. The bit corresponding to the partition , in which the XML fragment is inserted, is set to 1. Whenever the XML fragment is used to probe the hash partitions for the other XML streams, the bit corresponding to the hash partition is set to 1 when it can be used to join with at least one XML fragment in the partition. When all the bits of the *DoneBitmap* are set, the XML fragment is output as a result. Whenever a XML fragment f is used to probe a hash partition, a partial result is generated if the bits of the *DoneBitmap* for f are not completely set to 1. In contrast, a complete result is generated if all the bits of the *DoneBitmap* for f are set to 1.

Algorithm 1 describes the multi-way XML value join using RoP. In Line 1, the hash partitions are first sorted (ascending order) based on the number of partial results produced. *ProbeSequence* stores the information on the probing sequence for the hash partitions. In Line 2-6, we then probe the hash partitions in the order specified by *ProbeSequence*. In Line 5-6, we terminate the probing when one of the partitions do not produce any results.

Algorithm 1. MultiWayJoin with RoP

> **Data**   : n - Number of XML streams
> f_d - Newly Arrived XML Fragment,
> Hash Partitions Ht_i, where $0 \leq i < n$
>
> **Result** : R, Results
> **begin**
>
> 1 | *ProbeSequence* = SortHashPartitionsAsc() ;
> 2 | **for** *(i=0; i < n; i++)* **do**
> 3 | | idx = *ProbeSequence$_i$* ;
> 4 | | numResults = Ht_{idx}.probe(f_d) ;
> 5 | | **if** *(numResults == 0)* **then**
> 6 | | └ break;
>
> **end**

---

4   Performance Evaluation

We implemented all the algorithms in C++, and conduct the experiments on a Pentium 4 2.4 Ghz PC (1GB RAM). Similar to [8], we make use of the SAX Parser - Expat [10]. Whenever memory is full, we flush 10% of the in-memory XML Fragments flushed to disk. We compare the performance of Twig-RPJ (naive extension to RPJ [2] for XML), Twig'n Join (TnJ). In addition, we also included a Random method as a baseline. Whenever memory is full, the Random method randomly selects a partition (containing XML Fragments) to be flushed to disk. We evaluated the performance of all the algorithms using both synthetic datasets as well as several real-life datasets.

4.1   XMark

In this section, we evaluate the performance of Twig'n Join and Twig-RPJ using synthetic datasets generated using XMark [12]. XMark generates a single XML document consisting of information on the annotation, person, category, closed auction, open auction and the items. For the purpose of the experiments, we extracted out the details of the items and closed auctions into 2 separate XML

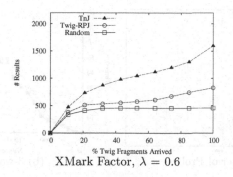

XMark Factor, $\lambda = 0.6$

Fig. 2. XMark

files. This is used to simulate two XML data streams. In this experiment, we join the item IDs reference of the closed auctions with the items. The join attribute is *id* (string). The following twig queries are defined on the Item and Closed Auctions streams respectively: **Items**: //item[id][name] and **ClosedAuctions**: //closed_auction[itemref/id][price].

In the experiment, we vary the scaling factor of XMark, λ, between 0.2 and 2.0. As we observed similar trends for varying XMark factor, we present only the results for $\lambda = 0.6$ in Figure 2. In all cases, Twig'n Join outperforms Twig-RPJ and the random flushing strategy by a large margin.

4.2 Multi-way XML Join

In this section, we compare the performance of the multi-way join using various probing techniques. These includes: (1) RoP (2) Sequential and Apriori. RoP uses the dynamic probing sequence described in Section 3.1. In the Sequential probing strategy, we probe the hash partitions in the order in which the XML streams arrive. In the Apriori strategy, we assume that we know the join selectivity of each of the XML streams. We then probe the hash partitions in order of increasing join selectivity. Thus, hash partitions with lower join selectivity are probed first. We evaluate the performance based on two metrics. Firstly, we count the total number of probes on the hash partitions. Secondly, we measured the time taken to produce results.

The XML streams used in this experiment is generated as follows. We first extracted all the name of authors from SIGMOD Record [11]. Using the names of authors, we generated a reference XML stream in which consists of blog entries written by the authors. Next, we generated the other XML streams to be used in the multi-way join by controlling the selectivity, μ. μ determines the probability that a author from the reference XML stream is included in the stream to be generated. We vary μ between 0.0 to 1.0. When $\mu = 0.0$, none of the authors from the reference XML stream are included. When $\mu = 1.0$, all the authors from the reference XML streams are included. Various m-way joins are evaluated (m varies between 3 to 5). From Figure 3(a), we can observe that dynamic result-oriented probing (RoP) outperforms the Sequential probing technique.

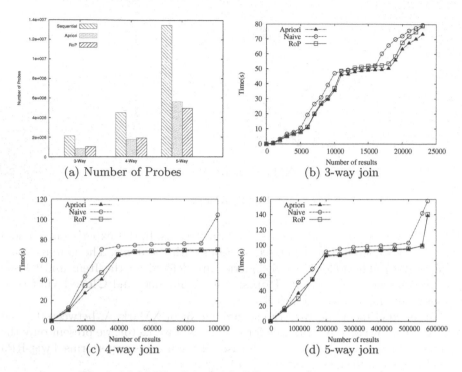

Fig. 3. Multi-Way Join (with different probing sequence)

In addition, RoP performs almost as well the Apriori strategy. This shows that the dynamic RoP technique is effective even without prior information on the join selectivities. From Figure 3(b)-(d), we can observe that RoP outperforms the Sequential probing technique for varying m. This is due to the significant reduction on the number of unnecessary probes.

5   Conclusion

We propose a practical approach for progressive processing of (FWR) XQuery queries on multiple XML streams, called Twig'n Join. We decompose a (FWR) XQuery query into a query plan consisting of twig queries and join processing. The twig queries are used for processing the structural constraints. Compared to conventional XQuery processing, the novelty of this approach lies in the decomposition of the XQuery queries into several independent components. This reduces the complexity for the design of XQuery query processing algorithms. Though we show this for XQuery queries involving joins, the technique can be applied to the the various type of (FWR) XQuery queries as well. Due to the large amount of streaming XML data, it is infeasible to keep all the XML data in-memory during join processing. We make use of the RRPJ method [3] to flush the XML data whenever memory is full.

In addition, we introduce a novel dynamic probing technique, called Result-Oriented Probing (RoP), which determines an optimal probing sequence for the multi-way join. This significantly reduces the amount of redundant probing for results. Experiment results show that Twig'n Join is indeed effective and efficient for the processing of both synthetic and real-life datasets.

References

1. Hong, M., Demers, A., Gehrke, J., Koch, C., Riedewald, M., White, W.: Massively multi-query join processing in publish/subscribe systems. In: SIGMOD, pp. 761–772 (2007)
2. Tao, Y., Yiu, M.L., Papadias, D., Hadjieleftheriou, M., Mamoulis, N.: RPJ: Producing fast join results on streams through rate-based optimization. In: SIGMOD, pp. 371–382 (2005)
3. Tok, W.H., Bressan, S., Lee, M.-L.: RRPJ: Result-rate based progressive relational join. In: Kotagiri, R., Radha Krishna, P., Mohania, M., Nantajeewarawat, E. (eds.) DASFAA 2007. LNCS, vol. 4443, pp. 43–54. Springer, Heidelberg (2007)
4. Tok, W.H., Bressan, S., Lee, M.L.: Twig'n join: Progressive query processing of multiple Xml streams. Technical Report TRA9/07, National University of Singapore (2007)
5. Florescu, D., Hillery, C., Kossmann, D., Lucas, P., Riccardi, F., Westmann, T., Carey, M.J., Sundararajan, A., Agrawal, G.: The bea/xqrl streaming xquery processor. In: Aberer, K., Koubarakis, M., Kalogeraki, V. (eds.) VLDB 2003. LNCS, vol. 2944, pp. 997–1008. Springer, Heidelberg (2004)
6. Peng, F., Chawathe, S.S.: Xpath queries on streaming data. In: SIGMOD, pp. 431–442 (2003)
7. Li, X., Agrawal, G.: Efficient evaluation of xquery over streaming data. In: VLDB, pp. 265–276 (2005)
8. Chen, Y., Davidson, S.B., Zheng, Y.: An efficient xpath query processor for Xml streams. In: ICDE, p. 79 (2006)
9. Viglas, S., Naughton, J.F., Burger, J.: Maximizing the output rate of multi-way join queries over streaming information sources. In: Aberer, K., Koubarakis, M., Kalogeraki, V. (eds.) VLDB 2003. LNCS, vol. 2944, pp. 285–296. Springer, Heidelberg (2004)
10. Clark, J.: The expat Xml parser (2003), http://expat.sourceforge.net
11. XML Data Repository (2002),
http://www.cs.washington.edu/research/xmldatasets/
12. Schmidt, A., Waas, F., Kersten, M.L., Carey, M.J., Manolescu, I., Busse, R.: Xmark: A benchmark for Xml data management. In: Bressan, S., Chaudhri, A.B., Li Lee, M., Yu, J.X., Lacroix, Z. (eds.) CAiSE 2002 and VLDB 2002. LNCS, vol. 2590, pp. 974–985. Springer, Heidelberg (2003)

TwigBuffer: Avoiding Useless Intermediate Solutions Completely in Twig Joins

Jiang Li and Junhu Wang

School of Information and Communication Technology
Griffith University, Gold Coast, Australia
Jiang.li@student.griffith.edu.au, J.Wang@griffith.edu.au

Abstract. Twig pattern matching plays a crucial role in XML data processing. TwigStack [2] is a holistic twig join algorithm that solves the problem in two steps: (1) finding potentially useful intermediate path solutions, (2) merging the intermediate solutions. The algorithm is optimal when the twig pattern has only //-edges, in the sense that no useless partial solutions are generated in the first step (thus expediting the second step and boosting the overall performance). However, when /-edges are present, a large set of useless partial solutions may be produced, which directly downgrades the overall performance. Recently, some improved versions of the algorithm (e.g., TwigStackList and iTwigJoin) have been proposed in an attempt to reduce the number of useless partial solutions when /-edges are involved. However, none of the algorithms can avoid useless partial solutions completely. In this paper, we propose a new algorithm, TwigBuffer, that is guaranteed to completely avoid the useless partial solutions. Our algorithm is based on a novel strategy to buffer and manipulate elements in stacks and lists. Experiments show that TwigBuffer significantly outperforms previous algorithms when arbitrary /-edges are present.

1   Introduction

The importance of fast processing of XML data is well known. *Twig pattern matching*, which is to find all matchings of a query tree pattern in an XML data tree, lies in the center of all XML processing languages. Therefore, finding efficient algorithms for twig pattern matching is an important research problem.

Over the last few years, many algorithms have been proposed to perform twig pattern matching. Most of these algorithms find twig pattern matching in two phases. In the first phase, a query tree is decomposed into smaller pieces, and solutions against these pieces are found. In the second phase, all of these partial solutions are joined together to generate the final results. Binary *structural join*(e.g.,[1]) and *holistic twig join*(e.g., [2,3,4,5]) are two important types of two-phase twig pattern matching algorithms. Holistic twig join algorithms have significantly reduced the number of useless intermediate path solutions compared with structural join algorithms. When only //-edges are present in a query tree,

J.R. Haritsa, R. Kotagiri, and V. Pudi (Eds.): DASFAA 2008, LNCS 4947, pp. 554–561, 2008.

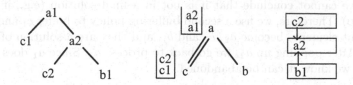

Fig. 1. An example to basic ideas of `TwigBuffer`

all of the intermediate paths will definitely appear in the final solutions. However, if /-edges are involved, none of them can completely eliminate the useless intermediate path solutions.

In this paper, we present a novel holistic twig join algorithm,`TwigBuffer`, that completely avoids the useless partial solutions for arbitrary twig patterns. With ingenious manipulation of data elements in *buffer stacks* and *result lists*, the algorithm also ensures the linear worst-case complexity in the first phase. Our experiments show that this algorithm significantly outperforms previous algorithms when arbitrary /-edges are present in the twig pattern.

The rest of the paper is organized as follows. `TwigBuffer` is presented in detail in Section 2. In Section 3, we discuss the correctness and time complexity of `TwigBuffer`.The experiment results are reported in Section 4. Finally, Section 5 concludes this paper.

2   TwigBuffer: Our Holistic Twig Join Algorithm

2.1   Overview of TwigBuffer

We explain the basic ideas used in `TwigBuffer` using the example in Fig. 1.

`TwigBuffer` avoids useless partial solutions by doing a *thorough* check of the P-C relationships, so that the current element of node n is regarded useful only if it has a descendant in T_j for each child j of n, and it has a child in T_i for each child i of n that has the P-C relationship with n, and every child of n recursively satisfies the above condition. Like `TwigStackList`, it buffers elements read from the streams in order to check the P-C relationships; but unlike `TwigStackList`, it uses two different buffering policies. The first buffering policy, *PCBuffering*, is used to buffer ancestors elements (and to some extent this is similar to the buffering in `TwigStackList`). To conduct thorough checking of the P-C relationships, when the buffer stack is not empty, it will check the top element only. To ensure no useful element is abandoned, a second type of buffering, *Sbuffering*, is used to buffer elements that are potentially descendants of elements in the stack, so that these elements can be checked later for A-D or P-C relationships with the elements in the parent stack.

In the example above, a_1 and a_2 will be buffered using the first buffering policy because they are ancestors of b_1. Now the current element of node a is a_2 (note: in `TwigStackList`, the current element is a_1).After this step, the current elements of query nodes become a_2, c_1 and b_1, but we cannot abandon c_1 yet, because we cannot conclude that it is not in a final solution (e.g., if a_1 has a b-child too). Therefore, we use a second buffering policy to buffer c_1 and c_2. Now the current elements become a_2, c_2 and b_1, and they are a solution of the twig pattern. After popping up a_2, we go back to process a_1. Since a_1 does not have a b-child, we know it can be abandoned.

2.2   Notation and Data Structures

Similar to other holistic twig pattern matching algorithms (e.g. TwigStack, TwigStackList), the Containment labels of the elements with the same value are stored or organized in one stream for access. For each stream T_n, there exists a pointer PT_n pointing to the current element. The function $Advance(T_n)$ can make the pointer PT_n point to the next element in the stream T_n. In addition, we can use $isEnd(T_n)$ to judge whether PT_n points to the position after the last element in the stream T_n.

There are two major types of data structures used in TwigBuffer. One is stacks for buffering elements read from the streams. The other is lists used for compactly storing and representing partial root-to-leaf solutions. Elements in a buffer stack are arranged in ascending order of the *start* value from bottom to top, but unlike TwigStackList, they may not be strictly nested. The elements in a result list are also in ascending order and strictly nested. An element is an ancestor of the elements after it.

For the buffer stacks, the basic operations are: *isEmpty, length, pop, push, top* and *bottom*. The functions built on the lists are: *head, tail, removeTail, isEmpty* and *insert*. The function *insert* should be noted. With it, an element can be inserted into a result list and the ascending order of elements is maintained. Similar to the linked stack in TwigStack, each element in the list contains a pointer pointing to an element in the parent list (See Fig. 1).

During query processing, the current element of each query node should be known. The basic rule is that if the buffer stack is not empty, the current element should stand at the top of the stack. Otherwise the current element will be the one in the stream pointed to by PT_n. Based on this rule, the function $getElement(n)$ will return the current element of node n for processing, and the function $proceed(n)$ will make the current element to be the next one:

$getElement(n)$:   returns $top(S_n)$ if S_n is not empty;
                  returns $getElementFromStream(n)$ otherwise.
$proceed(n)$: $pop(S_n)$ if S_n is not empty, $advance(T_n)$ otherwise.

For the nodes in twig pattern Q, the functions $isRoot(n)$ and $isLeaf(n)$ check whether node n is the root and is the leaf respectively, $parent(n)$ and $children(n)$ return the parent of n and the set of children of n, respectively, and $isPCChild(n)$ returns TRUE if n is not the root and n is connected to its parent by a /-edge.

For any two elements e_1 and e_2 in data tree t, the function $isAD(e_1, e_2)$ returns TRUE iff e_1 is an ancestor of e_2. The function $subtreeNodes(q)$ returns all of the roots of q's subtrees.

2.3 TwigBuffer

The *getNext(n)* function The function *getNext(n)*, shown in Algorithm 1, is a core function of `TwigBuffer`. The function takes a query node n as input , and returns a query node that may be n itself or a descendant of n.

Algorithm 1. getNext(n)

1: **if** $isLeaf(n)$ **then**
2:     **return** n
3: **for all** node n_i in $children(n)$ **do**
4:     **if** $isPCChild(n_i)$ **AND** $isAD(getElement(n), getElement(n_i))$ **then**
5:         $PCBuffering(n, n_i)$
6:     $r_i = getNext(n_i)$
7:     **if** $r_i \neq n_i$ **then**
8:         **return** r_i
9: $n_{min} = minarg_{n_i \in children(n)} getElement(n_i).start$
10: $n_{max} = maxarg_{n_i \in children(n)} getElement(n_i).start$
11: **if** $\neg isEmpty(BS_n)$ **AND** $\neg isEmpty(BS_{n_{min}})$ **then**
12:     **while** $getElement(n_{min}).end < getElement(n).start$ **do**
13:         $SBuffering(n, n_{min})$
14:         **if** $getElement(n_{min}).end < getElement(n).start$ **then**
15:             $Proceed(n)$
16: **if** $\neg isEmpty(BS_{n_{max}})$ **AND** $\neg isEmpty(BS_n)$ **then**
17:     **if** $getElement(n).end < getElement(n_{max}).start$ **then**
18:         **return** n_{max}
19: **while** $getElement(n).end < getElement(n_{max}).start$ **do**
20:     $Proceed(n)$
21: $PCBuffering(n, n_{max})$
22: $n_{min} = minarg_{n_i \in children(n)} getElement(n_i).start$
23: **if** $getElement(n_{min}).start < getElement(n).start$ **then**
24:     **return** n_{min}
25: **if** $ret := getNotSatisfyPC(n)$ **then**
26:     **return** ret
27: **else**
28:     **return** n

Before explaining the details of *getNext(n)*, we need to introduce the *buffering schemes*, which play an essential role in the matching process. Generally, there are two situations that buffering will happen:

First, if the query node n has the P-C relationship with its child n_i, and the current element in T_n is an ancestor of the current element in T_{n_i}, then buffering will occur and the general rule is:

PCBuffering: *Given a query node p and its P-C child c, suppose the current element of c is e_c. In the stream T_p, all of the elements that are ancestors of e_c will be buffered. The elements that are not ancestors of e_c will be skipped because they will not contribute to the final solutions. Apply this rule recursively on parent(p) if p is also a P-C child.*

Second, a different buffering occurs when the pointer of a stream needs to advance, but its parent's buffer stack is not empty. In other words, the current element in the stream can not be abandoned at current stage because it may be in the final solutions. The buffering rule is:

SBuffering: *Given a query node p and its child c, suppose some elements are buffered (through PCBuffering or SBuffering) in the buffer stack BS_p of the node p. In the stream T_c, all of the elements whose start value lies in the range (bottom(BS_p).start, top(BS_p).start) will be buffered in BS_c firstly, and then, the first set of strictly nested elements whose start value lies in the range (top(BS_p).start, top(BS_p).end) will also be buffered. After buffering, all of the elements in T_p on the left of the nested elements just mentioned above are skipped because they will not contribute to fianl solutions. The buffering process above may change the current element of node c, which means the PCBuffering on p may become invalid if c is a P-C child, so PCBuffering on p needs redo.*

In the explanation below, we use *current(n)* to denote the current element of node n.

Algorithm 2. Subroutines

1: **procedure** PCBUFFERING(p, c)
2:      $e_p = getElementfromStream(p)$
3:      $e_c = getElement(c)$
4:      **while** $e_p.start < e_c.start$ **do**
5:          **if** $e_p.end > e_c.end$ **then**
6:              $ClearBufferStack(p, e_p)$
7:              $MovetoBufferStack(p, e_p)$
8:          $Advance(T_p)$
9:      **if** $isPCChild(p)$ **AND** $isAD(getElement(parent(p)), getElement(p))$ **then**
10:         $PCBuffering(parent(p), p)$
11: **procedure** SBUFFERING(p, c)
12:     Buffer   all   the   elements   of   node c in   the   range   of $(Bottom(BS_p).start, Top(BS_p).start)$
13:     Buffer   the   first   set   of   nested   elements   of   node c in   the   range   of $(Top(BS_p).start, Top(BS_p).end)$
14:     **if** there are elements buffered AND $isPCChild(c)$ **then**
15:         $PCBuffering(p, c)$
16:     **while** $getElementfromStream(p).end < getElement(c).start$ **do**
17:         $Advance(T_p)$
18: **procedure** GETNOTSATISFYPC(n)
19:     **for all** node n_i in $children(n)$ **do**
20:         **if** $isPCChild(n_i)$ **then**
21:             **if** $getElement(n_i).level - getElement(n).level \neq 1$ **then**
22:                 **return** n_i

In Algorithm 1, lines 11 to 15 is an important step, which deals with the situation that both buffer stacks of a query node n and its child n_{min} are not empty. n_{min} is the child node whose current element has the minimum start value. The current element of node n should proceed if its start position is greater than the end position of the current element of node n_{min}, because it can not contribute to any useful path solutions in the future. Lines 16 to 18 deal with the situation that both buffer stacks of query node n and n_{max} are not empty.

Line 18 returns n_{max} because $current(n_{max})$ cannot contribute to the final solution. Lines 19 to 20 check whether $current(n)$ lies to the left of the current element of at least one of $n's$ children, and if so, it cannot contribute to the final solution, and will be skipped. Line 21 should be noted. If the node n proceeds in the last step, the PCBuffering on the node n needs redone. All the ancestors of current element of node n_{max} will be buffered. Line 22 is used for re-acquire the n_{min}, because n_{min} may change due to the buffering actions.

The main algorithm. Algorithm 3 presents the details of the main algorithm. It iteratively invokes $getNext(n)$ to get the appropriate query node for further processing. If ancestors or parents can be found in the parent result list, the current element of the returned node will be moved into the result list. Otherwise, the returned node can not contribute to final solutions in the future, so its current element should be abandoned.

Line 4 should be noted. It is used for cleaning self result list to guarantee the elements are strictly nested. Additionally, when an element is inserted into the result list, the ascendant order in start value should be always kept. It should be noted that the action of clean parent's result list does not exist in TwigBuffer, but it exists in TwigStack and TwigStackList. This change is mainly because TwigBuffer adopts more complex buffering schemes, clean parent's result list may cause some elements removed too early.

3   Correctness and Complexity of TwigBuffer

Due to the limit of space, the correctness proof is not included. This part can be found in the full paper. For **complexity**, since TwigBuffer uses more complicated buffering schemes to avoid useless partial solutions, one would naturally wonder whether the first phase of the algorithm has become computationally too expensive. We point out that although the element manipulation in TwigBuffer is more complex, the worse-case time complexity remains linear in the sum of the number of nodes in Q and the length of the output list.

4   Experiments

We implement TwigStack, TwigStackList and TwigBuffer in C programming language. The XML parser we used is Libxml2. All the experiments were performed on 1.6GHz Intel Centrino Duo processor with 1G RAM. The operating

---

Algorithm 3. TwigPatternMatching(Q)

1: **while** ¬$end(Q)$ **do**
2:     $q_{act} := getNext(root(Q))$
3:     **if** $isRoot(q_{act})$ **OR** ¬$isEmpty(RL_{parent(q_{act})})$ **then**
4:         $cleanSelfResultList(q_{act})$
5:         $MovetoResultList(q_{act})$
6:         **if** $isLeaf(q_{act})$ **then**
7:             $showSolutions(q_{act})$
8:             $removefromRL(q_{act})$
9:     **else**
10:         **if** $length(BS_{parent(q_{act})}) > 1$ **AND** $getElement(q_{act}).start >$ $bottom(BS_{parent(q_{act})}).start$ **then**
11:             $SBuffering(parent(q_{act}), q_{act})$
12:         **else**
13:             $Proceed(q_{act})$
14: **procedure** END(q)
15:     **return** $\forall q_i \in subtreeNodes(q) : isLeaf(q_i) \bigwedge isEmpty(BS_n) \bigwedge end(T_n)$
16: **procedure** CLEANSELFRESULTLIST(n)
17:     **if** $isEmpty(BS_n)$ **then**
18:         **while** $getElement(n).end < tail(RL_n).start$ **OR** $getElement(n).start > tail(RL_n).end$ **do**
19:             $RemoveTail(n)$
20: **procedure** MOVETORESULTLIST(n)
21:     p:= $getPointer(n)$
22:     e := $getElement(n)$
23:     Insert result node (e, p) to RL_n, the ascendant order in start value should be kept
24: **procedure** GETPOINTER(n)
25:     p points to the end of $RL_parent(n)$
26:     e := $getElement(n)$
27:     **while** $e.start < p.start$ **OR** $e.end > p.end$ **do**
28:         $p := previous(p)$
29:     **return** p

---

system is Windows XP. We used the following two data sets for evaluation: Tree-Bank and DBLP obtained from the University of Washington XML repository. The metrics of evaluation we selected is running time.

The queries on both data sets are presented in Table 1. It can be seen that all the queries have different twig structures. This consideration will make the comparisons more comprehensive.

The experimental results are illustrated in Fig.2. As shown, the performance of TwigStackList is nearly the same with TwigBuffer on the queries that do not have /-edges or /-edges happen under non-branching nodes. However, TwigBuffer performs better than TwigStackList when the queries have /-edges under branching nodes.

Table 1. Queries over TreeBank and DBLP

Data set	Query	XPath expression
TreeBank	Q1	//S[//MD]//ADJ
TreeBank	Q2	//S[//VP/IN]//NP
TreeBank	Q3	//S[/JJ]/NP
TreeBank	Q4	//S/VP/PP[/IN]/NP/VBN
TreeBank	Q5	//EMPTY[//VP/PP//NNP][/S[//PP//JJ]/VBN]//PP/NP
DBLP	Q1	//dblp/inproceedings[//title]/author
DBLP	Q2	//dblp/article[//author][//title]//year
DBLP	Q3	//dblp/inproceedings[//cite][//title]/author
DBLP	Q4	//dblp/article[//author][//title][//url][//ee]//year
DBLP	Q5	//article[//volume][//cite]//journal

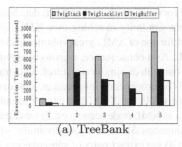
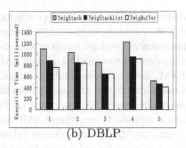

(a) TreeBank   (b) DBLP

Fig. 2. Experiment results

5   Conclusion

We presented a novel holistic twig join algorithm that efficiently finds root-
to-path matchings. Our algorithm completely avoids useless intermediate path
matchings for arbitrary twig patterns, and thereby improves the overall performance of previous two-phase twig join algorithms. The better overall performance of our algorithm has been substantiated in our experiments.

References

1. Al-Khalifa, S., Jagadish, H.V., Patel, J.M., Wu, Y., Koudas, N., Srivastava, D.:
   Structural joins: A primitive for efficient XML query pattern matching. In: ICDE,
   p. 141 (2002)
2. Bruno, N., Koudas, N., Srivastava, D.: Holistic twig joins: optimal XML pattern
   matching. In: SIGMOD Conference, pp. 310–321 (2002)
3. Chen, T., Lu, J., Ling, T.W.: On boosting holism in XML twig pattern matching
   using structural indexing techniques. In: SIGMOD Conference, pp. 455–466 (2005)
4. Lu, J., Chen, T., Ling, T.W.: Efficient processing of XML twig patterns with parent
   child edges: a look-ahead approach. In: CIKM, pp. 533–542 (2004)
5. Yu, T., Ling, T.W., Lu, J.: TwigStackList-: A holistic twig join algorithm for twig
   query with not-predicates on XML data. In: Li Lee, M., Tan, K.-L., Wuwongse, V.
   (eds.) DASFAA 2006. LNCS, vol. 3882, pp. 249–263. Springer, Heidelberg (2006)

An Approach for XML Similarity Join Using Tree Serialization

Lianzi Wen[1], Toshiyuki Amagasa[1,2], and Hiroyuki Kitagawa[1,2]

[1] Department of Computer Science, Graduate School of Systems and Information Engineering
[2] Center for Computational Sciences
University of Tsukuba
1-1-1 Tennodai, Tsukuba, Ibaraki 305-8573, Japan
moon@kde.cs.tsukuba.ac.jp,
{amagasa,kitagawa}@cs.tsukuba.ac.jp

Abstract. This paper proposes a scheme for similarity join over XML data based on XML data serialization and subsequent similarity matching over XML node subsequences. With the recent explosive diffusion of XML, great volumes of electronic data are now marked up with XML. As a consequence, a growing amount of XML data represents similar contents, but with dissimilar structures. To extract as much information as possible from this heterogeneous information, *similarity join* has been used. Our proposed similarity join for XML data can be summarized as follows: 1) we serialize XML data as XML node sequences; 2) we extract semantically/structurally coherent subsequences; 3) we filter out dissimilar subsequences using textual information; and 4) we extract pairs of subsequences as the final result by checking structural similarity. The above process is costly to execute. To make it scalable against large document sets, we use *Bloom filter* to speed up text similarity computation. We show the feasibility of the proposed scheme by experiments.

1 Introduction

XML (Extensible Markup Language) [1] is a general-purpose markup language for data exchange on the Internet and has been recommended by the World Wide Web Consortium (W3C). XML is currently used in many applications, including web data, business data, science data, logs, etc. It is popular because it can represent any kind of data from multiple sources. As a consequence, the use of XML is expected to continue growing enormously.

As XML becomes increasingly popular for data representation, a growing amount of XML data contains similar contents but with different structures. For example, DBLP Bibliography and ACM SIGMOD are published in XML on the Internet. The two XML data convey similar contents, but each may have its own extra information not found in the other. To extract as much information as possible from this heterogeneous information, we must efficiently measure similarity between XML data for integrating such data sources. Figures 1 and 2 show an example of two XML data from different databases. Figure 1 has additional information about the publisher and Figure 2 has web side information. It is important to integrate complementary information to extract more information from these databases.

J.R. Haritsa, R. Kotagiri, and V. Pudi (Eds.): DASFAA 2008, LNCS 4947, pp. 562–570, 2008.

```
<textbook>
 <title>Database Systems: The Complete
       Book</title>
 <authors>H. Garcia-Morina J. Ullman
         J. Widom</author>
 <bibinfo>
   <publisher>Prentice Hall</publisher>
   <year>2001</year>
   <isbn>123</isbn>
 </bibinfo>
</textbook>
```

```
<book year="2001">
 <title>Database Systems: The Complete
       Book</title>
 <authors>
   <author>
    <fname>Hector</fname>
    <lname>Garcia-Morina</lname>
   </author>
   <author>
    <fname>Jeffrey</fname>
    <lname>Ullman</lname>
   </author>
   <author>
    <fname>Jennifer</fname>
    <lname>Widom</lname>
   </author>
 </authors>
 <url>http://infolab.stanford.edu/
      ~ullman/dscb.html</url>
</book>
```

Fig. 1. Bibliographic data in XML (1)   **Fig. 2.** Bibliographic data in XML (2)

The operation used for this purpose is *similarity join*. So far, similarity join has been actively studied in relational databases, and research applied to XML data is underway. The importance of similarity join is expected to increase in the future as XML data continues to proliferate.

Several problems arise in XML similarity join. The tasks of dividing XML data into independent structures at a proper position and measuring the approximate similarity between XML data are not easy to perform. Figure 1 represents author information by the *authors* tag; Figure 2 shows the author name in more detail. In Figure 1, the publication year is expressed by the *year* element, but in Figure 2, it is expressed by the *year* attribute. Not only differences in markup vocabulary, but also the order in which elements appear and the tree structure of the XML data might differ. Thus, we have to accurately measure the approximate similarity between XML data, even if there are heterogeneities in content and structure. If the data is similar, similarity join processing should be possible. This paper focuses on dividing XML data and measuring similarity.

The tree structure representation of XML data makes it difficult to measure similarity. To solve this problem, serialization of XML tree structure has been proposed [2,3,4]. Serialization is the process of converting an XML tree into node sequences by traversing the tree in a particular order. The operation that extracts independent subtrees can be achieved by extracting node subsequences that correspond to subtrees.

The rest of this paper is organized as follows: Section 2 briefly describes related work. Section 3 discusses the proposed scheme. Effectiveness is evaluated in Section 4 and Section 5 concludes the paper with an outline of future work.

2   Related Work

Liang et al. [5] proposed a scheme of Approximate XML Join. In their approach, XML document trees are clustered into many subtrees that represent independent items. The approximate similarity between them is determined by calculating the similarity of text contents and path expressions. When calculating similarity, only identical text nodes

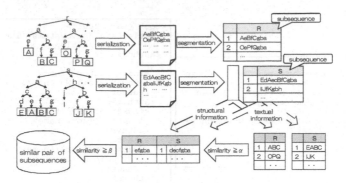

Fig. 3. An overview of the proposed scheme

are considered and the number of shared tag names are counted over two distinct path expressions.

It is well known that *tree edit distance* (TED) [6] is a metric for measuring the structural similarity in XML documents. It is defined as the minimum cost operations (insertions, deletions and substitutions) required to transform one tree to another. It is difficult for TED to be applied to massive data, because the computational cost is extremely high; in the worst case, it is an $O(n^4)$ operation for the document of size n.

Although the result is the same, our approach to extracting independent units from node sequences is faster than the approaches used in other research for extracting independent units from an XML tree structure. The approximate similarity between well-segmented subsequences can be effectively determined by computing the similarity both of textual information and structural information.

3   The Proposed Scheme

Figure 3 shows an overview of the proposed scheme. The main contributions of our proposal follow:

1. We serialize two given XML data as XML node sequences. In the serialization, we use postorder based on NoK (Next-of-Kin) [3] pattern.
2. We extract semantically/structurally coherent subsequences. This process can be executed automatically by giving parameters.
3. We measure similarity of the textual information and filter out dissimilar subsequences. In this measurement, we use Bloom filter for speed and effectiveness.
4. We extract pairs of subsequences as the final result by checking structural similarity.

3.1   Serialization of XML Data

We serialize the XML tree structure into XML node sequences by a tree-traversal order (e.g., pre-and postorder). Postorder is used in ViST [2] and Prüfer order is used in PRIX [7].

In general, information on structure is lost by serializing the tree structure to node sequences[1]. In Figure 4, the postorders are the same as "bca" although the two XML tree structures differ. Therefore, we see that structural information is lost.

To cope with this problem, the NoK pattern [3] is used in this serialization to maintain structural information. This paper uses postorder instead of the preorder used in the original Nok pattern. This is done because it is convenient to extract independent units by scanning the node sequence from its head. For example, "((b)(c)a)" is a string representation of the tree (Figure 4). The "(" preceding a indicates the beginning of a subtree rooted at "a"; its corresponding ")" indicates the end of

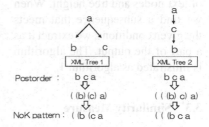

Fig. 4. Examples of serialization

the subtree. In the Nok pattern, it removes all closing parentheses and retains only open parentheses as in "((b(ca" because each node (a character in the string) actually implies a closing parenthesis. Even if a closing parenthesis is omitted, tree structural uniqueness is kept as we see in Figure 4.

Figure 5 shows the string and graphical representations of an XML tree. It is created according to the following rule: if an open parenthesis comes, the line will go down one step, and if a node comes, the line will go up one step while a NoK pattern is scanned. The result shows the depth of the XML tree. We can observe some properties. First, text nodes always appear in the lowest representation. Suppose we

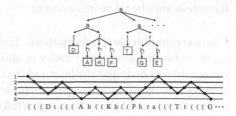

Fig. 5. Post-order NoK pattern

want to find the parent of node "t" in the sequence. Since the nodes are postordered, the parents of "t" must be presented after "t" and the level is one step higher. Therefore, parents of "t" are "a" because it is the first node on the right side that exists one step higher than "t" excluding "(". In this way, for a given set of nodes, we can find nodes that have particular relationships with the given nodes.

3.2 Extraction of Subsequences

After serialization, the XML data becomes one long node sequence and we have to extract independent subsequences corresponding to the XML subtree for subsequent similarity join. At this time, it is preferable to extract semantically/structurally coherent subsequences; this is not easily done automatically.

The basic idea here is as follows: We give only three parameters t_{min}, t_{max} and h, where t_{min} is the least number of the text node that the subsequence should have, t_{max} is the maximum number of the text node that the subsequence can have, and h is the least height from the leaf node that a subsequence should have. To extract an

---

[1] Prüfer order can serialize with structural information maintained.

independent unit, from the starting point, we try to read the sequence one by one by counting the number of text nodes and tree height. When we find a subsequence that meets the given conditions, we extract it as a part of the output. The algorithm is illustrated as algorithm 1.

3.3 Similarity Measure

Our technique to detect similar subsequences that correspond to an XML subtree has two steps. We measure textual information first and then check structural informa-

---

Algorithm 1. GetSubsequences(s, t_{max}, t_{min}, h)

Input: N, t_{max}, t_{min}, h {NoK pattern, max and min # of text nodes, and least height.}

Output: $S = \{s_1, s_2, \ldots, s_n\}$ {Subsequences.}

$S := \emptyset$; $s =$ "";
curheight = 0; textnum = 0;
for $i = 0$ to N.size **do**
  s.append(N[i]);
  if N[i].type == TEXT **then**
    textnum++;
  else if N[i].type == ELEMENT **then**
    curheight−−;
    if $t_{min} \leq$ textnum $\leq t_{max} \wedge h \leq$ curheight **then**
      $S := S \cup s$; {Output current subsequence.}
      $s :=$ ""; textnum := 0; {Initialize related variables.}
      continue;
    end if
  else if N[i] == "(" **then**
    curheight++;
  end if
end for
return S;

---

tion only for the pair of subsequences having similar textual information. We do this because XML data comprises textual and structural information, and both types of information must be taken into account.

Comparison of textual information. Textual information in this work is defined as a set of words included in text nodes or attribute nodes being considered; therefore, we extract the text node and attribute node from the subsequence in which the comparison is being attempted. For example, suppose we want to calculate two pieces of textual information t_1 and t_2, which are extracted from subsequences s_1 and s_2. t_1="database systems Hector Jennifer", t_2 ="database systems Jennifer Jeffrey". In this work, we use Jaccard similarity to calculate the ratio of words common to both t_1 and t_2 among all words in t_1 and t_2. Jaccard similarity between t_1 and t_2, $JR(t_1, t_2)$, is defined as

$$JR(t_1, t_2) = \frac{wt(t_1 \cap t_2)}{wt(t_1 \cup t_2)}$$

From this formula, the similarity of textual information $JR(t_1, t_2) = \frac{3}{5} = 0.6$. We can say that the textual information is similar if the calculated similarity is above the given threshold α. Notice that it is quite time consuming if we compute Jaccard similarity for all possible textual information over the subsequences. To cope with the problem, we use the idea of the *Bloom filter* [8] to accelerate the processing. *Bloom filter* provides a probabilistic way to determine if an element is a member of a given set.

An empty *Bloom filter* is a bit array of m bits, all set to 0. There must also be k different hash functions defined, each of which maps a key value to one of the m array positions. In this work, every word is hashed to give k results of bit arrays by using each of k hash functions. Then, the entire results of bit arrays are combined using "OR" operations. For example, suppose there are "database", "system", and "book" in a subsequence, and we use two hash functions. We get the bit array shown in Figure 6.

$h_1(database)$	0010000000000000
$h_2(database)$	0000000001000000
$h_1(system)$	0100000000000000
$h_2(system)$	0000000000000010
$h_1(book)$	0000000001000000
$h_2(book)$	0000001000000000
Result	0110001001100010

Fig. 6. Example of bit array

An important point to note here is that we can estimate the similarity of textual information by calculating the similarity of bit arrays. We can calculate the similarity of bit arrays with the following formula:

$$Sim(t_1.sig, t_2.sig) = \frac{count(t_1.sig \; AND \; t_2.sig)}{count(t_1.sig \; OR \; t_2.sig)}$$

where $t_1.sig$ and $t_2.sig$ denote bit arrays that are hashed from textual information of subsequence t_1 and t_2, respectively, and $count()$ is a function that counts the number of "1" in the bit array. Because the similarity of the bit arrays is guaranteed to be larger than the real value, $Sim(t_1.sig, t_2.sig) \geq JR(t_1, t_2)$, we can use it to filter out unnecessary candidates. Finally, we refine the resulting candidates by computing real Jaccard similarities to get rid of false positives.

Comparison of structural information. We must check structural information of the pair having textual similarity. Structural information here is a subsequence of the element node that extracts the text node from the NoK pattern. If the subsequences of element nodes are similar, it can be said that structural information is also similar, since the subsequences made by postorder NoK pattern maintains the tree structure. Therefore, we calculate similarity of subsequences of an element for structural information.

For this measure, we use *edit similarity*, which is based on edit distance. Edit distance is a similarity measure between two strings. It is defined as the minimum number of point mutations required to change one string to another, where a point mutation is change, insertion, or deletion of a letter. Several variations are possible, depending on how point mutations are defined and weighted. In this work, to maintain brevity, we use Levenshtein distance, where change, insertion, and deletion are permitted and are equally weighted.

From the definition, for a given two strings, the edit distance between them is affected by their lengths, which is inappropriate in similarity measure. To cancel those effects, we use edit similarity [9], which can be obtained from edit distance by the following formula:

$$ES(\sigma_1, \sigma_2) = 1.0 - \frac{ED(\sigma_1, \sigma_2)}{max(|\sigma_1, \sigma_2|)}$$

where σ_1 and σ_2 denote strings being compared, and ED and ES denote edit distance and edit similarity, respectively.

When comparing a given two XML subtrees, we need only to regard each location step (tag name) as an alphabet. Suppose that we attempt to compute the edit similarity between d_1 and d_2 and the NoK pattern based sequences are d_1: ((y(t(ab and d_2: ((t(a(ub, respectively. The edit distance can be computed as follows:

From the result that $ED(d_1, d_2) = 4$, we have $ES(d_1, d_2) = 1.0 - \frac{4}{8} = 0.5$. Finally, we judge that the node subsequences are similar if the calculated similarity is more than given threshold β.

d_1	(	(	y	(	t	(	a			b
d_2	(	(			t	(	a	(	u	b
$Cost$	0	0	1	1	0	0	0	1	1	0

4 Experimental Evaluation

We used a 2-way Dual Core AMD Opteron(TM) processor (2.4GHz) with 16GB memory running Sun OS 5.10. The program was implemented using J2SE 1.5, and we used PostgreSQL 8.1.0 as the underlying RDBMS.

We used XML versions of SIGMOD Record (464 KB) [2] and DBLP bibliography (400,170 KB) [3].

4.1 Subsequence Extraction

Table 1 shows the number of subsequences extracted from each XML data and processing time.

Table 1. Subsequence extraction

XML data	Size (KB)	#subsequence	Time (sec)
SigmodRecord.xml	464	1530	2.5
DBLP.xml	400,170	904,732	555.6

In our experiment, we extracted semantically/structurally coherent subtrees by setting appropriate parameters (three parameters $t_{min} = 3$, $t_{max}=15$ and $h = 2$) by taking into account the features of XML data.

4.2 Textual Similarity Computation

We tested the effectiveness of the proposed textual similarity computation method. We measured elapsed time for processing 1,384,239,960 (= $1,530 \times 904,732$) pairs of subsequences with different textual similarity thresholds. For Bloom filter, we used three hash functions of 256 bits. Table 2 shows 1) number of pairs remaining after Bloom filter, 2) number of correct pairs according to Jaccard similarity, and 3) elapsed time for the two processes. From the result, we observe that the method is successful in filtering dissimilar pairs, and the selectivity of Bloom filter is more than 99% on average.

Table 2. Textual similarity

Textual Similarity	# pairs by BF	# correct pairs	Time (sec.)
$T_{sim} \geq 0.30$	45,299	977	2,367
$T_{sim} \geq 0.40$	837	475	1,559
$T_{sim} \geq 0.50$	328	181	1,558
$T_{sim} \geq 0.60$	103	64	1,558
$T_{sim} \geq 0.70$	26	11	1,559
$T_{sim} \geq 0.80$	2	0	1,558
$T_{sim} \geq 0.90$	0	0	1,559

Table 3. Structural similarity

Structural similarity	Time (ms)	#pairs
$S_{sim} \geq 0.05$	882	436
$S_{sim} \geq 0.1$	882	323
$S_{sim} \geq 0.2$	883	173
$S_{sim} \geq 0.3$	885	22
$S_{sim} \geq 0.4$	883	2
$S_{sim} \geq 0.5$	885	0

We additionally investigated the efficiency of the scheme, in particular to see the benefit of Bloom filter, by comparing processing time of the proposed method and that without the filtering step by Bloom filter. That is, for the baseline, we attempted to compute Jaccard similarity for each pair. Figure 7 illustrates the elapsed time. The similarity threshold is 0.5. We observe that the baseline is much slower than the proposed scheme, thereby showing the effectiveness of the filtering step by Bloom filter.

4.3 Structural Similarity Computation

We checked the structural information of 475 pairs whose textual information is similar. Table 3 lists the number of pairs with structural similarity above the given threshold, and shows the elapsed time to calculate edit similarity.

To evaluate the efficiency of our algorithm based on XML serialization, we compared the processing time with that of the tree edit distance (TED) for 10,000 (= 100×100), 50,000 (= 100×500), 100,000 (= 100×1000), and 1,000,000 (= $1,000 \times 1,000$) pairs

---

[2] http://www.sigmod.org/record/xml/
[3] http://dblp.uni-trier.de/xml/

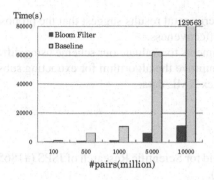

Fig. 7. Elapsed time (textual similarity)

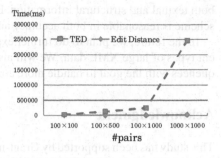

Fig. 8. Elapsed time (structural similarity)

from SigmodRecord.xml and DBLP.xml. Figure 8 shows the result. The figure shows that our approach is much faster than TED, which is known to be costly to process. With our approach, we can achieve good performance even for large XML datasets.

4.4 Extracted Pairs

Figures 9 and 10 show examples of similar pairs extracted by our proposed method. It appears that the proposed scheme is successful in finding similar pairs.

```
T_simi = 0.77, S_simi = 0.25

<article>
  <title articleCode="264376">The Five-Minute Rule
        Ten Years Later, and Other Computer
        Storage Rules of Thumb</title>
  <authors>
    <author AuthorPosition="00">Jim Gray</author>
    <author AuthorPosition="01">Goetz Graefe</author>
  </authors>
</article>
- - - - - - - - - - - - - - - - -
<article mdate="2004-06-09"
         key="journals/corr/cs-DB-9809005">
  <author>Jim Gray</author>
  <author>Goetz Graefe</author>
  <title>The Five-Minute Rule Ten Years Later,
   and Other Computer Storage Rules of Thumb</title>
  <ee>http://arxiv.org/abs/cs.DB/9809005</ee>
  <year>1998</year>
  <journal>CoRR</journal>
  <volume>cs.DB/9809005</volume>
  <url>db/journals/corr/corr9809.html</url>
</article>
```

Fig. 9. Similar subdocuments (1)

```
T_simi = 0.42, S_simi = 0.22

<article>
  <title articleCode="304029">Data Mining-based
    Intrusion Detectors: An Overview of the Columbia
    IDS Project</title>
  <authors>
    <author AuthorPosition="01">S. J. Stolfo</author>
    <author AuthorPosition="02">W. Lee</author>
    <author AuthorPosition="03">P. K. Chan</author>
    <author AuthorPosition="04">W. Fan</author>
    <author AuthorPosition="05">E. Eskin</author>
  </authors>
</article>
- - - - - - - - - - - - - - - - - - -
<article mdate="2006-04-03" key="journals/StolfoLCFE01">
  <author>Salvatore J. Stolfo</author>
  <author>Wenke Lee</author>
  <author>Philip K. Chan</author>
  <author>Wei Fan</author>
  <author>Eleazar Eskin</author>
  <title>Data Mining-based Intrusion Detectors: An
    Overview of the Columbia IDS Project.</title>
  <pages>5-14</pages>
  <year>2001</year>
  <volume>30</volume>
  <journal>SIGMOD Record</journal>
  <number>4</number>
  <ee>http://doi.acm.org/10.1145/604264.604267</ee>
  <ee>http://www.acm.org/sigmod/record/issues/0112/
      SPECIAL/1.pdf</ee>
  <url>db/journals/sigmod/sigmod30.html</url>
</article>
```

Fig. 10. Similar subdocuments (2)

5   Conclusions

In this paper, we proposed a scheme for similarity join over XML data based on XML data serialization. We focus on extracting many independent subsequences and measuring similarity of pairs of subsequences. In the similarity measure, we take into account both textual and structural information. Experimental results suggest that the proposed scheme is reasonable in both accuracy and effectiveness.

In the future, we plan to do further experiments to evaluate our scheme using different types of large XML data. We will also improve the algorithm for extracting subsequences with the goal to handle more complex XML data.

Acknowledgment

This study has been supported by Grant-in-Aid for Scientific Research of JSPS (#18650018 and #19700083) and of MEXT (#19024006).

References

1. W3C: Extensible Markup Language (XML) 1.0 (Fourth Edition) (August 2006)Recommendation, http://www.w3.org/TR/REC-xml/
2. Wang, H., Park, S., Fan, W., Yu, P.S.: ViST: A Dynamic Index Method for Querying XML Data by Tree Structures. In: Proc. the 2003 ACM-SIGMOD Conference (SIGMOD), pp. 110–121 (2003)
3. Zhang, N., Kacholia, V., Ozsu, M.T.: A succinct physical storage scheme for efficient evaluation of path queries in Xml. In: Proc. ICDE 2004, p. 54 (2004)
4. Li, Q., Moon, B.: Indexing and Querying XML Data for Regular Path Expressions. In: Proc. VLDB 2001, pp. 361–370 (2001)
5. Liang, W., Yokota, H.: A Path-sequence Based Discrimination for Subtree Matching in Approximate XML Joins. In: Proc. The 2nd Int'l Special Workshop on Databases for Next-Generation Researchers (SWOD) 2006, p. 116 (2006)
6. Zhang, K., Shasha, D.: Tree pattern matching. Pattern Matching Algorithms. In: Tree pattern matching. Pattern Matching Algorithms, vol. 11, Oxford University Press, Oxford (1997)
7. Rao, P., Moon, B.: PRIX: Indexing And Querying XML Using Prufer Sequences. In: Proc. ICDE 2004, p. 288 (2004)
8. Gong, X., Qian, W., Yan, Y., Zhou, A.: Bloom Filter based XML Packets Filtering for Millions of Path Queries. In: Proc. ICDE 2005, pp. 890–901 (2005)
9. Chaudhuri, S., Ganti, V., Kaushik, R.: A Primitive Operator for Similarity Joins in Data Cleaning. In: Proc. ICDE 2006, p. 5 (2006)

A Holistic Algorithm for
Efficiently Evaluating Xtwig Joins

Bo Ning[1], Guoren Wang[1], and Jeffrey Xu Yu[2]

[1] Northeastern University, Shenyang Liaoning 110004, China
ningyibai@hotmail.com, wanggr@mail.neu.edu.cn
[2] Chinese University of Hong Kong, China
yu@se.cuhk.edu.hk

Abstract. More and more XML data have been generated and used in the data exchange. XML employs a tree-structure data model, but lots of queries submitted by users are not like the tree-structure. Those queries contain ancestor axis in predicates, and specify the pattern of selection predicates on multiple elements from descendants to ancestors. Efficiently finding all occurrences of such an xtwig pattern in an XML database is crucial for XML query processing. A straightforward method is to rewrite an xtwig pattern to equivalent reverse-axis-free one. However, this method needs scanning the element streams several times and is rather expensive to evaluate. In this paper, we study the xtwig pattern, and propose two basic decomposing methods, *VertiDec* and *HoriDec*, and a holistic processing method, *XtwigStack*, for processing xtwig queries. The experiments show that the holistic algorithm is much more efficient than the rewriting and decomposition approaches.

1 Introduction

XML has become a de-facto standard for exchanging and representing information on the Web. XML data may be very complex and deeply nested, and queries typically specify patterns of selection predicates on multiple elements. In real applications, XML queries may contain predicates with ancestor axes. Consider the following XPath[9] expression.

$$Q_1 : book\,[\,ancestor :: Computer][ancestor :: Addison - Wesley]$$
$$[\,title = ``XML"]//author[name = ``John"]$$

The expression matches book elements that (i) have a child subelement *title* containing 'XML', (ii) have a subelement *author* whose name is 'John', (iii) are descendants of *Addison-Wesley* elements and (iv) are descendants of *computer* elements. This expression can be represented as a complex node-labeled pattern in which elements and string values are regarded as node labels, as shown in Figure 1(c). Q_1 not only cares for the information of the book's descendant nodes, but also constrains the condition of the book's ancestor nodes in XML document, by using the reverse axis 'ancestor' in the predicates.

J.R. Haritsa, R. Kotagiri, and V. Pudi (Eds.): DASFAA 2008, LNCS 4947, pp. 571–579, 2008.
© Springer-Verlag Berlin Heidelberg 2008

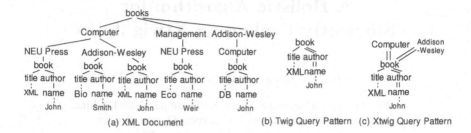

Fig. 1. Twig and xtwig patterns

Olteanu et al.[6] propose some rewriting rules to transform absolute XPath location paths with reverse axes into equivalent reverse-axis-free ones. There are lots of work of evaluating queries with descendant and child axes, such as path pattern and twig pattern. By the rewriting rules, we may transform the Q_1 to a twig query, and evaluate the query using classic algorithms of evaluating ancestor-axis-free query patterns, such as *PathStack*, *TwigStack*, *TJ-Fast*[3], *Extended Dewey*[2], *Twig2Stack*[4], *TwigList*[5] and so on. But the transformed query patterns contain equality joins of node identity inevitably, and remain expensive to evaluate. Specially, when there are multiple ancestor axis in query pattern, the performance is dissatisfactory. In this paper, we develop a new holistic processing algorithm to match xtwig queries without unnecessary joins.

2   Background and Definitions

2.1   Twig Pattern and TwigStack

A twig pattern has selection predicates with 'descendant' or 'child' on multiple elements in an XML document. Holistic twig join algorithm *TwigStack* is the first one which regards the query of twig as an indivisible entity, without breaking the query into pairs of nodes. It has two phases, *TwigStack* and Merge-join. In the first phase, some solutions to individual query root-to-leaf paths are computed. In the other phase, these solutions are merge-joined to compute the answers to the query twig pattern.

2.2   Xtwig Pattern

Definition 1. *An **xtwig pattern** is a query that has parent or ancestor axis in its predicates. In the pattern tree, the node whose in-edge is parent or ancestor axis is called **vbranchnode**, and vbranchNode's parent is called **vnode**. The branch rooted at vnode is called **vbranch**. The reverse twig rooted at the vbranchnode is called **vpattern**. If each vbranch of a vpattern is a vnode, we call it a **simple vpattern**. The number of the vnode on a vbranchnode is called **Reverse Fan-out Degree(RFOD)** of it.*

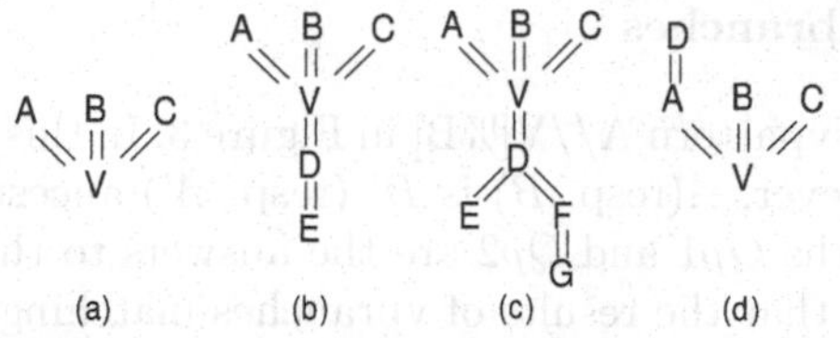

Fig. 2. Xtwig patterns

In Figure 2, (a) is a simple vpattern, and node V is a vbranchnode, and nodes A, B and C are vnodes. Each branch of a simple vpattern can be a path pattern. A vbranchnode can be the root node of path patterns or twig patterns, so grows the xtwig, as shown in Figure 2.

3 Xtwig Pattern Matching

3.1 Limitations of Basic Methods

There are three schemes to compute answers to an xtwig query pattern. The straightforward one is, at each vbranchnode, to vertically decompose the xtwig into multiple twig patterns, use the traditional twig algorithm to identify solutions to each individual twig pattern, then merge-join these solutions via the vbranchnodes to get the final results of the query. This scheme is named *VertiDec*. Querying each decomposed twig pattern needs scanning the streams of twig part in xtwig pattern once, so evaluating the xtwig pattern using *VertiDec* has several times scanning of streams. Furthermore, many intermediate results may not be part of the final results, e.g. if *John* published another *XML* book at another publisher. The second approach is, at each vbranchnode, to horizontally decompose the xtwig pattern into vpatterns and twig patterns, find all occurrences of twig patterns and vpatterns separately, and merge-join them via the vbranchnodes to get the final results. This approach is named *HoriDec*. Compared with *VertiDec*, *HoriDec* scans fewer element streams, but still has to scan the vbranchnode tag stream twice. Moreover, it faces the same problem of unnecessary intermediate results as *VertiDec*. The third approach is to solve the xtwig pattern matching using the rewriting rules[6,7] which can transform an xtwig pattern to a twig pattern. The idea of rewriting is that, instead of looking back from the context node, one can look forward from the beginning of the document for matching the tag T in the predicate with ancestor, then look for a descendant node identical to the context node. For example, Q_2 can be transformed to Q_3:

Q_2 : *book* [*ancestor* :: *Computer*]

Q_3 : *book* [/*descendant* :: *Computer*/*book* :: *node*() == *self* :: *node*()]

Although there is no ancestor axis in Q_3, evaluating Q_3 is rather expensive, since the streams have to be scanned twice, and there is an equality join caused by the rewriting. For an xtwig pattern containing multiple predicates with ancestor axis, the evaluation cost of the rewriting way becomes larger.

3.2   Matching Vbranches

Consider the simple vpattern $A//V[\%B]$ in Figure 3. In this pattern, A and B are V's ancestors. However, A(resp. B) is B' (resp. A') ancestor. Therefore all the answers to both paths $Qp1$ and $Qp2$ are the answers to the simple vpattern Q. So we can conclude that the results of vbranches matching are the permutation of the vnodes' elements which have the a-d relationships and satisfy respective vbranch pattern.

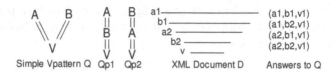

Fig. 3. Example of Vbranches Matching

Algorithm 1 gives the computation of matching those vbranches. We associate each node q in the vbranch pattern with a stream T_q containing all the elements of tag q and each stream has an imaginary cursor, which can either move to the next element or read the current element. The operations over streams are **eof, advance, next, nextL** and **nextR**. The last two operations return the startpos and endpos of element in the stream. We also associate a single stack S for all the vnodes, and associate a separate stack S_T for each node which is the ancestor of respective vnode to maintain their relations in the path. The operations over two kinds of stacks are: **empty, popStack, pushStack, topL** and **topR**. All the elements of vnodes will be pushed into stack S to maintain the a-d relationships of all the vnodes' elements, while the other tags' elements will be pushed into respective stack to preserve the answers of tag itself. In the algorithm 1, **getMin(v)** returns the tag, of which stream's next element has minimal startpos among the v's vnodes' tags. **cleanStack(S, T_q)** will pop the stack S, until the top of S is the ancestor of next(T_q). **RFOD(v)** is the reverse fan-out degree of v. When the algorithm runs to line 14, the next element is not the descendant of top of stack S, then the algorithm has to determinate whether the current stack S contains elements of all the vnodes (line 15).

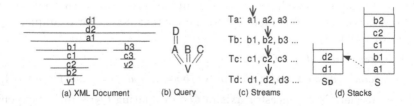

Fig. 4. Example of Algorithm VBStack

Algorithm 1. VBStack(v)

1: **while** not eof(v) **do**
2: $q_{act} = getMin(v)$;
3: **if** (q_{act} is *vnode*) **then**
4:   **if** not isRoot(q_{act}) **then**
5:    cleanStack(parent(q_{act}), nextL(q_{act}));
6:   **end if**
7:   **if** (isRoot(q_{act}) or not empty($S_{parent(q_{act})}$)) **then**
8:    **if** empty(S) **then**
9:     PushStack(next($T_{q_{act}}$), S);
10:   **else**
11:    **if** (NextL($T_{q_{act}}$)>topL(S) and nextR($T_{q_{act}}$)<topR(S)) **then**
12:     PushStack(next($T_{q_{act}}$), S));
13:     Advance($T_{q_{act}}$);
14:    **else**
15:     **if** S.KeySet.sizeof equals RFOD(v) **then**
16:      //Vbranches pattern is matched.
17:      cleanStack(S, nextL($T_{q_{act}}$));
18:      PushStack(next($T_{q_{act}}$), S);
19:     **end if**
20:    **end if**
21:   **end if**
22: **else**
23:   advance($T_{q_{act}}$);
24: **end if**
25: **end if**
26: **if** (q_{act} is not *vnode*) **then**
27:   **if** not isRoot(q_{act}) **then**
28:    cleanStack(parent(q_{act}), nextL(q_{act}));
29:   **end if**
30:   **if** (isRoot(q_{act}) or not empty($S_{parent(q_{act})}$)) **then**
31:    cleanStack(q_{act},next(q_{act}));
32:    moveStreamToStack($T_{q_{act}}$, $S_{q_{act}}$, pointer to top($S_{parent(q_{act})}$)));
33:   **else**
34:    advance($T_{q_{act}}$);
35:   **end if**
36: **end if**
37: **end while**

 Procedure moveStreamToStack(T_q; S_q; p)
  push(Sq; (next(T_q); p));
  advance(T_q);
 Function getMin(v)
  return $q_i \in Tags\ of\ Vbranches$ such that nextL(T_{q_i}) is minimal

---

For example, in Figure 4, we evaluate the query (b) on the XML document (a). There are three branches in the query pattern, $D//A$, B, and C. We associate node D with a stack S_D, as node D is the ancestor of vnode A, and associate all the vnodes with a stack S. The first two elements d_1 and d_2 are both pushed into S_D, because d_2 is the child of d_1. The next element a_1 is pushed to stack S, because a_1 is from vnode streams T_A, and a_1 points his parent pointer to d_2, which is the top of stack S_D. Then b_1, c_1, c_2 and b_2 are pushed into S too. When the next element b_3 comes, we confirm that the current stack S contains elements from all the vnodes, so we find the matchings of vbranches pattern, and pop the stack S until the top of S is the ancestor of b_3. Note that the algorithm does not pop S_D, until the next element of tag D is not the descendant of S_D's top.

3.3 XtwigStack: Matching Xtwig Pattern

To solve the xtwig pattern matching, we have to deal with the complex sub-tree rooted at the vbranchnode, such as patterns (b) and (c) in Figure 2.

We propose algorithm *XtwigStack*, which apply *VBStack* algorithm to *Path-Stack* or *TwigStack*. We define function *VBStackStep*, which is invoked once, to get a matching of the vbranch pattern. Since we may regard the set of vbranches as a virtual node, we need to change the function *advance* in *Path-Stack* or *TwigStack*. In function *advance*, if the tag node is a virtual node composed by many vbranches, we invoke the *VBStackStep* to move the relative streams for finding the next answer set for the virtual node. In this case, each stream is scanned only once, and algorithm *XtwigStack* is therefore I/O and CPU optimal.

4   Performance Evaluation

4.1 Experiment Setup

The following real-world and synthetic data sets are used for our experiments: (1) XMark, the XML data set is synthetic and is generated by an XML data generator. It contains information about an auction site. It is "information oriented", and has many repetitive structures and fewer recursions. (2) a synthetic XML data set which contains some deep recursions.

To do the performance evaluation, we implement two basic methods *VertiDec* and *HoriDec*, and the rewriting method (short for *RRM*). There is no abbreviated syntax for reverse axis ancestor, so in this section, we abbreviate **ancestor::** as symbol %. All the algorithms are implemented in Java 1.5 and performed on a system with 2.4GHz Pentium processor and 512MB RAM running on windows XP.

4.2 Performance Study

Firstly we select some queries on the synthetic XML documents and XMark to compare the elapsed time of *XtwigStack*, *VeriDec*, and *RRM* vs reversed fan-out degree. On the data set XMark, we evaluate queries 1 to 3 below, whose RFOD varies from 2 to 4. On the synthetic XML data, we evaluate queries 4 to 8, whose RFOD varies from 2 to 6.

$Query1: /regions//from[\%asia]$

$Query2: /regions//from[\%asia][\%mailbox]$

$Query3: /regions//from[\%asia][\%mailbox][\%mail]$

$Query4: /A//F[\%B]$

$Query5: /A//F[\%B][\%C]$

$Query6: /A//F[\%B][\%C][\%D]$

$Query7: /A//F[\%B][\%C][\%D][\%E]$

$Query8: /A//F[\%B][\%C][\%D][\%E][\%F]$

As shown in Figures 5 and 6, Algorithm *XtwigStack* performs better than *VeriDec*, and is independent of the reverse fan-out degree, the *VeriDec*'s performance gets worse as RFOD increases. *RRM* is not independent of the reverse fan-out degree and its performance is the worst. This is because *VertiDec* has to scan more elements as RFOD increases and merges the more intermediate results, while *RRM* has to scan streams from the beginning several times, and conduct multiple joins.

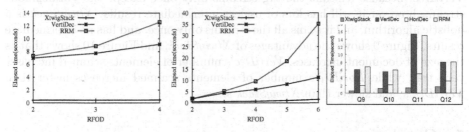

Fig. 5. XMark   **Fig. 6.** Synthetic data   **Fig. 7.** Elapsed time

On the data set XMark, we evaluate the following queries, to evaluate the performance. Those xtwig patterns have paths with different length:

$Query9 : /regions//from[\%mailbox]$

$Query10: /people//person//street[\%address]$

$Query11: /open\_auctions//open\_auction//time[\%bidder]$

$Query12: /closed\_auctions//closed\_auction//description[\%annotation]$

From Figure 7 we can see that the performances of *VertiDec* and *RRM* are similar, and are the worst, while *HoriDec* is slower than *XtwigStack*. In Figure 8, *XtwigStack* scans the fewest elements, while *VertiDec* and *RRM* scan the most. *VertiDec* scans many times the streams of the nodes under vbranches when matching respective twig patterns, and scans lots of intermediate results for merge-joining. While *HoriDec* avoid the scanning of the streams of the nodes under vbranches, so it performs better. *XtwigStack* scans the streams only once, and have not intermediate results, so it performs best.

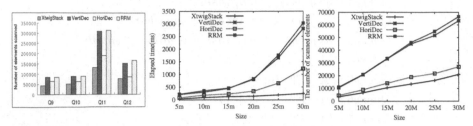

Fig. 8. Scanned Count   **Fig. 9.** Elapsed time   **Fig. 10.** Scanned Count

Finally, we use the following query Query13 to compare the performances of *XtwigStack*, *VertiDec*, *HoriDec* and *RRM* on the XMark data set.

Query13:/regions//item[%Asia]//name

As explained in section 3, *VertiDec* scans some streams many times and has a lot of intermediate results, so it performances worst. *HoriDec* is better than *VertiDec*, because it scans the tag streams only once except the vbranchnode stream, however it still has lots of useless intermediate results. *XtwigStack* is a holistic algorithm, and it scans all the streams only once, and has no intermediate results. Figure 9 shows the advantages of *XtwigStack*, and Figure 10 shows that as the size of document increases, *VertiDec*'s number of elements scanned increases very fast, while *HoriDec*'s number of elements scanned increases faster than *VertiDec*'s, and slower than *XtwigStack*'s.

5   Conclusions

In this paper, we address the issues of using ancestor axis in XPath expressions. In real applications, XML queries are more complex and may contain predicates with ancestor axis. In this paper, we propose a new query pattern xtwig to meet the requirement. We analyze the characteristics of the xtwig pattern matching, and develop an *XtwigStack* algorithm, which works efficiently and effectively, to match the queries with ancestor axis in predicates. Finally we present experimental results on a range of real and synthetic data, to complement our analytical results. The performance study shows *XtwigStack* is efficient, and every stream is scanned only once, and there is not useless intermediate results.

Acknowledgements

This work is supported by the National Natural Science Foundation of China (Grant No. 60773221 and 60773219) and National 863 Plans Projects of China (Grant No. 2006AA09Z139).

References

1. Bruno, N., Srivastava, D., Koudas, N.: Holistic twig joins: Optimal XML pattern matching. In: Proc of SIGMOD, pp. 310–321 (2002)
2. Lu, J., Ling, T.W., Chan, C., Chen, T.: From region encoding to extended dewey: On efficient processing of XML twig pattern matching. In: Proc. of VLDB (2005)
3. Lu, J., Chen, T., Ling, T.W.: TJFast: Effcient processing of XML twig pattern matching. Technical report, National university of Singapore, https://dl.comp.nus.edu.sg/dspace/handle/1900.100/1516
4. Chen, S., Li, H., Tatemura, J., Hsiung, W., Agrawal, D., Candan, K.S.: Twig2Stack: Bottom-up Processing of Generalized TreePattern Queries over XML Documents. In: Proc of VLDB, pp. 283–294 (2006)
5. Qin, L., Yu, J.X., Ding, B.: TwigList: Make Twig Pattern Matching Fast. In: Kotagiri, R., Radha Krishna, P., Mohania, M., Nantajeewarawat, E. (eds.) DASFAA 2007. LNCS, vol. 4443, pp. 850–862. Springer, Heidelberg (2007)
6. Olteanu, D., Meuss, H., Furche, T., Bry, F.: XPath: Looking Foward. In: Jensen, C.S., Jeffery, K.G., Pokorný, J., Šaltenis, S., Bertino, E., Böhm, K., Jarke, M. (eds.) EDBT 2002. LNCS, vol. 2287, pp. 109–127. Springer, Heidelberg (2002)
7. Barton, C., Charles, P., Goyal, D., Raghavachari, M., Fontoura, M., Josifovski, V.: Streaming XPath Processing with Forward and Backward Axes. In: Proc of ICDE, pp. 455–466 (2003)
8. Tatarinov, I., Viglas, S., Beyer, K., Shekita, E., Shanmugasundaram, J., Zhang, C.: Storing and querying ordered XML using a relational database system. In: Proc of SIGMOD, pp. 204–215 (2002)
9. XPath, http://www.w3.org/TR/xpath
10. XMARK, http://monetdb.cwi.nl/xml

Redundant Array of Inexpensive Nodes for DWS

Jorge Vieira[1], Marco Vieira[2], Marco Costa[1], and Henrique Madeira[2]

[1] Critical Software SA
Coimbra, Portugal
{jvieira,mcosta}@criticalsoftware.com
[2] CISUC, Department of Informatics Engineering, University of Coimbra
Coimbra, Portugal
{mvieira,henrique}@dei.uc.pt

Abstract. The DWS (Data Warehouse Striping) technique is a round-robin data partitioning approach especially designed for distributed data warehousing environments. In DWS the fact tables are distributed by an arbitrary number of low-cost computers and the queries are executed in parallel by all the computers, guarantying a nearly optimal speed up and scale up. However, the use of a large number of inexpensive nodes increases the risk of having node failures that impair the computation of queries. This paper proposes an approach that provides Data Warehouse Striping with the capability of answering to queries even in the presence of node failures. This approach is based on the selective replication of data over the cluster nodes, which guarantees full availability when one or more nodes fail. The proposal was evaluated using the newly TPC-DS benchmark and the results show that the approach is quite effective.

Keywords: Data warehousing, redundancy, replication, recovery, availability.

1   Introduction

A data warehouse (DW) is an integrated and centralized repository that offers high capabilities for data analysis and manipulation [8]. Data warehouses represent nowadays an essential source of strategic information for many enterprises. In fact, as competition among enterprises increases, the availability of tailored information that helps decision makers during decision support processes is of utmost importance.

Data warehouses are repositories that usually contain high volumes of data integrated from different operational sources. Thus, the data stored in a DW can range from some hundreds of Gigabytes to the dozens of Terabytes [7]. Obviously, this scenario raises two important challenges. The first is related to the storage of the data, which requires large and highly-available storage devices. The second concerns accessing and processing the data in due time, as the goal is to provide low response times for the decision support queries issued by the users.

In order to properly handle large volumes of data, allowing performing complex data manipulation operations, enterprises normally use high performance systems to host their data warehouses. The most common choice is systems that offer massive

J.R. Haritsa, R. Kotagiri, and V. Pudi (Eds.): DASFAA 2008, LNCS 4947, pp. 580–587, 2008.

parallel processing capabilities [1], [10], as Massive Parallel Processing (MPP) systems or Symmetric MultiProcessing (SMP) systems. Due to the high price of this type of systems, some less expensive alternatives have already been proposed [5], [6], [9]. One of these alternatives is the Data Warehouse Stripping (DWS) technique [2], [4].

In a simplified view, the DWS technique distributes the data of a data warehouse by a cluster of computers, providing near linear speedup and scale up when new nodes are added to the cluster [2]. However, adding nodes to the cluster also increases the probability of node failure, which in turn leads to a decrease in the cluster MTBF (Mean Time Between Failures). It is worth mentioning that a failure of a single DWS node is enough to render the whole DWS cluster out of service, as the execution of queries requires the availability of all data partitions [2].

Data redundancy is used in several applications to tolerate failures [12], [13]. In this paper we propose an approach that provides DWS clusters with high-availability even in the presence of node failures. The proposed approach, named RAIN (Redundant Array of Inexpensive Nodes), allows DWS to deliver exact query answers and is able to tolerate failures of several cluster nodes (the number of node failures tolerated depends on the configuration used). The RAIN technique is based on the selective replication of data over a cluster of low-cost nodes and comprises two types of replication: simple redundancy (RAIN-0) and striped redundancy (RAIN-S).

The proposal is illustrated using the newly TPC-DS benchmark [11]. Several configurations tolerating failures of different numbers of nodes have been tested. Results show that the performance in the presence of node failures is quite good.

The structure of the paper is as follows: section 2 briefly presents the DWS technique; section 3 describes the RAIN technique; section 4 presents the experimental evaluation; and Section 5 concludes the paper.

2 The DWS Technique

In the DWS technique the data of each star schema [2], [3] of a data warehouse is distributed over an arbitrary number of nodes having the same star schema (which is equal to the schema of the equivalent centralized version). The data of the dimension tables is replicated in each node of the cluster (i.e., each dimension has exactly the same rows in all the nodes) and the data of the fact tables is distributed over the fact tables of the several nodes using strict row-by-row round-robin partitioning or hash partitioning (see Fig. 1).

DWS data partitioning for star schemas balances the workload by all computers in the cluster, supporting parallel query processing as well as load balancing for disks and processors. The experimental results presented in [3] show that a DWS cluster can provide an almost linear speed up and scale up.

In a DWS cluster typical OLAP (OnLine Analytical Processing) queries are executed in parallel by all the nodes available. If a node of the cluster fails, the system is still capable of computing approximated answers to queries, which are computed by applying statistical formulas to the partial results obtained from the available nodes. The system is also capable of providing confidence intervals for the provided answers. Obviously a degradation of the quality of answers is observed as the number of failed nodes increases.

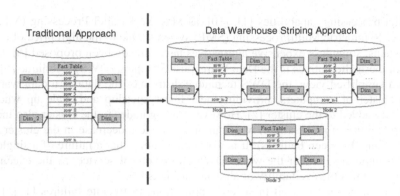

Fig. 1. Data Warehouse Striping Technique

The approximate answers provided by DWS systems are normally useful for decision support, especially because the confidence intervals provide the user with a measure of the accuracy of the query results. However, in some business areas the approximated answers are not enough, as it is required that the system always provides exact answers to queries, even in the presence of node failures.

An important aspect is that the DWS technique allows enterprises to build large data warehouses at low cost. In fact, DWS can be built using inexpensive hardware and software and still achieve very high performance.

3   Redundant Array of DWS Nodes

In the DWS when a node fails the fact tables data stored in that node is no longer available, preventing the system to compute exact answers to queries. In order to overcome this problem, allowing the system to provide exact answers to queries in the presence of node failures, it is necessary to guarantee that the fact table data from a given node is still available in the cluster, even when that node fails.

The RAIN technique is based on the replication of fact table data across the cluster nodes. Two distinct approaches are proposed for the implementation of this technique: simple redundancy (RAIN-0) and stripped redundancy (RAIN-S).

Although the technique requires more disk space in each node, as the data of the fact tables of each node will be replicated in the other nodes (see details further on), the current disk sizes in inexpensive computers (the ones used in the DWS clusters) are large enough to accommodate the extra data space required. From our experience, an inexpensive machine (a typical PC) can process queries over a star schema up to 20 GBytes of fact data with acceptable response time. However, a typical PC has disks of 200 GBytes or more, which is more than enough to store the facts data of the node plus the portions of data replicated from other nodes (and the typical DW materialized views and indexes). That is, it really makes sense to trade disk space by system availability, as proposed in the RAIN technique.

3.1 Simple Redundant Array of Inexpensive Nodes

The simple redundancy approach consists of replicating the facts data from each node in other nodes of the cluster. Depending on the configuration used RAIN-0 can tolerate the failure of one or more nodes. The following notation is used to refer to a RAIN-0 cluster with N nodes that tolerates the failure of Y nodes: **RAIN-0(N, Y)**.

Fig. 2 presents the facts data replication schema in a cluster of five nodes designed to tolerate the failure of a single node: RAIN-0(5, 1). As shown, each node stores its own facts data (i.e., 1/5 of the total facts data) plus the facts data from another node (i.e., each node stores a total of 2/5 of the total facts data). This way if one of the nodes temporarily fails it is still possibly to compute exact answers, since the data from that node also exists in another node.

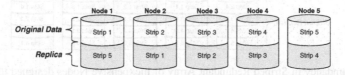

Fig. 2. Redundancy in Simple Redundant Array of Inexpensive Nodes designed to tolerate the failure of a single node: RAIN-0(5, 1)

It is important to emphasize that replicated data is used only in the case of node failures and no additional nodes are added to the cluster as that would represent a waste of computational resources in the absence of faults. Clearly, if more nodes are available they should be used during normal operation to increase the cluster performance and not only when there are failed nodes. To tolerate failures of several nodes the facts data must be replicated several times in different nodes. In fact, to tolerate the failure of N nodes the data must exist in N+1 nodes.

An important aspect is the overhead during the execution of queries in the presence of node failures. As the data from each node is replicated in at least another node, when a node fails one of the remaining nodes will have to process two times more data (i.e, a total of 2/N of all the data stored in the cluster, where N is the total number of nodes). For example, considering the RAIN-0(5, 1) cluster (see Fig. 2), when node 1 fails then node 2 must compute 2/5 of the facts data while nodes 3, 4, and 5 have to compute 1/5 of the facts data (as if there was no failure).

A basic rule is that to tolerate the failure of Y nodes a RAIN-0 cluster must comprise a minimum of Y+1 nodes. However, using the minimum number of nodes required implies that in the worst case scenario (failure of Y nodes) the single remaining node has to process all the facts data, which makes the system to act as a single server machine. Obviously, this is not the best approach and does not take advantage of the DWS data partitioning. This way, the minimum number of nodes recommended for a cluster should be equal to the number of simultaneous node failures to be tolerated multiplied by two: N=Y*2. This rule limits the maximum amount of data processed by any node to 2/N regardless of the number of failed nodes (that must obviously be inferior or equal to N/2).

3.2 Stripped Redundant Array of Inexpensive Nodes

The stripped replication is an evolution of the simple replication that intends to reduce the overhead caused on the queries execution when a node of the cluster fails. In this approach the facts data from each node is randomly distributed in N-1 sub-partitions (where N is the number of nodes) and each sub-partition is replicated in at least one of the other nodes (Fig. 3). As shown, each node stores $1/N + 1/N * 1/(N-1)$ of the facts data. The following notation is used to refer to a RAIN-S cluster with N nodes that tolerates the failure of Y nodes: **RAIN-S(N, Y)**.

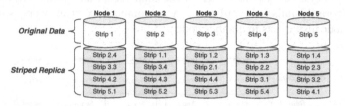

Fig. 3. Redundancy in Striped Redundant Array of Inexpensive Nodes designed to tolerate the failure of one node: RAIN-S(5, 1). Records from each partition are randomly distributed across four sub-partitions that are stored in the remaining four nodes.

Using RAIN-S replication strategy reduces drastically the overhead caused in the execution of a query when a node fails (when compared to RAIN-0). As the data from each node is evenly distributed by all the other nodes, when a node fails processing is distributed by all the remaining nodes. In other words, the remaining nodes have to compute only $(1/N) * 1/(N-1)$ more data (i.e., a total of $1/N + 1/N * 1/(N-1)$ of the all facts data stored in the cluster, where N is the total number of nodes). For example in a RAIN-S(20, 1) cluster each node will have to process approximately 0,2631% more data than it normally would: $(1/20) * 1/(20-1) = 0.2631\%$.

RAIN-S can also be configured to tolerate failures in several nodes. In this case, to tolerate the failure on Y nodes, each sub-partition must be replicated in at least Y nodes. An important aspect is that a RAIN-S cluster must have at least three nodes (with two nodes it is equal to a RAIN-0 cluster). Depending on the replication used, a cluster of 3 nodes is able to tolerate the failure of 1 or 2 nodes. Thus, the basic rule is that to tolerate the failure of Y nodes we need a RAIN-S cluster with Y+1 nodes. However, using the minimum number required nodes would mean that in the worst case scenario (failure of Y nodes) the single remaining node would have to process all the facts data, which is not the best approach as it does not take advantage of the DWS data partitioning. This way, the minimum number of nodes in a RAIN-S cluster should be equal to the number of simultaneous nodes failures to be tolerated multiplied by two: N=Y*2.

3.3 Execution Middleware and Node Recovery

Query execution in DWS is enabled through the use of a middleware [2] that allow client applications (e.g., Oracle Discoverer, JPivot, Crystal Reports) to connect to the system without knowing the cluster implementation details. This middleware receives queries from the clients, analyses the queries to be distributed by the nodes, submits the queries to the nodes, receives the partial results from the nodes and constructs the final result that is sent to the client application.

The DWS middleware was adapted to the RAIN technique, which means that it is prepared to transparently allow the system to continue providing exact queries answers when nodes fail. In fact, the middleware is able to detect failed nodes and uses the node's replicas to compute the queries answers.

Node recovery is one problematic aspect when considering typical DWS systems. In fact, node recovery is not possible when stored data gets corrupt. The only way to recover a node is to recollect the data from the operational sources and rebuild the entire cluster. The use of RAIN eases the node recovery, as the recovery process can be accomplished using the data existing in the failed node replicas, without need of cluster downtime.

4 Experimental Results and Discussion

The goal of the experiments presented in this section was to measure the impact of node failures in the query execution when using RAIN-0 and RAIN-S approaches.

Table 1 presents the set of configurations tested. The basic platform used consists of seven Intel Pentium IV servers with 2GB of memory, a 120GB SATA hard disk, and running PostgreSQL 8.2 database engine over the Debian Linux Etch operating system. The servers were connected through a dedicated fast-Ethernet network.

Table 1. Set of cluster configurations used. All configurations include RAIN-0 and RAIN-S. The benchmark was executed two times for each configuration (a total of 24 runs).

RAIN	System	# Nodes	# Node Failures Tolerated
RAIN-0 & RAIN-S	(5, 0)	5	0
RAIN-0 & RAIN-S	(5, 1)	5	1
RAIN-0 & RAIN-S	(5, 2)	5	2
RAIN-0 & RAIN-S	(7, 0)	7	0
RAIN-0 & RAIN-S	(7, 1)	7	1
RAIN-0 & RAIN-S	(7, 2)	7	2

The PostgreSQL database is one of the most complete open-source databases available and is frequently used to support non-critical applications. For these reasons, we have chosen this DBMS as case-study for the validation of the RAIN technique.

The TPC Benchmark™ DS (TPC-DS) [11] is a performance benchmark for decision support systems. This benchmark evaluates the essential features of decision support systems, including queries execution and data load. The benchmark includes a set of seven stars with many dimensions and using snow-flakes and mini-dimensions. A scale factor must be defined, which allows the evaluation of systems of different sizes. In our experiments we used a database with 10GB (scale factor 10 in TPC-DS benchmark). Due to time constraints, a representative subset of the TPC-DS queries was used to perform the experiments. The queries were selected based on their intrinsic characteristics and taking into account the changes needed for the queries to be supported by the PostgreSQL DBMS. Note that, as the goal is to evaluate the RAIN technique and not to compare the performance of the system with other systems, the subset of queries used is sufficient. The queries used are the following (see TPC-DS specification [11] for more details): 06, 07, 13, 15, 26, 27, 30, 37, 40, 45, 48, 52, 55, 75, 90, 91, 92, 96, 97, and 98.

Fig. 4 presents the execution times observed. As we can see, the RAIN-0(7, 0) presents an average execution time of 11.42 minutes, which is approximately 69% of the execution time observed for the RAIN-0(5, 0) configuration. This indicates a linear speedup during normal operation (without failures) of the cluster. Note that RAIN-0 and RAIN-S present the same execution time when the same number of nodes is considered (e.g., RAIN-0(5, 0) and RAIN-S(5, 0) have an execution time of 16,5 minutes). This is due to the fact that in these situations the cluster is operating in the absence of faults and thus the replication strategy does not influence the results.

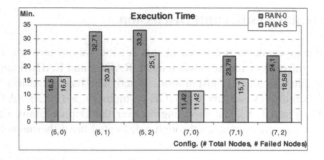

Fig. 4. Average execution time observed during the three executions of the benchmark in the several configurations tested

As expected the cluster with seven nodes presents better results when compared to the cluster with five nodes. Comparing RAIN-0 and RAIN-S execution times, we can observe that the latter presents much better results. For example, for the cluster with 5 nodes the execution time when one of the nodes is down is 32.71 minutes for RAIN-0 and 20.3 minutes for RAIN-S. As explained before, this is due to the fact that for RAIN-0 one of the remaining nodes has to process all the data from the failed node while for RAIN-S the processing is distributed across the several nodes.

An important observation is that for RAIN-S the execution time increases slightly as the number of failed nodes increases.

As mentioned before, the main objective of DWS is to allow the construction of low cost data warehouse systems with acceptable performance. From our experience a DWS cluster with 10 nodes can process with acceptable performance queries for a data warehouse with a data volume of 200 GB. Considering off-the-shelf computers with the Intel Core 2 Quad Processor, 4GB of DDRII memory and two 160 GB SA-TAII hard drives, this cluster will have a retail price of approximately 11.000€.

The introduction of RAIN in a DWS cluster has a very low impact on the final price of the cluster. The cluster presented above has enough disk space to allow the use of RAIN technique without needing further hard drive devices, as each node has 320 GB of storage available and needs to store less than 40GB of data. Nevertheless we can add two extra hard drives on each cluster node for a cost of less than 1.000€. This represents an increase in price of less than 10%, and allows using RAIN technique with a lower performance impact.

5 Conclusion

This paper proposes a technique that endows DWS clusters with the capability of providing exact queries answers even in the presence of node failures. The RAIN technique is based on the selective replication of data over the cluster nodes, which guarantees full availability when one or more nodes fail. Two types of replication are considered: simple redundancy (RAIN-0) and striped redundancy (RAIN-S).

The proposal was evaluated using the newly TPC-DS benchmark running on top of PostgreSQL and Debian Linux Etch operating system. The experimental results obtained were analyzed and discussed in detail. Results show that replication has no impact during the normal operation of the cluster. In addition, the performance impact in the presence of node failures is quite low.

Results show that by spending some disk space it is possible to achieve high availability with low performance degradation in the presence of node failures. This way, we believe that that the RAIN technique can be successfully applied to DWS clusters.

References

1. Agosta, L.: Data Warehousing Lessons Learned: SMP or MPP for Data Warehousing, DM Review Magazine (2002)
2. Bernardino, J., Madeira, H.: A New Technique to Speedup Queries in Data Warehousing. In: ABDIS-DASFA, Symp. on Advances in DB and Information Systems, Prague (2001)
3. Bernardino, J., Madeira, H.: Experimental Evaluation of a New Distributed Partitioning Technique for Data Warehouses. In: IDEAS 2001, Grenoble, France (2001)
4. Critical Software SA, DWS, http://www.criticalsoftware.com/
5. DATAllegro, DATAllegro v3™, http://www.datallegro.com/
6. ExtenDB, ExtenDB Parallel Server for Data Warehousing, http://www.extendb.com/
7. IDC, Survey-Based Segmentation of the Market by Data Warehouse Size and Number of Data Sources (2004)
8. Kimball, R., Ross, M.: The Data Warehouse Toolkit: The Complete Guide to Dimensional Modeling, 2nd edn. J. Wiley & Sons, Inc, Chichester (2002)
9. Netezza, The Netezza Performance Server® Data Warehouse Appliance, http://www.netezza.com/
10. Sun Microsystems, Data Warehousing Performance with SMP and MPP Architectures, White Paper (1998)
11. Transaction Processing Performance Council, TPC BenchmarkTM DS (Decision Support) Standard Specification, Draft Version 32 (2007), available at: http://www.tpc.org/tpcds/
12. Lin, Y., et al.: Middleware based Data Replication providing Snapshot Isolation. In: ACM SIGMOD Int. Conf. on Management of Data, Baltimore, Maryland, USA (2005)
13. Patino-Martinez, M., Jimenez-Peris, R., Alonso, G.: Scalable Replication in Database Clusters. In: Herlihy, M.P. (ed.) DISC 2000. LNCS, vol. 1914, Springer, Heidelberg (2000)

Load-Balancing for WAN Warehouses

Pedro Furtado

Universidade de Coimbra
Departamento Engenharia Informatica
Polo II, 3000 Coimbra, Portugal
+351239790000
pnf@dei.uc.pt

Abstract. Although the basic Data Warehouse schema concept is centralized, there are increasingly application domains in which there is the need to have several sites or computers input and analyze the data, therefore distributed data placement and processing is necessary. Given that sites may have different amounts of data and different processing capacities, how can we conform to the placement requirements of the context and balance such a system effectively? In WAN environments the network speed is a very relevant factor and there are application requirements concerning the place where each piece of data stays, based on who produced the data (ownership). We propose a new strategy that accepts the placement requirements of the desired context and uses an effective automatic approach to determine fixed-sized chunks and to balance and process those chunks efficiently. Our experimental results show the validity of the approach and how to minimize the context limitations.

Keywords: Data Warehouses, Grid.

1 Introduction

The Data Warehouse is a generic concept for storing, managing and analyzing data, typically used both in businesses and scientific domains. Some current and future applications of distributed data warehouses, from business historical information to molecular simulations or measurements of sensor data, can produce large amounts of data and many data warehouse applications have a distributed nature. One of the most significant challenges is maintaining an efficient data warehouse while abiding to the requirements of the data layout, and to build a distributed query processor for the data warehouse that is partitioned. Previous work on load-balancing considered only single site parallel databases, therefore did not have to deal with the requisites of having a WAN environment, different amounts of data in different sites and application-determined placement constraints. Our first main new contribution is to propose a basic schema partitioning that is the basis for the inter-site data warehouse while abiding to placement requirements in what concerns determinants such as ownership, organization or geographical issues that are present in the intended application contexts.

Another important contribution of this work is the proposal of an efficient load balancing approach tailored for the context, with a new automatic chunk formation

J.R. Haritsa, R. Kotagiri, and V. Pudi (Eds.): DASFAA 2008, LNCS 4947, pp. 588–595, 2008.

and replication approach that is crucial for the efficiency of the approach. Finally, we propose the load-balanced chunk processing strategy and evaluate the throughput of the solution against the non-load balanced alternative. We also analyze how to avoid imbalance in the partial replication that is required for load-balancing.

The paper is organized as follows: in section 2 we discuss related work. Section 3 discusses our proposal: we first describe data placement designed for the approach and context, chunk-based placement and processing and then load-balanced query processing. Section 4 presents some experimental results and section 5 concludes the paper.

2 Related Work

Data warehouse schemas usually have multidimensional characteristics [10], with large central fact relations containing several measurements (e.g. the amount of sales) and a size of up to hundreds or thousands of gigabytes, and dimensions (e.g. shop, client, product, supplier). These benefit from deployment in a parallel setting and conventional shared-nothing architectures such as reviewed in [2]. One of the major concerns in those environments is to decide how to partition or cluster relations into nodes, and query workload-based partitioning [17, 12] uses the query workload to determine the most appropriate partitioning attributes. WAN-based systems such as a "grid" of sites sharing a common data schema pose different requirements. Frequently, the data is organized according to ownership or other geographical and workload-related attributes.

Another relevant issue for WAN-based data warehouses is availability and load-balancing, and these can be accommodated by replication and replacing the processing of non-available or slower nodes by available or faster ones. Replica placement has been studied in different contexts, from RAID disks [11, 13] to chained [6, 7, 8] or interleaved declustering [14]. The use of chained declustering in the shared-nothing data warehouse context was first proposed in [5] and applied in [1]. In [5] we argued that, using a simple chained-declustering approach, we were able to have efficient availability guarantees (with small efficiency loss) in the event of node failures. In [1] we proposed the multi-site data warehouse grid with load-balancing and availability, but it assumed inter-site replication of the whole data warehouse for practical reasons (so that the data warehouse would be available both locally and globally if connections were off). In practice, loading and maintaining full replication over all sites is expensive, as in many typical environments (e.g. scientific labs) it would be too costly to maintain full inter-site data warehouse replication, and both ownership and organization policies may disallow such an approach. Load balancing is also discussed in [16], but in the context of a shared-storage architecture.

3 Placement and Load-Balancing over the Sites

The central concept of any Data Warehouse, including inter-site data warehouses, is a logically centralized schema deployed in relational databases as a star schema. These schemas are comprised of a few huge central fact relations and several smaller dimension relations. Each fact references a tuple in each dimension. For instance, a fact may consist of sales measure information and occupy anything from a few megabytes to 100 GB, while much smaller dimensions may include time, shop, customer, product.

The basic inter-site architecture assumes a set of sites that contain the data and offer query processing capabilities over that data – Executor Sites. Executor sites must have a database engine that is able to store and query the data using SQL and a simple middleware daemon (a Java application) that links to the remaining sites. It also assumes one or more controller sites, which have metadata and a middleware daemon that includes a global query processor. The metadata in the controller stores layout information concerning the data in the sites, and the global query processor takes queries as input, decomposes them into chunk queries (queries to be processed within executor sites) and schedules those chunk queries to sites for efficient execution. In order to have efficient load-balancing functionalities, we need to take into account how we may place and partially replicate the data in the system.

3.1 Data Layout and Replication

The inter-site warehouse is made of a set of sites, where each site is responsible for a fraction of the data in the data warehouse. The data must be partitioned not according to load-balancing objectives but to the global application requirements of the scientific or business data. Dimensions, which contain the relevant descriptive elements for analysis, are replicated into every site. Maintaining dimensions replicated into every site is reasonably inexpensive, because they are typically small. Facts are partitioned into all sites, but following a policy that depends on the application context – Partitioning Policy (PP). For instance, in scientific environments, facts generated by a specific research group may be placed at that research groups' site – Owner PP. The placement could also be geographical for instance – company sales of each region are placed in a site for that region. This means that it may not be feasible to follow a placement that would promote a balance of the amount of data among sites – Balanced PP. For the placement to happen, the system administrator must indicate which tables are to be partitioned and by what attribute, and which ones are to be replicated.

Figure 1 shows an example of how we partitioned the TPC-H schema from [15] in our experiments with inter-site data warehousing.

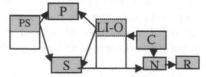

LI-lineitem, O-orders, PS-partsupp, P-part, S-supplier, C-customer, N-nation, R-region

Fig. 1. Layout for TPC-H data warehouse over Sites

The schema represents ordering and selling activity. Each node received the data depicted in gray in Figure 1, where partially filled rectangles represent only parts of tables and completely filled rectangles represent fully-replicated dimensions. In the case shown in this Figure, large fact relations (O, LI, PS) are partitioned (LI and O are partitioned by the common order_key attribute) and the remaining (smaller) dimension relations are replicated into all sites. This is similar to the approach in [3], but while in [3] the facts were hash-partitioned for efficient inter-node processing of joins, here facts are partitioned as dictated by each specific application context (e.g. ownership).

We also needed to find a way to divide the partial node data into smaller amounts that would be easy to form and would lend themselves to efficient load-balancing by partial replication. This was done as follows: when data is loaded into a table in a site, the system evaluates if the current chunk being filled reached a pre-defined chunk size. If so, it creates a new chunk and places the subsequent data into that new chunk. This approach is valid whether the data is placed using the Origin PP, Balanced PP or any other policy. The chunk size is a configurable parameter with default value of 250 MB (this should be a value such that queries involving the data in a chunk are processed reasonably fast, but it should also not be too small, to avoid the need for excessive chunk results merging overhead). With this simple approach, each site builds its facts as sets of "chunks", and queries are divided into chunk-sized queries by the global query processor - chunk queries.

Neighbour replicas (N-R): Chunks are also created/placed redundantly for load balancing. After a chunk is filled, it is copied into neighbouring sites. This way the system is able to replace the processing of a broken or slow site by an available or faster one. To enable this, sites are numbered from 1 to N (considering N sites) and, within a site, chunks are also numbered. They are then replicated into neighbouring sites, so that chunk j from site i is replicated into neighbouring site $(i + j) \, mod(N)$. For better availability and load-balancing, multiple replicas of each chunk can be created by replicating the same chunk into consecutive sites as well. For instance, chunk j can be replicated into site $i+j \, mod(N)$ and $i+j+1 \, mod(N)$. Figure 2 illustrates the approach with two of n sites (sites 1 and 2). Dimension relations C, P and S are dimensions replicated into all nodes. Relations LI-O are partitioned between the nodes and within each site they are sub-divided into chunks. The chunk replication approach means that site 2 will hold replicated chunk LI-O 1.1 from site 1. Likewise, site 1 will have replicated chunks n.1 from site n.

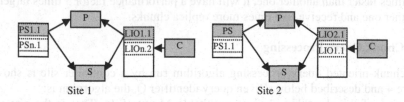

Fig. 2. Illustration of Site with replicated chunks (white dashed rectangles)

Given this replication approach, survival to site failures and load-balancing can both be guaranteed by an on-demand chunk processing approach, as processing can be diverted into other sites that contain replicated chunks. The initial query is decomposed into chunk queries and the controller maintains a queue of chunk queries for each site in the system, while simultaneously knowing that each chunk is also available as replicated in neighbour sites (in practice, only an identification of a generic chunk query template, together with a chunk ID and an enumeration of sites that hold the chunk are necessary in each queue position, because the same chunk query (chunk query template) is applied to all chunks by modifying only the chunk name termination to reflect the chunk identification). The queues are depicted in Figure 3.

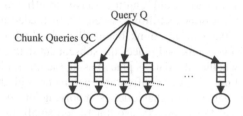

Fig. 3. Chunk processing in the architecture

When the query is submitted, the global query processor decomposes it into the chunk queries, places those chunk queries into the queues and notifies sites to start demanding chunks for processing. As sites end their work, they ask for further chunk queries to help deplete the amount of chunk queries that stand on the queues (there is a heartbeat function to evaluate site availability and the on-demand policy takes care of chunks that were not processed by unavailable sites). As soon as a site depletes all its chunk queries and asks for more work, the controller dispatches to it chunk queries from neighbouring queues that correspond to chunks it has replicated (this way, when faster sites end their own work, they help slower neighbouring sites).

Least-Used-Site Replicas (LUS-R): Placing chunk replicas into predefined neighbours does not provide guarantees of a balanced placement. The LUS-R approach is to determine the sites for replicas based on some performance factor, so that faster sites will have more chunks. We are are currently testing an automatic procedure to determine the performance factors, but in this work we consider that a manually configured performance factor is given for each node. For instance, if a site is 5 times faster than another one, it will have a performance factor 5 times larger than the other one and receive five times more replica chunks.

3.2 Chunk Query Processing

The chunk-oriented query processing algorithm run by a controller site is shown in Figure 4 and described below. Given query identifier Q, the algorithm is:

Step 1 of the algorithm chooses a suitable Merger Site. That is the site which receives results coming from all sites and applies a merge query. The Merger Site can be any executor site and in our implementation we balance the merge load by assigning query merging roles in a round-robin fashion to a set of sites that is manually indicated (sites with best performance features). Steps 2 and 3 of the algorithm command each site to start demanding chunk queries for the local execution step. From then on and until all work has finished (step 4. condition), the controller receives requests from sites. As soon as it receives a request, it tests if there is a chunk query in the corresponding chunk query queue (step 6.) and sends it for processing by the site. If there is none, the controller checks in the neighbor sites queues (for N-R, in all sites for LU-R) if there is some chunk query that could be processed by the site, because it has replicas of the corresponding chunk (steps 9., 10. and 11.). Step 12 of the algorithm represents the case in which there was no chunk the site could process, meaning that the site ended all useful chunk processing. In this case the controller simply notifies the site that its work ended and the site sends the result set for that operation to the Merger node (step 14).

Chunk-Oriented query processing: (n sites)
// command all sites to start demanding chunk queries
1. Elect a Merger Site
2. FOR (each Site S_i holding the schema)
3.   Notify S_i to start demanding chunk queries for Q
          Site[S_i]=STARTED
4. WHILE (there is at least one site with Site[S_i]=STARTED)
.   Wait for demand received from any site S_d
6.   IF(at least 1 query waiting in S_d chunk queries queue)
7.   Send a chunk query to S_d
8.   ELSE
9.   FOR(i=1;i<Number of Chunk Replicas;i++)
10.   IF(at least 1 chunk query S_d could do in the chunk queries queue of the i-th replica)
11.   Send the chunk query to S_d and exit loop
12.   IF (there was no chunk available for the site S_d to process)
13.   Site[S_i]=ENDED
14.   Command the site to merge partial chunk results and send the partial result to Merger Site
15 . IF all Sites have sent all partial chunk result to Merger Site
16.   Command Merger Site to merge all partial site results into the final answer

Fig. 4. Chunk-oriented Query Processing

4   Experimental Results

We show, using an heterogeneous setup, that if we do not use our load balancing approach, the performance of the whole system is dictated by the slowest elements, while with load balancing the processing replacement allows the system to be much faster. The computer nodes used are in Figure 5. The benchmark TPC-H with scale factor 10 was used [15] and we report on Queries-per-Hour (QpH).

Site Classification	Number of nodes	Processor	Memory	Disk	Network Bandwidth
High performance sites (HP)	20	P D 3.4 GHz	1GB	100GB	1Gbps
Slow sites (SS)	20	Celeron 1.8 GHz	100MB	30GB	100Mbps

Fig. 5. Characteristics of Nodes used in the Experiment

In Figure 6a we use different configurations to assess the advantage of our load-balancing approach (we used LU-R replica placement variant in this experiment) over 20 nodes, when compared with no load balancing. The results show that, in an heterogeneous environment (SS+HP), the load balancing approach improved the throughput significantly. These are the configurations that have 10 very fast sites and 10 slow sites. We can also see from the Figure that load balancing was crucial to that improvement, because without it (no load balancing) the heterogeneous configurations (SS+HP) did not improve much over the corresponding lower performance alternative.

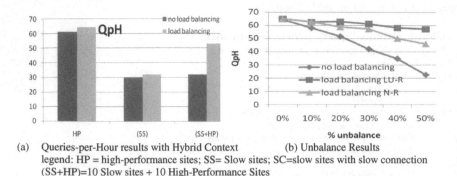

(a) Queries-per-Hour results with Hybrid Context   (b) Unbalance Results
legend: HP = high-performance sites; SS= Slow sites; SC=slow sites with slow connection
(SS+HP)=10 Slow sites + 10 High-Performance Sites

Fig. 6. Experimental Results

For the next experiment, we took 20 HP sites and placed the data sets with increasing unbalance by displacing an increasing number of the chunks from 10 of the sites into the other 10 sites. Then we compared the use of load balancing with no load balancing in those scenarios. This simulates data unbalance that can result from each site having its owned data - the Owner-PP policy we discussed in section 3.1. We included the two replica placement alternatives – Neighbor Replica (N-R) and Least-Used Replica (LU-R). Figure 6b shows the results: without load-balancing, the data unbalance resulted in a drastic drop in performance, the larger the unbalance, the larger the slope of the drop. The load balanced versions maintained a higher throughput despite the data unbalance, because by executing replicated versions of chunks, the sites with a lower amount of work were able to undo the data unbalance (in exchange for only a small increase in merging overhead). The N-R approach could not undo the imbalance as well as the LU-R approach, as LU-R was able to place the replica chunks in a manner that minimized imbalance.

5 Conclusions

In this paper we proposed a solution to the problem of placing and load-balancing a data warehouse that is built as a global inter-site structure. Our objective was to propose and test a highly functional, efficient and simple to handle approach that took typical placement requirements of the context into consideration. The approach is both fault tolerant and load-balanced, as it uses chunk-based processing, partial replication of the load into multiple sites and dynamic on-demand based allocation of chunk processing to sites. Our future work includes applying and testing this approach for fault-tolerance.

Acknowledgments. This work was partially developed under project Auto-DWPA (POSC/EIA/ 7974/2004), financed by "Fundação para a Ciência e Tecnologia" FCT.

References

1. Costa, R., Furtado, P.: Data Warehouses in Grids with High QoS. In: Proceedings of International Conference on Data Warehousing and Knowledge Discovery, Krakow, Poland, September 2006. LNCS, Springer, Heidelberg (2006)
2. DeWitt, D., Gray, J.: The Future of High Performance Database Processing. Communications of the ACM 3(6)
3. Furtado, P.: Experimental Evidence on Partitioning in Parallel Data Warehouses. In: Proceedings of the ACM DOLAP 04 - Workshop of the International Conference on Information and Knowledge Management, Washington USA (November 2004)
4. Furtado, P.: Efficiently Processing Query-Intensive Databases over a Non-dedicated Local Network. In: Proceedings of the 19th International Parallel and Distributed Processing Symposium, Denver, Colorado, USA (May 2006)
5. Furtado, P.: Replication in Node Partitioned Data Warehouses. In: Proceedings of the VLDB Ws. on Design, Implementation, and Deployment of Database Replication, Trondheim, Norway (August 2006)
6. Hsiao, H., DeWitt, D.: Chained Declustering: A New Availability Strategy for Multiprocessor Database Machines. In: International Conference on Data Engineering (1990)
7. Hsiao, H., DeWitt, D.: Replicated Data Management in the Gamma Database Machine. In: Workshop on the Management of Replicated Data (1990)
8. Hsiao, H., DeWitt, D.J.: A Performance Study of Three High Availability Data Replication Strategies. In: Proceedings of the Parallel and Distributed Systems (1991)
9. Hua, K.A., Lee, C.: An Adaptive Data Placement Scheme for Parallel Database Computer Systems. In: Proceedings of the Sixteenth Very Large Data Bases Conference, Brisbane, Queensland, Australia, August 1990, pp. 493–496 (1990)
10. Kimball, R., Reeves, L., Ross, M., Thornthwaite, W.: The Data Warehouse Life Cycle Toolkit. John Wiley & Sons, Chichester (1998)
11. Patterson, D.A., Gibson, G., Katz, R.H.: A case for redundant arrays of inexpensive disks (raid). In: Proceedings of the International Conference on Management of Data, Chicago, USA, June 1998, pp. 109–116 (1998)
12. Rao, J., Zhang, C., Megiddo, N., Lohman, G.: Automating Physical Database Design in a Parallel Database. In: Proceedings of the ACM International Conference on Management of Data, Madison, Wisconsin, USA, June 2002, pp. 8–69 (2002)
13. Tandem: NonStop SQL, A Distributed, High-Performance, High-Reliability Implementation of SQL. In: Workshop on High Performance Transactional Systems, CA, USA (September 1987)
14. Teradata: Teradata DBC/1012, Database Computer System Manual 2.0, C10-0001-02, Teradata, November
15. TPC. TPC Benchmark H, Transaction Processing Council (June 1999), Available at: http://www.tpc.org/
16. Raman, V., Han, W., Narang, I.: Parallel Querying with Non-Dedicated Computers. In: Proceedings of the 31st VLDB Conference, Trondheim, Norway (2005)
17. Zilio, D.C., Jhingran, A., Padmanabhan, S.: Partitioning Key Selection for a Shared-Nothing Parallel Database System. IBM Research Report RC 19820 (87739) (1994)

Enabling Privacy-Preserving e-Payment Processing

Mafruz Zaman Ashrafi and See Kiong Ng

Institute for Infocomm Research
21 Heng Mui Keng Terrace, Singapore 119613
{mashrafi,skng}@i2r.a-start.edyu.sg

Abstract. The alarming increase in the number of data breaching incidents from high profile companies reflects that buying goods or services from online merchants can pose a serious risk of customers' privacy and the merchants' business reputation. The conventional approach of encrypting customer data at merchant side using the merchant's secret key is no longer adequate for preserving customer privacy. An e-payment scheme that can guarantee customer authenticity while keeping the customer's sensitive details secret from the various parties involved in the online transaction is needed. We propose here an online protocol for processing e-payments that minimizes the customer's privacy as well as merchant business risks. Using a non-reusable password-based authentication approach, the proposed protocol allows consumers to purchase goods or services from an online merchant anonymously, thus achieving the ideal privacy environment in which to shop. The payment details sent to a merchant will become obsolete after the first use, thereby preventing any subsequent fraudulent transactions by a third party. Such protocol can be easily deployed in an e-commerce environment to strengthen the integrity of the electronic payment system.

1 Introduction

Online purchasing is poised to be on the rise as more and more consumers become internet-savvy, fueled by the rapid emergence of ubiquitous high speed broadband and wireless internet connections. As consumers increasingly use cell phones, personal digital assistants, PCs, and television set-top boxes to shop for goods and services anywhere in the world, there is a growing associated concern about the security and privacy issues of online transactions and electronic payments. Unlike the traditional way of selling goods and services, the online merchant must process transactions in not only a non-cash fashion but also a card-not-present environment. This makes it a highly vulnerable target for fraudsters [1].

Merchants in the e-commerce environment process transactions in a virtual and card-not-present environment. For verification purposes, an online customer will typically be asked more information than compared to when conducting a conventional transaction in a physical environment. Yet, online customers possess no control over the information flow, nor do they have the authority to manage their sensitive payment information. Obviously, the online merchants are very concerned about customer privacy and often implement database access control and/or encryption techniques.

J.R. Haritsa, R. Kotagiri, and V. Pudi (Eds.): DASFAA 2008, LNCS 4947, pp. 596–603, 2008.

However, recent incidents [8, 15] suggest that applying these conventional techniques no longer protect customers' privacy adequately. For example, recently TJX Companies Inc. (which runs more than 2500 shops world wide) found that intruders or hackers have stolen more than 45 million credit and debit card numbers from their computer system [8]. Data files containing encrypted sensitive customers' personal details were also stolen. This was not a standalone case as similar privacy breaches have also been found to occur with many other companies [11].

Such incidents pose serious threat for the healthy growth of electronic commerce. When customers disclose their payment details and related sensitive information to online merchants, not only are they exposing themselves to the costly risk of jeopardizing their privacy [1], the online merchants may also suffer immeasurable financial and reputation consequences if data breaches occur on their sites. To protect customer privacy, and mitigate the merchant's business risk, many online merchants now employ online payment services from third party payment gateways. However, this gives a false sense of security to the customer, as sending payment details to a third-party payment gateway merely shifts the target of attack. For example, recently a payment gateway *CardSystem* admitted that it has lost customer details [11].

The ease and scale of recent data breaches suggests that the conventional approach of encrypting customer data at merchant side using merchant secret key is no longer sufficient for preserving customer privacy. We need a privacy-preserving e-payment scheme that can guarantee customer authenticity while keeping the customer's sensitive details secret from the respective parties involved in the online transaction. In this paper, we propose an online payment protocol that minimizes the customer's privacy threat as well as the merchant's business risks. The proposed protocol allows consumers to anonymously purchase goods or services from an online merchant, thus achieving a highly private environment in which to shop. Therefore, the payment details sent to a merchant will become obsolete after the first use. In this way, should a merchant lose the payment information from their proprietary database, or any attacker ever obtained the information, no third party can execute any fraudulent transactions by using the obsolete information.

1.1 Our Contributions

In this paper, we address the issues of online customers' *information privacy* in terms of the secrecy and ownership of their details, and how to mitigate privacy risks when they are shopping from an online merchant. The main contributions of this paper can be summarized as follows:

1. *Anonymity.* Our proposed payment protocol offers customers anonymity while shopping online. At the same time, it also mitigates the merchant's risks by minimizing the possibility of the fraudulent transactions without incurring any extra cost.

2. *Authenticity.* We also proposed an efficient authentication technique for this payment protocol, requiring the issuing card company to not only verify the credit or debit card details, but also the customer's card password. This reduces the chances of subsequent illegitimate transactions should a customer lose his/her wallet or if someone were able to discover his/her card details.

3. *Non-reusability.* Our proposed payment protocol also utilized a random session key and a time stamp for encrypting the payment details; this guarantees that the

customer payment details will be unique every time they make a request for payment. As such, the same payment request cannot be used more than once, thereby minimizing identity theft risk.

2 Related Works

Several well-known protocols have been proposed by the industry to address the potential privacy breaches associated with the use of electronic payments to conduct commerce. The Secure Socket Layer (SSL) protocol [9] proposed in 1994 has now become the de facto standard adopted by most (if not all) merchants for secure online transactions. However, this protocol only protects e-commerce customers against external eavesdropping attackers and assumes that the merchant can be trusted to keep customers' details private [2, 6].

To address the rising concerns on customers' information privacy, a consortium of large credit card companies, including Visa and MasterCard come up with the Secure Electronic Transaction (SET) protocol more than a decade ago, so that each customer can sign and hide the details of their purchase [14]. SET is a very comprehensive protocol that maintains the confidentiality of the sensitive information, insures the integrity of the payment, and verifies the identities of all parties, including both the buyer and seller involved with any ecommerce payment process. Unfortunately, the computational complexity and overhead of this protocol has rendered it impractical. [2-5, 16].

Alternative identity verification techniques based on two-factor authentication have been proposed recently [15]. These techniques typically require additional hardware devices that every credit or debit cardholder has to carry in order to complete their online shopping transactions. The obvious drawback of this approach is the cumbersome requirement for customers to carry additional devices, which may not always be available when the customer wants to make an online purchase. Besides employing additional hardware authentication devices, another technique is to offer customers a one-time credit or debit card number that becomes obsolete after the customer finishes the transaction [7, 16]. However, at least one of the prominent financial institutes has been known to dispose of this technique after initiating this service for their customers and this technique has yet to flourish [16].

3 Proposed Method

We propose a protocol that uses a non-reusable password-based authentication approach for anonymous online e-payments. The aim of our protocol is not only to preserve the privacy of the customer but also to keep the merchant's existing payment handling process. The use of a password-based approach provides the necessary convenience for the customers for practical adoption. For encryption, we use both random number and timestamp so that the cipher text of the customer's payment details will be unique each time a payment request is submitted to the merchant. Generating different cipher text ensures that if the merchant retains a copy of that cipher text and an attacker or hacker were able to retrieve that information, he will be unable to find the credit card details and replay the cipher text to carry out fraudulent transactions, subsequently minimizing the threat of *man-in-the-middle* attack.

Let us now describe our privacy-preserving protocol for e-payment that keeps the customer's account details secret from the merchant and his choice of goods secret from the bank, and yet allowing the customer, merchant, and bank to authenticate the relevant details of a transaction even when some of those details are kept from them. Together with the protocol, we will describe a method that will achieve a one-time cipher text for each message to ensure message freshness and make data breaches useless exercises for a hacker or attacker.

3.1 Protocol

The proposed e-payment protocol has eight phases as described below:

Initialization Request: When a customer indicates (by pressing the submit button) that he/she is interested to buy goods/services from a merchant, he/she requests an invoice and the public keys of the merchant and credit card issuing company.

Initialization Response: The merchant replies with a signed message that includes an invoice consisting of a transaction identifier, and two certificates; customer's software (i.e. browser) then verifies the two certificate details.

Payment Response: The customer then generates two packets of a message: (*i*) payment and (*ii*) common order information separately. The payment information includes the hashcode of card details, the hashcode of the cardholder's password, the hashcode of common order details, and also a timestamp and validity period. The common order information includes the transaction ID, the purchase amount agreed with the merchant, and also the timestamp and its validity, which will be encrypted using the merchant's public key. The Time-to-valid parameter included in both messages indicates message freshness (i.e. how long the payment message will be valid); the default or generic value of this field would be the average processing time, say less than 60 seconds. The public key of card issuing company and merchant are used to encrypt the payment and common order information respectively.

Payment Processing: Upon receiving the payment response from the customer, the merchant decrypts the common order details using the merchant's private key. If it finds any discrepancy in the information that was sent earlier to the customer as an invoice, it immediately stops processing and notifies the customer. Otherwise, it generates a unique payment ID for this transaction and sends the customer's payment and common order details packet to the payment gateway for authorization, along with pre-generated authorization requests. Note that the payment gateway is an optional service; some merchants may choose to communicate directly with the card issuing company, in which case we can bypass following step *authorization request-I*.

Authorization Request– I: Similar to the merchant, the payment gateway first verifies the message, including the pre-generated merchant's authorization message, and then sends it to the credit card issuing company if it finds all messages are tamper-free.

Authorization Request– II: After receiving all messages from the payment gateway (or the merchant directly), the card issuing company verifies the decrypts the payment information using its own private key, checking the timestamp and its expiration. The company also checks the hashcode of the customer's password as we shall see detail in next section, which it stores in its local server. Finally, it sends messages to the issuing bank, who verifies whether sufficient funds are available for this transaction and sends back an approval or denial message to the credit card issuing company.

Authorization status: The credit card issuing company then sends back the authorization status to the payment gateway, which then organizes the payment, if approved by the cardholder's issuing bank. After that, it sends the approval/denial status along with the common order to the merchant, who can and finally updates the order status in its own server and sends it to the customer to let him/her know about the success of the purchase request.

3.2 Encryption

To protect the customer's privacy, we require that the cryptosystem that we used in the above protocol (i.e. sending payment details) always generates a unique cipher text every time the customer sends the payment details. As described earlier, the payment information message payment-request-details contain numerous items of information including a random number, customer card details, etc. As such, the length of the entire packet may be bigger than the key size. In this case, we will apply the block-cipher technique. We shall now discuss the cryptosystem of our protocol.

Let us assume the above protocol uses the RSA cryptosystem. To construct a one-way cipher text we add some random noise to each block of the original message:

$$c_i \rightarrow Crypt_{pubEK(cardissuing)} [m_i \oplus r] \tag{1}$$

where r is a random number, possibly half of the length of *pubEK*. While the customer adds the random noise r to his/her payment details, the card issuing company has no idea about the exact value of that random number at this stage, but to discover the exact payment details, the card issuing company requires that random number. Thus, the cipher text of that random number r is embedded at the beginning of the payment-request-details message.

$$C \rightarrow [c_1, c_2 \dots c_i] \tag{2}$$

In order to make sure that each of the hashcode values (except the card number) that the customer has included in the payment-request-details is dynamic, we incorporate the timestamp value. For simplicity, let us name these hash values *timestamp embedded dynamic hash*. Intuitively, these values will ensure that the customer's payment details cannot be reused, regardless of whether a dishonest merchant is able to separate the timestamp block from the payment cipher text or not. The timestamp value will be incorporated in the following manner:

$$h_1 \rightarrow Hash[cardholder\_name] ; h_2 \rightarrow Hash[timestamp]$$

$$H_1 \rightarrow Hash[h_1 \parallel h_2] \tag{3}$$

where '$\parallel$' represents an operator for concatenating the string h_1 and h_2.

3.3 Decryption

Once the card issuing company receives the cipher text, it reads the first block and finds the exact value of random number r that the customer embedded. Then the subsequent blocks are decrypted in the following way:

$$m_i \rightarrow (Decryp_{priEK(cardissuing)} [c] - r) \tag{4}$$

As mentioned earlier, every hashcode value of the customer's payment details (except the card number) are timestamp-embedded dynamic hash coded. Therefore, the card issuing company will still be unable to verify the cardholder details immediately after decryption. In this case, the card-issuing company will first discover the cardholder's information from their local database that is mapped to the decrypted card number. Let us assume that the card issuing company stores its cardholders' information in hashed forms in its database. In this case, the card issuing company will incorporate the timestamp value using formula (3) with its local hashed value of the customer details to verify the customer's identity. If the card issuing company stores its cardholders' information as unhashed plain text, then it will need to hash the information first before incorporating the timestamp value.

The exact time required to process a transaction generally relies on various participating parties and the communication speed. The validity period attributes are thus set with some default value, say equal to the average time for completing a customer transaction. However, the card issuing company needs to pay special attention to ensure there should not be more than one approved transaction within the validity period time window. To do so, every time the card issuing company receives an approval status from the card issuing bank, it will put that payment information to a table and remove it from that table once the validity is over. Every time that the card issuing company receives a payment request, it can now verify whether it has already processed the same request within the valid period or not.

4 Analysis

We have proposed a unique e-payment scheme that focuses on preserving the customer's information privacy. Table 1 below shows the comparisons between our method and the well known existing methods such as SET, SSL, and Visa 3D-Secure [9, 10 and 14]. Our method uses the card issuing company's public key and signed message using the customers' password. Let us now analyze the unique characteristics of our method in further details.

Our method can offer full *anonymity* for customers but still allow the merchant to be able to have full trust in their customers because every customer will verify with the card issuing company in two ways with (*i*) what they possess, i.e. the card number and the security code, and (*ii*) what they know, i.e. the password. As the customer payment details will always be authorized by the card issuing company, when the merchant sends the customer's payment details, the corresponding card issuing company can verify both the card number as well as the password, and will notify the merchant whether that particular customer actually possesses that card or not. The notification directly from the card issuing company allows the merchant to know the trustworthiness of their customers even though the payment details are concealed.

Secondly, the encryption scheme of our method permits the maintenance of *confidentiality* and enhances *efficiency*. As the payment information is encrypted using the issuing card company's public key, only the card issuing company can decrypt it. No eavesdropper or dishonest party in the communication network can decrypt that message at any given time as long as the private key of the issuing card company remains secure. Moreover, unlike SET [14], our method only uses the card issuing company's public key, which reduces the number of encryption/decryption operations.

Table 1. Comparison Proposed method vs. Other well known Methods

		Proposed Method	SET	SSL	3- D Secure
Anonymity		*Full*	*Limited/No*	*Limited/No*	*No*
Reusability		*No*	*Yes*	*Yes*	*No*
Verification	**Pass**	*Yes*	*No*	*No*	*Yes*
	Key	*No*	*Yes*	*No*	*No*
Additional Log In		*No*	*No*	*No*	*Yes*
Complexity		*Low*	*High*	*Low*	*Low*
Registration		CIC^1	CA^2	Nc	CIC^1

^{1} *CIC*- Card Issuing Company; ^{2}Certificate Authority.

Thirdly, our proposed protocol uses a timestamp and a random session key to encrypt the payment details, which cannot be *reused*. The information that the customer provided to their merchant while shopping becomes obsolete after the first use, thus preventing a man-in-middle or replay attack as long as the private key of the issuing card company remains secure. As the payment details become obsolete after the first use, the ownership of that information is effectively returned to the customer after use, giving back the customers full control of their payment information.

Fourthly, we have shown an efficient authentication technique that verifies each customer against their password. Unlike other methods [7, 10, 12 and 13], our authentication method is efficient because it does not require customers to log in another web page in order to verify themselves. Although the customer's password is sent along with the payment details, our protocol is also secure as the dynamic hashing ensures that any eavesdropper will be unable to steal it from the payment details.

Last but not least, our proposed protocol can easily be *deployed* in an ecommerce environment without requiring great changes to the current processes. As the cardholder registers with the card issuing company, the customer can set his/her card password with the company. This means that each customer can utilize the proposed protocol as soon as they receive their card, as there is no need for any further registration with a third party certificate authority, as required by many other current proposed solutions. Moreover, the merchant can process the transaction with the same process that they are currently employing; in other words, without the need to modify the current interface of their payment portal.

5   Conclusion

The use of electronic payments to conduct commerce is here to stay as buyers and sellers recognize the significant benefits in purchasing goods and services anywhere in the world. However, the virtual and card-not-present online environment makes it especially vulnerable to fraudsters. Indeed, data breaches have become more frequent; at the same time, it is also getting more costly for all the parties involved. Beyond the immediate financial and reputational consequences, repeated occurrence of data breaches is certain to undermine the public's confidence in electronic commerce.

Current electronic transaction systems have focused mostly on providing a secure communication environment but an oft-overlooked issue is the protection of customer's information privacy. In this paper, we have presented an electronic payment protocol that ensures that the customer is able to minimize their privacy and identity theft risk. The proposed protocol allows consumers to anonymously purchase goods or services from an online merchant, thus achieving the ideal privacy environment in which to shop. Moreover, the payment details they send to a merchant become obsolete after the first use. Therefore, if a merchant loses their payment information, or any attacker ever obtains the information, any third party will be inhibited from executing any fraudulent transaction. By enabling privacy-preserving e-payment processing, we can help strengthen the integrity of online payment systems to protect the healthy growth of electronic commerce.

References

1. Federal Trade Commission, Consumer Fraud and Identity Theft Complaint Data, available electronically at: http://www.consumer.gov/sentinel/pubs/Top10Fraud2005.pdf
2. Bella, G., Massacci, F., Paulson, L.C.: The verification of an industrial payment protocol: the SET purchase phase. In: Proc. of the 9th ACM CCS, pp. 12–20 (2002)
3. Bella, G., Massacci, F., Paulson, L.C.: Verifying the SET Registration Protocols. IEEE Journal of Selected Areas in Communications 21(1), 77–87 (2003)
4. Bella, G., Massacci, F., Paulson, L.C.: Verifying the SET Purchase Protocols. Journal of Automated Reasoning 36(1-2), 5–37 (2006)
5. Ruiz, C.M., Cazorla, D., Cuartero, F., Pardo, J.J.: Analysis of the SET e-commerce protocol using a true concurrency process algebra. In: Proce. ACM SAC, pp. 879–886 (2006)
6. Wagner, D., Schneier, B.: Analysis of the SSL 3.0 Protocol. In: Proc. of the 2nd USENIX Workshop on Electronic Commerce, pp. 29–40 (1996)
7. Citibank Virtual Account Number, available at: http://www.citicards.com/cards/wv/detail.do?screenID=700
8. Boston Globe, Breach of data at TJX is called the biggest ever, available at: http://www.privacy.org/archives/2007_03.html
9. Netscape Communication, The SSL Protocol Version 3.0, available electronically: http://wp.netscape.com/eng/ssl3/ssl-toc.html
10. Visa Verified By Visa, available at: https://usa.visa.com/personal/security/vbv/index.html
11. Schneier, B.: CardSystems Exposes 40 Million Identities (July 2005) available electronically at: http://www.schneier.com/blog/archives/2005/06/cardsystems_exp.html
12. Samos, M.H.: Electronic Payment Systems (20-763), Official Course Web, available electronically at: http://euro.ecom.cmu.edu/program/courses/tcr763/2002pgh/cards7.ppt
13. Discover Card, Secure Online Account Number available electronically at: http://www.discovercard.com/discover/data/faq/soan.shtml
14. Mastercard & VISA. SET Secure Electronic Transaction: External Interface Guide (1997)
15. VeriSign Unified Authentication, available electronically at: http://www.verisign.com/products-services/security-services/unified-authentication/index.html
16. MSN Money Online, credit cards are the only way to buy, available electronically at: http://moneycentral.msn.com/content/Banking/creditcardsmarts/P114591.asp

Managing and Correlating Historical Events Using an Event Timeline Datatype

Abhishek Biswas[1], B.V. Sagar[1], and Jagannathan Srinivasan[2]

[1] Amrita School of Engineering, Kasavanahalli, Bellandur Post, Bangalore 560037, India
[2] Oracle, One Oracle Drive, Nashua, NH 03062, USA

Abstract. An event is occurrence of an incident at a particular instant of time with a set of attributes describing it. A collection of similar events ordered by time is called an *event timeline[1]*. This paper presents the notion and design of an event timeline datatype, which can be used to store, maintain and operate on a collection of events as a single abstraction. The datatype represents event occurrences as bitmaps with methods for accessing and updating the timeline. Two additional pair-wise operators union (OR) and intersection (AND) allow correlating timelines to discover new information. Semi-structuring event attributes of a timeline as (name, value) pairs provides a basis for search and indexing. The datatype can be used either to develop simple applications requiring timelines, such as genealogy, patient history, etc. or to develop more involved applications tracing the evolution of a field based on timelines depicting developments in related areas.

Keywords: Event, Timeline, Datatype, Bitmaps, Indexing.

1 Introduction

Chronologically ordering events is something generally referred to as history. Such an event map may sketch the progress of an organization, highlight advances in a technological field, or simply depict lifespan of a person, or a product. Furthermore, chronologically ordered events can grouped by certain common attributes to form interesting event timelines, like, a record of births, deaths or marriages in a family.

An event in general can have *temporal*, *spatial* and *schematic* attributes associated with it. For example, a birth event has date of birth, place of birth, and name of parents associated with it. If we assume that the time when the event occurred is always specified, then we can formulate the concept of *event timeline* as events grouped by one or more common attributes and ordered by time. For example, an event timeline of a company may contain events such as its launch, major product releases, acquisitions, etc. Each event may contain a different set of attributes but they can be filtered out from a general pool of events by the common "organization name" attribute.

This paper addresses the issue of managing such event timelines with a relational database system by creating an event timeline datatype with suitable representation of

---

[1] Timelines can be classified as discrete (event) or periodic timelines. The focus of this paper is on discrete event timelines, which is referred to as timelines in the rest of the paper.

J.R. Haritsa, R. Kotagiri, and V. Pudi (Eds.): DASFAA 2008, LNCS 4947, pp. 604–612, 2008.
© Springer-Verlag Berlin Heidelberg 2008

event occurrences and its attributes. This abstraction allows timelines to be a fundamental part of a schema design as timeline type columns and variables, which can be accessed by SQL queries through methods.

For the timeline (EVENT_TIMELINE_ty) datatype, the paper proposes:

- Bitmaps to record event occurrences with a one for occurrence of an event, zero, otherwise. Bitmaps can represent a timescale of any required granularity consuming minimal space and can be compressed further [7]. It also simplifies logical operation on timeline type variables.
- Semi-structured representation of event attributes as (name, value) pairs is flexible and facilitates search and filter operations.
- Two logical operators, Union (OR) and Intersection (AND), for correlating two event timeline type variables. These operators aim to discover obscure associations between timelines.
- APIs for use in DML commands and queries on tables with timelines.
- Three granularities (day, month, year) to model different granularity timelines.

Applications: Supporting event timelines in DBMS would enable building interesting database applications. Historical events can now be stored as a single timeline, for example, events pertaining to a country, or an organization, and used for pedagogical purposes. The ability to correlate event timelines would allow tracking the evolution of a field based on timelines depicting developments in related areas. For example, an application can trace the evolution of Computer Science by exploring and correlating Hardware, Software and Internet timelines [11].

History of events that occur in a database application itself could be interesting and worth recording. For example, in an order processing application, one may want to maintain purchase order history of a customer. Another interesting usage is to record the order status modification history itself. This could be correlated to an event timeline of inventory replenishment events. Similarly, event timelines could also be useful for auditing applications where modification history of one or more of sensitive data tables is maintained, which can be used for investigation purposes.

Sample Example: Let us consider the history of book publication and painting exhibition events. Consider a books_and_arts events table with a column of the EVENT_TIMELINE_ty datatype, recording timelines of famous books published or timelines of celebrated paintings exhibited. Let each row represent the timeline of a different country. The table can be created as follows:

```
CREATE TABLE books_and_arts (country VARCHAR2(64),
                             timeline EVENT_TIMELINE_ty);
```

To instantiate the datatype the set_timeline() constructor is invoked. The following command inserts a book publication event timeline of India:

```
INSERT INTO books_and_arts VALUES('India',
  EVENT_TIMELINE_ty.set_timeline ('book_timeline',
   '(27-May-1912, "title=Gitanjali",
     "author=Rabindra Nath Tagore")
   (24-Oct-1935, "title=Swami and Friends",
     "author=R. K. Narayan")', 1900, 1950));
```

The first argument to set_timeline() method specifies the name of timeline, whereas the last two arguments specify the time interval in units of year.

In general, an event has temporal, spatial, and schematic attributes associated with it. The mandatory temporal attribute forms the timeline and is stored as a bitmap. The optional spatial and schematic attributes are stored as (name, value) pairs. In addition, one can specify other structured fields in the relational table, which are associated with the entire timeline (e.g., the country field).

To display events in a timeline (as a string), the following query can be used:

```
SELECT ba.timeline.get_all_events()
FROM books_and_arts ba WHERE ba.country='India';
```

Methods for searching and filtering events in a timeline are also provided. For example, a simple search query, using a Table function, looking for events with a certain (name, value) pair, is as follows:

```
SELECT t.event_date, t.event_attributes
FROM books_and_arts ba,
   TABLE (ba.timeline.contains('author',
   "Rabindra Nath Tagore')) t WHERE a.country='India';
```

In addition, pair-wise timeline correlation operations accept two event timeline type variables and return the resulting event timeline as result. To illustrate this, we insert a timeline of paintings exhibited in USA:

```
INSERT INTO books_and_arts VALUES ('USA',
   EVENT_TIMELINE_ty.set_timeline ('painting_timeline',
   '(24-Oct-1935, "title=Thanksgiving",
       "painter=Doris Lee")', 1900, 1950));
```

Now, a query correlating the timelines of books in India and paintings from USA can be obtained by performing an intersection (AND) operation as follows:

```
SELECT AND_timelines (a.timeline,
                      b.timeline).get_all_events()
AS Result FROM
(SELECT timeline FROM books_and_arts
   WHERE country='India') a,
(SELECT timeline FROM books_and_arts
   WHERE country='USA') b;
```

Result
___

```
(24-Oct-1935, "title=Swami and Friends", "author=R.K. Narayan")
(24-Oct-1935, "title=Thanksgiving", "painter=Doris Lee")
```

The operation brings out the interesting fact that publication of the R. K. Narayan's book coincided with the exhibition of Doris Lee's painting. Furthermore, by typecasting timelines to higher granularity (say month, or year) and then performing logical operations, user can find co-occurrences in broader time interval.

The event timeline datatype is implemented in Oracle Database Release 9iR2 using Oracle's Extensibility Framework [10] and includes an indexing scheme to support efficient filtering based on timelines operators. We conducted experiments with synthesized genealogy dataset that characterizes the performance of timeline datatype.

Related Work: There are quite a few software packages dealing with creation, maintenance, and visual display of timelines for legal cases, law enforcement, project management, and health care systems [1, 2, 3, 4]. This paper however focuses on timelines as a datatype, which can be part of a database system as a built in type.

Bitmaps have been used for bitmap indexes supported in DBMS like Oracle, DB2, and Sybase. Concept of a bitmap datatype [5] has been used for RFID item tracking. We however use bitmaps to represent timescales and identify event occurrences, which enables performing logical AND & OR operations on timelines.

Semi–structured data are similar attributed objects grouped together in arbitrary order. Length of an attribute can vary and presence of every attribute in an object is not necessary. XML is typically used to identify structural units within a document [6]. This idea has been adopted to represent event attributes as (name, value) pairs.

Time series analysis [9] focuses on identifying the pattern in the observed set of data points so that one can understand the nature of phenomenon represented by the time series or predict its future values. Event timelines can be viewed to contain a time series. However, the time series involved is irregular as historical events rarely occur at uniform intervals. Also, the focus here is on events described by attributes with the ability to discover trends and correlate multiple timelines. Also, event timelines allows representing more than one event at a given instant of time.

Temporal databases [8] address the problem of handling time-qualified data. In contrast, our focus is to support event timelines in a traditional DBMS with date type.

2 Key Concepts

An event is defined by a set of event attributes. Event attributes can be classified into temporal, spatial and schematic attributes. Usually, one or more of the event attributes (such as time or location) may be common across the events thereby facilitating their organization. For example, if time attribute is common across all events then we can chronologically order events to form an event timeline (Fig. 1a). Similarly, if location attribute is common across all events then we can form an event space map (Fig. 1b).

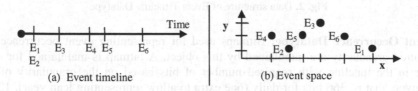

Fig. 1. Organizing events on temporal and spatial dimensions

This paper focuses on event timelines with logical correlation operations on chronologically ordered event sets. The same idea can be explored for events organized on spatial or schematic dimensions but is beyond the scope of this paper.

The event timeline datatype represents historical events as a single internally managed timeline entity in a relational database. To correlate timelines, the logical AND and OR operations are provided. The *closure* property of these pair-wise operations is maintained, that is, an operation on two timeline variables results in another timeline variable, allowing multilevel recursively nested operations on timelines. The datatype form of timelines allows managing a timeline individually (as a variable) or as collections (stored in a column of a table). Indexing schemes can be implemented for efficient querying of timeline type collections.

To support varying precision of dates, timelines with different granularity (daily, monthly, or yearly) is supported. A lower granularity timeline type instance may be type casted to a higher granularity type for performing inter-type logical operations.

Regarding semantics of pair-wise logical operations, for *exact overlap* of timelines the resulting timeline would inherit the same span. For *partial overlapping* and *disjoint* timelines, the result timelines spans from earliest start year to latest end year.

Event Timeline Datatype: The EVENT_TIMELINE_ty provides APIs for timeline maintenance and correlation operations (Table 1). It includes attributes `timeline ID`, `timeline name`, `start` and `end year` of the timeline and a single instance each of `event occurrence` type and `event details` type (Fig. 2).

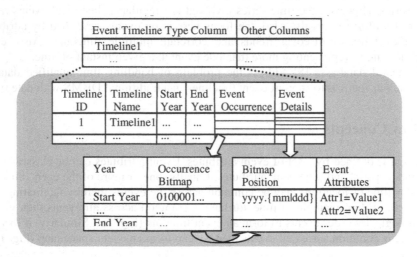

Fig. 2. Data structure of Event Timeline Datatype

Event Occurrence Datatype: Bitmaps used for representing event occurrences of required granularity are maintained by this object. A bitmap is maintained for each year in the timeline with required number of bits based on the granularity of the datatype, that is, 366 bits for daily (one extra to allow representing leap year), 12 for monthly, and one for yearly. A bit set in the bitmap represents occurrence of one or more events. The first day of the Julian calendar (January 1, 4712 BC) is used as a common reference point for dates. Event occurrences are often sparse so the bitmaps representing lower granularity can be compressed. Methods for bitmap creation and logical operations (AND & OR) on event occurrence bitmaps are provided.

Event Details Datatype: Event attributes defining the event are stored as (name, value) pairs and managed by this datatype. The event data is referenced by a position qualifier of the form *yyyy* (in case of yearly bitmap) or *yyyy.{mm|ddd}* (in case monthly or daily bitmap) where *yyyy* is the year offset relative to the start year of the timeline and *mm* (*ddd*) is the month (day) offset within the year. This bitmap position uniquely identifies a date in the timeline. The basic format can be easily extended to support hour, minute, or second granularity timelines.

Table 1. Key Methods of Event Timeline Datatype

Method	Description
set_timeline(timeline_name event_list, start_year,end_year)	returns event timeline; The event_list is of the form <(event-date, "Name=Val" [,"Name=Val"] ...) ...>
get_events_table()	returns EVENTS_TABLE(event_date, event_attributes) as a table
total_event_count()	returns total event count of the timeline
{yearly|monthly}_event_count()	returns ({yearl| month}, count) as a table
{add|delete}_event(event_date event_attributes)	add/deletes an event and returns modified timeline
{AND|OR}_timelines(tm1, tm2)	returns resulting (ANDed or ORed) timeline
filter_timeline(name, value)	generates a event timeline of events containing the specified (name,value) pair
contains_named_value(timeline_column, name, value)	returns "1" if the timeline has an event matching the (name, value) pair or else "0".
has_event_on(timeline_col, ev_date [,ev_date2])	returns "1" if timeline has an event on the specified date (date range) or else "0".
to_yearly(), to_monthly()	converts from monthly or daily timeline

The design of the datatype and the methods supported is covered in detail in [11].

3 Implementation

The timeline datatype has been implemented in Oracle 9iR2 Database using Oracle's Extensibility Framework [10], which allows creating complex object types, table functions, user-defined operators, and indexing schemes.

Different granularity timelines (daily, monthly, yearly) are implemented as three object types. Each of these types is composed of occurrence bitmap and event details, which in turn are implemented as object types (Fig. 2). Bitmaps representing the timeline are maintained as a VARRAY of Oracle RAW type. The timeline attributes are stored in nested table collections referenced by bitmap offset positions, allowing the timeline objects to be self-sufficient and maintain the closure property for timeline operations. Logical operations on the bitmaps are performed using the UTL_RAW package routines. CAST operations for transforming from finer to coarser granularity involve aggregating the bitmaps and the corresponding set of events.

A timeline datatype specific indexing scheme is created for efficient filtering timelines using CONTAINS_NAMED_VALUE (HAS_EVENT_ON) operator. For named value base filtering, an index_data_nv(name, value, rid) table is created where name and value pairs are extracted from the timeline event data fields of the indexed column and rid is the corresponding row identifier (Fig. 3).

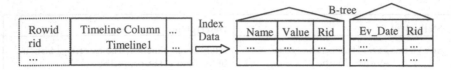

Fig. 3. Indexing Timeline Column

Similarly, for event date based filtering, index_data_ev(ev_date, rid) table is created. The index manipulation routines translate operations on the timeline column into operations on the index data table. For example, when the indexed timeline column is queried using the CONTAINS_NAMED_VALUE operator, an appropriate query is implicitly issued against the index_data_nv table to retrieve all the satisfying rows. This executes efficiently by using B-tree index structure on the index_data_nv table. Similarly, queries involving HAS_EVENT_ON operator leverages index_data_ev table and its B-tree index.

Performance Study: The experiments comparing the three timeline datatypes are performed using Oracle 9iR2 DBMS and MS Windows XP Professional on a Pentium 4 PC with 256 MB RAM, 128 MB database cache, and 8KB block size.

The *data set* consists of 50,000 randomly generated *genealogy* events of births and corresponding deaths spanning over 700 years with a raw size of 3.2 MB. The birth events have five attributes, whereas the death event has two attributes. They are stored as *birth* or *death* timelines in an event_history(year, ev_type, ev_timeline) table with each row holding a timeline of one year duration. Three versions of this table with daily, monthly, and yearly timelines are used.

Table 2 shows the load time and storage costs for different granularity timelines. The coarser granularity timelines have smaller load time because multiple events get grouped together thereby reducing the number of distinct bitmap offsets, which need to be computed for event detail rows. However, the storage costs across timelines are similar as the significant portion of storage (3.24 MB) is incurred by event details, which needs to be maintained across all timelines.

Table 2. Performance of timeline datatypes for three different granularities

Datatype Granularity	Load Time (sec.)	Space Occupied (MB)	AND of birth and death event timelines (sec.)			
			12,500 events	25,000 events	37,500 events	50,000 events
Yearly	153.68	3.33	0.88	1.02	1.92	2.39
Monthly	164.28	3.33	3.14	7.27	11.54	16.87
Daily	182.42	3.42	3.95	8.15	13.37	19.74
Daily$_{10year}$	N/A	3.42	1.16	1.88	3.02	3.97

Next, the query response time (averaged over 20 runs, standard deviation <= 0.4) of performing AND operation on birth and death timelines of different event counts was measured (Table 2). The response time of the AND operation increases (almost linearly) as the number of events increases. Also, finer granularity timelines have more overhead (compared to coarser granularity timeline) as they have to process larger number of event sets (of lower cardinality). However, maintaining decade timelines of ten year duration and repeating AND query shows significant reduction in response time ($Daily_{10year}$) as the method invocation overhead is reduced.

Next, we examine filtering operator query performances on an indexed vs. non-indexed daily timeline type column storing the data set described above. The index creation time is 126.65 seconds and index data table size is 3.45 MB. Table 3 shows the query performances for three CONTAINS_NAMED_VALUE queries (abbreviated as CNV in the table) averaged over 20 runs and with standard deviation <= 0.1. Index-based execution outperforms non-indexed based execution. Query 3 shows use of index even when the CNV operator is used with a scalar column predicate.

Table 3. Query Processing time with indexed vs. non-indexed timeline column

Query	WHERE Clause condition	Result size	Non-Indexed	Indexed
1	CNV('name', 'abc165342') = 1	1	18.70 sec.	0.46 sec.
2	CNV('place', 'city3028') = 1	9	20.03 sec.	1.53 sec.
3	CNV('place', 'city3028') = 1 AND ev_type='Death'	3	10.21 sec.	0.69 sec.

4   Conclusion and Future Work

This paper proposed a novel datatype for managing and correlating historical events. Special timeline operations (union and intersection) allow correlating two timelines and the closure property of timelines operations enables recursively nested timeline queries. The three granularities (yearly, monthly, and daily) of timelines allow handling data with varying precision of dates. The timeline datatype is implemented on Oracle DBMS, 9iR2. The experimental study using a genealogy dataset characterizes the performance of the datatype. In future, we plan to support timelines of finer granularity (minutes, hour, and seconds), and develop indexing schemes to speed up other common timeline operations.

Acknowledgments. We thank Sayanka Banerjee for help with data collection, and A. Srinivas, Ramesh Babu, and Jay Banerjee for their encouragement and support.

References

1. Timeline Maker Professional 2, http://www.timelinemaker.com
2. CareVoyant, http://www.infosysusa.com/products_physicain_home.asp
3. MinuteMan Project Management Software, http://www.minuteman-systems.com
4. Plaisant, C., Mushlin, R., Snyder, A., Li, J., Heller, D., Shneiderman, B.: LifeLines: Using Visualization to Enhance Navigation and Analysis of Patient Records. In: AMIA, pp. 76–80 (1998)

5. Hu, Y., Sundara, S., Chorma, T., Srinivasan, J.: Supporting RFID-based Item Tracking Applications in Oracle DBMS Using a Bitmap Datatype. In: VLDB 2005, pp. 1140–1151 (2005)
6. Sperberg-McQueen, C.M.: XML and Semi-structured data. ACM Queue 3(8), 34–41 (2005)
7. Wu, K., Otoo, E.J., Shoshani, A.: Optimizing Bitmap Indices with Efficient Compression. ACM Trans. Database Syst. 31(1), 1–38 (2006)
8. Etzion, O., Jajodia, S., Sripada, S. (eds.) Temporal Databases: Research and Practice, Dagstuhl Seminar 1997. LNCS, vol. 1399, Springer, Heidelberg (1998)
9. Time Series Analysis, http://www.statsoft.com/textbook/sttimser.html
10. Oracle9i Data Cartridge Developer's Guide Release 2 (9.2), Part No. A96595-01 (2002)
11. Biswas, A., Sagar, B.V., Banerjee, S.: Event Timeline Datatype: Final Semester Project, Amrita School of Engineering (affiliated with VTU), Karnataka, India (2007)

Online Collaborative Stock Control and Selling Among E-Retailers

Horng-Ren Tsai and Toly Chen*

Department of Information Technology, Ling Tung University
*Department of Industrial Engineering and Systems Management, Feng Chia University,
100, Wenhwa Road, Seatwen, Taichung City, Taiwan
tolychen@ms37.hinet.net
http://www.geocities.com/tinchihchen/

Abstract. Zwass claimed that in e-commerce there are five areas (commerce, collaboration, communication, connection, and computation) in which innovation opportunities could be found. This research is devoted to investigating one way of innovative collaboration among e-retailers and applies the concept of virtual warehousing under an online inter-organizational framework, so as to facilitate selling and to accelerate inventory depletion by resources sharing and mutual assistance. An online collaborative stock control and selling mechanism among e-retailers is therefore constructed. To evaluate the effectiveness of the constructed mechanism, two experimental B2C websites are also built for simulation analyses. According to experimental results, the constructed mechanism could indeed improve the efficiency of selling products for the participating e-retailers by shortening the average time to stock depletion up to 14%.

1 Introduction

In the past, without Internet a retailer used telephone or fax to contact its wholesaler for stock replenishment. Such a stock replenishment action had to be accompanied with at least a fixed number of products, which increased the risk on the retailer's side, because the replenishment action might lead to the accumulation of stock. With the widespread development of Internet, the communication channels between a retailer and its wholesaler are becoming more and more diverse and convenient; nevertheless, a mutually beneficial plan of order quantity negotiation is still needed. Many mechanisms have therefore been constructed, e.g. quick response (QR), efficient customer response (ECR), collaborative planning, forecasting, and replenishment (CPFR) (including vendor managed inventory (VMI) and jointly managed inventory (JMI)), etc. This study considers from another point of view, i.e. retailers themselves can also communicate with each other (not with their wholesalers) very conveniently through Internet. As a result, it becomes possible to exchange stock among retailers, especially e-retailers, to satisfy transient and unexpected demand, so as to promote sales, which provides the motive for constructing an online collaborative stock control and selling mechanism among e-retailers. Besides, due to the diversification of the trading environment that happened recently, inter-organization collaboration is

J.R. Haritsa, R. Kotagiri, and V. Pudi (Eds.): DASFAA 2008, LNCS 4947, pp. 613–620, 2008.
© Springer-Verlag Berlin Heidelberg 2008

becoming an important topic in organization theory fields. In addition, innovative and evolving technologies, dynamic environmental changes, and rising competition pressure force an organization to look for the opportunity of cooperating with another, so as to maintain the global competitive advantage, which also holds for an e-retailer. The concept of collaborative commerce (e-commerce) is therefore proposed. According to Park et al. [5], the stakeholders of a c-commerce alliance may collaborate on product designs, procurement plans, demand forecasts, manufacturing schedules, and distribution activities.

An e-retailer collaborates with another to satisfy customer requirements in various ways. For example, in affiliate marketing the affiliate (associate) e-retailer promotes the products and services of the merchant e-retailer on the affiliate's website for a commission. In addition, the affiliate's website content can therefore be more fresh and dynamic. Many famous websites have provided affiliate marketing programs, e.g. the Amazon.com Associates program and the Yahoo Affiliates program [1]. In addition, Ito et al. [4] proposed a cooperative exchanging mechanism in which an e-retailer can online exchange for a product if he does not have enough of the product for sale, with the aid of software agents. Affiliate marketing is basically unidirectional stock provision and commissioned selling, not bi-directional stock sharing and collaborative selling. Besides, Ito et al.'s model only considers the stock exchange between two e-retailers, and lacks the function of stock aggregation among many e-retailers. To solve these problems, an online collaborative stock control and selling mechanism (OCSSM) among e-retailers is constructed for this purpose. In OCSSM, the stock levels of a product on multiple websites are monitored, virtually aggregated, and shared among these websites. This is one of the most important functions of c-commerce: aggregating information from a variety of sources to provide an integrated view of these resources [5]. They can therefore share stock and exchange products among each other. Unlike an affiliate marketing program in which the merchant e-retailer determines the price, is actually responsible for the delivery of the product, and occupies the total profit, every e-retailer in OCSSM can determine a different price for the same product and makes a different profit. In other words, every e-retailer plays both affiliate and merchant roles. In this way, the channels of selling a product will be significantly increased. Selling can therefore be enhanced and inventory depletion will be accelerated. Namely, the community-oriented strategy is to shorten the average time to stock depletion, and we expect that the outcome will be a win-win situation for participating e-retailers. Besides, to enhance the efficiency of OCSSM, related operations (e.g. stock aggregation and sharing) will be automated, and do not need human intervention.

2 Literature Review

This section reviews several concepts and technologies that are related with the construction of OCSSM. The characteristics of e-commerce are different with different trading objects or application fields. In application aspect, e-commerce can be classified into business-to-business (B2B), business-to-consumer (B2C), business-to-business-to-customer (B2B2C), consumer-to-consumer (C2C), peer-to-peer (P2P), mobile commerce, intra-organizational, business-to-employee (B2E), collaborative commerce, non-profit-making, digital learning, exchange-to-exchange (E2E), and e-government categories.

In Zwass's opinion, e-commerce comprises five aspects including commerce, collaboration, communication, connection, and computation. These aspects can be exploited to find innovational opportunities to organize and address marketplaces, offer innovative products, collaborate with business partners, transform business processes, and organize the delivery of information-system services. If the five innovational opportunities are mapped to these categories, then a matrix showing the innovational opportunities in these e-commerce categories can be constructed in Table 1.

Table 1. Innovative opportunities in e-commerce categories

	B2B	B2C	C2C	C2B	Others
Organize and address marketplaces					
Offer innovative products					
Collaborate with business partners					
Transform business processes					
Organize the delivery of information-system services					

This study is focused on the innovative coordination and cooperation among e-retailers, which belongs to "collaborate with business partners" and B2B EC. However, all participants are sellers, which is different from traditional B2B EC in which usually one is buyer and the other is seller. Nevertheless, the ultimate goal is also to satisfy a consumer's (final customer's) demand. Therefore, it belongs to B2B2C EC, which is usually involved in B2C EC.

2.1 Collaborative Stock Control and Selling Among Business Partners

The mechanism constructed in this study is one way of cooperation among multiple e-retailers in the aspect of stock management. For this reason, this section firstly reviews the concepts of stock and stock management. Then how to aggregate the stock is investigated.

Owing to the uncertainties in the time and quantities of demand and supply, the purpose of holding stock is to balance such uncertainties [2]. From the supply point of view, stock has the functions of preventing material shortage and maintaining the continuity of production. From the demand viewpoint, stock has to satisfy predicted demand, periodic demand, and fluctuating demand. Every enterprise holds more or less stock. However, too much stock leads to the accumulation of fund and increases the difficulties in financial operations. Conversely, insufficient stock increases the times of ordering and increases the associated expenses. Further, production might be paused because of the shortage of materials. Besides, the risk of shortage and losing orders in selling is also increased. Therefore, the purpose of stock management is to seek the balance between the stock level and times of ordering, so as to enhance productivity or sales benefits. According to Davis et al. [2], holding stock has the following advantages: satisfying predicted demand, reducing ordering costs, reducing shortage costs, maintaining the independence of operations, making production operations smooth and flexible, gaining protection from the increase in the price. Conversely, there are also disadvantages associated with holding stock:

1. Holding costs are increased.
2. It becomes less flexible to respond to customer's transient requirement changes.
3. Causing the waste of capacity.

Following the trend of e-business, how to apply information technologies to optimize stock management is one of the challenges facing an enterprise. On the other hand, in business models of global logistics, stock is no longer treated as a necessary asset, but a burden that increases the costs for the enterprise. Therefore, if the stock level in the warehouse can be lowered, or the stock in the supplier's warehouse can be flexibly utilized, then the costs of holding stock can be reduced and the efficiency of stock management can be elevated, from which the concept of "virtual stock" originates.

Most inventory systems place stock at the same location. Companies usually construct one or few warehouses to store stock. The fewer the number of warehouses is, the easier the management becomes. The major advantages of such inventory systems include the easiness to obtain stock related information, and the convenience in stock management and control. However, due to the rapid development of information technologies and internet, obtaining the stock information of a distant warehouse becomes as easy as that of a nearby one. As a result, even the inventory is spread everywhere, the company can also control effectively.

3   The OCSSM Mechanism

This section firstly states the basic assumptions in constructing the OCSSM mechanism. Then the properties of collaboratively sold products and the construction and operation procedures of OCSSM are introduced. To enhance the applicability of the OCSSM mechanism in practical environment, some basic assumptions are made in this study:

1. An consensus can be achieved among participating shopping websites about the trading prices or commission of products: In this study, a shopping website has to input the price and commission percentage of a product. Then the trading interval is specified as [price * (1 – commission percentage), price * (1 + commission percentage)]. Product exchange only happens between two websites which trading intervals overlaps, and the exchange price is equal to the center of the overlapped intervals.
2. All participating websites will cooperate in shifting orders, paying commission, and delivery of goods.
3. The prosperity of market is good. Namely, all stock will be sold out eventually. Otherwise, it is not necessary to exchange the stock.

A product sold with the OCSSM mechanism has the following characteristics:

1. The product can be sold from any participating website.
2. The product can be simultaneously for sale on all participating websites, but will eventually be sold from a single website.

3. The stock of the product on any participating website is limited. Therefore there is necessity for stock exchange.

With the OCSSM mechanism, the stock of the product on all participating websites are monitored, virtually aggregated, and shared among these websites. There are the following roles in the mechanism:

1. The original seller of a product, which is the website that sells the product and inputs the basic data of the product into the system database.
2. The collaborative seller of the product: In addition to the original seller of the product, other websites joining the collaborative selling of the product afterward are called the collaborative sellers of the product.
3. The system manager: The system manager is a software agent composed of three parts - a register dealing with the registration of websites and products, a stock monitor continuously recording and aggregating the stock levels on all participating websites, and a request sender sending requests to all participating websites to get or update the current stock levels.

Besides, the total online quantity obtained by OCSSM is in fact a virtual stock level, which contains the stock of every website joining the mechanism, and can be sold from any website. For this reason, to prevent from instantaneous stock shortage, the interval of reading (and updating) stock levels has to be very small. The most important consideration in constructing OCSSM is not to disturb the normal operations of the shopping website. The operation procedure of OCSSM is shown in Fig. 1, and is detailed as follows:

1. An e-retailer goes to the system register to register its basic data.
2. After checking the validity of the registration data, the e-retailer is informed of a set of username and password to register new products that will be sold with the mechanism. The e-retailer can also join the collaborative sale of other products that have been involved in the mechanism.
3. The e-retailer has to download a stock reader and a stock updater that are simple script files used only to get and update the stock levels of products involved in OCSSM, respectively. The default values of parameters in the script files have to be modified.
4. The request sender of the system manager sends a request to run the stock reader on each participating website every a small time interval, and the stock reader will respond the current stock levels of products involved in OCSSM on the website. Then the stock levels are aggregated to obtain the total stock levels (the total quantities that still can be sold online). The calculation process is explained as follows. Assume there are totally m products involved in the mechanism and denoted by P_i, $i = 1 \sim m$. There are n websites selling product P_i and indicated by $W_{ij}, j = 1 \sim n$. The current stock level of product P_i on website W_{ij} is S_{ij}. The previously obtained total stock level of product P_i is PTS_i. Then the current total stock level of product P_i can be obtained as:

$$CTS_i = PTS_i - \sum_{j=1}^{n}(PTS_i - S_{ij}). \tag{1}$$

5. The request sender of the system manager sends a request to run the stock updater on each participating website to update the stock level of product P_i on the website to the current total stock level CTS_i. The PTS_i of product P_i is also updated to the same value.

$$PTS_i = CTS_i. \tag{2}$$

4 Simulation Experiments

Two experimental shopping websites and the OCSSM system server have been constructed to simulate the practical operations of the mechanism. The experimental environment is described:

1. Platform: Microsoft Windows XP and IIS server installed on a PC.
2. Shopping websites: constructed using TimeShop shopping system.
3. A series of computer programs simulating different kinds of users' accesses to the website, which have been developed with BCB programming language

The procedure of doing simulation experiments is detailed:

1. Construct the two experimental EC websites.
2. Construct the OCSSM server with VBScript and HTML programming languages.
3. Develop a series of computer programs to simulate the accesses by various kinds of users (including visitors and customers) to the EC website. The sequence of browsing documents by a simulated user is randomized. Backward accesses are also allowed, but have to conform to the limitations in the website structure. The browsing path of a customer must contain the payment page. Simulated browsing paths might not be straight. Simulated users might retrace steps, linger, take side paths and come back.
4. Go to the OCSSM system server to register the two websites.
5. Go to the OCSSM system server to register collaboratively sold products.
6. Download the stock level reading and updating scripts for the two websites.
7. Simulate customers' browsing and purchasing behaviors on the two websites.

The OCSSM system server periodically (every 10 minutes) calls the stock reading scripts on the two websites to get the current stock levels. After aggregation, the OCSSM server calls the stock updating scripts to show the total online stock levels, which are only virtual stock levels. A series of simulation experiments have also been performed to compare the average time to stock depletion in collaborative selling and in separate selling. Firstly, assuming demand cannot be partially satisfied, which means that if the stock level is less than the demand, then the customer will not purchase. The results are shown in Fig. 2. Obviously the time to stock depletion is shorter in collaborative selling than in separate selling. The average advantage of collaborative selling is 13.7%.

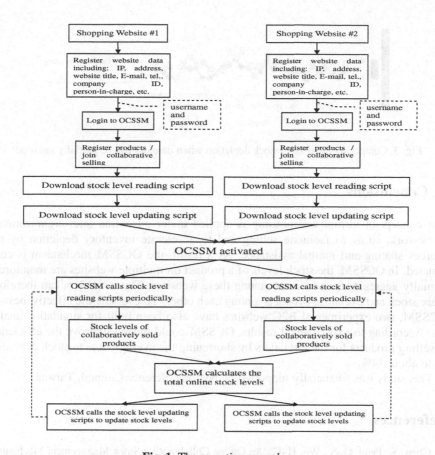

Fig. 1. The operation procedure

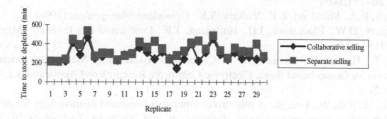

Fig. 2. Comparing the time to stock depletion when demand cannot be partially satisfied

Subsequently, assuming demand can be partially satisfied, which means that if the stock level is less than the demand, then the customer will reduce the demand and then purchase. The results are shown in Fig. 3. The time to stock depletion is also shorter in collaborative selling than in separate selling. The average advantage of collaborative selling is 14.5%. In other words, the superiority of collaborative selling is consistent.

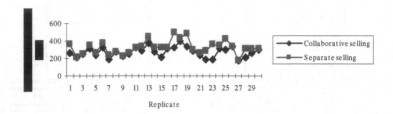

Fig. 3. Comparing the time to stock depletion when demand can be partially satisfied

5 Conclusions

The concept of virtual warehousing is applied under an online inter-organizational framework, so as to facilitate selling and to accelerate inventory depletion by resources sharing and mutual assistance. As a result, the OCSSM mechanism is constructed. In OCSSM, the stock levels of a product on multiple websites are monitored, virtually aggregated, and shared among these websites. The e-retailers can therefore share stock and exchange products among each other. To evaluate the effectiveness of OCSSM, two experimental B2C websites have also been built for simulation analyses. According to experimental results, OCSSM could indeed improve the efficiency in selling products for the e-retailers by shortening the average time to stock depletion up to about 14%.

This study was financially supported by National Science Council, Taiwan.

References

1. Chen, T., Peng, G.-S., Wu, H.-C.: An Online Collaborative Stock Management Mechanism for Multiple E-retailers. Lecture Series on Computer and Computational Sciences, vol. 2, pp. 20–23 (2005)
2. Davis, R.A., Markland, R.E., Vickery, S.K.: Operations Management (1998)
3. Fogarty, D.W., Blackstone, J.H., Hoffmann, T.R.: Production and Inventory Management (1991)
4. Ito, T., Hattori, H., Shintani, T.: A Cooperative Exchanging Mechanism among Seller Agents for Group-based Sales. Electronic Commerce Research and Applications 1, 138–149 (2002)
5. Park, H., Suh, W., Lee, H.: A role-driven component-oriented methodology for developing collaborative commerce systems. Information and Software Technology 46, 819–837 (2004)
6. Zwass, V.: Electronic Commerce and Organizational Innovation: Aspects and Opportunities. International Journal of Electronic Commerce 7(3), 7–37 (2003)

Mining Automotive Warranty Claims Data for Effective Root Cause Analysis

Ashish Sureka, Sudripto De, and Kishore Varma

Infosys Technologies Limited,
Bangalore 560100, India
{Ashish_Sureka,Sudripto_De,Kishore_Varma}@infosys.com

Abstract. We present an application of text analytics in automotive industry and describe a research prototype for extracting named-entities in textual data recorded in automotive warranty claim forms. We describe an application for gaining useful insights about products defect reported to the dealer during the warranty period of vehicles. The prototype is developed for air-conditioning subsystem and consists of two main components: a text tagging and annotation engine a query engine. We present some real world examples with sample output and share our design and implementation experiences.

Keywords: Text analytics, product warranty data analysis, text tagging and annotation.

1 Introduction

Product defects and warranty claims results in heavy costs to manufacturers. The top 50 U.S.-based warranty providers together reported $23.0 billion in warranty claims during 2006, up 5.1% from 2005. It is interesting to note that auto manufacturing companies in USA such as General Motors Corporation and Ford Motors Corporation are amongst the top 50 U.S.-based warranty providers of product warranties in terms of the total dollar amounts they reported in warranty claims during calendar 2006. GM and Ford together spent $8.6 billion on claims during 2006 and auto manufacturers grab the bulk of the pie in the overall extended warranty market. Auto manufacturing companies are spending a huge percentage (around 2.5%-3.0%) of their sales revenue fixing vehicles under warranty which puts a tremendous pressure on auto manufacturing companies to come up with innovative ways to reduce the overall cost to the company on warranty claims by reducing the detection to correction time of a product failure and also to increase customer satisfaction and brand value [7][9][11].

More importantly, the famous Bridgestone and Firestone's recall of 6.5 million tires used primarily on Ford Explorer vehicles, and the deaths of more than 100 people in accidents blamed on the failure of those tires triggered legislators to introduce a new U.S. law called as Transportation Recall, Enhancement, Accountability and Documentation (TREAD) Act which makes it mandatory for automakers to compile quarterly reports on consumer complaints and warranty claims. Recall of

J.R. Haritsa, R. Kotagiri, and V. Pudi (Eds.): DASFAA 2008, LNCS 4947, pp. 621–626, 2008.
© Springer-Verlag Berlin Heidelberg 2008

vehicles due to safety related product defects costs a huge amount of money to auto makers and has become a very serious issue because of the loss of human life and injuries as a result of those defects. Another famous example is recalls of around 800,000 Jeep Liberty vehicles because of a defective steering part. The defect caused drivers to lose control of their vehicle and suffer serious injuries in road accidents. The US Consumer Product Safety Commission provides public access to data on recalls under various product types, companies and date [4][6][8][10][12].

The pressure to reduce spending on warranty claims and compliance to legal regulations has prompted automakers to look more carefully into warranty claims data. The business driver for the work presented in this paper is to build tools and techniques that can help discover defects early in the product life-cycle and enable a warranty analyst to identify root causes of failures by leveraging textual data in conjunction with the structured data recorded in claim forms.

1.1 Textual Data in Product Warranty Claim Forms

Automotive warranty data is generally gathered by filling a claim form by a customer and a technician. The form is either filled on a paper which is later scanned and imported into a database or the information is directly entered online. A form can contain many fields to be entered by a customer or a technician. Some of the fields require information such as the product code, model number, date and time-stamp and customer id. This information falls into the category of structured data in the sense that the information has a well defined format and requires close-ended answers i.e. there are finite choices from which a selection can be made. Usually the form also contains a comments section where a customer or a technician can provide detailed information about the problem. The comment section is provided as it is not possible to capture details about a defect using the structured data fields alone. This is the section where information is entered in the form of a natural language text or a free-form text. The data entered in comment sections is a key element in diagnosing and understanding the problem. Following are some of the examples of the customer complaint, technician comments and action taken field data.

Customer complaint
- Air conditioning not working
- Poor performance from a/c system
- Water ingress into passenger foot well
- Room lamp flickering when switched ON

Technician Comments
- Found expansion valve defective
- Thermostat by-pass valve not working
- No engine cranking noise. Solenoid check.
- AC Knob found broken

Action Taken
- Removed and replaced the seals on the elbow joints
- Cable set properly & refitted

- AC cable replaced resulting smooth movement
- Connected the coupler to compressor

If the auto manufacturer suspects a recurring problem they sift through the claims data and manually go through the customer and technician comments to see if they can find any kind of patterns or clue that can help them finding the cause of the problem. A high level analysis of a defect can be done from data stored in structured data fields. However, a drill down analysis or an in depth analysis requires a warranty claim analyst to read the free-form textual data fields also. The main challenge is that the manual process of reading each and every comment is impractical and time consuming. Hence there is a strong need in automating the process of analyzing the natural language text data stored in claim forms for an efficient data analytics.

2 Solution Approach

The end user of the system that we developed is an automotive warranty analyst who is primarily a domain expert belonging to the quality department of the automaker. The system has been designed keeping in mind the requirements outlined by a warranty analyst. One of the primary requirements of the warranty analyst was to have a graphical user interface based system where he can query the unstructured data (customer complaints and technician diagnosis expressed in free-form textual format) using high-level or natural language queries and generate reports. Hence, the system was designed to have the following two main components.

1. Text Tagging and Annotation Engine
2. Query Engine

We divided the process of mining warranty claims data into two phases. Phase 1 consists of converting free-form text data in warranty claim forms into structured data using a natural language processing technique called as named-entity extraction. Phase 2 is of reporting and analytics where a warranty analyst queries and analyzes the structured data obtained from Phase 1 process using high level queries. Figure 1 presents the high level architectural diagram illustrating the data flow and the two phases.

Text Tagging and Annotation also called as Named Entity extraction forms an important component of many language processing tasks such as text mining, information extraction and information retrieval. Named Entity extraction consists of identifying the names of entities in free-form or unstructured text. Some of the common types of entities are proper nouns such as person names, products, organization, location, email addresses, vehicle, computer parts and currency, temporal entities such as dates, time, day, year, month and week, numerical entities such as measurements, percentages and monetary values. There can be numerous domain specific entities also [1][2][3][5].

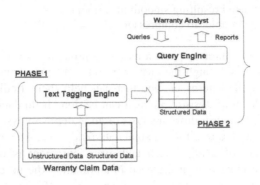

Fig. 1. High-level architectural diagram illustrating the two stages of the analysis

We developed a rule-based system for extracting named entities from customer complaint, technician comments and action taken field of the warranty claim forms. Some of the named-entities that we identified are technician action, car part location of a defect, reason of failure, effect of failure, defect type, condition under which defect occurred and customer action that caused the defect. Table 1 and 2 give examples of some customer complaints and the named entities extracted by our tool. Table 1 and 2 are for illustration purposes as it is not possible to present all the named entities with examples due to limited space in the paper. We made use of lookup tables to increase the accuracy of our system. The tagging and annotation engine is based on hand-crafted rules and lookup tables containing domain terms and clue words or phrases. Following is a simple example of a rule to illustrate the technique. Action taken by a technician is an entity that we wanted to extract from the action

Table 1. Customer complaints in warranty claim forms

Customer Complaints	
ID	**Complaint**
WCF01	Vehicle does not start when cranked
WCF02	Vehicle causes noise from below during turning
WCF03	Horn is not working
WCF04	When vehicle is stopped, lot of noise from engine
WCF05	Room lamp flickering when switched ON

Table 2. Output of tagging and annotation

Tagged Data for Customer Complaints			
ID	**Component**	**Problem**	**Customer Action**
WCF01	-	does not start	when cranked
WCF02	-	causes noise from below	during turning
WCF03	Horn	not working	-
WCF04	Engine	lot of noise	When vehicle is stopped
WCF05	Room Lamp	flickering	when switched ON

taken field in warranty claim forms. Some of the examples of technician action are replaced (replaced fuse or horn relay replaced), removed (removed evaporator unit), refitted (refitted AC able) and cleaned (cleaned the duct of AC system) etc.

It is practically not possible and scalable to create a lookup table of all the actions that a technician can take and hence we implemented a rule which scans each word in a sentence and checks if it ends with "ed" or "ing" If the word ends with "ed" and "ing" in technician action taken field then there are good chances that it is an entity of type action. However, just this rule is not enough as the word connected ends with "ed" but may not be an action in the context of the sentence "replaces belt connected to pulley". Hence some more levels of check or a chain of rules is required to identify an entity of correct type and disambiguate it from other entities. For instance, in the example "replaces belt connected to pulley", connected is not a technician action as it is also preceded by a word "to". The complexity of the rule depends on the type of entity that needs to be extracted and also depends on the writing style.

We did an evaluation of two popular text mining toolkits, GATE (General Architecture for Text Engineering) and LingPipe to see their fitment with the problem at hand [3]. To perform named-entity detection, LingPipe requires a supervised training of a statistical model and once a model is built it can be used to detect named-entities on unseen data of the similar nature as the training data. The training data must be labeled with all of the entities of interest and their types. We had around 250 sample data points which was not enough for us to select a machine learning based approach. The dataset available to us was limited, but all the 250 data points we had were of different types without any duplicates. Machine learning based approach is successful when there is a good quantity of annotated corpora to train a model. We tried using LingPipe but we were not getting good results due to insufficient training data. We will again evaluate LingPipe in the future when more sample data becomes available to us. GATE was another alternate and provides a mechanism to write hand-crafted rules and regular expressions in the form of pattern specification language called as JAPE (Java Annotations Pattern Engine) grammar. We tried writing rules using JAPE but realized that the heuristics required for extracting entities are easier to code in a programming language like Java as it required operations like finding the presence of a substring in a string, usage of features from previous annotations, usage of *if-then-else* statements and nested *for* loops. GATE provides functionality to call Java code from JAPE rules but we realized that the majority of our rules require flexibility of a programming language like Java and we found it easier to write our own custom code rather than calling our Java code from within GATE environment. Moreover, textual data in claim forms contain language which has lots of spelling mistakes, grammatical errors and short forms and that is why we chose to implement a custom rule-based system. There are two kinds of approaches to named entity recognition. One approach is based on using statistical modeling or machine learning whereas the other approach is based on developing rules and heuristics. We evaluated both the approaches in the form of LingPipe and GATE. However, our requirements were such that we finally decided to implement a custom rule-based named entity recognition system. We also created our own gazetteer list and pattern matching rules. Our gazetteer consists of two types of text tokens. One type of gazetteer list consisted commonly occurring terms such as vehicle parts (condenser, cooling coil etc) and technician actions (removed, replaced, adjusted). The other type of gazetteer list consisted of trigger words. Trigger words are text tokens that provide indication or clue for an entity occurrence such as the presence of token "on" for a location entity.

In the prevailing circumstances where million claims being filed per annum, it becomes practically impossible for any warranty analyst to go through the text of the claims manually. As a result most of the info reported in the text goes unnoticed and undecipherable. Text analytics will help integrate and automate the process of deciphering information from text resulting in more effective defect discovery.

References

1. Tan, A.-H.: Text Mining: The state of the art and the challenges. In: Zhong, N., Zhou, L. (eds.) PAKDD 1999. LNCS (LNAI), vol. 1574, pp. 65–70. Springer, Heidelberg (1999)
2. McCallum, A.: Information Extraction: Distilling Structured Data from Unstructured Text. Social Computing 3(9), 48–57 (2005)
3. Cunningham, H., Maynard, D., Bontcheva, K., Tablan, V.: GATE: A Framework and Graphical Development Environment for Robust NLP Tools and Applications. In: Proceedings of the 40th Anniversary Meeting of the Association for Computational Linguistics (ACL 2002), Philadelphia (July 2002)
4. Batesa, H., Holwegb, M., Lewisc, M., Oliverd, N.: Motor vehicle recalls: Trends, patterns and emerging issues. OMEGA: International Journal of Management Science 35(2), 202–210 (2007)
5. Zhang, L., Pan, Y., Zhang, T.: Focused named entity recognition using machine learning. In: Proceedings of the 27th annual international ACM SIGIR conference on Research and development in information retrieval, pp. 281–288 (2004)
6. Fournier, R., Shovelton, T., Stolle, L.: Walking the automotive industry tightrope: Keeping customer and your brand safe, An Executive strategy report of IBM Global Services published on (April 14, 2003)
7. Teret, S.P., Vernick, J., Mair, J.S., Sapsin, J.W.: Role of Litigation in Preventing Product-Related Injuries. Epidemiological Reviews 25, 90–98 (2003)
8. Automotive Warranty Management: Paying the Bill and Solving the Problem by Kevin Prouty, A report on Manufacturing, AMR Research(November 01, 2000)
9. Recalls and Product Safety News, U.S. Consumer Product Safety Commission, http://www.cpsc.gov/cpscpub/prerel/prerel.html
10. The Warranty Process Flow within the Automotive Industry: An Investigation of Automotive Warranty Processes and Issues. Center for Automotive Research (August 2005)
11. Warranty Week, The Newsletter for Warranty Management Professionals, http://www.warrantyweek.com/

A Similarity Search of Trajectory Data Using Textual Information Retrieval Techniques

Yu Suzuki, Jun Ishizuka, and Kyoji Kawagoe

Ritsumeikan University, 1-1-1 Noji-Higashi, Kusatsu, Shiga 5258577, Japan

Abstract. In this paper, we propose a novel similarity measure between trajectory data and geometry data using textual information retrieval techniques. Currently, many trajectory data are generated and used for sightseeing. When users search trajectory data at sightseeing, if a user's current position is similar to a retrieval target trajectory datum, this trajectory datum should be useful. However, even if the euclidean distances between the moving points in trajectory data and the user's position have small values, these trajectory data are not always relevant to the user's interests. In this paper, we deal with textual information retrieval method to measure the similarity values between retrieval target trajectory data and user's current position. Using our proposed method, users can gain relevant sightseeing spots as retrieval results. In our experimental evaluation, we confirmed that our proposed method can retrieve intuitively relevant trajectory data.

1 Introduction

Recently, many global information systems (GIS), such as car navigation systems and sightseeing navigation systems have been proposed and widely used. Because, Global Positioning System (GPS) is developed and embedded to many devices, such as cell phones, or PDAs. When these devices take users' current positions, the devices can provide several information related to the user's position. In this paper, we develop the user's sightseeing navigation system as one of examples of global information systems.

In this paper, we propose a trajectory data retrieval system which returns trajectory data related to the user's current position. For example, when users are in a temple, and the temple is relevant to the user's interests, the system gives several trajectory data which consist of temples with similar backgrounds, or similar building periods. As a result, the user can understand the background or building periods deeply by sightseeing these temples.

Our proposed method has two contributions: one contribution is the method to input queries by users. In our method, users only input user's current position, such as latitude, and longitude, to the system. This is because, if users do not know about the retrieval target sightseeing spots, the users can easily input the users' queries.

Another contribution is that, for calculating similarity values, we deal with textual information about trajectory data and sightseeing spot instead of the actual geometrical position. The aim of our proposed method is to retrieve intuitively relevant sightseeing spots. Current similarity searches methods for trajectory data are based on geometrical

J.R. Haritsa, R. Kotagiri, and V. Pudi (Eds.): DASFAA 2008, LNCS 4947, pp. 627–634, 2008.
© Springer-Verlag Berlin Heidelberg 2008

position [1,2]. However, even if geometrical positions in trajectory data and user's current position are close, these trajectory data are not always intuitively relevant. To solve this issue, we deals with textual information instead of geometrical position.

2   Our Proposed Similarity Measure for Trajectory Data

First, we introduce the overview of our proposed system. In this section, we do not concern the order of the sightseeing spots in the users' trajectory data.

Our proposed system has the following five steps.

1. **A user inputs a current position to our proposed system**
   Using GPS (Global Positioning System), a user inputs user's current position, such as latitude, and longitude values, to our proposed system.
2. **The system extracts textual features from user's current position**
   The system gain textual information about user's current position. For example, if a user is at Kyoto station, the system gain textual information about Kyoto station. In our system, we use semantic location extracting technique [3].
3. **The system calculates similarity values between user's current position and retrieval target trajectory data**
   The system extracts textual information from retrieval target trajectory data using the same method for user's current position. Using these textual information, the system calculates similarity values between user's current point and retrieval trajectory data. Textual information retrieval techniques are used to calculate similarity values.
4. **The system outputs retrieval target trajectory data as result list**
   The system outputs several trajectory data which have high similarity values.

2.1   Extract Textual Information from Retrieval Target Trajectory Data

In this section, we describe the method to extract feature values from retrieval target trajectory data. In our system, we deal with the same method to extract feature values from user's current position.

Before describe our proposed method, we describe about trajectory data. We assume that a system administrator prepares many trajectory data, and s/he stores these data into the retrieval system. These trajectory data G are a set of users' geometric position data. Using converting techniques such as [3], we convert from these geometric position data to the time series data G', which are a set of sightseeing spot data. For example, we have a trajectory data G as follows:

$$G = [(154.2, 128.3), (154, 7, 128.2), (154.9, 128.5)] \tag{1}$$

We denote that the elements of G have the order. Each element of G describes the position; the first number is a latitude, and the next number denotes a longitude.

Convert Position into Spot Name. We convert raw position data into sightseeing spot name using converting techniques.

$$G' = [\text{Kyoto station}, \text{Honganji temple}, \text{the Imperial Palace}] \tag{2}$$

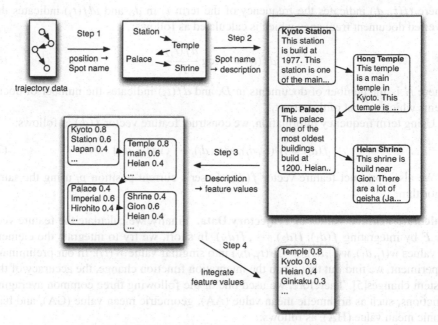

Fig. 1. Extract feature values from trajectory data

Generally, G' consists of the keywords of the trajectory data as follows.

$$G' = [k_1, k_2, \cdots, k_N] \tag{3}$$

where k_i ($i = 1, 2, \cdots, N$) is a keyword of sightseeing spot.

Retrieve Description about Spot. The system retrieves the description of these spot names. In our experiment, we deal with Wikipedia[1] to retrieve the descriptions of sightseeing spots. Using these encyclopedia data, we generate the description array D as follows:

$$D = [d_1, d_2, \cdots, d_N] \tag{4}$$

where d_i ($i = 1, 2, \cdots, N$) is a description of the sightseeing spot k_i.

Calculate Feature Values from the Description. From the description d_i, we extract terms, and count term frequency and inverted document frequency (tfidf) of the term. As the article [4] describes, the weight $w(t_j, d_i)$ of the term t_j ($j = 1, 2, \cdots, M$) are calculated as follows:

$$w(t_j, d_i) = tf(t_j, d_i) \cdot idf(t_j) \tag{5}$$

---

[1] http://en.wikipedia.org/

where $tf(t_j, d_i)$ indicates the frequency of the term t_j in d_i, and $idf(t_j)$ indicates the inverted document frequency which is calculated as follows:

$$idf(t_j) = -\log \frac{df(t_j)}{N} \tag{6}$$

where N is the number of documents in D, and $df(t_j)$ indicates the number of documents which consist of the term t_j.

Using term frequency information, we construct feature vector $f(d_i)$ as follows:

$$f(d_i) = [w(t_1, d_i), w(t_2, d_i), \cdots, w(t_M, d_i)] \tag{7}$$

We also construct feature vector $f(d_i)$ of user's current position p_i using the same algorithm.

Calculate Feature Values of Trajectory Data. Finally, we calculate the feature vector F by integrating $f(d_1), f(d_2), \cdots, f(d_N)$. In short, we try to integrate the element of values $w(t_j, d_1), w(t_j, d_2), \cdots, w(t_j, d_N)$ into singular value $W(t_j)$. In our preliminary experiment, we find out that when the integration function change, the accuracy of the system changes[5]. Therefore, we used one of the following three common averaging functions, such as arithmetic mean value (AA), geometric mean value (GA), and harmonic mean value (HA), as follows:

AA(Arithmetic mean value)

$$W_{AA}(t_j) = \frac{\sum_{i=1}^{N} w(t_j, d_i)}{M} \tag{8}$$

GA(Geometric mean value)

$$W_{GA}(t_j) = \sqrt[N]{\Pi_{i=1}^{N} w(t_j, d_i)} \tag{9}$$

HA(Harmonic mean value)

$$W_{HA}(t_j) = \frac{\frac{1}{M}}{\sum_{i=1}^{N} \left(\frac{1}{w(t_j, d_i)} \right)} \tag{10}$$

Using the above elements, we construct the trajectory data feature vector F as follows.

$$F = [W(t_1), W(t_2), \cdots, W(t_M)] \tag{11}$$

where $W(t_j)$ is replaced with $W_{AA}(t_j)$, $W_{GA}(t_j)$, or $W_{HA}(t_j)$.

2.2   Similarity Values between Retrieval Target Trajectory Data and User's Current Position

In this section, we indicate the method to calculate the similarity value $S(p_i)$ between users' feature vector F and the feature vector of sightseeing spot $g(p_i)$ is defined using the following function.

$$S(p_i) = \cos(F, g(p_i))$$
$$= \frac{F \cdot g(p_i)}{|F| \cdot |g(p_i)|} \tag{12}$$

where $|F|$ and $|g(p_i)|$ are the lengths of the feature vector F and $|g(p_i)|$, respectively.

3 Gradient Score: The Order of Trajectory Data

In section 2.1, the system extracts feature vector from retrieval target trajectory data. In this method, we do not consider the order of sightseeing spots in trajectory data. Hence, if two trajectory data have same sightseeing spots using different pathway, the system extracts the same features from the trajectory data. However, frequently, if the trajectory data have different pathway, we should consider that these trajectory data have different features.

To solve this problem, we introduce the similarity measure about the order of trajectory data, and add this similarity values to our original similarity values.

3.1 Calculation of the Gradient Score

In this section, we describe a method to calculate the gradient score $G(p_i)$, which is a score which considers the order of trajectory data.

We describe the method of calculating the gradient score in detail. The gradient score is calculated using summation of differences between retrieval target sightseeing spot and each sightseeing spot of trajectory data as follows:

$$G(p_i) = \sum_{j=1}^{M} v(p_i, d_j) \tag{13}$$

where $v(p_i, d_i)$ is calculated as follows:

$$v(p_i, d_i) = \frac{f(p_i) \cdot f(d_j)}{|f(p_i)| \cdot |f(d_j)|} \tag{14}$$

3.2 Integration of Textual Score and Gradient Score

We calculate the similarity value $S'(p_i)$ using $S(p_i)$, the similarity value of textual information, and $G(p_i)$, the gradient score, as follows:

$$S'(p_i) = (1 - \alpha) \cdot S(p_i) + \alpha \cdot G(p_i) \tag{15}$$

where α is a parameter of the degree of considering the gradient score. Using this score instead of function 12, the accuracy of our proposed system increase.

4 Experimental Evaluation

To confirm the accuracy of our proposed method, we did two experiments.

1. **Comparison for three averaging functions**

   In section 2, we calculate the trajectory data feature vectors using three different mathematical functions. However, we cannot decide which mathematical function is suitable. Therefore, we deal with these mathematical functions for our proposed method, and compare the accuracy of our proposed system using recall and precision ratio.

2. **Effectiveness using the order of trajectory data**

   In section 3, we describe the calculation method of the gradient scores to consider the order of sightseeing spots. Therefore, in this experiment, we confirm the effectiveness of the gradient scores using recall and precision ratio. In our method, we use the gradient scores with the scores calculated by textual retrieval techniques. To integrate these two scores, we used the parameter α. In this experiment, we also try to decide suitable values of α.

4.1   Experiment 1: Deciding Suitable Averaging Functions

In this section, we try to select the three averaging function such as averaging mean value, geometric mean value, and harmonic mean value, using experimental evaluation. In this experiment, we do not use the gradient score.

Experimental setup.   In our experiment, we did our experiment using the following steps.

1. **The user input trajectory data as a query to our proposed system**

   We prepared the following two trajectory data Query (a) and (b).
   - Query (a) - The users who are interested in "Japanese gardens."
   - Query (b) - The users who are interested in "Buddhist statue."

   We prepared 5 observers. Each observer decides one user's position which satisfies Query (a), and s/he also decides another user's position which satisfies Query (b).

2. **We prepared several suitable trajectory data**

   We prepared 10 sets of query-answer sets. Each query-answer sets consists of from 8 to 10 trajectory data as collect retrieval results, and one user's current position as user's query. In our experiment, if these trajectory data are retrieved, we treat that our proposed system can retrieve suitable trajectory data.

3. **Our proposed system outputs several trajectory data as ranked lists**

   We prepared 108 trajectory data and their metadata from Wikipedia, and input these data into our proposed system. Using retrieval target trajectory data, the system outputs severals sightseeing points as ranked lists.

4. **We calculate a recall and precision ratio**

   Using the 10 sets of query-answer sets, we calculate a recall and precision ratio, and draw 11-pt interpolated recall-precision graphs.

Experimental result.   Fig. 2 shows the 11-pt interpolated recall-precision graphs. From this figure, we find out that, using averaging mean value, we can gain better precision than geometric and harmonic mean value. We also unveil that when the system deals with AA, the accuracy of our proposed system is higher than the system using the other averaging functions. The reason of this result is that if very small values are integrated using either GA or HA, the results are very small even if the other integrated values are large.

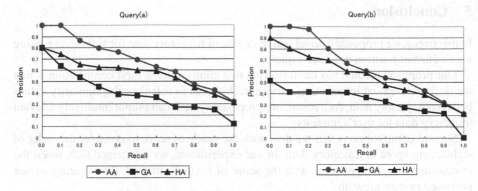

Fig. 2. The recall-precision graphs of AA, GA, and HA

4.2   Experiment 2: Effectiveness of the Gradient Score

Experimental setup.   In this experiment, we did our experiment as the same way as experiment 1. We used the similarity measure $S'(p_i)$ in section 3 instead of $S(p_i)$ described in section 2. We changed the value α from 0.0001 to 0.0020 for 0.0001 steps, and tried to find the most suitable values of α for each query. We used arithmetic mean value as the averaging function. Because, in experiment 1., when we use the function AA with our proposed method, we can gain the best performance.

Experimental result.   Table 1. shows 11-pt averaging precision of our proposed system using either arithmetic mean value or gradient. In this table, "Observer" means an ID of observers, "AA" means the averaging 11-pt precision ratio of the system using only $S(p_i)$, "gradient" means the averaging 11-pt precision ratio of the system using $S'(p_i)$ instead of $S(p_i)$, "AA - gradient" means the difference ratio between AA and gradient, and "α" means the most suitable value of the parameter α.

From this result, we unveil that, in any case, when the system use the gradient score, the accuracy of the system improved. However, we also find that the appropriate parameters are different in case of the users' interest and the users' query trajectory data. There is no rule to find adequate parameter.

Table 1. The comparison of averaging presioon ratio between average mean value and gradient score

Observer	AA(%)	gradient(%)	AA-gradient(%)	α
1	76.39	77.79	1.39	0.010
2	74.02	76.94	2.92	0.005
3	66.95	66.95	0.00	0.000
4	80.35	81.39	1.04	0.005
5	62.11	63.79	1.68	0.002

5   Conclusion

In this paper, we proposed a novel search engine of trajectory data for sightseeing, using user's current position as a user's query.

Our proposed method has the following two contributions: First contribution is that we deal with textual information retrieval techniques to calculate the similarity values between trajectory data. As a result, our proposed system can output intuitively relevant trajectory data for user's interests.

Second contribution is that we deal with the gradient score to consider the order of sightseeing spots in trajectory data. In our experiments, we confirmed that, when the system use the gradient score with the score of textual retrieval, the accuracy of our proposed system grew up.

Finally, we describe an open problem as a future work. In this paper, we assume that a user inputs one position data, and retrieve trajectory data using our proposed system. However, if a user can input multiple position data, our system can receive more accurate retrieval result trajectory data. Here, we will face the problem about calculation cost of scores. Therefore, to realize this method, we should solve this calculation cost problem as a future work.

References

1. Agrawal, R., Faloutsos, C., Swami, A.: Efficient similarity search in sequence databases. In: Proceedings of the 4th International Conference on Foundations of Data Organization and Algorithm, pp. 69–84 (1993)
2. Faloutsos, C., Ranganathan, M., Manolopoulos, Y.: Fast subsequence matching in time-series databases. In: Proceedings of the ACM SIGMOD Conference on Management of Data, pp. 419–429 (1994)
3. Liu, J., Wolfson, O., Yin, H.: Extracting Semantic Location from Outdoor Positioning Systems. In: Proceedings of International Workshop on Managing Context Information and Semantics in Mobile Environments (MCISME 2006), pp. 34–41 (2006)
4. Salton, G., McGill, M.J.: Introduction to Modern Information Retrieval. McGraw-Hill Book Company, New York (1983)
5. Suzuki, Y., Hatano, K., Uemura, S.: A calculation method of document scores for multimedia document retrieval method. In: Proceedings of the IASTED International Conference on Information Systems and Databases (ISDB 2002), pp. 178–183 (2002)

Constrained k-Nearest Neighbor Query Processing over Moving Object Trajectories

Yunjun Gao[1], Gencai Chen[1], Qing Li[2], Chun Li[1], and Chun Chen[1]

[1] College of Computer Science, Zhejiang University, Hangzhou 310027, P.R. China
{gaoyj,chengc,lichun,chenc}@cs.zju.edu.cn
[2] Department of Computer Science, City University of Hong Kong, Hong Kong, P.R. China
itqli@cityu.edu.hk

Abstract. Given a set D of trajectories, a query object (point or trajectory) q, a time interval T, and a constrained region CR, a constrained k-nearest neighbor (CkNN) query over moving object trajectories retrieves from D within T, the k (≥ 1) trajectories that lie closest to q and intersect (or are enclosed by) CR. In this paper, we propose several algorithms for efficiently processing CkNN search on moving object trajectories. In particular, we thoroughly investigate two types of CkNN queries, viz. CkNN$_P$ and CkNN$_T$ queries, which are defined w.r.t. stationary query points and moving query trajectories, respectively. The performance of our algorithms is evaluated with extensive experiments using both real and synthetic datasets.

1 Introduction

With the advances of wireless communication, mobile computing, and positioning technologies, it has become possible to obtain and manage (e.g., model, query, etc.) the trajectories of moving objects in real life. A number of interesting applications are being developed based on the analysis of trajectories. For instance, it is very useful for zoologists to determine the living habits and migration patterns of certain groups of animals by mining the motion trajectories of animals in a large natural protection area. Therein, an important type of queries surely useful for the above applications is the so-called k-nearest neighbor (kNN) search, which retrieves the k (≥ 1) trajectories that are closest to a given query object from a dataset within a predefined time extent.

In some realistic cases, however, users may enforce constrained regions on kNN queries. In other words, they require only finding the nearest neighbors (NNs) in a portion of the data space rather than in the whole data space. As an example, assuming that the trajectories of animals over a long history are known in advance, the zoologists may pose the following query: *find the two closest animal trajectories in a restricted region (e.g., the rectangle area that locates at the east of the lab) to a given query object (e.g., lab, food source, etc.) within a specified time interval*. This query can be thought of as the kNN retrieval with a region constraint for trajectory data. The final query result must satisfy the given constraint. This application example motivates the constrained k-nearest neighbor (CkNN) query processing for moving object trajectories, the subject of this paper.

J.R. Haritsa, R. Kotagiri, and V. Pudi (Eds.): DASFAA 2008, LNCS 4947, pp. 635–643, 2008.
© Springer-Verlag Berlin Heidelberg 2008

Given a set D of trajectories, a query object (point or trajectory) q, a time interval T, and a constrained region CR, a CkNN query over trajectories retrieves from D within T, the k trajectories that lie closest to q and cross (or fully fall in) the area defined by CR. In spite of the huge bibliography in unconstrained kNN queries for spatial and spatio-temporal objects [1, 3, 4, 5, 6, 8, 9, 10], to our knowledge, none of the existing work has examined the CkNN query problem on moving object trajectories. In this paper, we study CkNN queries over moving object trajectories and propose several algorithms to handle such queries. Specifically, we thoroughly investigate two types of CkNN queries, termed as CkNN$_P$ and CkNN$_T$ queries which are defined w.r.t. stationary query points and moving query trajectories, respectively. Our methods are based on existing R-tree-like structures storing historical information about trajectories (e.g., TB-tree [7], etc.) in order to achieve low I/O cost (i.e., number of node/page accesses) and CPU overhead. Finally, we evaluate the efficiency and scalability of our proposed algorithms with extensive experiments, using both real and synthetic datasets.

The rest of the paper is organized as follows. Section 2 gives the problem statement and briefly surveys the related work. Sections 3-4 describe the algorithms for CkNN$_P$ and CkNN$_T$ queries, respectively. Experimental evaluation is presented in Section 5. The last section concludes the paper.

2 Background Information

In this section, we first state the problem and then give a brief review of related work. The notations used frequently in this paper are summarized in Table 1.

2.1 Problem Statement

The problem of CkNN search on moving object trajectories is formalized as follows.

Definition 1 (CkNN Query over Moving Object Trajectory). *Given D, q (including Q_P and Q_T), T, CR, and k, a CkNN query retrieves from D within T, a set S_{rslt} of k trajectories such that all the objects in S_{rslt} are closest to q and intersect with (or are enclosed by) CR. Note that when $|S_{CRT}| \leq k$, $S_{rslt} = S_{CRT}$ (i.e., S_{CRT} is the solution).*

We refer to the CkNN retrieval w.r.t. Q_P and w.r.t. Q_T as CkNN$_P$ query and CkNN$_T$ query, respectively. Moreover, for simplicity, we assume that CR is a rectangle, although arbitrary query shapes can be used as well. Consider, for example, Figure 1 illustrating a CkNN$_P$ ($k = 2$)

Table 1. Symbols used in this paper

Symbol	Description
D	a set of moving object trajectories
k	number of requested nearest neighbors
q	query object (involving Q_P or Q_T)
Q_P	query point
Q_T	query trajectory
T	the query time interval of the form $(T.t_s, T.t_e)$
CR	constrained region
S_{CRT}	the set of moving object trajectories that cross (or are contained in) CR within T
S_{rslt}	the set of query result for CkNN search

query on $D = \{Tr_1, Tr_2, ..., Tr_6\}$ within $T (= [t_1, t_3])$. Then, $S_{rslt} = \{Tr_2, Tr_3\}$, intersecting CR (shadowed area in Figure 1) during T, is the final query result. Notice that in this case, Tr_1 is not an answer, although it is the nearest trajectory to Q_P inside T for

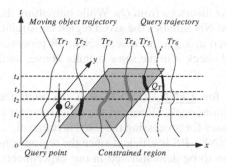

Fig. 1. Example of C*k*NN$_P$ and C*k*NN$_T$ queries over moving object trajectories for $k = 2$

the unconstrained *k*NN search, because Tr_1 does not cross *CR*. Similarly, in Figure 1 the result of a C*k*NN$_T$ ($k = 2$) query over *D* within T (= $[t_2, t_4]$) contains both Tr_4 and Tr_5.

2.2 Related Work

One area of related work concerns indexing of moving object trajectories. The trajectory of a moving object is the path taken by it across time. Thus, trajectories can be considered as 2D or 3D time slice data. Here, we only focus on R-tree-like structures that store historical information about trajectories (e.g., TB-tree [7], STR-tree [7], etc.). In our study, we assume that the dataset is indexed by the TB-tree due to its high efficiency in *trajectory-based queries*.

Our work in this paper is also related to *k*NN queries in spatial and spatio-temporal databases. In the past decade, numerous algorithms for *k*NN (and NN) queries have been proposed in the database literature [1, 3, 4, 5, 6, 8, 9, 10]. Recently, Frentzos *et al.* [3] and Gao *et al.* [4, 5] investigated the problem of processing *k*NN search for trajectories on R-tree-like structures. Another area of related work is C*k*NN search in spatial databases. Ferhatosmanoglu *et al.* [2] introduced and solved the C*k*NN retrieval for spatial objects, by finding the NNs in a specified spatial region. Nevertheless, this problem differs from our work in this paper, since it does not consider the temporal information of objects. Due to space limitation, we omit the details of these techniques.

3 C*k*NN$_P$ Query Processing

In this section, we propose our algorithms for handling C*k*NN$_P$ queries on moving object trajectories. Section 3.1 briefly describes the two-step algorithms for C*k*NN$_P$ queries, by combining range query and unconstrained *k*NN search. Sections 3.2 and 3.3 present one-step algorithms for C*k*NN$_P$ queries, based on *depth-first* and *best-first* traversal paradigms, respectively.

3.1 Two-Step Algorithms

C*k*NN queries naturally contain both range and *k*NN queries. Motivated by this fact, a simple approach for C*k*NN$_P$ queries can comprise two phases incorporating, in sequence, these two types of queries. Different orders of the phases may bring forth specific properties and advantages. Next, we outline two straightforward methods for tackling C*k*NN$_P$ search.

NN Search Followed by a Range Query for C*k*NN$_P$ (C*k*NN$_P$-SR). C*k*NN$_P$-SR follows a two-phase framework. The first stage retrieves the unconstrained NNs (i.e., closest trajectories) of Q_P within T, using our previously proposed BFP*k*NN algorithm in [4] to find the NNs that are ranked by their distances from Q_P. While outputting the NNs, CkNN$_P$-SR verifies whether the current NN satisfies the given region constraint. If so, the NN is indeed an answer and is stored in S_{rslt}; otherwise, CkNN$_P$-SR proceeds to retrieve the next non-constrained NN and check it in the same manner above, until the final query result is returned.

Range Query Followed by NN Search for CkNN$_P$ (CkNN$_P$-RS). As with the CkNN$_P$-SR, CkNN$_P$-RS adopts a two-step framework. However, the difference w.r.t. the CkNN$_P$-SR is that it reverses the two phases for processing the CkNN$_P$ retrieval.

In summary, neither of CkNN$_P$-SR and CkNN$_P$-RS is efficient since the region constraint is applied at a very late stage, as also to be demonstrated in our experiments. Based on this observation, we present two single-phase approaches below, called Depth-First based CkNN$_P$ query (CkNN$_P$-DF) algorithm and Best-First based CkNN$_P$ query (CkNN$_P$-BF) algorithm respectively, which interleave the range constraints with NN conditions efficiently.

3.2 CkNN$_P$-DF Algorithm

CkNN$_P$-DF provides the ability to process the CkNN$_P$ search w.r.t. Q_P during T, as shown in Algorithm 1. In fact, CkNN$_P$-DF adapts the Point$k$NNSearch algorithm proposed in [3] by merging the region constraint into the algorithm. The details of the CkNN$_P$-DF algorithm are as follows.

Algorithm 1. Depth-First based CkNN$_P$ query algorithm (CkNN$_P$-DF)

Algorithm CkNN$_P$-DF (N, Q_P, T, CR, $kNearest$)
1: Initialize $kNearest.MaxDist$ to ∞
2: **If** N is a leaf node
3:      **For** each leaf entry E in N
4:          **If** GetEntryInConstraint (E, E', T, CR)
5:              $Dist = Euclidean\_Dist\_2D$ (Q_P, E')   // compute entry's actual distance from Q_P
6:              **If** $Dist < kNearest.MaxDist$
7:                  Add E' with $Dist$ to $kNearest$ and update $kNearest.MaxDist$ value if necessary
8:              **Endif**
9:          **Endif**
10:     **Next**
11: **Else**   // N is an intermediate (i.e., non-leaf) node
12:     $BranchList$ = GenBranchList (N, Q_P, T, CR)
13:     SortBranchList ($BranchList$)   // sort active branch list by $mindist$
14:     **For** each child node entry E in $BranchList$
15:         CkNN$_P$-DF (E, Q_P, T, CR, $kNearest$)
16:         PruneBranchList ($BranchList$) // use pruning heuristics in [2, 12] to prune $BranchList$
17:     **Next**
18: **Endif**

Function GenBranchList (N, Q_P, T, CR)
1: **For** each entry E in N
2:      **If** GetNodeInConstraint (E, E', T, CR)
3:          $Dist = Mindist$ (Q_P, E') : Add E' to $list$ together with its $Dist$
4:      **Endif**
5: **Next**
6: **Return** $list$

At the leaf level of the tree structure that indexes trajectory data, CkNN$_P$-DF iteratively accesses every leaf entry E in the leaf node N visited currently (Lines 3-10). In particular, CkNN$_P$-DF utilizes a function **GetEntryInConstraint** to check whether E intersects (or is enclosed by) CR during T (Line 4). If so, the **GetEntryInConstraint** interpolates E to produce E' (i.e., a portion of E) whose temporal component is within T and spatial component is contained in CR, and returns TRUE; otherwise, it returns FALSE. Without loss of generality, we suppose that **GetEntryInConstraint** returns TRUE at this time. Subsequently, CkNN$_P$-DF calculates the actual Euclidean distance *Dist* between Q_P and E' (Line 5). Notice that, here the *Euclidean_Dist_2D* between Q_P and E' is computed in the same way as [3]. After this point, CkNN$_P$-DF determines if *Dist* is smaller than *kNearest.MaxDist* (set to infinity in Line 1) which specifies the maximum distance in the *kNearest* structure, the latter is used to store the final result of the CkNN$_P$-DF query. If *Dist* < *kNearest.MaxDist* holds, then CkNN$_P$-DF adds E' to *kNearest* together with its *Dist* and updates the value of *kNearest.MaxDist* if necessary (Lines 6-8). At the non-leaf level of the tree structure, CkNN$_P$-DF recursively visits every child entry of the intermediate (i.e., non-leaf) node (Lines 12-17). When a potential candidate is retrieved, the algorithm, backtracking to the upper level, prunes the nodes in the active branch list (Line 16) using the pruning heuristics developed in [8]. Note that when CkNN$_P$-DF invokes the function GenBranchList to generate a node's branch list using its entries that intersect or are enclosed by CR within T (Line 12), we also combine region constraint into the GenBranchList. For this purpose, we use another function **GetNodeInConstraint**, which can verify whether or not the time interval of a given non-leaf node entry N overlaps with T, and if N's spatial extent intersects (or is contained in) CR.

3.3   CkNN$_P$-BF Algorithm

CkNN$_P$-BF is actually a variant of our proposed BFPkNN algorithm in [4] which can efficiently process the unconstrained kNN search over moving object trajectories. CkNN$_P$-BF, like BFPkNN, follows the *best-first* traversal fashion as well as enables efficient pruning strategies to discard all non-qualifying entries, but (unlike BFPkNN) it integrates the region constraint into the algorithm. Below, we show the CkNN$_P$-BF algorithm in Algorithm 2.

By starting from the root in the tree R on the set of trajectory data (Line 2), CkNN$_P$-BF recursively traverses R in a *best-first* manner (Lines 3-30). Specifically, CkNN$_P$-BF first de-heaps the top entry E from the heap *hp* (Line 4). If E is an actual entry of trajectory segment and its identifier *id* is not included in S_{rslt}, then CkNN$_P$-BF inserts E as a final answer into S_{rslt} (Line 7) provided that the number of elements in S_{rslt} is smaller than k; otherwise, the algorithm returns S_{rslt} (Line 9) because the final result of the CkNN$_P$-BF query has already been found. When E is a node entry, there are two possible cases as follows: (i) If E is a leaf node, then CkNN$_P$-BF chooses the entry *NearestE* in E with the smallest distance to Q_P and inserts *NearestE* into *hp* (Lines 12-21). Here, we also employ our proposed pruning heuristics in [4] to prune away the unnecessary entries that do not contribute to the query result, in order to reduce the number of node accesses and speed up the algorithm. (ii) If E is a non-leaf node, then CkNN$_P$-BF visits only the qualifying nodes that may contain the actual answers, and inserts them into *hp* (Lines 23-27). Note that CkNN$_P$-BF first employs

---

Algorithm 2. Best-First based CkNN$_P$ query algorithm (CkNN$_P$-BF)

Algorithm CkNN$_P$-BF $(R, Q_P, T, k, CR, S_{rslt})$
 1: Create and initialize heap hp
 2: Insert all entries of the root in R into hp
 3: **Do While** $hp.count > 0$
 4:      De-heap the top entry E in hp
 5:      **If** E is an actual trajectory segment entry and its identifier id is not in S_{rslt}
 6:          **If** $|S_{rslt}| < k$ // the cardinality of S_{rslt} is smaller than k
 7:              Insert E as an actual result into S_{rslt}
 8:          **Else**
 9:              **Return** S_{rslt} // report the final query result
10:          **Endif**
11:      **ElseIf** E is a leaf node
12:          $MinimalDist = \infty$
13:          **For** each entry e in E
14:              **If** GetEntryInConstraint (e, e', T, CR)
15:                  $Dist = Euclidean\_Dist\_2D\ (Q_P, e')$
16:                  **If** $Dist < MinimalDist$
17:                      $MinimalDist = Dist : NearestE = e'$
18:                  **Endif**
19:              **Endif**
20:          **Next**
21:          Insert $NearestE$ with $MinimalDist$ into hp
22:      **Else**   // E is an intermediate (i.e., non-leaf) node
23:          **For** each entry e in E
24:              **If** GetNodeInConstraint (e, e', T, CR)
25:                  $MinDistE = Mindist\ (Q_P, e') :$ Insert e' with $MinDistE$ into hp
26:              **Endif**
27:          **Next**
28:      **Endif**
29: **Loop**
30: **Return** S_{rslt}

---

the **GetEntryInConstraint** (**GetNodeInConstraint**) function to determine if every entry e in the leaf (non-leaf) node E satisfies both the temporal and spatial constraints in Line 14 (24) before it visits e. This check is necessary because some entries in E may not meet the given constraints (therefore they do not need to be visited).

4 CkNN$_T$ Query Processing

So far we have discussed CkNN query processing algorithms for moving object trajectories w.r.t. the stationary query **point** Q_P. In this section, we extend our approaches to address the CkNN retrieval over moving object trajectories w.r.t. a given query **trajectory** Q_T. Such a CkNN query is termed as CkNN$_T$, in contrast to CkNN$_P$ retrieval.

The CkNN$_T$ query is a variation of the CkNN$_P$ query with the following difference: a CkNN$_T$ search takes a query trajectory instead of a query point as one of its inputs. Therefore, our previously proposed CkNN$_P$-SR, CkNN$_P$-RS, CkNN$_P$-DF, and CkNN$_P$-BF algorithms for dealing with CkNN$_P$ queries (presented in Section 3) can be easily adapted for processing the CkNN$_T$ search after slight modification. In view of this, we have developed corresponding algorithms for CkNN$_T$ queries, called CkNN$_T$-SR, CkNN$_T$-RS, CkNN$_T$-DF, and CkNN$_T$-BF, respectively. The details of these algorithms are omitted here due to space limitation.

5 Performance Evaluation

This section experimentally evaluates the efficiency and scalability of our proposed algorithms with both real and synthetic datasets. We compare only the I/O cost since the CPU cost leads to similar results following the same trend. All algorithms were coded in Visual Basic and run on a PC with Pentium IV 3.0GHz CPU and 1GB main memory. Notice that both CkNN$_P$-RS and CkNN$_T$-RS are always worse than the other algorithms by several orders of magnitude. Hence, they are omitted in our experimental results reported below.

We use two real datasets that consist of a fleet of trucks containing 276 trajectories and a fleet of school buses containing 145 trajectories. Both of them are available at the *R-tree Portal*, http://www.rtreeportal.org/. We also deploy several synthetic datasets generated by a GSTD data generator [11] to examine the scalability of the algorithms. Specifically, the synthetic data correspond to 100, 200, 400, 800, and 1600 moving objects, with the position of each object being sampled approximately 1500 times. Furthermore, the initial distribution of moving objects is *Gaussian* while their movement is ruled by a random distribution.

Each dataset is indexed by a TB-tree [7], using a page size of 4 KB and a (variable size) buffer fitting the 10% of the tree size with the maximal capacity of 1000 pages. The experiment studies four factors involving *CR*, *k*, temporal extent (*TE*), and the number of moving objects (*#MO*), which affect the performance of the algorithms. Performance is measured by executing workloads, each comprising of 100 queries. To achieve stable statistics, each measurement is obtained by averaging the results from the last 50 queries, after warming up the buffer with the first 50 queries. Moreover, the query points used in the CkNN$_P$ query algorithms utilize random ones in 2D space. Towards the algorithms for CkNN$_T$ queries on trucks dataset, we take random parts of random trajectories belonging to the school buses dataset as the query trajectory collection; while on GSTD datasets, the query sets of trajectories are also created by the GSTD data generator.

We first study the performance of the algorithms for CkNN$_P$ queries under a variety of settings, and present the experimental results in Figures 2-5. Clearly, CkNN$_P$-BF consistently outperforms the other algorithms significantly in all cases. A crucial observation is that when *CR* is small (e.g., 20%), CkNN$_P$-DF is always more effective than CkNN$_P$-SR, whereas the latter exceeds consistently the former when *CR* becomes large (e.g., 60%). Subsequently, we evaluate the efficiency and scalability of the algorithms for CkNN$_T$ queries. As expected, CkNN$_T$-BF performs the best consistently for all settings; CkNN$_T$-DF is over CkNN$_T$-SR when *CR* is small, but CkNN$_T$-SR outperforms CkNN$_T$-DF when *CR* is large.

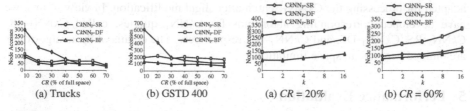

Fig. 2. I/O cost vs. *CR* (k = 4, *TE* = 6%)   **Fig. 3.** I/O cost vs. k (*TE* = 6%, GSTD 400)

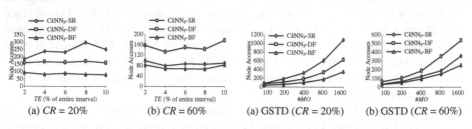

Fig. 4. I/O cost vs. *TE* (k = 4, GSTD 400)   **Fig. 5.** I/O cost vs. *#MO* (k = 4, *TE* = 6%)

6 Conclusion

In this paper, we have first introduced the concept of the CkNN retrieval and proposed several algorithms for efficiently processing CkNN queries over the trajectories of moving objects. In particular, we have thoroughly investigated two types of CkNN queries, termed as CkNN$_P$ and CkNN$_T$ queries w.r.t. stationary query points and moving query trajectories, respectively. We have conducted extensive experiments with both real and synthetic datasets, the results of which demonstrate the efficiency and scalability of our proposed algorithms.

References

1. Benetis, R., Jensen, C.S., Karciauskas, G., Saltenis, S.: Nearest Neighbor and Reverse Nearest Neighbor Queries for Moving Objects. In: Proc. of IDEAS, pp. 44–53 (2002)
2. Ferhatosmanoglu, H., Stanoi, I., Agrawal, D., Abbadi, A.: Constrained Nearest Neighbor Queries. In: Jensen, C.S., Schneider, M., Seeger, B., Tsotras, V.J. (eds.) SSTD 2001. LNCS, vol. 2121, pp. 257–278. Springer, Heidelberg (2001)
3. Frentzos, E., Gratsias, K., Pelekis, N., Theodoridis, Y.: Algorithms for Nearest Neighbor Search on Moving Object Trajectories. GeoInformatica 11, 159–193 (2007)
4. Gao, Y., Li, C., Chen, G., Chen, L., Jiang, X., Chen, C.: Efficient k-Nearest-Neighbor Search Algorithms for Historical Moving Object Trajectories. JCST 22, 232–244 (2007)
5. Gao, Y., Li, C., Chen, G., Li, Q., Chen, C.: Efficient Algorithms for Historical Continuous kNN Query Processing over Moving Object Trajectories. In: Dong, G., Lin, X., Wang, W., Yang, Y., Yu, J.X. (eds.) APWeb/WAIM 2007. LNCS, vol. 4505, pp. 188–199. Springer, Heidelberg (2007)
6. Hjaltason, G.R., Samet, H.: Distance Browsing in Spatial Databases. ACM TODS 24, 265–318 (1999)

7. Pfoser, D., Jensen, C.S., Theodoridis, Y.: Novel Approaches in Query Processing for Moving Object Trajectories. In: Proc. of VLDB, pp. 395–406 (2000)
8. Roussopoulos, N., Kelley, S., Vincent, F.: Nearest Neighbor Queries. In: Proc. of SIGMOD, pp. 71–79 (1995)
9. Song, Z., Roussopoulos, N.: K-Nearest Neighbor Search for Moving Query Point. In: Jensen, C.S., Schneider, M., Seeger, B., Tsotras, V.J. (eds.) SSTD 2001. LNCS, vol. 2121, pp. 79–96. Springer, Heidelberg (2001)
10. Tao, Y., Papadias, D., Shen, Q.: Continuous Nearest Neighbor Search. In: Proc. of VLDB. pp. 287–298 (2002)
11. Theodoridis, Y., Silva, J.R.O., Nascimento, M.A.: On the Generation of Spatiotemporal Datasets. In: Güting, R.H., Papadias, D., Lochovsky, F.H. (eds.) SSD 1999. LNCS, vol. 1651, pp. 147–164. Springer, Heidelberg (1999)

Location Update Strategies for Network-Constrained Moving Objects

Zhiming Ding[1] and Xiaofang Zhou[2]

[1] Institute of Software, Chinese Academy of Sciences,
South-Fourth-Street 4, Zhong-Guan-Cun, Beijing 100080, P.R.China
zhiming@iscas.ac.cn
[2] School of Information Technology & Electrical Engineering
The University of Queensland, Brisbane, QLD 4072 Australia
zxf@itee.uq.edu.au

Abstract. Location update strategy is one of the most important factors that affect the performance of moving objects databases. In this paper, a new location update mechanism, Location Update Mechanism for Network-Constrained Moving Objects (Net-LUM), is proposed. Through active-motion-vector-based network-matching and special treatment with junctions, Net-LUM can achieve better performances in terms of communication costs and location tracking accuracy, which is confirmed by the experimental results.

Keywords: Database, Spatiotemporal, Moving Object, Location Update.

1 Introduction

Moving Objects Database (MOD) is the database that can track and manage the locations of moving objects such as cars, ships, flights, and pedestrians. One of the key problems with MOD is to continuously track the locations of moving objects with minimum communication and computation costs. To answer location related queries, the server has to know the current and the historical location information of all registered moving objects, which means that the location information at the server has to be refreshed through location update messages sent by the moving objects.

There exist a large number of location-tracking mechanisms in the literature, and there are also several practical (and simpler) solutions adopted by real-world applications [4, 5, 6, 1]. In general, these methods can be divided into three categories: Fixed-Time Location Update Mechanisms (FTLU), Fixed-Distance Location Update Mechanisms (FDLU), and Motion-Vector-Based Location Update Mechanisms (MVBLU). Currently, FTLU and FDLU are widely used in real-world systems because of their simplicity. However, in terms of efficiency and accuracy, FTLU and FDLU are not the best choices, and as a result, MVBLU has become increasingly influential in the MOD area [4, 6].

In [5, 6], Wolfson et. al. have proposed a location update mechanism which can utilize the geometry of routes and the topology of the traffic network. In their method, however, only distance/deviation triggered location updates are defined so that when the moving object transfers from one route to another the change can not be reported

J.R. Haritsa, R. Kotagiri, and V. Pudi (Eds.): DASFAA 2008, LNCS 4947, pp. 644–652, 2008.

to the server promptly, which can affect the accuracy of location tracking. In [1], Civilis and Jensen et al. have proposed a road-network-based location tracking mechanism for moving objects. However, their method heavily depends on GPS logs in linking segments into routes and in generating accelerating profiles, which limits its usability in real-world applications. In [2], Ding and Güting have proposed an MODTN model and provided some rough location update principles for network constrained moving objects. However, the method is based on mile-meters and a lot of key problems, such as network-matching, junction treatment, detailed algorithms, and performance evaluation, remain unsolved.

Except the above limitations, another important problem with all the existing FTLU and MVBLU location update methods is that junctions are not well treated so that location update costs around junctions can be a big problem in real-world applications. For instance, in a lot of Chinese cities with heavy traffic, say in Beijing, nearly no junction allows vehicles to pass through smoothly. Moving objects have to speed up and gear down for several times in the junction area so that a lot of location updates will be triggered. However, these location updates are not necessary since the junction area is relatively small and the increased location update messages will not help to improve the location tracking accuracy.

To solve these problems, we propose a new location tracking mechanism, Location Update Mechanism for Network Constrained Moving Objects (Net-LUM), in this paper. The remaining part of this paper is organized as follows. Section 2 defines the network constrained moving objects database model, Section 3 describes the location update strategies, and Section 4 provides performance evaluation results and finally concludes the paper.

2 Data Model for Network-Constrained Moving Objects

In this section, we present the data model for network-constrained moving objects. The model is an improvement to the MODTN model proposed in [2]. The main improvements are as follows: (1) The geometry of a junction is expressed by a point plus a radius so that it's considered as an area instead of as a point; (2) A "graph point" value can be expressed either by a (rid, pos) pair, or by a junction ID, to accommodate the situation when the moving object is inside a junction area; and (3) The "motion vector" and the "moving graph point" definitions are extended accordingly to meet the situation when the moving object is inside a junction.

In the following discussion, we suppose that moving objects are uniquely identified and each of them is equipped with a portable computing platform and some other integrated location tracking equipments (such as GPS and wireless interface). Let $junct(jid)$ and $route(rid)$ be functions which return the junction and the route corresponding to the specified identifiers respectively.

Definition 1 (Graph). A transportation graph (or graph) G is defined as a pair:

$G = (Routes, Juncts)$

where $Routes$ is a set of routes and $Juncts$ is a set of junctions.

Definition 2 (Route). A route of graph G, denote by r, is defined as follows:

$r = (rid, geo, len, fd)$

where *rid* is the identifier of *r*, *geo* is a polygon-line (or polyline) which describes the geometry of *r* (the beginning point and the end point of *geo* are called "0-end" and "1-end" respectively), *len* is the length of *r*, and $fd \in \{+, -, \pm\}$ is the traffic flow directions allowed in *r*.

Definition 3 (Junction). A junction of graph *G*, denoted by *j*, is defined as follows:

$$j = (jid, loc, ((rid_i, pos_i))_{i=1}^{n}, \gamma, m)$$

where *jid* is the identifier of *j*, *loc* is the location of *j* which can be presented as a point value in the $X \times Y$ plane, $((rid_i, pos_i))_{i=1}^{n}$ describes the routes connected by *j*, γ is the radius of the junction area, and *m* is the connectivity matrix [2] of *j*.

The radius γ can describe the size of the junction. As a result, the junction is no longer viewed as a point in the traffic network model. Instead, it is viewed as a junction area.

Definition 4 (Graph Point). A graph point is a point residing in the graph. The set of graph point of graph *G* is defined as follows:

$$GP = \{ jid \mid junct(jid) \in G.Juncts \} \cup \{(rid, pos) \mid route(rid) \in G.Routes, pos \in [0,1]\}$$

The position of a graph point *gp* can have two possibilities: either residing in a junction (we say *gp* is "in junction" in this case), or reside in a route (in this case we say *gp* is "in route"). For every route, we suppose that its total length is 1, so that any location inside the route can be presented by a real number $p \in [0, 1]$. We define two Boolean functions, IsinJunct(*gp*) and IsinRoute(*gp*), to check whether *gp* is in junction or in route.

The dynamic position of a network-constrained moving object is modeled as a moving graph point, which is a function from time to graph point. Discretely, a moving graph point is expressed as a sequence of motion vectors, and each motion vector describes the movement of the moving object at a certain time instant.

Definition 5 (Motion Vector). A motion vector, *mv*, is a snapshot of the moving object's movement and is generated by location updates. *mv* is defined as follows:

$$mv = (t, gp, \vec{v})$$

where *t* is a time instant, $gp \in GP$ is a graph point describing the location of the moving object at time *t*, and $\vec{v}$ is the speed measure of the moving object at time *t*. $\vec{v}$ contains both speed and direction information. If *gp* is in route, then $\vec{v}$ is a real number value. Its absolute value is equal to the speed of the moving object at time *t*, while its sign (either positive or negative) indicates the traffic flow direction the moving object belongs to. If *gp* is in junction, then $\vec{v} = \perp$ ($\perp$ means "undefined").

Definition 6 (Moving Graph Point). A moving graph point *mgp* is defined as:

$$dmgp = (mv_i)_{i=1}^{n}$$

where $mv_i = (t_i, gp_i, \vec{v}_i)$ $(1 \le i \le n)$ is the *i*th motion vector of the moving graph point, and for $\forall i \in \{1, ...n-1\}$ we have: $t_i < t_{i+1}$.

For a running moving object *mo*, its last motion vector, $mv_n = (t_n, gp_n, \vec{v}_n)$, contains key information for prediction and location updates, and we call it "active motion vector". Through the active motion vector, we can derive the computed location of the moving object (see Section 3). We call the route on which the moving object is running at time t_n "active route". If gp_n is in route and $gp_n = (rid_n, pos_n)$, then the active route is route(rid_n). If gp_n in in junction and $gp_n = jid_n$, then the active route is the route on which *mo* is running before it reaches junct(jid_n).

3 Location Update Strategies for Network Constrained Moving Objects

In this section, we propose a new location update mechanism, Location Update Mechanism for Network-Constrained Moving Objects (Net-LUM). In Net-LUM, both the moving objects and the database server need to store the graph of the traffic network. The basic idea of Net-LUM is the "Inertia Principle". That is, the system assumes that the moving object will continue to move along the current route at roughly steady speed for some more time, and whenever this assumption becomes invalid, the moving object will launch a location update so that the up-to-date information of the moving object can be reported to the database server to reflect the new situations.

3.1 Transformation of Euclidean Position to Network Position

When a moving object *mo* is running inside the network, it will repeatedly transform the Euclidean position measured from GPS of the (x, y) form to the network position of the *jid* or (rid, pos) form. This process is called "network-matching". If (x, y) is inside a junction area, then the junction ID is returned as the network-matching result. Otherwise, (x, y) will be matched to a route and the graph point value in the route will be returned as the result.

During the network-matching process, data errors from multiple sources have to be dealt with. In an MOD system, there can be two kinds of errors. First, since we use polylines to describe the geometry of the routes, the digitalized route is only an approximation of the real-world traffic road no matter how fine the granularity is. Second, GPS can also incur sampling errors. In current commercial GPS products for civil use, the sampling error can amount to 15 meters or more. These two kinds of errors together can amount to 30 meters or more in an MOD system so that they should not be ignored. We call these two kinds of errors together "data error" in the following discussion. To deal with data errors, we can utilize the active motion vector and the active route to improve the network matching accuracy, as shown in Figure 1.

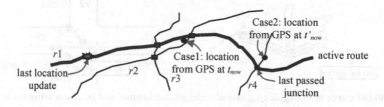

Fig. 1. Transformation of Euclidean Position to Network Position

As illustrated in Figure 1, suppose that the moving object's active motion vector is $(t_n, gp_n, \vec{v}_n)$, and the active route is route $r1$. When mo is running, in most cases we only need to check the active route $r1$, since mo can not "jump" to another route without going through a junction. Therefore, in Case 1 of Figure 1, even though the Euclidean position from the GPS is very close to r2, the Algorithm can still spot the moving object to route $r1$ correctly. Only when mo has just passed through a junction we need to check the other routes connected by the junction, as shown in Case 2 of Figure 1. As a result, the network-matching accuracy and efficiency can be improved.

3.2 Location Update Mechanism

When a moving object mo is running inside the transportation network, it continuously compares its current moving parameters (for instance route identifier, location, speed, and direction) with the active motion vector it has submitted at last location update. Whenever certain conditions are met, a new location update will be triggered to send the current motion vector to the server so that the location of the moving object can be tracked.

In Net-LUM, we define three kinds of location updates, ID-Triggered Locations Updates (IDTLU), Distance-Threshold-Triggered Location Updates (DTTLU), and Speed-Threshold-Triggered Location Updates (STTLU). These three kinds of location updates work together to fulfill the location tracking of moving objects (see Algorithm 1). Among them, only IDTLU and DTTLU are basic ones while STTLU is optional and is needed only when uncertainty management [3] is involved.

Definition 7 (ID-Triggered Location Update (IDTLU)). For a running moving object, whenever it transfers from one route to another, a location update will be triggered to reflect the change of route identifiers. We call this kind of location updates ID-Triggered Location Updates (IDTLU).

Definition 8 (Distance-Threshold-Triggered Location Update (DTTLU)). When the moving object mo is running along a certain route, it repeatedly compares its actual position measured from GPS (denoted as gp_{gps}) with the computed position derived from the active motion vector (denoted as gp_{cmp}). If one of the following two conditions is met: (1) the distance between gp_{gps} and gp_{cmp} exceeds a certain predefined threshold ξ (for instance, 0.5 kilometer), (2) gp_{cmp} is in junction while gp_{gps} is in the active route, a new location update will be triggered to report the actual location of the moving object. This kind of location updates is called Distance-Threshold-Triggered Location Updates (DTTLU). Figure 2 illustrates two cases of DTTLU.

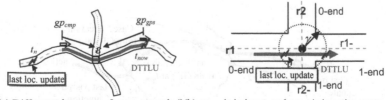

(a) Difference btw. gp_{cmp} & gp_{gps} exceeds ξ (b) gp_{cmp} is in junct. and gp_{gps} is in active route

Fig. 2. Distance-Threshold-Triggered Location Update (DTTLU)

Let's consider how to derive the computed location gp_{cmp} from the active motion vector $mv_n = (t_n, gp_n, \vec{v}_n)$. If gp_n is in route (suppose $gp_n = (rid_n, pos_n)$), then the computed position at the current time t_{now} is $gp_{cmp} = (rid_n, pos_{cmp})$, where pos_{cmp} can be computed with the following formula:

$$pos_{now} = pos_n + \frac{vm_n \times (t_{now} - t_n)}{route(rid_n).length} \quad (route(rid_n).length \text{ is the length of } route(rid_n))$$

If gp_n is in junction (suppose $gp_n = jid_n$), then the computed position at time t_{now} is still in the junction, that is, $gp_{cmp} = jid_n$.

When evaluating the computed position, a special case should be considered when the moving object is near the end of the route and the actual speed is lower than $|\vec{v}_n|$ ($|\vec{v}_n|$ is the absolute value of $\vec{v}_n$). In this case, the computed position can exceed the scope of $[0, 1]$ and we can interpret the extra value as the distance covered by the moving object in other routes after it finishes the current route, so that the location update policy does not need to be changed.

Definition 9 (Speed-Threshold-Triggered Location Update (STTLU)). Suppose that the active motion vector of the moving object mo is $mv_n = (t_n, gp_n, \vec{v}_n)$. If $\vec{v}_n$ is defined, then mo will repeatedly compare its actual speed measure $\vec{v}_{gps}$ with $\vec{v}_n$ during its move. If one of the following two conditions is met: (1) the difference between $|\vec{v}_{gps}|$ and $|\vec{v}_n|$ exceeds a certain predefined threshold ψ (for instance, 10 kilometer/hour); (2) $\vec{v}_{gps}$ and $\vec{v}_n$ are in different flow directions, a location update is triggered. We call this kind of location updates Speed-Threshold-Triggered Location Updates (STTLU).

Through STTLU, we can be assured that between any two consecutive location updates (suppose that the corresponding motion vectors are $mv_i = (t_i, gp_i, \vec{v}_i)$ and $mv_{i+1} = (t_{i+1}, gp_{i+1}, \vec{v}_{i+1})$, and gp_i is in route), the speed of the moving object is between $(|\vec{v}_i| - \psi)$ and $(|\vec{v}_i| + \psi)$.

From the above definitions we can have the following two inferences (proofs omitted).

Theorem 1. Location update can happen at most one time inside a junction area.

Theorem 2. If moving object mo triggers a location update inside a junction, then when it drives out of the junction, it will launch another location update immediately.

From the above analysis we can see that, Net-LUM can dramatically reduce location update costs around junctions. In real-world traffic systems, moving objects often run most irregularly around junctions. If junctions are not treated separately, there would be a lot of location updates, as illustrated in Figure 3(a). However, in Net-LUM, moving objects, as shown in Figure 3(b), will trigger only few location updates around a junction (mo1), or trigger no location updates at all if it moves roughly in a steady speed along a certain route (mo2). As an ideal situation, mo may drive through the whole route without triggering any location update, even though it passes through multiple junctions.

In Net-LUM, IDTLU, DTTLU, and STTLU work together to provide a complete location tracking mechanism for moving objects. The overall location update algorithm is shown in Algorithm 1 (we suppose that the algorithm is called frequently enough so that key location update chances will not be missed).

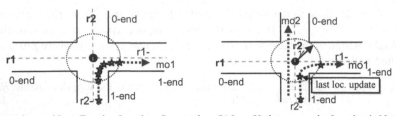

(a) Loc. Updates without Treating Junctions Separately   (b) Loc. Updates around a Junction in Net-LUM

Fig. 3. Location Updates around a Junction

Algorithm 1. Location Update Algorithm Running at the Moving Object End

General Arguments:

    $G; \xi; \psi;$ //the graph of the traffic network, the distance threshold, and the speed threshold

1.   **WHILE** (*mo* is active) **DO**
2.   Read GPS signal, and get Euclidean position $p = (x, y)$, speed v, and direction d;
3.   Transform (p, d, gp, fd);   //*network-matching*
4.   Let $\vec{v} := fd * v$;
5.   **LET** $mv_n = (t_n, gp_n, \vec{v}_n)$ be the active motion vector sent at the last location update;
6.   **LET** $actv\_rid$ be the ID of the active route;
7.   **IF** (IsinJunct(gp_n))
8.   **IF NOT**(IsinJunct(gp) **AND** ($gp=gp_n$)) **THEN**
9.   SendtoSVR(CreateLUM($mid, t_{now}, gp, \perp$));   //*DTTLU or IDTLU*;
10.   **ENDIF**;
11.   **ELSE** // gp_n is in route, suppose that $gp_n = (rid_n, pos_n)$
12.   **IF** IsinJunct(gp) **THEN** TransformtoRoute($gp, gp\_inactvroute, actv\_rid$);
13.   **ELSE** $gp\_inactvroute := gp$;
14.   **ENDIF**;
15.   // $gp\_inactvroute$ is in route. suppose $gp\_inactvroute = (rid, pos)$;
16.   **IF** ($rid \neq actv\_rid$) **THEN**
17.   SendtoSVR(CreateLUM(($mid, t_{now}, gp, \vec{v}$));   //*IDTLU*
18.   **ELSE**
19.   **LET** gp_{cmp} be the computed position derived from mv_n;
20.   **IF** (one of the DTTLU conditions is met) **THEN**
21.   SendtoSVR(CreateLUM(($mid, t_{now}, gp, \vec{v}$));   //*DTTLU*
22.   **ELSE IF** (one of the STTLU conditions is met) **THEN**
23.   SendtoSVR(CreateLUM(($mid, t_{now}, gp, \vec{v}$)); //*STTLU*
24.   **ENDIF**;
25.   **ENDIF**;
26.   **ENDIF**;
27.   **ENDIF**;
28.   **ENDWHILE**;

In Algorithm 1, the function TransformtoRoute($gp, gp*, rid$) transforms a graph point value from the *jid* form to the (rid, pos) form inside route(rid); CreateLUM() and SendtoSVR() create and send a location update message respectively.

4   Performance Evaluation and Conclusion

To evaluate the performance of the Net-LUM model, we have implemented an experimental system and conducted a series of experiments. We choose two typical location update policies proposed by Wolfson *et al.* in [6], Eu_Deviation_Avg and Eu_Distance, as the controls of the experiments.

In the experiments, we mainly focus on two factors, (1) average number of location updates per moving object (φ); and (2) error introduced by location updates (δ), which reflects the deviation of the computed position and the actual position. The experimental results are shown in Figure 4.

From Figure 4 we can see that Net-LUM performs better than Eu_Deviation_Avg and Eu_Distance in terms of location update costs and location tracking accuracy. This is because that Net-LUM can better deal with data errors and changes of route identifiers. Besides, Net-LUM can reduce location update costs around junctions.

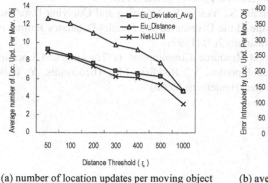
(a) number of location updates per moving object

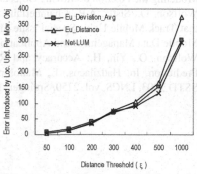
(b) average error introduced by location updates

Fig. 4. Location update costs and introduced errors

To sum up, compared with previously proposed location update mechanisms, Net-LUM has the following features.

(1) In Net-LUM, moving objects can utilize the active motion vectors and also the topology of traffic network so that the transformation from the GPS position to the network position can be fulfilled more accurately and efficiently;

(2) Junctions are specially considered for location updates in Net-LUM so that the location update costs are reduced around junctions;

(3) Net-LUM can report the changes in route identifiers, directions, and speeds to the server more promptly so that the accuracy can be improved;

Net-LUM is a "complete" network-based location update mechanism, which means that in Net-LUM, not only the location update policy itself, but also the underlying data model, are network-based. Therefore, it can be seamlessly integrated into the network-base MOD model such as MODTN [2] as the data sampling module without any further transformation involved.

Acknowledgments. The work was partially supported by NSFC project under grant number 60573164, and by SRF for ROCS, SEM.

References

1. Civilis, A., Jensen, C.S., Pakalnis, S.: Techniques for Efficient Road-Network-Based Tracking of Moving Objects. IEEE Trans. Knowl. Data Eng. 17(5) (2005)
2. Ding, Z., Güting, R.H.: Managing Moving Objects on Dynamic Transportation Networks. In: Proc. of the 16th International Conference on Science and Statistical Database Management (SSDBM 2004), Santorini, Greece (June 2004)
3. Ding, Z., Güting, R.H.: Uncertainty Management for Network Constrained Moving Objects. In: Galindo, F., Takizawa, M., Traunmüller, R. (eds.) DEXA 2004. LNCS, vol. 3180, Springer, Heidelberg (2004)
4. Wolfson, O., Chamberlain, S., Dao, S., Jiang, L., Mendez, G.: Cost and Imprecision in Modeling the Position of Moving Objects. In: Proc. of ICDE (1998)
5. Wolfson, O., Sistla, A.P., Chamberlain, S., Yesha, Y.: Updating and Querying Databases that Track Mobile Units. Special issue of the Distributed and Parallel Databases Journal on Mobile Data Management and Applications 7(3) (1999)
6. Wolfson, O., Yin, H.: Accuracy and Resource Consumption in Tracking and Location Prediction. In: Hadzilacos, T., Manolopoulos, Y., Roddick, J.F., Theodoridis, Y. (eds.) SSTD 2003. LNCS, vol. 2750, Springer, Heidelberg (2003)

A P2P Meta-index for Spatio-temporal Moving Object Databases

Cecilia Hernandez[1], M. Andrea Rodriguez[1,3], and Mauricio Marin[2,3]

[1] Dept. of Computer Science, Universidad de Concepción, Chile
{cecihernandez,andrea}@udec.cl
[2] Yahoo! Research, Chile
mmarin@yahoo-inc.com
[3] Center for Web Research, Universidad de Chile, Chile

Abstract. In this paper we propose a distributed meta-index using a peer-to-peer protocol to allow spatio-temporal queries of moving objects on a large set of distributed database servers. We present distribution and fault tolerance strategies of a meta-index that combines partial trace of object movements with aggregated data about the number of objects in the database servers. The degrees of distribution and fault tolerance were compared by using discrete-event simulators with demanding spatio-temporal workloads. The results show that the distributed meta-index using Chord provides good performance, scalability and fault tolerance.

1 Introduction and Related Work

For query processing, an important aspect of distributed moving object applications is the distribution of the index structure. Advances in this sense are studies that distribute spatial index structures, such as the Quadtree and R-tree structures, in peer-to-peer (p2p) networks [4,7]. In similar way, studies in distributed spatio-temporal databases have typically considered a distributed global index structure, which organizes spatio-temporal information in a structure defined in terms of space and temporal partitions [2,5]. These studies have addressed window queries (i.e., time-instant and time-interval queries) and not queries about locations of particular objects or aggregated queries.

In this paper, we describe an alternative approach to distributed spatio-temporal database servers. First, and unlike previous works, we design a structure to support different types of queries including coordinate-based queries (i.e., time-slice and time interval queries), aggregated queries (top-K servers with the largest number of objects and geographic extent), and combined queries of the form "Where was an object o at a time instant t or time interval $[t_1, t_2]$?" We argue that it is impractical to think of solving these types of queries with a global index when the system is composed of a large and dynamic number of servers. Second, instead of having a global and distributed spatio-temporal structure, we propose to handle independent local indexes in database servers and a global and distributed meta-index structure. This could be seen as a two-level architecture;

J.R. Haritsa, R. Kotagiri, and V. Pudi (Eds.): DASFAA 2008, LNCS 4947, pp. 653–660, 2008.
© Springer-Verlag Berlin Heidelberg 2008

however, we are not forcing any coordination between the meta-index and the local indexes.

A preliminary study for this meta-index [3] analyzes the viability of creating and updating a centralized structure in a highly dynamic environment of moving objects. This paper continues with our previous work and describes the distributed meta-index structure using Chord [6]. In this paper we present experimental evaluation that shows the good performance, scalability and fault tolerance properties of the proposed system.

The structure of the paper is as follows. Section 2 describes the meta-index, whose distribution and fault tolerance strategies are presented in section 3. Section 4 presents experimental evaluation supporting our claims on performance, scalability and fault tolerance. Finally, we outline our conclusions in Section 5.

2   Meta-index Description

A basic component of the system is the meta-index structure that guides the search process to local servers or gives approximated answers to different queries. In particular, we studied the following types of queries: (1) time-slice or time-interval queries that return objects or trajectories within a query window and time instant or interval, (2) aggregated queries concerning the top-K servers with largest number of objects or trajectories and largest extent area, (3) nearest neighbors to a specific object and time instant, and (4) queries about the location of a particular object at a specific time instant.

The meta-index stores partial data about the time-varying location of objects. These data include: time-varying number of objects per server (statistical or aggregated information), time-varying geographic extent including the location of objects in a server, and coarse traces of object visits across servers. Coarse traces of objects in the meta-index are not "real" sparse trajectories, but lists of servers that objects have visited sorted by the time of the data collection.

Among all queries of interest, this paper concentrates only on queries about the location of specific objects at a particular time instant, since they represent a challenging and not previously addressed query in the context of distributed spatio-temporal databases. This type of queries uses the object traces and number of objects per server at different time stamps in the meta-index. Using coarse traces for answering this type of queries is novel since we do not use the classical space partition to organize or distribute spatio-temporal information.

The search algorithm, based on object traces, finds the first location of the object at a time instant t' such that $|t' - t|$ is minimum, with t the query time. If the trace of an object is not found in the meta-index, the algorithm starts search on servers with largest number of objects at query time. When a first location of the object is found, the algorithm follows the object path in the corresponding servers until finding the location of the desired object at query time.

To maintain the meta-index, we define a crawling strategy that collects data from database servers asynchronously [3]. In every data collection from a server, a crawler transfers to the meta-index aggregated data and the ids of objects which have been in the server but not transfered in previous visits. Consequently, some objects might no longer be present in the server at the time of the data collection.

3   Meta-index Distribution

We base the distribution of the meta-index on a p2p network using Chord protocol [6]. Chord defines a common address space for nodes and data keys and provides a lookup algorithm. This algorithm enables distribution by mapping data keys among a changing set of nodes with $O(\log n)$ routing cost.

In order to use bandwidth efficiently, we propose two strategies for distributing the meta-index data. First, we introduce the concept of *MSB (Most Significant Bits)* to map moving object trace data to Chord keys. Crawlers use the MSB to group object trace data in *Composite Objects*. The number of bits defined by the MSB constitutes a threshold for the update message maximum size performed by crawlers' robots. Second, since the amount of aggregated data per update message is very small we piggy it back to composite objects. This scheme allows crawlers to send fewer update messages into the p2p network avoiding overloading the network. Figure 1 shows an example of how a robot groups object traces into two composite objects. In this case the MSB is 28 which means that the maximum number of moving object ids in a composite object is 16 (2^4). Here, composite objects are identified by *compositeObj₁* and *compositeObj₂* with their corresponding moving object ids and aggregated data (SD1).

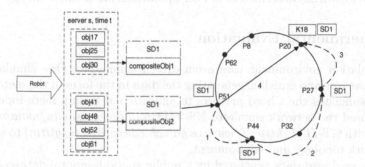

Fig. 1. Meta-index distribution

The meta-index distribution strategy requires the administration of two types of messages. *S-messages* (messages 1-3 in Figure 1) and *T-messages* (message 4 in Figure 1). *S-messages* carry control and aggregated data but they do not carry composite object content. Aggregated data are stored in all peers visited until finding the destination peers. *T-messages* transfer composite object content from the entry peer to the peer that will store the composite object.

The distribution of the meta-index supports fault tolerance in two ways: by the partial replication of aggregated data in all peers, and by distributing the object traces uniformly at random in p peers. The replication of aggregated data is useful when no data about the trace of the desired object is found in the distributed meta-index. In such case, the number of objects stored in any of the peers is used to rank the servers where to most likely find the desired objects. Second, data collected by a crawler are grouped into *sub-composite* object ids based on MSB and a random number between 1 and p. Then, object traces are associated randomly with one of the p possible sub-composite objects and stored in the corresponding peer. When a client looks for a specific object at a given time, it starts the search at a random peer, this peer looks up the p sub-composite objects in parallel and sends back the best answer to the client, where best here means the estimated location of the object at the closest time to the query. If one of the peers holding the requested object is down, another peer among the p ones may have it. Even though the reply might not be the best, it would serve as starting search point for the client.

Operations on our meta-index system architecture are decentralized. First, crawlers collect data from database servers and update the meta-index in the p2p network in parallel. Crawlers build composite objects based on MSB and p, and send them to the peer-to-peer network through $Update()$ messages. Second, peers receive and delegate $Update()$ operations (Algorithm 1) to other peers in order to find the destination peer. We allow peers to receive queries from any node on the network through the $RecvClientQuery()$ procedure. This procedure takes in a moving object and query time and finds at most p peers that resolve the query. The procedure $DoQuery()$ looks up recursively the composite object id in the peer-to-peer network and then the query resolving involves applying the search algorithm described in 2. Peer operations are defined in Algorithm 1.

4   Experimental Evaluation

Our simulation environment uses event-driven simulators. One simulates the database servers and crawling generating the data in the form of the meta-index. Another simulates the Chord protocol to allocate and lookup meta-index data. We also used the network simulator NS-2 ($http://www.isi.edu/nsnam/ns$) in tandem with GT-ITM ($http://www.cc.gatech.edu/projects/gtitm$) to simulate the network topology and environment.

We use workload data generated by a public spatio-temporal dataset generator; the Network-based Generator of Moving Objects (NGMO) [1]. The data set contains 50,000 initial moving objects, existing around 150,000 along the simulation time. In similar ways as seen in [7], we run the experiments with 4, 8, 16, and 32 crawlers' robots and 16, 32, 64 and 128 peers. We chose peers and clients randomly from stub nodes defined in a transit-stub network of 588 nodes created with NS-2/GT-ITM.

Search performance of centralized versus distributed meta-index. We first evaluate the performance of the search process for a centralized versus distributed meta-index by using the number of visits to local servers as a performance metric to answer a query of type "find object o at time stamp t," (Figure 2 (a)) . We used three types of benchmarks, each with 200 queries: (b1) random

---

Algorithm 1. Algorithms at peers

```
 1: procedure Update (peer entrypeer, composite object id coid, aggregated data sd)
 2: if aggregated data sd is not in thispeer then
 3:     Add sd into this thispeer
 4: end if
 5: if coid must be stored in thispeer then
 6:     Get composite object co with co.id = coid from entrypeer
 7:     Add composite object co with co.id = coid into thispeer
 8: else
 9:     call Update message to nextpeer (based on Chord protocol)
10: end if
11: end procedure

 1: procedure RecvClientRequest (moving object id moid, timestamp queryTime)
 2: Get all composite object ids coidList using moid, MSB and P
 3: for each coid in coidList in parallel do
 4:     call DoQuery(coid) on a random peer
 5: end for
 6: Resolve best reply to client
 7: end procedure

 1: procedure DoQuery (composite object id coid)
 2: if coid is in thispeer then
 3:     call Algorithm1 defined in section 2
 4:     return
 5: else
 6:     call DoQuery(coid) on nextpeer (based on Chord protocol)
 7: end if
 8: end procedure
```

---

objects (average trajectory length of 32 servers), (b2) objects with largest trajectories (average trajectory length of 175 servers), and (b3) objects with shortest trajectories (average trajectory length of 2 servers). Here, the meta-index contains traces for 71% of the total number of moving objects (150,000) that exist during simulation time; that is, approximately 30% of queries for the whole data set would need to use statistical data from the meta-index.

Figure 2 (b) shows the performance of the distributed meta-index for 200 random queries using different percentages of searches that require the statistical data. Here, using statistical data in a search is important because this data is partially recovered from peers, which can affect the performance with respect to the number of visits to local servers during searches. The quality of the statistical data recovery would improve when using p greater than 1, since query solving would combine statistical data from p peers. We present the ratio between a centralized and distributed meta-index search using 8 robots and statistical data recovery from only one peer. In this figure, values close to 1.0 mean that the distributed and centralized meta-index have similar performance with respect to the number of visits to local servers.

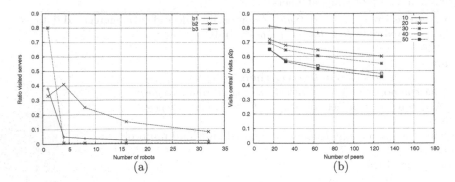

Fig. 2. Performance of the meta-index: (a) The ratio number of servers visited using the meta-index to the number of servers visited using the server random selection strategy (The x-axis indicates the different numbers of robots updating the meta-index), (b) Search performance with a distributed meta-index with respect to a centralized meta-index (Different curves indicate percentages of use of aggregated data in the search)

Communication Overhead. In order to evaluate the communication overhead associated with the distribution of the meta-index in the p2p network, we run an experiment that shows the number of bytes transmitted in update operations (i.e., bytes in S-messages and T-message). This number of bytes is normalized by the total number of bytes of the meta index, such that we can obtain an overhead with respect to the original data content we want to distribute. Figure 3(a) shows the normalized number of bytes with respect to different numbers of peers and different setting of MSB.

Elasticity. We run an experiment to compare the scalability of the distributed meta-index with the centralized system. We measured the response time for a protocol that use MSB=20, P=1, 128 peers and different rates of query concurrency, from 10 to 1000 queries per second given to the system. Figure 3 (b) indicates that over 200 queries per second, the performance of the centralized system starts to be affected significantly due to the workload. In this case, the response time in the centralized scheme does not scale because of network congestion. The P2P system has a great amount of available bandwidth allowing it to scale gracefully with increasing query rate.

Fault Tolerance. We analyzed the fault tolerance in terms of the strategies described in 3. Figure 3(c) shows the efficiency of the distribution of the aggregated data among peers. We define efficiency in a strict hand-to-hand manner by averaging over all servers and peers on the ratio average number of objects to the maximum number of objects observed in any peer for the same server. This is a demanding distribution test for both the degree of completeness and precision of the samples kept in each real p2p node. There is a trade-off with MSB, since fewer bits lead to a small number of composite objects circulating among the peers.

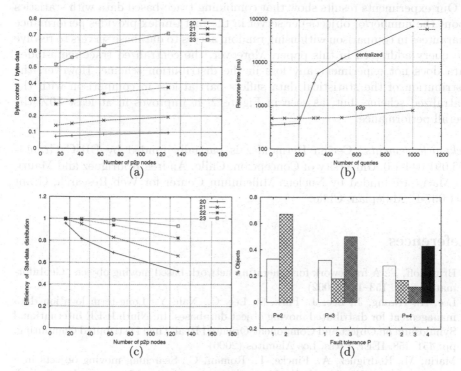

Fig. 3. Communication overhead, elasticity, and fault tolerance of the meta-index: (a) relative number of bytes transmitted in update operations, (b) response time for different query rates given to the system for centralized and p2p scheme (MSB=20, $p=1$ and 128 peers), (c) efficiency of distribution of statistical data in peers, and (d) quality of fault tolerance distribution for different values of p

We also measured the distribution of an object trace in the participating p peers. We show results about the percentage of objects that maintain traces between 1 and p peers. In Figure 3(d), the first bar for each p represents the percentage of objects whose traces are only in one peer (objects with shortest traces), the second bar represents the percentage of objects whose traces are in two peers, and so on. The results indicate that our strategy is able to provide fault tolerance for object traces for 70% of the objects in the system.

5   Conclusions

Overall our strategy allows the two main processes, namely crawling and searching to make efficient use of the p2p network. Crawlers can store their payload in any peer and user queries can also start up in any peer. Thus the scheme can accommodate many crawlers, which improve the probability of registering at least one trace per moving object. Also, user queries throughput can be increased by evenly distributing the start of searches across all of the p2p nodes.

Our experiments results show that combining trace-based data with statistics about the number of objects per server in the meta-index provides performance guarantees in comparison with using random access to database servers to resolve the query addressed in this paper. Moreover, the centralized trace meta-index data does not experiment any loss in the distribution scheme. However, the distribution of the statistical data suffers partial loss in comparison with the centralized scheme, but, as seen in Figure 2, it improves in at least 50% the overall performance.

Acknowledgements. Cecilia Hernandez is partially funded by DIUC Grant - 206.091.044-1.0, University of Concepcion, Chile. Andrea Rodriguez and Mauricio Marin are funded by Nucleus Millennium Center for Web Research, Grant P04-067-F, Mideplan, Chile.

References

1. Brinkhoff, T.: A framework for generating network-based moving objects. GeoInformatica 6(2), 153–180 (2002)
2. Lee, H., Hwang, J., Lee, J., Park, S., Lee, C., Nah, Y.: Long-term location data management for distributed moving object databases. In: Ninth IEEE International Symposium on Object and Component-Oriented Real-Time Distributed Computing, pp. 451–458. IEEE Press, Los Alamitos (2006)
3. Marín, M., Rodríguez, A., Fincke, T., Román, C.: Searching moving objects in a spatio-temporal distributed database servers system. In: Meersman, R., Tari, Z. (eds.) OTM 2006. LNCS, vol. 4276, pp. 1388–1401. Springer, Heidelberg (2006)
4. Mondal, A., Lifu, Y., Kitsuregawa, M.: P2pr-tree: An r-tree-based spatial index for peer-to-peer environments. In: Lindner, W., Mesiti, M., Türker, C., Tzitzikas, Y., Vakali, A.I. (eds.) EDBT 2004. LNCS, vol. 3268, pp. 516–525. Springer, Heidelberg (2004)
5. Nah, Y., Lee, J., Lee, W.J., Le, H., Kim, M.H., Han, K.J.: Distributed scalable location data management system based on the GALIS architecture. In: Tenth IEEE International Workshop on Object-Oriented Real Time Dependable Systems, pp. 397–404. IEEE Press, Los Alamitos (2005)
6. Stoica, I., Morris, R., Liben-Nowell, D., Karger, D.R., Kaashoek, M.F., Dabek, F., Balakrishnan, H.: Chord: a scalable peer-to-peer lookup protocol for internet applications. IEEE/ACM Trans. Netw. 11(1), 17–32 (2003)
7. Tanin, E., Harwood, A., Samet, H.: Using a distributed quadtree index in peer-to-peer networks. VLDB J. 16(2), 165–178 (2007)

An Update Propagation Strategy Considering Access Frequency in Peer-to-Peer Networks

Toshiki Watanabe, Akimitsu Kanzaki, Takahiro Hara, and Shojiro Nishio

Dept. of Multimedia Eng., Graduate School of Information Science and Technology,
Osaka University, 1-5 Yamadaoka, Suita, Osaka 565-0871, Japan
{watanabe.toshiki,kanzaki,hara,nishio}@ist.osaka-u.ac.jp

Abstract. In a P2P network, a data update occurred on a particular peer should be immediately propagated to other peers holding its replicas. In our previous work, we proposed a novel update propagation strategy using a tree structure for delay reduction and node failure tolerance. This strategy propagates the updated data according to the tree. In this paper, we extend our previous strategy to selectively propagate each updated data considering the data access frequency of each peer. The extended strategy propagates the updated data to peers which frequently access the data, whereas only a small message informing that the replica has become invalid is propagated to peers which rarely access the data.

Keywords: peer-to-peer, replication, update propagation, access frequency.

1 Introduction

Recently, peer-to-peer (P2P) networks and applications have been becoming popular and a large number of research projects on P2P systems are ongoing. In a P2P network, it is common that data items are replicated on multiple peers for efficient data retrieval and load balancing [2,3,4,5]. In many systems such as a distributed file system in a P2P network, shared data items are updated by particular peers and other peers might read old replicas that are not the latest versions. In this paper, we assume an application where such dirty reads to old replicas are not allowed and an updated data has to be immediately propagated to all peers that hold the replicas.

In our previous work [6], we proposed the UPT-FT (Update Propagation Tree with Fault Tolerance) method, which is an update propagation strategy for delay reduction and node failure tolerance using a tree structure. UPT-FT creates an n-ary tree, which we call the *UP (Update Propagation)* tree, for each data item in a P2P network and propagates the updated data according to the tree.

Here, among replica holders of a data item, some of them frequently access the data, whereas others do not. In UPT-FT, however, the updated data is necessarily propagated to all replica holders regardless of their access frequencies. This may cause redundant load for update propagation.

In this paper, we extend our previous strategy and propose the UPT-HL (UPT with H L peers) method which reduces such redundant load for update propagation. UPT-HL propagates the updated data to peers which frequently access the data, whereas it propagates only a small message (invalidation report: IR) informing that the replica has

J.R. Haritsa, R. Kotagiri, and V. Pudi (Eds.): DASFAA 2008, LNCS 4947, pp. 661–669, 2008.

become invalid to peers which rarely access the data. This approach further reduces the load and delay for update propagation since the number of peers that receive the updated data decreases.

The remainder of this paper is organized as follows. In Section 2, we introduce some related works. In Section 3, we explain our proposed UPT-HL. In Section 4, we show the simulation results regarding the performance evaluation of our strategy. Finally, in Section 5, we summarize this paper and discuss future work.

2   Related Works

In the proposed in [3], a peer who updates its replica propagates the update information based on a gossiping approach, in which each peer that receives the update information continuously forwards it to a limited number of the peer's neighbors. However, this method does not guarantee that all peers holding the replica can receive the update information. In the method proposed in [5], each data item has a logical replica chain composed of all the replica holders. Each peer on the chain acquires partial knowledge of the bi-directional chain by keeping a list of information of m nearest peers, called *probe peers*, in each direction. When a peer updates the data item, it pushes the update information to all active probe peers through the chain. In each direction, the farthest probe peer that received the update information further pushes the update information to its probe peers through the chain. This process continues until all peers on the chain receive the update information. However, the propagation delay is still long, i.e., $O(p)$, where p is the number of replica holders.

In [6], we proposed an update propagation method for delay reduction and node failure tolerance. As mentioned in Section 1, this method creates an n-ary tree, which we call the UP tree, for each data item and propagates the updated data according to the tree. The root of the UP tree is the owner of the data item, and the other nodes are replica holders. However, as mentioned in Section 1, in this method, the updated data must be propagated to all replica holders and causes heavy traffic for update propagation.

3   UPT-HL Method

In this Section, we explain the detailed behaviors in UPT-HL.

3.1   Structure of an UP Tree

Figure 1 shows the structure of an UP tree in UPT-HL. UPT-HL classifies the peers that have the same replica into two types of peers. One is *H peer* that frequently accesses the data and constantly needs the latest version. The other is *L peer* that rarely accesses the data and does not always need the latest version. In addition, UPT-HL creates one tree that consists of only H peers (H tree) and some trees that consist of only L peers (L trees) for each data item. Each L tree follows an H peer, and both types of peers can transit to another type according to their own access frequencies for the data. In UPT-HL, the root node in the H tree (*original node*) has the original data and can update the data.

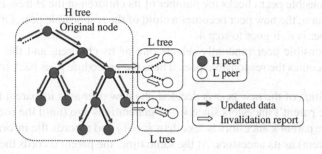

Fig. 1. Structure of an UP tree and update propagation

Each H peer records the information on the L peer (the root of the L tree) that follows the H peer as well as the information on its ancestors since k generations ago and its children on the H tree. On the other hand, each L peer records the information on its parent and children on the L tree, and the H peer that the L tree follows (we call this H peer *responsible H peer*).

3.2 Propagation of Update Information

Figure 1 shows how to propagate the update information in UPT-HL. The root node in the H tree updates the data and propagates the updated data according to the H tree. H peers that receive the updated data send it to their children until the data reaches leaf nodes in the H tree. On the other hand, if there exists an L tree that follows an H peer, the H peer propagates an IR to the L tree. L peers that receive the IR invalidate their own replica and forward it to their children until the IR reaches leaf nodes in the L tree. When an L peer holding invalid replica needs the latest version of the data, it requests the data to its responsible H peer.

3.3 Participation of a New Peer

When a peer newly creates a replica of a data item, it participates in the UP tree (n-ary tree) of the data item. We call this peer a *new peer*. In UPT-HL, a new peer firstly participates in the H tree as an H peer. We also call a peer that determines the position of the new peer in the tree a *responsible peer*. The detailed procedure is as follows:

1. If the peer that responded to a data request from the new peer is an H peer, the H peer becomes the responsible peer. Otherwise, i.e., the responded peer is an L peer, it returns the information on its responsible H peer to the new peer and the H peer becomes the responsible peer.
2. If the responsible peer is the root node in the H tree (*original node*), the procedure goes to step 3. Otherwise, the responsible peer firstly sends a participation request to its ancestor since k generations ago (the root node if the number of ancestors is less than k). The ancestor which receives the request becomes the responsible peer and the procedure goes to step 3.

3. The responsible peer checks the number of its children in the H tree. If the number is less than n, the new peer becomes a child of the responsible peer. Otherwise, i.e., the number is n, it goes to step 4.
4. The responsible peer randomly selects one of its children, and the selected one newly becomes the responsible peer. Then, the procedure goes back to step 3.

After the position of the new peer is decided, the new peer asks its parent for the information on the parent's ancestors since $k - 1$ generations ago (until the root node if the number of the parent's ancestors is less than $k - 1$) and records the information (with that of the parent) as its ancestors. At the same time, the parent records the information on the new peer as its child.

3.4 Exit of a Peer

When a peer deletes a replica of a certain data item, it exits from the UP tree of the data item. We call this peer an *exit peer*. When a peer exits, a procedure for repairing the UP tree is needed in order to avoid the partition of the UP tree. The same repairing procedure can be applied to both cases that an H peer and an L peer exit. The detailed procedure is as follows:

1. If the exit peer is a leaf node in the UP tree, it informs its parent of that it will exit from the UP tree. Then, the parent peer deletes the exit peer from the information on its children. The exit peer also deletes the k ancestor nodes from the information on its ancestors and the procedure finishes. Otherwise, it goes to step 2.
2. If the exit peer has children, it randomly selects one of its children and informs the selected child of that it will exit from the UP tree (i.e., sends an exit request). Then, if the child peer is not a leaf node, it selects one of its children and forwards the received exit request to the selected child. This process continues until the exit request reaches a leaf node. If the selected peer is a leaf node, it is exchanged with the exit peer. Then, the parent of the leaf node (actually, it is no longer a leaf) deletes the leaf node from the information on its children, and the procedure goes to step 3.
3. The exchanged peer (originally, a leaf node) informs the new parent and new descendants within k generations of that the peer has been exchanged with the exit peer. The parent and the descendants update the information on its children and their ancestors by replacing the exit peer with the exchanged peer.
4. If the exit peer is an H peer that has an L tree following it, it sends the information on the root of the L tree to the exchanged peer (the parent node if the exit peer is a leaf node in the H tree). If the H peer that receives the information has no L tree, the L tree following the exit peer is changed to follow the H peer. Otherwise, the H peer sends a participation request to the root of the L tree that follows the H peer. Then, the root L peer becomes the responsible peer, and the L tree following the exit peer is changed to attach to an L peer in the same way as described in Section 3.3. Finally, the exit peer deletes all the ancestors and children from the information on its ancestors and children, and the procedure finishes.

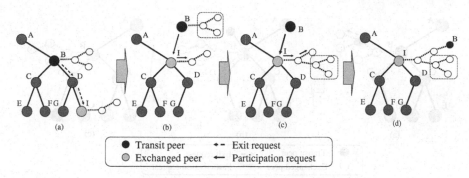

| ● | Transit peer | ← - | Exit request |
| ○ | Exchanged peer | ← | Participation request |

Fig. 2. Transition from H peer to L peer

3.5 Transition of the Peers

As described in Section 3.1, in UPT-HL, each replica holder checks its data access frequency to the corresponding data and changes its peer type (H or L) if necessary.

Determination of the transition: Each replica holder counts the number of its own accesses to the replica (*access count*) and that of receptions of the update information (*update count*). Each peer decides whether it changes its type based on the above two values every time it receives the update information. In particular, if the access count is larger than the update count at a peer whose peer type is L, the peer transits to the H peer. On the other hand, if the update count is larger than the access count at a peer whose peer type is H, the peer transits to the L peer.

Transition from H peer to L peer: When an H peer decides to transit to an L peer (we call this peer a *transit peer*), it firstly checks whether an L tree follows it. If an L tree follows, the H peer transits to L peer after moving the L tree to another H peer. In what follows, we explain the detailed procedure using Fig. 2:

1. If the transit peer is a leaf node, it informs its parent of its exit. Otherwise, a transit peer randomly selects one of its children on the H tree and informs the selected child of its exits. The child that receives this information (an exit request) randomly selects one of its children and forwards the received exit request to the selected child. This process repeats until the exit request reaches a leaf node as shown in Fig. 2(a). Then, the leaf node that receives the exit request is exchanged with the transit peer. This procedure is similar to that of step 2 in Section 3.4.
2. If there exists an L tree following the transit peer, it is moved to another H peer as shown in Fig. 2(b). This procedure is similar to that of step 4 in Section 3.4.
3. The transit peer sends a participation request for L tree to the exchanged peer (peer I in Fig. 3). If the exchanged peer has no L tree, the transit peer follows the H peer directly as an L peer. Otherwise, the H peer forwards the request to the root of the L tree that follows the H peer. Then, the root of the L tree becomes the responsible peer, and the transit peer participates in the L tree according to the procedure described in Section 3.3 (as shown in Fig. 2(c) and (d)).

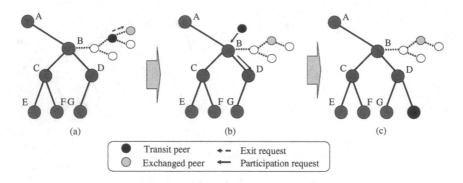

Fig. 3. Transition from L peer to H peer

Transition L peer to H peer: When an L peer decides to transit to an H peer, it forwards an exit request to one of its children until the request reaches a leaf node as shown in Fig. 3(a), and the L peer exits from the L tree in the same way as step 2 in Section 3.4. Then, the transit peer sends a participation request to its responsible H peer, and it participates in the H tree. This procedure is similar to that of steps 2 to 4 in Section 3.3 (as shown in Fig. 3(b) and (c)).

4 Simulation Experiments

In this section, we show the results of simulation experiments to evaluate the performance of our proposed method.

4.1 Simulation Model

There are 1,000 peers in the entire system. These peers construct an unstructured P2P network of a Power-Law Random Graph (PLRG) [1], where the i-th peer (i: peer identifier) has $d_i = \lfloor 20 \cdot i^{-0.4} \rfloor$ neighboring peers. By doing so, peers with smaller identifiers have more neighbors.

A hundred peers of identifiers 1 to 100 hold original data of identifiers 1 to 100, respectively. At every peer, the query-issuing frequency at each time slot is constantly 0.1. When issuing a query, the target data of each query is determined based on the Zipf distribution [7], where the probability of requesting j-th data, q_j, is given by the following equation:

$$q_j = \frac{j^{-\alpha}}{\sum_{k=1}^{100} k^{-\alpha}}. \tag{1}$$

This represents that the smaller the data identifier is, the more often the data is requested. Here, α is called the Zipf coefficient, which determines the skew of the query frequency. In the simulations, α is set as 0.5.

Queries are processed by expanding ring model [4], which starts with TTL=1 and expands the TTL linearly by 1 until finding the target data item. Data replication is

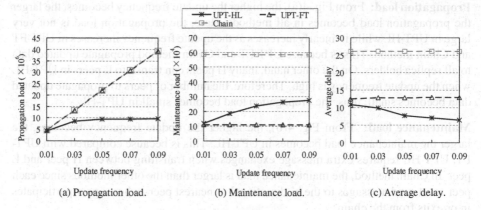

(a) Propagation load.   (b) Maintenance load.   (c) Average delay.

Fig. 4. Simulation results

performed based on owner replication, where the target data item is replicated on the query issuing peer. It is assumed that the UP tree is a binary tree, i.e., $n = 2$, and each peer in the tree records the information on its parent, i.e., $k = 1$.

All data items are of the same size and each peer has data storage sufficient to create 10 replicas excluding the space for the original data. If a peer whose data storage is full wishes to create a new replica, it selects the least recently used replica in its data storage and deletes the selected replica to prepare a space for the new replica. Here, the original data that a peer holds is not deleted whatever happens. The original data holders update their own original data item based on the constant probability (update ratio).

Based on this simulation model, we ran the simulation during 10,000 time slots. In the following, we show the simulation results by focusing on the 1st data item (data identifier is 1).

4.2 Simulation Results

We compare the performances of UPT-HL, UPT-FT, and update propagation using chain structure [5] ($m = 5$) described in Section 2 (we call it Chain method). In the experiments, we examine the following three criteria when varying the update frequency from 0.01 to 0.09 (by 0.02).

- Propagation load: The total load (volume of transmitted data) that H peers propagate the updated data to their children and that H peers or L peers propagate the IRs. In these simulation experiments, the sizes of updated data and IR are set as 100 and 1, respectively.
- Maintenance load: The total volume of transmitted messages that are exchanged to construct or repair the overlay network (tree or chain structure) and to transit to H peer or L peer. The size of the exchanged messages is set as 1 (same as that of an IR).
- Average delay: The average hopcount between the original node (the peer that updated the data) and other peers in the H tree.

Fig. 4 shows the simulation results.

Propagation load: From Fig. 4(a), the higher the update frequency becomes, the larger the propagation load becomes in all methods. Here, the propagation load is not very large in UPT-HL while it linearly increases as the update frequency increases in UPT-FT and Chain method. This is because UPT-FT and Chain method propagate updated data to all replica holders. On the other hand, many H peers can transit to L peers in UPT-HL when the update frequency is high. Therefore, the number of peers to propagate updated data becomes small, thus, the propagation load becomes small in UPT-HL.

Maintenance load: From Fig. 4(b), the higher the update frequency becomes, the larger the maintenance load becomes in UPT-HL. This is because compared with UPT-FT, UPT-HL requires extra message exchanges when transiting between H peer and L peer. In Chain method, the maintenance load is larger than the other methods since each peer must send messages to the bi-directional m nearest peers when a peer participates in or exits from the chain.

From these results, we can see that the maintenance load becomes large in UPT-HL when the update frequency is high. However, since the amount of the decrease of the propagation load is very large as shown in Fig. 4(a), the increase of the maintenance load is not so serious.

Average delay: From Fig. 4(c), the delay is smaller in UPT-HL than in UPT-FT. This is because UPT-HL propagates updated data to only H peers while UPT-FT must propagate it to all replica holders. Moreover, in UPT-HL, the average delay becomes smaller as the update frequency gets higher. This is because as the update frequency gets higher, the update counts of most peers become larger than the access counts, and thus, the number of H peers decreases. In Chain method, the average delay becomes large since the updated data is propagated linearly to bi-directional m peers.

5 Conclusions

In this paper, we have proposed a new update propagation strategy, UPT-HL method, by extending our previous method proposed in [6]. This strategy propagates the updated data to peers which frequently access the data, whereas it propagates only a small invalidation message to peers which rarely access the data. The simulation results show that this approach further reduces the load and delay for update propagation compared with our previous method.

In UPT-HL, we assume an environment where the update frequency is static. As part of our future work, we plan to consider an efficient approach for adaptively changing peer types (H and L) in an environment where the update frequency dynamically changes.

Acknowledgement

This research was partially supported by Special Coordination Funds for Promoting Science and Technology:"Yuragi Project", and Grant-in-Aid for Scientific Research (18049050) of the Ministry of Education, Culture, Sports, Science and Technology, Japan.

References

1. Admic, L.A., Lukose, R.M., Puniyani, A.R., Huberman, B.A.: Search in power-law networks. Physical Review E 64(4), 046135 (2001)
2. Cohen, E., Shenker, S.: Replication strategies in unstructured peer-to-peer networks. In: Proc. SIGCOMM 2002, pp. 177–190 (2002)
3. Datta, A., Hauswirth, M., Aberer, K.: Updates in highly unreliable, replicated peer-to-peer systems. In: Proc. ICDCS 2003, pp. 76–85 (2003)
4. Lv, Q., Cao, P., Cohen, E., Li, K., Shenker, S.: Search and replication in unstructured peer-to-peer networks. In: Proc. ICS 2002, pp. 84–95 (2002)
5. Wang, Z., Das, S.K., Kumar, M., Shen, H.: An Efficient Update Propagation Algorithm for P2P Systems. Computer Communications, 1106–1115 (2007)
6. Watanabe, T., Kanzaki, A., Hara, T., Nishio, S.: An Update Propagation Strategy for Delay Reduction and Node Failure Tolerance in Peer-to-Peer Networks. In: Proc. FINA 2007, pp. 103–108 (2007)
7. Zipf, G.K.: Human Behavior and the Principle of Least Effort. Addison-Wesley, Reading (1949)

Towards Automated Analysis
of Connections Network
in Distributed Stream Processing System

Marcin Gorawski and Pawel Marks

Silesian University of Technology,
Institute of Computer Science,
Akademicka 16,
44-100 Gliwice, Poland
{Marcin.Gorawski,Pawel.Marks}@polsl.pl

Abstract. Not so long ago data warehouses were used to process data sets loaded periodically during ETL process (*Extraction, Transformation and Loading*). We could distinguish two kinds of ETL processes: full and incremental. Now we often have to process real-time data and analyse them almost on-the-fly, so the analyses are always up to date. There are many possible applications for real-time data warehouses. In most cases two features are important: delivering data to the warehouse as quick as possible, and not losing any tuple in case of failures. In this paper we describe an architecture for gathering and processing data from geographically distributed data sources and we define a method for analysing properties of the connections structure, finding the weakest points in case of single and multiple node failures. At the end of the paper our future plans are described briefly.

Keywords: stream processing, fault-tolerance, analysis algorithm.

1 Introduction

These days it becomes more common to process continuous data streams. It may have application in many domains of our life such as: computer networks (e.g. intrusion detection), financial services, medical information systems (e.g. patient monitoring), civil engineering (e.g. highway monitoring) and more.

Thousands or even millions of energy meters located in households or factories can be sources of meter readings streams. Continuous analysis of power consumption may be crucial to efficient electricity production. Unlike other media such as water or gas, electricity is hard to store for further use. That is why a prediction of energy consumption may be very important. Real-time analysis of the media meter readings may help to manage the process of energy production in the most efficient way.

There are many systems for processing continuous data streams and they are still developed [1]. In [2] there is presented a fault tolerant Borealis system. This is a dedicated solution for applications where a low latency criterion is essential. Another system facing infinite data streams is described in [3]. Authors of

J.R. Haritsa, R. Kotagiri, and V. Pudi (Eds.): DASFAA 2008, LNCS 4947, pp. 670–677, 2008.

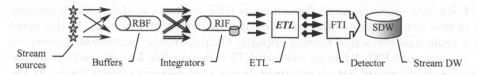

Fig. 1. Layered structure of the distributed system

the work deal with sensors producing data continuously, transferring the measured data asynchronously without having been explicitly asked for that. They proposed a *Framework in Java for Operators on Remote Data Streams* (Fjords).

In our research we have focused on processing data originating from a radio-based measurement system [4]. We carried research on efficient recovery of interrupted ETL jobs and proposed a few approaches [5,6] based on the Design-Resume algorithm [7].

Basing on the previous experience, we have focused on fault-tolerance and high availability in a distributed stream processing environment. In [8] we proposed a new set of modules increasing the probability that a failure of one or more modules will not interrupt the processing of endless data streams. Then we prepared a model [9,10] of data sources to estimate the amounts of data to be processed useful in the configuration of the environment. In this paper we want to address the problem of the weakest point analysis of the modules connection network structure.

In section 2 we define the problem we want to solve. Section 3 contains a detailed description of the analysis we propose and in the last section we summarize the paper.

2 Definition of the Problem

We work on a telemetric system [4] for remote and automatic reading of media consumption meters (Fig. 1). Data from particular meters are transferred via radio to so called collecting nodes (further called *data stream sources*) and then to a stream data warehouse where the data are further processed. A collecting node can be compared to a LAN switch, which only transfers data from one point to another. Receiving data from sources is analogous to transferring data from transactional databases to a data warehouse during an ETL process.

Our goal is to assure the continuity of the ETL process reading data from stream sources, or unless it is possible, to enable recovery of the interrupted processing without any stream data loss, no matter how long a failure lasts. The designed system is a geographically distributed set of modules providing various functionality and communicating with one another.

The system is comprised of a few layers of modules (Fig. 1): stream data sources, remote stream buffers (RBF), remote persistent stream integrators (RIF), ETL modules, module for error detection and stream integration (FTI).

Each data source transmit data to many RBF buffering modules, which communicate with RIF integrating modules offering persistent buffering. At this stage a replicated extraction process appears. Outputs of the extraction process are connected to FTI detecting module. FTI is responsible for not loading the improperly processed data (malformed during processing or transmitting). At the end of the modules chain there is a stream data warehouse and the systems using it. A complete description of the tasks of the particular system layers can be found in [8,9]. In this paper we want to focus on the four first layers only, namely: data sources, remote buffers, remote integrators and ETL modules.

One of the objectives of the RBF and RIF layers is to deliver data from data sources to ETL modules. As the sources may be located far away from ETL modules the data need to be transmitted via intermediate devices such as network routers. The sources are unable to buffer any data itself, it is a result of the hardware limitations, and a tuple sent to ETL modules is no longer remembered in the source. It is assumed that such a tuple will be luckily delivered to the destination. Unfortunately, tuples may be lost during the transmission, the connections may be broken so the tuples should be retransmitted. This cannot be done because these tuples are no longer available in the source module. To avoid data loses we introduced two additional layers to our system [8].

Our aim is to analyse the structure of connections between data sources via RBFs and RIFs to ETL modules and find the weakest points in it. The weakest point is the module in one of the intermediate layers whose failure is the most detrimental to the system. We are interested in a single module failure, and also in longer scenarios of multiple nodes failures. Detection of the weak points lets us to improve the structure, add new connections or new modules if needed.

3   System Analysis

The entry point to the modules connections analysis is the definition of the connections. To define the connections in a universal form we chose a connection matrix built as follows:

Definition 1. *Connection matrix CM_{XY} between two module layers X and Y is a matrix having $|X|$ rows and $|Y|$ columns, where $x_i \in X$ and $y_j \in Y$ and*
$$\forall i \in [1, |X|] \, \forall j \in [1, |Y|] \, p_{i,j} = \begin{cases} 0 \text{ if there is no connection between } x_i \text{ and } y_j \\ 1 \text{ if there exists a connection between } x_i \text{ and } y_j \end{cases}$$

In our system three connections matrices are required to describeconnections between the four module layers: GB matrix for connections between the sources (also called **G**enerators) and RBF modules, BI matrix for RBF-RIF connections, and EI matrix for connections between RIF layer and **E**TL layer. Consider the structure presented in the figure 2. Generators G_1 and G_2 transmit data to three RBFs each. Only the RBF_2 module transmits data to all available RIF integrators, whereas the other RBFs are connected to only two RIFs each.

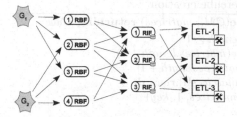

Fig. 2. Example of a modules connections network structure

Finally each RIF is connected with two of three ETL modules. Such a connections network is described by the following connection matrices:

$$CM_{GB} = \begin{bmatrix} 1 & 1 & 1 & 0 \\ 0 & 1 & 1 & 1 \end{bmatrix} \quad CM_{BI} = \begin{bmatrix} 1 & 1 & 0 \\ 1 & 1 & 1 \\ 0 & 1 & 1 \\ 1 & 0 & 1 \end{bmatrix} \quad CM_{IE} = \begin{bmatrix} 1 & 0 & 1 \\ 1 & 1 & 0 \\ 0 & 1 & 1 \end{bmatrix}$$

3.1 Connections Hypercube Creation

To answer the question which node in the RBF or RIF layer is the most critical in terms of the fault-tolerance, we need to express the connections described by CM_{GB}, CM_{BI} and CM_{IE} matrices in a form from which we can easily extract information about complete processing paths. We decided to use an n-dimensional hypercube representation. The cube has as many dimensions as many layers we want to analyse. Continuing our example we have four module layers: G, B, I and E.

Definition 2. *The n-dimensional hypercube $C_{L_1 L_2 \ldots L_n}$ is a representation of an n-dimensional analysis space. Cube cells are addressed as a multidimensional array using the notation $C[l_{1i}, l_{2j}, l_{3k}, \ldots, l_{nz}]$ to denote a path starting in the module $l_{1i} \in L_1$, via modules $l_{2j} \in L_2$ and $l_{3k} \in L_3$ and ended in the module $l_{nz} \in L_n$, where $\forall i \in [1..n] \exists j \in [1..|L_i|]$ such that l_{ij} corresponds to the jth module in the ith module layer.*

Definition 3. *The connection key k_n for n-dimensional hypercube $C_{L_1 L_2 \ldots L_n}$ is an n-elements array of indices such that $\forall i \in [1..n]$ $k_n[i] \in L_i$. The key uniquely identifies a connection path in the cube.*

The hypercube defined above gathers all connection information in a single data structure and makes it easy to analyse. Each cell of the cube can contain one of two values: 1 or 0 to indicate if respectively there exists a connection for the addressed path or not. But before we can start analysis of the cube contents it has to be built basing on the connection matrices. To do it we designed an algorithm which can create a hypercube from any number of the connection matrices (Alg. 1).

Algorithm 1. Hypercube creation

Function buildCube(*CM*[] *matrices*) **returning** *Cube*
Require: *matrices.length* $\geq$ 1
 1: *cube* $\Leftarrow$ **new** Cube(*matrices.length* + 1)
 2: *key* $\Leftarrow$ **new** int[*matrices.length* + 1]
 3: **for all** *key*[1] **in** *matrices*[1].*rows* **do**
 4:     nextDim(*cube, matrices,* 1, *key*)
 5: **end for**
 6: **return** cube
Procedure nextDim (*Cube cube, CM*[] *matrices, int dim, int*[] *key*)
 1: **for all** *key*[*dim* + 1] **in** *matrices*[*dim*].*columns* **do**
 2:     **if** *matrices*[*dim*][*key*[*dim*], *key*[*dim* + 1]] $\neq$ 0 **then**
 3:         **if** *dim* = (*matrices.length*) **then**
 4:             *cube*[*key*] $\Leftarrow$ 1
 5:         **else**
 6:             *nextDim*(*cube, matrices, dim* + 1, *key*)
 7:         **end if**
 8:     **end if**
 9: **end for**

The algorithm input is a sorted array of connection matrices. The algorithm consists of two units: the `buildCube` function and the private recursive procedure `nextDim`. The `buildCube` function creates an empty cube by instantiating an abstract object `Cube` (line 1). The cube has as many dimensions as the layers are covered by the matrices (number of matrices + 1). After initial creation the cube contains no defined connections, all cells contain zeros. There is also created a key object instance as an array of integers representing dimension indices (line 2). Its task is cube cells indexing. Iterating through the connection matrices cells, complete connection paths are discovered and remembered in the hypercube.

The `nextDim` procedure traverses recursively consecutive dimensions. In line 1 we force the value in the `key` array corresponding to the current dimension to be set to all possible columns values of the current matrix. In line 2 it is checked whether there is a connection (value 1) between rows and columns of the connection matrix. If so a recursive call takes place in line 6 or if it is the last dimension, the cube cell indicated by the current value of the key is set to 1 in line 4.

3.2   Module Failure Influence Computation

Having a hypercube with all declared paths defined we can start the analysis of the connections structure. We are interested in finding the weakest points of the structure. As the weakest point we understand the node in any intermediate layer, whose failure (loss) causes the most damage to the structure, and increases the most the probability that the modules network stops working. First we define some basic terms:

Definition 4. *The connection key mask m_n for the cube $C_{L_1 L_2 \ldots L_n}$ is an extension of the key k_n such that $\forall i \in [1..n]\ m_n[i] \in L_i \lor m_n[i] = *$, where $*$ is a wildcard denoting any value.*

Definition 5. *The connection key set K_n for the cube $C_{L_1 L_2 \ldots L_n}$ is a set of single k_n keys.*

Definition 6. *The cube intersection $\cap_c$ is a two-argument operator selecting from the cube $C_{L_1 L_2 \ldots L_n}$ a set of connection keys K_n defined by a given connection key mask m_n.*
$$K_n = C_{L_1 L_2 \ldots L_n} \cap_c m_n \Rightarrow \forall k_n \in K_n\ \forall i \in [1..n]\ k_n[i] = m_n[i] \lor m_n[i] = *$$

The first step in the cube analysis is computation of some fault-tolerance properties for each module from the intermediate layers:

- $ftLostPaths_{i,j}$ - the number of paths to be broken if the jth module from the ith layer fails
- $ftLostRatio_{i,j}$ - percentage of all paths to be broken

The properties express what is the influence of a given node on the overall condition of the system. They are computed as follows:

```
1:  Mask m ⇐ cube.EmptyMask
2:  totalConnections ⇐ |cube ∩c m|
3:  for all dim in cube.dimensions \ {1, n} do
4:      for all node in cube.dimensions[dim] do
5:          Mask nodeMask ⇐ m
6:          nodeMask[dim] ⇐ node
7:          ftLostPath_dim,node ⇐ |cube ∩c nodeMask|
8:          ftLostRatio_dim,node ⇐ ftLostPath_dim,node ÷ totalConnections
9:      end for
10: end for
```

In line 1 an empty mask is declared using predefined constant $EmptyMask$ of a *cube* object. It contains "$*$" on each index position and covers all the connections declared in the cube. Using the mask in line 2 the total number of defined connections is computed by counting the connection keys which fit the mask. Then the iteration over the intermediate layers starts. The *cube.dimensions* expression is a set of all indices of all dimensions in the cube. For an n-dimensional hypercube it is a set of values from 1 to n. As can be noticed in line 3, the first (1) and the last (n) layers are omitted. The second loop in line 4 iterates over all modules in each selected layer. For each module the mask of connections affected by the module is prepared (lines 5 and 6). The number of paths that passes the module is the number of paths to be lost if the module fails (line 7). In line 8 the computation of lost paths ratio takes place.

Unfortunately basing only on these properties we cannot point which nodes are critical. It may happen that removing node A breaks only 1 path, but it is the only one between two modules, and removing node B breaks 10 of 50 paths. It forces us to compute the influence of each node on the connection paths affected by the node failure:

- $ftpiAllPaths_{s,d}$ - total number of paths between the sth module in the sources layer and the dth module in the destination (ETLs) layer
- $ftpiNodeLostPaths_{s,d,i,j}$ - number of paths described above passing the jth module from the ith layer
- $ftpiNodeLostRatio_{s,d,i,j}$ - percentage of lost paths
- $ftpiRemainingNodesInLayer_{s,d,i,j}$ - number of remaining redundant nodes

The computation goes as follows:

```
 1: for all dim in cube.dimensions \ {1, n} do
 2:     for all node in cube.dimensions[dim] do
 3:         for all src in cube.dimensions[1] do
 4:             for all dest in cube.dimensions[n] do
 5:                 Mask mask ⇐ cube.EmptyMask
 6:                 mask[1] ⇐ src
 7:                 mask[n] ⇐ dest
 8:                 mask[dim] ⇐ node
 9:                 ftpiNodeLostPaths_{src,dest,dim,node} ⇐ |cube ∩_c mask|
10:                 if ftpiNodeLostPaths_{src,dest,dim,node} > 0 then
11:                     Mask nodeMask ⇐ mask
12:                     nodeMask[dim] ⇐ *
13:                     ftpiAllPaths_{src,dest} ⇐ |cube ∩_c nodeMask|
14:                     ftpiNodeLostRatio_{src,dest,dim,node} ⇐
15:                         ftpiNodeLostPaths_{src,dest,dim,node} ÷ ftpiAllPaths_{src,dest}
16:                     ftpiRemainingNodesInLayer_{s,d,i,j} ⇐
17:                         countRemainingNodes(cube, mask, nodeMask)
18:                 end if
19:             end for
20:         end for
21:     end for
22: end for
```

In lines 1-2 there is an iteration through all nodes in the intermediate layers. Lines 3-4 generate all combinations of transfers from the source layer to the destination layer. It is examining all possible transfers via particular nodes. First, in lines 6-9 we compute how many connections goes from the selected source to the selected destination module via the current node. If there is no connections, there is nothing to analyse. Otherwise, we compute how many connections are going via any intermediate nodes (lines 11-13), and the lost connections ratio is calculated. In lines 16-17 we compute how many nodes connecting current source and destination remain in the current layer after a failure of the current node.

Now we can compute how many paths were broken ($brokenPaths_{i,j}$) by counting the paths having $ftpiRemainingNodesInLayer = 0$ and how many paths became risky ($riskyPaths_{i,j}$) by counting paths with $ftpiRemainingNodes InLayer = 1$. Having computed all these properties we can start analysis of the data we obtained. To find the most critical module we sort the computed values of the system nodes in descending order: $brokenPaths$, $riskyPaths$, $ftLostRatio$.

4   Conclusions

In this paper we described a part of our research on the processing of data streams in geographically distributed environment. To avoid data loss in the case of transmission failures we proposed to use a network of modules offering various functionality, such as simple low-cost buffering modules, persistent stream integrators, ETL machines, and an error detection module. The proper modules configuration assures that even in case of failures of a few network components, the data transmission and processing will not be affected. To obtain high availability the connections network must be appropriately designed. Presented analysis method supports the designer of the connections network. It detects the weakest points in the structure and gives the chance to improve the structure prior to failures causing processing interruption. In future we plan to continue research on the presented system and extend the algorithm. Our plan is to create an algorithm which can automatically modify the network and eliminate its weaknesses.

References

1. Arasu, A., Babcock, B., Babu, S., Datar, M., Ito, K., Motwani, R., Nishizawa, I., Srivastava, U., Thomas, D., Varma, R., Widom, J.: Stream: The stanford stream data manager. IEEE Data Eng. Bull. 26(1), 19–26 (2003)
2. Balazinska, M., Balakrishnan, H., Madden, S., Stonebraker, M.: Fault-Tolerance in the Borealis Distributed Stream Processing System. In: ACM SIGMOD Conf., Baltimore, MD (2005)
3. Madden, S., Franklin, M.J.: Fjording the stream: An architecture for queries over streaming sensor data. In: ICDE, pp. 555–566. IEEE Computer Society, Los Alamitos (2002)
4. Gorawski, M., Malczok, R.: Distributed spatial data warehouse indexed with virtual memory aggregation tree. In: Sander, J., Nascimento, M.A. (eds.) STDBM, pp. 25–32 (2004)
5. Gorawski, M., Marks, P.: High efficiency of hybrid resumption in distributed data warehouses. In: DEXA Workshops, pp. 323–327. IEEE Computer Society, Los Alamitos (2005)
6. Gorawski, M., Marks, P.: Checkpoint-based resumption in data warehouses. In: Socha, K. (ed.) IFIP International Federation for Information Processing, Warsaw. Software Engineering Techniques: Design for Quality, vol. 227, pp. 313–323 (2006)
7. Labio, W., Wiener, J.L., Garcia-Molina, H., Gorelik, V.: Efficient resumption of interrupted warehouse loads. In: Chen, W., Naughton, J.F., Bernstein, P.A. (eds.) SIGMOD Conference, pp. 46–57. ACM, New York (2000)
8. Gorawski, M., Marks, P.: Fault-tolerant distributed stream processing system. In: DEXA Workshops, pp. 395–399. IEEE Computer Society, Los Alamitos (2006)
9. Gorawski, M., Marks, P.: Towards reliability and fault-tolerance of distributed stream processing system. In: DepCoS-RELCOMEX, pp. 246–253. IEEE Computer Society, Los Alamitos (2007)
10. Gorawski, M., Marks, P.: Distributed stream processing analysis in high availability context. In: ARES 2007: Proceedings of the The Second International Conference on Availability, Reliability and Security, Washington, DC, USA, pp. 61–68. IEEE Computer Society Press, Los Alamitos (2007)

RAIN: Always on Data Warehousing

Jorge Vieira[1], Marco Vieira[2], Marco Costa[1], and Henrique Madeira[2]

[1] Critical Software SA
Coimbra, Portugal
{jvieira,mcosta}@criticalsoftware.com
[2] CISUC, Department of Informatics Engineering, University of Coimbra
Coimbra, Portugal
{mvieira,henrique}@dei.uc.pt

Abstract. The Redundant Arrays of Inexpensive DWS Nodes (RAIN) technique is a node-level data replication approach that introduces failover capabilities to DWS (Data Warehouse Striping) clusters. RAIN is based on the selective replication of fact tables' data across the cluster nodes and endows DWS clusters with the capability of providing query answers even when one or more nodes are unavailable. Two distinct replication modes are supported: simple redundancy (RAIN-0) and stripped redundancy (RAIN-S). In this demo we are going to show a DWS cluster using the RAIN technique, focusing on the execution of queries in the presence of nodes failures and on the process of recovering failed nodes.

Keywords: Data warehousing, redundancy, replication, recovery, availability.

1 Introduction

The success of organizations depends more and more on the information they have and, consequently, on the systems used to store and manage that information. In fact, markets globalization, with the consequent increase of the competitiveness, lead organizations to regard information as one of their most valuable resources.

The availability of tailored information to help decision makers during decision support processes is of utmost importance. Data warehouses (DW) are becoming one of the main assets for enterprise information analysis and manipulation [3]. Enterprises continuously create huge amounts of operational data that is typically integrated in a centralized data warehouse.

A data warehouse can store data ranging from hundreds of Gigabytes to the dozens of Terabytes [2]. Obviously, this requires large and highly-available storage devices and the capability of accessing and processing the data in due time. A low response time for the decision support queries issued by the users is of utmost importance.

DWS is a low-cost approach that distributes the data of a data warehouse by a cluster of computers, providing near linear speedup and scale up as new nodes are added to the cluster [1]. However, adding nodes to the cluster also increases the probability of node failure, which in turn leads to data availability problems.

J.R. Haritsa, R. Kotagiri, and V. Pudi (Eds.): DASFAA 2008, LNCS 4947, pp. 678–681, 2008.
© Springer-Verlag Berlin Heidelberg 2008

In [5] we propose an approach that provides DWS clusters with high-availability even in the presence of node failures. The RAIN (Redundant Array of Inexpensive Nodes) technique is based on the selective data replication over a cluster of low-cost nodes and includes two types of replication: simple redundancy (RAIN-0) and striped redundancy (RAIN-S). RAIN-0 consists of replicating the facts data from each node in other nodes of the cluster. In RAIN-S facts data from each node is randomly distributed in N-1 sub-partitions (where N is the number of nodes in the cluster) and each sub-partition is replicated in at least one of the other nodes. See [5] for more details on RAIN.

The goal of this demo is to show the always on data warehousing capabilities of DWS clusters based on the RAIN approach. During the demo we will exhibit the advanced potential of DWS and RAIN, namely in what concerns to: high-performance using low cost hardware and software; always on data warehousing even in the presence of node failures; very easy and fast node recovery; and non-stop data loading.

2 The Demo

Fig. 1 depicts the setup that will be used for the demo. The basic platform consists of a small cluster using four heterogeneous machines and running PostgreSQL 8.2 database engine over the Debian Linux Etch operating system. The machines are connected using a dedicated Fast Ethernet Network.

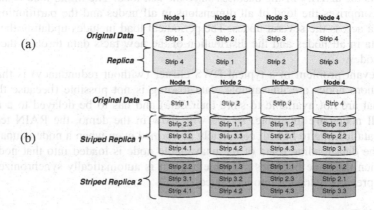

Fig. 1. Cluster setup for the demo. Two configurations will be used: a) a RAIN-0 cluster designed to tolerate the failure of one node (RAIN 0(4, 1)) and b) a RAIN-S cluster configured to tolerate the failure of two nodes (RAIN S(4, 2)).

The newly TPC-DS™ Benchmark (TPC-DS) [4] is used as case study. TPC-DS is a performance benchmark for decision support systems. This benchmark evaluates the essential features of decision support systems, including queries execution and data load. In the demo we use a small size database with 10GB (scale factor 10 in TPC-DS).

2.1 Queries Execution

Query execution in DWS is enabled through the use of a middleware that allow client applications (e.g., Oracle Discoverer, JPivot, Crystal Reports) to connect to the system without knowing the cluster implementation details. The DWS middleware was adapted to the RAIN technique, which means that it is prepared to transparently allow the system to continue providing exact queries answers when nodes fail.

When a node failure is detected in a RAIN-0 configuration the DWS middleware automatically redirects the corresponding query to one of the nodes that contains the replica of the failed node data.

In a RAIN-S configuration, when a node fails the DWS middleware automatically forwards a set of queries (equivalent to the query that should be sent to the failed node) to all other nodes. If Y nodes fail at the same time in a configuration able to tolerate Y node failures or more the DWS middleware always distributes the queries in such way that the final nodes workload is as uniform as possible.

As mentioned before, both RAIN-0 and RAIN-S cluster operation will be shown in the demo, including operation in the absence of node failures and operation in the presence of one or two node failures.

2.2 Cluster Data Loading

In typical data warehousing systems data loading occurs in two different moments: 1) initial load when the system starts to be used; 2) periodical load of the new data that represents the activity of the business since the last load. The initial load of the DWS system comprises the load of all dimensions in all nodes and the partitioning of the facts data across the several nodes. The periodical load includes updating the dimensions data in all nodes and the distribution of the new facts data through the several cluster nodes.

A relevant problem in a typical DWS cluster (without redundancy) is that when one or more nodes are unavailable, data loading is not possible (because there are nodes that are not available to store their data) and has to be delayed to a moment where all nodes are available. As we will show in the demo, the RAIN technique allows data loading to be still completely performed even when a node is unavailable during the data loading (data of the unavailable node is loaded into that node replicas). When the node becomes available again it is automatically synchronized using the data previously loaded into its replicas.

2.3 Node Recovery

Node recovery is another important aspect that will be demonstrated. In fact, in a typical DWS cluster node recovery is not possible when stored data gets corrupt. The only way to recover a node is to recollect the data from the operational sources and rebuild the entire cluster. The use of RAIN eases the node recovery, as the recovery process can be accomplished using the data existing in the failed node replicas, without need of cluster downtime. After a node failure, two kinds of recovery can be needed: complete recovery or partial recovery. Complete recovery is needed when the node failure results in lost of node data. Partial recovery is required when the node still contains all its data, with exception to data loaded during the node unavailability.

Both complete recovery and partial recovery will be demonstrated. When partial recovery is performed the missing data is copied to the node from its replicas. When using RAIN-S the impact on other nodes is very small, as only a small amount of data has to be copied from each cluster node. When using RAIN-0, the impact is slightly high (all data is obtained from a single node), but still not very high, as the data to be copied corresponds to only a small period of time.

A complete recovery can be performed using the same approach used for the partial recovery, as all the data needed for a given node is distributed among the other nodes. However this has a higher impact in the performance as more data has to be loaded.

An important aspect we will show is that during node recovery the system continues to process query requests by using the replicas to process them. Nodes being recovered are kept offline until all its data is updated. Note that, node recovery is performed while the cluster is idle (or with minimum load) which minimizes the overhead caused by nodes recovery in the queries execution. In fact, our mechanism is able to pause or slowdown the recovery process when queries are being executed.

3 Conclusion

In this demo we present the Redundant Arrays of Inexpensive DWS Nodes (RAIN) technique, which is a node-level data replication approach that introduces failover capabilities to DWS (Data Warehouse Striping) clusters. This technique endows DWS clusters with the capability of providing exact queries answers even in the presence of node failures. The advanced capabilities of DWS and RAIN are demonstrated using a 4 nodes cluster designed to tolerate failures of one and two nodes. The goal of the demo is to show that high-performance is possible even using low cost hardware and software; always on data warehousing is feasible even in the presence of node failures; node recovery is very easy and fast; and non-stop data loading is achievable. This guarantees high data availability, a key characteristic for future data warehouses.

References

1. Bernardino, J., Madeira, H.: A New Technique to Speedup Queries in Data Warehousing. In: ABDIS-DASFA, Symp. on Advances in DB and Information Systems, Prague (2001)
2. IDC, Survey-Based Segmentation of the Market by Data Warehouse Size and Number of Data Sources (2004)
3. Kimball, R., Ross, M.: The Data Warehouse Toolkit: The Complete Guide to Dimensional Modeling, 2nd edn. J. Wiley & Sons, Inc., Chichester (2002)
4. Transaction Processing Performance Council, TPC BenchmarkTM DS (Decision Support) Standard Specification, Draft Version 32 (2007), available at, http://www.tpc.org/tpcds/
5. Vieira, J., Vieira, M., Costa, M., Madeira, H.: Redundant Array of Inexpensive Nodes for DWS. In: Haritsa, et al. (eds.) DASFAA 2008. LNCS, vol. 4947, pp. 580–587, Springer, Heidelberg (to appear, 2008)

Data Compression for Incremental Data Cube Maintenance

Tatsuo Tsuji, Dong Jin, and Ken Higuchi

Graduate School of Engineering, University of Fukui
{tsuji,jindong,higuchi}@pear.fuis.fukui-u.ac.jp

Abstract. We have proposed an incremental maintenance scheme of data cubes employing extendible multidimensional array model. Such an array enables incremental cube maintenance without relocating any data dumped at an earlier time, while computing the data cube efficiently by utilizing the fast random accessing capability of arrays. But in practice, most multidimensional arrays for data cube are large but sparse. In this paper, we describe a data compression scheme for our proposed cube maintenance method, and demonstrate the physical refreshing algorithm working on the data structure thus compressed.

1 Introduction

In [1] we have presented an incremental maintenance scheme of MOLAP data cubes. It allows efficient construction of a data cube from a source relation in the frontend OLTP database. Here, the incremental maintenance of a data cube means the propagation of only its changes[3]. When the amount of changes during the specified time period are much smaller than the size of the source relation, computing only the changes of the source relation and reflecting them to the original data cube is usually much cheaper than recomputing from scratch. We have employed *extendible multidimensional arrays* to realize the capability of incremental data cube construction. An extendible multidimensional array enables incremental cube maintenance without relocating any data dumped at an earlier time even if array size is extended along any dimension, while computing the data cube efficiently by utilizing the fast random accessing capability of arrays.

But in practice, it is common that most multidimensional arrays for data cube are large but sparse. Multidimensional arrays are good containers to store dense data, but for sparse data cubes huge memory will be wasted because a large number of array cells are empty and thus are very hard to use in actual implementation. In particular, the sparseness problem becomes serious for *delta data cube* [1] whose logical array size can be the same as that of the original data cube, but usually much more sparse than that.

HOMD (*History Offset implementation scheme for Multidimensional Datasets*) model presented in [2] seems to be one of the efficient storage schemes to store such sparse data. It employs extendible multidimensional arrays as its underlying logical data structure. In this paper, we will describe a data compression scheme for our proposed incremental data cube maintenance strategy. Also, we will demonstrate the subarray-based algorithm in [1] working on the data structure thus compressed.

J.R. Haritsa, R. Kotagiri, and V. Pudi (Eds.): DASFAA 2008, LNCS 4947, pp. 682–685, 2008.

2 Data Compression Scheme

The HOMD model is based on the extendible array explained in Section 2 in [1]. Each dimension of a data cube corresponds to a dimension of the extendible array and each dimension value of the data cube is uniquely mapped to a subscript value of the array dimension. A subarray is constructed for each distinct dimension value. Fig. 1 shows the HOMD implementation of the two dimensional data cube in Fig. 3 in [1]. The readers should consult Section 3 in [1] for more details of our incremental maintenance scheme.

For an n-dimensional data cube Q, the corresponding logical structure of HOMD is the pair (M, A). A is an n dimensional extendible array created for Q and M is a set of mappings. Each m_i $(1 \leq i \leq n)$ in M maps i-th dimension values of Q to subscript values of the dimension i of A. A will be often called as a logical extendible array.

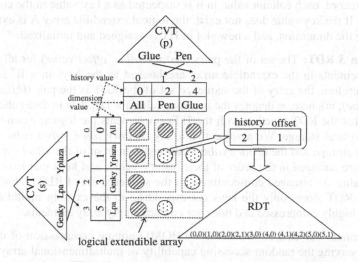

Fig. 1. Physical structure for a data cube implemented by HOMD

Each element of an n dimensional extendible array can be specified by its n dimensional coordinate. In HOMD model, we have directed our attention to that each element can be specified by using the pair of history value and offset value. Note that since each history value h is unique and has a one-to-one correspondence with its corresponding subarray S, S is specified uniquely by h. Moreover, the offset value of each element in S can be computed as in Section 2 in [1] and this is also unique in the subarray. Therefore each element of an n dimensional extendible array can be referenced by the pair (*history value, offsetvalue*). In the HOMD logical structure (M, A), each mapping m_i in M is implemented using a single B^+ tree called CVT (key subscript ConVersion Tree), and the logical extendible array A is implemented using a single B^+ tree called RDT (Real Data Tree) and n HOMD tables, each of which is an extension of the three auxiliary tables of an extendible array.

Definition 1 CVT: CVT_k for the k-th dimension of an n dimensional data cube is defined as a structure of B^+ tree with each distinct dimension value v as a key value and its associated data value is subscript i of the k-th dimension of the logical extendible array A. So the entry of the sequence set of the B^+ tree is the pair (v, i). i references to the corresponding entry of the HOMD table in the next definition.

Note that the special value *All* is unnecessary to be mapped in CVT as it is always the first subscript value 0 in each array dimension.

Definition 2 HT: HT (HOMD Table) corresponds to the auxiliary tables explained in Section 2 in [1]. It includes the history table and the coefficient table. Note that the address table can be void in our HOMD physical implementation.

HT is arranged according to the insertion order. For example, the dimension value "Genky" is mapped to the subscript 3 as the insertion order, though in the sequence set of CVT, the key "Genky" is in position 1 due to the property of B^+ tree. At insertion of a record, each column value in it is inspected as a key value in the corresponding CVT. If the key value does not exist, the logical extendible array A is extended by one along the dimension, and a new slot in HT is assigned and initialized.

Definition 3 RDT: The set of the pairs (*history value, offset value*) for all of the effective elements in the extendible array are housed as the keys in a B^+ tree called RDT. Therefore, the entry of the sequence set of the RDT is the pair ((*history value, offset value*), m), here, m denotes the measure value for fact data in data cube.

Note that the RDT together with the HTs implements the logical extendible array on the physical storage. We assume that a key (*history value, offset value*) occupies fixed size storage and the *history value* is arranged in front of the *offset value*. Hence the keys are arranged in the order of their history values and keys that have the same history value are arranged consecutively in the sequence set of RDT. Note also that since the RDT stores only the keys corresponding to the existing multidimensional data, it is highly compressed and does not contain empty array elements.

We implement our data cube scheme by HOMD aiming compression of data cubes while preserving the random accessing capability of multidimensional arrays. For an n dimensional data cube in our data cube scheme, its HOMD implementation is the set of n CVTs, n HTs and RDT. For our shared dimension method in [1], HOMD implementation of original data cube Q and delta cube ΔQ share one set of n CVTs and n HTs, while each data cube has independent RDT to store fact data; RDT for Q and ΔRDT for ΔQ.

3   Refreshing of Data Cubes

We implement subarray-based method proposed in [1] by HOMD. We will use a B^+ tree called IRT(Intermediate Result Tree) in main memory to contain the intermediate result for the dependent cells instead of n intermediate result arrays. The data structure of IRT is the same as RDT. Note that we can easily delete those intermediate results which are not needed in later computation by IRT. In Fig. 2, we will describe the physical refreshing algorithm for subarray-based method corresponding to the logical one described in [1].

Inputs:
ΔRDT for the delta cube ΔQ (only include base cuboid);
RDT for the original data cube Q ;
Output:
The updated RDT for data cube Q ;
Method:
Initialize IRT on main memory to be empty;
For each history value h from max history value in ΔRDT to 1 **do**
   let d = the extended dimension on history value h;
   For each element e with history value h in IRT
      insert or update corresponding element e_d in IRT by aggregating with
                                                  value of e along dimension d;
   End for
   load all the elements with history value h in ΔRDT into main memory;
   For each element e with history value h in ΔRDT **do**
      insert or update corresponding element e_d in IRT by aggregating
                                       with value of e along dimension d;
   End for
   get all the dependent cells of S[h] in ΔQ by aggregating with all the elements
                                       with history h in ΔRDT & IRT;
   delete all the elements with history value h in IRT;
   Refresh RDT with the result calculated for S[h] in ΔQ in main memory;
End for
Refresh (0,0) in RDT by aggregating with (0,0) in IRT

Fig. 2. Physical refreshing algorithm for subarray-based method

4   Conclusion

We have described a data compression scheme for incremental maintenance of data cubes without sacrificing the random accessing capability of arrays, and demonstrated the physical refreshing algorithm working on the data structure thus compressed.

References

1. Jin, D., Tsuji, T., Tsuchida, M., Higuchi, K.: An Incremental Maintenance Scheme of Data Cubes. In: Haritsa, et al. (eds.) DASFAA 2008. LNCS, vol. 4947. Springer, Heidelberg (to appear, 2008)
2. Hasan, K.M.A., Kuroda, M., Azuma, N., Tsuji, T., Higuchi, K.: An Extendible Array Based Implementation of Relational Tables for Multidimensional Databases. In: Tjoa, A.M., Trujillo, J. (eds.) DaWaK 2005. LNCS, vol. 3589, pp. 233–242. Springer, Heidelberg (2005)
3. Lee, K.Y., Kim, M.H.: Efficient Incremental Maintenance of Data Cubes. In: Proc. of VLDB, pp. 823–833 (2006)

A Bilingual Dictionary Extracted from the Wikipedia Link Structure

Maike Erdmann, Kotaro Nakayama, Takahiro Hara, and Shojiro Nishio

Dept. of Multimedia Engineering,
Graduate School of Information Science and Technology,
Osaka University
1-5 Yamadaoka, Suita, Osaka 565-0871, Japan
{erdmann.maike, nakayama.kotaro, hara, nishio}@ist.osaka-u.ac.jp

Abstract. A lot of bilingual dictionaries have been released on the WWW. However, these dictionaries insufficiently cover new and domain-specific terminology. In our demonstration, we present a dictionary constructed by analyzing the link structure of Wikipedia, a huge scale encyclopedia containing a large amount of links between articles in different languages. We analyzed not only these interlanguage links but extracted even more translation candidates from redirect page and link text information. In an experiment, we already proved the advantages of our dictionary compared to manually created dictionaries as well as to extracting bilingual terminology from parallel corpora.

1 Introduction

Bilingual dictionaries are required in many research areas, for instance to enhance existing dictionaries with technical terms, as seed dictionaries to improve machine translation results, in cross-language information retrieval or for second language teaching and learning. Unfortunately, the manual creation of bilingual dictionaries is not efficient since linguistic knowledge is expensive and new or highly specialized domain specific terms are difficult to cover.

In recent years, a lot of research has been conducted on the automatic extraction of bilingual dictionaries. Dictionary extraction from large amounts of bilingual text corpora is an emerging research area. For English-Japanese dictionary extraction, e.g. corpora of paper abstracts [1] or software documentations [2] have been exploited. However, that approach faces several issues. Particularly, for very different languages such as English and Japanese, and for domains where sufficiently large text corpora are not available, accuracy and coverage of extracted bilingual terminology is rather low.

Therefore, in order to provide a high accuracy and high coverage dictionary, we extracted bilingual terminology from Wikipedia. Wikipedia is a very promising resource since the continuously growing encyclopedia already contains more than 5 million articles in several hundred languages and a broad variety of topics. We already proved that because of its dense link structure Wikipedia can be used to create an accurate association thesaurus [3][4].

J.R. Haritsa, R. Kotagiri, and V. Pudi (Eds.): DASFAA 2008, LNCS 4947, pp. 686–689, 2008.

In this demonstration, we present an English-Japanese dictionary which we constructed from our proposed methods. The dictionary has a Web-based interface and can be accessed freely.

2   Proposed Methods

We extracted bilingual terminology by analyzing the link structure of Wikipedia. Wikipedia contains a lot of links between articles, not only within the articles of the same language but also between articles of different languages. As opposed to the plain text in bilingual corpora, Wikipedia links contain to some extent semantic information. For instance, an interlanguage link usually indicates that one page title is the translation of the other. This can decrease difficulties of dictionary creation caused by natural language processing issues.

On the other hand, an article in the source language has usually at most one interlanguage link to an article in the target language. Thus, in order to extend the coverage of our dictionary, we analyzed other kinds of link information as well. In the following, we briefly introduce our extraction methods. They are explained more comprehensively in [5].

Interlanguage Link Method
An interlanguage link in Wikipedia is a link between two articles in different languages. We assume that in most cases, the titles of two articles connected by an interlanguage link are translations of each other. Therefore, we analyze all interlanguage links in Wikipedia to create a baseline dictionary.

Redirect Page Method
A redirect page in Wikipedia contains no content but a link to another article (target page) in order to facilitate the access to Wikipedia content. Redirect pages are usually strongly related to the concept of the target page, thus we can enhance the number of translation candidates by the titles of all redirect pages. Furthermore, we assign a score to all extracted translation candidates and filter unsuitable terms through a threshold.

Link Text Method
A link text is the text part of a link, i.e. the text that is presented to the user in the browser where he clicks to reach the target page. We already realized that link texts are usually strongly related to the target page title [3][4]. Therefore, we can enhance our baseline dictionary with the link texts of all backward links of a page. After that, we filter unsuitable terms by calculating a score and setting a threshold.

Redirect Page and Link Text Method
At last, we combine the RP (redirect page) method with the LT (link text) method. The score of each translation candidate thus becomes the weighted sum of the RP method score and the LT method score.

Translation	Score
欧州連合	0.704818
EU	0.136269787
ヨーロッパ連合	0.04821555
EU	0.046526315
歴史	0.008258775
共通外交・安全保障政策	0.008258775
機構	0.008258775
3つの柱	0.008258775
欧 州 連 合	0.007570544
欧州連合 (EU)	0.002064694
EU圏	0.001376462
EU(欧州連合)	0.001376462
EU(ヨーロッパ連合)	0.001376462
欧州連合(EU)	0.000688231
欧州連合#公用語	0.000688231
欧州	0.000688231
ヨーロッパ連合(EU)	0.000688231
ヨーロッパ統合	0.000688231
EU諸国	0.000688231
EU圏内	0.000688231

[European Union]:

Fig. 1. Wikipedia Bilingual Dictionary

3   Implementation

We downloaded the English and Japanese Wikipedia database dump data from
November/December 2006 [6] containing 3,068,118 English and 455,524 Japanese
pages (including redirect pages). From that data, we extracted all interlanguage
links, link texts and redirect pages as well as the number of backward links
for each page. In total, we extracted 103,374 interlanguage links from English
to Japanese, 108,086 interlanguage links from Japanese to English, 1,345,318
English and 91,898 Japanese redirect pages, 7,215,301 different English and
2,019,874 different Japanese link texts.

From that data, we created the English-Japanese dictionary. All translations
are being created in real-time from our extracted data stored in MySQL 5.0
tables. The tables are equipped with indices in order to accelerate the analysis
of large amounts of data.

Figure 1 shows a screen shot of the developed system with translation candidates for the term "EU". First, using redirect page information, the system
detects that "EU" stands for "European Union." In the second step, translation

candidates for "European Union" are extracted from interlanguage link, redirect page and link text information. Some of the listed translation candidates have different meanings such as "history", "organization" or "Europe." However, since their scores are lower than those of the correct translation candidates, they can easily be filtered. Our dictionary can be accessed under the following URL.

http://wikipedia-lab.org:8080/WikipediaBilingualDictionary

We already conducted an experiment in which we compared the translations of 200 terms extracted by our methods to the translations extracted from a parallel corpus. We used the two standard criteria precision and recall to compare accuracy and coverage of our methods and the parallel corpus approach. The experiment confirmed our conviction that Wikipedia is an invaluable resource for bilingual dictionary extraction and that redirect pages and link texts are helpful to enhance a dictionary constructed from interlanguage links. A detailed description of the experiment is given in [5].

We believe that Wikipedia will become much more comprehensive in near future which will result in an even better coverage. For general terms it is also promising to integrate manually constructed dictionaries. For context-sensitive translations (e.g. machine translation), we can benefit from combining our approach with the parallel corpus approach.

Acknowledgment

This research was supported in part by Grant-in-Aid on Priority Areas (18049050), and by the Microsoft Research IJARC CORE project.

References

1. Tsuji, K., Kageura, K.: Automatic generation of japanese-english bilingual thesauri based on bilingual corpora. Journal of the American Society for Information Science and Technology 57(7), 891–906 (2006)
2. Fung, P., McKeown, K.: A technical word- and term-translation aid using noisy parallel corpora across language groups. Machine Translation 12(1-2), 53–87 (1997)
3. Nakayama, K., Hara, T., Nishio, S.: A thesaurus construction method from large scale web dictionaries. In: Proc. of IEEE International Conference on Advanced Information Networking and Applications (AINA 2007), pp. 932–939 (2007)
4. Nakayama, K., Hara, T., Nishio, S.: Wikipedia mining for an association web thesaurus construction. In: Benatallah, B., Casati, F., Georgakopoulos, D., Bartolini, C., Sadiq, W., Godart, C. (eds.) WISE 2007. LNCS, vol. 4831, Springer, Heidelberg (2007)
5. Erdmann, M., Nakayama, K., Hara, T., Nishio, S.: An approach for extracting bilingual terminology from wikipedia. In: Haritsa, et al. (eds.) DASFAA 2008. LNCS, vol. 4947. Springer, Heidelberg (to appear, 2008)
6. Wikimedia Foundation: Wikimedia downloads, http://download.wikimedia.org/

A Search Engine for Browsing the Wikipedia Thesaurus

Kotaro Nakayama, Takahiro Hara, and Shojiro Nishio

Dept. of Multimedia Eng., Graduate School of Information Science and Technology
Osaka University, 1-5 Yamadaoka, Suita, Osaka 565-0871, Japan
TEL.: +81-6-6879-4513 FAX: +81-6-6879-4514
{nakayama.kotaro,hara,nishio}@ist.osaka-u.ac.jp

Abstract. Wikipedia has become a huge phenomenon on the WWW. As
a corpus for knowledge extraction, it has various impressive characteristics such as a huge amount of articles, live updates, a dense link structure,
brief link texts and URL identification for concepts. In our previous work,
we proposed link structure mining algorithms to extract a huge scale and
accurate association thesaurus from Wikipedia. The association thesaurus
covers almost 1.3 million concepts and the significant accuracy is proved in
detailed experiments. To prove its practicality, we implemented three features on the association thesaurus; a search engine for browsing Wikipedia
Thesaurus, an XML Web service for the thesaurus and a Semantic Web
support feature. We show these features in this demonstration.

1 Introduction

A thesaurus is a kind of dictionary that defines semantic relatedness among
words. Although the effectiveness is widely proved in various research areas,
automated thesaurus dictionary construction (esp. machine-understandable) is
one of the difficult issues. Since it is difficult to maintain huge scale thesauri,
they do not support new concepts in most cases. Therefore, a large number of
studies have been made on automated thesaurus construction based on NLP.
However, issues due to the complexity of natural language, for instance the
ambiguous/synonym term problems still remain on NLP.

We noticed that Wikipedia, a collaborative wiki-based encyclopedia, is a
promising corpus for thesaurus construction. According to statistics of Nature,
Wikipedia is about as accurate in covering scientific topics as the Encyclopedia
Britannica. It covers concepts of various fields such as Arts, Geography, History,
Science, Sports and Games. It contains more than 2 million articles (Dec. 2007,
English only) and it is becoming larger day by day. Because of the huge scale
concept network with a wide-range topic coverage, it is natural that Wikipedia
can be used as a knowledge extraction corpus. In fact, we already proved that it
can be used for accurate association thesaurus construction[1,2]. In this demonstration, we describe the overview of our thesaurus construction method and
show three features which we developed for the thesaurus; a search engine for
browsing Wikipedia Thesaurus, an XML Web service for the thesaurus and a
Semantic Web (RDF) support feature.

J.R. Haritsa, R. Kotagiri, and V. Pudi (Eds.): DASFAA 2008, LNCS 4947, pp. 690–693, 2008.
© Springer-Verlag Berlin Heidelberg 2008

2   pfibf

pfibf (Path Frequency - Inversed Backward link Frequency), an association thesaurus construction method we proposed, is a link structure mining method which is optimized for Wikipedia mining. The relativity between any pair of articles (v_i, v_j) is assumed to be strongly affected by the following two factors:

- the number of paths from article v_i to v_j,
- the length of each path from article v_i to v_j.

The relativity is strong if there are many paths (sharing of many intermediate articles) between two articles. In addition, the relativity is affected by the path length. In other words, if the articles are placed closely together in the concept graph and sharing hyperlinks to articles, the relativity is estimated to be higher than further ones.

The number of backward links on articles is also estimated as a factor of relativity because general/popular articles have a lot of backward links and these articles easily have high relativity to many articles. Therefore, we must consider the inversed backward link frequency *ibf* in addition to the two factors above. The relativity becomes stronger if there are many paths (sharing many intermediate articles) between them. In addition, the relativity becomes stronger according to the path length.

Therefore, if all paths from v_i to v_j are given as $T = \{t_1, t_2, ..., t_n\}$, the relativity *pf* (path frequency) between them can be expressed as follows:

$$pfibf(v_i, v_j) = pf(v_i, v_j) \cdot ibf(v_j), \tag{1}$$

$$pf(v_i, v_j) = \sum_{k=1}^{n} \frac{1}{d(|t_k|)}, \tag{2}$$

$$ibf(v_j) = \log \frac{N}{bf(v_j)}. \tag{3}$$

$d()$ denotes a function which increases the value according to the length of path t_k. A monotonically increasing function such as a logarithm function can be used for $d()$. N denotes the total number of articles and $bf(v_j)$ denotes the number of backward links of v_j. This means a page which shares hyperlinks with a specific page but not with other pages, has a high *pfibf*.

However, counting all paths between all pairs of articles in a huge graph is a computational resource consuming work. Therefore, we proposed an efficient data structure named "Dual binary tree" (DBT) and a multiplication algorithm for the DBT[2].

3   Architecture and Applications

We constructed a huge scale association thesaurus named Wikipedia Thesaurus by using *pfibf* described above. We used the Wikipedia dump data created in Sept. 2006. It contains more than 1.3 million concepts and 243 million association relations among the concepts. The thesaurus is stored in a database (MySQL 5.0)

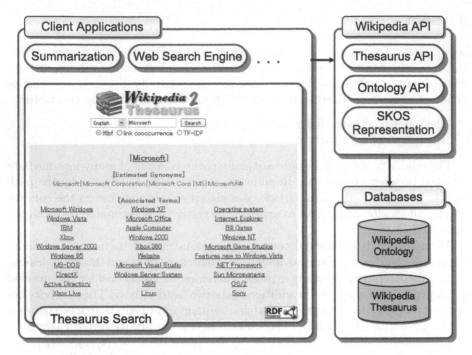

Fig. 1. Search engine for Wikipedia Thesaurus and architecture

and indexed to provide practical APIs for other applications such as information retrieval and text summarization. The APIs are provided by XML Web services to achieve high interoperability among different environments. They allow developers to add association term extraction capability to their own applications.

We developed a thesaurus browsing system by using the APIs to prove the capability of this approach and the accuracy of our association thesaurus. We are also working on a huge scale Web ontology construction from Wikipedia. The ontology will be provided as an XML Web service. Figure 1 shows the architecture of our system and a screen shot of the thesaurus browsing system.

The concept search engine provides SKOS[3] (an application of RDF) representation function. SKOS is a developing specification and standard to support knowledge organization systems (KOS) such as thesauri, classification schemes, subject heading systems and taxonomies within the Semantic Web. SKOS supports several basic relations between two concepts such as "broader," "narrower" and just "related." Further, since $pfibf$ extracts relatedness between two concepts, we extended SKOS in order to express the relatedness. We defined an extension for SKOS as the "The Wikipedia Thesaurus Vocabulary (WTV)." Currently, WTV supports some simple relations not in SKOS such as relatedness. The URLs for concepts in the Wikipedia Thesaurus correspond to the URLs of Wikipedia articles. For example, "http://wikipedia-lab.org/concept/Music" in the thesaurus corresponds to the article "http://en.wikipedia.org/wiki/Music" in Wikipedia. Figure 2 show an example of a SKOS representation of "Music."

```
-<rdf:RDF>
  -<skos:Concept rdf:about="http://wikipedia-lab.org/concepts/Music" rdfs:seeAlso="http://en.wikipedia.org/wiki/Music">
    <skos:prefLabel>Music</skos:prefLabel>
    <skos:altLabel>musicaK</skos:altLabel>
    <skos:altLabel>musician</skos:altLabel>
    <skos:altLabel>musicians</skos:altLabel>
    <skos:altLabel>genres</skos:altLabel>
    <skos:related rdf:resource="http://wikipedia-lab.org/concepts/Performance" wtv:relatedness="0.0579156"/>
    <skos:related rdf:resource="http://wikipedia-lab.org/concepts/Music_venue" wtv:relatedness="0.0564693"/>
    <skos:related rdf:resource="http://wikipedia-lab.org/concepts/Publication" wtv:relatedness="0.0230024"/>
    <skos:related rdf:resource="http://wikipedia-lab.org/concepts/Dance" wtv:relatedness="0.0170662"/>
    <skos:related rdf:resource="http://wikipedia-lab.org/concepts/Organization" wtv:relatedness="0.0111082"/>
    <skos:related rdf:resource="http://wikipedia-lab.org/concepts/Art" wtv:relatedness="0.0104583"/>
    <skos:related rdf:resource="http://wikipedia-lab.org/concepts/Drama" wtv:relatedness="0.00829775"/>
    <skos:related rdf:resource="http://wikipedia-lab.org/concepts/Theatre" wtv:relatedness="0.00791607"/>
    <skos:related rdf:resource="http://wikipedia-lab.org/concepts/Philosophy" wtv:relatedness="0.00704391"/>
    <skos:related rdf:resource="http://wikipedia-lab.org/concepts/Poetry" wtv:relatedness="0.00689856"/>
    <skos:related rdf:resource="http://wikipedia-lab.org/concepts/Music_theory" wtv:relatedness="0.00669445"/>
    <skos:related rdf:resource="http://wikipedia-lab.org/concepts/Pitch_(music)" wtv:relatedness="0.00668502"/>
    <skos:related rdf:resource="http://wikipedia-lab.org/concepts/Painting" wtv:relatedness="0.00658941"/>
    <skos:related rdf:resource="http://wikipedia-lab.org/concepts/Literature" wtv:relatedness="0.00651561"/>
    <skos:related rdf:resource="http://wikipedia-lab.org/concepts/Theater" wtv:relatedness="0.00628334"/>
    <skos:related rdf:resource="http://wikipedia-lab.org/concepts/Piano" wtv:relatedness="0.00589881"/>
    <skos:related rdf:resource="http://wikipedia-lab.org/concepts/Lyrics" wtv:relatedness="0.00566608"/>
    <skos:related rdf:resource="http://wikipedia-lab.org/concepts/Melody" wtv:relatedness="0.00563164"/>
    <skos:related rdf:resource="http://wikipedia-lab.org/concepts/Entertainment" wtv:relatedness="0.00556938"/>
    <skos:related rdf:resource="http://wikipedia-lab.org/concepts/Composer" wtv:relatedness="0.00548984"/>
```

Fig. 2. Sample of SKOS (RDF) representation

Our thesaurus browsing system and XML Web services are available under the following URLs.

- Wikipedia Thesaurus:
  http://wikipedia-lab.org:8080/WikipediaThesaurusV2
- Wikipedia API:
  http://wikipedia-lab.org/en/index.php/Wikipedia_API

We are going to extract much more complicated relations from Wikipedia by using NLP techniques. Relatedness is just a first step, but it will support much more relation types in the future.

Acknowledgment. This research was supported in part by Grant-in-Aid on Priority Areas (18049050), and by the Microsoft Research IJARC Core Project.

References

1. Nakayama, K., Hara, T., Nishio, S.: A thesaurus construction method from large scale web dictionaries. In: Proc. of IEEE International Conference on Advanced Information Networking and Applications (AINA 2007), pp. 932–939 (2007)
2. Nakayama, K., Hara, T., Nishio, S.: Wikipedia mining for an association web thesaurus construction. In: Benatallah, B., Casati, F., Georgakopoulos, D., Bartolini, C., Sadiq, W., Godart, C. (eds.) WISE 2007. LNCS, vol. 4831, Springer, Heidelberg (2007)
3. World Wide Web Consortium: Simple knowledge organisation systems (skos) (2004), http://www.w3.org/2004/02/skos/

An Interactive Predictive Data Mining System for Informed Decision

Esther Ge and Richi Nayak

Faculty of Information Technology
Queensland University of Technology,
Brisbane, Australia
t.ge@student.qut.edu.au, r.nayak@qut.edu.au

Abstract. There exists a need to utilize the predictive data mining models for querying to obtain the predicted outcome based on user provided inputs in its real use. This demo illustrates a real-world situation in which the trained predictive data mining system is being deployed and now users can interact with the model for informed decision.

Keywords: Data Mining, predictive model, interactive model.

1 Introduction

The training of a predictive data mining model and then understanding rules and patterns inferred by the mining model should not be the end of the prediction task. The data mining model should allow the user to query the system for future cases. The created data mining model needs to be deployed in practice as a user-driven prediction system so that the data mining system can be used for querying to obtain the predicted outcome based on user provided inputs. In some real-world applications, the training set contains relatively complete attribute information while the unseen cases (user queries) do contain many missing attribute values. Consider a predictive data mining model that is built to predict the "Service Life" of building components based on the input attributes such as "Location", "Component", "Material", "Salt Deposition", and "Mass Loss". Suppose a builder (a typical user of the predictive model/tool/system) wants to know the service life of a "Gutter" with "Galvanized Steel" at a particular location. The user does not explicitly know the "Salt Deposition" and "Mass Loss" in that location. The user query will include two missing values. In such a case, the predicted service life by the predictive data mining tool will not be as accurate as tested in the evaluation phase of the predictive model, especially when the missing attributes play key roles in predicting the outcome. On the other hand, if the "Salt Deposition" and "Mass Loss" features are excluded from the model building, the performance of the model may not be acceptable. Hence, a major problem that needs to be solved is how to select the appropriate attributes to build the model for real-world situations where the users can not provide all the inputs for querying to the system. In other words, how to deal with the missing attribute values in user queries (unseen cases). We developed an interactive data mining model for predicting the

J.R. Haritsa, R. Kotagiri, and V. Pudi (Eds.): DASFAA 2008, LNCS 4947, pp. 694–697, 2008.

service life of metallic components in buildings which allows the user to input the queries based on their limited knowledge, while maintaining the accuracy of the predicted outcome.

2 An Interactive Predictive Data Mining System

The proposed interactive predictive data mining system consists of nine phases structured as sequences of predefined steps (as shown in Figure 1). The system includes the standard data pre-processing, data analysis and result post-processing phases. Additionally, it includes the phase of Query Based Feature Selection (QBFS) separated from the data pre-processing step. The QBFS phase has the involvement of users or domain experts and hence is different from the usual feature selection. The Results Post-processing and the Use of Model phase are added into the modeling process in order to ensure the predictive data mining system is being used in practice by users. An external domain knowledge base is involved in results post-processing and missing inputs pre-processing in the Use of Model phase.

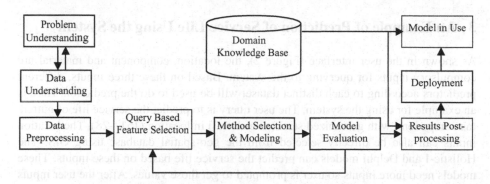

Fig. 1. An Interactive Predictive Data Mining System

In order to select the appropriate attributes to build the predictive model, the Query Based Feature Selection (QBFS) algorithm [1] is applied to datasets. The QBFS algorithm allows selecting the attributes according to the interest of a user/domain expert. This algorithm first divides the attributes available for training into three categories according to their accessibility for querying. The first group contains attributes that are easily accessible to users. The second group contains attributes that are difficult for users to access. In other words, users cannot directly provide the values of these attributes while querying, but it is still feasible to get those values via some indirect sources. The third group contains attributes that users can never provide values for querying. The attributes belonging to third groups are not included in model building. The QBFS algorithm selects a minimum subset of features which can be provided by users or obtained from domain knowledge based on the groups 1 and 2, while maintaining the acceptable accuracy of the model as well. The QBFS has been proven [1] to provide good generalization accuracy.

The datasets used in this system are from real-life including four different sources of service life information of building components in Queensland schools [1]. The features selected by the QBFS include some which can be provided by users such as "longitude", "latitude", "component", "material" and others which can not be provided by users such as "Salt Deposition" and "Rainfall". Therefore, the domain knowledge is used to get these two attribute values in user queries. The knowledge is represented as items in the database. For example, an item for Salt Deposition knowledge is a tuple with longitude, latitude, Salt Deposition. Once the user inputs the location (longitude, latitude), the SQL query language is used to search the knowledge base to find the same location or the nearest location and accordingly the values of Salt Deposition and the Rainfall are obtained. As the predictors are built from different datasets, the predicted results may not be consistent. The domain knowledge base also includes some generalised rules to post-process the inconsistent results. One example of the generalised rules is (Component, Environment, Material, Min years, Max years). These generalised rules give a reasonable range of service life for matched user inputs.

3   An Example of Prediction of Service Life Using the System

As shown in the user interface (Figure 2), the location, component and material are compulsory inputs for querying to the system. Based on these three inputs, different predictors according to each distinct dataset will be used to do the prediction. Here is an example for using the system. The user query is to predict the service life of gutters (as component) with galvanized steel (as material) in location (151, -28). The location inputs can also be directly selected from the geo-spatial database using GIS. The Holistic-I and Delphi models can predict the service life based on these inputs. These models need more inputs so user is prompted to get those values. After the user inputs the gutter position, maintenance and environment information, the system automatically gets values from domain knowledge for other features needed by the predictors. For example, the Holistic-I predictor requires salt deposition in this location as an input as well. The system gets the salt deposition from the salt database and predicts the service life to be 14.5 years from the Holistic-I predictor. A similar process is done for the Delphi predictor and the predicted service life is predicted as 14.4 years.

Sometimes the results of predictors conflict with each other. An example of such a case is the service life of roof with Zincalume in location (153.0310, -27.4315). The predicted result of the Delphi model is 51.8 years while of the Holistic-III predictor is only 29.9 years. In such a case, the system consults the "domain knowledge base" that includes the generalized rule set. For example, in this case, the rule that matches is "The range of service life for roofs with Zincalume in benign environment is greater than 50 years". Based on this rule, the system finally displays the service life to be 51.8 and discards the life value of 29.9 based on Holistics-III predictor.

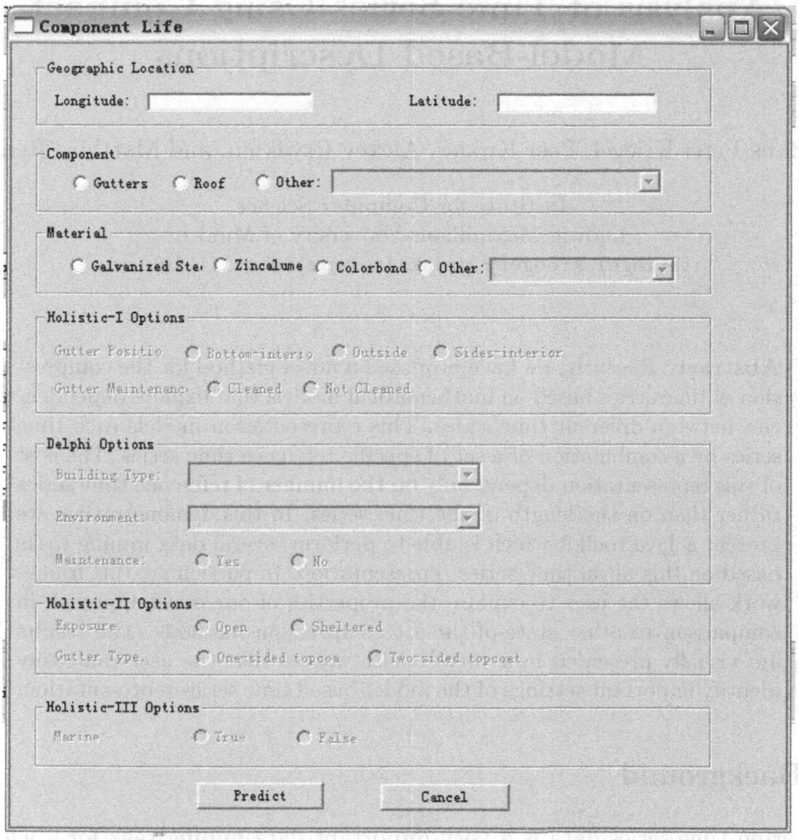

Fig. 2. User Interface of the Interactive Predictive Data Mining System

4   Conclusions

This paper presents a real-world situation in which the learned predictive data mining model is deployed as a user-oriented prediction system. The developed system is easy to use for people with little expertise in data mining and domain non-experts. It provides accurate prediction where not all inputs are available for querying to the system with incorporation of query based feature selection algorithm.

References

1. Ge, E., Nayak, R., Xu, Y., Li, Y.: A User Driven Data Mining Process Model and Learning System. In: Haritsa, et al. (eds.) DASFAA 2008. LNCS, vol. 4947. Springer, Heidelberg (to appear, 2008)

Analysis of Time Series Using Compact Model-Based Descriptions

Hans-Peter Kriegel, Peer Kröger, Alexey Pryakhin, and Matthias Renz

Institute for Computer Science
Ludwig-Maximilians-University of Munich
{kriegel,kroegerp,pryakhin,renz}@dbs.ifi.lmu.de

Abstract. Recently, we have proposed a novel method for the compression of time series based on mathematical models that explore dependencies between different time series. This representation models each time series by a combination of a set of specific reference time series. The cost of this representation depend only on the number of reference time series rather than on the length of the time series. In this demonstration, we present a Java toolkit which is able to perform several data mining tasks based on this novel time series representation. In particular, this framework allows the user to explore the properties of our novel approach in comparison to other state-of-the-art compression methods. The results are visually presented in a very concise way so that the user can easily identify important settings of the model-based time series representation.

1 Background

Clustering time series data is a very important data mining task for a wide variety of applications. In many scenarios, the set of time series that need to be analyzed is very large and each time series of this set has an enormous length. Both aspects have obviously a negative influence on the runtime of the clustering process. As a consequence, in many applications, the user is happy with an approximative clustering result, as far as the approximation is not too coarse compared to the original result (effectivity) and the approximative clustering is considerably faster (efficiency) tentatively enabling an interactive analysis. Beside standard compression techniques for similarity search (e.g. [1]), an approach for fast approximative clustering of time series is presented in [2].

Recently, we proposed a novel method for the compression of time series based on mathematical models that explore dependencies between different time series [3]. The resulting representation consists of some low-dimensional feature vector that can easily be indexed by means of any Euclidean index structure. The similarity distance used for the clustering is computed by applying the parameters that specify the combination. Consequently, — in contrast to other existing methods — the cost of the clustering process depend only on the number of reference time series rather than on the length of the time series.

Figure 1 illustrates our approach. A set of reference time series (marked as "T1", "T2", and "T3" on the left hand side of Figure 1) is used to approximate an

J.R. Haritsa, R. Kotagiri, and V. Pudi (Eds.): DASFAA 2008, LNCS 4947, pp. 698–701, 2008.
© Springer-Verlag Berlin Heidelberg 2008

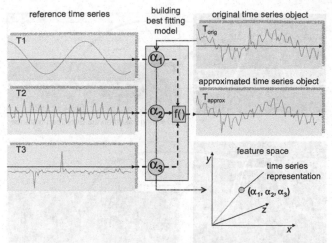

Fig. 1. Model-based time series representation

original time series T_{orig} (shown on the upper right hand side of Figure 1) by an arbitrary complex combination T_{approx}. In case of Figure 1 this is a combination of the coefficients $\alpha_1, \ldots, \alpha_3$ representing the three input time series using a function f. The resulting approximated time series (marked as "output" in the middle of the right hand side of Figure 1) is similar to the original time series. For clustering, the approximation is represented by a feature vector of the coefficients of the combination (cf. lower right hand side of Figure 1). Obviously, the choice of a suitable set of reference time series is crucial.

2   A Visual Data Mining Framework for Time Series

The application presented in this work implements a visual data mining approach that analyzes and describes the data in a clear and user-friendly way in order to enable interactive data exploration, in particular cluster analysis. Basically, our framework is designed for the following purposes:

1. It allows visually inspecting the results of comparative data mining, i.e. the results of different compression methods can be evaluated interactively.
2. It assists the user in analyzing the properties of different compression methods, in particular of our model-based technique proposed in [3]. Thus, e.g. the impact of the choice of the reference time series can be evaluated interactively.
3. It is a key tool for a novel step-wise procedure for identifying potentially interesting reference time series for our model-based compression approach.

The concept of our application supports the extraction of novel insights in supervised as well as in unsupervised settings. If class labels are available, the user can evaluate different settings of reference time series based on classification accuracies in cross-validation experiments. But even in an unsupervised

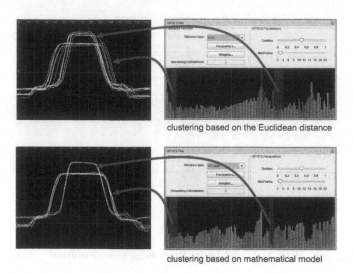

clustering based on the Euclidean distance

clustering based on mathematical model

Fig. 2. Comparative time series mining

situation where no pre-classified time series are available, our framework can be very helpful. By a quick visual inspection of several clustering results derived for example by OPTICS [4] it is possible to discover important and interesting reference time series based on their ability to form distinct cluster structures.

Comparative Data Mining. In our framework we included several time series similarity measures such as DTW and the Euclidean distance for the purpose of comparing different similarity notions. The collection of implemented distance functions can easily be extended when need arises. Furthermore, most of the existing compression techniques for time series have been implemented including dimensionality reduction techniques as well as our novel model-based compression approach. Our framework enables to visually inspect and comparatively evaluate several clusterings that may be generated using different distance measures or compression models. An illustrative example is visualized in Figure 2 where we compared a clustering based on the Euclidean distance using no compression with the clustering based on our new mathematical model (model-based compression).

Evaluation of Model-based Time Series Compression. The visual clustering output can be used to understand the properties of our novel model-based compression method. For example, it assists in evaluating the impact of the choice of reference time series on the quality of the compression (and, therefore, on the quality of the clustering results) in an interactive way. In a supervised scenario, the user can inspect precision and recall values of different clusterings obtained using different reference time series in order to evaluate the quality of the respective results. In an unsupervised scenario, the user can inspect the cluster hierarchies obtained using different reference time series in order to evaluate the

(a) Supervised scenario

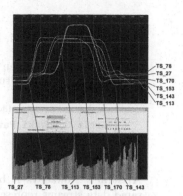

(b) Unsupervised scenario

Fig. 3. Evaluation of model-based time series compression

quality of the respective results. Screenshots of our framework for both scenarios are visualized in Figure 3.

Determination of Reference Time Series for Model-based Compression. As discussed above, the choice of reference time series is crucial for the quality of data compression and, thus, for the quality of the approximative clustering result. Here, we propose a step-wise procedure for interactively optimizing the set of reference time series based on our novel framework. Starting from an initial clustering of the time series, we can iteratively refine the clustering and the set of reference time series by inspecting the clustering results and updating the set of reference time series accordingly.

References

1. Faloutsos, C., Ranganathan, M., Maolopoulos, Y.: Fast Subsequence Matching in Time-series Databases. In: Proc. SIGMOD Conference (1994)
2. Ratanamahatana, C.A., Keogh, E., Bagnall, A.J., Lonardi, S.: A Novel Bit Level Time Series Representation with Implication for Similarity Search and Clustering. In: Ho, T.-B., Cheung, D., Liu, H. (eds.) PAKDD 2005. LNCS (LNAI), vol. 3518, Springer, Heidelberg (2005)
3. Kriegel, H.-P.: Approximate Clustering of Time Series Using Compact Model-based Descriptions. In: Haritsa, et al. (eds.) DASFAA 2008. LNCS, vol. 4947. Springer, Heidelberg (to appear, 2008)
4. Ankerst, M., Breunig, M.M., Kriegel, H.P., Sander, J.: OPTICS: Ordering Points to Identify the Clustering Structure. In: Proc. SIGMOD Conference (1999)

Collecting Data Streams from a Distributed Radio-Based Measurement System

Marcin Gorawski, Pawel Marks, and Michal Gorawski

Silesian University of Technology,
Institute of Computer Science,
Akademicka 16,
44-100 Gliwice, Poland
{Marcin.Gorawski,Pawel.Marks,Michal.Gorawski}@polsl.pl

Abstract. Nowadays it becomes more and more popular to process rapid data streams representing real-time events, such as large scale financial transfers, road or network traffic, sensor data. Analysis of data streams enables new capabilities. It is possible to perform intrusion detection while it is happening, it is possible to predict road traffic basing on the analysis of the past and current vehicle flow. We addressed the problem of real-time analysis of the stream data from a radio-based measurement system. The system consists of large number of water, gas and electricity meters. Our work is focused on data delivery from meters to the stream data warehouse as quick as possible even if transmission failures occur. The system we designed is intended to increase significantly system reliability and availability. During this demonstration we want to present an example of the system capabilities.

Keywords: stream processing, fault-tolerance, sensor networks.

1 Introduction

These days it becomes more common to process continuous data streams. It may have application in many domains of our life such as: computer networks (e.g. intrusion detection), financial services, medical information systems (e.g. patient monitoring), civil engineering (e.g. highway monitoring) and more.

Thousands or even millions of energy meters located in households or factories can be sources of meter readings streams. Continuous analysis of power consumption may be crucial to efficient electricity production. Unlike other media such as water or gas, electricity is hard to store for further use. That is why a prediction of energy consumption may be very important. Real-time analysis of the media meter readings may help to manage the process of energy production in the most efficient way.

There are many systems for processing continuous data streams and they are still developed [1]. In [2] there is presented a fault tolerant Borealis system. This is a dedicated solution for applications where a low latency criterion is essential. Another system facing infinite data streams is described in [3]. Authors of

J.R. Haritsa, R. Kotagiri, and V. Pudi (Eds.): DASFAA 2008, LNCS 4947, pp. 702–705, 2008.

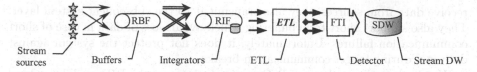

Fig. 1. Layered structure of the distributed telemetric system

the work deal with sensors producing data continuously, transferring the measured data asynchronously without having been explicitly asked for that. They proposed a *Framework in Java for Operators on Remote Data Streams* (Fjords).

In our research we have focused on processing data originating from a radio-based measurement system [4]. We carried research on efficient recovery of interrupted ETL jobs and proposed a few approaches [5,6] based on the Design-Resume algorithm [7].

Basing on the previous experience, we have focused on fault-tolerance and high availability in a distributed stream processing environment. In [8] we proposed a new set of modules increasing the probability that a failure of one or more modules will not interrupt the processing of endless data streams. Then we prepared a model [9,10] of data sources to estimate the amounts of data to be processed useful in the configuration of the environment. In this paper we want to present how the system works in practice and how it has been implemented.

2   Research System

Our research is based on a telemetric system [4] for remote and automatic reading of media consumption meters. It's main task is transferring data from particular meters using wireless communication to local collecting nodes (further called *data stream sources*). Next, the data streams need to be transferred to a *stream data warehouse* (SDW) in which the data can be processed analysed.

Our goal is to assure the continuity and correctness of the data streams being transmitted into a stream data warehouse. We want the process to run continuously, and we want it to be failure resistant. There are two goals to be achieved: (a) transferring all the data, (b) transferring data as fast as possible, minimizing the delay between measurement and loading into SDW. To achieve the above mentioned goals we introduce three additional layers to the system structure [8] (Fig. 1): RBF remote buffers, RIF persistent integrators, FTI detector.

Data sources (collecting nodes) are very simple devices comparable to LAN routers. Their task is to receive data from meters and send it further to the destination. The most important feature of the data sources is their inability to buffer data. If the outgoing connection is lost, the outgoing data is lost also. To avoid such situations each data source transmit data to many RBF buffering modules. If one connection is lost, the others remain.

RBF buffers are simple and low-cost buffering modules. Because they are cheap, they can be used in many copies. The more RBFs receive data from a data source, the lower is the probability of data loss. The RBFs not only receive data from sources, they also transmit it to the subsequent system layer. They also support short-term buffering which avoids loss of data in case of short communication failures. Unfortunately, it does not protect the system against data loss during longer communication breaks.

We also introduce a layer of RIF persistent integrators. RIF modules extend the RBF functionality. They offer persistent buffering based on a persistent storage (e.g. hard disk). Another task of an RIF module is integration of stream parts from many RBFs into a single data stream. It is necessary when one RBF fails, and the RIF starts to receive data from another one. The data safely received can be processed now.

Behind the RIF layer we placed the layer of ETL modules. ETL stands for *Extraction, Transformation and Loading*. In this layer the data streams received from RIF modules are processed (filtered, transformed, recalculated, joined, aggregated). The ETL process is described using *Directed Acyclic Graph*. Graph nodes are responsible for data processing, whereas graph edges define tuple flow directions. Graph nodes belong to one of three node classes: extractors responsible to retrieving data to be processed, transformations in which tuples are processed, and inserters which save data in a destination. The ETL process supports three resumption algorithms: Design-Resume (DR) [7], hybrid DR-based resumption [5] and checkpoint-based algorithm [6]. In stream processing only checkpointing is applicable.

The FTI layer is responsible for merging redundant data stream from replicated ETL processes. It also checks the correctness of received data comparing all received stream copies. Finally the data is stored in the data warehouse.

The use of checkpointing enables replication and migration of ETL process. Saved ETL process state can be easily copied and used to restart ETL on another machine or network node. The checkpointing algorithm uses filtration known from DR algorithm. It enables synchronization with streams incoming from RIFs and delivered to the last FTI layer.

3 Technical Details

For research purposes the system has been implemented in Java. We have built the stream data generator which works similarly to the real collecting nodes. RBF and RIF modules are also implemented in Java, although in real world RBFs should be a small and simple devices based on microcontrollers e.g. AVR or ARM equipped with necessary communication interfaces. The RIF module can be built similarly to RBF using microcontroller; however, it is not to be used in so many copies as RBFs, so its functionality can be realized in a PC running necessary software.

ETL module is a set of Java classes implementing ETL process, communication with RIF and FTI layers and resumption algorithms. Each instance of an ETL process is started in a separate *Java Virtual Machine* (JVM).

Each system module (data generator, RBFs, RIFs, ETLs, FTI) runs on a separate instance of JVM. This references the complete independence of modules in real world. The communication between layers uses RMI and TCP/IP connections. RMI interfaces of modules are required to locate needed services and initialize the communication. When the modules agree for the connection options after exchange of host IPs and port number, they exchange data via TCP/IP connections, which are much more efficient than RMI.

During experiments failures are simulated in two ways: by "unexpected" terminating of random system modules or by breaking the connections between system modules. In both cases the system is expected to handle the failure correctly and continue the processing without any data loss.

Our demonstration is going to be preceded by a short slideshow clarifying how the system works. During the demonstration we want to show the working system using the software simulator running on a PC-class machine. We are going to simulate module failures and observe the system reaction.

References

1. Arasu, A., Babcock, B., Babu, S., Datar, M., Ito, K., Motwani, R., Nishizawa, I., Srivastava, U., Thomas, D., Varma, R., Widom, J.: Stream: The stanford stream data manager. IEEE Data Eng. Bull. 26(1), 19–26 (2003)
2. Balazinska, M., Balakrishnan, H., Madden, S., Stonebraker, M.: Fault-Tolerance in the Borealis Distributed Stream Processing System. In: ACM SIGMOD Conf., Baltimore, MD (2005)
3. Madden, S., Franklin, M.J.: Fjording the stream: An architecture for queries over streaming sensor data. In: ICDE, pp. 555–566. IEEE Computer Society, Los Alamitos (2002)
4. Gorawski, M., Malczok, R.: Distributed spatial data warehouse indexed with virtual memory aggregation tree. In: Sander, J., Nascimento, M.A. (eds.) STDBM, pp. 25–32 (2004)
5. Gorawski, M., Marks, P.: High efficiency of hybrid resumption in distributed data warehouses. In: DEXA Workshops, pp. 323–327. IEEE Computer Society, Los Alamitos (2005)
6. Gorawski, M., Marks, P.: Checkpoint-based resumption in data warehouses. In: Socha, K. (ed.) IFIP International Federation for Information Processing, Warsaw. Software Engineering Techniques: Design for Quality, vol. 227, pp. 313–323 (2006)
7. Labio, W., Wiener, J.L., Garcia-Molina, H., Gorelik, V.: Efficient resumption of interrupted warehouse loads. In: Chen, W., Naughton, J.F., Bernstein, P.A. (eds.) SIGMOD Conference, pp. 46–57. ACM, New York (2000)
8. Gorawski, M., Marks, P.: Fault-tolerant distributed stream processing system. In: DEXA Workshops, pp. 395–399. IEEE Computer Society, Los Alamitos (2006)
9. Gorawski, M., Marks, P.: Towards reliability and fault-tolerance of distributed stream processing system. In: DepCoS-RELCOMEX, pp. 246–253. IEEE Computer Society, Los Alamitos (2007)
10. Gorawski, M., Marks, P.: Distributed stream processing analysis in high availability context. In: ARES 2007: Proceedings of the The Second International Conference on Availability, Reliability and Security, Washington, DC, USA, pp. 61–68. IEEE Computer Society, Los Alamitos (2007)

A Web Visualization Tool for Historical Analysis of Geo-referenced Multidimensional Data

Sonia Fernandes Silva

Etruria Telematica, Strada di Basciano 22 Monteriggioni 53035, Siena Italy
sonia.silva@etelnet.it, sonia.silva@tiscali.it

Abstract. This paper describes the recent visual features of an interactive web-based tool that couples geographic map and visual diagrams in order to promote its use for users who need to remotely explore large multidimensional datasets in a spatio-temporal context for decision-making. The tool puts together useful techniques for data exploration and optimization, by enabling the user to create and explore several interactive web-based visual reports of summarized data almost instantaneously. We describe the new contributions in a particular tourist Datawarehouse.

1 Introduction

With the recent advances in the area of WebGIS and *Spatial OLAP* (SOLAP) technology [4], state of the art solutions propose web visualization tools that support geographic display and navigation during the explorative analysis of geo-referenced multidimensional data, including the spatio-temporal analysis [1].

Following this emerging area, our research has focused in developing a web interactive tool for analysts (decision-makers, statisticians) who need to remotely explore large geo-referenced datasets from a DataWarehouse (DW) in a spatio-temporal context, in particular, the tourist DW from the Public Administration in the province of Siena (Italy), whose specific requirement about the remote data exploration include: fast information about trend analysis and spatio-temporal distribution of summarized thematic indices at the various levels of detail.

Two issues were considered to address such requirement. The first concerns aspects of internal representation and manipulation of spatio-temporal aggregations from a DW. The transactional architecture of the DBMS and GIS tools that currently handle the non-spatial and spatial data from the tourist DW are not suitable to meet the requirement above. In spite of the variety of SOLAP proposals, we decided to adopt an architecture that take advantage of the current tools and that would not impose restrictions on the flexibility of the web interface design for including new exploratory visualization techniques.

Therefore, we adopted an approach that puts together some existing techniques in the literature for data exploration and optimization. For a high interactivity with the maps on the web, we adopted the image-based technique [6], where the geographic knowledge is encoded into the raster images which are delivered to the web client. For a quick query processing against a DW, we

J.R. Haritsa, R. Kotagiri, and V. Pudi (Eds.): DASFAA 2008, LNCS 4947, pp. 706–709, 2008.
© Springer-Verlag Berlin Heidelberg 2008

extended the technique which efficiently encodes multidimensional data cubes [3] with an additional indexing mechanism, in order to facilitate exploratory spatio-temporal analysis. This integration enables the user to remotely create and explore several interactive visual reports of summarized data in sub-seconds [5]. The second issue is how the geo-referenced multidimensional data should be displayed and manipulated to a user. The main purpose of the paper is to focus only on this issue (to address in detail the first one is out of the scope of this paper, the reader may refer to [3], [5], [6] for more details). In the following we briefly describes the new improvements of the web tool that give additional significant insights and shows its use in the tourist DW.

2 Visually Exploring Geo-referenced Data Cubes

The tourist DW contains a meaningful volume of daily historical series of tourist's presence in the receptive structures in the province of Siena. The raw data is structured into *multidimensional cubes* in which a *fact* is associated with numerical *measures* and *dimensions* that characterize the fact. Decision-makers analyze the facts representing the arrivals and stays measures aggregated over the range of the dimensions *location* of the receptive structures (spatial dimension), *period* of arrivals and stays (temporal dimension), *structure typology* and *tourist origins* (thematic dimensions). The spatial dimension include geometric shapes that are geo-referenced. Figure 1 illustrates the dimensional hierarchy.

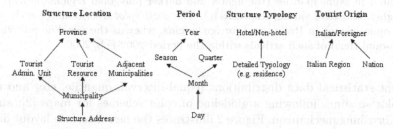

Fig. 1. Dimensional hierarchy of the tourist DW with multiple hierarchy in some dimensions

The spatial and temporal dimensions are fully exploited in order to rapidly reveal spatio-temporal patterns hidden within the raw data. Such analysis is carried out in two phases: the user specifies incoming queries in a web page (by selecting the measures and corresponding aggregation function, the hierarchy level of the thematic dimensions, the spatial and temporal coverages) for a successive remote interaction with the query result. The user continuously explores the resulting dataset by using visual interactions and making progressive spatial and temporal *drill-down/roll-up* interactions, i.e. descending from coarser aggregates to more detailed ones (e.g. *quarter → month*) or vice-versa.

The original web user interface [5] used in the second phase has been improved with new interactive features such as: a visual filtering panel; the availability of

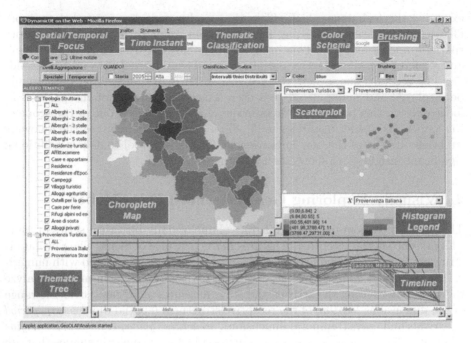

Fig. 2. Snapshot of the spatial distribution (*choropleth map* and its *histogram legend*) of the foreign tourist's arrivals in the low cost accommodations at high season 2005, aggregated in Municipalities (the lighter and darker polygonal regions indicate lower and higher statistical values, respectively). The *scatterplot* discriminates such arrivals in the same instant in Italian and foreign origins, whereas the *timeline plot* visualize the seasonal trend of such arrivals within the period 2002-Feb/2006.

different statistical data distributions (equal-interval, quantile, etc) and different color shadings following a guideline of color schemes for maps [2]; and the visual brushing mechanism. Figure 2 illustrates the new interface layout divided into three main areas. The top and left areas contain the interactive widgets that allows the user to accomplish exploratory visual interactions. In particular, the left area implements the *slice-and-dice* operator applied on the thematic dimensions (each one at a specific hierarchy level). It displays the content of such dimensions in form of a file-browser (*thematic tree*) for visual data filtering.

In the main area the results of the user interactions are visualized in interactive data displays (map display and visual plots) that are sensitive to the mouse movement. Such displays are tightly coupled meaning that by moving the mouse on a geographic object in a specific display (the polygon on the map, the mark in scatterplot, the line in timeline plot), there is an immediate identification of its statistical information (as illustrated in Figure 2 in timeline plot), and the same object is highlighted (in violet color) in the other displays. As users manipulate the widgets, the map and visual plots animate to give immediate feedback in order to facilitate the exploratory data analysis on the web.

A new visual interaction of the interface include the *dynamic time-oriented classification*, where the choropleth map is colored according to specific statistic method, time instant and color shade. An instant is specified by manipulating temporal spinners at a specific temporal aggregation level, including the *all history* option to visualize statistical values accumulated along the history. Other new visual interaction is the *visual brushing*, that involves the interactive selection of a subset of geographic objects (by dragging a box on the diagrams) in order to highlight the objects inside the box in all visualizations (the filtered elements are evidenced in color red). Dragging a box on the scatterplot and the timeline visually specifies a 2-dimensional and temporal filtering respectively.

3 Conclusion

Many proposals in the literature combine map with visual plots in order to gain insights (that is, to rapidly reveal patterns, trends or anomalies) into the multidimensional data for decision-making. In this paper we presented a web visualization tool that apply this combination in the context of a highly interactive web exploration of geo-referenced multidimensional data cubes, and show its use in a particular tourist DW. The great benefit of this tool is that it enables the user to remotely create and explore several visual reports of summarized data in a nearly real-time interactivity. It is worth noting that it is not our intention to propose a new SOLAP tool but to demonstrate that integrating different techniques for data exploration and optimization was suitable to meet our specific requirements, without imposing the total implementation of an integrated multidimensional and geographic environment.

Acknowledgements. This work is supported by the Public Administration of Siena and the Monte dei Paschi di Siena Foundation.

References

1. Andrienko, N., Andrienko, G.: Interactive Visual Tools to Explore Spatio-Temporal Variation. In: Proc. of ACM Symposium on Advanced Visual Interfaces, pp. 417–420 (2004)
2. Brewer, C., Harrower, M.: ColorBrewer Online Tool,
   http://www.personal.psu.edu/cab38/ColorBrewer/ColorBrewer.html
3. Fu, L., Hammer, J.: CubiST++: Evaluating Ad-Hoc CUBE Queries Using Statistics Trees. Journal of Distributed and Parallel Databases 14, 221–254 (2003)
4. Rivest, S., et al.: SOLAP technology: Merging business intelligence with geospatial technology for interactive spatio-temporal exploration and analysis of data. ISPRS Journal of Photogrammetry & Remote Sensing 60, 17–33 (2005)
5. Silva, S.F., Menicori, P.: Uno Strumento Web di Esplorazione Interattive Multidimensionale dei Dati Geo-Riferiti. In: Proc. of 15th Italian Symposium on Advanced Database Systems (SEBD 2007), Torre Canne (Fasano, BR), Italy, pp. 136–147 (2007)
6. Zhao, H., Shneiderman, B.: Colour-coded pixel-based highly interactive Web mapping for geo-referenced data exploration. International Journal of Geographical Information Science 4, 413–428 (2005)

Is There *really* Anything Beyond Frequent Patterns, Classification and Clustering in Data Mining?

Vikram Pudi

IIIT Hyderabad, India
vikram@iiit.ac.in

Data mining, the automatic discovery of interesting patterns from large databases, has become a mainstream database research topic. Various kinds of patterns have been identified by researchers including classification and regression models, clusters, association rules, frequent patterns, time-series patterns, summaries, etc. Among these different data mining tasks, frequent pattern mining, classification and clustering have received most attention from the database research community. This is perhaps justified by the fact that most other data mining tasks can be reduced to these three. Here are some examples:

1. Association rules may be considered as merely another way to represent frequent itemsets.
2. Time-series are a special case of sequence databases and mining them reduces to mining frequent sequences.
3. Regression can be reduced to classification if the dependent attribute (i.e. the attribute whose values are to be estimated) is discretized into small ranges.
4. Outlier mining can be reduced to clustering, and then identifying data points that do not lie in any cluster.

In this context, the question naturally arises as to whether there is *really* anything more in data mining than frequent pattern mining, classification and clustering. This is an important question to be answered because if the question is answered in the affirmative, then future research may rather focus on the aspects of data mining that are beyond these basic tasks. On the other hand, if the question is answered in the negative, then we may well consider data mining as a "closed problem" because of the availability of very efficient algorithms for these three basic tasks.

The objective of this panel is to discuss this high impact and provocative question, which is basically a disguised way of asking: Do any fundamental research issues remain to be solved in data mining?

J.R. Haritsa, R. Kotagiri, and V. Pudi (Eds.): DASFAA 2008, LNCS 4947, p. 710, 2008.
© Springer-Verlag Berlin Heidelberg 2008

Author Index

Lecture Notes in Computer Science

Sublibrary 3: Information Systems and Application, incl. Internet/Web and HCI

For information about Vols. 1– 4557
please contact your bookseller or Springer

Vol. 4796: M. Lew, N. Sebe, T.S. Huang, E.M. Bakker (Eds.), Human–Computer Interaction. X, 157 pages. 2007.

Vol. 4794: B. Schiele, A.K. Dey, H. Gellersen, B. de Ruyter, M. Tscheligi, R. Wichert, E. Aarts, A. Buchmann (Eds.), Ambient Intelligence. XV, 375 pages. 2007.

Vol. 4777: S. Bhalla (Ed.), Databases in Networked Information Systems. X, 329 pages. 2007.

Vol. 4761: R. Obermaisser, Y. Nah, P. Puschner, F.J. Rammig (Eds.), Software Technologies for Embedded and Ubiquitous Systems. XIV, 563 pages. 2007.

Vol. 4747: S. Džeroski, J. Struyf (Eds.), Knowledge Discovery in Inductive Databases. X, 301 pages. 2007.

Vol. 4744: Y. de Kort, W. IJsselsteijn, C. Midden, B. Eggen, B.J. Fogg (Eds.), Persuasive Technology. XIV, 316 pages. 2007.

Vol. 4740: L. Ma, M. Rauterberg, R. Nakatsu (Eds.), Entertainment Computing – ICEC 2007. XXX, 480 pages. 2007.

Vol. 4730: C. Peters, P. Clough, F.C. Gey, J. Karlgren, B. Magnini, D.W. Oard, M. de Rijke, M. Stempfhuber (Eds.), Evaluation of Multilingual and Multi-modal Information Retrieval. XXIV, 998 pages. 2007.

Vol. 4723: M. R. Berthold, J. Shawe-Taylor, N. Lavrač (Eds.), Advances in Intelligent Data Analysis VII. XIV, 380 pages. 2007.

Vol. 4721: W. Jonker, M. Petković (Eds.), Secure Data Management. X, 213 pages. 2007.

Vol. 4718: J. Hightower, B. Schiele, T. Strang (Eds.), Location- and Context-Awareness. X, 297 pages. 2007.

Vol. 4717: J. Krumm, G.D. Abowd, A. Seneviratne, T. Strang (Eds.), UbiComp 2007: Ubiquitous Computing. XIX, 520 pages. 2007.

Vol. 4715: J.M. Haake, S.F. Ochoa, A. Cechich (Eds.), Groupware: Design, Implementation, and Use. XIII, 355 pages. 2007.

Vol. 4714: G. Alonso, P. Dadam, M. Rosemann (Eds.), Business Process Management. XIII, 418 pages. 2007.

Vol. 4704: D. Barbosa, A. Bonifati, Z. Bellahsène, E. Hunt, R. Unland (Eds.), Database and XML Technologies. X, 141 pages. 2007.

Vol. 4690: Y. Ioannidis, B. Novikov, B. Rachev (Eds.), Advances in Databases and Information Systems. XIII, 377 pages. 2007.

Vol. 4675: L. Kovács, N. Fuhr, C. Meghini (Eds.), Research and Advanced Technology for Digital Libraries. XVII, 585 pages. 2007.

Vol. 4674: Y. Luo (Ed.), Cooperative Design, Visualization, and Engineering. XIII, 431 pages. 2007.

Vol. 4663: C. Baranauskas, P. Palanque, J. Abascal, S.D.J. Barbosa (Eds.), Human-Computer Interaction – INTERACT 2007, Part II. XXXIII, 735 pages. 2007.

Vol. 4662: C. Baranauskas, P. Palanque, J. Abascal, S.D.J. Barbosa (Eds.), Human-Computer Interaction – INTERACT 2007, Part I. XXXIII, 637 pages. 2007.

Vol. 4658: T. Enokido, L. Barolli, M. Takizawa (Eds.), Network-Based Information Systems. XIII, 544 pages. 2007.

Vol. 4656: M.A. Wimmer, J. Scholl, Å. Grönlund (Eds.), Electronic Government. XIV, 450 pages. 2007.

Vol. 4655: G. Psaila, R. Wagner (Eds.), E-Commerce and Web Technologies. VII, 229 pages. 2007.

Vol. 4654: I.-Y. Song, J. Eder, T.M. Nguyen (Eds.), Data Warehousing and Knowledge Discovery. XVI, 482 pages. 2007.

Vol. 4653: R. Wagner, N. Revell, G. Pernul (Eds.), Database and Expert Systems Applications. XXII, 907 pages. 2007.

Vol. 4636: G. Antoniou, U. Aßmann, C. Baroglio, S. Decker, N. Henze, P.-L. Patranjan, R. Tolksdorf (Eds.), Reasoning Web. IX, 345 pages. 2007.

Vol. 4611: J. Indulska, J. Ma, L.T. Yang, T. Ungerer, J. Cao (Eds.), Ubiquitous Intelligence and Computing. XXIII, 1257 pages. 2007.

Vol. 4607: L. Baresi, P. Fraternali, G.-J. Houben (Eds.), Web Engineering. XVI, 576 pages. 2007.

Vol. 4606: A. Pras, M. van Sinderen (Eds.), Dependable and Adaptable Networks and Services. XIV, 149 pages. 2007.

Vol. 4605: D. Papadias, D. Zhang, G. Kollios (Eds.), Advances in Spatial and Temporal Databases. X, 479 pages. 2007.

Vol. 4602: S. Barker, G.-J. Ahn (Eds.), Data and Applications Security XXI. X, 291 pages. 2007.

Vol. 4601: S. Spaccapietra, P. Atzeni, F. Fages, M.-S. Hacid, M. Kifer, J. Mylopoulos, B. Pernici, P. Shvaiko, J. Trujillo, I. Zaihrayeu (Eds.), Journal on Data Semantics IX. XV, 197 pages. 2007.

Vol. 4592: Z. Kedad, N. Lammari, E. Métais, F. Meziane, Y. Rezgui (Eds.), Natural Language Processing and Information Systems. XIV, 442 pages. 2007.

Vol. 4587: R. Cooper, J. Kennedy (Eds.), Data Management. XIII, 259 pages. 2007.

Vol. 4577: N. Sebe, Y. Liu, Y.-t. Zhuang, T.S. Huang (Eds.), Multimedia Content Analysis and Mining. XIII, 513 pages. 2007.

Vol. 4568: T. Ishida, S. R. Fussell, P. T. J. M. Vossen (Eds.), Intercultural Collaboration. XIII, 395 pages. 2007.

Vol. 4566: M.J. Dainoff (Ed.), Ergonomics and Health Aspects of Work with Computers. XVIII, 390 pages. 2007.

Vol. 4564: D. Schuler (Ed.), Online Communities and Social Computing. XVII, 520 pages. 2007.

Vol. 4563: R. Shumaker (Ed.), Virtual Reality. XXII, 762 pages. 2007.

Vol. 4561: V.G. Duffy (Ed.), Digital Human Modeling. XXIII, 1068 pages. 2007.

Vol. 4560: N. Aykin (Ed.), Usability and Internationalization, Part II. XVIII, 576 pages. 2007.

Vol. 4559: N. Aykin (Ed.), Usability and Internationalization, Part I. XVIII, 661 pages. 2007.

Vol. 4558: M.J. Smith, G. Salvendy (Eds.), Human Interface and the Management of Information, Part II. XXIII, 1162 pages. 2007.